Animal Sonar Systems

NATO ADVANCED STUDY INSTITUTES SERIES

A series of edited volumes comprising multifaceted studies of contemporary scientific issues by some of the best scientific minds in the world, assembled in cooperation with NATO Scientific Affairs Division.

Series A: Life Sciences

Recent Volumes in this Series

The series is published by an international board of publishers in conjunction with NATO Scientific Affairs Division

A Life Sciences B Physics	Plenum Publishing Corporation London and New York
C Mathematical and Physical Sciences	D. Reidel Publishing Company Dordrecht, Boston and London
D Behavioral and Social Sciences E Applied Sciences	Sijthoff & Noordhoff International Publishers Alphen aan den Rijn, The Netherlands, and Germantown U.S.A.

Animal Sonar Systems

Edited by
René-Guy Busnel
Ecole Pratique des Hautes Etudes
Jouy-en-Josas, France

and

James F. Fish
Sonatech, Inc.
Goleta, California

PLENUM PRESS • NEW YORK AND LONDON
Published in cooperation with NATO Scientific Affairs Division

Library of Congress Cataloging in Publication Data

International Interdisciplinary Symposium on Animal Sonar Systems, 2d, Jersey, 1979.
  Animal sonar systems.

  (NATO advanced study institutes series: Series A, Life sciences; v. 28)
  Symposium sponsored by the North Atlantic Treaty Organization and others.
  Bibliography: p.
  Includes indexes.
  1. Echolocation (Physiology) – Congresses. I. Busnel, René Guy. II. Fish, James F.
III. North Atlantic Treaty Organization. IV. Title. V. Series.
QP469.I57 1979   599'.01'88   79-23074
ISBN 0-306-40327-7

This work relates to Department of the Navy Grant N00014-79-G-0006
issued by the Office of Naval Research. The United States Government
has a royalty-free license throughout the world in all copyrightable
material contained herein.

Proceedings of the Second International Interdisciplinary Symposium on
Animal Sonar Systems, held in Jersey, Channel Islands, April 1–8, 1979.

© 1980 Plenum Press, New York
A Division of Plenum Publishing Corporation
227 West 17th Street, New York, N.Y. 10011

ORGANIZING COMMITTEE
-=-=-=-=-=-=-=-=-=-=-=-=-=-=-=-=-=-=-

- R.G. Busnel
  Laboratoire d'Acoustique Animale
  E.P.H.E. - I.N.R.A. - C.N.R.Z.
  78350 Jouy-en-Josas, France

- J.F. Fish
  U.S. Navy
  Naval Ocean Systems Center
  Kailua, Hawaii 96734, U.S.A.

- G. Neuweiler
  Department of Zoology
  University of Frankfurt
  D-6000 Frankfurt, F.R.Germany

- J.A. Simmons
  Department of Psychology
  Washington University
  St. Louis, Mo. 63130, U.S.A.

- H.E. Von Gierke
  Aerospace Medical Research Lab.
  Wright Patterson Air Force Base
  Dayton, Ohio 45433, U.S.A.

A C K N O W L E D G M E N T S

-=-=-=-=-=-=-=-=-=-=-=-=-=-=-=-

This symposium was sponsored by various Organizations, National and International. The organizing Committee would like to thank them and their representatives :

- North Atlantic Treaty Organization, N.A.T.O.

- Advanced Study Institutes Programme

- Office of Naval Research of the U.S.A.

- United States Air Force

- Volkswagen-Stiftung from F.R. of Germany

- Ecole Pratique des Hautes-Etudes, Laboratoire d'Acoustique Animale, I.N.R.A. - C.N.R.Z. - 78350 JOUY-en-JOSAS, France

Preface

Thirteen years have gone by since the first international meeting on Animal Sonar Systems was held in Frascati, Italy, in 1966. Since that time, almost 900 papers have been published on its theme. The first symposium was vital as it was the starting point for new research lines whose goal was to design and develop technological systems with properties approaching optimal biological systems.

There have been highly significant developments since then in all domains related to biological sonar systems and in their applications to the engineering field. The time had therefore come for a multidisciplinary integration of the information gathered, not only on the evolution of systems used in animal echolocation, but on systems theory, behavior and neurobiology, signal-to-noise ratio, masking, signal processing, and measures observed in certain species against animal sonar systems.

Modern electronics technology and systems theory which have been developed only since 1974 now allow designing sophisticated sonar and radar systems applying principles derived from biological systems. At the time of the Frascati meeting, integrated circuits and technologies exploiting computer science were not well enough developed to yield advantages now possible through use of real-time analysis, leading to, among other things, a definition of target temporal characteristics, as biological sonar systems are able to do.

All of these new technical developments necessitate close cooperation between engineers and biologists within the framework of new experiments which have been designed, particularly in the past five years.

The scientists who have been working on these problems in various fields (electronics experts, signal processors, biologists, physiologists, psychologists) have produced new and original results, and this second symposium furnished the opportunity of cross-disciplinary contacts permitting an evaluation of the state of present research.

The Jersey meeting in April, 1979, brought together more than 70 participants from 8 different countries. This meeting was particularly necessary considering the number of new research groups that have appeared in various fields. I mention in particular: the Federal Republic of Germany, where two young scientists who participated at the Frascati meeting, Gerhard Neuweiler and Hans-Ulrich Schnitzler, have since become professors and have founded two schools of highly productive research; the United States, where funded by the U.S. Navy, studies on dolphins have spread to San Diego and Hawaii, given impetus by Bill Powell, Forrest Wood, Sam Ridgway, C. Scott Johnson, Bill Evans, and Ron Schusterman, and where certain of Donald Griffin's most gifted students, such as Alan Grinnell and Jim Simmons, are continuing his work on bats at various universities; Canada, where Brock Fenton is performing outstanding research.

Although much research has been carried out in the Soviet Union since 1969-1970, it is most unfortunate that, for reasons independent of their wishes, our colleagues from this country, who have moreover published several excellent reviews of their work, were not able to participate in our discussions.

The tendancy which appeared at Frascati towards a certain zoological isolation corresponding to a form of animal specialization dominated by either dolphins or bats, has partially regressed, thanks to several physicists who use both bat and dolphin signals in their theoretical approaches. While this segregation by field still remains a dominant behavior made obviously necessary, up to a certain point, by the different biological natures of these two groups of animals, the phenomenon is aggravated by the use of semantics specific to each group. Nevertheless, common interests and attempts at mutual understanding which appeared are encouraging and should be congratulated.

The proceedings of the Jersey meeting demonstrate the extent to which technology has advanced in the past decade, in performances of transducers, various microphones, hydrophones, as well as in recording apparata, analytical methods, particularly in the use of signal processing techniques, and in the application of new ideas such as time-domain, auditory processing, frequency-domain, Doppler compensation, target-acoustic imaging, and so on.

It is also interesting to note to what extent experimental strategies have been refined, and one can only admire the elegance of certain demonstrations carried out on dolphins as well as on bats. Many aspects of the performances of diverse species continue, however, to intrigue us, as they reveal sensory abilities whose fine analysis still eludes us, particularly the central mechanisms which control and regulate them. As an example of this, I would particularly like to mention how enriching was the experience that we were able to have using Leslie Kay's apparatus for the blind,

which gave spatiodirectional sensations analogous to those of airborne animal sonar systems. I do not doubt in the least that those
several minutes will lead to a totally new concept of biosonar problems.

As Henning von Gierke pointed out during the last session and
in a personal communication, the trends appearing at the Jersey
meeting indicate that pattern recognition theory is becoming more
and more important to biosonar research, replacing range finding
and echo theories as the promising research areas of the future.
Pattern recognition specialists should be included in future meetings as well as experts on the spatial frequency analysis of acoustic
and visual perception. Although animal sonar might be used to a
large extent for acoustic imaging of space, we know from ultrasono-
graphy that acoustic images are different from optical images.
Acoustic space perception therefore differs from visual space perception. Since the acoustic space is scanned sequentially, the
total "acoustic image" depends primarily on memory capability, which
is the major difficulty encountered by Leslie Kay in his device for
the blind. The findings of this Symposium may have a major impact
on general auditory physiology regarding, on one hand, the example
of sharp filter (acoustic fovea) and on the other, peripheral processing.

A reading of the present set of volumes presenting the current
state of research will bring out at the same time the unknowns of
the problem, the uncertainties, the hypotheses, and will allow vet-
erans of Frascati to measure the progress made since then.

I am most happy to thank here my American colleagues who agreed
to respond to my call in 1977, particularly Henning von Gierke, who
was our referee and support for N.A.T.O. as well as the U.S. Air
Force, and Bill Powell and Forrest Wood for the U.S. Navy, who later
recommended to me Jim Fish, a young associate, dynamic and efficient,
who held a preponderant place in the Organizing Committee.

I also would particularly like to express my gratitude to Gerhard
Neuweiler for his constructive participation in our planning group.
It is thanks to his outstanding reputation that our Symposium was
able to obtain funding from Volkswagen for the active and highly
productive German delegation. He assumed as well the heavy respon-
sibility of financial management of our funds.

I also wish to thank, on behalf of our Committee and myself,
the diverse individuals from my laboratory who, with devotedness,
took on countless tasks, often thankless and lowly: Michèle Bihouée,
Annick Brézault, Marie-Claire Busnel, Sophie Duclos, Diana Reiss,
and Sylvie Venla. Leslie Wheeler, who assisted me in editing these
volumes, deserves special mention, as without her help I would not
have been able to publish them so rapidly.

During the last plenary session in Jersey, the Organizing Committee and the Co-chairmen of the different sessions decided on the publication format of the proceedings, and suggested to the Symposium participants to dedicate this book to Donald R. Griffin. Our colleagues unanimously rendered homage to the spiritual father, the inventor, of echolocation.

The scientific wealth brought out during the three half-day poster sessions, bears witness to the interest and importance of the work of numerous young scientists and makes me optimistic for the future of the field of animal sonar systems. For this reason, and thanks to the two experiences which the majority of you have considered successful, I wish good luck to the future organizer of the Symposium of the next millenium.

                                    René-Guy Busnel
                                    Jouy-en-Josas
                                    France

Contents

CONTENTS

On behalf of the members of the organizing Committee and all the Symposium participants, we dedicate this book to

Donald R. GRIFFIN

in homage to his pioneering work in the
field of echolocation.

Donald R. Griffin

DEDICATION

by Alan D. Grinnell

Just over 40 years ago, a Harvard undergraduate persuaded a physics professor to train his crystal receiver and parabolic horn at active bats. They detected ultrasonic pulses, and the contemporary field of echolocation research was born. The undergraduate was Donald R. Griffin, the professor, G. W. Pierce, and their first report appeared in 1938. In recognition of his founding role and his countless important contributions to echolocation research this volume is dedicated to Don Griffin. Fortuitously, this coincides with the approach of his 65th birthday, a time when particularly popular and influential figures in a field are often honored with a "Festschrift" volume.

We were fortunate, at this meeting, that Don Griffin was persuaded to add a few recollections of his experiences during the infancy of the field. These are included in this volume. Additional perspective on some of the early years has been volunteered by his partner in the first experiments demonstrating echolocation, Dr. Robert Galambos, then a Harvard graduate student, now a well-known auditory neurophysiologist and Professor of Neuroscience at UCSD.

"In early 1939 Don found out that Hallowell Davis at the Harvard Medical School was teaching me how to record electric responses from guinea pig cochleas and asked if I could slip in a few bats on the side. He wanted to know whether their ears responded to the "supersonic notes" he and Professor G. W. Pierce had just discovered. So I asked Dr. Davis, who said "go ahead" and thus Don found himself a collaborator in some unforgettable adventures. (This was not the only time Don enlisted me in one of his enterprises--I remember spending 3 days helping him build a bird blind on Penekeese Island in weather so foul the Coast Guard had to come rescue us.)

During that spring I worked out the high frequency respon-
sivity of the bat cochlea, and in the fall we assessed the
obstacle avoidance capabilities of the 4 common New England
species (using an array of wires hanging from the ceiling to
divide a laboratory room into halves). We recorded the inaudible cries bats make in flight and demonstrated (by inserting
earplugs or tying the mouth shut) that they must both produce
and perceive them if their obstacle avoidance is to be successful. (We also made a sound movie of all this which nobody can
find.) That research yielded my Ph.D. thesis and launched Don
on the career this book honors.

All the crucial new measurements we made used the unique
instruments devised by Professor Pierce. A physics professor
who, like Don, seemed to me really a naturalist at heart, Pierce
had designed his "supersonic" receivers in order to listen to
the insects singing around his summer place in Vermont (or was
it New Hampshire?). How Don found Professor Pierce and then
talked him into letting us use his apparatus I do not know.

In the spring of 1940, when we felt we understood how bats
avoided obstacles in the laboratory, it occurred to us to test
our ideas in the field. So we made an expedition to a cave
Don knew in New York State taking along the portable version of
Pierce's supersonic receiver shown in the accompanying picture
(Don was a skillful photographer even in those days).

The cave in question opened just beyond the bank of a
mountain brook. Once inside you first climbed down a ways and
then up. We had no trouble getting in and, after the small
upward climb, found ourselves looking along a straight tunnel
as far as our flashlights could penetrate. We set up the
equipment and Don went further on to the gallery where the bats
roosted. He sent several of them flying, one by one, down the
tunnel in my direction and as they approached Professor Pierce's
machine emitted the chattering clamor we hoped for. Then Don
and I changed places so he could hear the noise too. Since
the readout of Pierce's portable device came only via earphones,
the sole record of the first bat ultrasonic cries ever heard
outside a laboratory is the one engraved in our memories.

Throughout all of this Don kept urging me not to waste
time, and I thought he closed off the exercise and moved us
toward the entrance after an almost indecently short interval.
He explained his behavior once we had scrambled down and then
up and then out to cross the brook: the spring sun was rapidly
melting the snow, the brook was rising fast, and he had been
worried from the moment we arrived that the water might fill

the entrance to the cave and trap us inside. My captain, I
thought to myself, always looks out for the safety of his crew."

.During the nearly four decades since that first demonstration
of echolocation, the field has flourished. This is evident from
the quality and diversity of work described in this volume. It can
also be seen, quantitatively, in the accompanying graph, which shows
the number of publications on echolocation in microchiropteran bats
during each 3-year period since 1938. This graph, kindly prepared
by Uli Schnitzler, shows several surges in productivity - one about
1961, when the second generation of U.S. scientists began publishing,
and a very large one in 1967, when, following Frascati, the second
generation of German scientists and the Russian group entered the
field. Now, with the Jersey Symposium, the third generation is beginning to contribute importantly. Throughout this entire period,
while moving professionally from Harvard to Cornell, back to Harvard,
and then to the Rockefeller University, and while making major contributions to a number of other fields, Don Griffin has continued to
guide development of the field of echolocation with imaginative, incisive experiments: demonstrating the usefulness of echoes for orientation in the lab and in the field, showing that echolocation was
used for insect capture, documenting the sensitivity of the system
and its resistance to jamming, developing techniques to show how
accurately targets can be discriminated, and analyzing the laryngeal

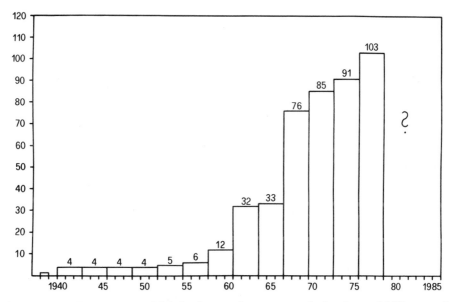

Histogram of papers published per 3-year period since 1938 on echolocation in microchiropteran bats.

D. R. Griffin at Simla, Trinidad, studying capture of food by
<u>Noctilio</u> (1960).

mechanisms involved in ultrasonic orientation. It was in his lab,
with his encouragement, that the first major advances were made in
understanding the neural adaptations for echolocation, an aspect of
the field that has grown to enormous and impressive proportions. In
association with Al Novick, he established the great value of comparative approaches to echolocation, traveling the world not only to
document the variety of sounds and skills shown by bats, but to do
the first careful studies of echolocating birds, as well. Out of
early interests in the feeding habits, home ranges, and seasonal mi-
grations of bats (not to mention a long-standing fascination with
homing and migration in birds) came pilot studies on the role of
echolocation, passive hearing and vision in homing behavior. Indeed,
in countless instances, he has broken new ground and set standards
of rigor and experimental elegance that have served the field well.

A major milestone in the field of animal behavior was the publication, in 1958, of <u>Listening in the Dark</u>, Griffin's monograph on
his experiments and thoughts on echolocation, and winner of The Daniel
Giraud Elliot Medal of the National Academy of Sciences. With the
perspective of 20 years subsequent work in the field, it is astonishing to anyone re-reading this classic to realize how fully Griffin
already understood the phenomenon of echolocation, how many critical
discoveries he had already made, and how profound were his insights.
This was one of five books Don Griffin has written, the most recent

of which is "The Question of Animal Awareness: Evolutionary Con-
tinuity of Mental Experience" (1976).

It was not only for his incisive early work that Don Griffin
deserves recognition as "father" of the field (or Godfather as he
was also described during the meeting); he has also been the academic
father, or colleague, of a high percentage of those who have become
active in the field. Most of the contributors to this volume who
have worked on bats have felt the imprint of Griffin's personality
and experimental approach directly, as graduate students or postdoctoral research associates working in his laboratory, or as their
students, and it has been a powerful influence on their lives and
careers.

Don is a great storyteller, as his chapter in this volume at-
tests. More than that, however, his approach to science is guar-
anteed to never leave a dull moment. Whether it is fighting to stay
aloft after releasing birds from a small plane piloted by Alex Forbes,
or trying to convince suspicious authorities that he had valid reason
for clamboring around the fire escapes of a mental hospital, or barely
escaping a flooded cave, or covering for a student who had fallen
through the ceiling of the Mashpee Church, or searching the tombs of
an ancient Italian cemetery for Rhinolophus, his approach to experimentation is as direct and audacious as his ideas. Many of the
breakthroughs in echolocation research were the result of his introduction of new or unfamiliar technology, from home-made ultrasonic
microphones to high speed tape recorders, sonagraphs, and information
theory. This applies in all of his other fields, as well. In recent
years, for example, much of his energy has been devoted to radar
identification and tracking of migrating birds. One of his collab-
orators in this research, Charles Walcott, tells of just one of the
complications that this has gotten them into, and of Don's characteristic aplomb in solving the problem:

"That season Don arrived in Stony Brook with a trailer
carrying what looked like a giant coffin. Actually the box
contained a helium filled kitoon - a sort of hybrid balloon -
kite combination that could be flown up to several thousand
feet in the air. Its purpose was to place instruments in the
same air mass that the radar showed birds to be flying in.
Indeed on its trials in Stony Brook the kitoon worked splen-
didly - we were able to get it up at least a thousand feet
or more - amply high enough to be where birds flew. Unfortunately they aren't the only things that fly there - a
helicopter soon appeared flying significantly lower than the
kitoon. As it came close and inspected the kitoon we noticed
that the helicopter bore the inscription "Police" in large
letters. We rapidly hauled down the kitoon and anxiously
awaited the arrival of sirens and patrol cars; fortunately
they never came. But on other occasions, the law has become

involved and Don had been read the Federal Air Regulations
which prohibit the flying of balloons, kites or other such
devices more than a few feet above ground. The State Police
in Millbrook, New York next to The Rockefeller field station
had alerted the FAA about Don's activity. Don promised the
FAA representative that it wouldn't happen again. It was the
following Easter Sunday when Don and his colleagues were
following a kitoon with radar that the kitoon's string broke.
Following the kitoons progress with the radar, they saw that
it suddenly stopped drifting in the wind and became stationary.
It was hanging only a few hundred feet in the air directly over
the State Police headquarters. The string had caught in the
upper branches of a small sapling. With a typical display of
ingenuity, Don rigged a second kitoon with a grappling iron
and managed to snag the string of the escaped kitoon retrieving
both just before dawn on Easter Sunday!"

To those of us who have had the privilege of working with Don
Griffin, perhaps the greatest lesson has been his emphasis on rigor
in experimentation. No one is more acutely aware of the ambiguities
of an experiment, the difficulties involved in demonstrating something convincingly. Nor does he ever jump to conclusions. His intellectual vigilance has been characterized by the claim that if he
were in a car passing a flock of sheep in a field, and a travelling
companion commented on the fact that among the sheep were two that
were black, he would reply, "They're black on the side facing us,
anyway." (Attributed to Don Kennedy.) On the other hand, he is not
afraid to consider some of the most complex formsof animal behavior,
and to take on established dogma, decrying the excessive use of "simplicity filters" in interpreting behavior, a position he argues elo-
quently in his recent writing on animal awareness.

This mixture of insistence on careful unbiased observation of
what is really there and rigorous proof of any conclusions, combined
with a brilliant imagination and willingness to adapt new technology
to biological problems, has done much to develop the field of animal
behavior research. It has served as particular inspiration for all
of us.

We are honored to dedicate this volume to Donald R. Griffin.

THE EARLY HISTORY OF RESEARCH ON ECHOLOCATION

Donald R. Griffin

The Rockefeller University
New York, N.Y. 10021
U.S.A.

It has been very gratifying to hear and read about the exciting new discoveries concerning echolocation by both bats and cetaceans from so many active laboratories on both sides of the Atlantic. I am particularly impressed because from about 1940 to 1960 very few scientists were actively investigating these fascinating problems, despite what seemed to me their obvious significance. The Frascati and Jersey conferences in 1966 and 1979 have not only confirmed but exceeded my expectations.

I was asked to describe the early history of echolocation, but unfortunately I cannot bring you firsthand recollections of Lazzaro Spallanzani or his contemporaries. As you know he was the true pioneer who first discovered that there was a problem, and that bats could navigate dexterously without the use of their eyes. This history has been reviewed, both in my book Listening in the Dark (1958) and by Dijkgraaf (1960). I can only recount experiences during the Twentieth Century, but it is sobering to realize that forty-five years has elapsed since I first began to study the orientation and navigation of bats. Another forty-five years before that time would take us back well into the "Dark Ages" when the brilliant early work of Spallanzani and Jurine had been virtually forgotten, a time when Cuvier's tactile theory was generally accepted.

As an undergraduate in the 1930s my interest in bat migration and homing led to banding bats in substantial numbers. Several friends at Harvard kept telling me that I really should visit a physics professor, George Washington Pierce, who had developed apparatus to analyze what were then called supersonic sounds. His was probably the only apparatus in the entire world which could detect and analyze sounds over a broad frequency range extending from the upper limit of human hearing to about 100 kHz. He was already deeply engaged in studying the high frequency sounds of insects (Pierce, 1948), and once I visited his laboratory he enthusiastically invited me to bring some bats to his apparatus. A cage full of little brown bats (Myotis lucifugus) set the apparatus buzzing and clicking. But when bats were allowed to fly their high frequency sounds were picked up only occasionally. Therefore our first paper (Pierce and Griffin, 1938) now seems absurdly cautious, for it then seemed quite possible that these newly discovered high frequency sounds had nothing to do with orientation in the dark.

Discussions with a fellow student, Robert Galambos, led to much more thorough investigations. Turning somewhat reluctantly from my program of banding bats, I plunged into what was completely joint work with Galambos during the years from 1938-1940. He was already an expert in auditory physiology, and I was familiar with the natural history of bats, knew where to catch them, how to handle them, and what questions in their natural behavior were of primary interest. One of our first steps was to repeat and extend the type of quantitative experiments described by Hahn (1908) to measure obstacle avoidance ability under various conditions. We began by repeating with minor variations the experiments pioneered by Jurine and extensively replicated by Hahn, and confirmed that covering the ears of bats produced almost total disorientation. It was one of Galambos' major contributions to suggest that we should also interfere with the production of high frequency sounds; and indeed this had almost the same effect as plugging the ears. We soon found that the high frequency sounds could be picked up from flying bats provided we located Pierce's microphone, which was equipped with a parabolic horn, in front of the flying bat. When we took care to do this the brief pulses of high frequency sound proved to be invariable accompaniments of flight and to increase in repetition rate whenever the bat faced a difficult problem such as dodging small obstacles.

This is now so rudimentary that it is difficult to realize how unexpected was the discovery that bats actually orient their flight by hearing echoes of sounds generated for this purpose. For example, the distinguished physiologist Selig Hecht was so incredulous when he heard our report of these experiments at a meeting of the American Zoological Society around Christmas, 1940 that he seized Galambos by the shoulders and shook him while complaining that we could not possibly mean such an outrageous suggestion. Radar and sonar were still highly classified developments in military technology, and the notion that bats might do anything even remotely analogous to the latest triumphs of electronic engineering struck most people as not only implausible but emotionally repugnant. It was partly to overcome this sort of resistance that I suggested the general term echolocation to cover the wide variety of orientation mechanisms, natural and artificial, based upon the emission of probing signals and the location of distant objects by means of echoes (Griffin, 1944).

One amusing incident occurred when Galambos borrowed Pierce's apparatus for the production of high frequency sounds in order to stimulate bats from which he was recording cochlear microphonics. Such recordings were still relatively new and exciting aspects of auditory physiology and the only available apparatus was in the laboratory of Hallowell Davis at the Harvard Medical School. It is also necessary to realize that Pierce's apparatus was absolutely unique, as far as we knew in the entire world, both for detecting high frequency sounds and also for generating them under anything approaching controlled conditions. Thus the only way in which Galambos could look for cochlear microphonics in bats at high frequencies was to take one of Pierce's devices across Boston to the Harvard Medical School. We were somewhat hesitant to ask for this privilege, but after the importance of the enterprise was explained to Professor Pierce he readily agreed.

Unfortunately Galambos did not realize that some of the innocent-looking power outlets in Davis' laboratory supplied DC rather than AC power, and when Pierce's sound producing apparatus was first plugged in the result was a puff of white smoke and a serious diplomatic problem. I remember Galambos saying "Don, I think I can fix the thing, but won't you please be the one who tells Pierce what happened." This was an obviously expedient division of labor, and to our great relief Pierce was quite sympathetic and reasonable, worrying only that the repair work be done properly.

I worked during World War II in the psychoacoustic laboratory at Harvard directed by S. S. Stevens and learned a great deal about acoustic apparatus and measurements which I had never heard of during my graduate studies in the biology department. Francis Wiener was especially helpful and arranged to let me borrow a Western Electric 640AA microphone, preamplifier, and oscilloscope for improved measurements of the acoustic properties of the high frequency sounds we had been able to detect but only crudely to characterize with Pierce's apparatus. It was during one of these unofficial sessions that a Myotis lucifugus escaped into Leo Beranek's enormous anechoic chamber. The bat had no difficulty detecting the supposedly anechoic walls, and indeed it was quite a problem to retrieve it since the chamber was so large that direct approach to the walls was impossible. Later when Beranek asked to borrow a bat in order to demonstrate to the press the total lack of sound reflection from the walls of this chamber, I was able to spare him the embarrassment of a totally unconvincing demonstration.

As I think you all know, Sven Dijkgraaf in Utrecht independently discovered bat echolocation without the benefit of any electronic apparatus (Dijkgraaf, 1943, 1946). By listening carefully as bats flew about in quiet surroundings he was able to hear the faint audible component which accompanies the ultrasonic orientation sounds. The Ticklaut as he named this audible sound, and what Galambos and I called the audible click accompanying the physically much more intense ultrasonic signals were clearly one and the same. After the war we straightened out these matters by cordial correspondence, and at one point I even sent Dijkgraaf a live Myotis lucifugus to compare with the very closely related European species he had studied.

Something I should like to emphasize, because it may have implications for the future, is that after these basic facts had been generally accepted there was what now seems in retrospect an incredible lack of interest in further studies of echolocation. A typical reaction was "Well, Griffin you've convinced us that bats really do detect stalactites in caves by hearing echoes of these high frequency sounds, and that's all very well, but why do you want to spend any more time on this sort of thing? Who cares whether they can detect slightly smaller wires or whether one species uses a slightly different kind of sound from another? Why do you want to go to the tropics just because some of the bats there have funny looking excrescences on their faces? Isn't it time you turned your attention to something really important and forgot about all those silly bats."

This attitude was so pervasive that I found it difficult to maintain my own motivation. It seemed silly to worry about how bats might catch their insect prey, and almost unthinkable that they might use echolocation. It was really only because of the failure of some other, indoor experiments during the summer of 1950 that I persuaded myself to drag all of the necessary apparatus out-of-doors in order to study the high frequency sounds of bats when they were catching insects under natural conditions. I cannot overemphasize the intellectual inertia and difficulty in justifying the necessary effort to continue studies of echolocation and inquire what further ramifications might develop after a wider variety of species had been studied under a wider range of natural conditions. We should be alert in the future for similar mental blocks that may restrict imagination and thus retard progress.

One exception was Harold Trapido who had been working for several years at the Gorgas Laboratories in Panama and who persuaded me after rather extensive correspondence to come to Panama with the necessary apparatus to inquire whether echolocation might be different in any of the neotropical bats conveniently available there for further study. My apparatus filled 15 or 20 large packing cases, and the first of my two months in Panama was consumed in tracking the air freight shipment which had gone astray.   Nevertheless during that trip, and a follow-up trip the next year by Alvin Novick, it was possible to sketch out something of the enormous range of orientation sounds used by various families and genera of neotropical bats (Griffin and Novick, 1955).   Yet highly distinguished experts had attempted to discourage me from wasting the time and funds necessary for this trip to Panama.   "Why" they said "a bat is a bat and what makes you think there will be any significant differences just because they live in the tropics and have funny looking faces?"   This sort of resistance extended to arguing that alleged differences in frequency patterns were of no significance, and that all these high frequency sounds were simply noise bursts.   This argument was advanced even after Mohres had clearly demonstrated that the European horseshoe bats used sounds of much longer duration and of almost constant frequency in contrast to all bats previously studied which had been members of the family Vespertilionidae or Molossidae.

One episode from the early 1950s is especially appropriate to relate at this second International Conference on Animal Sonar Systems.   Among the very first European visitors who came to see something of our experiments on echolocation in bats were Rene-Guy and Marie-Claire Busnel who spent several hours with me and Novick at the Harvard Biological Laboratories.   Wishing also to show them how we study bats under natural conditions I drove to Kenneth Roeder's home in Concord, Massachusetts where he was studying the hearing of moths.   A colony of Eptesicus fuscus emerged every evening from a neighboring house and many of them flew over Roeder's yard where he often set up moth preparations and recorded action potentials in response to the passing bats.   In order to elicit insect pursuit maneuvers I had for some time been using a slingshot to throw pebbles into the air 2 or 3 meters in front of an approaching bat.   This very often elicited a buzz and often a spectacular pursuit maneuver.

As so often happens in such cases we arrived a little late and the bats were already emerging for the evening.   Hastily assembling in Roeder's yard the apparatus which then consisted of a 640AA microphone, battery operated amplifiers, and oscilloscope I took careful aim with my slingshot for a point three meters in front of an approaching bat.   To the surprise of all concerned (except the bat itself) there was a sharp "plunk" as the pebble collided with one of the bat's flight membranes.   The Busnels obviously thought that I was so proficient with my slingshot that I could hit a flying bat at a range of 10 or 15 meters.   I was happy to make excuses about not using the slingshot again so as to leave the Busnels with the glowing impression that a typical American field biologist was such a crack shot that he could hit bats with pebbles from a slingshot. What none of us realized at the time was that bats actively reach out with their wing membranes to catch insects. Undoubtedly this Eptesicus had done so, but it was only some years later thanks to the photographic talents of Frederic Webster that we learned about one more unsuspected aspect of bat echolocation.

A long time elapsed after Galambos' original studies of cochlear microphonics before anyone else took up the neurophysiology of audition in bats. I can only explain this in terms of the same sort of conservatism and lack of interest in unexpected new comparative studies which I mentioned earlier. The man who opened up the neurophysiology of bat hearing came to me first as a freshman student asking to be excused from taking the elementary course I was then teaching. He also expressed interest in whatever I might be doing, and soon became actively involved in studies of obstacle avoidance and sensitivity of echolocation. During this interval an amusing episode occurred which, while it does not relate directly to echolocation, shows how throughly involved in studies of bat behavior this undergraduate student became. Approximately twenty years earlier I had banded a large number of bats in certain caves in Vermont. John Hall of the University of Massachusetts recaptured one of these which must have had a minimum age of about 19 years. I therefore returned to these caves with several colleagues equipped with records of all band numbers so that suspected longevity records could be examined with maximum care before the bat was released. We found three or four individual Myotis lucifugus that were 19 or 20 years old. When one of them was positively identified the man who was holding it was heard to remark "Why it's older than I am."

But bat banding and obstacle avoidance tests were not sufficiently challenging to hold the full attention of this student, and while still an undergraduate he wanted to record action potentials from the brains of bats. I discouraged this for some time, explaining patiently that recording from the central nervous system was a complex and difficult business which no one in the Harvard Biological Laboratories was equipped to undertake. But finally he wore down my resistance and we anesthetized a Myotis lucifugus with Nembutal and exposed the dorsal surface of the brain. Using a piece of copper wire (or perhaps a paper clip) as an electrode we displayed on an oscilloscope whatever potentials might result when the anesthetized bat was stimulated by short duration high frequency sounds or by the orientation sounds of another bat. To my amazement beautiful action potentials could easily be recorded from anywhere on the bat's brain -- from cerebellum to olfactory lobes. This was the beginning of Alan Grinnell's studies of N_4 and later other auditory evoked potentials. This first crude experiment certainly served to hook us both, and Grinnell only escaped with difficulty several years later for postdoctoral work with Bernard Katz in London while I sought to compensate for his

loss by writing Professor Katsuki in Tokyo to ask whether he had
a young postdoc who would like to extend Grinnell's studies of
auditory responses in bat brains. Fortunately he did and the
result was the long and brilliant series of studies by Nobuo Suga
and later his several students and colleagues. The latest of their
many extremely significant results have been presented at the Jersey
meetings.

Dr. Suga had occasion to learn that studying bat echolocation
leads one into a variety of challenging situations and opportunities.
Shortly after he had arrived from Japan we were approaching a
dilapidated old building in rural Massachusetts to collect bats
when he saw what he thought was a cat with an unusual pattern of
black and white markings. He was rescued barely in time from too
close an approach to a skunk (Mephitis). Later, with his charming
wife, he was introduced to the tropics at the William Beebe
Tropical Research Station in Trinidad. This former estate had well
equipped laboratory buildings and very pleasant gardens where one
could investigate bats and other animals under ideal conditions.
For instance, the director, my wife Jocelyn Crane, saw to it that
the food did not make any of the northern visitors sick. The high
point of the day usually was the evening cocktail hour on the
terrace over which flew more species of bats than we could ever
identify. But in listening to them with a bat detector, usually
with a rum punch known as a "Simla special" in hand, we often heard
the high frequency sounds of many kinds of insects. Most were also
audible to the unaided ear, but one in particular was not.
Frequently this insect-like sound came from small potted bushes
which could be approached closely from all sides. By quietly
approaching with the bat detector it was easy to ascertain that
the sound came from the bush in question, but close examination
of every part of the bush failed to disclose any signs of the animal
which had been making the sound before it was disturbed by our
close approach.

At times, especially after a second Simla special, we even
entertained the possibility that the sound might be generated by
the plant! The Sugas were especially intrigued by this unidentified
sound source called the "Snarley Buzz." I suspect that Nobuo
secretly hoped to discover the first echolocating plant. But for
whatever reason he and his wife worked long and hard to discover
the actual source, often climbing into tall trees on rickety ladders,
being attacked by biting ants and stinging bees, but in the end
locating the insect responsible -- a small long horned grasshopper
of the genus Pflugus (Suga, 1966).

In short, echolocation has been a most rewarding subject to
investigate, both in the laboratory and in the field. It is now
flourishing and ramifying in many significant directions. I am

confident that this happy state of affairs will continue well into
the future, and I hope all students of echolocation will have as
satisfying experiences as I have been privileged to enjoy.

REFERENCES

Dijkgraaf, S., 1943, Over een merkwaardige functie van den gehoorsin
     bij vleermuizen, Verslagen Nederlandische Akademie van
     Wetenschappen Afd. Naturkunde, 52:622-627.
Dijkgraaf, S., 1946, Die Sinneswelt der Fledermäuse, Experientia,
     2:438-448.
Dijkgraaf, S., 1960, On Spallanzani's unpublished experiments on
     the sensory basis of object perception in bats, Isis, 51:9-20.
Griffin, D. R., 1944, Echolocation by blind men, bats, and radar,
     Science, 100:589-590.
Griffin, D. R., 1958, "Listening in the Dark," Yale University
     Press, New Haven, Conn. (reprinted 1974 by Dover Publications,
     New York).
Griffin, D. R., and Novick, A., 1955, Acoustic orientation of
     neotropical bats, J. Exptl. Zool., 130:251-300.
Hahn, W. L., 1908, Some habits and sensory adaptations of cave-
     inhabiting bats, Biol. Bull., 15:135-193.
Pierce, G. W., 1948, "The Songs of Insects," Harvard University
     Press, Cambridge, Mass.
Pierce, G. W., and Griffin, D. R., 1938, Experimental determination
     of supersonic notes emitted by bats, J. Mammalogy, 19:454-455.
Suga, N., 1966, Ultrasonic production and its reception in some
     neotropical Tettigoniidae, J. Insect Physiol., 12:1039-1050.

Chapter I
Performances of Animal Sonar Systems

Chairman : F.G. Wood

co-chairmen

R.J. Schusterman and H.U. Schnitzler

- Underwater

    Behavioral methodology in echolocation by marine
    mammals.
    R.J. Schusterman

    Detection range and range resolution of echolocating
    bottlenose porpoise.
    A.E. Murchison

    Odontocete echolocation performance on object size,
    shape and material.
    P.E. Nachtigall

    Cetacean obstacle avoidance.
    P.W.B. Moore

- Airborne

    Performance of airborne animal sonar systems
    I. MICROCHIROPTERA.
    H.U. Schnitzler and O.W. Henson, Jr.

    Performance of airborne animal sonar systems
    II. VERTEBRATES other than MICROCHIROPTERA
    O.W. Henson, Jr. and H.U. Schnitzler

BEHAVIORAL METHODOLOGY IN ECHOLOCATION BY MARINE MAMMALS

Ronald J. Schusterman
Department of Psychology and Biology
California State University
Hayward, California 94542

INTRODUCTION

Despite the fact that a preponderance of the investigations dealing with echolocation by marine mammals has been done with intact animals performing some characteristic behavior, only a single paper written some fifteen years ago (Turner, 1964) has ever dealt in any profound way with problems of behavioral methodology. The 1964 paper by R. N. Turner addressed itself primarily to conditioned responses and included response measures, techniques of target or signal presentation, threshold measurement and reinforcement contingencies. When Turner's paper was published, there had only been a few pioneering studies done on porpoise echolocation and hearing, and by today's standards, these studies were, for the most part, relatively unsophisticated. A recent Russian publication (Lekomtsev and Titov, 1974) attempted to update Turner's work, but in reality the paper served primarily to highlight Russian behavioral procedures to the study of porpoise sonar abilites.

Even though numerous advances in the behavioral analysis of sensory and perceptual capabilities of a wide variety of animal species have been made since the mid 1960's, there has been no attempt to review and evaluate procedural variables in the behavioral study of animal sonar systems. In particular, modern operant conditioning techniques (Skinner, 1961) have become very efficient, and methodological advances have been applied to many problems and many animal species. With these advances has come a marked interest in a relatively new field of animal psychology which has been termed "animal psychophysics" (Blough and Blough, 1978; Stebbins, 1970).

The purpose of this paper is to review and evaluate behavioral techniques used to study echolocation by marine mammals, particularly dolphins, within the context of this new field of animal psychophysics. The goal of these techniques is to establish and maintain stimulus control of an animal's behavior. Stimulus control is essentially a convenient expression for saying that a stimulus change brings about a change in some measurable aspect of behavior. The degree to which a stimulus exerts control over an animal's behavior may be measured in a variety of ways including probability, latency, amplitude, or rate of some specified response and it is usually assumed that a controlling stimulus either signals the animal that a particular class of responses will be reinforced or it signals the animal that a particular class of responses will not be reinforced. Once stimulus control has been established, the experimenter may want to identify the manner in which the animal classifies or dimensionalizes the stimulus. This has been particularly true in recent echolocation experiments in which the controlling aspects of the stimulus targets were interpreted in one way by the experimenter and in quite a different way by the porpoise (Schusterman and Kersting, 1978). In animal psychophysics, the answer to the question about what the porpoise "sees" when it is stimulated with suprathreshold echoes depends on a series of perceptual investigations in which tests of equivalence, transfer or generalization are given. Thus, the porpoise is placed in experimental situations in which different stimulus configurations, as defined by the experimenter, are responded to in the same fashion.

The emphasis in this paper will be on problems associated with the types of indicator responses used by echolocating animals and the ways in which stimulus targets have been presented. In this regard, it should be pointed out that the major difference between a passive hearing task and an active sonar task is that in the former task it is not necessary to preclude the animal's dependence on vision unless visual stimuli are correlated with acoustic stimuli in signalling reinforcement. However, in echolocation tasks the targets invariably have a visual analogue, and if the animal can use visual cues, it may not depend on its acoustic auto-communication cues, i.e., on echoes which return following its own sound emissions. When using marine mammals in echolocation tasks, enucleation has rarely, if ever, been used and the traditional way of occluding or eliminating vision has been to either have the animal work in turbid water or in water where the light level was very low, place a blindfold or eye cups on the animal, use targets which are quite different acoustically but appear visually to be the same, or place the targets behind a visually opaque but acoustically transparent screen. All of these methods have costs and benefits and the technique used will frequently depend on the species under investigation and the type of problem being studied.

This paper will not discuss some procedural problems of echolocation experiments which have been reviewed elsewhere (Turner, 1964; Lekomtsev and Titov, 1974). These include randomization of stimulus presentation and motivation or food deprivation variables.

UNCONDITIONED RESPONSES

In procedures with unconditioned responses, the indicator response appears to require no special training, i.e., it is reflex-like in its action, being directly elicited by a variety of submerged targets and sounds.

Perhaps the first systematic use of reflexes as indicator responses were the experiments by Kellogg (1961) with two Atlantic bottlenose dolphins (<u>Tursiops truncatus</u>) which were conducted in a pool with very limited visibility. Kellogg's targets included: (a) polelike sheet-metal devices, triangular in cross section; (b) BB shot; (c) wooden streamlined shapes which could be silently submerged; (d) food fish, and (e) human swimmers. Kellogg recorded two classes of behavior: (a) bursts of sonar clicks, and (b) changes in movement or swimming patterns. The latter behavior was subdivided into either approach or avoidance responses. These experiments suggested that almost any target "noisely" submerged would elicit echolocation pulses by <u>Tursiops</u> sometimes followed by approach (e.g. fish) or avoidance (human swimmer or long metal pole). Oscillating head movements were frequently noted as the porpoise, emitting clicks, swam toward a hand-held fish. If the water was splashed without submerging a target, then the animals did not sustain their clicks. However, when reflecting targets were submerged, sustained clicking occurred, and it was assumed by the experimenter that the porpoises detected targets and differentiated between them on the basis of the sound signals emitted and the type of swimming movement (approach or avoidance) observed.

In retrospect, it is likely that one of the most vexing problems in Kellogg's experiments was a lack of an objective and quantitative evaluation of the porpoises's behavior under these extremely variable stimulus conditions. On occasion, the experimenter must have been forced to make a subjective judgment as to whether there was indeed an avoidance response or an approach response-- particularly if the porpoise did not take the hand-held fish. The fact that Kellogg did not present quantitative data in reporting these unconditioned responses to submerged targets and water splashing supports the notion that the use of unconditioned responses is of limited value in the study of echolocation by marine mammals. Another confounding factor in these experiments

is the possibility that these reflex-like responses habituated
with repeated stimulation by similar sounds or similar targets.
Habituation of sudden changes in the swimming movements detected
visually by the experimenter when relatively loud pure tones were
presented may have led to an initial underestimate of the upper
frequency limit of hearing in Tursiops (Kellogg and Kohler, 1952;
Kellogg, 1953).

CONDITIONED RESPONSES

Unconditioned responses depend on a relatively simple and
direct stimulus-response relationship in which echoes from the
target or other auditory stimuli elicit the indicator response,
which may not be easy to define and quantify. When conditioned
responses are used, the experimenter usually chooses some carefully
defined motor response which can be readily measured. In classical
or Pavlovian procedures, the unconditioned stimulus (US) or
reinforcer is contingent on the conditioned stimulus (CS) i.e.,
the CS and US are paired. In operant or Skinnerian procedures,
reinforcement is made contingent on some class of responses. If
a stimulus sets the occasion for reinforcing a class of responses,
then the stimulus becomes a discriminative stimulus (S^D) and
serves as a signal very much as the CS serves as a signal in
classical conditioning procedures. Following repeated pairings
on the CS and US or the S^D and reinforcement, a highly stereotyped
indicator response is elicited or emitted whenever the CS or S^D
is presented.

Operant procedures may involve the use of positive or negative
reinforcers. When positive reinforcement is used, if the animal
makes a response in the presence of an S^D, it receives a reward.
When negative reinforcement is used, if the animal makes a response
in the presence of an S^D, it avoids an aversive stimulus event.

Classical conditioning in which the US is shock and the CS
is an acoustic signal has been used extensively by the Russians
to study the sound reception by echolocating porpoises. For
example, sound detection thresholds were obtained in the common
porpoise (Phocaena phocaena) by pairing auditory stimuli (CS)
with shock (US) and measuring a restrained animal's galvanic skin
response (Supin and Sukhorunchenko, 1970). In another experiment
on restrained bottlenose dolphins the characteristics of directional
sound signals in the horizontal plane were investigated by measuring
the cardiac component of what the Russians call "motor-defensive
conditioned reflexes" (Ayrapet'yants, et al., 1973). All of
these experiments have involved passive hearing in porpoises and
I do not know of a single experiment that has used classical
conditioning procedures to determine the sonar capabilities of

marine mammals. It seems that only operant conditioning procedures
have been used to investigate the performance of echolocating
marine mammals.

RESPONSE MEASURES

Examination of a variety of response measures and their
related reinforcement contingencies reveals remarkable similarity
among different experimenters in different laboratories around
the world in their use of operant procedures to study echolocation
in both marine mammals and in bats. Investigators from the
Soviet Union and United States have used very different terminology
in describing essentially the same procedure. For example, when
a Russian researcher says he or she used the "method of motor-
food conditioned reflexes in unrestrained dolphins" this is
nearly the equivalent of an American researcher saying he or she
used "a two-alternative forced-choice procedure",

Retrieval

Typically, this procedure involves the arrangement of reinforcement contingencies such that if the porpoise picks up a single
submerged target with its mouth or rostrum and returns it to the
experimenter or trainer, then the animal is positively reinforced
with a food fish. In theory, at least, the retrieval technique
could be used as a way of determining the echo dimensions along
which auditory quality is represented in the echolocating porpoise.
This could be done by repeatedly presenting a blindfolded Tursiops
with a variety of targets differing in some aspect of fine target
structure and differentially reinforcing the animal for retrieving
targets differing in this single aspect. In fact, however, the
retrieval procedure has been used infrequently to study the sonar
abilities of marine mammals, and its use has generally been
limited to simple detection tasks.

An experiment by C. Scott Johnson (1967) used the retrieval
procedure in an attempt to determine the relationship between
pulse rate and target range. The targets used in the experiment
included a weighted air-filled foam rubber ring and single vitamin
pill. Scronce and Ridgway (this volume) failed in their attempt
to demonstrate echolocation in a blindfolded gray seal (Halichaerus
gyrpus) by requiring the seal to retrieve an air-filled plastic
ring 20 cm in diameter situated randomly in one large section of
a 10 m diameter redwood tank. Most recently Dziedzic and Alcuri
(this volume) studied changes in the click burst of T. truncatus
when the task was to retrieve different shaped objects.

Obstacle Avoidance

The most definitive early demonstrations of echolocation in
bats were those reported by Griffin and Galambos (1941) in which
enucleated bats were required to navigate through a maze of wires.
Similar experiments have been done with free-swimming T. truncatus
and P. phocaena under a variety of conditions (Busnel et al., 1965;
Busnel and Dziedzic, 1967; Kellogg, 1961). The complexity of these
obstacle avoidance tasks makes it extremely difficult to determine
what aspects of the targets were important for the occurrence of the
complex swimming patterns of porpoises which included short and
difficult turns in order to avoid colliding wth poles, wires and
nets. The obstacle avoidance technique is extremely limited from
this standpoint and is rarely used today in echolocation experiments
with marine mammals where the emphasis is the determination of a
well-defined response and a carefully controlled target or signal
dimension.

Go/No-Go

Surprisingly, the "go/no-go" procedure has rarely been chosen
to measure the echolocation performance of marine mammals despite
the fact that it is, perhaps, the most simple and straight forward
procedure to study discriminative echolocation. There are several
variations of the go/no-go paradigm. In the study of marine mammal
bioacoustics, the paradigm operates within the context of a discrete
trial procedure in which target or signal presentation is relatively
brief and is contingent on the animal's maintaining a reasonable
fixed position between target presentations. During a trial, indicated by a sound or a light, the animal orients from its starting
position towards the target area. Then it either performs a carefully defined action, e.g. pressing a paddle (a "yes" response) in
the presence of a standard target, or it inhibits or withholds the
action (a "no" response) in the presence of a comparison target or
in the absence of a target. Following the indicator response the
animal stays or returns to its starting position. Response measures
include the likelihood of a response occurring on each trial and the
latencies of "yes" responses. Reinforcement may be programmed so
that "yes" responses in the presence of a target are positively
reinforced and are extinguished, i.e., are not reinforced in the
absence of a target, or reinforcement may be programmed symmetri-
cally so that it follows "yes" responses in the presence of a target
("hits") and "no" responses in the absence of a target ("correct
rejections"). Errors (which may be punished with "time-outs" or
other aversive events) are of two types: a "yes" response in the
absence of a target ("false alarms") or a "no" response in the
presence of a target ("misses").

The go/no go procedure is usually employed when targets or

stimuli are presented sequentially and is thus used most extensively
in the earlier stages of training when only a single standard
target is presented on repeated trials (Lekomstev and Titov, 1974).

One type of go/no-go procedure is exemplified by an experiment
on discriminative echolocation (Ayrapet'yants, et al., 1969) in
three Black Sea dolphins (Tursiops truncatus), under high and
very low illumination. The dolphins were required to tug at a
ring in the presence of a standard target (a cylinder 25 mm in
length and 110 mm in diameter) and to withhold this response when
presented with comparison targets (cylinders 28, 30, 35, 45, 55
and 75 mm in length and 110 mm in diameter). A 5 kHz tone
signalled the dolphin to swim "from the starting position at a
distance of 20 meters to the experimenter who immersed different
targets into the water...." (Ayrapet'yants, et al., 1969).
Tugging the ring when the standard target was presented resulted
in a fish reward. The manner in which this paper was written
makes it difficult to be sure about the reinforcement contingencies,
but in my interpretation the contingencies were not symmetrical
which means that withholding the "ring-tug response" in the
presence of the comparison targets did not result in a fish
reward. Following each response, the dolphin was required to
swim back to the start position. Eventually, all dolphins accurately
differentiated all but the 28 and 30 mm comparison cylinders from
the standard 25 mm standard cylinder. The authors state that "it
was more difficult to differentiate the 30 mm cylinder which was
only 5 mm taller than the positive one. Such differentiation was
not absolute (correct answers constituted 70%). Presentation of
the 28 mm cylinder, which was 3 mm taller than the positive one,
elicited no recognition. Thus, the threshold of differentiation
of targets according to height was 5-10 mm" (Ayrapet'yants, et
al., 1969).

The Russian experiment highlights some of the least desirable
aspects of the go/no-go procedure. With only a single well-
defined response, it is sometimes difficult to separate "no"
reports from a failure to respond, and there is no measurable
latency for the "no" response. Moreover, it would seem that
unless the reinforcement contingencies or what signal detection
theorists call the payoff matrix was symmetrical there might be a
preference for the "yes" response, thereby driving the threshold
estimate down. Furthermore, the statistical definition of a
"threshold" is sorely missing in this particular paper.

Multiple-Response Forced-Choice Procedures

The multi-response technique which has probably been used
most extensively for doing experiments on marine mammal biosonar

Figure 1.   Schematic of a testing tank illustrating the use of the
            two-response forced-choice procedure with simultaneous
            target presentation.   Prior to the onset of an 8 kHz
            tone, the dolphin positioned itself in front of the
            emitter.   The targets were lowered silently into the
            water at two points 2 m apart.   In response to the
            audio signal the dolphin emerged from the enclosure
            through the passageway and headed for the platform
            where it detected the targets from a certain distance
            (adapted from Lekomtsev and Titov, 1974).

is the two-response forced-choice procedure.   The major variation
on this procedure depends on the simultaneous or successive
presentation of targets.

    Simultaneous Target Presentation.   Following the stationing
of an animal each of two spatially separated targets are presented
near their associated response manipulanda and the animal is
reinforced for responding to the "correct" stimulus.   Frequently,
a vertical partition or separation screen has been placed between
the targets so the animal is forced to make its choice at some
minimum specified distance (see Figure 1).

    Variables affecting the echolocation performance of several
species of dolphins, including T. truncatus and D. delphis, have
frequently been investigated by the two-response forced-choice
procedure using a simultaneous presentation technique.   These
factors include object detection with noise disturbance (Titov,
1975), angular and range resolution (Bel'kovich et al, 1970;
Lekomtsev and Titov, 1974; Murchison, this volume), size discrimination of planometric and stereometric figures (Kellogg, 1961;

Turner and Norris, 1966; Barta, unpublished ms.; Ayrapet'yants et
al, 1969), shape discrimination of planometric and stereometric
figures made of metal or plastic (Gurevich, 1969; Barta, unpublished
ms.; Nachtigall and Murchison, this volume; Bagdonas et al.,
1970) and material compostiion of plates and solid elastic spheres
(Evans and Powell, 1967; Dubroskiy et al, 1971).

Perhaps the only disadvantage of the two-alternative forced-
choice procedure with a simultaneous presentation of targets is
that even with a partition between the targets it is extremely
difficult to force a dolphin or seal without eye cups to make its
choice at some fixed distance from the targets when the experimenter
has difficulty visually tracking the animal in turbid water or at
very low levels of light (see Schusterman, 1967).

Successive Target Presentation. A group of American investigators at the Naval Ocean Systems Center (NOSC) in Hawaii have
done a wide variety of echolocation experiments by presenting
targets on successive trials and requiring the dolphin to respond
to one of two spatially separated response manipulanda. In
"presence-absence" experiments, a response to one manipulandum
(e.g. to the animal's right) is reinforced if a target is present
("yes" response) and a response to the other manipulandum (e.g.
to the animal's left) is reinforced if the target is absent ("no"
response). In so-called discrimination or recognition experiments,
targets usually differing multi-dimensionally control differential
responding and the porpoise is reinfoced for responding to the
manipulanda associated with a specific target. Despite spatially
disparate response manipulanda, targets are presented in the same
location. Thus, unlike the two-alternative forced-choice method
with simultaneous presentation, where stimulus targets and associated
response manipulanda have approximately the same relative spatial
location (so that different reflected sounds come from different
places), the two-alternative forced-choice technique with successive
stimulus presentation requires the animal to make a bilateral or
spatially different response to different sounding reflected
echoes emanating from the same locus in space.

An illustration of this procedure within the context of a
detection experiment has been reported by Penner and Murchison
(1970), who used it to determine the smallest diameter copper
wire that could be detected by an Amazon River dolphin, Inia
geoffrensis. The Inia stationed between the response manipulanda
(mounted on the opposite side of the pool from the stimulus
display) until a trial began with the onset of a 1-sec burst of a
30 kHz tone. Then the Inia swam across the 7.3 diameter tank
toward the target area emitting echolocation pulses, and returned
to the response manipulanda area. Activation of the "yes" manipulandum in the presence of a target was a hit and the dolphin was
reinforced. In the absence of a target activation of the "no"

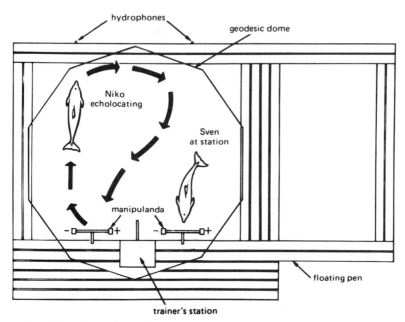

Figure 2.   Testing situation showing the use of the two-response
            forced-choice procedure with successive presentation
            of targets.   Presence of a target was reported on the
            "yes" manipulandum and the absence of a target was
            reported on the "no" manipulandum.   For further
            details of this procedure see Murchison (this
            volume).

manipulandum was a correct rejection and the dolphin was reinforced.
Errors (false alarms as well as misses) were never reinforced .

    An interesting variation of this procedure was used to
calculate the furthest distance at which solid and water filled
spheres could be detected by Tursiops (Murchison, this volume).
The study used two Tursiops working alternately on the same task
(see Figure 2).   In an attempt to determine whether cooperative
or cueing behavior occurred in this type of paired-porpoise range
detection task, Penner (1977) conducted a similar experiment
comparing performance when animals worked on alternate trials,
simultaneously or alone, and he found that echolocation performance
appeared unaffected by these conditions.

    Finally, Hammer (1978) and Au and Hammer (this volume)
applied the two-alternative forced-choice procedure with successive
target presentation to the study of salient target characteristics
of different-size cylinders.   One set of cylinders (A) controlled
A reports and the other set of cylinders (B) controlled B reports
(see Figure 3).

Figure 3. Experimental setting and apparatus showing porpoise
           stationing, A and B manipulanda, experimenter's
           shack and the lowering of hollow cylindrical targets
           attached to monofilament lines mounted on I-beam
           support.

Collateral Behavior

     As previously indicated, a response-contingent reinforcement
procedure within the context of a discrete trial procedure is the
method by which echolocation is studied in marine mammals. For
example, following the intake of a fish reward, the beginning of
the next trial depends upon the dolphin resuming its stationing
or starting position. At the start of the trial, the dolphin
makes an "observing response" by emitting echolocation pulses at
the target, and on the basis of its echo perception, the dolphin
makes a choice (e.g. hitting one of two rubber balls mounted on
flexible stalks as shown in Figure 3) which results in either a
fish reward, no reward, or a time-out period (negative reinforcement).
In this type of response-contingent procedure, the form or topography
as well as the temporal patterning of each response--stationing →
observing → choice → feeding or time-out--are all likely to
influence the relationship between the "choice" response, the
target variables and the reinforcement in either unknown or at
least in rather imprecisely measured ways. Thus, at the start of
a trial the dolphin leaves its starting position, swims to its
right, begins to emit pulses while swinging its head from side to
side as well as up and down, turns its entire body around, swims
back to the manipulandum, breaks the water surface by spinning on
its vertical axis and terminates the behavioral sequence by
striking one of the rubber balls with its rostrum while simultaneously emitting squeaking sounds. Then the dolphin immediately
opens its mouth, swallows the fish reward, begins swimming along

the right wall of the tank before returning to station where the
animal subsequently begins bobbing its melon and rostrum up and
down prior to the start of the next trial. This bewildering
array of responses, their sequential dependencies and their
duration are, to some extent, important regarding an interpretation of the dolphin's auditory perception of echo complexes.

In a discriminative echolocation experiment with different
sized spheres presented simultaneously to an Atlantic bottlenose
dolphin, Turner and Norris (1966) called some of these behaviors
"collateral behavior" and found a relationship between "disruption
of a stable pattern of collateral behavior" and a substantial
increase in errors. Moreover, these experimenters could accurately
predict incorrect choices or errors on the basis of the dolphin's
swimming pattern. Evans and Powell (1967) also used the term
"collateral behavior" to refer to all behaviors except target
pressing and showed that there was an inverse correlation between
"performance scores" (correct target presses) and "scanning
rate". They also found that choices of a target (metal plates)
were more rapid for easy discriminations as compared to difficult
ones, but that this was largely due to the greater head scanning
at distances of less than one meter from the plates for the
difficult comparisons. Here again we see that responses occurring
before the originally conditioned operant were predictive of the
animal's performance.

Although Evans and Powell (1967) emphasized various aspects
of the observing response in an echolocating dolphin and showed
correlations between this behavior and the "terminal" response,
Penner and Murchison (1970) noted that an echolocating Inia
geoffrensis sometimes swam upside down, jaw gapped, jaw snapped
and yawned. In terms of predicitng whether their fresh water
dolphin would hit the "yes" or "no" manipulandum, Penner and
Murchison found that Inia terminated its observing response
sooner or later depending on the absence or presence of a target.
Similar observations have been reported in echolocation experiments
dealing with the effects of distance and reverberation on object
detection in Tursiops truncatus (Murchison, 1979).

Collateral behavior in echolocating porpoises may be viewed
in the larger context of environmentally induced sequential
patterns of behavior. Falk (1961; 1969; 1971) coined the term
"adjunctive behavior" in reference to excessive drinking in a
hungry rat when the animal was lever pressing on a schedule of
intermittent food reinforcement. The distinction between adjunctive
or interim activities and terminal responses was first made by
Staddon and Simmelhag (1971) who showed that during interfood or
intertrial intervals predictable sequential behavior patterns
increased and decreased with terminal responses (e.g. a conditioned
operant or Pavlovian response) occurring just before the delivery

of food. In a recent review Staddon (1978) has shown that temporal
sequences of behavior depend on their interactions with other
behaviors that are induced in the situation. Interim activities
for an echolocation dolphin would include all aspects of the
observing response (pulse emission, head scanning, etc.) as well
as head bobbing, jaw clapping, jaw gapping, side swimming, etc.
Various measures of the terminal or indicator response might
include its form or topography, its duration, and its strength.
Any and all of these response measures may be correlated with the
"choice" response in an echolocation task and are likely to help
in our understanding of the behavior of a porpoise following its
analysis of complex echo inputs in relation to its previously
stored information.

       To illustrate the interaction between interim activities and
terminal responses, I will describe one of a series of discriminative
echolocation experiments which I conducted at the Naval Ocean
Systems Center in Kanoehe Bay, Hawaii. Hollow steel and hollow
bronze targets were presented successively to an experienced
adult male porpoise (Sven) in a two-alternative forced-choice
procedure. Figure 3 shows a schematic diagram of the experimental
setting. The dolphin's task was to differentiate echoes from
hollow steel and hollow bronze targets--each having two different-
sized outer diameters (OD) and wall thicknesses. The targets
were described in detail by Hammer (1978). All targets were 17.8
cm long. The smaller OD's were 3.81 cm with a wall thickness of
0.32 cm and the larger OD's were 7.62 cm with a wall thickness of
0.40 cm. Both the large and small steel targets were to be
reported on the $\underline{A}$ or left manipulandum and the large and small
bronze targets were to be reported on the $\underline{B}$ or right manipulandum.
In earlier experiments Sven had been trained to report the presence
of a target by hitting the $\underline{A}$ manipulandum and the absence of a
target by hitting the $\underline{B}$ manipulandum. To take advantage of those
previously established response contingencies, Sven was first
trained to make an $\underline{A}$ report when presented with either a large or
a small hollow steel target submerged about 1.14 m from the
surface of the water and to make a $\underline{B}$ report in the absence of a
target. Once Sven perfected this differentiation, both the large
and small bronze targets were submerged gradually, i.e., the
vertical positions of the bronze targets were initially always
closer to the surface of the water than the steel targets. The
vertical position of the bronze targets gradually approached that
of the steel targets until eventually all targets were presented
in the same position. In general, this type of <u>stimulus</u> <u>fading</u>
procedure minimizes errors (see below) during the acquisition
phase of a learned discrimintation.

       Figure 4 shows the acquisition of the discrimination between
steel and bronze targets regardless of size. Because a <u>fading</u>
procedure was used, the number of errors was minimized even when

Figure 4.   Acquisition of a hollow bronze and hollow steel
            differentiation regardless of the size of the outer
            diameter of the targets. "Long" observing responses
            occurred only in the presence of the small hollow
            steel target ("A" reports) and were a precursor
            of a significant increase in total correct responses.

the porpoise was first confronted with the discrimination.
Nevertheless, the percentage of total correct responses, i.e.,
for all four targets, rose significantly from the first 14 test
sessions to the last 14 sessions ($\underline{p}$ < .01; sign test). In addition
to measuring the terminal choice responses as a function of the
type of target presented, we also recorded response latencies and
classified observing responses into two categories. The most
frequently occurring response was one in which the turn from
station toward the target area at the beginning of a trial was
quickly followed by another abrupt turn to the manipulanda.
During the course of these rapid observing responses, Sven never
swam more than two meters from station and the response latencies
were invariably less than four seconds. This observing response
occurred most frequently in the presence of both bronze targets
and the large steel target. However, about one-quarter of the
way through the experiment, Sven began to make a different observing
response. When he made a complete turn from station toward the
target area at the start of a trial, he rolled onto his left side
and with his head tucked slightly he faced the target area.
Following a pause of between one to two seconds, after swimming
at least three meters from station, Sven swung his entire body
around in a very deliberate fashion before returning to the
manipulanda. The response latencies for these observing responses
were usually more than four seconds. These "long" observing
responses occurred primarily in the presence of the small steel
target. As Figure 4 shows, "long" observing responses were very

infrequent during the first seven test sessions and then increased dramatically. The increased probability of "long" observing responses in the presence of the small steel target was a precursor to the significant increase in total correct responses. Thus, in this experiment, there was a strong interaction between interim activities and terminal responses as a function of specific target stimuli. Moreover, a measure of the interim responses showed that the porpoise was actually differentiating between the large and small hollow steel targets even though the terminal response (A-report) was the same for both echo complexes.

METHOD OF TARGET PRESENTATION

The development and maintenance of stimulus control in echolocation experiments makes it absolutely necessary to ensure that the animal is appropriately oriented to the target area and that targets are chosen and presented in such a way that target parameters cover a wide range of values including threshold values. Target detectability must be assured if reinforcement contingencies are to produce the desired results of establishing and maintaining stable responding. The spacing of target parameter values around threshold and the sequence of presentation are also crucial considerations in training and maintaining stimulus control. Once established, the degree of stimulus control by an echo complex can be tested or probed by presenting novel targets and finding out the degree to which targets contain similar echo configurations (Hammer, 1978). In general, experimenters using porpoises in studies of echolocation have used traditional psychophysical methods to arrive at threshold estimates.

Observing Response

In an early experiment on discriminative echolocation with different sized spheres (Turner and Norris, 1966), a blindfolded dolphin had to deflect a lever corresponding to the standard sphere in order to collect a fish which was thrown to a specific "zone" or area of the pool. Once the dolphin collected the fish, it oriented to the apparatus to make another response. Obviously, this type of control of the animal's positioning while it echolocates is crude and makes it difficult to specify the properties of the echoes returned to the dolphin which controlled the animal's final choice response. Without controlling the position from which a discrimination is made, it is not even possible to be certain that the intensity of the echo return, depending as it does on the distance between the porpoise and the targets and the manner in which the porpoise faced the targets, is the critical cue in differentiating objects of different target strength (e.g. see Evans and Powell, 1967). Much of the early Russian research

Figure 5.   Schematic of a dolphin assuming a fixed stationing posi-
             tion in a chin cup behind an acoustic screen during a
             range resolution task. The animal's task was to report
             whether the left or right target was closer and the
             report was made by striking the appropriate manipulan-
             dum. On this trial the correct report is on the left.
             The two-alternative forced-choice procedure with
             simultaneous target presentation was used.

suffers from the same problem, i.e., the ambiguous nature of the
echo return in relation to the position of the dolphin and the
targets at the time the discrimination is made.

    During the past few years, much more refined and unambiguous
control of an animal's orientation with respect to the targets
has been accomplished by making the observing response contingent
upon target presentation. In a range resolution task, Murchison
(1979) made target presentation (consisting of two identical 7.6
cm diameter polyurethane spheres) contingent on "chin-cup" stationing
(see Figure 5). Once the porpoise was stationed in the fiberglass
chin cup and the targets were in position, an acoustic screen was
dropped and the target display was presented to the dolphin.
Thus, the acoustic screen prevented the animal from prematurely
scanning the target display and has the potential of allowing the
experimenter to make rather precise measurements of pulse emission
and the echo return at the relatively precise position of pulse
emission (Au, Floyd and Haun, 1978).

HOOP STATION

SIDE VIEW   **3/4 VIEW**

Figure 6.   Diagram of an Atlantic bottlenose dolphin (<u>T. truncatus</u>)
maintaining a fixed position in a water-filled plastic
hoop. The dolphin remained in the hoop up to its
flippers and when given an audio signal it emitted
a burst of echolocating clicks at a target 6 m
directly in front of its melon. Following the termination of the last pulse within a burst of pulses,
the dolphin backed out of the hoop and reported the
presence of a sphere or cylinder by hitting the "A"
or "B" response paddle or manipulandum with its rostrum.

Another technique of controlling not only the postural
aspect but also the pulse emission aspect of the observing response
in an echolocation task has been developed by Schusterman, Kersting
and Au (this volume). In a sphere-cylinder discrimination task
the porpoise had to position itself in a 41 cm water-filled
plastic hoop with its rostrum directly in line with the center of
the target located 6 m away and inhibit the emission of echolocation
pulses until an audio cue was given (see Figure 6). The preciseness
of the observing response just described permitted a determination
of the type of signal return the dolphin used to make its differentiation (Au, Schusterman, and Kersting, this volume).

Most recently, Penner and Kadane, in an evaluation of biosonar
detection in noise (this volume), were able to record pulse train
latencies and response latencies following the <u>last</u> pulse of a
train from two bottlenose dolphins who were trained to make a
precise postural observing response. Thus, these investigators
were and are capable of making relatively exact measurements of

the moment in time in which the dolphins make their decision
regarding the presence or absence of a target.

Fading Procedures

Perhaps the most powerful technique currently available
which minimizes the number of incorrect responses during the
acquisition phase of a detection or discrimination task via
echolocation is the technique of fading. In general terms the
technique depends on initiating training or stimulus control
procedures with an easy detection or discrimination (high signal/
noise ratio) and gradually changing some feature of the targets
on successive trials so that stimulus control of responding is
transferred from one feature or characteristic to another. For
example, in the visual mode Schusterman (1965; 1966) showed that
sea lions could repeatedly reverse their responses to different
visual shapes (circles and triangles) errorlessly if the animals
were transferred from a previously well-established size discrimination. These and other experiments suggest that, in a wide
variety of species using a wide variety of stimuli, interdimensional
training frequently results in less errors than intradimensional
training probably because of the greater differences between
stimulus configurations in the interdimensional condition (see
Castillo and Pinto-Hamuy, 1978, for a recent review of the use of
fading procedures).

In, perhaps, the best controlled experiment on discriminative
echolocation of planometric shapes, Simmons and Vernon (1971)
found that enucleated big brown bats (Eptesicus fuscus) initially
could not be readily trained to distinguish between different
shaped triangles (made of polystyrene) having the same surface
area. However, when the experimenters presented the bats with
two triangular targets differing both in size and shape, the
discrimination was acquired rapidly. Thereafter, the size dimension
was gradually reduced so that eventually only differences in
triangular shape controlled responding by the bats.

In a dolphin counterpart to the bat echolocation experiment
on fading from a size discrimination to a shape discrimination,
Barta (unpublished ms.) and Evans (personal communication) used
the type of fading technique first described by Schusterman
(1965; 1966). After the successful discrimination of planometric
circular targets (made of neoprene cemented to aluminum) differing
in size, a blindfolded Tursiops truncatus readily distinguished
between equal-size triangles and circles following the fading or
gradual elimination of the size dimension.

Several studies on echo ranging in bottlenose dolphins have
shown that thresholds were considerably lowered by presenting

progressively more difficult detections or discriminations, i.e.,
by incorporating fading procedures. Lekomtsev and Titov (1974)
report that the detection of a steel sphere 1.1 cm in diameter
occurred at a significantly greater range if the separation
screen was advanced by successively smaller steps (0.5 m) than if
it was advanced by successively large steps (1.0 m). Murchison
(1979) showed that in a range resolution study daily sessions of
small incremental changes of distance in the near and far targets
(ΔR) had to be preceded by large ΔR trials and extensive fading
of the ΔR in order to maintain stable responding and the obtaining
of reliable estimates of ΔR thresholds.

Probe Trials

     In virtually all previously described studies, the behavioral
methodologies allowed for interpretation of the dolphin's echolocation capabilities primarily in terms of the systematic covariation
between values of target parameters (i.e., size, material, shape,
etc.) and performance. These methods are unsuitable for identifying
the manner in which dolphins classify target or echo parameters.
In animal psychophysics, many investigators depend on generalization
tests or test of response equivalence to determine the dimensions
along which some sensory quality is represented in a given species
or population. One variation on the generalization test which
has been used in echolocation experiments is to establish an
extremely stable discrimination between two different targets or
sets of targets and then, interspersed with trials in which
targets from the original or baseline discrimination are presented,
the investigator presents novel targets on the "probe" or test
trials.

     Recent research by Hammer (1978) and Au (Au and Hammer, this
volume) serves as a fine illustration of the probe-trial technique.
These investigators applied the two-alternative forced-choice procedure with successive target presentation to the study of discriminative echolocation in which the object of the study was to have the
porpoise classify echo complexes in terms of response equivalence.
An adult male Tursiops, who had been well-trained on a range detection task, was presented with two sets of different-sized cylindrical targets. One set of cylinders (A) controlled responses to the
left response manipulandum (A) and the other set of cylinders (B)
controlled responses to the right response manipulandum (B) (see
Figure 3). In one very interesting experiment, the porpoise was
first trained to differentiate between hollow aluminum cylinders and
solid coral rock. The animal was then presented with probe trials
of different-sized hollow cylinders of glass, bronze and steel and
the probability of trials on which the porpoise gave the aluminum
response was calculated. On the basis of these ranked probabilities,
Hammer and Au found that the porpoise classified glass most like
aluminum and steel least like aluminum, regardless of target

size. The results of synthesized broadband dolphin-like signals
bounced off the same targets and subjected to a matched-filter
analysis indicated a relatively close resemblance to the dolphin's
performance on the task.

Psychophysical Procedures

    Once maximal stimulus control is attained, then an estimate
of the animal's absolute or differential sensitivity is sought.
Regardless of the signal or noise parameters involved, the threshold
is usually taken as some interpolated stimulus value associated
with some arbitrary criterion of detectability or discriminability.
Most threshold estimates are arrived at by one of several graphic
methods in which a broad range of target or noise values yield a
broad range of correct choices extending from chance to perfect
performance.

    As previously noted, fading procedures are usually incorporated
in the application of psychophysical methods to the study of
echolocation parameters. Frequently, the dolphin is initially
confronted with detections of discriminations with high signal/noise
ratios followed by decreasingly lower signal/noise ratios. Thus,
in general, the most widely used traditional psychophysical
procedure for studying the echolocation abilities of dolphins has
been some variation of the descending "method of limits". In
order to maintain reliable performance, investigators have frequently
used blocks of several trials prior to decreasing the signal/noise
ratio.

    Titration methods have been used on occasion in order to
maintain tight stimulus control in an efficient manner. In
titrating (also known as "stair-case" or "tracking"), the signal/
noise ratio is decreased (either in terms of single trials or in
blocks of trials) until the animal makes an error or reaches some
arbitrary error criterion; then the signal/noise ratio is stepped
up. Thus, in titration methods, the signal/noise ratio is either
decreased or increased, depending on the animal's performance.
Murchison (1979) incorporated a titration procedure as a warm-up
technique in his study of range resolution in Tursiops truncatus
and Barta (unpublished ms.) used a titration procedure to obtain
a size discrimination threshold in the same species.

    In the psychophysical method of constant stimuli, a set of
stimulus values occurs in random order, either in terms of single
trials or in blocks or trials. Interestingly, Barta (unpublished
ms.), in an attempt to obtain a size discrimination threshold,
began with the method of constant stimuli by using comparison
target values considerably above threshold and when an asymptotic
level of performance had not been attained, he shifted to a

titration method. A recent exception to this trend of using the descending method of limits in psychophysical studies of dolphin echolocation performance is illustrated by an experiment by Penner and Kadane (this volume) on the effects of noise on the detection of a 7.65 cm steel water-filled sphere at a distance of 16.5 m. These researchers, in using a modified method of constant stimuli, randomly presented five levels of white noise (ranging from 67 to 87 dB re 1 µPa) in 10-trial blocks and generated a very predictable relationship between noise levels and detectability in two experienced Atlantic bottlenose dolphins.

PROCEDURAL PROBLEMS

I have already focused in on some fairly important problems related to the behavioral methods used in the study of echolocation in marine mammals. These have included the desirability of explicitly controlling in precise measurable ways various aspects of observing responses, indicator or terminal responses and ways of measuring what the animal "sees" when it is stimulated with suprathreshold echoes. There are other related problems that I would like to deal with in my concluding remarks, and these include learning, response bias, and the subtle difference between the recognition or discrimination of targets and the detection of targets.

Baseline Behavior

There have been several measurements of thresholds related to such target parameters as distance, size, resolution, wall thickness, etc. which have resulted in considerably detailed data on the echolocation capabilities of several species of porpoises (see the papers by Moore, Murchison and Nachtigall in this chapter). In all the psychophysical experiments reviewed, it usually takes months before stable asymptote or baseline is achieved. Several psychophysical studies have reported learning effects over a relatively long period of time (e.g. see Turner and Norris, 1966, and Lekmotsev and Titov, 1974).

The continued learning effect on the echolocation capability of a bottlenose dolphin is clearly illustrated in an experiment reported by Lekmotsev and Titov (1974). Following several days of training, the dolphin could reliably detect a lead sphere 5.0 cm in diameter at a distance of 11 m. After two additional months of training, the dolphin reliably detected an 0.8 cm sphere of the same material at the same distance, a greater than fourfold increase in detection acuity.

In perceptual experiments of echolocation in which generaliza-

tion techniques are used, it is extremely important, as Hammer
(1978) has pointed out, that probe trials be "superimposed" on
unchanging baseline behavior. It is meaningless to attempt to
interpret responses to novel targets either in terms of sound
emission parameters or echo return parameters if response to the
original training targets are no longer reliable.

Detection, Recognition and Discrimination

In the literature of animal sonar systems, sometimes the
terms "detection", "discrimination", and "recognition" are used
interchangeably. However, traditional, as well as operational,
use of these concepts in the psychophysical literature suggests a
distinction between these concepts (Galanter, 1962). Classically
speaking, detection (in echolocation tasks) should refer to the
problem of how much echo strength is necessary in order for an
animal to reliably differentiate the target or signal from the
ambient noise. Recognition and discrimination, on the other
hand, assume detectability and refer to the question of how
distinctive or how similar echoes emanating from two or more
targets are from each other. Even a cursory review of the literature
leads me to believe that relatively few investigators working on
animal sonar systems have realized that behavioral methods,
particularly in terms of a whole host of variables dealing with
the animal's previous training experience, critically determine
the way in which an animal perceives the common as well as the
distinctive or salient features of different target-reflected
echoes.

In an experiment on shape discrimination in bats (Myotis
oxygnathus), the animals were first food reinforced for approaching
squares and not triangles (presented simultaneously) of equal
surface area (Konstantinov and Akhmarova, 1968). After learning
this discrimination to a high degree of accuracy, when the nonreinforced triangle was replaced with a circle, the probability of
the bats still approaching the square remained unchanged. Later,
when the square was presented simultaneously with both the circle
and the triangle, performance still remained relatively unchanged.
A similar experiment was conducted under water with T. truncatus
and the results were essentially the same, i.e, the introduction
of a new target presented simultaneously with the previously
reinforced target did not effect performance deleteriously (Barta,
unpublished ms.). These results might be expected if during or
following original learning it was primarily the presence of
distinctive echo characteristics from the positive or reinforced
target which controlled responding and not the distinctive echo
characteristics of the negative or nonreinforced targets or the
echo characteristics which both positive and negative targets had
in common. However, the results are surprising if it is understood

that during and following original training both positive and
negative targets were equally or nearly equally effective in
controlling discriminative performance (Mackintosh, 1974). The
question of what would have happened in these and other similar
experiments if the positive target were replaced and a novel
target was paired with the negative target seems never to really
have been considered. My guess is that performance would have
deteriorated. It would appear, therefore, that the animals in
these experiments were listening for distinctive echoes from the
positive target and that echoes from other targets acquired the
status of background or ambient noise similar to the situation
that prevails in detection experiments.

Perhaps the clearest example of a dolphin being trained
inadvertently to listen for distinctive echoes reflected from one
set of targets and not another comes from the work of Hammer
(1978). As previously mentioned, Hammer's Atlantic bottlenose
dolphin (Sven) had previously been trained in a series of detection
tasks to report the presence of a target to its left and the
absence of a target (ambient noise) to its right, and was then
trained by Hammer in a two-alternative forced-choice procedure to
report the presence of small and large hollow aluminum cylinders
(A targets) to its left (A response manipulandum) and to report
the presence of small and large solid coral rock (B targets) to
its right (B response manipulandum). After the porpoise had
perfected this baseline differentiation, Hammer conducted a
series of experiments in which he used probe trials to determine
what characteristics the probe targets had in common with the
baseline targets of rock and aluminum. In almost all instances,
probe targets--even hollow aluminum targets with a wall thickness
of only 0.16 cm difference from the standard hollow aluminum
targets--were classified as the standard solid rock targets,
i.e., as B targets or ambient noise. These results suggest that
most targets were classified as ambient noise because the dolphin,
Sven, was listening for a salient portion of the echo complex to
make its differentiation and that in the absence of the salient
or distinctive feature, Sven reported on the B manipulandum.
Since Sven's previous training probably played a crucial role in
the way the dolphin reported on the A and B manipulandum, I
suspect that if Hammer had trained Sven to report solid rock
cylinders on A and hollow aluminum cylinder on B, most of the
probe targets would have been classified as aluminum, or, in
other words, as ambient noise. Thus, Hammer's results and his
interpretations may have strongly depended simply on the way he
constructed the original or baseline reinforcement contingencies.

I tested the hypothesis that Sven made his classification on
the basis of a distinguishing feature or features emanating from
the A targets and on the basis of irrelevant features or ambient
noise emanating from the B targets in the following way. First,

I trained the dolphin to differentiate between successively
presented small and large hollow steel cylinders (A targets) and
small and large hollow bronze cylinders (B targets). The dolphin
perfected and continued this baseline behavior during each test
session for 56 of the 64 trials, but during the other eight
trials, "compound probe" trials were presented. On these compound
probe trials, instead of presenting a single A or B target, both
the large and small A targets were presented side by side approximately 4–8 cm apart. If Sven was listening for a specific echo
feature from a single A target (large or small) in order to
report A, this feature, according to my hypothesis, should be
blocked or in some way masked or negated by the presence of both
the large and small A targets and Sven should classify such
reflected echoes as irrelevant noise and report on the B manipulandum. If, on the other hand, Sven was indeed distinguishing
the A target features from the B target features, then reflected
echoes from two A targets should be classified as A and reported
as such. During eight compound probe trials with both A targets,
Sven classified all such presentations as B (p < .01, binomial
test); thus confirming the hypothesis. A control procedure in
which compound probes consisted of large and small B targets
(hollow bronze) resulted in 7 of 8 B reports (p < .05, binomial
test).

These experiments clearly demonstrate the importance of
considering both the previous training and the current behavioral
techniques in use in interpreting results from so-called "discriminative" echolocation tasks. Apparently the way echolocating
dolphins perceived and classify echo complexes from specified
targets depends, not only on previous experience, but also on the
manner in which these targets are presented in relation to the
specific responses required by the experimenter, i.e., the response-
contingent reinforcement procedures. It might very well be the
case that bilateral responding to different echo complexes emanating
from the same location in space (as is the case in the two-
alternative forced-choice procedure with successive presentations)
has a different effect on the processing of echo information by
dolphins than does bilateral responding to different echo complexes
emanating from different locations.

Response Bias

When sonar targets are presented successively in order to
determine a threshold, e.g., the maximum range at which a given
sized object is detected by a dolphin on 50 percent of the trials
(see Murchison, this volume), response bias may affect the threshold estimate. The bias can be manifested either as a position
preference in a forced choice procedure or as a preference for
responding or not responding in a go/no-go procedure. A bias in

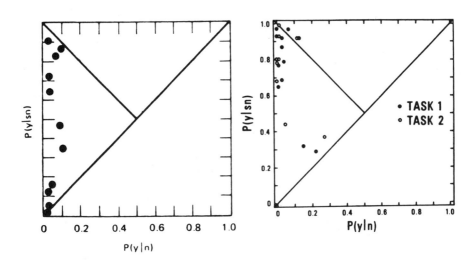

Figure 7. Hits or P(y/sn) plotted against false alarms or P(y/n)
for echolocating Amazon River porpoise in a size
detection task (left panel) and two echolocating
Atlantic bottlenose dolphins (T. truncatus) in a
maximum detection range task (right panel).

favor of a "no" response would result in an increased threshold
while a bias in favor of a "yes" response would result in a
decreased estimated threshold. The theory of signal detectability
is a relatively recent approach which permits one to evaluate
independently the contributions to discriminative behavior of an
animal's sensitivity and its response bias (Swets, 1973). Schusterman (1974) has pointed out that in some echolocation experiments
biases introduced by training conditions and reinforcement contingencies have led to underestimates of sensitivity.

Figure 7 presents conventional signal detection plots with
"hits" (y/sn) on the ordinate and "false alarms" (y/n) on the
abscissa. The data were originally plotted as ogives and involved
an Amazon River dolphin in a size detection task (Penner and
Murchison, 1970) and two Atlantic bottlenose dolphins performing
on two range detection tasks using different sonar targets (Murchison, this volume). The data points lie between the upper left-
hand corner (perfect detection) and the major diagonal (chance
detection). Points lying along the minor diagonal (the line

drawn from the upper left-hand corner to the major diagonal)
would represent "zero" bias since the two possible types of error
(false alarms and misses) would be equally probable. Both graphs
show that all three echolocating dolphins (regardless of the
task) have a strong and relatively consistent bias against making
false alarms. This is particularly true of the Amazon River
dolphin. The sensitivity index, called d', is a measure of
sensory effects uncontaminated by bias and represents the distance
of a data point from the major diagonal. The signal detection
arrangement of plotting hits in relation to false alarms is a
convenient way to display response bias and to estimate its
interaction with apparent sensitivity to stimulus events. A
careful inspection of the right panel of Figure 7 shows that the
smallest d' values also yielded the weakest bias against making
false alarms. This analysis suggests that for the two echo
ranging detection tasks by Tursiops there is an interaction
between response bias and apparent sensitivity to the strength of
the returning echo.

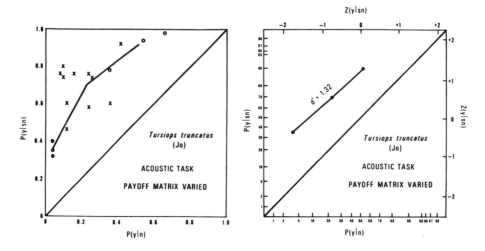

Figure 8.   ROC functions for an Atlantic bottlenose dolphin
             (T. truncatus). The animal attempted to detect
             an 8-kHz signal varying between -9 to -11 dB re 1 μbar
             with ambient noise varying by as much as 15 dB. Each
             data point in the left plot is based on 100 trials
             The x's represent ratios of number of fish rewards for
             hits; correct rejections of 1:1. The open circles
             represent ratios of 4:1 and the solid circles
             represent ratios of 1:4. The right plot shows the
             same data plotted on Z-score coordinates.

Several recent studies with marine mammals have shown that in underwater passive hearing tasks with acoustical signals having a relatively low signal/noise ratio, variation in such nonsensory variables as signal probability, amount of reinforcement and probability of reinforcement, resulted in highly predictable changes in response bias without appreciably altering their ability to detect the acoustic signal (Schusterman, 1976; Schusterman, 1974; Schusterman and Johnson, 1975; Schusterman, Barrett and Moore, 1975). If signals of similar strength are repeatedly presented while the amount of reinforcement for "yes" and "no" responses is varied, then a series of data points along a "receiver operating characteristic" (ROC) or "isosensitivity" curve are produced. Figure 8 shows ROC curves for a dolphin. The curves were generated by holding signal strength constant (the ambient noise levels were relatively unstable) and varying the number of fish for hits and correct rejections. According to the theory of signal detectability, when the averaged data points in the left panel of Figure 8 are transformed into standard deviation units or Z-scores, they should yield an ROC curve parallel to the major diagonal as shown in the right panel of Figure 8 (Green and Swets, 1966).

In order to obtain a threshold estimate uncontaminated by response bias, one could extrapolate a d' value from a family of ROC curves where the curves cross the minor diagonal, i.e., at the point of zero bias. Unfortunately, ROC curves have not, as yet, been generated from dolphin performance on echolocation tasks. It seems that a signal detection approach would certainly further the understanding of signal processing in echolocating marine mammals by disentangling the effects of bias (e.g. motivation and attention) from psychophysical data dealing with their sonar receiver, i.e., the peripheral and central mechanisms of their auditory system.

ACKNOWLEDGEMENTS

There is not enough space to thank all those people who made my sabbatical (1977-78) at the Naval Ocean Systems Center, Kailua, Hawaii, so successful from almost every standpoint. However, I am especially indebted to Whitlow Au, Bob Floyd, Jim Fish, Earl Murchison, Paul Nachtigall, and Ralph Penner for their stimulating ideas and helpful criticisms which ultimately led to the writing of this paper. My productivity would have been nil without the help of Debbie Kersting in Hawaii and Patrick Moore at the California State University, Hayward. I acknowledge the financial support of ONR Contract N00014-77-C-0185, and I am forever grateful to those two purposeful porpoises - Sven and Ekahi.

REFERENCES

Aronsohn, S, Castíllo, O., and Pinto-Hamuy, T., 1978, Fading pro-
    cedure effects on a visual pattern discrimination reversal
    in the albino rat, Anim. Learn. and Behav., 6:72.
Au, W. W. L., Floyd, R. W., and Haun, J. E., 1978, Propagation of
    Atlantic bottlenose dolphin echolocation signals, J. Acoust.
    Soc. Am., 64:411.
Au, W. W. L., and Hammer, C., 1979, Target recognition via echo-
    location by Tursiops truncatus, this volume.
Au, W. W. L., Schusterman, R. J., and Kersting, D. A., 1979,
    Sphere-cylinder discrimination via echolocation by Tursiops
    truncatus, this volume.
Ayrapet'yants, E. Sh., Golubkov, A. G., Yershova, I. V., Zhezherin,
    A. R., Zvorkin, V. N., and Korollev, V. I., 1969, Echo-
    location differentiation and characteristics of radiated
    pulses in dolphins, Report of the Academy of Science of the
    USSR, 188:1197 (English translation JPRS 49479).
Ayrapet'yants, E. Sh., Voronov, V. A., Ivanenko, Y. V., Ivanov,
    M. P., Ordovskii, D. L., Sergeev, B. F., and Chilingiris,
    V. I., 1973, The physiology of the sonar system of Black
    Sea dolphins, J. Evol. Biochem. Physiol., 9:364. (English
    translation JPRS 60298).
Bagdonas, A., Bel'kovich, V. M., and Krushinskaya, N. L., 1970,
    Interaction of analyzers in dolphins during discrimination
    of geometrical figures under water, J. Higher Neural Act.,
    20:1070 (English translation in: "A Collection of Trans-
    lations of Foreign Language Papers on the Subject of Bio-
    logical Sonar Systems", K. J. Diercks, 1974, ed., Applied
    Research Laboratories, U. of Texas, Austin, Tech. Rept.
    74-9).
Barta, R. E., 1969, Acoustical pattern discrimination by an At-
    lantic bottlenosed dolphin, unpublished manuscript, Naval
    Undersea Center, San Diego, Ca.
Bel'kovich, V. M., Borisov, V. I., Gurevich, V. S., 1970, Angular
    resolution by echolocation in Delphinus delphis, in: "Proc-
    eedings of Scientific Technical Conference, Ministry of
    Higher and Secondary Specialized Education RSFSR, Leningrad"
    (English translation in: "A Collection of Translations of
    Foreign Language Papers on the Subject of Biological Sonar
    Systems", K. J. Diercks, 1974, ed., Applied Research Lab-
    oratories, U. of Texas, Austin, Tech. Rept. 74-9).
Blough, D. and Blough, P., 1978, Animal psychophysics, in: "Hand-
    book of Operant Behavior", W. K. Honig and J. E. R. Staddon,
    eds., Prentice-Hall, Englewood Cliffs, N. J.
Busnel, R.-G., and Dziedzic, A., 1967, Résultats métrologiques
    expérimentaux de l'écholocation chez le Phocoena phocoena
    et leur comparaison avec ceux de certaines chauves-souris,
    in: "Animal Sonar Systems, Biology and Bionics", R.-G. Bus-
    nel, ed., Laboratoire de Physiologie Acoustique,

Jouy-en-Josas, France.

Busnel, R.-G., Dziedzic, A., and Andersen, S., 1965, Seuils de perception du système sonar du Marsouin Phocoena phocoena, en fonction du diamètre d'un obstacle filiforme, C. R. Acad. Sci., 260:295.

Dubrovskiy, N. A. and Krasnov, P. S., 1971, Discrimination of elastic spheres according to material and size by the bottle-nose dolphin, Trudy Akusticheskogo Institute, 17, as cited in: Bel'kovich, V. M., and Dubrovskiy, N. A., 1976, "Sensory Bases of Cetacean Orientation", Nauka, Leningrad (English translation JPRS L/7157.).

Dubrovskiy, N. A., Krasnov, P. S., and Titov, A. A., 1971, On the question of the emission of ultrasonic ranging signals by the common porpoise, Akusticheskiy Zhurnal, 16:521 (English translation JPRS 52291).

Dziedzic, A., and Alcuri, G., 1979, Variations in characteristics of the pulsed emissions of a T. truncatus during the approach process and acoustic identification of different polygonal shapes, this volume.

Evans, W. E., and Powell, B. A., 1967, Discrimination of different metallic plates by an echolocating delphinid, in: "Animal Sonar Systems, Biology and Bionics", R.-G. Busnel, ed., Laboratoire de Physiologie Acoustique, Jouy-en-Josas, France.

Falk, J. L., 1961, Production of polydipsia in normal rats by an intermittent food schedule, Science, 133:195.

Falk, J. L., 1969, Conditions producing psychogenic polydipsia in animals, Ann. N. Y. Acad. Sci., 157:569.

Falk, J. L., 1971, The nature and determinants of adjunctive behavior, Physiol. and Behavior, 6:577.

Galanter, E., 1962, Contemporary psychophysics, in: "New Directions in Psychology", Holt, Rinehard, and Winston, New York.

Green, D. M., and Swets, J. A., 1966, "Signal Detection Theory and Psychophysics", John Wiley and Sons, New York.

Griffin, D. R., and Galambos, R., 1941, The sensory basis of obstacle avoidance by flying bats, J. Exp. Zool., 86:481.

Gurevich, V. S., 1969, Echolocation discrimination of geometric figures in the dolphin, Delphinus delphis, Moscow, Vestnik Moskovskoga Universiteta, Biologiya, Pochovedeniye, 3:109. (Translation JPRS 49281).

Hammer, C. E., 1978, Echo-recognition in the porpoise (Tursiops truncatus): An experimental analysis of salient target characteristics, Naval Ocean Systems Center, San Diego, Tech. Rep. 192.

Johnson, C. S., 1967, Discussion to paper by Evans and Powell, in: "Animal Sonar Systems, Biology and Bionics", R.-G. Busnel, ed., Laboratoire de Physiologie Acoustique, Jouy-en-Josas, France.

Johnson, C. S., 1967, Sound detection thresholds in marine mammals, in: "Marine Bio-Acoustics, Proc. of the Second Symposium on Marine Bio-Acoustics, New York", W. N. Tavolga, ed.,

Pergamon Press, New York.

Kellogg, W. N., 1953, Ultrasonic hearing in the porpoise, J. Comp. Physiol. Psychol., 46:446.

Kellogg, W. N., 1960, Auditory scanning in the dolphin, Psych. Rec., 10:25.

Kellogg, W. N., and Kohler, R., Responses of the porpoise to ultrasonic frequencies, Science, 116:250.

Konstantinov, A. I., and Akhmarova, N. I., 1968, Object discrimination with the aid of echolocation by bats (Myotis oxygnathus), Nauchnye Doklady Vysshei Shkoly, Biol. Naak., 11:22.

Lekomtsev, V. M., and Titov, A. A., 1974, Procedures for studying the echolocation apparatus of dolphins, Bionika, 8:83, (Translation JPRS 63492).

Mackintosh, N. J., 1974, "The Psychology of Learning", Academic Press, New York.

Murchison, A. E., 1979, "Maximum Detection Range and Range Resolution in Echolocating Tursiops truncatus (Montague), Ph. D. Dissertation, U. of Calif. at Santa Cruz, Ca.

Nachtigall, P. E., Murchison, A. E., and Au W. W. L., Cylinder and cube shape discrimination by an echolocating blind-folded bottlenose dolphin, this volume.

Penner, R. H., 1977, Paired simultaneous echo ranging by Tursiops truncatus, Proc. (Abstracts) 2nd Conf. on the Bio. of Mar. Mammals, 2:38.

Penner, R. H., and Kadane, J., 1979, Tursiops biosonar detection in noise, this volume.

Penner, R. H., and Murshison, A. E., 1970, Experimentally demonstrated echolocation in the Amazon river porpoise Inia geoffrensis (Blainville), Proc. 7th Ann. Conf. Bio. Sonar and Diving Mammals, 7:17.

Schusterman, R. J., 1965, Errorless reversal learning in a California sea lion, Proc. A. Conf. Am. Psychol. Ass., 73(1): 139.

Schusterman, R. J., 1967, Perception and determinants of underwater vocalization in the California sea lion, in: "Animal Sonar Systems, Biology and Bionics", R.-G. Busnel, ed., Laboratoire de Physiologie Acoustique, Jouy-en-Josas, France.

Schusterman, R. J., 1974, Low false-alarm rates in signal detection by marine mammals, J. Acoust. Soc. Am., 55:845.

Schustermnan R. J., 1976, California sea lion auditory detection and variation of reinforcement schedules, J. Acoust Soc. Am., 59:997.

Schusterman, R. J., Barrett, B., and Moore, P. W. B., 1975, Detection of underwater signals by a California sea lion and a bottlenose porpoise: variation in the payoff matrix, J. Acoust. Soc. Am., 57:1526.

Schusterman, R. J., Johnson, B. W., 1975, Signal probability and response bias in California sea lions, Psych. Rec., 25:39.

Schusterman, R. J., and Kersting, D., 1978, Selective attention in discriminative echolocation by the porpoise (Tursiops truncatus), Paper read at the Animal Behavior Society Annual Meeting, June 19-23, U. of Wash., Seattle.

Schusterman, R. J., Kersting, D. A., and Au, W. W. L., 1979, Stimulus control of echolocation pulses in Tursiops truncatus, this volume.

Scronce, B. L., and Ridgway, S. H., 1979, Grey seal, Halichoerus grypus: echolocation not demonstrated, this volume.

Simmons, J. A., and Vernon, J. A., 1971, Echolocation: discrimination of targets by the bat, Eptesicus fuscus, J. Exp. Zool., 176:315.

Skinner, B. F., 1961, "Cumulative Record", Appleton-Century-Crofts, New York.

Staddon, J. E. R., 1978, Schedule-induced behavior, in: "Handbook of Operant Behaviour", W. K. Honig and J. E. R. Staddon, eds., Prentice-Hall, Englewood Cliffs, N. J.

Staddon, J. F. R., and Simmelhag, V. L., 1971, The "superstition" experiment: A re-examination of its implications for the principles of adaptive behavior, Psychol. Rev., 78:3.

Stebbins, W. C. (ed.), 1970, "Animal Psychophysics", Prentice-Hall, Englewood, N. J.

Supin, A. Ya., Sukhoruchenko, M. N., 1970, The determination of auditory threshold on Phocoena phocoena by the method of skin-galvanic reaction (Russian), Trudy Akousticheskogi Inst., 12:194.

Suthers, R. A., 1965, Acoustic orientation by fish-catching bats, J. Exp. Zool., 158:319.

Swets, J. A., 1973, The relative operating characteristic in psychology, Science, 182:990.

Titov, A. A., 1975, Recognition of spherical targets by the bottlenose dolphin in the presence of sonic interference, in: "Marine Mammals, Proceedings of the Sixth All-Union Conference on the Study of Marine Mammals", G. B. Agarkov, ed., Naukova Dumka, Kiev (English translation JPRS L/6049).

Turner, R. N., 1964, Methodological problems in the study of behavior, in: "Marine Bio-Acoustics", W. N. Tavolga, ed., Pergamon Press, Oxford.

Turner, R. N., and Norris, K. S., 1966, Discriminative echolocation in a porpoise, J. of the Exp. Anal. of Beh., 9:535.

DETECTION RANGE AND RANGE RESOLUTION OF ECHOLOCATING BOTTLENOSE PORPOISE (TURSIOPS TRUNCATUS)

A. Earl Murchison

Naval Ocean Systems Center

Kailua, Hawaii   96734

MAXIMUM DETECTION RANGE

Most of the quantifiable information available on object detection by echolocating odontocetes has resulted from studies conducted in tanks, and only a few of them have had maximum range of detection as their primary goal (Evans, 1973). The results of Zaslavskiy et al. (1969) as reported by Ayrapet'yants and Konstantinov (1974), indicated detection ranges (90% correct response) of 11 m for a metal cylinder (75 mm in height and diameter), 7.3 m for a plastic cylinder (115 mm high, diameter not given) and 6.8 m for a wooden cylinder (115 mm high, diameter not given). The subject of this experiment was a Tursiops truncatus. No other information was given.

Morozov et al. (1972), in their study of the pulse repetition rate of Tursiops truncatus as a function of food object distance, reported a maximum "starting position" of 30 m for approach and consumption of horse mackeral (species not given) of "about 20 cm in size." Although their assumption of detection by echolocation of the food fish for 100% of the only 33 runs at 30 m by the animal may have been justified, their vaguely defined target strength, and small number of trials makes their results only prefigurative as they relate to maximum echolocation range of Tursiops truncatus.

Titov (1972 – in Bel'kovich and Dubrovskiy, 1976) reported 50-mm spherical target detection by a bottlenose porpoise. The maximum detection ranges given were 5.9 m for rubber targets, 7.8 m for wax targets, 11.1 m for lead targets, and 12.4 m for steel targets.

Bel'kovich and Dubrovskiy (1976) also presented the bottlenose porpoise maximum detection range results reported by Babkin
and Dubrovskiy (1971). This experiment determined the maximum
detection ranges of several sizes (diameters from 4.0 mm, to 12.3
mm) of steel (presumably solid) spheres and lead spheres (diameters from 3.0 mm, to 12.7 mm) by two <u>Tursiops truncatus</u>. The
maximum detection range for the smallest steel sphere was 3.5 m,
and for the largest it was 7.5 m. The maximum detection range
for the smallest lead sphere was 3.0 m, and 9.0 m for the largest.
Bel'kovich and Dubrovskiy (1976) reported that the 75% correct
detection level of all of the steel spheres was near "5' [minutes
of arc] angular dimension." This refers to the expression:

$$\theta = \frac{d}{R}$$

where θ is the angular dimension (in radians), d = sphere diameter,
and R = maximum detection range. They reported 2.5' for lead
spheres.

The Babkin and Dubrovskiy (1971) study also reported (according to Bel'kovich and Dubrovskiy, 1976) a 9.8 m maximum detection
range by a bottlenose porpoise for a mackerel (species not given)
13.5 +1 cm long, 3.8 +0.5 cm wide and 1.8 +0.3 cm thick with the
swim bladder removed. They reported that the porpoise-mackerel
aspect relationship was very important with the maximum detection
range being attained when the fish was presented "lateral" to the
porpoise. For multiple fish targets, Bel'kovich and Dubrovskiy
applied an echo magnitude increase of $\sqrt{n}$ (n = number of fish) to
predict a 100 m maximum detection range for a school of 4000
mackerel, 19 cm long, by <u>Tursiops truncatus</u>. Assuming a live
mackerel could be detected from 12 m, and utilizing the same
school echo magnitude ($\sqrt{n}$) estimate, Ayrapet'yants and Konstantinov (1974) predicted a range of 350 meters for a porpoise's
detection of a 4000-fish school.

Titov (1972 - as reported in Bel'kovich and Dubrovskiy,
1976) conducted experiments to determine the effect of reverberation resulting from other objects near a target on maximum
detection range by a <u>Tursiops truncatus</u>. Smooth round rocks
between 5 and 30 mm in diameter were strewn on the bottom of the
test tank in circles 40 cm in diameter. The targets to be
detected were presented at the center of these rock circles, on
the tank bottom, amid the rocks, as well as various distances
above the rock circles. The 75% correct response detection range
for a lead sphere (50 mm) and a steel sphere (33 mm) was reported
to be "over 11 m" with the targets "30 cm beneath the surface"
(depth of tank not given), but the porpoise could detect both
targets "5 m away only if they extended at least 2.0 cm above the
largest stones." More complete results were not given. However,

the wording suggests that the porpoise could <u>not</u> detect the target
on the bottom, among the stones.

The only other report in the literature giving any information on maximum echolocation detection range of <u>Tursiops</u> is a
field observation. Eberhardt and Evans (1962) and Evans and
Dreher (1962) observed five Pacific bottlenose porpoises behaving
in a manner that strongly suggested their detection, from approximately 366 m, of 15 air-filled aluminum tubes, 5.1 cm in diameter
and 4.6 m long placed vertically 3.7 m deep at 15.2 m intervals.

Detection Range Experiments in Open Waters

Two male <u>Tursiops truncatus</u> (Sven and Niko) were the subjects
of three maximum detection range experiments conducted in a
floating porpoise pen (Steele, 1971) at the Naval Ocean Systems
Center's (NOSC) facility, at Kaneohe Bay, Hawaii. Two of these
psychophysical experiments were designed to determine the maximum
detection range of two different spherical targets, and the third
was designed to determine the effects of target depth (or nearness
to bottom) on maximum detection range. The rectangular, wire-
mesh, floating pen (Figure 1) had an aperture (2 m x 1.2 m)

Figure 1. Floating pen used in the maximum detection range
experiments with cutaway view through the geodesic
dome.

that was opened during echolocation trials to afford the echolocating porpoises an unobstructed "view" of the target range. The
targets were presented at various distances from the pen aperture
by means of a "skyhook" cable and pully arrangement between two
metal poles 158 m apart (Figure 2). The mean water depth along
the range was 6.3 m. Mean water visibility (Secchi disk) was 3.2
m. The bottom was covered with soft sandsilt up to 2 m deep along
the range. The reef slope at the end of the range consists of
coral rubble, live coral and benthic algae. The measured mean
sound velocity was 1542 m/sec and the transmission loss was 0.044
dB/m (Au et al., 1974).

Two detection versus distance ogives were determined utilizing
a 2.54-cm diameter solid steel sphere (target strength (TS) = −41.6
dB) and a 7.62-cm diameter water-filled stainless steel sphere
(TS = −28.3 dB). A third target, used to determine the effects of
depth or nearness to the bottom on range of detection, was a 6.35-
cm diameter solid steel sphere (TS = −34.8 dB). Target strengths
were determined by the method of Au and Hammer (1978).

Each animal was conditioned to hold position at a station
area near the side of the pen opposite the pen's aperture; each
station had two response manipulanda. Sixty-trial sessions (30
trials/animal) were conducted, and target-present and target-
absent trials were randomly presented. Trials were alternated
between animals during sessions. While one animal's trial was
being conducted, the other animal remained at its station. During
the first two experiments targets were presented at 1.22 ± 0.15 m
depth at all distances. One or two sessions were conducted per
day, preceded by warm-up trials. Details of the operant conditioning
techniques used are presented by Murchison (1979).

In the third experiment (6.35-cm solid steel sphere), the
same procedure was repeated. Sessions were conducted with this

Figure 2. Skyhook detection range (not to scale).

target suspended at depths of 1.22 ± 0.15 m, 2.43 ± 0.15 m, 3.66 ± 0.15 m, 4.88 ± 0.15 m, and on the bottom 6.3 ± 0.76 m. Separate detection distance ogives were determined for each of the 5 target depth increments.

The mean hit rates for each distance increment and target size, computed by the method of Woodworth and Schlosberg (1954), is presented in Figure 3 as a composite curve for both animals. Utilizing the same method, mean hit rates were plotted as a function of distance and target depth for the third experiment. The distances at which the 0.50 probabilities of detection occurred for each depth is shown in Figure 4, as a function of depth, distance and bottom contour.

Noise Limited or Reverberation Limited?   Although the echolocation systems of both bats and odontocetes have evolved to reduce the effects of noise (Grinnell, 1967; Evans, 1967; Bel'ko-

Figure 3.   Composite (both Tursiops) detection performance (hit rate) as a function of distance for the 2.54-cm and 7.62-cm targets.

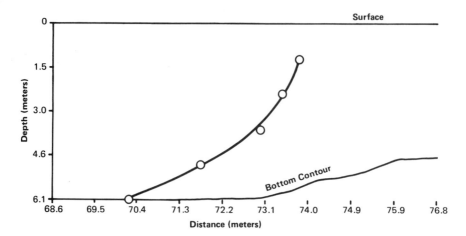

Figure 4. Detection thresholds (0.50 hit rate) for the 6.35-cm
           target for each depth increment tested as a function
           of distance and bottom contour.

vich and Dubrovskiy, 1976) on their detection and discrimination
of objects, they have not completely eliminated it.

    The two noise dimensions important to odontocetes are the
ambient noise and reverberation. The echo level from a sonar
signal decreases with target range but the ambient noise is
independent of target range. On the other hand, reverberation
level varies as a function of complex interactions of target
range, oceanographic conditions, and objects and boundaries
(surface, bottom, reefs) (Urick, 1975). The Tursiops used in
these experiments produced echolocation pulses with maximum
energy between 120 and 130 kHz. This is much higher in frequency
than previously reported for Tursiops and may have been an
adjustment by the animals that optimized the signal-to-noise
ratio in the ambient noise environment of Kaneohe Bay (Au et al.,
1974).

    Reverberation from the surface, bottom, or water volume can
mask, distort or otherwise interfere with target echoes. Additionally, echoes from objects other than the target may be distinctly
separable by an echolocating porpoise, yet affect target detection
by increasing the difficulty of recognition. That is, detection
becomes confounded by discrimination (see Schusterman, this
volume).

    The detection threshold (DT) of an active sonar occurs when
the returning echo level equals the masking level of the noise or
reverberation in the environment. For the noise-limited case,
the appropriate sonar equation is (Urick 1975):

$$SL - 2TL + TS = NL - DI + DT$$
$$\text{(Echo Level)} = \text{(Noise Masking Level)}$$

Assuming that during the skyhook experiments the porpoises maintained a constant DT and that the source levels (SL) of their pulses, their receiver directivity index (DI) and the mean ambient noise level (NL) were the same during the sessions with the 2.54 cm and 7.62 cm targets, the only parameters of the equation different for the two targets were transmission loss (TL) and target strength (TS), if the porpoises were noise limited. Transmission loss involves range—dependent losses due to spherical spreading of acoustic waves and to absorption (TL = 20 log R + αR; R = range, α = absorption coefficient) (Urick, 1975). Au et al. (1974) measured the TL from absorption for simulated porpoise pulses with peak energy at 120 kHz in the Kaneohe Bay test area as 0.044 dB/meter. Therefore, for the 2.54—cm target, two—way transmission loss was 80.7 dB when the target was at the 0.50 hit rate range. The target strength of the 7.62—cm sphere was 13.3 dB greater than that of the 2.54—cm sphere and, therefore, if the detection level is the same for both targets, the 7.62—cm target should have a 121 m detection range. That is 44.4 m farther than the mean detection range actually attained during the 7.62—cm target sessions (Figure 3). Because the noise—limited form of the sonar equation does not predict the porpoises' maximum detection range for the 7.62—cm target based on their performance with the 2.54—cm target, it is likely that the porpoises' detection of the larger target was limited by reverberation rather than ambient noise.

It is not surprising that the porpoises' biosonar systems were probably reverberation limited at their maximum, 7.62—cm sphere, detection range. Porpoise hearing is highly directional (Bullock et al., 1968; Ayrapet'yants et al., 1969; Ayrapet'yants and Konstantinov, 1974) and resistant to masking by noise from directions outside of their cone of best reception (Morozov et al., 1975). Their biosonar pulses are also very directional (Norris et al., 1961; Norris and Evans, 1967; Evans et al., 1964; Romanenko et al., 1965; Zaslavskiy, 1974 – according to Bel'kovich and Dubrovskiy, 1976; Au et al., 1978; Au, this volume). Thus, the greatest amount of potential masking noise will be created by backscatter from objects and surfaces that are in the animal's cone of inspection.

The effects of reverberation from the reef at the end of the skyhook range is further indicated by calculating the predict-ed detection range of the 6.35—cm target. Based on results from the 1.2 m (depth) sessions only, the detection range of the 2.54—cm target predicts a noise limited 0.50 hit rate for 6.45—cm target at 95 m. However, the two animals' mean detection range

was 21 m less than that (Figure 4). Thus, based on the animals'
detection range for the 2.54-cm target, the TL + TS model predicts
maximum detection ranges that are 22% greater for the 6.35-cm
target and 40% greater for the 7.62-cm target than were attained.
The degradation of detection ability with range suggests that the
closer the targets were to the reef, the greater were the possible
effects of reverberation. However, it is not possible to determine to what extent reverberation from the reef was affecting the
animals' maximum detection ranges for the 2.54-cm target or even
whether its detection range was reverberation limited at all. It
may be that reverberation limiting occurred only at greater
ranges, when the targets were closer to the reef.

Odontocetes may have "two functionally (and possibly to some
extent morphologically) distinct subsystems" of hearing, one
functional for the perception of low frequencies and the other
for the perception of higher frequencies (Vel'min and Dubrovskiy,
1975; 1976; Bel'lovich and Dubrovskiy, 1976). Vel'min and
Dubrovskiy propose that the lower frequency "passive" subsystem
is important for emotional and/or communicative sound perception,
and the other, "active hearing", subsystem is especially adapted
for echolocation. An important function of this "active hearing"
subsystem may be the reduction of the effects of reverberation
(Vel'min, 1975; Vel'min and Dubrovskiy, 1975; 1976; Bel'kovich
and Dubrovskiy, 1976). Giro and Dubrovskiy (1972 - according to
Vel'min, 1975) proposed an odontocete reverberation-limiting
mechanism of temporary gating or strobing of hearing sensitivity
whereby the auditory system is blocked during periods of potential
reverberation and operates with increased sensitivity during echo
reception. This mechanism implies a timing capability for the
interval between pulse emission and the "appearance of the
strobe" (Vel'min, 1975). Webster (1967) proposed similar predictive time gates for bats, describing them as "the anticipated
interval(s) of echo reception". Norris (Norris and Evans, 1967)
also mentioned the possibility of gating in a <u>Tursiops</u>.

Vel'min and Dubrovskiy (1975; 1976) reported the results of
several experiments and concluded there is an interval within
which odontocetes fuse "the indiviudal pulses comprising an echo
into the complete acoustic shape of an object." They conclude
that 300 microsec echo interpulse interval is the time that
determines the maximum size of an object perceived by porpoises
as an "integral echoranging object". These experiments are also
discussed by Vel'min and Titov (1975), Vel'min (1975) and Bel'kovich and Dubrovskiy (1976).

<u>Variable Target Depth Experiment.</u> The results of the
variable target depth experiment, as summarized in Figure 4, also
show the effects of reverberation. The closer the target was to

the bottom, the shorter the animals' maximum detection range was.
This strongly suggests that the reef and bottom were a source of
detection-limiting reverberation. It should be noted that
although the maximum range of target detection was reduced when
the target was on the bottom, the porpoises still detected it at
a mean distance of 70.2 m. This is considerably greater than
would be supposed from the results of Titov's (1972) bottom-
clutter experiment. However, in his experiment the bottom
composition was quite different from the skyhook range bottom.
Reverberation level in any location is influenced by complex
physical, environmental and acoustical factors and predicting it
for any particular area or time is very difficult (Urick, 1975).

There was considerable temporal variability in the detection
ranges of Sven and Niko. It was suspected that this was due, in
part, to factors that resulted in fluctuating reverberation
levels (e.g., fish, entrained gas bubbles, volume reverberation)
in any area, particularly shallow, inshore areas. Future psychophysical studies conducted in a field setting such as the one
described here, should attempt to quantify possible environmental
effects on echolocation and search for convarience between them
and the porpoises' detection and/or discrimination performance.

If these environmentally influenced differences and fluctuations in biosonar capabilities also occur in the porpoises'
natural habitats, the accuracy of the predictions made here (and
those made by Babkin and Dubrovskiy, 1971, and Ayrapet'yants and
Konstantinov, 1974) of odontocete detection ranges in those
habitats will vary considerably with time and place. Ecologically,
this means that the availability of prey species might not be the
only factor affecting the hunting success of porpoises and that
bioacoustic parameters are also important. Just as bats avoid
flying in fog (Pye, 1971; Sales and Pye, 1974), porpoises may
avoid areas during high bubble entrainment or that have other
conditions adverse to their biosonar capabilities.

With these reservations concerning differences and fluctuations in detection capabilities and assuming that the detection
range for the 2.54-cm target was not reverberation limited, rough
predictions of detection ranges of fish by Tursiops can be made
from these experiments. Using Love's (1971) formula for the side
aspect target strength of a 19-cm fish, at the pulse frequencies
produced by Sven and Niko, a detection range of 98 meters is
predicted for a single fish and 292 m for a school of 4000 fish.
This is almost three times the maximum detection range predicted
by Bel'kovich and Dubrovskiy (1976) for the same school and 17%
less than that predicted by Ayrapet'yants and Konstantinov
(1974), for a 25-30 kHz biosonar pulse. The net-detection capabilities of Sven and Niko predicted from their 2.54-cm target detection

performance and the net-target-strength measurements given by
Leatherwood et al. (1977), produce considerable variability in
the detection ranges predicted for different nets. The extreme
examples are the ranges of 130 m predicted for a 11.4-cm, twisted
net (TS = -26dB at 120 kHz) and 310 m for a 5.1-cm knotless net
joined in the middle to a 2-inch (5.1-cm) braided net (TS = 4 dB
at 120 kHz). Little wonder that Norris and Brocato had such a
difficult time catching Tursiops (Norris, 1974). Although it may
be safe to assume that Niko and Sven might detect such nets at at
least these ranges in quieter, less reverberant, open ocean
waters, the detection ranges of the species (Stenella attenuata,
S. longirostris, Delphinus delphis) of concern to Leatherwood et
al. may have different detection ranges. However, if one net
were predicted as more than twice as detectable as another for
the two Tursiops used in the present experiments, similar ratios
might be expected for other odontocetes.

Performance predictions based on the sonar equations should
be considered as only "best-guess time average(s)." (Urick,
1975). The detection ranges predicted above should be considered
as rough estimates of the potential detection ranges of the two
Tursiops under conditions similar to Kaneohe Bay, but possibly
with less bottom and reef reverberation. They do, in general,
support Ayrapet'yants' and Konstantinov's (1974) prediction and
statement that the delphinid biosonar system has detection ranges
not exceeding several hundred meters.

Bel'kovich and Dubrovskiy (1976) reported that the 75%
correct response detection level for Tursiops truncatus was 5' of
arc for steel spheres. This relationship predicts a 17.5 m
detection range for a 2.54-cm sphere such as used in the skyhook
experiments. Although the experimental design and methodology
underlying the relationship was not given by Bel'kovich and
Dubrovskiy, environmental (particularly reverberation) or other
test differences could account for my measured difference with
their prediction.

Target Range Inspection Behavior. Body movements and head-
tucking behaviors were observed during Sven and Niko's detection
trials and were probably related to their acoustic inspection of
the target range. These head and body movements may facilitate
internal scanning (Bel'kovich and Nestereno, 1972; Evans, 1973)
or may optimize echo reception (Norris, 1968). These behaviors
occurred more often when targets were at greater distances (and
closer to the reef), and may have helped reduce the effects of
reverberation, or aided detection at longer ranges.

The utility of head scanning movements observed in Tursiops
by a number of authors (Schevill and Lawrence, 1956; Kellogg,

1960; Norris, 1968, Norris et al., 1961) toward reducing the
effects of reverberation is indirectly indicated by an observation
by Evans (1973). He points out that Tursiops and Inia geoffrensis
(the other species reported to have head scanning movements like
Tursiops; Penner and Murchison, 1970) have much less fusion of
their cervical vertebrae than most pelagic delphinids. Both of
these species live in relatively shallow, inshore or riverine
environments where reverberation levels are high and multisourced.
Marine mammals' echolocation behaviors in open deep water may be
quite different than in "tidal flats and muddy bays" (Norris,
1968). It would be interesting to learn if the other odontocete
species that live in reverberant environments exhibit more distinctive posturings or elaborate scanning movements than pelagic
species. The experience of Norris (1968) with Delphinus delphis,
Lagenorhynchus obliquidens, and Tursiops truncatus gilli that were
unable to echolocate with sufficient efficiency to keep from
hitting tank walls or find objects when they were first placed in
reverberant tank environments, suggested that these pelagic
animals were ill equipped and/or had not learned to cope with high
reverberation levels. It is tempting to speculate that because of
their inshore ecology that Tursiops truncatus may have a biosonar
system that is among the better adapted relative to reverberation,
particularly bottom and volume reverberation. The role of learning
in scanning behaviors was illustrated by a gradual changes in the
scanning strategies used by Niko and Sven during the initial
training of their target detection and reporting behaviors.

 Pulse Repetition Rate. A number of studies have reported
odontocete echolocation pulse repetition rates that increase with
decreasing animal-to-target range (Busnel and Dziedzic, 1966;
Evans and Powell, 1967; Norris et al., 1967; Johnson, 1967;
Morozov et al., 1972). However, the pulse repetition rate does
not always increase in the systematic fashion that calculations of
animal-to-target sound travel time would predict (Norris, 1964;
Johnson, 1967). Thus, Norris (1964) suggested that repetition
rate may be, at least in part, related to the degree of discrimination desired by the porpoise at any particular time.

 In their study of pulse repetition rate as a function of
target distance with five Tursiops truncatus, Morozov et al.
(1972) found that the interpulse interval varied between 20
milliseconds and 3 milliseconds longer than the time necessary for
the pulse to travel to the target and the echo to return. They
expressed their results by the formula:

$$T_o = (2L/c) + 3\text{-}20 \text{ milliseconds}$$

(T_o = interpulse interval; L = the distance from the target to the
animal; c = speed of sound in seawater). Au et al. (1974),

recording animals echolocating at greater ranges than the Morozov
et al. animals, also found that the interpulse interval was
always just greater than the calculated two-way transit time (30-
50 milliseconds). Morozov et al. concluded that the time difference between echo reception and production of the following pulse
represented processing time, and "that a dolphin during echo
ranging emits each successive pulse only after (italics theirs)
reception of the echo signal from the preceeding pulse." Diercks
(1972) in a discussion of the Morozov et al. conclusions suggests
that pulse generation by porpoises may be triggered by feedback
from the animal's receiver system; however, he qualifies this
suggestion with the observation that test animals seem to have
some control of their repetition rate.

If, as Morozov et al. (1972) suggests, odontocete pulses are
produced only upon the reception of the echo from the previous
pulses, the interpulse interval would always be reflective of the
nearest object in the porpoise projecting-receiving field, not
necessarily the salient target object being tested. Their model
could be modified to include some sort of "neural template" (Simmons,
1969), or "spectral copy" (Mermoz, 1967), thus allowing optimal
reception of echoes from a particular target object while ignoring
echoes from other objects. Even so modified, the target-echo
pulse triggering model of Morozov et al. does not account for the
interpulse intervals recorded from the Tursiops during the maximum
detection range experiments reported here. During the skyhook
experiments, the animals continued to produce pulses during trials
when no target was present in the test range and interpulse intervals during target-present trials and target-absent trials were
essentially the same (Au, this volume, Figure 9). The important
meaning of those target-absent trials' repetition rates and intertrial intervals is that the animals maintained intertrial intervals that allowed sufficient time for echoes to return from an
expected target, at an expected approximate range. If an echo was
not received 30-50 milliseconds beyond that interval, another
pulse was produced without waiting for an echo. The animals'
target-absent range expectancies in those experiments were set
(learned) by the perceived target range during target-present
trials of the same and recent detection sessions.

The important point here is that the interpulse interval for
an echolocating porpoise is not automatically controlled by the
distance of a target, but it is a measurable function of range of
attention as indicated by measurable interpulse intervals. The
relationships between echolocation signal parameters and target
dimensions may offer an ideal sensory vehicle with which to study
selective attention and expectancy in echolocating animals. The
Schusterman et al. (this volume) study illustrates that potential.

In a discussion of the "search image" concept of von Uexküll

(1934) and L. Tinbergen (1960), Hinde (1970) points out that such
a selective attention concept implies the existence of a mechanism
in the absence of an activating stimulus, yet the evidence for
such mechanisms has all been based on results obtained during the
presence of activating stimuli. The interpulse intervals maintained by Sven and Niko in the maximum range study, (Figure 9, Au,
this volume) during target-stimulus-absent trials, might be
considered to support the "search image" concept. The porpoises
had an expected or "searched-for" echo "image," at least along the
pulse-to-echo delay-time dimension they possibly perceived as
target-stimulus distance. This all assumes that interpulse intervals
are indeed a function of the distance of the target or area of
attention.

    The adoption of the word "target" into biosonar terminology
may have brought a subtle bit of inappropriate conceptual simplicity with it because it implies that the experimenter knows the
one stimulus, among all the stimuli available, that the echolocating animal is attending to. This implies that the stimulus
defined as target by the experimenter is, at all times during the
experiment, defined as target by the echolocating animal. Analysis of repetition rate/target-to-animal distance relationships,
might be affected by these assumptions and implications.   As
Schusterman (1968) has pointed out, echolocation cannot be understood as a phenomenon isolated from the developmental processes,
learning and memory characteristics, attention mechanisms, and
other sensory modalities of the echolocating animal. The only
thing that defines a biosonar target as such is the selective
attention of the echolocating animal.

    If the interpulse intervals of odontocete echolocation
pulses are behaviorally controlled by the animal and indicative of
its area of attention or search (distance and/or "depth of field")
the variability reported for Tursiops truncatus pulse repetition
rates by Johnson (1967) may simply reflect momentary attention
shifts by that animal as it swam toward the target. If viewed in
this light, Johnson's statement that pulse repetition rate is not
used for range determination may need re-examination.

    The variable relationships between pulse repetition rates and
target distances found by Johnson may have been the results of
momentary attention shifts as reflected in interpulse intervals.
This variability would have been confounded if there were variability in the time lags between echo reception and production of the
following pulse as reported by Morozov et al. (1972) and Au et
al. (1974 this volume).

    Experiments designed to compare the pulse repetition rate of
relatively static porpoises in a search mode (i.e., the animals
know the target, its direction and depth, but not its distance)

with those of porpoises detecting targets at known distances might
clarify the relationship of repetition rate and detected object
distance.

RANGE RESOLUTION

Although Ayrapet'yants and Konstantinov (1974) present the
results of a number of Soviet experiments designed to determine
porpoises' abilities to discriminate target differences in azimuth
(horizontal separation), elevation (vertical separation), and
range, there is insufficient information available in the English
translation of their work to make a meaningful evaluation of these
results as they relate to the range resolution experiment reported
here. They report range resolution for spheres of various sizes
and materials that were presumably 5 to 7 cm apart, but the absolute
porpoise-target distances were not given. Range resolution results,
without absolute target range, are of little value and the discrimination of target separation in studies with targets so close
together could be based on differences in echo spectra of the
multiple target complex rather than the determination of range
differences between individually perceived targets.

There are two sets of competing hypotheses of echo-information processing by bats in the determination of target distance.
The hypothesis that bats process their FM signal echoes in a
manner that allows a <u>direct</u> measure of echo travel time (Hartridge, 1920), led Strother (1961), McCue (1966), and others to
postulate a receiving system incorporating a neural equivalent of
a matched filter or template of the outgoing signal. The competing models of animal echolocation all propose some <u>indirect</u>
determination of distance by the perception of neural or acoustic
interactions between outgoing signals and echoes. Trains of brief
acoustic pulses produce the sensation of an <u>apparent</u> tone. The
frequency of this tone is determined by the temporal distance
between the pulses of the train (Thurlow and Small, 1955). Most
of the indirect biosonar models of distance determination are
based on some variation of this time-separation-pitch (TSP)
phenomenon, or frequency-difference beat tones. In general, these
models propose that the echolocating animal determines target
distance by perceiving these time-separation-pitch or frequency
beat tones resulting from the interaction of the echoes with the
outgoing signal, or its replica (Nordmark, 1961; Pye, 1961; Kay,
1962; Stewart, 1968; Johnson and Titlebaum, 1976). The Johnson
and Titlebaum model emphasizes spectrum analysis of the echolocation
pulse plus its echo as the proposed processing mechanism and is
related to other TSP models by their conclusion that TSP can be
explained by frequency domain, correlation-equivalent, spectrum
analysis; however, their model does not necessitate the assumption
that the animal perceive a pitch. Simmons (1969) proposed that

the autocorrelation function of a bat's echolocation signal approximates the crosscorrelation function of emitted and received signals
and can be used to predict the animal's abilities to discriminate
distances between objects (i.e., range resolution). The results
of his experiments with Eptesicus fuscus supported his matched-
filter model, and later experiments (Simmons, 1971a, 1971b, 1973)
included other species and distances.

Caine (1976) reported the results of pulsepair interpulse
interval discrimination by a Tursiops truncatus. She interpreted
her results according to a TSP processing model of distance
determination. The interpulse-interval difference thresholds she
reported were 100 microsec (standard = 0.65 millisec), 170 microsec (standard = 1.35 millisec), and 130 microsec (standard = 4.0
millisec). These temporal difference thresholds predict echolocation range resolutions of 7 cm (absolute range = 0.49 m), 13 cm
(absolute range = 1.01 m), and 10 cm (absolute range = 3.00 m).

Range Resolution Experiments

A Tursiops truncatus (Heptuna) was the subject of three
experiments designed to determine the range resolution capabilities
of an echolocating porpoise at three different distances.

The experiments were conducted in the same area of Kaneohe
Bay in a pen (Figure 5) similar to the one used for the maximum
range experiments. Two identical spherical, polyurethane foam
targets, 7.62-cm in diameter (TS = -33.8 dB), were used as

Figure 5. Floating pen used in the range resolution experiments.

discriminative stimuli. Target stability was enhanced by suspending each of them on 4 monofilament lines 40 cm below the water (Figure 6). The targets were attached to aluminum travelers, and the relative target distances could be adjusted by line and pulley arrangements from the operator's position in an instrument shack (Figure 5).

The porpoise was conditioned to place the anterior part of his lower jaw snugly into a molded fiberglass "chin-cup" (Figure 6). An acoustically opaque panel obscured the targets until their relative distances were adjusted. It was then dropped, presenting the target display to the porpoise. The fiberglass chin-cup was mounted 50 cm beneath the surface in a plexiglass swivel and could rotate horizontally to allow the porpoise to remain positioned in the chin-cup while scanning back and forth between the two target spheres. The targets were always 40° from one another relative to the center of the chin-cup. The porpoise was conditioned to indicate his choice of which of the two targets was the closer by striking one of two manipulanda located at the right and left of the chin-cup position. His choice was indicated by striking the manipulandum on the same side as the closer (right or left) of the two spheres. Details of the operant conditioning techniques used are given by Murchison (1979).

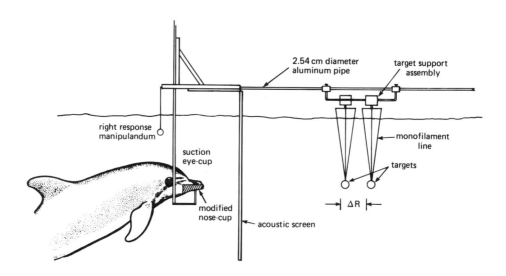

Figure 6.   Diagramatic view of the porpoise's position during the range resolution trials. The acoustic screen was removed (dropped) and targets presented to the porpoise after ΔR was adjusted.

Sessions were conducted in blocks of 10 trials (limited random right-left, 5 trials each) utilizing a modified, descending method of limits. One-meter absolute range means that the target closest to the porpoise during any particular trial (and therefore the "correct" response on that trial) was always one meter from the tip of the chin-cup. The same is true for the 3-meter and 7-meter sessions.

Heptuna's target range discrimination performance results are summarized by Figures 7, 8 and 9. The three absolute range experiments were plotted separately, and the 75% correct response level ΔR's (range resolution) estimated from them (Woodworth and Schlosberg, 1954). The theoretical curve in Figure 7 was calculated from the envelope of the autocorrelation function from one of Heptuna's echolocation pulses in the manner proposed by Simmons (1969, 1971a, 1971b, 1973); the pulse and its autocorrelation function is shown by Figure 10.

The 75% correct ΔR's were plotted in Figure 9 with the absolute ranges and ΔR's converted to two-way pulse-echo travel

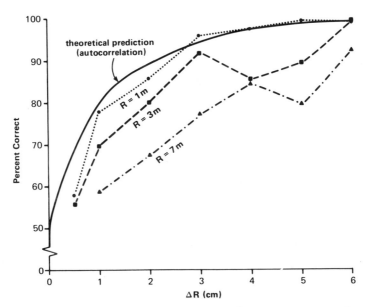

Figure 7.   Range resolution performance (percent correct) as a
            function of ΔR (range differences) for absolute
            ranges (R) of 1, 3 and 7 meters. The theoretical
            prediction curve was taken from the envelope of the
            autocorrelation function of the echolocation
            pulse shown in Figure 10 in the manner given by
            Simmons (1969).

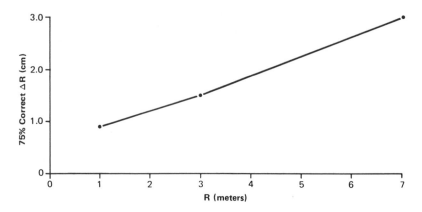

Figure 8.   Range difference (ΔR) thresholds (75 percent correct)
as a function of absolute ranges (R) of 1, 3 and 7
meters.

time. This figure also shows the results of R. Johnson's (this
volume) study of time-separation-pitch (TSP) discrimination in
humans. The abscissa is the standard pulse-pair time separations
(T) and the ordinate is discrimination performance (75% correct
level) of subjects, converted to the pulse-pair time-separation
differences (ΔT). Figure 9 also includes the reported results of
the range resolution experiments of Simmons (1969, 1971b, 1973)
with bats (<u>Eptesicus fuscus</u>). Simmons' range resolution results
were also converted to two-way pulse-echo travel time using a two-
way travel time of 5.8 microsec/1 mm of range (Simmons, 1973).
The dashed line on Figure 9 is Heptuna's performance predicted by
Simmons' autocorrelation technique.

The most accurate prediction of porpoise range resolution was
made by Kenshalo (1967) who predicted that the "resulting function
would resemble the standard Weber-Fechner function of ΔD/D=K or
conform to the Stevens Power Law." Figures 8 and 9 confirm his
prediction.

The comparisons of the porpoises' range resolution performance
with that predicted by Simmons' model (1969, 1971a, 1971b, 1973),
(Figures 8 and 9) argue strongly against the same type of signal
processing by porpoises and bats. The range resolution performance
of <u>Eptesicus fuscus</u> reported by Simmons (1969, 1971b, 1973) and
shown in Figure 9, seems puzzling in light of Kenshalo's prediction
and the results of the porpoise experiments reported here.
However, porpoises and bats are very different creatures. About
the only two things they have in common are mammalian ancestry and
the ability to echolocate. As Pye (1967) pointed out, it may

Figure 9.   Two-way pulse travel time differences (ΔT) thresholds
            (75 percent correct) as a function of two-way pulse
            travel time to the nearest target (T) at 1, 3 and 7
            meters.   ΔT is in μsec and T is in millisec.   The
            Tursiops curve is from the present study, Eptesicus
            curve is from data reported by Simmons (1969, 1971b
            and 1973), and human curve is from R. Johnson's
            (this volume) study of time-separation-pitch
            discrimination.

not be valid to assume that there is only one signal processing
mechanism used by all species of echolocating animals.   On the
other hand, Johnson and Titlebaum (1976) argue that it would be
"convenient" to have one model for all echolocators.   The function
generated by human TSP experiments and the one generated by
Heptuna's range resolution performance (Figure 9) show greater
similarity than any other proposed predictive models or evidence.
Although Simmons et al. (1976) and Johnson and Titlebaum (1976)
each caution that their particular model may not uniquely explain
the results of range resolution experiments like the one reported
here, comparison of their results indicates that neither of their
models predicts or explains the other's results (Figure 9).

     The ideal detector (matched filter) model of echolocation
range determination proposed by Simmons (1969, 1971a, 1971b, 1973)
relies heavily on the envelope of the calculated autocorrelation
function of the produced pulse to predict the range resolution
capabilities of an echolocator.   However, as Diercks (1972) has
emphasized, no proposed sonar signal processing technique can
yield a meaningful analog in the absence of considerations of the
biological, ear-brain processor.   The matched filter, model
proposed by Simmons (1969, 1971a, 1971b, 1973) and others, seems
to have left out an important consideration in this regard.   The
abcissas of calculated, predictive, autocorrelations are based on the
physical continua, i.e. measured time intervals.   However, the
output of any type of time-domain processing by echolocating

animals is going to be based on a psychological continua, i.e.
sensed time intervals. (See Stevens, 1975 for a discussion of the
non-linearity of the relationship between the two). As Green and
Swets (1966) have pointed out, the unrealistic assumption of ideal
detector models is that they "measure with infinite accuracy with
no computational errors."

If Heptuna's sensation of duration is assumed to be nonlinear; that is, if he does not have a perfect internal clock his
judgments of relative pulse-to-echo delay times would be different
at different absolute ranges. If the differences between his
sensation of duration and the calculated abcissa of the autocorrelation function predicted by Simmons model are considered, the
difference in resolution at the different absolute ranges might
still be accounted for by a time (perceived) domain correlation
similar to that proposed by Simmon's model. That is, Heptuna
could have been utilizing pulse-to-echo correlation-equivalent
processing with an imperfect timing mechanisms, and the ΔT as a
function of T plot could be a measure of that imperfection.

These may have been the kinds of considerations that prompted
Kenshalo (1967) to predict Weber-Fechner or Stevens Power Law

Figure 10. Echolocation pulse produced by Heptuna during range
            resolution experiment and its Fourier transform
            and autocorrelation function. (Pulse recorded and
            Fourier transform and autocorrelation function
            calculated by Whitlow W. L. Au, Naval Ocean
            Systems Center, Hawaii Laboratory).

results from such a range resolution experiment as the one reported
here. This still leaves the puzzle of Simmons' (1969, 1971a,
1971b, 1973) results, as shown in Figure 9. Do bats have perfect
clocks?

If they don't, perhaps the pulse-to-echo delay times that the
bats in Simmons' experiments were discriminating were in a region
of optimal and/or constant time judgment. Just as the frequency
difference thresholds (Δf) plots of many vertebrates show frequency
regions of relatively constant difference thresholds (Fay, 1974),
there may be regions of relatively constant duration (pulse-to-
echo) difference thresholds (ΔT). It could be that the plot of E.
fuscus' time resolution, shown in Figure 9, represents such a
region. If this were the case, and the porpoise has such a region,
it might be at shorter (or longer) pulse-to-echo delay times than
represented by the ranges tested.

The only similarity between Caine's (1976) predicted Tursiops
truncatus range resolution as a function of absolute distance and
the actual performance of Heptuna was that she predicted an
increase in ΔR between the first two absolute ranges (0.49m to
1.01m). However, the similarity ended there. All of the predicted
ΔR's were much larger (less acuity) than Heptuna's actual performance.
Caine predicted range resolutions of 7 cm at R = 0.49m, 13 cm at
R = 1.35m, and 10 cm at R = 3m. As she points out, the reverbera-
tions of her experimental tank may have affected the porpoise's
resolution of the pulse time separations presented to him.

Vel'min and Titov (1975) concluded in their report of their
pulse-pair interval difference threshold experiment that Tursiops
truncatus differential threshold "was approximately 10%." This
predicts T. truncatus' range resolution of 10 cm at R = 1m, 30 cm
at R = 3m, and 70 cm at R = 7m. As Figure 8 shows, Heptuna's
range resolution was more acute than that. However, Vel'min and
Titov worked with pulse-pair intervals much smaller than the
pulse-to-echo delay times that Heptuna was discriminating. They
stated that they could not train their porpoise to discriminate
pulse pair intervals longer than 500 microsec. It was for this
reason that they concluded that their results supported the
critical-interval hearing hypothesis of Vel'min and Dubrovskiy
(1975, 1976).

The fact that Heptuna was obviously able to discriminate
pulse-to-echo delay intervals longer than 500 microsec does not
necessarily invalidate Vel'min's and Dubrovskiy's critical interval
hypothesis. It could indicate that the Tursiops in the Soviet
experiments had learned to discriminate pulse-pair intervals
with delay times less than 500 microsec (Vel'min and Titov, 1975)
or 300 microsec (Vel'min and Dubrovskiy, 1976) utilizing signal
processing that was optimal at those delay times (but not longer

delay times) and, therefore, was unable to discriminate longer
pulse delay-times because they necessitated a different signal
analysis that the animal may have been capable of but had not
<u>learned</u> (or been trained) to utilize to accomplish the discrimina-
tion task. This is consistent with that part of the critical-
interval hypothesis that proposes different signal processing of
pulse-to-echo delay times less than 300 microsec than for intervals
greater than that. The ability of Heptuna to obviously discriminate
pulse-to-echo delay times considerably greater than 300 microsec,
and the reported inability of the Soviet porpoises to discriminate
pulse intervals greater than 300-500 microsec might also indicate
that the open hearing-sensitivity window ("temporarily gating")
hypothesis of Giro and Dubrovskiy (1972), proposed as one method
that odontocetes use to combat reverberation, is variable in
duration according to the acoustic environment, especially reverberation. That is, the 300-microsec interval that Vel'min and
Dubrovskiy (1975, 1976) proposed might have been the duration of
the strobed sensitivity interval their animals <u>learned</u> in order to
optimize their performance in the reverberant tank environment.
The duration of the proposed range-gating, hearing sensitivity
window may also vary as a function of target distance. In the
Vel'min and Dubrovskiy (1975) 40mm sphere detection experiment
with variable artificial noise-pulse delay times, the animals-to-
target distance was the same for all noise-to-pulse delay times;
therefore, this possibility was not eliminated.

SPECULATIONS

     Since echolocation is a part of the behavior of the animals
that have that ability, perhaps behavioral as well as mammalian
hearing and sonar engineering models should be included for
consideration. An open acoustic sensitivity window, as proposed
by Giro and Dubrovskiy, (1972) could be opened for any expected
target distance. Echolocation range of attention, regulated by
interpulse intervals, and a time gating of echo sensitivity
regulated by expectancy (or search depth), would reduce reverberation resulting from echoes from objects or surfaces behind or in
front of the target range of interest.

     The critical interval hypothesis (Vel'min and Dubrovskiy,
1975) can be combined with the strobed-hearing hypothesis (Giro
and Dubrovskiy, 1972) in the following way. Heptuna's range
resolution performance showed that a porpoise can discriminate
pulse-to-echo delay times greater than 300 microsec. Whatever
mechanism he utilized to do that could be considered the timing
mechanism necessary for the Giro and Dubrovskiy (1972) strobed
sensitivity window. If it is assumed that the delay time, from
pulse production to the opening of the hearing sensitivity window,
is a function of the animal's attention or expectancy, it could be

opened at any time after the production of the pulse and the proposed critical-interval processing could then apply to all echoes received during the time that the sensitivity window is open. The Vel'min and Dubrovskiy (1975) conclusion that pulses closer together than 300 microsec result in an "integral echo-ranging object" is complementary to the spectrum analysis models of construction, size, and materials proposed by Dubrovskiy et al. (1971) and Au and Hammer (1978), because the secondary echoes that resulted from the material and structure of the targets in those experiments were all less than 300 microsec apart and would be, therefore, coherently added to the processed spectrum.

A visual analogy emerges from this speculation and that on the relationship between interpulse intervals and target range. Funcationally, interpulse interval control may be roughly analogous to visual focus and the strobed sensitivity window may function somewhat like visual depth of field with all echoes received during the period of open sensitivity (possibly adjustable) that are closer in time than the 300 microsec critical interval perceived as an integrated, coherent "image".

Regardless of whether this particular analogy is meaningful or not, it does seem that echolocating Tursiops do have the ability to attent ("focus") to one particular area or object at a chosen distance.

REFERENCES

Au, W. W. L, 1979, Echolocation signals of the Atlantic bottlenose dolphin Tursiops truncatus in open waters, this volume.

Au, W. W. L., Floyd, R. W., and Haun, J. E., 1978, Propagation of Atlantic bottlenose dolphin echolocation signals, J. Acoust. Soc. Am., 64:411.

Au, W. W. L., and Hammer, C., 1978, Analysis of target recognition via echolocation by an Atlantic bottlenose porpoise (Tursiops truncatus), J. Acoust. Soc. Am., 60:587.

Au, W. W. L., Floyd, R. W., Penner, R. H., and Murchison, A. E., 1974, Measurement of echolocation signals of the Atlantic bottlenose dolphin, Tursiops truncatus (Montagu), in open waters, J. Acoust. Soc. Am., 56:1280.

Aryapet'yants, E. Sh., Golubkov, A. G., Yershova, I. V., Zhezherin, A. R., Zvorkin, V. N., and Korolev, V. I., 1969, Echolocation differentiation and characteristics of radiated pulses in dolphins, Report of the Academy of Science of the USSR, 188:1197 (English translation JPRS 49479).

Ayrapet'yants, E. Sh., and Konstantinov, 1974, "Echolocation in Nature", Nauka, Leningrad (English translation JPRS 63328).

Babkin, V. P., and Dubrovskiy, N. A., 1971, Range of action and
     noise stability of the echolocation system of the bottle-
     nose dolphin in detection of various targets, Tr. Akust.
     Inst., Moscow, 17:29, as cited in: "Sensory Bases of Ceta-
     cean Orientation", Nauka, Leningrad (English translation
     JPRS L/7157).
Bel'kovich, V. M., and Dubrovskiy, N. A., 1976, "Sensory Bases of
     Cetacean Orientation", Nauka, Leningrad (English translation
     JPRS L/7157).
Bel'kovich, V. M., and Nesterenko, I., 1971, How the dolphin's
     locator operates, Priroda, 7:71 (German translation by W.
     Petri, 1974, Naturwissenschaftliche Rundschau, 0004:0143).
Bullock, T. H., Grinnell, A. D., Ikezono, E., Kameda, K., Katsuki,
     Y., Nomoto, M., Sato, O., Suga, N., Yanagisawa, K., 1968,
     Electrophysiological studies of central auditory mechanisms
     in cetaceans, Zeitschrift für vergleichende Physiol., 59:117.
Busnel, R.-G., Dziedzic, A., 1966, Acoustic signals of the pilot
     Whale Globicephala melaena and the porpoises Delphinus
     delphis and Phocoena phocoena, in: "Whales, Dolphins, and·
     Porpoises", K. S. Norris, ed., U. of California Press,
     Berkeley.
Caine, N. G., 1976, "Time Separation Pitch and the Dolphin's Sonar
     Discrimination of Distance", Master's Dissertation, San
     Diego State University.
Diercks, K. J., 1972, Biological Sonar Systems: A Bionics Survey,
     Applied Research Laboratories, ARL-TR-72-34, Austin, Texas.
Dubrovskiy, N. A., Krasnov, P. S., and Titov, A. A., 1971, Dis-
     crimination of solid elastic spheres by an echolocating
     porpoise, Tursiops truncatus, in: "Proc. 7th International
     Acoust. Conf.", Budapest.
Eberhardt, R. L., and Evans, W. E., 1962, Sound activity of the
     California gray whale, Eschrichtius glaucus, J. Aud. Eng.
     Soc., 10:324.
Evans, W. E., 1967, Discussion, in: "Animal Sonar Systems,
     Biology and Bionics", R.-G.Busnel, ed., Laboratoire de
     Physiologie Acoustique, Jouy-en-Josas, France.
Evans, W. E., 1973, Echolocation by marine delphinids and one
     species of fresh-water dolphin, J. Acoust. Soc. Am., 54:191.
Evans, W. E., and Dreher, J. J., 1962, Observations on scouting
     behavior and associated sound production by the Pacific
     bottlenosed porpoise (Tursiops gilli Dall), Bull. Soc.
     Calif. Acad. Sci., 61:217.
Evans, W. E., and Powell, B. A., 1967, Discrimination.of different
     metallic plates by an echolocating delphinid, in: "Animal
     Sonar Systems, Biology and Bionics", R.-G. Busnel, ed.,
     Laboratoire de Physiologie Acoustique, Jouy-en-Josas, France.
Evans, W. E., Sutherland, W. W., and Beil, R. G., 1964, The dir-
     ectional characteristics of delphinid sounds, in: "Marine
     Bio-Acoustics, Proceedings of the First Symposium on Marine
     Bio-Acoustics, Bimini", W. N. Tavolga, ed., Pergamon Press,

New York.

Fay, R. R., 1974, Auditory frequency discrimination in vertebrates,
    J. Acoust. Soc. Am., 56:206.

Giro, L. R., and Dubrovskiy, N. A., 1972, Correlation between
    recurrence frequency of delphinid echolocation signals and
    difficulty of echoranging problem, in: Marine Instrument
    Building Series, 2:84, as cited in: "Marine Mammals, Proc-
    eedings of the Sixth All-Union Conference on the Study of
    Marine Mammals," G. B. Agarkov, ed., Nauka Dumka, Kiev
    (English translation JPRS L/6049).

Green, D. M., and Swets, J. A., 1966, "Signal Detection Theory
    and Psychophysics", John Wiley and Sons, New York.

Grinnell, A. D., 1967, Mechanisms of overcoming interference in
    echolocating animals, in: "Animal Sonar Systems, Biology
    and Bionics", R.-G. Busnel, ed., Laboratoire de Physiologie
    Acoustique, Jouy-en-Josas, France.

Hartridge, H., 1920, The avoidance of objects by bats in flight,
    J. Physiol., 54:54.

Hinde, R. A., 1970, "Animal Behavior, A Synthesis of Ethology and
    Comparative Psychology", McGraw-Hill, Palo Alto.

Johnson, C. S., 1967, Discussion to paper by Evans and Powell,
    in: "Animal Sonar Systems, Biology and Bionics", R.-G. Bus-
    nel, ed., Laboratoire de Physiologie Acoustique, Jouy-en-
    Josas, France.

Johnson, C. S., 1967, Sound detection thresholds in marine mammals,
    in: "Marine Bio-Acoustics, Proc. of the Second Symposium
    on Marine Bio-Acoustics, New York", W. N. Tavolga, ed.,
    Pergamon Press, New York.

Johnson, R. A., 1979, Energy spectrum analysis in echolocation,
    this volume.

Johnson, R. A., and Titlebaum, E. L., 1976, Energy spectrum anal-
    ysis: a model of echolocation processing, J. Acoust. Soc.
    Am., 60:484.

Kay, L., 1962, A plausible explanation of the bat's echolocation
    acuity, Animal Behaviour, 10:34.

Kellogg, W. N., 1960, Auditory scanning in the dolphin, Psych.
    Rec., 10:25.

Kenshalo, D. R., 1967, Discussion, in: "Animal Sonar Systems,
    Biology and Bionics", R.-G. Busnel, ed., Laboratoire de
    Physiologie Acoustique, Jouy-en-Josas, France.

Leatherwood, J. S., Johnson, R. A., Ljungblad, D. K., and Evans,
    W. E., 1977, "Broadband measurements of underwater acoustic
    target strengths of panels of tuna nets", NOSC TR 126, San
    Diego.

Love, R. H., 1971, Measurements of fish target strength: a re-
    view, Fish. Bull., 69:703.

McCue, J. J. G., 1966, Aural pulse compression by bats and humans,
    J. Acoust. Soc. Amer., 40:545.

Mermoz, H., 1967, Discussion, in: "Animal Sonar Systems, Biology
    and Bionics", R.-G., Busnel, ed., Laboratoire de Physiologie
    Acoustique, Jouy-en-Josas, France.

Morozov, V. P., Akopian, A. I., Burdin, V. I., Donskov, A. A.,
    Zaytseva, K. A., and Sokovykh, Tu. A., 1971, Delphinid
    audiogram, Sechenov Physiol. Jour., of the USSR, 57(6):843.

Morozov, V. P., Akopian, A. I., Burdin, V. I., Zaytseva, K. A.,
    and Sokovykh, Yu. A., 1972, Tracking frequency of the lo-
    cation signals of dolphins as a function of distance to the
    target, Biofizika, 17:139 (English translation JPRS 55729).

Morozov, V. P., Akopian, A. I., Zaytseva, K. A., and Titov, A. A.,
    1975, On the characteristics of spatial directivity of the
    dolphin acoustic system in signal perception against the
    background of noise, in: "Marine Mammals, Proceedings of the
    Sixth All-Union Conference on the Study of Marine Mammals,"
    G. B. Agarkov, ed., Naukova Dumka, Kiev (English translation
    JPRS L/6049).

Murchison, A. E., 1979, "Maximum Detection Range and Range Reso-
    lution in Echolocating Tursiops truncatus (Montague), Ph. D.
    Dissertation, U. of California at Santa Cruz, California.
    (Manuscript).

Nordmark, J., 1961, Perception of distance in animal echolocation,
    Nature, 190:363.

Norris, K. S., 1964, Some problems of echolocation in cetaceans,
    in: "Marine Bio-Acoustics, Proceedings of the First Sym-
    posium on Marine Bio-Acoustics, Bimini", W. N. Tavolga, ed.,
    Pergamon Press, New York.

Norris, K. S., 1968, The evolution of acoustic mechanisms in odont-
    ocete cetaceans, in: "Evolution and Environment", E. T.
    Drake, ed., Yale Univ. Press, New Haven.

Norris, K. S., 1974, "The Porpoise Watcher", George J. McLeod,
    Limited, Toronto.

Norris, K. S., and Evans, W. E., 1967, Directionality of echo-
    location clicks in the rough-tooth porpoise Steno bredan-
    ensis (Lesson), in: "Marine Bio-Acoustics, Proceedings
    Second Symposium on Marine Bio-Acoustics, New York", W. N.
    Tavolga, ed., Pergamon Press, New York.

Norris, K. S., Evans, W. E., and Turner, R. N., 1967, Echolocation
    of an Atlantic bottlenose porpoise during discrimination,
    in: "Animal Sonar Systems, Biology and Bionics", R.-G. Bus-
    nel, ed., Laboratoire de Physiologie Acoustique, Jouy-en-
    Josas, France.

Norris, K. S., Prescott, J. H., Asa-Dorian, P. V. and Perkins, P.,
    1961, An experimental demonstration of echolocation behav-
    ior in the porpoise, Tursiops truncatus (Montagu), Biol.
    Bull., 120:163.

Penner, R. and Murchison, A. E., 1970, Experimentally demonstrated
    echolocation in the Amazon river porpoise Inia geoffrensis
    (Blainville), Proc. 7th Ann. Conf. Biol. Sonar and Diving
    Mammals, 7:17.

Pye, J. D., 1961, Perception of distance in animal echolocation, Nature, 190:362.

Pye, J. D., 1967, Discussion, in: "Animal Sonar Systems, Biology and Bionics", R.-G. Busnel, ed., Laboratoire de Physiologie Acoustique, Jouy-en-Josas, France.

Pye, J. D., 1971, Bats and fog, Nature, 229:572.

Romanenko, Ye. V., Tomilin, A. G., and Artemenko, B. A., 1965, in: "Bionica", M. G. Gaaze-Rapoport and V. E. Yakobi, eds., Nauka, Moscow (English translation JPRS 35125).

Sales, G., and Pye, D., 1974, "Ultrasonic Communication by Animals", Chapman and Hall, London.

Schevill, W. E., and Lawrence, B., 1956, Food-finding by a captive porpoise (Tursiops truncatus), Breviora Museum Comp. Zool., 53:1.

Schusterman, R. J., 1968, Experimental laboratory studies of pinniped behaviour, in: "The Behaviour and Physiology of Pinnipeds", R. J. Harrison, R. C. Hubbard, R. S. Peterson, C. E. Rice, and R. J. Schusterman, eds., Appleton-Century-Crofts, New York.

Schusterman, R. J., 1979, Behavioral methodology in echolocation by marine mammals, this volume.

Simmons, J. A., 1969, "Depth Perception by Sonar in the Bat Eptesicus fuscus", Doctoral dissertation, Princeton University, University Microfilms, Inc., Ann Arbor.

Simmons, J. A., 1971a, The sonar receiver of the bat, Ann. N. Y. Acad. Sci., 188:167.

Simmons, J. A., 1971b, Echolocation in bats: Signal processing of echoes for target range, Science, 171:925.

Simmons, J. A., 1973, The resolution of target range by echolocating bats, J. Acoust. Soc. Am., 54:157.

Simmons, J. A., Altes, R. A., Beuter, K. J., Bullock, T. H., Capranica, R. R., Goldstein, J. L., Griffin, D. R., Konishi, M., Neff, W. D., Neuweiler, G., Schnitzler, H. U., Schuller, G., Sovijarvi, A. R. A., and Suga, N., 1977. Localization and identification of acoustic signals, with reference to echolocation, group report, in: "Dahlem Workshop on Recognition of Complex Acoustic Signals, Berlin, 1976", T. H. Bullock, ed., Abakon Verlagsgesellschaft, Berlin.

Steele, J. W., 1971, Marine-environment cetacean holding and training enclosures, NUC TP, 227, San Diego.

Stevens, S. S., 1975, "Psychophysics: Introduction to its Perceptual, Neural, and Social Prospects," G. Stevens, ed., John Wiley and Sons, New York.

Stewart, J. L., 1968, Analog simulation studies in echoranging, T. D. R. N°. AMRL-TR-68-40, Aerospace Medical Div., U. S. A. F. Systems Command.

Strother, G. K., 1961, Note on the possible use of ultrasonic pulse compression by bats, J. Acoust. Soc. Am., 33:696.

Thurlow, W. R., and Small, A. H., 1955, Pitch perception for
    certain periodic auditory stimuli, J. Acoust. Soc. Am.,
    27:132.
Tinbergen, L., 1960, The natural control of insects in pinewoods,
    I. Factors influencing the intensity of predation by song
    birds, Arch. neerl. Zool., 13:265.
Titov, A. A., 1972, Investigation of sonic activity and phenomeno-
    logical characteristics of the echolocation analyzer of
    Black Sea delphinids. Canditorial dissertation, Karadag, as
    cited in: Bel'kovich, V. M., and Dubrovskiy, N. A., 1976,
    "Sensory Bases of Cetacean Orientation", Nauka, Leningrad
    (English translation JPRS L/7157).
von Uexküll, J., 1934, Streifzuge durch die Umwelten von Tieren
    und Menschen, Translated in: "Instinctive Behaviour", 1957,
    C. H. Schiller, ed., Methuen, London.
Urick, R. J., 1975, "Principles of Underwater Sound", McGraw-Hill
    Co., New York.
Vel'min, V. A., 1975, Target detection by the bottlenose dolphin
    under artificial reverberation conditions, in: "Marine
    Mammals, Proceedings of the Sixth All-Union Conference on
    the Study of Marine Mammals", G. B. Agarkov, ed., Naukova,
    Dumka, Kiev, (English translation JPRS L/6049).
Vel'min, V. A., and Dubrovskiy, N. A., 1975, On the auditory anal-
    ysis of pulsed sounds by dolphins, Doklady Akad. Nauk.,
    SSSR, 225(2):470.
Vel'min, V. A., and Dubrovskiy, N. A., 1976, The critical interval
    of active hearing in dolphins, Sov. Phys. Acoust., 22(4):
    351.
Vel'min, V. A., and Titov, A. A., 1975, Auditory discrimination of
    interpulse intervals by bottlenose dolphin, in: "Marine
    Mammals, Proceedings of the Sixth All-Union Conference on
    the Study of Marine Mammals", G. B. Agarkov, ed., Naukova
    Dumka, Kiev (English translation JPRS L/6049).
Webster, F. A., 1967, Interception performance of echolocating
    bats in the presence of interference, in: "Animal Sonar
    Systems, Biology and Bionics", R.-G. Busnel, ed., Labora-
    toire de Physiologie Acoustique, Jouy-en-Josas, France.
Woodworth, R. S., and Scholsberg, H., 1954, "Experimental Psychol-
    ogy", revised edition, Holt, Rinehart, and Winston, Inc.
    New York.
Zaslavskiy, G. L., 1974, Experimental study of time and space
    structure of the dolphin's echolocating signals, candidat-
    orial dissertation, Karadag, as cited in: Bel'kovich V. M.,
    Dubrovskiy, N. A., 1976, "Sensory Bases of Cetacean Orient-
    ation", Nauka, Leningrad (English translation JPRS L/7157).
Zaslavskiy, G. L., Titov, A. A., and Lekomtsev, V. M., 1969, Re-
    search on the sonar abilities of the bottlenose dolphin,
    Trudy Akusticheskogo Instituta, 8:134, as cited in: Ayra-
    pet'yants, E. Sh., and Konstantinov, A. I., 1974, "Echoloc-
    ation in Nature", Nauka, Leningrad (English trans. JPRS 63328)

ODONTOCETE ECHOLOCATION PERFORMANCE ON OBJECT SIZE, SHAPE AND MATERIAL

Paul E. Nachtigall

Naval Ocean Systems Center

Kailua, Hawaii 96734

A cetacean swimming in its natural environment must frequently encounter a variety of sources for returning echos. An echolocation system which only allowed for the detection of objects would be insufficient for gathering information necessary to meet needs like prey location and predator avoidance. The ability of cetaceans to differentiate and recognize characteristics of objects via echolocation has an obvious biological benefit. The purpose of this paper is to review the behavioral experiments that have examined delphinid abilities to differentiate between objects differing in size, shape, or material.

Since the last Animal System Sonar Conference was held in 1966, a high percentage of the experiments examining delphinid discrimination have been conducted within the Soviet Union. Two fine reviews of the Soviet work, one by Ayrapet'yants and Konstantinov (1974) and the other by Bel'kovich and Dubrovskiy (1976), have been written. Because of the thoroughness of those reviews, the Soviet work reviewed within this paper will be confined to those studies that are applicable or comparable to recent work done outside of the USSR. The tables presented within this review were prepared to be comparable extensions of the thorough tables presented in both Soviet reviews.

The emphasis within this review is placed on animal performance. The procedures used in animal discrimination experiments generally require the animal to choose between at least two stimuli presented either simultaneously or successively. The stimuli normally differ along only one dimension and the limits of the animal's discrimination are frequently tested by making the differences progressively smaller. The point at which the animal can no

longer effectively make the discrimination is termed the threshold.
For comparative purposes, threshold levels for this review are
arbitrarily set at 75% correct. Although direct comparison
between performance levels and thresholds will be made within
this review, the reader must bear in mind that comparisons between
the results of differing experiments must be considered as approximate. Different animals, training procedures, environments, and
experimental procedures frequently lead to variability in performance levels independent of the actual sensitivities of the
cetacean subjects.

SIZE DISCRIMINATION

Kellogg's (1958) initial work on echolocation demonstrated
that a bottlenosed dolphin (Tursiops truncatus) could choose
between food fish of different sizes. Since that early work, a
number of experiments have demonstrated that a variety of delphinids
are quite capable of discriminating targets differing in size.
The first tests of the limits of delphinid size discrimination
ability were both presented at the previous animal sonar symposium
(Busnel, 1967). Busnel and Dziedzic (1967), using an unconditioned
obstacle avoidance technique, demonstrated that a blindfolded
harbor porpoise (Phocaena phocaena) could easily avoid wires made
of copper, iron, and steel .5 mm in diameter but frequently ran
into wires .2 mm in diameter. While Norris, Evans and Turner
(1967) demonstrated that a blindfolded Tursiops truncatus could
choose the larger of two simultaneously presented nickel-steel
ball bearings. The results of the experiment (Turner and Norris,
1966) indicated that the animal could discriminate between a 5.71
cm in diameter standard sphere and spheres larger than 6.35 cm in
diameter.

Shortly after the publication of the last animal sonar
symposium, scientists within the Soviet Union began publishing
work on the size discrimination ability of cetaceans. Ayrapet'yants,
Golobkov, Yershova, Zhezherin, Zvorykin and Korolev (1969) investigated the ability of bottlenosed dolphins to discriminate steel
disc shaped cylinders of varying heights. The animals were
trained to swim a distance of 20 meters and pull a ring to receive
a fish when a standard 25 mm high 110 mm in diameter disc-shaped
cylinder was presented and to withhold the response when a cylinder
of the same diameter but of a greater height was presented.
While Schusterman (this volume) is rightfully unsure about the
balancing of the reinforcement for "yes" or "no" responses, in
that the animals may only have received reinforcement in the
presence of the standard cylinder, the data indicate that the
animal did discriminate a 30 mm high cylinder from the 25 mm high
standard at the 70% level. That threshold estimate may be considered
as an underestimate if the animal was, in fact, biased by the

unbalanced reinforcement contingencies.

Zaslavskiy, Titov, and Lekomstev (1969) conducted an experiment using the harbor porpoise (Phocaena phocaena) discriminating the heights of a similar series of targets. Rather than using the ring-pull technique used by Ayrapet'yants et al., targets were apparently presented simultaneously in a two alternative forced choice technique. Echolocating from a distance of 2.5 meters, the porpoise was able to differentiate 75 mm in diameter cylinders 75 mm high from cylinders 95 mm high at the 80% level. Ayrapet'yants and Konstantinov (1974) indicate that the results of these two experiments (even with the disparity of species) and other experiments on cylinder height discrimination conducted within the Soviet Union, generally agree and show that the threshold height ratios are within 1.1 to 1 and 1.3 to 1.

While Soviet scientists were comparing species on cylinder height discrimination, American scientists were examining two very different species on cylinder diameter discrimination. Performance of an Atlantic bottlenosed dolphin (Tursiops truncatus), trained by Evans and Hall in San Diego, and an Amazon River dolphin (Inia geoffrensis), trained by Penner and Murchison in Hawaii, was compared using solid chloroprene cylinders varying in target strength in 1 dB increments. Both animals were trained to wear blindfolding devices and to choose the standard cylinder (see Table 1 for target size details). Once this was achieved, targets were paired with the others in the series in a two alternative forced choice simultaneous presentation procedure with position of targets varied in a modified random order. The results of the comparison with targets having a greater target strength than standard indicated that both Inia and Tursiops could discriminate targets varying by one dB with performance levels above 70% (Evans, 1973).

Sphere Diameter Discrimination

According to Ayrapet'yants and Konstantinov (1974) Soviet Scientists were unsatisfied with the spherical size discrimination work presented by Norris, Evans, and Turner (1967):

It should be kept in mind that the experiment was not set up entirely properly: The balls were connected together by a metal rod, and the distance between them exceeded the diameter of the large ball by only two times. Thus acoustic interaction through the water was not excluded. Therefore, it would be more accurate to refer not to discrimination between balls but to recognition of asymmetry in the acoustic field scattered by a system of two balls and a connecting rod.

The reviews by both Ayrapet'yants and Konstantinov (1974) and Bel'kovich and Dubrovskiy (1976) indicate that Soviet scientists conducted a series of experiments to further examine delphinid sphere size discrimination using both steel and lead balls. Unfortunately neither review paper provides sufficient methodological detail to know how the targets were presented to overcome the reviewer's objection to Norris et al.'s procedure. In the first of these experiments (Dubrovskiy, 1972), bottlenosed dolphins were trained to discriminate a 5 cm in diameter standard lead ball from balls 10, 6 and 5.5 cm in diameter. Ayrapet'yants and Konstantinov (1974) indicate that discrimination precision "dropped regularly in almost all cases as the distance to the target was increased". The animal, at a range of 8 meters from the target, discriminated a 5.5 cm in diameter sphere from the 5.0 cm in diameter standard at the 78% level. Bel'kovich and Dubrovskiy (1976) indicate that additional work with lead spheres by Dubrovskiy and Krasnov (1971) and Fadeyeva (1973) showed that bottlenosed dolphins could discriminate a 1.4 cm in diameter sphere from 1.6 cm sphere at the 70% level at a distance of 4.8 meters from the targets; and a 1.02 cm in diameter lead sphere from a 1.17 in diameter sphere at the 75% level at a distance of 3 meters. Additional tabular data presented in Ayrapet'yants and Konstantinov (1974) indicate that at a distance between 2 and 6 m to the targets, lead spheres varying in diameter from 10.2 to 14.3 cm were compared. The results indicate an orderly relationship between target size and overall percentage correct with a threshold of 75% correct being achieved when comparing 10.2 and 11.7 cm lead spheres.

A considerable amount of Soviet work was also done by Dubrovskiy et al. (1973) on the discrimination of the size of steel spheres. The size of steel spheres was varied and presented in a two alternative forced choice procedure at distances varying between 2 and 6 meters. These data were collected within the same experiments and with the same animals previously used in the lead ball diameter discrimination presented above. Tabular data presented within Ayrapet'yants and Konstantinov (1974) indicate that with steel spheres in a graded series between 10.0 and 20.7 cm the bottlenosed dolphin could discriminate between a steel sphere 10.4 cm in diameter and a steel sphere 14.3 cm in diameter at the 75% level. Further data with larger targets indicate that, with sphere radii varying between 50.8 and 63.4 cm, the animal's 77% threshold level was reached with a comparison of spheres with radii of 57.1 and 63.5 cm.

A comparison of lead vs steel extrapolated from the above data indicates a 1.5 cm threshold difference (10.2 vs 11.7) in radii for lead balls and a 3.9 cm (10.4 vs 14.3) threshold difference in radii for steel balls of similar sizes. Ayrapet'yants and Konstantinov (1974) indicate that this interaction between size

and materials was the most important discovery in the series of
experiments.

Planometric Targets

Barta (unpublished manuscript) conducted the first Western
size discrimination experiment using flat targets. Targets were
constructed of 3.18 mm neoprene (celltite) foam cemented to a .79
mm aluminum sheet metal backing. The circular targets ranged
from 15.2 cm to 25 cm in diameter. The dolphin (Tursiops truncatus)
was trained to choose the smaller of two simultaneously presented
targets. A divider placed between the targets assured that the
animal was not closer than .7 m from the targets during discrimination. The results indicated that the animal discriminated a 16.1
cm in diameter target from the standard 15.2 cm target at a 75%
correct level. Subsequent reflectivity measures indicated that
this size difference threshold corresponded to a 1 dB difference
in target strength.

Bel'kovich, Borisov, Gurevich and Krushinskaya (1969) determined the difference threshold for the discrimination of flat
plastic foam squares of different sizes. The animal was trained
to choose the larger of two simultaneously presented targets.
The standard was 100 cm^2 while the comparison targets measured
50, 81, and 90.25 cm^2. The common porpoise (Delphinus delphis)
showed an orderly decline in performance with decreasing size
differences, with performance failing to the 77% level on comparisons between the 100 cm^2 and 90.25 cm^2 targets. The size difference threshold was therefore reported to be around 10%.

Target or Wall Thickness

The first evidence that cetaceans could discriminate between
objects varying in thickness of the material was presented by
Evans and Powell (1967) at the initial Animal Sonar Systems
Meeting. Circular shaped 20 cm in diameter metal discs varying
in both type of metal and thickness were mounted within brass
backed sponge neoprene 30 x 30 cm square target holders. The
Tursiops truncatus was trained to wear eyecups and swim 7 m to
choose a .22 cm thick copper standard target. The standard and
one comparison target were presented on each trial using a two
alternative forced choice technique and the psychophysical method
of constant stimuli. The animal was unable to discriminate .16
and .27 cm thick targets from the .22 cm standard but the .32 and
.64 cm copper discs were discriminated at the 75% and 90% levels.

Although few details are available, both Ayrapet'yants and
Konstantinov (1974) and Bel'kovich and Dubrovskiy (1976) indicate

that Titov (1972) examined the bottlenosed dolphin's ability to
discriminate differences in the wall thickness of hollow cylinders.
The steel cylinders measured 50 mm high with an outer diameter of
50 mm and were presented 5 m away from the animal (presumably
using the two alternative forced choice procedure). The animal
was trained to choose the thinner of two cylinders with target
wall thickness varying from .1 to 2 mm (according to Ayrapet'yants
and Konstantinov, 1974) and was able to discriminate a wall thickness
difference of .2 mm at the 75% level.

Hammer (1978) conducted an interesting series of experiments
using hollow and solid cylinders. In the first study, he examined
a discrimination of hollow aluminum cylinders and solid cylinders
made of coral rock chunks embedded in epoxy resin (see materials
section for further details). The results of that first experiment
led him to believe that the salient cue for target "recognition"
was the thickness of the wall of the hollow aluminum cylinders.
In order to determine the limits of wall thickness, a series of
cylinders varying in wall thickness was constructed. The animal
continued the same basic task as that of the first experiment.
Four targets were presented to maintain the animal's baseline
performance. Only one target was presented on each trial. If
either of two standard aluminum hollow cylinders (see Table 1 for
description of target dimensions) was presented the animal was
required to hit a manipulandum on the left. If either of two
solid coral rock molded in epoxy targets (with the same outside
measurements as the aluminum targets) was presented the animal
was required to hit a manipulandum on the right. During each
session the animal continued to perform this task with performance
levels near 100%. Occasionally a hollow aluminum target with a
different wall thickness was inserted. Responses to either
manipulandum were not reinforced. The wall thickness of these
"probe" targets varied around the two hollow aluminum standard
targets. Those around the smaller standard varied in .16 cm
increments while those around the large standard varied in .32 cm
increments. Each session was made up of 76 trials. The first 12
were "warm-up" trials during which only standard aluminum and
coral rock targets were presented. In the following 64 trials,
56 were continued as standard baseline trials while 8 were randomly
chosen as trials. The aluminum targets varying in wall thickness
were presented within these probe trials.

Data were collected in three phases. During the first and
third phases, targets were presented 6 m from the animal. During
the second phase targets were presented 16 m from the animal.
The data from the first phase indicated that the targets most
similar in wall thickness to the hollow aluminum standards were
infrequently "classified" as those standard targets, but the data
from the second and third phases indicated that all of the hollow
aluminum targets with wall thicknesses differing from the standards

were <u>classified</u> <u>along</u> <u>with</u> <u>the</u> <u>solid</u> <u>coral</u> <u>rock</u> <u>in</u> <u>epoxy</u> <u>targets</u>.
These results would appear to be unexpected but see Schusterman's
(this volume) chapter on the importance of behavioral methodology
in the interpretation of psychophysical data.

The data indicate that the animal could classify a hollow
aluminum target with a .16 cm wall thickness difference as being
different from the standard on 100% of the trials. A difference
threshold was not obtained but that is not surprising considering
Titov's data indicating wall thickness difference thresholds as
small as .2 mm with steel hollow cylinders.

SHAPE DISCRIMINATION

Planometric Targets

Two experiments were conducted within the late 1960's examining
cetacean ability to discriminate differences between the shapes
of flat targets. Barta (unpublished manuscript) examined the
ability of a blindfolded <u>Tursiops</u> to choose between circles,
squares and triangles, while Bagdonas, Bel'kovich and Krushinskaya
(1970) tested <u>Delphinus'</u> ability to discriminate a large square
from a smaller triangle. Barta constructed triangles and squares
from the same materials previously reported in the planometric
size discrimination work. The same circular target standards
were used in comparison to triangles and squares with varying
overall surface areas (see Table 1 for target dimensions). Using
the same two alternative forced choice paradigm, (used in the
earlier reported size discrimination work) the animal reliably
chose the circular target.

Bagdonas et al. (1970) presented targets made from ebonite
10 mm thick. The animal (<u>Delphinus delphis</u>) was trained to swim
to a 100 cm^2 square as compared to a 50 cm^2 triangle. This
compounding of size and shape differences makes it impossible to
precisly know which stimulus dimension controlled the animal's
behavior. The animal was not blindfolded and initial presentations
were made in clear water with sessions conducted both day and
night. The fact that the animal was correct 82% of the time
during the day and 57% during the night sessions indicated to the
authors that the animal was relying on visual cues. After some
15 sessions, however, the experimenters began hearing echolocation
signals and the dolphin suddenly began to echolocate intensely.
On the 17th session the animal reached the 100% performance level
independent of visibility conditions.

An additional planometric shape discrimination experiment
was conducted in 1971. Bel'kovich and Borisov tested the common

dolphin's (Delphinus delphis) ability to differentiate flat
squares from similar sized squares with circular holes cut in the
center. The animal was initially trained to choose a foam rubber
covered plexiglass 10 x 10 cm square when it was simultaneously
presented with a 7 x 7 cm square. When the 7 x 7 cm square was
paired with a 7 x 7 cm square with a hole cut out of the center,
in unreinforced probe trials, the animal chose the (previously
negative) solid square 89.5% of the time.

A subsequent series of comparisons of the hole-cut and solid
squares indicated that the animal could distinguish squares with
6.5% of area cut out. A previous study, mentioned earlier in
this review, (Bel'kovich et al., 1969) with discrimination based
solely on the size of the squares, had indicated an area difference
threshold of 10%. The authors concluded, therefore, that the
animal did not make the discrimination solely on the basis of
overall reflectivity but must have been provided with other cues,
such as spectrum differences of the return echo, on targets with
the cut out hole.

Planometric and Three Dimensional Targets

The first experiment designed to determine the delphinid's
ability to discriminate three-dimensional from flat targets was
conducted by Bel'kovich, Borisov, Gurevich and Krushinskaya
(1969). Two Delphinus delphis were trained to choose a three-
stepped pyramid plastic foam target when compared to a simultaneously
presented flat plastic foam triangle in a two alternative forced
choice procedure. Once this discrimination was established, flat
squares with size equal to that of the base of the three dimensional
pyramid were also easily differentiated.

Bel'kovich et al. further examined the dolphin's ability to
discriminate stepped pyramids. The standard pyramid was constructed
of foam plastic squares layered in three 12 mm thick steps. The
base step measured 100 cm^2, the middle step measured 49 cm^2,
while the top step was only 9 cm^2. This standard was compared to
pyramids in which: (1) the top step was omitted (a two-step
pyramid), (2) The size of the top step was made smaller, and (3)
the thickness of the top two steps was changed.

When the pyramid with the top step omitted was compared to
the standard 3 step pyramid, the animals made the discrimination
easily at the 95% level. When area differences of the top step
were made smaller (surface areas of 6.25 and 8.4 cm^2 as compared
to the standard 9 cm^2) performance fell to near 70% for both
target comparisons.

The thickness of the top two steps was also varied in order to "ascertain the ability of the dolphin to detect any shift of the echo signal in time". One comparison pyramid was varied by reducing the thickness of each of the top two steps from 12 mm to 6 mm. The animal discriminated this target from the standard easily (at the 100% level). A second comparison pyramid varied only by changing the thickness of the middle step to 6 mm. This target was discriminated at the 86.7% level. The authors concluded that "it is reasonable to assume that the dolphins distinguished between the figures (step thickness differences) by using the change in spectral composition of the reflected echo signals, including change in the relationship between the time of their return and the elements constituting the truncated stepped pyramid".

A bottlenosed dolphin's ability to discriminate the difference between three dimensional solid cylinders and cubes made of syntatic foam was examined by Nachtigall, Murchison and Au (this volume). The blindfolded animal was required to examine two simultaneously presented targets located 38 cm apart, and two meters from the tip of its rostrum, and to press one of two adjacent manipulanda indicating the randomly predetermined left or right position of a cylinder as compared to a cube. Sizes of both the cylinders and cubes were varied to control for possible differences in overall reflectivity (see Table 2). Each of three different sized cylinders was repeatedly paired with each of three different sized cubes. Performance data, presented in Table 2, indicate an interesting interaction of shape discrimina-bility and target size. Comparisons of cubes and cylinders of the same sizes show increased discriminability of shape as target sizes increase.

Once the ability of the animal to differentiate cubes from cylinders was well established, the aspect of the targets presented to the animal was systematically varied using a probe technique. Baseline performance with targets presented upright was maintained within 56 of the 63 trials per session. On the other 7 probe trials one of the targets was either rotated or layed down horizon-tally. During the first two probe sessions the cube was rotated so that the edge, rather than a flat surface, faced toward the animal. This change did not disrupt the animal's performance. During the next two sessions, both the cubes and the cylinders were changed during probe trials. The cubes were once again presented with the edge forward but the cylinders were layed down horizontally. Overall baseline performance remained high but performance during probe trials declined to 71%. During the final two sessions, both the cubes and the cylinders were presented with the flat surface (i.e., top of the cylinder) facing the animal. Performance on these probes declined to near chance.

Following the completion of the experiment the targets were

examined acoustically with a monostatic sonar measurement system
that projected simulated dolphin echolocation signals. The echo
returns for the various targets were processed, and resulting
frequency spectra were obtained, by taking a 1024 point fast
fourier transform. Comparisons of the target spectra failed to
reveal consistent and obvious cues for shape discrimination.

Fifteen measures of target strength for each target, however,
revealed an interesting possible cue for shape discrimination.
Standard deviations around the mean of the measures indicated
high variability in measures for both the cube and the cylinder
with the flat face forward and low variability with the cylinder
straight up. These differences in variability paralleled the
performance given by the animal. The animal most likely received
repeated pulse echos varying in amplitude when scanning across
cubes and relatively uniform pulse echos when scanning across the
curved portion of cylinders.

An animal discriminating the shapes of objects has more than
one acoustic cue (dimension) available to it. Barta's work
indicated that Tursiops could readily discriminate flat squares
from flat circles, yet in Nachtigall et al.'s experiment the
dolphin did not discriminate the difference between the flat
square surface of a cube and the flat circular cylinder top. It
seems very likely that the primary discriminative cue for differentiating flat objects of differing shapes is not the same as the
primary cue for differentiating three dimensional objects. One
may conclude that the dolphin in Nachtigall et al.'s experiment
was quite capable of making the discrimination of flat circles
and flat squares, but had not been trained to make that particular
discrimination. This is an important point. The interpretation
of the results of a behavioral experiment depend heavily upon
both the type of training that the animal received and the type
of experimental procedure used to gather the data. Had the three
dimensional targets been rotated during training, prior to presenting
them within probe trials, the animal would have most likely
learned both discriminative cues and easily made the discrimination,
but the hypothesis of differing discriminative cues for the two
tasks would not have been suggested.

MATERIALS DISCRIMINATION

Targets made of differing types of material will reflect
differing amounts of acoustic energy. A target composed of air
filled styrofoam, for instance, will return a higher amplitude
echo under water than a similarly shaped target composed of
plexiglas. Thus an animal might easily differentiate targets
made up of these differing materials solely on the amplitude
differences of the echos. Evans and Powell (1967) predetermined

the theoretical differences in the amplitude of the return of
metal plates of varying thickness and composition. They found
that when copper plate targets of varying thickness were presented
to a Tursiops truncatus, the animal could not discriminate between
copper discs differing in calculated reflectivity of +.04 and −.06.
However, when an aluminum disc, with a calculated reflectivity
difference of only .02, was compared to the standard copper disc,
the animal easily made the discrimination. This was the first
solid evidence indicating that a delphinid could discriminate
differences in target materials independent of their overall
reflectivity. According to Evans (1973) this experiment was
replicated with a second Tursiops and a Lagenorhynchus obliquidens
with comparable results for both species. Johnson (1967) stated
that the animal may have been making this discrimination based on
phase differences.

    Following the initial Evans and Powell work, Soviet scientists
began a lengthy series of experiments examining delphinid ability
to discriminate spheres and cylinders made up of different materials.
These experiments are more than adequately reviewed by both
Ayrapet'yants and Konstantinov (1974) and Bel'kovich and Dubrovskiy
(1976). Zaslavskiy, Titov and Lekomtsev (1969) investigated the
common porpoise's (Phocaena phocaena) ability to discriminate
11.5 cm high by 7.5 cm in diameter cylinders composed of steel as
compared to other materials. Using an interesting technique of
determining the range at which 75% correct thresholds were obtained,
they found wood and plastic to be more discriminable from steel
than was glass. Three experiments (Babkin, Dubrovskiy, Krasnov,
and Titov, 1971; Abramov, Ayrapet'yants, Burdin, Golubkov,
Yershova, Zhezherin, Koroleve, Malyshev, Ulyanov and Fradkin,
1971; and Titov, 1972) examined the bottlenosed dolphin's ability
to discriminate 5 cm in diameter solid spheres made of various
materials (steel, aluminum, brass, ebonite, textolite, flouroplastic,
rubber, wax, paraffin, plexiglas, and lead) using the two alternative
forced choice technique. Performance on most comparisons was
well above 90% but the discrimination of steel and aluminum was
shown to be particularly difficult in all three studies.
Dubrovskiy, Krasnov and Titov (1971) noted that this difficulty
existed in spite of the fact that the specific impedance of steel
is about 2.6 times more than that for aluminum, meaning that the
"specific impedance (as well as density) of (the) material is not
an important factor in the discrimination of solid spheres".
They found, however, that a parameter Δ, which is monotonically
related to the speed of the shear waves within the target material,
was very similar for steel and aluminum targets. Dubrovskiy et
al. hypothesized that target materials were discernable as long
as differences in Δ were discriminable. According to the hypothesis
"the discrimination is performed by means of an auditory analysis
of the intensity and the spectrum of a so-called secondary echo
which contains information about free vibrations of the target

forced to vibrate by a porpoise signal". Dubrovskiy, Krasnov and
Titov (1971) indicated, however, that they were "far from the
opinion" that this difference in the secondary echo spectrum was
the only cue for discrimination. They assumed discrimination
could also be made based on: (1) the fine structure of the power
spectrum, (2) differences in average intensity of echos, and (3)
phase shifts.

Although no work on the discrimination of spheres made up of
differing materials has been done outside the Soviet Union,
Hammer (1978) conducted two experiments on the classification of
cylinders made up of various materials. Using the procedure
outlined in an earlier section of this paper, Hammer first trained
the dolphin (Tursiops truncatus) to press a paddle on its left
when either of two standard hollow aluminum targets was presented
and to press a paddle on its right if either of two solid cylinders
composed of coral rock chunks embedded in epoxy resin was presented.
The animal continued on this basic task for 56 out of 64 trials
per session. On the other 8 randomly dispersed probe trials, any
one of 8 different cylinders was presented. Four of these cylinders
were composed of aluminum and the other four were composed of
various nonmetallic materials (see Table 3, Experiment 1).

The data, once again, were collected in three phases.
During the first and third phases the targets were presented 6 m
from the animal. During the second phase the distance was
increased to 16 m. Data collected during the first phase indicated
that the animal classified a small and a large aluminum hollow
cylinder and a small solid solid aluminum cylinder along with the
standard hollow aluminum cylinders. But, in the next two phases,
all cylinders not exactly like the standard aluminum cylinders
were classified along with the coral rock epoxy cylinders 100% of
the time. These data indicate that with continued training, an
animal may narrow its classification criterion down to a particular
target type, and that, without specific training, dolphins do not
necessarily categorize stimuli along a dimension of target material.

In order to more precisely examine the animal's ability to
classify targets of differing materials, Hammer (1978) conducted
another experiment using the same basic procedure but different
targets. During this experiment the comparison probe targets,
made of bronze, glass, and stainless steel, matched the standard
aluminum targets in all dimensions other than target materials.
The data on the stainless steel targets indicated clearly and
consistently that the animal did not classify them as the same as
aluminum standards. Responses on the bronze targets were mixed
while the glass targets were consistenly classified the same as
the aluminum.

Using the same animal, experimental set up, and glass and

aluminum targets (but not the same experimental procedure) that
Hammer had previously used, Schusterman, Kersting and Au (this
volume) continued training the animal on the aluminum-glass
discrimination. They found that the animal could eventually
discriminate the smaller glass target from the two aluminum
standards but that the animal could not discriminate the large
glass cylinder from the aluminum standards. In fact, sessions,
during which the animal was presented the "insolvable" problem,
biased the animal and made the solvable problem more difficult.

INTERACTIONS

       Up to this point the experiments on cetacean discrimination
have been classified into sections based on differences in target
size, shape, or material. Although we may assume these categories
to be broadly defined, each experiment is usually limited to a
particular target type. An experimenter usually varies only one
of the three dimensions. An experiment on size discrimination,
for instance, frequently contains targets of only one shape and
only one material. This procedure tends to limit the experimental
findings because the interactions between the three dimensions
are frequently not explored. Experiments which have examined
more than one dimension have frequently produced fascinating
results. The fact that Evans and Powell (1967), for example,
used targets of different materials in their experiment on the
bottlenose dolphin's ability to discriminate flat plates of
different thicknesses, led to the discovery that the animal was
capable of making the discrimination on some basis other than
reflectivity.

       It is also possible to examine interactions by comparing
data across experiments. A comparison of the Soviet work on the
discrimination of solid aluminum and steel spheres and the Hammer
(Experiment 3) work on the classification of aluminum and steel
hollow cylinders, provides an interesting example of the interaction
of target materials and target shape. In the three Soviet experiments bottlenosed dolphins were shown to have great difficulty
discriminating solid aluminum and solid steel spheres, yet in
Hammer's work aluminum and steel hollow cylinders were distinguished
with comparative ease. It seems apparent that the cues that an
echolocating dolphin uses to differentiate target materials may
be particular to the target's shape or structure. Aluminum
hollow cylinders are apparently easily discriminated from steel
hollow cylinders but solid steel spheres are not easily discriminated
from solid aluminum spheres. The dolphin's ability to discriminate
steel from aluminum may depend on the shape of the targets or it
may depend on whether the targets are solid or hollow.

       A similar interaction may be observed when comparing Evans

and Powell's (1967) thickness discrimination work with that
conducted by Titov (1972). Where Evans and Powell found that
copper discs were 75% discriminable at thickness differences of
.1 cm using solid flat discs, Titov found that the hollow steel
cylinder wall thickness difference threshold for a bottlenosed
dolphin was .02 cm. Once again there appears to be some sort of
interaction occurring. The better performance reported by Titov
may have been due to the difference in materials or the difference
in shape. Steel may yield more discriminable thickness difference
return echoes than does copper, or the echo returns from solid
flat plates may not be as discriminable as the echo returns from
hollow cylinders.

The fact that these interactions occur makes it difficult to
speak strictly about a single target dimension. All three dimensions
which form the division of this paper - size, shape, and material -
must be considered as relevant to any one particular discriminative
echolocation task. Further understanding of the process of
discriminative echolocation may well be enhanced by examining the
causes of the interactions.

Table 1. Size Discrimination

SHAPE	TARGET MATERIAL	VARYING TARGET SIZE (cm) POS.	VARYING TARGET SIZE (cm) NEG.	SIZE DIMEN.	METHOD	THRESHOLD	% CORR.	MIN. DIST.	SPECIES	LIT. SOURCE
Solid Spheres	Nickel-Steel	6.35	5.72 5.40 5.08 4.76 4.45 4.13 3.97 3.81 3.18	Dia.	2 Alt. Force Choice	.63 cm	77 90 93 98 100 100 100 100 100	Ø	Tursiops Truncatus	Turner & Norris (1966) Norris, Evans & Turner (1967)
Wires	Iron Copper and Steel	—	.35 .15 .075 .05 .035 .02	Dia.	Obstacle Avoidance	.035 cm	98.9 98.9 94.7 90.9 78.9 46	Ø	Phocaena Phocaena	Busnel, Dziedzic & Anderson (1964) Busnel & Dziedzic (1967)
	Nylon		.15				72.5			

Table 1 (Cont'd). Size Discrimination

| TARGET | | VARYING TARGET SIZE (cm) | | SIZE DIMEN. | METHOD | THRES-HOLD | % CORR. | MIN. DIST. | SPECIES | LIT. SOURCE |
SHAPE	MATERIAL	POS.	NEG.							
Solid Cylinders H=17.78 cm	Chloroprene (DC-100)	1.64	2.07 2.62 3.28 4.16 5.20	Dia. & dB -18dB -17dB -16dB -15dB -14dB	2 Alt. Forced Choice	.43 cm	85 94 89 96 90	.77 m	Tursiops Truncatus	Evans (1973)
		1.64	2.07 2.62 3.28 4.16 5.20	-18dB -17dB -16dB -15dB -14dB		.98 cm	70 77 89 90 90	.77 m	Inia Geof-frenis	
Circular Disks	Neoprene (Cell-tite) Back With Aluminum	15.2	25.0 20.0 18.0 17.2 16.1 15.7 15.2	Dia.	2 Alt. Forced Choice	.9 cm	98.3 98.5 97 94.1 74.8 57.5 50.7	.7 m	Tursiops Truncatus	Barta & Evans (Unpublished Manuscript)

Table 1 (Cont'd). Size Discrimination

| TARGET | | VARYING TARGET SIZE (cm) | | SIZE DIMEN. | METHOD | THRES- HOLD | CORR. | MIN. DIST. | SPECIES | LIT. SOURCE |
SHAPE	MATERIAL	POS.	NEG.							
Circular Disks D=20 cm	Copper	.22	.16 .27 .32 .64	Thick- ness	2 Alt Forced Choice	.10 cm	50 60 75 90	.7 m	Tursiops Truncatus	Evans & Powell (1976)
Hollow Cylin- ders H=17.78 cm	Aluminum	OD= 7.62 .95	.31 .63 1.27 1.59	Wall Thick- ness	2 Alt Succes- ive Probe	–	100 100 100 100	6 m 16 m	Tursiops Truncatus	Hammer (1978)
		OD= 3.18 .64	.32 .48 .80 .96				100 100 100 100			

Table 2. Shape Discrimination

TARGET SHAPE & MATERIAL		TARGET SIZE CM		METHOD	DISTANCE TO TARGET	PERCENT CORRECT	SPECIES	SOURCE
POSITIVE	NEGATIVE	POSITIVE	NEGATIVE					
Circular Neoprene Disks	Flat Squares	182 cm^2	428 cm^2 269 cm^2 182 cm^2 87 cm^2	2 Alt Forced Choice	.7 m	100 98.6 93.1 100	Tursiops Truncatus	Barta & Evans (Unpublished Manuscript)
Circular Neoprene Disks	Flat Triangles	182 cm^2	346 cm^2 187 cm^2 48 cm^2			97.9 92.3 100		
	Circular Disks	182 cm^2	182 cm^2			49.6		
Solid Styrofoam Cylinders (Straight Up)	Solid Styrofoam Cubes (Flat Face Forward)	H 4 cm D 4 cm H 5 cm D 5 cm H 6 cm	H 4 cm D 4 cm H 5 cm D 5 cm H 6 cm D 6 cm H 4 cm D 4 cm H 5 cm D 5 cm H 6 cm D 6 cm H 4 cm	2 Alt Forced Choice	2 m	75 94 91 87 87 98	Tursiops Truncatus	Nachtigall, Murchison, And Au (This Volume)

Table 2 (Cont'd). Shape Discrimination

| TARGET SHAPE & MATERIAL | | TARGET SIZE CM | | METHOD | DISTANCE TO TARGET | PERCENT CORRECT | SPECIES | SOURCE |
POSITIVE	NEGATIVE	POSITIVE	NEGATIVE					
		D 6 cm	D 4 cm H 5 cm D 5 cm H 6 cm D 6 cm		2 m	91 94 96		
Cylinders Straight Up	Cubes Edge Forward	All Sizes	All Sizes	2 Alt Forced Choice Probe		93		
Cylinders Layed Horizontal	Cubes Edge Forward	All Sizes	All Sizes	2 Alt Forced Choice Probe		71		
Cylinders Flat Top Forward	Cubes Flat Face Forward	All Sizes	All Sizes	2 Alt Forced Choice Probe		57		

Table 3. Materials Discrimination

SHAPE	TARGET MATERIAL & SIZE (CM) POSITIVE	NEGATIVE	METHOD	PERCENT CORRECT	MIN. DIST. TO TARGETS	SPECIES	SOURCE
Circular Disks	Copper D-20 T.22	Aluminum D-20 T.32 D-20 T.64 D-20 T.79 Brass D-20 T.32 D-20 T.64	2 Alt Forced Choice	100 98 93 55 100	∅	Tursiops Truncatus	Evans & Powell (1967)
Cylinders H-17.78 cm	Hollow Aluminum D-3.81 WT-.64 D-7.62 WT-.95	Solid Coral Rock/ Epoxy Resin D-3.81 D-7.62 D-3.81 D-7.62 Solid Aluminum D-3.81 D-7.62 Hollow Aluminum D-6.35 WT-.48 D-11.43 WT-.64	2 Alt Successive Presentations (Baseline) 2 Alt. Successive Presentations Probe	 100 100 100 100 100 100 100 100	6 m 16 m	Tursiops Truncatus	Hammer (1978) Experiment I

Table 3 (Cont'd). Materials Discrimination

SHAPE	TARGET MATERIAL & SIZE (CM) POSITIVE	NEGATIVE	METHOD	PERCENT CORRECT	MIN. DIST. TO TARGETS	SPECIES	SOURCE
		Solid Coral Rock/Epoxy Resin D-11.43		100			
		Solid Chloroprene D-6.35		100			
		D-4.06		100			
		Polyvinyl-Chloride D-7.62 WT-.79		100			
Cylinders H-17.78 cm	Hollow Aluminum	Solid Coral Rock/Epoxy Resin	2 Alt Successive Presentations (Baseline)			Tursiops Truncatus	Hammer (1978) Experiment III
	D-3.81 WT-.32	D-3.81		100			
		D-7.62		100			
	D-7.62 WT-.40	D-3.81		100			
		D-7.62		100			
		Hollow Bronze	2 Alt Successive Presentations Probe				
		D-3.81 WT-.32		75	6 m*		
		D-3.81 WT-.32		37.5	16 m		
		D-7.62 WT-.40		100	6 m*		
		D-7.62 WT-.40		62.5	16 m		

Table 3 (Cont'd).   Materials Discrimination

TARGET					MIN.		
SHAPE	MATERIAL & SIZE (CM)		METHOD	PERCENT CORRECT	DIST. TO TARGETS	SPECIES	SOURCE
	POSITIVE	NEGATIVE					
		Hollow Glass					
		D-3.81 WT-.32		∅	6 m*		
		D-3.81 WT-.32		∅	16 m		
		D-7.62 WT-.40		12.5	6 m*		
		D-7.62 WT-.40		12.5	16 m		
		Hollow Steel					
		D-3.81 WT-.32		87.5	6 m*		
		D-3.81 WT-.32		100	16 m		
		D-7.62 WT-.40		100	6 m*		
		D-7.62 WT-.40		100	16 m		

KEY:
 D = Diameter
 T = Thickness
 WT = Wall Thickness
 H = Height
 * = 6 m data from the second series
     following the 16 meter data collection.

REFERENCES

Abramov, A. P., Ayrapet'yants, E. Sh., Burdin, V. I., Golubkov,
    A. G., Yershova, I. V., Zhezherin, A. R., Korolev, V. I.,
    Malyshev, Yu. A., Ul'yanov, G. K., and Fradkin, V. G., 1971,
    Investigations of delphinid capacity to differentiate be-
    tween three-dimensional objects according to linear size and
    material, "Report from the 7th All-Union Acoustical Confer-
    ence", Leningrad, as cited in: Bel'kovich, V. M., and Du-
    brovskiy, N. A., 1976, "Sensory Bases of Cetacean Orient-
    ation", Nauka, Leningrad (English translation JPRS L/7157).
Ayrapet'yants, E. Sh., Golubkov, A. G., Yershova, I. V., Zhezherin,
    A. R., Zvorkin, V. N., and Korolev, V. I., 1969, Echoloca-
    tion differentiation and characteristics of radiated echo-
    location pulses in dolphins, Report of the Academy of Sci-
    ence of the USSR, 188:1197 (English translation JPRS 49479).
Ayrapet'yants, E. Sh., and Konstantinov, 1974, "Echolocation in
    Nature", Nauka, Leningrad (English translation JPRS 63328).
Babkin, V. P., Dubrovskiy, N. A., Krasnov, P. S., and Titov, A. A.,
    1971, Discrimination of material of spherical targets by
    the bottlenose dolphin, in: "Report from the 7th All-Union
    Acoustical Conference, Leningrad".
Bagdonas, A., Bel'kovich, V. M., and Krushinskaya, N. L., 1970,
    Interaction of analyzers in dolphins during discrimination
    of geometrical figures under water, J. Higher Neural Act.,
    20:1070 (English translation in: "A Collection of Trans-
    lations of Foreign Language Papers on the Subject of Bio-
    logical Sonar Systems", K. J. Diercks, 1974, ed., Applied
    Research Lab, U. of Texas, Austin, Tech. Rept. 74-9).
Barta, R. E., 1969, Acoustical pattern discrimination by an Atlan-
    tic bottlenosed dolphin, unpublished manuscript, Naval
    Undersea Center, San Diego, Ca.
Bel'kovich, V. M., and Borisov, V. I., 1971, Locational discrimi-
    nation of figures of complex configuration by dolphins,
    Trudy Akusticheskogo Institute, 17:19.
Bel'kovich, V. M., Borisov, I. V., Gurevich, V. S., and Krushin-
    skaya, N. L., 1969, Echolocating capabilities of the common
    dolphin (Delphinus delphis), Zoologicheskiy Zhurnal, 48:876.
Bel'kovich, V. M., and Dubrovskiy, N. A., 1976, "Sensory Bases of
    Cetacean Orientation", Nauka, Leningrad, (English transla-
    tion JPRS L/7157).
Busnel, R.-G., and Dziedzic, A., 1967, Résultats métrologiques
    expérimentaux de l'écholocation chez le Phocoena phocoena
    et leur comparaison avec ceux de certaines chauves-souris.
    in: "Animal Sonar Systems, Biology and Bionics", R.-G. Bus-
    nel, ed., Laboratoire de Physiologie Acoustique, Jouy-en-
    Josas, France.
Busnel, R.-G., Dziedzic, A., and Andersen, S., 1965, Seuils de
    perception du système sonar du Marsouin Phocoena phonoeca
    en fonction du diamètre d'un obstacle filiforme, C. R. Acad.

<u>Sci.</u>, 260:295.

Dubrovskiy, N. A., 1972, Discrimination of objects by dolphins
using echolocation, "Report of the 5th All-Union Conference
on Studies of Marine Mammals, Part 2, Makhachkala", as cited
<u>in</u>: Ayrapet'yants, E. Sh. and Konstantinov, A. I., 1974,
"Echolocation in Nature", Nauka, Leningrad, (English Translation JPRS 63328-2).

Dubrovskiy, N. A., and Krasnov, P. S., 1971, Discrimination of
elastic spheres according to material and size by the bottlenose dolphin, <u>Trudy Akusticheskogo Institute</u>, 17, as cited
<u>in</u>: Bel'kovich, V. M., and Dubrovskiy, N. A., 1976, "Sens-
ory Bases of Cetacean Orientation", Nauka, Leningrad (English translation JPRS L/7157).

Dubrovskiy, N. A., Krasnov, P. S., and Titov, A. A., 1971, On the
question of the emission of ultrasonic ranging signals
by the common porpoise, <u>Akusticheskiy Zhurnal</u>, 16:521 (English translation JPRS 52291).

Dubrovskiy, N. A., Krasnov, P. S., and Titov, A. A., 1971, Discrimination of solid elastic spheres by an echolocating
porpoise, <u>Tursiops truncatus</u>, <u>in</u>: "Proc. 7th International
Conf. Acoust.", Budapest.

Evans, W. E., 1973, Echolocation by marine delphinids and one
species of fresh-water dolphin, <u>J. Acoust. Soc. Am.</u>, 54:191.

Evans, W. E., and Powell, B. A., 1967, Discrimination of different
metallic plates by an echolocating delphinid, <u>in</u>:"Animal
Sonar Systems, Biology and Bionics", R.-G. Busnel, ed.,
Laboratoire de Physiologie Acoustique, Jouy-en-Josas, France.

Fadeyeva, L. M., 1973, Discrimination of spherical targets with
different echo signal structure by the dolphin, <u>in</u>: "Report
from the 8th All-Union Acoustical Conference, Moscow, 1"
As cited <u>in</u>: Bel'kovich, V. M., and Dubrovskiy, N. A., 1976,
"Sensory Bases of Cetacean Orientation", Nauka, Leningrad.
(English translation JPRS L/7157).

Hammer, C. E., 1978, Echo-recognition in the porpoise (<u>Tursiops</u>
<u>truncatus</u>): an experimental analysis of salient target
characteristics, Naval Ocean Systems Center, San Diego,
Tech. Rep. 192.

Johnson, C. S., 1967, Discussion to paper by Evans and Powell,
<u>in</u>: "Animal Sonar Systems, Biology and Bionics", R.-G. Bus-
nel, ed., Laboratoire de Physiologie Acoustique, Jouy-en-
Josas, France.

Johnson, C. S., 1967, Sound detection thresholds in marine mammals,
<u>in</u>: "Marine Bio-Acoustics, Proc. of the Second Symposium
on Marine Bio-Acoustics, New York", W. N. Tavolga, ed.,
Pergamon Press, New York.

Kellogg, W. N., 1958, Echo ranging in the porpoise, <u>Science</u>, 128:982.

Norris, K. S., Evans, W. E., and Turner, R. N., 1967, Echolocation
of an Atlantic bottlenose porpoise during discrimination,
<u>in</u>: "Animal Sonar Systems, Biology and Bionics", R.-G. Bus-
nel, ed., Laboratoire de Physiologie Acoustique, Jouy-en-

Josas, France.

Schusterman, R. J., 1979, Behavioral methodology in echolocation
    by marine mammals, this volume.

Schusterman, R. J., Kersting, D. A., and Au, W. W. L., 1979, Stim-
    ulus control of echolocation pulses in Tursiops truncatus,
    this volume.

Titov, A. A., 1972, Investigation of sonic activity and phenomeno-
    logical characteristics of the echolocation analyzer of
    Black Sea delphinids. Canditorial dissertation, Karadag,
    as cited in: "Sensory Bases of Cetacean Orientation", Nauka,
    Leningrad (English translation JPRS L/7157).

Turner, R. N. and Norris, K. S., 1966, Discriminative echolocation
    in a porpoise, J. Exp. Anal. Behav., 9:535.

Zaslavskiy, G. L., Titov, A. A., and Lekomtsev, V. M., 1969,
    Research on the sonar abilities of the bottlenose dolphin,
    Trudy Akusticheskogo Instituta, 8:134, as cited in:
    Ayrapet'yants, E. Sh., and Konstantinov, A. I., 1974,
    "Echolocation in Nature", Nauka, Leningrad (English trans-
    lation JPRS 63328).

CETACEAN OBSTACLE AVOIDANCE

Patrick W. B. Moore

SEACO, Inc.
146 Hekili St.
Kailua, Hawaii 96734

As Schevill (introduction to McBride, 1956) has pointed out, the original ideas about cetacean echolocation were based on observations of the porpoises' ability to avoid obstacles in turbid waters. McBride (1956) noted that although porpoises (Tursiops truncatus) avoided fine mesh nets, by either rolling over the corkline or swimming directly through any openings in the net, they would charge larger mesh nets (24 cm) and were easily caught.

W. N. Kellogg in 1958 performed the first experimental obstacle avoidance task involving cetaceans. Two bottle-nosed dolphins (T. truncatus) were placed in a specially dredged pool, 16 x 21 m and 1.6 to 2.1 m deep (depending on tide level). The water was turbid (visibility less than 50 cm). Using suspended barriers of triangular metal poles or posts 1.32 m long, 5.08 cm wide and spaced 2.44 m apart, the number of animal collisions were recorded in twenty minute sessions. Only four such collisions occurred in the first session by the two animals. In those instances the animal had passed the obstacle but brushed it with its tail flukes. Sessions were also conducted during dark, moonless nights with no resulting collisions.

Kellogg (1958) also conducted experiments involving glass and plexiglass barriers. The experimenter required the porpoise to select the one of two openings in a suspended net which did not contain a barrier in order to receive a fish. Additional incentive to pass through the selected opening was generated by slowly crowding the animal with a larger net. A combined 98% correct performance level was demonstrated in the plexiglass experiment and 100% correct level using plate glass.

Interestingly enough, it is now known that even though porpoises can detect the presence of a plexiglass (or acrylic) barrier, a thin sheet of this material is sufficiently transparent to the echolocation pulses and return echoes of the animal that it can detect objects through it (Penner and Murchison, 1970). The visual analogy would be looking through a glass window; one knows it is there, but it is still transparent.

Norris, Prescott, Asa-Dorian and Perkins (1961) demonstrated echolocation unequivocably by blindfolding a female porpoise (T. truncatus) with soft rubber eye cups and requiring her to swim through a series of mazes formed by two rows of iron pipes. The rows were 3.05 m apart and pipes were suspended at 1.22 to 2.13 m intervals. The animal swam through various maze patterns several times a day over a 60-day period. Norris et al. (1961) reported that only one collision occurred during this time, and they considered that collision to be deliberate.

Busnel and Dziedzic (1967) were the first to attempt to make a defined and quantifiable evaluation of the performance of an underwater biosonar system other than T. truncatus, using a harbor porpoise (Phocoena phocoena) in an obstacle avoidance experiment with various diameter iron, steel, or copper wires (diameters of 92 mm to 0.2 mm). These obstacles were vertically suspended in a tank 1 m apart in alternate rows. The animal was either blindfolded or sighted. When blindfolds were used the animal showed "practically 100%" avoidance for wire diameters between 4 and 2.8 mm, 90% for diameters of 2.8 and 0.5 mm, 79% for diameters of 0.35 mm and finally reached 46% for diameters of 0.2 mm.

When the animal was sighted the results indicated "practically 100%" performance for wire diameters down to 0.75 mm; 90% for diameters of 0.6 mm and 0.5 mm; and 79% for diameters of 0.35 and 0.2 mm. The authors reported that "there appeared to be no correlation between the nature of the metal thread and ability of the animal to avoid it". Measurement of the acoustic signals emitted by this animal was reported to have a maximum intensity at 2 kHz.

Two problems seem to exist with Busnel and Dziedzic's original experiments. First, the reported peak frequencies around 2 kHz would indicate the detection of wire diameters of extremely small fractional sizes compared to the wavelength of the echolocating pulse. Schevill, Watkins, and Ray (1969) have pointed out that the large number of harmonic frequencies found by Busnel and Dziedzic were probably a result of their use of recordings made with systems of limited response. Schevill et al. (1969) reported occasional frequencies as high as 10 kHz in the clicks of P. phocoena with the "favored click" generally

narrow band and less than 2 kHz. Anderson's (1970) audiogram of
this species indicates their best hearing range is between 4-40
kHz, with upper frequency limits near 150 kHz. Supin and Sukhoruchenko (1974) found Phocoena hearing as high as 190 kHz, with the
best frequency range between 64 and 128 kHz; and Sukhoruchenko
(1973) says they hear from 3-190 kHz. Comparison of these Phocoena
hearing studies with the echolocation pulse frequencies reported
by Busnel and Dziedzic, and Schevill et al. indicates some discrepancy, as previous studies have indicated that other echolocating
animals produce signals that are matched to their hearing abilities
(Johnson, 1966; Diercks et al., 1973).

        The second problem in the Busnel and Dziedzic obstacle
avoidance study was related to the 0.35 mm detection threshold.
Penner and Murchison (1970) during a study of Inia geoffrensis
wire detection thresholds found that data fluctuated widely on
wire diameters below 1.0 mm. The confounding variable turned out
to be minute bubble formation on the wires which occurred within
60 sec. following submersion. An elaborate procedure involving
special polishing of target wires, wetting agent treatment between
trials, and greatly restricted immersion times (less than 15 sec)
was developed to eliminate this confounding variable. However,
Busnel and Dziedzic apparently did not eliminate bubble formation
on the wires used in the Phocoena experiments (see Schusterman,
this volume).

FIELD OBSERVATIONS

        Field recordings of cetacean sound and associated behavior
have been conducted in an attempt to discover the extent of
echolocation throughout this group of animals. Recordings of
blue whale (Balaenoptera musculus) clicks and Minke whale
(B. acutorostrata) clicks by Beamish and Mitchell (1971, 1973)
and gray whale (Eschrichtius robustus) "clicks" by Wenz (1964),
Gales (1966), Poulter (1968) and Fish, Sumich and Lingle (1974)
suggested that mysticete cetaceans may also be possible echolocators.

        However, Eberhardt and Evans (1962) observed gray whales in
their calving ground in Baja, California, to collide with an
acoustically reflective barrier. Recordings were also made of
the associated sound activity of the gray whales. Eberhardt and
Evans erected an acoustic barrier consisting of a string of 5.08
cm diameter, 4.57 m aluminum tubes, 6.09 m apart. This barrier
was placed part way across the Piedra channel. Gray whale sounds
were recorded as were the "sounds caused by spar buoys hitting
together as whales crashed through the barrier". They noted that
"no biological sounds from the whales were evident" (Eberhardt
and Evans, 1962).

During this same research cruise, five Tursiops gilli were
observed moving toward the channel barrier and echolocation
clicks were recorded. As the animals approached the barrier they
moved over to shoal water and schooled up. After five minutes
one animal made a "sonar run" on the barrier and returned to the
group. Several more single animal runs were observed, followed
by the entire group slowly proceeding up to and through the
barrier with no collisions reported (Evans and Dreher, 1962).

The reported collisions of the gray whales do not completely
eliminate the possibility of echolocation in an animal producing
sounds of long wavelengths. It is very possible that these
animals could use the various sounds they produce to obtain echo
information about water depth, gross bottom contours and large
obstacles in much the same fashion that blind persons detect
walls and estimate room size based on the timbre and other echo
characteristics produced by their footsteps, fingersnaps or their
own voices (Rice, C. E., Feinstein, S. H., Schusterman, R. J.,
1965). Although it may not be the primary function of their
"singing", water depth information is certainly available in the
echoes of humpback whale (Megaptera novaeangliae) "songs".

Evidence for echolocation among the mysticeti may remain in
question for some time, as their size would seem to eliminate the
possibility of the kinds of captive experiments necessary to
prove echolocation.

THE TUNA-PORPOISE PROBLEM

In order to reduce the annual porpoise mortality rate due to
purse seine fishing of yellowfin tuna, several lines of investigation have been pursued that are directly related to obstacle
avoidance by the various species of porpoise (Stenella attenuata,
S. longirostris, Delphinus delphis) which are associated with the
tuna catch. The basic question is how the porpoise group associated
with the tuna school can be released from the net confines
without significant loss of the tuna catch.

According to Norris et al. (1978), the tuna catch procedure
is as follows: After the porpoise group has been spotted, the
seiner "tends to turn continuously counterclockwise in an ever
diminishing spiral". At the onset this spiral is quite large and
speedboats "work the area where the wake is weakest and reinforce
the wake by turning in arcs and circles at high speed". The
porpoise, approaching the bubble wake, turn and avoid the wake.
The seiner drops the net skiff "when the diameter of the spiral
reaches the appropriate size and when the wind is in the proper
quadrant", and starts to deploy the net in a circle-like tract,

using the wake (an acoustic obstacle) and the net (an acoustic obstacle and a physical barrier) to surround the animals. Entrapment is complete when the seiner and net skiff rejoin.

Norris et al. (1978) has clearly described the general behavior of the porpoise group, saying that during the capture sequence the animals stopped in the spiral-shaped bow wake of the seiner, about 137 m from the wake. This suggests biosonar detection of the bubble wake. They do not penetrate the wake or approach it. Some do, however, get past the speed boats attempting to herd the animals, passing behind the smaller boats. The majority of the animals form an aggregate inside the net. Within this aggregate a compression of natural behavior occurs with associated increases of aggressivity and acoustic behavior. A few animals dive below a rafting group (animals hanging quietly in the water) down to approximately 60 feet.

W. F. Perrin and J. R. Hunter (1972) performed a number of experiments to assess the escape behavior of the spinner porpoise (S. longirostris). Perrin and Hunter tried to discover what configuration, if any, of a rescue gate in a purse-seine net could be best recognized and utilized by the entrapped porpoise. The subjects were five spinner porpoises, three of which were trained animals and two naive animals. The animals were tested in a 24.7 m circular crowding tank, 4 meters deep at the center. Nets were placed around the edge to create a 20 m diameter circular crowding chamber. Two radial crowding nets, one fixed and one pivoting, extended from a central aluminum mast to the outer edge of the chamber. The moving net wall was used to simulate a slowly closing net, crowding the animals closer to the fixed net which was modified for various types of escape openings. The net was tuna purse-seine webbing 10.8 cm, #42 thread, knotted nylon. Flotation was by purse-seine corkline with 15.0 cm diameter x 9 cm sponge plastic floats. The basic 5.5 m wide x 1.8 m deep escape opening was modified via a purse-seine net flap to decrease the width and/or depth. A cork line was strung across the surface of the opening to test responses to an overhead surface barrier, and various low acoustic reflective panel materials were attached for tests of responses to an occluded opening. The panels consisted of: 8.6 cm #12 mono filament webbing, a .38 mm or 1.04 mm thick polyvinyl sheet. Results of the various experiments were expressed as a function of area of chamber, at escape, in m vs. trials.

The effect of width of opening was tested and, generally, the results suggested that the opening width was not a significant factor effecting escape if it exceeded 1.5 m. Interesting behavior occurred with failure to escape in these experiments; occasionally, the test animal would expel air and sink passively

to the bottom of the chamber and would not move even when the net
was completely collapsed. Similar behavior was seen in the tuna
net at sea (Norris K., Personal Comm.).

   The effect of depth was tested by presenting various hole
depths of 0.15 m to 1.8 m and holding hole width at 5.5 m. The
authors' tentative conclusion was that a depth of opening of 1 m
or greater is "optimal" for escape. When a corkline was placed
across the opening, initial escape performance always decreased.
However, with greater depth of opening, the effect of the corkline
on escape performance was lessened.

   When a monofilament web panel was inserted into the opening
all animals swam into the webbing "as if it did not exist"
(Perrin & Hunter, 1972), and the animals had to be unentangled.
With a clear polyvinyl sheet placed in the opening (only for two
trials with the same animal) the porpoise hit the panel and slid
over the top as the panel buckled. Noteworthy was the fact that
with the polyvinyl sheet the animal passed back and forth over
the panel several times while the enclosure was constantly being
reduced in size.

   These series of experiments were repeated with two naive
porpoises beginning the day after capture. Both these animals
showed erratic behavior in the experimental area due to what was
believed to be shock or extreme fright and in the depth experiment one of the animals when confronted with a 0.61 m deep
opening dove and became entangled in the net. Width of opening
and corkline experiments with the non-naive animals (captive for
three years) showed the same general trends as with the naive
animals.

   The authors generally suggest that an escape gate in a purse
seine net should be no less than 1.5 m wide and 1.0 m deep.
There should be no corkline; self-actuating release ports, if
used, should be made of acoustically transparent materials.

   This study suggests that an entrapped spinner porpoise, when
faced with a net barrier obstacle and stressed by crowding, can
locate, either visually or acoustically, a near surface void and
manage an escape, given that the void meets some general size and
configurational criteria. Also, the results of the acoustically
transparent panels and net material suggest that echolocation was
used to access the escape route.

   Leatherwood, Johnson, Ljungblad and Evans (1977) measured
the target strength of various (1¼, 2 and 4½ inches) tuna net
panels at 0° and 35° aspect to access their "acoustic" visibility
to echolocating porpoises. Using a broadband sound source
representing a porpoise echolocation signal, Leatherwood et al.

reported that, "porpoises utilizing their echolocation system, should be able to "see" any of the mesh sizes measured regardless of the angles at which they are oriented and regardless of the degree of stretch in the nets". However, the degree of acoustic visibility varied 17dB between net types. They note that this 17 dB difference in target strength may have significant advantage in the complex acoustic environment of the purse seine net (see Murchison, this volume). Of course, to what extent the animals can or do echolocate inside the purse seine net is unknown.

Awbrey (in press) investigated the general acoustic environ of the purse seining operation during October 1976 to discover to what extent, if any, the purse seining operation would mask porpoise echolocation or communication signals. Recordings were made during one "water set" (no animals present) and nine actual porpoise-tuna sets.

Measurements of noise produced by net boat, the speed boats, and the sound of the net itself were analyzed. (Measurements reported were in dBre 1 μpa.) When the net boat approaches the tuna-porpoise aggregate, 100–200 m from the animals, 91% of the total noise energy was below 6 kHz peaking at 360 Hz. Also, low frequency pulsations were caused by propeller beats with sound pressure levels (SPL) of 120–125 dB at 200 m from the ship, 85% of the speed boat noise (Peak = 125 dB for short periods) during the set was below 6 kHz peaking at 4 kHz (SPL of 117 dB). Once the net is out and pursing begins high frequency clashing sounds of the pursing rings at the bottom of the net are produced. These transient sounds peaked at 5–6 kHz and contained energy at frequencies above 30 kHz (no SPL given). After the net is out and the pursing operation is completed the general net ambient noise is considerably reduced and noise is primarily generated from the hauling in of the net (SPL of 105–115 dB with major energy below 3 kHz). Measurements of animal whistles during the hauling operation suggest that vessel noise had little effect on the general characteristics of the vocalization. Lastly, during the "backing down" operation some low frequency noise bursts are produced by the net boat (70% energy below 6 kHz) at an SPL of 125 dB.

Awbrey (in press) summarizes his report by stating that "reducing vessel noise would probably contribute very little to reduction of porpoise mortality", in that the majority of the purse seining noise lies in a frequency range below the animal's vocalizations and echolocation clicks and probably does not prevent the animals from hearing their own whistle or echolocating on the net. However, the noise can be considered an aversive stimulus to the animals because loud transients do elicit fright responses.

Perryman and Au (1977) observed porpoises in the Commission Yellowfin Regulatory Area beginning to avoid the seiners at distances of 5-6 km. Awbrey (in press) reports that both spinner and spotters now began avoidance behavior (running or lying quietly) at greater distances, suggesting that with repeated exposure to purse seining the animals gain sophistication with respect to avoiding the tuna fishing operation. This notion is also supported by observations made by Pryor and Norris (1978). During the initial chase the animals are herded by the speed boats and are eventually stopped by the vast bubble curtain created by the bow wake of the large seiner, the net skif is set free and while the speed boats herd the animals, the net is set around the tuna-porpoise pod; however, some animals have "learned.. ...to go under the bubble curtain of the ship" (Pryor and Norris, 1978).

Ingrid Kang, (Personal Comm.) an observer on the vessel Queen Mary, reported spinners occasionally jumping and flipping inside   the net. Spotters, on the other hand, tended to raft. When a small dingy was inside the net with the animals, both spotters and spinners would avoid the dingy, but if divers entered the water, the animals did not seem to try to avoid them, and the rafting animals could easily be approached. Additionally, the tighter the net confines, the less difficult it was to approach the animals. Also, at no time was an animal observed to charge the net. During some net sets on combined spinners and spotters, large amounts of small bubbles were present throughout the animal aggregate, generated by the spinners whistling vocalizations; obviously these bubbles would have reduced the acoustic transmission quality of the surrounding water and hamper the animal's echo-location signals when it would seem to have been needed the most. Results from the Perrin and Hunter (1972) study and the Leatherwood et al. (1977) study coupled with Awbrey's (in press) noise measurements suggests that the porpoises could have used echolocation to find and exit the tuna net. The extent to which these animals utilize information, gathered via the various sensory modalities of passive hearing, echolocation, and vision, to evaluate and react to the tuna net operation is unknown. Also, the relative dominance of these sensory information channels, if any, is a purely speculative matter at this point.

In general, the entire tuna fishing operation is a complex of avoidance behaviors for the porpoise, the exact meaning of which has yet to be completely explained, but certainly relates to the acoustic abilities and obstacle avoidance strategies used by these animals. Also, to what degree the visual, echolocation and passive hearing systems share prominance in information acquisition is a question of current interest (see Schusterman, this volume).

Presently, Japan is involved in deep sea salmon fishing, using 16 km x 6 m monofilament gill nets. This fishing operation accounts for a seasonal mortality of Dall's porpoise of 10-20 thousand when animals became entangled and drowned over night in the nets. Various other small cetaceans are also subject to gill net fishing operations from Greenland to Northern Australia (Evans, W. E. and Norris, J., Personnal Comm.). The Perrin and Hunter (1972) study, using a monofilament panel, shows similar non-avoidance to these presumably acoustically transparent nets.

CETACEAN STRANDING

W. H. Dudock Van Heel (1962, 1966) put forth the hypothesis that mass strandings of Cetaceans were due to the inability of the animal to correctly navigate or identify a gently shelving shoreline due to either the physical properties of the returned echo (i.e. reduced strength, reverberation) or the animals not "attending" to the low frequency components of the return pulse which would, presumably, give an indication of a gently shelving shoreline. Other explanations, such as group sickness or parasitic infestations coupled with ideas of a disoriented dominant or leader animal stranding a generally healthy following, also abound (Wood, 1978).

F. G. Wood (1978) suggested a non-acoustic stress hypothesis to explain mass strandings. The stress hypothesis suggests that highly stressful situations would cause basic and primitive behavioral responses, and the stranding act is seen as a stress or fear response, which is a primitive and evolutionarily regressive response, to seek safety on land. Stress may also act as the catalyst to produce other behaviors. The observed behaviors of rafting inside the tuna net or the passive sinking of the spinner porpoise noted in the Perrin and Hunter (1972) study may also be stress induced.

CONCLUSION

The use of the obstacle avoidance paradigm in cetacean echolocation research has declined in recent years, since the technique does not yield the rich and precise results of more rigorous experimental designs (see Schusterman, this volume). It does, however, offer the advantage of less animal training.

It should be noted, however, that to date the only behavioral evidence relating to echolocation capabilities in mysticetes has come from field-conducted obstacle avoidance studies (Eberhardt and Evans, 1962) suggesting that obstacle avoidance has not yet

outlived its usefulness as a field observational technique
intended to glean echolocation behavior in less well documented
species.

ACKNOWLEDGEMENTS

I would like to thank Ronald Schusterman, Paul Nachtigall
and Earl Murchison for the helpful advice and support in the
preparation of this manuscript, Phyllis Johnson for her cooperation and understanding in typing this subchapter, Jack Harmon of
SEACO, Inc., for his financial support and the Naval Ocean Systems Center for providing this opportunity.

REFERENCES

Andersen, S., 1970, Directional hearing in the harbor porpoise,
        Phocoena phocoena, in: "Investigation on Cetacea, Vol. II",
        G. Pilleri, ed., Benteli Ag., Berne.
Awbrey, F. T., (in press), Background study of acoustical and
        bioacoustical factors in tuna fishing productivity and
        associated porpoise mortality, Fish. Bull.
Beamish, P., and Mitchell, E., 1971, Ultrasonic sounds recorded
        in the presence of a blue whale Balaenoptera musculus, Deep-
        Sea Res., 18:803.
Beamish, P., and Mitchell, E., 1973, Short pulse length audio
        frequency sounds recorded in the presence of a Minke whale
        Balaenoptera acutorostrate, Deep-Sea Res., Abstra., 20:375.
Busnel, R.-G., and Dziedzic, A., 1967, Résultats métrologiques
        expérimentaux de l'écholocation chez le Phocoena phocoena
        et leur comparaison avec ceux de certaines chauves-souris,
        in: "Animal Sonar Systems, Biology and Bionics", R.-G. Bus-
        nel, ed., Laboratoire de Physiologie Acoustique, Jouy-en-
        Josas, France.
Diercks, K. J., Trochta, R. T., and Evans, W. E., 1973, Delphinid
        sonar: measurement and analysis, J. Acoust. Soc. Amer.,
        54:200.
Dudok van Heel, W. H., 1962, Sound and Cetacea, Neth. Jour. Sea
        Res., 1:407.
Dudok van Heel, W. H., 1966, Navigation in Cetacea, in: "Whales,
        Dolphins, and Porpoises", K. S. Norris, ed., U. of Calif.
        Press, Berkeley.
Eberhardt, R. L., and Evans, W. E., 1962, Sound activity of the
        California gray whale, Eschrichtius glaucus, J. Aud. Eng.
        Soc., 10:324.
Evans, W. E., and Dreher, J. J., 1962, Observations on scouting
        behavior and associated sound production by the Pacific
        bottlenosed porpoise (Tursiops gilli Dall), Bull. Soc.
        Calif. Acad. Sci., 61:217.

Fish, J. F., Sumich, J. L., and Lingle, G. L., 1974, Sounds pro-
    duced by the gray whale, Eschrichtius robustus, Mar. Fish
    Rev., 36:38.
Gales, R. S., 1966, Pickup analysis and interpretation of under-
    water acoustic data, in "Whales, Dolphins, and Porpoises",
    K. S. Norris, ed., U. of Calif. Press, Berkeley.
Johnson, C. S., 1966, Auditory thresholds of the bottlenosed
    porpoise, (Tursiops truncatus Montagu), N. O. T. S. TP 4178.
Kellogg, W. N., 1958, Echo ranging in the porpoise, Science,
    128:982.
Leatherwood, J. S., Johnson, R. A., Ljungblad, D. K., and Evans,
    W. E., 1977, "Broadband measurements of underwater acoustic
    target strengths of panels of tuna nets", NOSC TR 126, San
    Diego.
McBride, A. F., 1956, Evidence for echolocation by cetaceans,
    Deep-Sea Res., 3:153.
Murchison, A. E., 1979, Detection range and range resolution of
    echolocating bottlenose porpoise (Tursiops truncatus),
    this volume.
Norris, K. S., Prescott, J. H., Asa-Dorian, P. V., and Perkins, P.,
    1961, An experimental demonstration of echolocation behav-
    ior in the porpoise, Tursiops truncatus (Montagu), Biol.
    Bull., 120:163.
Norris, K. S., Stuntz, W. E., and Rogers, W., 1978, The behavior
    of porpoises in the eastern tropical Pacific yellowfin tuna
    fishery - Preliminary Report, U.S. Marine Mammal Comm.,
    Report PB-283 970, Washington, D.C.
Penner, R., and Murchison, A. E., 1970, Experimentally demonstrated
    echolocation in the Amazon river porpoise Inia geoffrensis
    (Blainville), Proc. 7th Ann. Conf. Biol. Sonar and Diving
    Mammals, 7:17.
Perrin, W. F., and Hunter, J. R., 1972, Escape behavior of the
    Hawaiian spinner porpoise (Stenella cf. S. longirostris),
    U.S. Nat. Mar. Fish. Bull., 70:49.
Perryman, W., and Au, D., 1977, Aerial observations of evasive
    behavior of dolphin schools, Proc. 2nd Conf. Biol. Mar.
    Mammals, Abstr.
Poulter, T. C., 1968, Vocalization of the gray whales in Laguna
    Oja de Liebre (Scammon's Lagoon), Baja, California, Mexico,
    Norsk Hvalfangsttid, 58:53.
Pryor, K. and Norris, K. S., 1978, The tuna porpoise problem:
    Behavioral aspects, Oceanus, 21:31.
Rice, C. E., Feinstein, S. H., and Schusterman, R. J., 1965, Echo-
    detection ability of the blind: size and distance factors,
    J. Exp. Psych., 70:246.
Schevill, W. E., 1956, Evidence for echolocation by cetaceans,
    Deep-Sea Res., 3:153.
Schevill, W. E., Watkins, W. A., and Ray, C., 1969, Click structure
    in the porpoise, Phocoena phocoena, J. Mamm., 50:721.

Schusterman, R. J., 1979, Behavioral methodology in echolocation
        by marine mammals, this volume.
Sukhoruchenko, M. N., 1973, Frequency discrimination in dolphins
        (Phocoena phocoena), Sechenov Physiological Jour. of the
        U.S.S.R., 59(8):1205.
Supin, A. Ya., and Sukhoruchenko, M. N., 1974, Characteristics of
        acoustic analyzer of the Phocoena phocoena L. dolphin, in:
        "Morphology, Physiology, and Acoustics of Marine Mammals",
        V. Ye. Sokolov, ed., Nauka, Moscow (English translation
        JPRS 65139).
Wenz, G. M., 1964, Curious noises and the sonic environment in
        the ocean, in: "Marine Bio-Acoustics, Proceedings of the
        First Symposium on Marine Bio-Acoustics, Bimini", W. N.
        Tavolga, ed., Pergamon Press, New York.
Wood, F. G., 1978, The cetacean stranding phenomenon: an hypothesis,
        in: "Report on the Marine Mammal Stranding Workshop",
        N.T.I.S., Washington, D.C.

PERFORMANCE OF AIRBORNE ANIMAL SONAR SYSTEMS:

I. MICROCHIROPTERA

Hans-Ulrich Schnitzler and O'Dell W. Henson, Jr.

Fachbereich Biologie, Philipps-Universität
355 Marburg/Lahn BRD
Department of Anatomy, U. of North Carolina
Chapel Hill, N.C. 27514 U.S.A.

INTRODUCTION

Animals living in the dark are faced with the problem that vision is only of limited use for orientation in the environment and for the detection and identification of relevant targets. In order to cope with this situation, some animals, during evolution, developed eyes which were especially adapted for good visual efficiency at low light intensities whereas others improved the remaining sensory systems.

Echolocation is one of the adaptations that has evolved for a life in the dark. Echolocating animals emit sounds and analyze the returning echoes in order to detect the presence and character of the reflecting targets. In contrast to other sensory modes, echolocation is an active system which uses its own controlled energy beam to illuminate interesting parts of the environment and to probe object characteristics.

Of the numerous nocturnal birds, only the cave-dwelling oil bird and cave swiftlets use echolocation. In mammals, echolocation systems of low efficiency have been reported for some species of rodents, insectivores and Megachiroptera, whereas Microchiroptera and cetaceans use very efficient and sophisticated systems.

In the first part of this article on airborne animal sonar systems we shall describe the operational principles and functions of the highly developed echolocation systems of Microchiroptera. In the second part, we shall summarize the rather poor echolocation

performance of other animals which use airborne sonar, including
Megachiroptera. Reviews dealing with these topics have been
published by Griffin (1958), Schnitzler (1973b; 1978), Airapetianz
and Konstantinov (1970; 1974) and Novick (1977).

INFORMATION AVAILABLE FROM THE ECHO

Bats analyze the acoustical parameters of their orientation
sound echoes in order to obtain information on the features of
targets in their surroundings. Echolocation is capable of delivering the following information.

Detection: For the detection of a target, the bat has to make
the decision whether an echo of its own orientation sound is present
or not. However, it is hard to conceptualize detection independent
of further information extraction.

Range: The time lag between sound emission and echo reception
is a measure of the distance between bat and target. The more
exact the measurement of this travel time, the more accurate is
the range measurement.

Angular direction: The angle at which the wavefront of an echo
returns to the bat determines the direction of a target. Bats use
two ears, which act as directional antennae, to measure the angle
of the echoes received.

Relative velocity: The relative velocity between bat and
target is encoded in Doppler frequency shifts contained in the
echoes. However, it can also be determined by consecutive range
measurements.

Target oscillations: Fluttering movements of the target, e.g.
insect wingbeats, produce amplitude and frequency modulations in
the echoes owing to changes in the exposed surface area of the target
and due to Doppler shifts.

Target size: The amplitude of an echo can be used as a rough
size indicator for targets. However, this relationship is not always
reliable. A good determination of target size is dependent on the
wavelengths of the signals and is only possible if an echolocation
system can resolve dimensions which are significantly less than
the target dimensions (Skolnik, 1970).

Target shape and surface properties: The illumination of a
target with sound leads to a three-dimensional, scattered field of
echoes. The information on the shape and surface properties of the
target is encoded in the complex spectral composition and temporal
structure of the whole echo field. The analysis of the field

section which returns in the bat's direction gives limited information with regard to the shape and the surface properties of the
target itself. More information may be gained if the target is
probed from different directions.

An echolocation signal that is ideal for the extraction of all
these types of information does not exist. This becomes evident,
for example, when one compares the restraints for range and velocity
determination. For high accuracy in range measurement, broadband,
short signals are needed whereas long signals of constant frequency
allow great precision in velocity determinations. The echolocation
signals of bats are therefore compromises adapted to the special
information needs of each species.

INTERFERENCES IN ECHOLOCATION

Echolocating bats may have problems in characterizing their
targets owing to the following interference factors.

Noise: If a target is very small or far away, the echo may
be so weak that it is buried in external and/or internal noise.
Higher signal energy or an improved noise rejection of the auditory
system may help to overcome this problem.

Clutter: Echoes from reflectors other than the interesting
target are called clutter. In order to obtain maximal information
regarding the interesting target it is necessary to separate the
target echo from the clutter. A proper signal design, high
directionality of the echolocation system and special strategies
in information processing may help to overcome this problem.
Extreme clutter problems arise in fog (Pye, 1971) which scatters
the orientation sounds in all directions and thus makes echolocation
difficult to impossible.

Signals from other bats: If other bats are close to an echolocating bat, it will receive their signals and the echoes from
their signals. In the field, interference from other bats may not
be of any significance owing to the high degree of directionality
of the echolocation systems and the rather low bat density. However
in roosts, or when leaving the roosts, high bat densities are
possible. It may be that in such cases, bats use their highly
developed spatial memory (Neuweiler and Möhres, 1967) to overcome
this problem.

Frequency dependence of directionality: The directionality of
sound emission and hearing is frequency dependent. Echoes from
different directions therefore have a different spectral composition.
In order to overcome this effect, which alters all information based

on spectral cues, bats would have to take the spectral changes typical for each direction into account. Another strategy would be to use spectral cues only after a target has been fixated.

Signal changes due to atmospheric influences: The atmospheric attenuation of sound depends on temperature, humidity, composition of the air and frequency of the propagated sound. High frequencies are attenuated more than low frequencies and high humidity is more effective in attenuation than low humidity (Griffin, 1971). Tropical conditions therefore lead to marked changes in the spectral composition of echoes. A countermeasure would be for bats to recalibrate their echolocation systems to every atmospheric condition.

An echolocating signal ideal for avoiding all these interference factors is theoretically impossible. The echolocation signals of different bats are therefore not only adapted to the information needs but also to the special interference problems encountered by each species.

INFORMATION GATHERING PROBLEMS IN BATS

The more than 800 different species of echolocating bats live in various ecological situations. Apart from problems common to all bats, such as finding their way and their roosts and avoiding obstacles in the dark, the different species must also solve different problems. Those that hunt insects in the open must have different strategies than those hunting in areas full of obstacles or those that hunt stationary insects. The bats that feed on fruit, pollen, nectar, small vertebrates, surface fishes or on blood must also solve a wide variety of different information gathering and interference problems. As a result, the emitter and receiver of the echolocation systems in different species of bats, i.e. the sound emission apparatus, the peripheral auditory receptors and the information processing areas of the auditory system, have been adapted during evolution for an optimal gathering of relevant information.

A comparison of the physical characteristics of the orientation sounds found in different situations (e.g. when looking for food or avoiding obstacles) with signal detection theories developed for radar and sonar systems, allows conclusions to be made regarding the information collection preferences of the echolocation systems in different species.

Neurophysiological studies with stimuli which simulate natural echolocative signal parameters help to reveal the information processing principles of the auditory system. Owing to the fact that behaviorally relevant information can be rather clearly defined in bats, they are ideal experimental animals for the investigation of neural information processing.

TYPES OF ECHOLOCATION SIGNALS

The echolocation signals of different bats have been developed, through evolutionary processes, as optimal adaptations for gathering behaviorally relevant information and suppressing unwanted interference. The evolved signals were developed with differences in frequency structure, sound pressure level and duration.

The basic echolocative signals of bats consist of frequency modulated (fm) components alone or a combination of constant frequency (cf) components and fm components. The fm components are rather short, with durations below 5-10 msec, whereas cf components lasting from a few msec up to more than 100 msec have been found. Apart from the pure fm sounds the most frequently observed combinations are the long cf-fm, the short cf-fm and the fm-short-cf sounds. Further variations in the frequency structure of echolocation signals can be due to different harmonic content.

The determination of the sound pressure levels of bat sonar signals is very difficult owing to the highly directional sound emission patterns and the high directionality of the ultrasonic microphones used to pick up the sounds. Nevertheless it is evident that different species of bats produce signals with different sound pressure levels. Griffin (1958) stated, as a general rule, — that bats which hunt for small targets (e.g. insects) produce loud sounds whereas bats which echolocate large targets (e.g. fruit or small vertebrates) use faint signals (whispering bats).

Variations in structure, duration and intensity are not only found in the signals from different species but also in sounds used by one species under different situations. These differences are also interpreted as adaptations to changing information-gathering and interference problems. Only sounds recorded under similar conditions should therefore be compared.

Enough detailed information for such a comparison is only available for a few species. Since another review at this conference will deal with the structure of echolocation sounds of bats in detail, we will only give an example for the frequency composition of free flight signals in nine bats from nine different families (fig. 1) and an example for situation dependent variation of signals in one species (fig. 19).

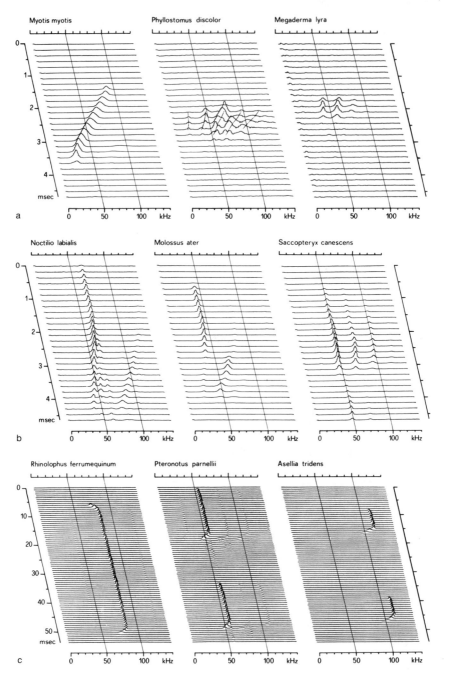

Fig. 1. Real-time-spectrograms of free flight orientation sounds of nine species of bats from different families (from Schnitzler, 1978).

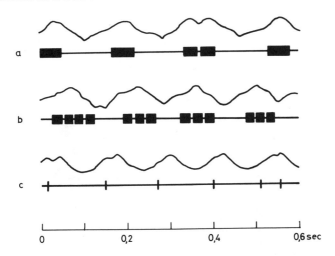

Fig. 2. Correlation between respiratory cycle and sound emission in resting Rhinolophus ferrumequinum (a), Rhinolophus euryale (b) and Phyllostomus hastatus (c) (a and b adapted from Schnitzler, 1968; c from Suthers, Thomas and Suthers, 1972).

PATTERNING OF ECHOLOCATION SIGNALS

Comparative studies of bat signal patterns reveal that different species arrange their sounds in a rather similar way in comparable situations.

All echolocating bats must adjust their pulse emission to the respiratory cycle. Recordings of the flank movements of resting rhinolophids (Schnitzler, 1968) demonstrated that, during one respiratory cycle, one pulse or a group of several pulses is emitted (fig. 2 a, b). Similar results were found in other bats by monitoring their exhalations and inhalations with a thermistor (Suthers, Thomas and Suthers, 1972; Roberts, 1972) (fig. 2c).

In flight, different species of bats also show a more or less distinct tendency to emit single pulses or groups of pulses. This was described for Myotis lucifugus (Grinnell and Griffin, 1958), rhinolophids (Schnitzler, 1967; 1968), Pteronotus parnellii (Schnitzler, 1970a) and hipposiderids (Pye and Roberts, 1970; Gustafson and Schnitzler, in press). Films of different species of bats flying against headwinds in a experimental wind tunnel showed that pulse emission is also correlated with wingbeat (Schnitzler 1971). Species of four different families emitted either a single pulse or a group of pulses per wingbeat (fig. 3 a-e). The

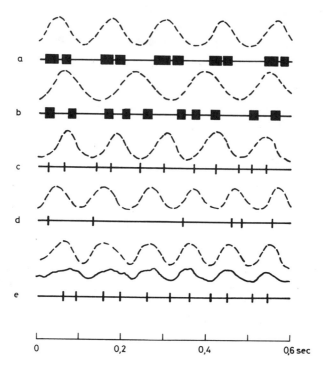

Fig. 3. Correlation between wing beat cycle and sound emission in *Rhinolophus ferrumequinum* (a), *Rhinolophus euryale* (b), *Myotis lucifugus* (c), *Carollia perspicillata* (d) and *Phyllostomus hastatus* (e) (a – d adapted from Schnitzler, 1971; e from Suthers, Thomas and Suthers, 1972). The additional curve in e describes the in and outward air flow during respiration.

correlation between sound emission and wing beat in flying bats led to the suggestion that flying bats make one wingbeat per respiratory cycle and, in so doing, either emit a single pulse or a group of pulses (Schnitzler, 1971). Suthers, Thomas and Suthers (1972) confirmed this with flying *Phyllostomus hastatus* (fig. 3 e).

The number of pulses which are produced in one respiratory cycle is dependent on the bat's orientation situation.

In resting bats, the number of pulses per respiratory cycle is determined by their interest in the surroundings. Undisturbed *Rhinolophus ferrumequinum* e.g. produce single pulses at every exhalation at a rate of 4 – 7 pulses/sec. If they are disturbed or become interested in a special target, groups with 2-8 pulses are emitted. *Rhinolophus euryale* and *Asellia tridens* echolocating a mealworm offered with forceps produce up to 10 pulses per group

(Schnitzler 1968; Gustafson and Schnitzler, in press). Resting
Phyllostomus hastatus produce either one pulse or groups of 2 - 4
pulses per respiratory cycle (Suthers, Thomas and Suthers, 1972).
The cited pulse repetition rates of 5 - 10 pulses/sec in resting
Myotis lucifugus (Grinnell and Griffin, 1958) also suggests the
emission of a single pulse per respiratory cycle. The rather sparse
data on echolocation behavior in resting bats reveal the following
tendencies: resting bats make approximately 4 - 10 respiratory
cycles/sec and, depending upon their interest in a target, per
cycle single sounds or groups of up to 10 sounds can be produced.

In flight different species of bats produce similar pulse
patterns. The three behavioral categories used by Griffin, Webster
and Michael (1960) to describe the pulse patterns of hunting vespertilionids can therefore be generalized for all bats. A "free"
or search flight category occurs when bats fly in the open in order
to search for food. When closing in on a target the pulse patterns
change. In this situation Griffin, Webster and Michael (1960)
distinguished between an approach phase and a terminal phase as
the animal neared the target.

Under laboratory conditions, a bat is considered as being in
free flight if there is no detectable target or obstacle in its
flight path within a range of about 2 m. In this situation
rhinolophids (Schnitzler, 1968), Pteronotus parnellii (formerly
Chilonycteris rubiginosa) (Schnitzler, 1970 a, b) and Phyllostomus
hastatus (Suthers, Thomas and Suthers, 1972) emit either single
sounds or groups of two sounds at a rate of approximately 8 - 12
per second (fig. 3e). The group duration, defined as the time
interval between the first sounds of consecutive groups, is normally
in the range of 80 - 120 msec. It is highly probable that these
free flight characteristics can be generalized for all other bats
(fig. 4). However, in bats with short pulses, the arrangement
of pulses into groups is not always clearly evident. When flying
outdoors in the open, most bats probably emit only one pulse per
wingbeat and respiratory cycle with a repetition rate which may
be lower than the 8 - 12 per second found in the laboratory.
Griffin (1958), for example, reports for high flying Eptesicus
fuscus a rate of only 4 pulses/sec.

A bat enters the approach phase when it produces more pulses
per group than in free flight (fig. 4). This increase in pulse
repetition rate indicates that a bat has detected and responded
to a specific target. The number of pulses per group increases
proportionally with decreasing distance to the target. The duration of the approach phase is situation dependant. When detecting
a small, unknown target, e.g. an insect, the approach phase only
contains 1 - 3 groups of sounds, while bats passing a well known

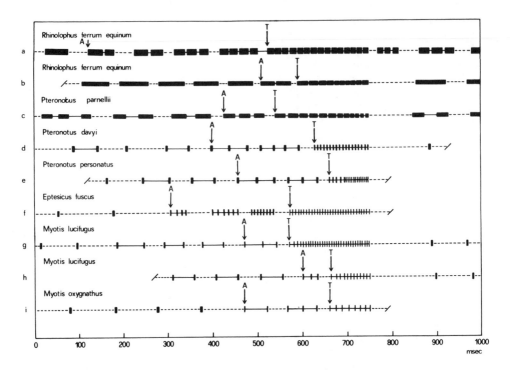

Fig. 4. Sound patterns of different species of bats when approaching a target. Pulses connected with a solid line belong, according to our interpretation, to a pulse group emitted per wingbeat and respiratory cycle. In (a) the bat was approaching a wire obstacle, in (b, c, d, e, h and i) the bats were persuing an insect and in (f and g) a ballistic target. The data are adapted from Schnitzler 1973b (a), Webster, 1967a (b), Novick and Vaisnys, 1964 (c), Novick, 1963 (d), 1965 (e), Griffin, 1962 (f), Webster as cited in Cahlander, 1967 (g), Griffin, Webster and Michael, 1960 (h) and Konstantinov, 1969 (i). Note the difference in the duration of the approach and terminal phases as found in Rhinolophus (a and b) and in Myotis (g and h). (A) determines the start of the approach phase and (T) the start of the terminal phase.

obstacle or landing on a familiar landing site produce up to 8 groups (fig. 4a, b).

Films of landing Rhinolophus ferrumequinum and Pteronotus parnellii demonstrated that the 1:1 correlation between wingbeat and arrangement of sounds in groups is also maintained during the approach phase (Schnitzler 1970 a,b). A similar arrangement of sounds in other species suggests that, in the approach phase, all bats

produce groups of sounds every 80 - 120 msec, these being correlated
with wingbeat and respiratory cycle.

It is not clear whether the start of the approach phase is always an indicator for a directly preceding initial detection of a
target. It can be calculated that, in the case of big targets, the
detection distance should be greater than the response distance,
characterized by the switching from free flight mode to the approach
phase. The bats appear to respond to targets only when they are
nearer than 2 - 3 m. For small targets, however, where the detection
occurs below this distance it is likely that detection is followed
immediately by the start of the approach phase (Grinnell and Griffin, 1958).

The terminal phase begins when a bat approaches the target
closely. It is characterized by a relatively long group of sounds
with rapidly repeated short pulses (fig. 4). The duration of the
terminal group of pulses may be twice as long as that of groups emitted in free flight and during the approach phase. The number of
pulses emitted in the terminal group depends on the echolocation
situation. Webster and Brazier (1965) found that, during the capture
of projected mealworms by Myotis lucifugus, the duration of the final
pursuit determines the duration of the terminal group. In extended
pursuits, where a bat has misjudged the position or the path of a
target, the terminal phase is prolonged and even a second terminal
group is sometimes emitted. During insect pursuit by Myotis
lucifugus, the terminal group contains 3 - 20 pulses (Griffin,
Webster and Michael, 1960) (fig. 4). In obstacle avoidance tests,
the number of pulses in the terminal group is reduced with decreasing
wire diameter (Griffin and Grinnell, 1958; Schnitzler, 1967, 1968)
(fig. 5).

The films of Webster and Griffin showing Myotis lucifugus
catching insects show that, during the terminal group, several
wingbeats are made which bring the bat into the final interception
position. For the terminal phase, therefore, the general rule of one
wing beat and one group of sounds per respiratory cycle, is not valid.

In the approach and terminal phases, the intervals between
pulses within a group also become reduced as the number of pulses per
group increases. The factor that determines the minimum interpulse
interval seems to be the distance to the target. In landing Pteronotus parnellii the duration of the interpulse interval is always
maintained at a length which allows the echo of the target to return
before the next pulse is emitted (Schnitzler, 1970a).

The approach to a target is also characterized by a more or
less pronounced decrease in the bat's flight speed. This was demonstrated for Lasiurus, Eptesicus and Myotis during insect catching

Fig. 5. Sound patterns of Rhinolophus ferrumequinum in obstacle experiments with different wire diameters. The arrows indicate the start of the approach phase. At 0 m the bats passed the obstacle (adapted from Schnitzler, 1968).

(Webster and Brazier, 1968) and also for obstacle avoidance and land-ing in rhinolophids (Schnitzler 1967, 1968) (fig. 5). This decrease in flight speed, which occurs together with the previously described increase in pulse repetition rate, results in a rapid increase in pulse density (number of pulses emitted per distance flown) when a bat closes in on a target (Schnitzler, 1967).

In all bats, the terminal group of pulses is often followed by a long interval with no sound emission, this being at least as long as individual group durations in free flight or during the approach phase (fig. 4). After this silent period, which may be a respiratory cycle without sound emission, the bats switch back to the pulse pattern of free flight with single pulses, or groups with two pulses.

In areas where Myotis, Eptesicus and Lasiurus have encountered targets in former flights, the pulse patterns are characterized by an increase in pulse repetition rate even if no targets are present (Webster and Brazier, 1968). This kind of 'target expectancy' has also been found in Rhinolophus ferrumequinum (Schnitzler, 1968) and Asellia tridens (Gustafson and Schnitzler, in press) accustomed to flying through obstacles. Trained Rhinolophus ferrumequinum, for

instance, emitted two pulses per wingbeat within a distance
of about 2 m in front of the place where the obstacle had previously
been positioned (fig. 5). Untrained bats, on the other hand, showed
typical free flight behavior with single pulses in the same situation.

DURATION OF ECHOLOCATION SIGNALS

Different species of bats show wide variation with respect to
their pulse duration but basic similarities in the way they change
this parameter in different situations.

In resting bats, sound duration is maximal in single pulses
and becomes reduced with an increase in the number of pulses per
group. In all bats, the longest pulses are normally emitted in free
flight. During the approach and terminal phases, the pulses are
progressively shortened as the number of pulses per group increases.
The shortest pulse durations are always found in the terminal phase.
In Pteronotus davyi, Pteronotus personatus (formerly Chilonycteris
psilotis), Pteronotus parnellii (Novick 1963, 1965; Novick and
Vaisnys, 1964), Noctilio leporinus (Suthers, 1965), Myotis lucifugus
(Cahlander, McCue and Webster, 1964) and Rhinolophus ferrumequinum
(Schnitzler, 1968) there is an approximately linear relationship
between the reductions in pulse durations and the target distance.

Under laboratory conditions, the size of the flight room probably influences sound duration in free flight. Griffin (1953) reports that Eptesicus fuscus flying in the open produce pulses of
10 - 15 msec whereas in the laboratory pulses with only 2 - 4 msec
durations were recorded. In Noctilio leporinus flying outdoors pulse
durations of 11.1 - 16.7 msec are produced but in a flight cage, the
pulse durations are 5.9 - 9.11 msec.

If the echo delay is shorter than the pulse duration, the returning echoes overlap the emitted pulse. In all vespertilionid
bats studied so far, the pulses are kept so short that no overlapping
occurs (Webster and Brazier, 1968). This is probably the case in all
bats emitting pure fm sounds. When catching mealworms Myotis lucifugus show, in the approach and terminal phases, a strong tendency to
maintain the pulse durations at about half the echo delay (Cahlander,
McCue and Webster, 1964). In bats which produce cf-fm sounds, the
returning echoes usually overlap the outgoing sounds. Detailed
studies of this problem in Rhinolophus ferrumequinum (Schnitzler,
1968) and in Pteronotus parnellii (Schnitzler, 1970 a,b) reveal,
however, that a decrease in the duration of the fm component during
the approach and terminal phases excludes any fm component pulse-
echo overlap. When the fm component of cf-fm sounds is considered
by itself, it emerges that this component is modified in a way
similar to that of the fm sounds in pure fm bats.

DETECTION OF TARGETS

     The detection limits of echolocation in bats have been studied
by measuring obstacle avoidance skills with vertically and horizontally stretched wires. The ability to detect the wires is usually
measured as the percentage of through flights without collisions.
The data for different species are not fully comparable owing to
the different methods used in the experiments. For example, a bat
which has been trained to fly through obstacles will attain a better
avoidance score than an untrained bat forced to fly the same route.
The results may also be influenced by differences in the distance
ratios of the wires and the wingspans of the animals being tested.
In order to obtain comparable data we suggest that, in future experiments, trained bats should be enticed to fly through the obstacles
with a food reward and a 2:3 ratio of distance between wires and
wingspan should be maintained.

     A score above the random avoidance level was reached at a
wire diameter of 0.26 mm in Myotis lucifugus (Curtis, 1952), of
0.175 mm in Carollia perspicillata, Glossophaga soricina and Artibeus jamaicensis (Griffin and Novick, 1955), of 0.19 mm in Macrotus
mexicanus (Griffin and Novick, 1955), of 0.2 mm in Myotis oxygnathus
and Plecotus auritus (Konstantinov, Sokolov and Stosman, 1967) and
of 0.21 mm in Noctilio leporinus and Pizonyx vivesi (Suthers, 1967).
The avoidance capacities of some other species are even better. The
detection limit of Rhinolophus ferrumequinum is between 0.08 and
0.05 mm (Schneider and Möhres, 1960; Schnitzler, 1967, 1968).
Rhinolophus euryale and Rhinolophus mehelyi are able to detect wires
with diameters of only 0.05 mm (Schnitzler, 1968; Konstantinov,
Sokolov and Stosman, 1967). Megaderma lyra can avoid wires of 0.06 -
0.08 mm (Möhres and Neuweiler, 1966). In Asellia tridens the avoidance score drops below the chance level when the wire diameter is
between 0.065 and 0.05 mm (Gustafson and Schnitzler, in press).

     These results demonstrate that wires with diameters far below
the orientation sound wavelength can be detected. In rhinolophids
the minimal wire diameter was 50 - 60 times smaller than the
wavelength of the pulse cf component.

     In the section on pulse patterns it was noted that with small
targets a change from the free flight to the approach phase is a
definite indication that a bat has just detected a target. The
analysis of the sound sequences produced during flights through
obstacles therefore allows the determination of the response distance
and also an evaluation of the detection distance. In addition, the
response to a target is very often characterized by a decrease in
pulse duration and flight speed (figs. 4 and 5). Response and detection distances have been determined for Myotis lucifugus (Grinnell
and Griffin, 1958), Rhinolophus ferrumequinum and Rhinolophus

euryale (Schnitzler, 1968), Myotis oxygnathus (Golubkov, Konstantinov
and Makarov, 1969) and Rhinolophus ferrumequinum and Rhinolophus
mehelyi (Sokolov, 1972). All these measurements revealed a general
tendency for the detection distance to diminish with decreasing wire
diameter. A rather rough comparison of the experimental data from
different species indicates that wires with diameters of about 2 mm
are detected at distances of approximately 2 m, whereas the detection
range for 0.2 mm diameter wires is about 1 m.

    Griffin (1958) studied the echolocation behavior of free-flying
Eptesicus fuscus in the field. He estimated that the bats detect
1 cm diameter objects thrown into their flight path at a distance
of about 2 m. Webster and Brazier (1965) investigated the detection abilities of Myotis lucifugus trained to catch airborne targets
such as mealworms and small spheres. They used two lines of
evidence to decide whether a bat had detected and responded to a
target. Apart from the increase in pulse repetition rate at the
start of the approach phase, they also found that, after detection,
the bats turned their heads toward a target and if they intended
to explore the target further they directed the flight path to the
target or to the place where they expected to intercept it. Using
both criteria, they determined that the response to spheres 2.1 mm
in diameter occurred at distances of about 60 cm and initial detection
occurred at about 90 cm. With 4.2 mm diameter spheres the response
range was between 90 - 105 cm and the detection range was judged to
be 120 - 135 cm. With spheres of greater diameter, however, the
response point was not further away although there should have
been an increase in echo strength.

    These findings and similar ones by Grinnell and Griffin (1958)
and Schnitzler (1968) in bats avoiding obstacles raise the question
as to whether objects large enough to reflect audible echoes are
also processed beyond the maximal response range indicated by the
pulse patterns. The observations of Webster and Brazier (1965), in
which bats turned away from a volley ball thrown into their flight
path at a distance of about 3 - 4 m without showing a change in the
pulse pattern suggest that these bats are able to detect targets
beyond the indicated range. Similar conclusions can be made in the
case of rhinolophids which respond with a fright response to big
obstacles moving towards them at distances of several meters. However, it is still possible that bats use a form of time or attention
gate which is mainly open for echoes coming back from targets within
a maximum response range of about 2 m. A time or attention gate
for the near field in front of a bat would allow the bats to concentrate on a few near-by targets and prevent overloading of the
echolocation system with echoes from more distant targets.

    Many species of bats collide with objects placed in an
accustomed flight path which, over a long period of time had

previously been obstacle-free. Although strong echoes returning from
these newly placed obstacles must be present, they are not detected
and avoided by the bats. Griffin called this situation, where
obviously present information is not used, the "Andrea Doria effect".
It has been shown by Neuweiler and Möhres (1967), that bats can
establish a very precise three-dimensional picture of the environment. We suggest that this spatial memory completely dominates the
echolocation system in cases where bats do not avoid new obstacles
in a well known environment. This raises the question as to
whether in this situation the "attention gate" is closed for obstacle information but open for prey detection. In this case bats,
after having established their spatial memory, would disregard
all echo information returning from the environment and only concentrate on the targets in which they are interested. Further experiments must be conducted to study and solve this question.

If one knows the detection range and the sound pressure level
of the emitted sounds, it is possible to calculate the sound pressure
level of the echoes used for detection of targets with known reflective properties. Griffin (1958) estimated that, in Eptesicus
fuscus, the echoes return with a SPL of 17 dB (relative to 0.0002
dyne/cm^2) from a spherical object detected at about 2 m distance.
For Myotis lucifugus catching fruitflies a detection level of about
30 dB at a range of 50 cm, and 15 - 20 dB at 100 cm was assumed
(Griffin, Webster and Michael, 1960). In bats avoiding obstacles
a detection level of 23 - 28 dB was calculated in the same
way for Myotis lucifugus (Griffin, 1958), of 25 - 34 dB for
Rhinolophus ferrumequinum (Sokolov, 1972) and 9.2 - 21.6 dB in
Myotis oxygnathus (Airapetianz and Konstantinov, 1974).

Novick suggested another criterion for the detection of targets
in mormoopids and other bats with cf-fm sounds. He assumed that detection first occurred when a returning echo overlapped the outgoing
pulse (Novick, 1963, 1965, 1971, 1973, 1977; Novick and Vaisnys,
1964). In Pteronotus personatus whose pulse durations are 4 msec
in free flight, the detection distance would be about 0.69 m
whereas Pteronotus parnellii with 20 msec sounds should detect
targets at 3.5 m. In Rhinolophus ferrumequinum, with its free
flight pulses of more than 50 msec duration, the detection should
occur at distances greater than 8.5 m. Since mormoopids produce
the same basic fm sound patterns in their free flight, approach and
terminal phases as other bats, it may be that the detection mechanisms in these bats do not differ from those of all other species.

DIRECTIONAL SENSITIVITY OF ECHOLOCATION SYSTEMS

The likelihood that a target attracts the attention of a bat decreases the more laterally the target is positioned relative to the flight path (Griffin, 1958). This observation can be explained by the directionality of sound emission and of hearing. The more lateral a target, the fainter are the echoes heard by the bats.

Measurements to determine the directional emission pattern have been made in several species of bats including Myotis lucifugus (Griffin, 1958; Griffin and Hollander, 1973; Shimozawa, Suga, Hand- ler and Schuetze, 1974), Myotis oxygnathus (Airapetianz and Kon- stantinov, 1974), Myotis grisescens (Shimozawa, Suga, Handler and Schuetze, 1974), Eptesicus fuscus (Simmons, 1969; Griffin and Hollander, 1973), Megaderma lyra (Möhres and Neuweiler, 1966), Pteronotus parnellii (Simmons, 1969) and Rhinolophus ferrumquinum (Möhres, 1953; Schnitzler, 1968; Sokolov and Makarov, 1971; Schnitzler and Grinnell, 1977).

In most cases, only a few angles in the horizontal and vertical axes were tested, but the results have demonstrated that the emission patterns are highly directional and frequency dependent. Some of these measurements show that the radiation patterns are determined by the structure of the nose leaf in bats which emit sounds through the nostrils and by the form of the mouth and lips in bats which emit sounds through the mouth. Complete measurements of the sound field are only available for 3 frequency components of the fm bat Myotis grisescens (Shimozawa, Suga, Handler and Schuetze, 1974) and of the cf components of the cf-fm sounds of Rhinolophus ferrumequinum (Schnitzler and Grinnell, 1977) (fig. 6B). Contour maps of the emission pattern of these bats reveal sound fields with a strong asymmetry in the vertical axis. In Myotis grisescens, the main beam was emitted 5 - 10° downward from the eye-nostril line, whereas Rhinolophus emitted the main energy perpendicular to the plane of the horseshoe-shaped nose leaf; this corresponds to the normal flight direction in horizontal flight. In both species rather wide and prominent downward-pointing side lobes were found.

The likelihood of target detection at different angles cannot be judged from the directionality pattern of sound emission alone. For instance, the presence of a downward pointing side lobe has led to the erroneous conclusion, that Rhinolophus has a form of ground sonar for monitoring the flight altitude (Sokolov and Makarov, 1971). However, to obtain a complete picture of the directional sensitivity of an echolocation system, the directionality of the hearing system also has to be taken into account.

The directionality of hearing has been measured in several species using neural responses as a criterion. This has the

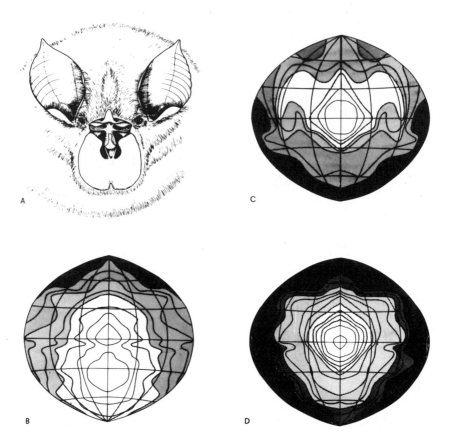

Fig. 6. Directional sensitivity of echolocation in Rhinolophus
ferrumequinum. (A) head of the Greater Horseshoe Bat as seen from
the direction perpendicular to the plane of the horseshoe. (B) contour map of the sound field in front of a bat. (C) contour map of
the behaviorally determined directionality of hearing. (D) contour
map of the directionality of the whole echolocation system as calculated by superimposing (B) and (C). Lines in the contour maps are
separated by 3 dB. Parallels of latitude and meridians of longi-
tude are shown at 30° intervals (adapted from Schnitzler and
Grinnell, 1977, and Grinnell and Schnitzler, 1977). Note in (D)
the symmetry of the pattern and the steep fall-off of echo effectiveness in all directions relative to the line of flight.

advantage that the effectiveness of sounds from any angle can be quickly measured. However, all these measurements have been made on one side of the brain only and do not provide a good means of assessing the capabilities of the hearing system as a whole. A method has been developed which allows the determination of behavioral hearing directionality in Rhinolophus ferrumequinum (Grinnell and Schnitzler, 1977). Contour plots of the hearing field of Rhinolophus reveal an asymmetry in the vertical axis with a sharp drop in sensitivity beneath the bat (fig. 6C). When the data on hearing are combined with the data on sound emission, a contour map of the sensitivity of the whole echolocation system can be constructed (fig. 6D). This map shows rather radially symmetrical contour lines with practically no side lobes. The sensitivity is reduced to half (- 6 dB) at about 15 degrees lateral to an axis perpendicular to the plane of the noseleaf. The prominent downward side lobes of the emission pattern therefore do not increase the chances of ground echo detection, since hearing sensitivity decreases rapidly at this angle.

The measurements in Rhinolophus ferrumequinum reveal that the asymmetries found in sound emission and hearing compensate each other. This results in a radially symmetrical and highly beamed sensitivity curve. If in Rhinolophus the chance to detect an echo only depends on echo strength the echoes returning at an angle 15 degrees lateral to the flight path should be about 6 dB louder and those from 30 - 40° approximately 20 dB louder than echoes returning from straight ahead to have the same chance of detection.

Unfortunately we only have complete measurements on directional sensitivity in one species. Further experiments have to be conducted in order to determine whether other species have similar radially symmetrical characteristics.

FIXATION, TRACKING AND CAPTURE OF TARGETS

Search Volume

The highly directional sensitivity of echolocation in bats suggests that their angular search volume may be rather limited and restricted to a cone-shaped area which extends outward from the mouth. Griffin, Webster and Michael (1960) have studied this problem in Myotis lucifugus during insect-catching in the laboratory. They estimated that Myotis is able to detect fruit flies in a spatial cone of about 120° diameter in front of the head at detection distances between 27 - 83 cm. There was, however, little correlation between angular position and detection distance. Webster and Brazier (1965) describe a case where Myotis was even able to detect a moth roughly 90° to one side and about 45 cm away.

Fig. 7. <u>Myotis lucifugus</u> selecting a mealworm out of a cluster of three discs. Image 5 and the third disc are omitted. Note the precise head aim of the bat (adapted from Webster and Durlach, 1963).

Fixation

        Immediately after detection, flying Myotis lucifugus,
Eptesicus fuscus and Rhinolophus ferrumequinum turn toward the
target and orient their head so as to bring the target into their
echolocation system's region of maximal sensitivity (Webster, 1963
a, b, 1967 a,b; Webster and Brazier, 1965, 1968). Webster (1967a)
estimated that this initial fixation of the target occurs within
about 1/10th sec. The precise aiming of the bat's mouth and ears
toward the target indicates that they are able to pinpoint it
exactly。 From photographs of Myotis lucifugus catching insects
Webster and Brazier (1965) judged the accuracy of head aim to be
± 5 degrees. Fig. 7 gives an example of the excellent head aim in
a Myotis lucifugus selecting a mealworm from a cluster of targets.

Tracking ⌐

        The tracking of targets, i.e. the ability to determine the
pathways of targets and to predict their future position, has been
studied in Myotis lucifugus Eptesicus fuscus, Lasiurus borealis
and Rhinolophus ferrumequinum (Webster, 1963 a,b, 1967 a,b,
Webster and Brazier, 1965, 1968).

        The analysis of the bat's flight path with respect to that of
a target reveals that bats are able to evaluate the course of
targets and to make predictions regarding their future position.
Experiments with ballistic targets (spheres, discs, etc.) demonstrate that trajectory evaluation contains a strong learning com-
ponent。 Bats initially misjudge the vertical position of targets
and try to capture objects at a position several cm above the
real one。 They are probably not able to appreciate acceleration
due to gravity. After a few trials, however, they learn to predict
the trajectories so precisely that they no longer need to correct
their flight path over the last half meter before target capture.

        Since flying insects are capable of maneuvering, tracking is
therefore far more difficult than in the case of ballistic targets,
especially when the insect is a moth making evasive movements.
Nevertheless remarkable predictions of flight paths have been
observed。 It may be that in this situation, bats make a type of
probabilistic estimation of the future target position. Well-timed
and sudden changes in flight direction, however, commonly result
in successful evasion of bats by insects.

        The tracking behaviour of Myotis, Eptesicus and Rhinolophus
is characterized by quick and precise following of the target with
the head。 Such precise head orientation was found at distances
up to about one meter in laboratory tests with Myotis and up to two
meters or more in Eptesicus hunting in the field. Even in situations

where the bats were no longer capable of bringing their bodies into
an interception position, the head aim still indicated that they had
located the target. A different tracking behavior was found in
Lasiurus borealis; this bat can accurately predict the path of its
prey without the precise head-following observed in other bats.

Capture Techniques in Bats

The following movements of the head, characteristic of target
flight-path evaluation, cease at close range and the final interception is made by simple deduction. The following target-capture
techniques for insects, projected mealworms or other ballistic
targets have been described for Myotis lucifugus, Eptesicus fuscus,
Lasiurus borealis and Rhinolophus ferrumequinum (Webster, 1962, 1963
a,b, 1967 a,b; Webster and Griffin, 1962; Webster and Durlach, 1963;
Webster and Brazier, 1965, 1968).

Mouth catch: In Myotis it was observed that small and slow
flying insects, such as fruit flies, may sometimes be seized directly with the mouth. That means that the insect must have been
localized with extremely high accuracy.

Tail membrane catch: Myotis, Eptesicus and sometimes Lasiurus
direct their flight path to the interception point and form a
pouch with the interfemoral membrane by forward flexion of the
hind limbs and tail. After the target is caught in this pouch, the
bat bends forward and seizes it with the jaws. This kind of catching
behavior was observed with insects as well as with projected mealworms and other ballistic targets.

Wing scoop catch: If the target is out of reach of the tail
membrane pouch, bats sometimes reach out for the target with the
wing, shovel it into the pouch and grasp it with the mouth. Even
targets which can be reached only with the tip of the wing may be
caught in this way. In Rhinolophus it is probable that insects
intercepted with the wing are brought to the mouth directly without
the use of the interfemoral membrane pouch.

Somersault catch: Lasiurus uses all the capture techniques
found in Myotis and Eptesicus and with an even higher predictive
skill. Sometimes, however, these bats fly underneath the target
and make a backward somersault by swinging the tail membrane
around in a complete 360° circle. During this action they
produce a form of funnel with the wings leading to the tail
membrane pouch. The bats come out of this maneuver with the target
in the pouch. After reorientation, they reach into the pouch and
seize the prey.

The use of the tail membrane and the wings for the capture of flying insects serves to increase the area within which a bat can seize its prey (Webster and Griffin, 1962).

The high capture efficiency observed in the different species indicates that the bats know exactly where their targets are. Webster (1963) estimated that targets are sometimes localized and intercepted with an accuracy of $\pm$ 0.5 cm. This means that bats are able to pinpoint their prey in space within about one cubic centimeter. The speed of action is also astonishing. The complete sequence with detection, initial fixation, tracking and capture may only last 300 - 500 msec.

The rate at which bats catch insects has been studied in Myotis lucifugus hunting insects in the open (Gould, 1955, 1959). During a 2369 sec observation time, the Myotis, on an average, pursued an insect every 3 sec and probably caught every second one. Similar results were presented by Griffin, Webster and Michael (1960). They estimated that, in the laboratory, Myotis caught fruit flies with an average rate of one every hour seconds.

DETERMINATION OF RANGE

Hartridge (1945 a,b) suggested that bats determine the range of targets by measuring the time between sound emission and echo reception. Griffin (1944) had already anticipated this principle by comparing echolocation with radar and sonar systems.

Beat Note Hypothesis

Pye (1960, 1961a, 1961b, 1963) and Kay (1961a, 1961b, 1962) suggested that in fm bats, range information might be contained in beat notes arising between the outgoing sound and the overlapping echo. Such a beat frequency would decrease linearly with decreasing range. Cahlander, McCue and Webster (1964), however, demonstrated that, in Myotis lucifugus during the approach and terminal phases, the sound duration decreases almost linearly with decreasing target range and in such a way that no overlap occurs. The beat note hypothesis can also be rejected for other species since the fm sounds in other pure fm bats and also the fm components in cf-fm bats do not overlap with their echoes in the approach and terminal phases.

Accuracy of Target Range Determination

When judging the ability of bats to measure distances by echolocation, it is important to distinguish between accuracy of range measurement and range resolution. Accuracy describes the acuity

with which the range of a single target can be measured whereas
resolution refers to the ability to distinguish between two or more
closely spaced targets.

In experiments to investigate range measurement performance
(Simmons, 1968, 1971, 1973; Simmons and Vernon, 1971; Airapetianz
and Konstantinov, 1974) bats had to decide which of two similar
targets offered in different directions and at different absolute
ranges was the closer. The discrimination performance was measured
as the percentage of correct responses and the threshold was set at
75%. In these experiments, the angle between the targets was so
great that the bat had no difficulty in resolving them. While
measuring and discriminating the distances, the bats scanned from
one target to the other, thus making it highly probable that determination of the target range was made successively and not simultaneously. The discrimination threshold is therefore a measure for
the accuracy of range measurement and not of range resolution.

Similar experiments have been conducted with Eptesicus fuscus,
Phyllostomus hastatus (both fm bats), Pteronotus suapurensis (short
cf/fm) and Rhinolophus ferrumequinum (long cf/fm); the bats were
trained to discriminate the range of triangular targets offered at
an angle of 40° to each other (Simmons, 1968, 1971, 1973; Simmons
and Vernon, 1971). In other experiments Myotis oxygnathus (fm bat)
and Rhinolophus ferrumequinum were trained to discriminate the
range of cylinders or corner reflectors offered at an angle of 13°
(Airapetianz and Konstantinov, 1974). Table 1 summarizes the results
of all bats tested to date and fig. 8 describes the discrimination
performance of Eptesicus fuscus. In general, range discrimination
performance diminishes with a decrease in the band widths of the
echolocation signals used for discrimination. In the bats tested,
thresholds between 8 - 45 mm were found. With a two channel target
simulator (two loudspeakers transmitting bat signals as echoes back to
the bat) where similar echoes differing only in arrival time were
presented, it was demonstrated that Eptesicus discriminates differences in echo arrival times with the same accuracy as comparable
difference in range (Simmons, 1973). This suggests that Eptesicus
and probably all other bats use time differences between signals for
the determination of ranges.

An equivalent discrimination performance was reached with the
two-channel simulator when the echoes were presented successively
instead of simultaneously (Simmons and Lavender, 1976). This also
indicates that absolute ranges (time delays) are compared and that
the range discrimination threshold is a measure for the accuracy of
range determination and not for range resolution.

An astonishing result of Simmons' measurements is that the discrimination threshold hardly changes when the targets are offered

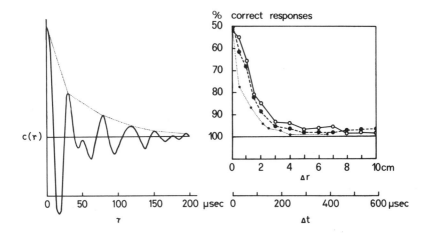

Fig. 8. Left: Autocorrelation function with envelope of an orientation sound of Eptesicus fuscus produced during a range discrimination experiment. Right: Average performance of 8 Eptesicus fuscus in a range discrimination experiment (solid line) compared with a predicted performance curve (dashed line) which was derived from the envelope curve of the autocorrelation function (dotted line). In order to obtain the prediction curve the envelope curve was corrected for the bats' head movements during discrimination (adapted from Simmons, 1968).

at different ranges (table 1). In Eptesicus, for example, the discrimination threshold was 12 mm and 14 mm when the target distance was 30 cm and 240 cm respectively. This indicates that range determination is independent of echo intensity since echoes returning from 240 cm must be far less intense than echoes from targets 30 cm away. Airapetianz and Konstantinov (1974) also discovered no difference in the range performance of Rhinolophus when the echo intensities differed between 71 dB and 64 dB. In Myotis, however, a threshold of 8 mm was measured with echo intensities of 25 dB, whereas only 23 mm was reached with 18 dB echoes.

Prediction of Range Determination Accuracy

Strother (1961, 1967) pointed out that the signals of fm bats are rather similar to the signals used in chirp or pulse compression radar systems. He therefore assumed that echo processing in bats might be comparable to that in the matched filter receiver of a pulse compression radar system; in these radar systems the returning

Table 1. Range discrimination threshold (dr) and corresponding
time measuring errors (dt) of different species of bats using
signals of different bandwidth (b) measured with targets positioned
at range (r). In one Eptesicus the measuring error was determined
directly with a two target simulator. The angle between targets
was 40° in the experiments of Simmons and 13° in the experiments
of the Russian scientists. The different thresholds obtained for
Myotis oxygnathus were explained by differences in echo intensity
(18 dB at dr = 43 mm and 25 dB at dr = 8mm).

Species	dr	dt	b	r	Author
Eptesicus fuscus	13 mm	75μsec	25 kHz	60 cm	Simmons 1971
Eptesicus fuscus	12 mm	70μsec	25 kHz	30 cm	Simmons 1973
Eptesicus fuscus	14 mm	81μsec	25 kHz	240 cm	Simmons 1973
Eptesicus fuscus	-----	60μsec	25 kHz	30 cm	Simmons 1973
Phyllostomus hast.	12 mm	70μsec	35 kHz	60 cm	Simmons 1971
Phyllostomus hast.	12 mm	70μsec	35 kHz	120 cm	Simmons 1973
Pteronotus suapur.	15 mm	87μsec	22.5kHz	30 cm	Simmons 1973
Pteronotus suapur.	17 mm	98μsec	22.5kHz	60 cm	Simmons 1973
Rhinolophus ferr.	25 mm	145μsec	15 kHz	30 cm	Simmons 1973
Rhinolophus ferr.	41 mm	240μsec	15 kHz	100 cm	Airapet.et al.1974
Myotis oxygnathus	8 mm	46μsec	85 kHz	100 cm	Airapet.et al.1974
Myotis oxygnathus	23 mm	133μsec	85 kHz	100 cm	Airapet.et al.1974

echo is cross-correlated with the transmitted signal by passing the
echo through a filter matched to the signal.

Various aspects of such optimal filtering in bats have been discussed in detail by Cahlander (1963, 1964, 1967), van Bergejk
(1964), McCue (1966, 1969), Altes and Titlebaum (1970), Altes
(1973, 1975, 1976), Glaser (1971a, 1971b, 1974) and Beuter (1977).
From all these theoretical studies, the only result indicating that
bats might use optimal filters was that the echolocation signals
of Myotis lucifugus would be optimally Doppler tolerant if they were
processed in a matched filter receiver.

A matched filter receiver (ideal receiver, optimal filter receiver) concentrates the signal energy within a small time interval. This compression increases the signal-to-noise ratio at the
filter output and allows an optimum estimate of the pulse arrival
time. Under the assumption that the postulated optimal filter in
bats is matched to the waveform of the signals used, the efficiency
of the receiver in compressing signals can be judged by comparing

the cross-correlation function as derived from the echo and the
emitted signal. Simmons therefore assumed in his range determination
experiments that the cross-correlation function reflects the ambiguity encountered by an optimal filter receiver in estimating the
arrival time of echoes. With the high frequencies used by bats, it
is unlikely that phase information can be extracted by the auditory
system. A postulated matched filter receiver would therefore be a
noncoherent receiver, i.e. the fine structure of the cross-correlation function is not available and only the envelope may be used to
make predictions regarding range measurement accuracy. Since the
echoes in the discrimination experiments are similar to the emitted
signals, the envelope of the cross-correlation function can be conveniently approximated to the envelope of the autocorrelation
function of the emitted signal.

In fig. 8, a predicted performance curve is derived from the
envelope of an echo's autocorrelation function in Eptesicus. It
is compared with the average range discrimination performance of
8 bats. After a correction of the prediction curve, necessary
to compensate for the bat's head movements, the measured performance
curve shows good correspondence with the corrected prediction curve.
Such a close fit between predicted curve and measured performance
was also found in Phyllostomus hastatus (fm) and Pteronotus
suapurensis (short cf-fm) when the entire sounds were used for
autocorrelation function derivation. In Rhinolophus (long cf-fm),
however, only the autocorrelation function of the fm component
produces a corresponding prediction curve.

These results led Simmons (1973) to suggest that there is a
neural equivalent of a matchedfilter, an ideal sonar receiver
which functionally cross-correlates a replica of the outgoing sound
with the returning echo to detect the echo and to determine the
arrival time. Whether this experimental approach allows this
far-reaching conclusion is a matter of conjecture.

The accuracy of time and therefore also of range measurements
in a matched filter receiver depends on the effective bandwidth (b)
of the signal and also on the signal energy and the noise conditions.
In the case of echolocation signals, the spectral bandwidth (b) is
a good approximation of the effective bandwidth (Glaser, 1974). The
maximum-peak-signal-to-mean-noise (power) ratio at the output of a
matched filter is equal to $2E/No$ where E is the signal energy and
N_O the noise power per cycle of bandwidth at the input. The error
of accuracy in time measurements δ_t for a coherent matched filter
receiver (Skolnik, 1962) was calculated as:

$$\delta_t = \frac{1}{\beta \, 2E/N_o}$$

This formula was used to predict time (range) error for the
signals with bandwidths between 10 - 100 kHz and for different
$2E/N_0$ ratios (fig. 9). If we assume constant noise conditions, the
following prediction can be made for a matched filter receiver which
uses phase information: The time error increases with decreasing
echo energy and decreasing echo bandwidth. This can also be generalized for a matched filter receiver with no phase information; in
this case, the time error must be greater, since the "fine grain"
information of the cross-correlation function is lost, and only the
envelope is available. However, the general concept that time
error depends on echo energy and bandwidth is the same.

In order to decide whether or not bats use a matched filter
receiver, the experimental data have to be compared with the predictions made for such a receiver.

One of Simmons' most important results is that the measured
discrimination thresholds do not depend on absolute target range
and therefore not on echo intensity. In Eptesicus, for instance,
the energy of the echoes returning from 240 cm must have been far
below that of echoes from 30 cm. Although great differences in the
$2E/N_0$ ratio must have been present, the measured accuracy of time
(range) determination was approximately the same at both distances.
This is in clear contradiction to the behavior of a matched filter
receiver, suggesting that it is improbable that bats use optimal
filtering.

For a second comparison, the experimental results of the range
discrimination experiments in different species of bats are shown
in fig. 9. At first glance one would assume that the measured time
errors of different species decrease with increasing bandwidth
in a predictable way i.e. parallel to a line of equal $2E/N_0$ ratio.
However, the energy of the echoes returning in the discrimination
experiments is very different in different bats. Simmons (1973),
for example, measured different signal durations and different sound
pressure levels of echoes ranging from 0.005 to 0.1 N/m^2. As the
noise conditions were approximately the same for all bats, the
differences in echo energy also mean differences in the $2E/N_0$
ratios. The experimental data from different species are therefore
not comparable and the observed tendency of decreasing time error
with increasing bandwidth cannot be used to conclude that bats use
a matched filter receiver.

Another argument against optimal filtering is the high variability of sound emissions in different echolocative phases and in
different species of bats. Should a matched filter receiver be
postulated, one would then have to assume that each sound had its
own matched filter.

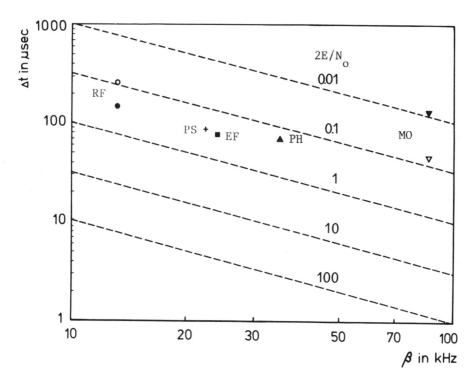

Fig. 9 Diagram which predicts for a coherent matched filter receiver the time measuring error for signals of different effective
bandwidth (β) and different $2E/N_O$ ratios. For comparison, experimental data on time measuring errors have also been included; these
were derived from range discrimination experiments with different
species. (RF) Rhinolophus ferrumequinum, (PS) Pteronotus suapurensis, (EF) Eptesicus fuscus, (PH) Phyllostomus hastatus, (MO)
Myotis oxygnathus. For further information see section on "prediction of range determination accuracy" and table 1.

          In summary, one can state that bats determine the distances of
targets separated in angle by measuring the delay between the outgoing sound and its returning echo. In bats with long cf-fm sounds
probably only the fm component is used for range determination
(Schnitzler, 1968; Simmons, 1973). This information processing in
the time domain allows a range determination accuracy of approximately 1 - 2 cm, this corresponds to a time accuracy of about
60 - 120μsec. The high degree of accuracy found suggests that some
kind of filtering improves the arrival time estimate. No experiment
to date, however, has been able to prove that this filtering is
optimal, as should be the case for a matched filter receiver. It

is probable that the range discrimination data can be explained on
the basis of suboptimal receiver types since the experiments were
performed under relatively favorable noise conditions.

Range Resolution

        Range resolution is the ability to distinguish between two
closely spaced targets. These two targets may be simulated by a
single target reflecting two wavefronts with a variable time
delay between them, as, for instance, a planar target with holes
of variable depth. In this case, the range resolution performance
could be measured as a percentage of correct choices for the two-
wavefront-target when discriminating it from a one-wavefront-target
(flat plate). Such an experiment has yet to be conducted. From
discrimination experiments with two-wavefront-targets made in another
context (Simmons et al., 1974; Habersetzer, Vogler and Leimer,
personal communication), it can be assumed that a bat's range resolution threshold is approximately 1 mm or less. How can this high
degree of resolution be explained?

        The echoes from a two-wavefront-target are distinctly different
from the echoes of a one-wavefront-target. The frequency spectrum
of the one-wavefront-target corresponds well with the spectrum of the
emitted sound. The spectrum of the two-wavefront-target, however,
has distinct minima which change their position in relation to the
time delay (hole depth) between the two wavefronts (fig. 18). These
minima reflect the interference patterns between the overlapping
echoes returning from the two reflecting planes of the two-wavefront-
target. Simmons et al. (1974) demonstrated that bats are able to
use such spectral cues to discriminate between two targets, each of
which produced two wavefronts but with different wavefront delays
between them. In this experiment, a discrimination threshold of
1 mm or less was measured. It is therefore highly probable that
this type of information processing in the frequency domain could
also be applied in the range resolution experiment suggested above
and that the range resolution threshold would be approximately the
same or even better.

DETERMINATION OF ANGULAR DIRECTION

Acoustical Cues

        For the determination of target direction, the angle at which
the echo's returning wavefront arrives at the bat has to be measured. For this measurement, the bats collect information with
their two ears which act as directional antennae. Such a system
could deliver information on interaural phase, time, and sound
pressure differences as well as monaural cues. Flying bats could

also use the angular dependence of the Doppler shift as additional
information for the directional localization of nonmoving targets
(Schnitzler, 1968). It is probable that various species of bats
use the available information in different ways. The following
suggestions have been forwarded to explain directional determination by bats.

Binaural Comparison of Interaural Differences in Arrival Time

The azimuth of target direction is encoded by differences
in the phase and arrival times of echoes. Phase differences,
however, do not allow unique direction determination, since the
half wavelength of ultrasonic sounds is shorter than the bat's
interaural distance. It is also improbable that phase information is preserved in the auditory system at frequencies above
5 kHz. It is therefore suggested that phase information is not
used for the determination of horizontal target angles.

The interaural differences in arrival time depend on the distance between the two ears. This distance is relatively short in
bats (8 - 22 mm). For reasonable accuracy in angle determination,
the barely detectable interaural time difference should also be
very small. Shimozawa et al., (1974) calculated for _Myotis_
grisescens that the interaural time difference changes 0.45 μsec
per degree. Behavioral data suggest that bats are able to determine
direction with an accuracy of about 5° (see below). Depending on
the interaural distance in different species of bats, the barely
detectable time difference would be in the range of 3 - 5 μsec.
This is comparable to the smallest detectable interaural time
difference of 5 μsec measured in man (cited in Shimozawa et al.,
1974). The fact that bats might be capable of using the interaural
time difference for the determination of a target's azimuth should
therefore still be considered a possibility. It is, however, more
probable that the interaural differences in the sound pressure levels
of different frequency components are the more important cues.

Binaural Comparison of Different Echo Frequency Components

Grinnell and Grinnell (1965) demonstrated for _Myotis lucifugus_
and _Plecotus townsendii_ with collicular evoked potentials that the
directionality of hearing is frequency dependent and that the directionality increases with frequency. They found that the measured
directionality depends mainly on the structure of the external ear
and demonstrated that every angle in space is uniquely characterized
by a family of binaural ratios of echo spectral components of different frequencies. They therefore suggested that bats with fm
sounds measure the binaural ratios of different frequency
components in the echo and use this information for determination of
direction. The conditions for this form of angle determination are

orientation sounds with different frequency components and a change
of ear directionality with different frequencies. These are not
only found in fm bats but also in bats with cf-fm sounds such as
Pteronotus and Saccopteryx (Grinnell, 1970; Grinnell and Hagiwara,
1972). Different frequencies not only occur in the fm components
but also in the harmonics of the cf components of the signals.

Binaural Comparison of cf Components in the Echo at Different Ear Positions

Bats with long cf-fm sounds such as rhinolophids, hipposiderids
and also Pteronotus parnellii often concentrate most of their sound
energy in the cf part. The terminal fm sweep is short and covers
only a small frequency range. This raises the question as to whether
these bats can use their cf part for the determination of angular
directions. The answer to this question might be determined if an
explanation could be found for the remarkable ear movements used by
bats with long cf-fm sounds (Rhinolophids: Schneider and Mohres,
1960; Griffin, Dunning, Cahlander and Webster, 1962; Pye, Flinn and
Pye, 1962; Webster, 1967a; Gorlinsky and Konstantinov, 1978; Hipposiderids: Pye and Roberts, 1970; Pteronotus parnellii: Schnitzler,
1970a). In order to explain the determination of direction with
cf components, Schnitzler (1973b) suggested that the ear movements
during echo reception change the binaural SPL ratio of an echo's
cf component in a way typical for each target direction. Since the
bats control their ear movements, they should be able to decode the
information concerning target direction contained in the changing
binaural SPL ratios. The dependence of hearing directionality on
the position of the external ear measured with collicular evoked
potentials in Rhinolophus (Neuweiler, 1970) supports this hypothesis.
Further evidence for the importance of ear movements in the determination of direction was given by Gorlinsky and Konstantinov (1978).
They showed that ablation or partial immobilization of the pinnae of
Rhinolophus prevents vertical and diminishes horizontal localization
of long cf signals played over a loudspeaker.

Directional Information from Doppler Shifts

In flying bats, the echo frequency is higher than the emission
frequency owing to Doppler shifts caused by the relative velocity
between bat and target. For stationary targets, the relative velocity is dependant on the angle between flight path and target. The
Doppler shifts decrease with the cosine of the increasing target
angles. It may be possible that, in bats with cf-fm sounds, Doppler
shifts might be a cue for a rough determination of target angle
(Schnitzler, 1968). In Rhinolophus flying 4.1 m/sec the maximal
Doppler shift is 2000 Hz, at 30° it is still 1732 Hz, at 60° 1000 Hz
and at 90° it would be 0 Hz. This cue would only be of aid in flight
and with stationary targets. It is therefore highly probable that it

may simply be an additional cue which can be used together with
other mechanisms for determination of direction。

Discrimination Experiments

     Peff and Simmons (1972) trained Eptesicus fuscus and Phyllosto-
mus hastatus to discriminate the horizontal angle between a spherical target offered at 0° and a fixed target at 19° from the angle
between the target at 0° and a movable target offered at variable
angles between 37° and 19°. The bats were able to discriminate the
two angles when the variable angle was 6° - 8° larger than the
fixed one in Eptesicus fuscus and 4° - 6° larger than in Phyllostomus hastatus (fig. 10). Airapetianz and Konstantinov (1974) determined the angular resolution in the following way: they measured
the ability of Rhinolophus ferrumequinum to discriminate a pair of
metal rods offered at an angle below the angular resolutation (24')
from a pair of similar rods offered at variable angles between
7° 50' and 2° 58'. The bats reached the threshold level of 75%
correct responses at an angle of 4° 30' (fig. 10).

Monaural Versus Binaural Localization of Targets by Echolocation

     Several experiments have been made to determine the influence
of auditory input attenuation for one ear on a bat's ability to
localize targets. All these experiments were made under the assumption that a decrease of localization acuity depends on a reduced
ability to determine direction. Griffin and Galambos (1941) showed
that monaural earplugs reduced the ability to avoid wire obstacles
in Myotis lucifugus Konstantinov, Sokolov and Stosman (1967) found
that, after unilateral perforation of the tympanic membrane, a reduction in obstacle avoidance ability occurred in Plecotus auritus
and Rhinolophus mehelyi but no effect was observed in Myotis oxygnathus. Möhres (1953) stated that obstacle avoidance in
Rhinolophus ferrumequinum is not influenced by plugging one ear.
Further experiments with Rhinolophus ferrumequinum (Flieger and
Schnitzler, 1972), however, showed that the obstacle avoidance
score was reduced from a normal value of 90% down to 60% when a
monaural earplug having a measured attenuation of 15 - 20 dB was
inserted. If the second ear was plugged in the same way, these
bats reached 90% again. This fact demonstrates that both ears
must be provided with a symmetrical input for optimal target
localization.

     These data show that monaural echo attenuation reduces the ob-
stacle avoidance performance in nearly all species of bats tested
to date。 This result and the restoration of the localization acuity
in Rhinolophus ferrumequinum with symmetrical attenuation in both
ears, leads to the conclusion that, in bats, the interaural comparison of echo intensity plays an important role in angular direction
determination。

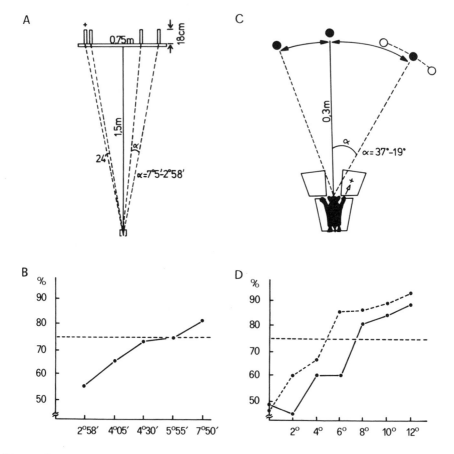

Fig. 10. Experimental set-up and results of discrimination experiments for the determination of the angular resolution threshold in Rhinolophus ferrumequinum (A and B) and of the angle measuring accuracy in Eptesicus fuscus (solid lines) and Phyllostomus hastatus (dashed lines) (C and D). Adapted from Airapetianz and Konstantinov, 1974 (A and B) and from Peff and Simmons, 1972 (C and D).

Localization of Sound Sources by Passive Hearing

The ability of bats to localize ultrasonic sound sources by passive hearing offers additional information on the acoustical cues required for the determination of angular direction.

Konstantinov, Stosman and Gorlinsky (1973) trained Rhinolophus ferrumequinum and Myotis sp. to discriminate which of two loud-speakers positioned at various angles (21° - 2°) 2.7 m away transmitted ultrasonic signals of constant frequency and with a sound duration of 5 msec. Both bats learned to make correct right-left decisions with large angles. The 75% level of correct responses was reached at an angle of 4° in Rhinolophus ferrumequinum and of 5° in Myotis sp. (at 83.5 and 40 kHz respectively). In further experiments the localization ability of Rhinolophus was also measured in the vertical plane (Gorlinsky and Konstantinov, 1978). With a signal duration of 50 msec and a frequency of 81 kHz, the bats reached the 75% discrimination level in both planes at angles of about 4°. With frequency modulated signals which imitated the terminal sweep of the echolocation sounds (2.5 msec duration, sweep of 81 to 62 kHz) the threshold angle of 19° in the vertical plane was far greater than the angle of 6° in the horizontal plane. The plugging of one ear prevented the discrimination of long cf signals in both planes even at angles of 25°. Partial immobilization or ablation of the pinnae prevented the vertical and reduced the horizontal discrimination ability with long cf signals at an angle of 25°. These data support the hypothesis (Schnitzler, 1973b) that, in Rhinolophus, direction determination is based upon evaluation of successive interaural intensity differences caused by the ear movements.

DETERMINATION OF RELATIVE VELOCITY BETWEEN BAT AND TARGET AND DOPPLER SHIFT COMPENSATION

Relative Velocity and Doppler Shifts

The relative velocity between bat and target can be determined either by measuring the decrease in distance between consecutive signals or by evaluating the Doppler shifts in echoes. No experiments have been conducted to date to determine a bat's efficiency in measuring relative velocity by range tracking. We will therefore concentrate in this chapter only on the Doppler shift evaluation.

Doppler shifts originate when a sound source or a receiver moves when transmitting or receiving sounds. In the case of a bat flying with a velocity v_B toward a target, which moves with a velocity v_T toward the bat, the echo frequency f_e can be calculated from the transmission frequency f_t (Schnitzler, 1968):

$$f_e = f_t \frac{c + v_B}{c - v_B} \times \frac{c + v_T}{c - v_T}$$

If v_B and v_T are small compared to the velocity of sound (c)
a good approximation for f_e is

$$f_e = f_t \frac{2 v_B + 2 v_T}{c}$$

As $v_B + v_T$ correspond to the relative velocity v_R between bat and
target, f_e can be expressed as

$$f_e = f_t + f_t \frac{2 v_R}{c}$$

The frequency shift caused by Doppler shifts is therefore

$$f_e - f_t = f_t \frac{2 v_R}{c} = \Delta f.$$

The Doppler shift thus increases linearly with frequency and relative velocity so that the ability of a bat to measure relative velocity is directly determined by its ability to measure the Doppler
shifts in echoes.

Accuracy of Frequency Measurement

Frequency cannot be measured with infinite accuracy as all
filters have finite time constants and bandwidths. A rule of thumb
in technical frequency analysis is that, with decreasing time for
analysis, the bandwidth of analysis filters must be increased so that

$$\Delta f \times \Delta t \approx 1$$

This means that the accuracy is already limited by the time a signal
is available for analysis. Accuracy is limited in addition by noise
and by the bandwidth of the analysis filters even when the analysis
time is long. If we use the same rule of thumb for bats, we can
assume for low noise levels that the accuracy in frequency determination is limited by the characteristics of the neuronal filters and
by the time each individual filter is excited.

In cf signals, the excitation time corresponds to the duration
of the echolocation signals (T). The theoretical upper limit of
frequency accuracy is therefore only limited by the duration of the
signal:

$$\Delta f \approx \frac{1}{T}$$

This theoretical limit cannot be attained when the bandwidth of the filtering neurons is larger than the predicted value.

In fm bats which use signals with bandwidths (B_s) larger than the bandwidth of the neurons (B_n) the signal sweeps through the neuronal filters. The signal duration therefore has to be reduced by the ratio B_n/B_s in order to determine the excitation time. The theoretical upper accuracy limit can thus be expressed as

$$\Delta f \approx \frac{1}{T} \times \frac{B_s}{B_n}$$

It must be emphasized that these calculations are made using a "rule of thumb" to obtain approximate predictions for the frequency measurement accuracy. It may well be that bats improve the accuracy of frequency measurement by additional informaion processing. There is no experimental evidence, however, to prove that this is the case.

From these considerations, it can be concluded that long signals of constant frequency are especially suited for Doppler shift evaluation whereas short fm signals with large sweeps are far less effective. Theoretically, frequency changes of 20 Hz could be detected with a cf sound of 50 msec duration.

Therefore it is probable that bats with long cf-fm sounds, such as rhinolophids, hipposiderids and the mormoopid bat Pteronotus parnellii, are very sensitive to Doppler shifts whereas bats using pure fm sounds do not avail themselves of Doppler information. This conclusion is supported by neurophysiological studies. In bats with long cf-fm sounds, many neurons are extremely sharply tuned, thus allowing good frequency determination whereas in fm bats, only widely tuned neurons have been found.

No systematic experiments have been made to date to investigate the accuracy of frequency determination in bats. In Rhinolophus ferrumequinum, however, two factors suggest that these bats are able to determine frequencies with an accuracy of 30 - 60 Hz: in the first place during Doppler compensation they have the ability to maintain the echo frequency in the cf component of consecutive sounds with a standard deviation of only 30 - 40 Hz; secondly, they have the ability to discriminate echoes with sinusoidal frequency modulations from non-modulated ones when the modulation depth is 60 Hz

(Schnitzler, 1978). Since the frequency of the cf component is
approximately 83 kHz, one can calculate that relative velocities
of 6 - 12 cm/sec could be detected by evaluating Doppler shifts.

Everyone who has worked with rhinolophids knows that these bats
are very sensitive to Doppler shifts. Even slow movements of the
hand or of a large target toward the bat cause a typical fright
response, characterized by an increase in scanning movements of the
head and a retraction of the body on the hind legs. More rapid movements immediately result in the bats flying off. Movements away from
the bat evoke no response at all. This indicates that the Doppler
shifts from large objects moving toward a horseshoe bat have a warning function.

Beat Note Hypothesis

In rhinolophids, hipposiderids, and Pteronotus parnellii the
cf component is so long that returning echoes often overlap the
transmitted signals. In the case of relative movements between bat
and target, the bats would hear two different frequencies, the
transmission frequency and the Doppler shifted echo frequency. Pye
(1960, 1961a, 1961b, 1963) and Kay (1961a, 1961b, 1962) suggested
therefore that cf bats might hear the frequency difference as low
frequency beat notes produced by a multiplicative process. This beat
frequency would correspond to the Doppler shifts and therefore allow
a precise determination of the relative velocity between bat and
target. Since bats have high auditory thresholds in the range of
possible beat frequencies (Neuweiler, Schuller and Schnitzler, 1971;
Long and Schnitzler, 1975) and since there are many results indicating that bats process their echo frequencies directly in the ultrasonic frequency range, this hypothesis is now only of historical
interest.

It should be noted, however, that the overlapping of signals
with two frequencies (f_1, f_2) sometimes produce amplitude modulations
with a modulation frequency ($f_2 - f_1$) and these modulations are
capable of evoking synchronized responses in the auditory system
(Grinnell, 1970; Suga, Simmons and Jen, 1975; Vasiliev, 1976).

Doppler Shift Compensation in Free Flight

At positive relative velocities between bat and target the echo
frequency is always higher than the transmission frequency owing to
Doppler shifts. For instance, in Rhinolophus ferrumequinum flying
with a speed of 4.1 m/sec toward a stationary target the difference
between transmission and echo frequency is as high as 2 kHz.

Laboratory studies with long cf-fm bats such as the rhinolophids
Rhinolophus ferrumequinum and Rhinolophus euryale (Schnitzler, 1968),

the mormoopid bat Pteronotus parnellii (Schnitzler 1970a) and the
hipposiderid bat Asellia tridens (Gustafson and Schnitzler, in
press) revealed that in flight, these bats lower the frequency of
the cf component in such a way that the Doppler shifts are compensated and the echo frequency is kept constant within a small frequency band about 200 Hz above the average cf frequency of resting
bats (resting frequency). This behavioral response to Doppler shifts
was termed "Doppler shift compensation". Russian scientists recently
repeated the experiments with flying Rhinolophus ferrumequinum and
confirmed the presence of Doppler shift compensation (Konstantinov,
Makarov, Sokolov and Sanotskaya, 1976; Konstantinov, Makarov and
Sokolov, 1978).

In order to decide whether horsehoe bats compensate Doppler
shifts in a control system or if they derive the necessary downward frequency shift from measurement of the air speed (perhaps with
their vibrissae), two crucial experiments with bats flying in a
wind tunnel or in a He-O_2 gas mixture were made (Schnitzler, 1973a).
The following results proved that bats use a feedback control system
to hold the echo frequency constant at a reference frequency. Bats
flying in head winds in a wind tunnel lowered the transmission
frequency to compensate for the Doppler shifts caused by the ground
speed and not with reference to the far higher air speed. In the
He-O_2 mixture, the bats also made no use of the flight speed information and compensated for the Doppler shifts with reference to the
different sound velocity in the gas mixture.

Bats also compensate for Doppler shifts originating from other
sources. A pendulum moving in front of a stationary horseshoe bat
will evoke the Doppler compensating response (Schnitzler, 1968).
Compensation was also found in Rhinolophus ferrumequinum sitting on
a moving platform (Sokolov and Lipmanova, 1977; Konstantinov, Makarov
and Sokolov, 1978) and Pteronotus parnellii sitting on a pendulum
(Henson, poster in this conference) or on a moving platform (O'Neill,
personal communication).

Doppler Shift Compensation with Simulated Doppler Shifts

With free flying bats, the performance of the feedback control
system for Doppler shift compensation cannot be investigated sufficiently well. In Rhinolophus, however, electronically simulated
Doppler shifts evoked the Doppler compensation response. It is thus
possible to determine the compensation range (Schuller, Beuter and
Schnitzler, 1974) and the dynamic properties of the compensation
system by this means (Schuller, Beuter and Rübsamen, 1975). Simmons
(1974) using a different method to simulate Doppler shifts in the
echoes also evoked Doppler compensation in a horseshoe bat.

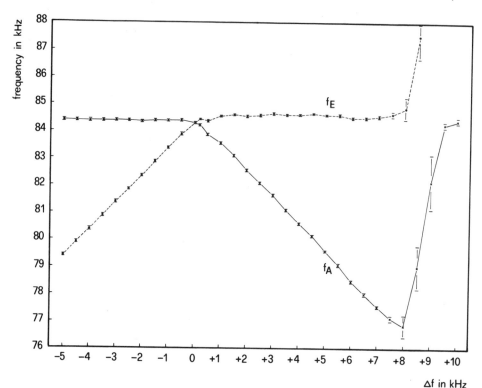

Fig. 11. Dynamic range of the Doppler shift compensation system of _Rhinolophus ferrumequinum_. For the cf components, the transmitted frequency (f_A – solid line) and the echo frequency (f_E – dashed line) was measured for simulated positive and negative Doppler shifts (Δf). Each point corresponds to the average frequency of 200 cf components with standard deviation of the mean. Adapted from Schnitzler (1978).

In the Doppler simulation experiments, bats were held in a restrainer in front of a microphone and a loudspeaker. The transmitted sounds were picked up with the microphone and electronically shifted in frequency upward or downward with the moving target simulator. The shifted signals were played back to the bats as artificial echoes thus simulating relative movements toward or away from the bat. Some bats responded to these simulated target movements by compensating positive Doppler shifts without paying attention to the fact that range did not change in this situation.

Fig. 11 gives an example for experiments with _Rhinolophus ferrumequinum_ in which positive and negative Doppler shifts were simulated over a wide range in order to determine the entire compensation ability.

With positive frequency shifts, horseshoe bats lower the transmission frequency (f_A) of their pulse cf component so that the
echo frequency (f_E) is kept constant at approximately the reference
frequency of the feedback control system. In Fig. 11 the reference
frequency is about 300 Hz above the average cf produced when there
is no frequency shift (resting frequency). In other species frequency
differences of 50 - 300 Hz between resting and reference frequency
have been observed (Schuller, Beuter and Schnitzler, 1974). The
horseshoe bat in our example (fig. 11) was able to compensate for
frequency shifts up to 8 kHz. At higher Doppler shifts, compensation
was incomplete and ceased above 10 kHz. The compensation range for
positive frequency shifts differs from bat to bat but normally lies
between 4 - 8 kHz. The normal flight speed of horseshoe bats is
seldom higher than 5 m/sec (Schnitzler, 1971). The Doppler shifts
this brings about do not exceed 2.5 kHz. Therefore in flying horseshoe bats sufficient compensation range remains to compensate for
additional Doppler shifts caused by targets moving toward the animal.

Negative Doppler shifts simulating relative movement away from
a bat are not compensated for. In this case, the transmission frequency is similar to the resting frequency. This behavior might
be of biological value since targets which move away do not indicate
danger or, in the case of prey, are rather difficult to catch since
the maximum flight speed of Rhinolophus ferrumequinum is low
(Schnitzler, 1971).

An astonishing result of the compensation experiments is the
high precision with which the echo frequency is kept constant. When
compensating positive Doppler shifts, some horseshoe bats reach
standard deviations of only 30 - 40 Hz. This indicates that the
accuracy of frequency determination must be equally good.

Insects changing their flight paths can cause rapid changes in
the relative velocity between a bat and its prey. It is therefore
of interest to know how bats respond to changing Doppler shifts. The
dynamic properties of Rhinolophus ferrumequinum's Doppler compensation
system has been studied in detail by Schuller, Beuter and Rübsamen
(1975). Fig. 12 summarizes some of their results. In their experiments, horseshoe bats had to compensate for positive frequency shifts
changing sinusoidally between +1 and +3 kHz with different modulation
frequencies.

At modulation frequencies below 0.1 Hz, the compensation system
is fast enough to compensate for the simulated Doppler shifts completely by lowering the transmission frequency for each signal (fig.
12 A). Therefore the echo frequency is always kept constant at the
reference frequency (fig. 12B). At a modulation frequency of 0.25
Hz the compensation system cannot respond quickly enough to compensate for the Doppler shifts for every sound. With increasing

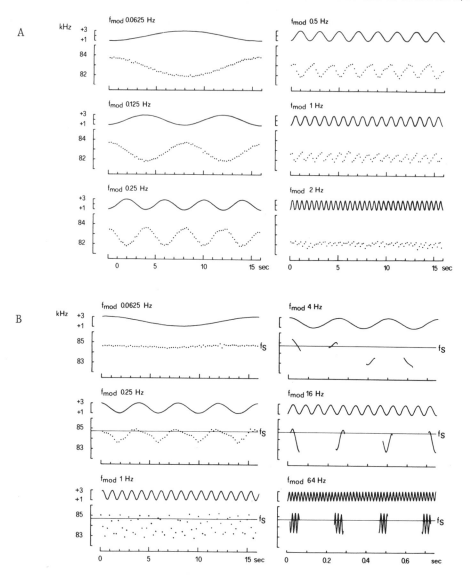

Fig. 12. Transmission frequency (A) and echo frequency (B) of the cf components of echolocation sounds in <u>Rhinolophus ferrumequinum</u> when compensating for simulated Doppler shifts changing sinusoidally between +1 and +3 kHz at different modulation frequencies (f_{mod}). In (A) and in the left half of (B) each point corresponds to the frequency of one cf component. In the right half of (B) the time base is changed to make the frequency structure of each individual echo visible. In (B) the echo frequency can be compared with the reference frequency (f_S). Part (B) adapted from Schnitzler (1978).

modulation frequencies only the highest Doppler shifts are completely
compensated for since the feedback system is faster when compensating
for increasing rather than for decreasing Doppler shifts. At modulation frequencies above 2 kHz, the transmission frequency remains at
the lower end of the corresponding compensation range the entire time
(fig. 12 A). In situations where the modulation frequency is so
high that one or more cycles fall within one echo (f_{mod}> 20 Hz), the
transmission frequency is also kept at the lower end of the compensation range so that the highest frequency components in the echoes
stay near the reference frequency.

The experiments with Doppler compensating bats have revealed
that these bats regulate their transmission frequency in order to
adjust the cf component of the echo to a specific reference
frequency. This raises the question as to whether the receiver,
i.e. the hearing system, shows a corresponding adaptation to the
reference frequency.

Hearing in Doppler Compensating Bats

Auditory thresholds of bats with fm signals are widely tuned
over the whole frequency range of their emissions. In bats with
long cf-fm sounds and Doppler shift compensation, however, a different threshold curve with a small minimum in the range of the reference frequency occurs. The hearing system of these bats is
further characterized by numerous neurons with extremely sharp
tuning and best frequencies near the reference frequency. These
specializations have been studied with neurophysiological methods
in rhinolophids (Neuweiler, 1970; Neuweiler, Schuller and Schnitzler,
1971; Schnitzler, Schuller and Neuweiler, 1971; Suga, Neuweiler and
Möller, 1976; Schnitzler, Suga and Simmons, 1976; Neuweiler and
Vater, 1977; Möller, Neuweiler and Zöller, 1978; Ostwald, 1978;
Airapetianz and Vasiliev, 1970, 1971; Vasiliev, 1971, 1975, 1976;
Vasiliev and Andreeva, 1971; Vasiliev and Timoshenko, 1973), in
Pteronotus parnellii (Grinnell, 1967, 1970; Pollak, Henson and
Novick, 1972; Suga, Simmons and Shimozawa, 1974; Suga, Simmons and
Jen, 1975; Suga and Jen, 1977; Suga, 1978; Suga and O'Neill, 1978)
and in hipposiderids (Grinnell and Hagiwara, 1972). The sharply
tuned threshold minumum near the reference frequency has also been
found in the behavioral audiogram of Rhinolophus ferrumequinum
(Long and Schnitzler, 1975).

Studies on the organization of the auditory cortex in Pteronotus parnellii (Suga and Jen, 1976; Suga, 1977; Suga, O'Neill and
Watanabe, 1978) and Rhinolophus ferrumequinum (Ostwald, 1978, and
poster at this conference) revealed, as a further adaptation, a disproportionate tonotopical organization with large areas specialized
for the processing of the cf and fm components of the echoes.

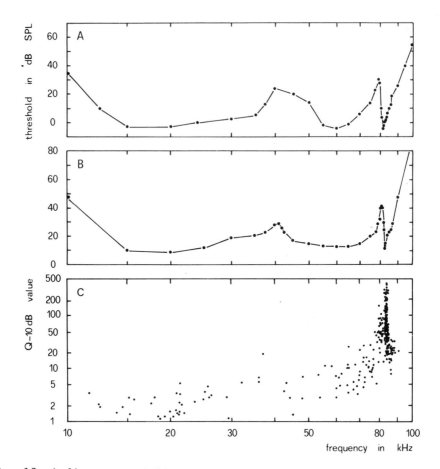

Fig. 13. Auditory specializations of Rhinolophus ferrumequinum.
(A) Behavioral audiogram (adapted from Long and Schnitzler, 1975);
(B) evoked potential threshold as measured in the colliculus
inferior (adapted from Schnitzler, 1978); (C) Q-values of primary
auditory cortex neurons (adapted from Ostwald, poster at this
conference).

Details on information processing in the auditory system of
Doppler compensating bats will be given in other reviews at this
conference. Therefore only a few basic data on hearing in Rhinolophus ferrumequinum, to understand the performance of such a specialized echolocation system, are summarized in fig. 13.

In Rhinolophus ferrumequinum, the behavioral audiogram (Long
and Schnitzler, 1975) is similar to the evoked potential threshold

curve (Neuweiler, 1970; Neuweiler, Schuller and Schnitzler, 1971).
Both curves are characterized by a narrowly tuned threshold minimum
in the reference frequency range. This minimum is separated from a
wide tuned area of low threshold which partly covers the frequency
range of the terminal fm sweeps (83 - 68 kHz) in the bat's cry. The
two areas where the threshold minimum is low are separated by an
insensitive region where the threshold is 25 - 30 dB higher. Another insensitivity peak occurs at the frequency of the orientation
sound's suppressed first harmonic (about 41.5 kHz) and this
separates the fm area from another area of low threshold which corresponds to the main frequency range of communication sounds. Below
15 kHz, the threshold is relatively high, indicating that horseshoe
bats have poor hearing at low frequencies.

The ability of the auditory system to measure frequencies is
determined by the sharpness of the tuning curves of single neurons.
The degree of tuning can be characterized by the Q_{10}-dB value (ratio
of best frequency to bandwidth at 10 dB above minimum threshold). The
sharper the tuning, the higher the Q_{10}-dB value and the better the
frequency resolution. In fig. 13 C the Q_{10}-dB values of auditory
neurons from the cf-processing area in the cortex of Rhinolophus
(Ostwald, poster at this conference) are compared with the auditory
threshold curves. This comparison shows that at frequencies below
the threshold minimum at the reference frequency, the Q-values are
low, indicating wide tuning. Similar values were also found in
fm bats (Suga, Neuweiler and Möller, 1976). However, near the reference frequency and a few kHz above it, the neurons are extremely
sharply tuned, reaching Q-values up to 400. This suggests a very
high accuracy in frequency measurement for this frequency range.

The described specializations of transmitters and receivers in
the echolocation systems of Doppler compensating bats are interpreted as adaptations for the detection and evaluation of movements.

It has already been pointed out that long cf-fm sounds are
advantageous for the determination of Doppler shifts and therefore
probably for the measurement of relative velocity between bat and
target. Doppler shift compensation and the sharply tuned threshold
minimum near the reference frequency may improve the determination
of the frequency differences between transmitted sounds and overlapping echoes. Faint echoes are adjusted in a frequency range of
low auditory threshold whereas loud outgoing sounds fall in frequency
ranges where auditory thresholds are high. The masking effect of the
loud transmitted sounds is therefore reduced (Neuweiler, 1970).

A further advantage of Doppler compensation combined with a
sharply tuned auditory system is that, in a situation where a bat
compensates for echoes from a target moving toward it, all echoes
from stationary or slowly moving targets are rejected since they

have lower frequencies and therefore fall in ranges of higher
auditory thresholds (Schnitzler, 1973 a, b; Simmons, 1974).

Long cf component sounds also have the potential to deliver
directional cues since they allow the measurement of the Doppler
shifts which are reduced laterally in accordance with the cosine of
target angle (Schnitzler, 1968).

The most important advantage of echolocation in Doppler compensating bats is that amplitude and frequency modulations in
echoes, brought about by the wingbeats of insects, can be detected.
This advantage is so important that it will be discussed in a
separate section.

DETECTION AND EVALUATION OF FLUTTERING TARGET MOVEMENTS

Hypothesis

The beating wings of insects frequency modulate and amplitude
modulate cf signals and insects are the prey of all Doppler-shift-
compensating bats studied to date. After the discovery of Doppler
shift compensation it was therefore suggested that these bats keep
the echo frequency constant at the reference frequency in order
to maintain a constant carrier frequency for such frequency and
amplitude modulations (Schnitzler, 1970a, 1970b). These modulations could aid in discriminating echoes from fluttering and nonfluttering targets. Since different species of insects cause
different modulations, it is even possible that bats use these
cues to classify or identify their insect prey (Schnitzler, 1978).

Behavioral Experiments

In order to answer the question as to whether bats utilize frequency modulations caused by fluttering target movements, the following experiment was conducted (Schnitzler, 1978; Schnitzler and
Flieger, in prep.). Several horseshoe bats were trained to discriminate an oscillating from a motionless target. By controlling the
oscillation amplitude and the oscillation frequency of the target,
frequency modulations of defined modulation depth (Δf) and modulation frequency (f_{mod}) were produced. It was therefore possible
to determine, for different modulation frequences, the threshold
modulation depth just necessary to discriminate the oscillating
target from the motionless target with a percentage of 75% correct
choices (fig. 14).

At oscillation rates of 30 Hz and below, the bats required a
modulation depth of at least 60 Hz in order to discriminate. Since
modulation depth corresponds to the difference between highest and

Fig. 14. Lower part: Performance of <u>Rhinolophus ferrumequinum</u> when discriminating a target oscillating at different rates (or modulation frequencies) from a nonoscillating target on the basis of Doppler shifts caused by the oscillations. The threshold curve indicates the echo modulation depth (maximal frequency difference in the sinusoidally modulated echoes) necessary to make the discrimination with 75% correct choices. Upper part: Comparison of echo spectra simulating threshold situations with the spectrum of a nonmodulated echo. For the simulation a carrier frequency of 84 kHz was used (adapted from Schnitzler, 1978).

lowest frequency in the sinusoidal changing echo frequency, this result suggests that bats have a frequency resolution of about 60 Hz in this situation. This value is comparable to the frequency determination accuracy of 30 - 40 Hz derived from Doppler shift compensation experiments. At oscillation rates above 30 Hz, however, the discrimination threshold decreases, finally reaching a value of only 12 Hz when the target's oscillating frequency is 200 Hz. This finding cannot be explained on the basis of the frequency discrimination performance of horseshoe bats.

A comparison of the spectra of echoes used at different modulation frequencies for threshold discrimination may help to explain these results (fig. 14). The unmodulated carrier frequency has a small single peak in the spectrum. At low modulation frequencies,

this peak oscillates up and down at a rate sufficient for the bats
to make normal frequency discriminations. A time averaged spectrum
of this situation shows a peak which is as wide as the modulation
depth. At oscillation frequencies above 50 Hz, the spectra are
characterized by side bands. The height of the side bands determined
by the modulation depth and the distance between the bands corresponds to the modulation frequency. Although modulation depth is
only 12 Hz at an oscillation rate of 200 Hz, the side bands are
clearly visible in the spectra and must be audible to the animals.
It was therefore suggested that, at low oscillation rates, bats
measure the frequency changes in the echoes directly, whereas the
high sensitivity to frequency modulations at high oscillation rates
can be explained on the basis of side band detection (Schnitzler,
1978). A similar mechanism is known in humans where tones with
high modulation rates and small modulation depth sound "rougher"
than pure tones.

Behavioral discrimination experiments support the hypothesis
that Doppler compensating bats utilize frequency modulations in
their echoes to obtain information on oscillating target movements.
These experiments proved, in addition, that this detection system
is extremely sensitive, e.g. a target oscillating with a frequency
of 40 Hz only requires an oscillation amplitude of 0.3 mm in order
to be distinguished from a motionless target.

The Doppler shifts produced by the wing movements of insects are
far above the discrimination thresholds measured in Rhinolophus
ferrumequinum, and it must be concluded that the frequency modulations in echos are detectable by Rhinolophus and other Doppler compensating bats.

The fact that wing movements are very important for the detection of insects can be derived from observations on horseshoe bats
hunting insects in a flight room. Only insects with beating wings
were pursued whereas moths whose wings were not moving did not
evoke the bats' interest even when they were in the acoustical beam
of the transmitted sounds (Schnitzler and Hollerbach, unpublished).
Russian scientists made similar observations with rhinolophids
(Sokolov, 1972; Airapetianz and Konstantinov, 1974) but interpreted
their results in a different way. They assumed that the bats hear
the wing movements of the moths and find their prey by passive acoustical localization. That the hearing of the wing sounds might act as
an additional cue for the detection of insects cannot be excluded.
We, however, suggest that the main information on flying insects is
derived from the frequency and amplitude modulations in the echoes,
since horseshoe bats still pursued fluttering insects when the sounds
produced by their wings were completely masked for a human listener,
by the loud noise of a ventilation fan (Schnitzler and Hollerbach,
unpublished).

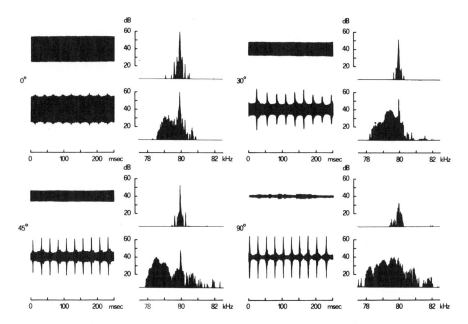

Fig. 15. Oscillograms and time averaged spectra of echoes returning from flying and nonflying sphingid moths <u>Daphnis nerii</u> when oriented at different angles in the acoustical beam of a loudspeaker transmitting an 80 kHz signal (adapted from Schnitzler, 1978).

The importance of wing beats has also been studied in <u>Pteronotus parnellii</u> by Goldman and Henson (1977). This species only pursued insects with flapping wings and it routinely attacked a stationary mechanical insect model when its wing-like parts were moving at rates of 15 Hz or higher. They found, in addition, that <u>Pteronotus</u> rejects some species of insects, thus demonstrating selective prey acquisition.

Echo Analysis

For the description of amplitude and frequency modulation caused by wing beats we have analyzed echoes from flying insects. Several species of moths were tethered in the beam of an ultrasonic loudspeaker which transmitted a carrier frequency of 80 kHz. The echoes returning from the insects were picked up with a microphone positioned near the loudspeaker, recorded on tape and afterwards analyzed (Schnitzler, 1978; Schnitzler and Flieger, in prep.). Figs. 15 and 16 summarize results from a sphingid moth (<u>Daphnis nerii</u>) flying and resting when oriented at different angles to the loudspeaker.

Echoes returning from a resting moth had the highest sound

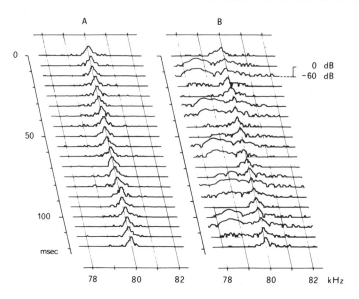

Fig. 16. Real-time spectrograms of echoes returning from a nonflying (A) and a flying (B) Daphnis nerii when oriented at an angle of 45° (adapted from Schnitzler, 1978).

pressure level (SPL) when the moth was oriented with the head directly toward the loudspeaker (60 dB re $2 \times 10^{-5} N/m^2$ at 0°). When the moth was turned to the side, the echo SPL decreased and only reached 35 dB at an orientation of 90°. At 0° the flapping wings caused only minor amplitude modulations whereas at 90°, a modulation ratio of 30 dB was reached. The number of amplitude peaks per second indicated a wing beat frequency of 38 - 40 Hz.

Frequency spectra averaged over several beats allow the degree of frequency changes caused by the wing beats in the echoes to be determined. The spectra of echoes from resting moths show a single small peak which decreases in height at lower echo intensities. In flying moths, the frequency and amplitude modulations in the echoes are represented by sidebands in the spectra. These sidebands are asymmetrical, the main energy being normally concentrated below the carrier frequency. The width of the sidebands and their fine structure differ for the moth's varying angles of orientation. At 30° the lower sideband is approximately 1.5 kHz wide whereas at 90°, widths of more than 2 kHz are reached.

Comparative studies with echoes from different species of moths also reveal species specific characteristics in amplitude and spectral cues. In this case, moths with large wings cause larger amplitude modulation ratios and wider sidebands than smaller moths. Also differences in wing beat frequencies were found.

Time averaged spectra do not show the time relationship between frequency changes in the spectrum and wing beat rhythm. In order to obtain this information, real time spectra with time intervals of 6.25 msec were made (fig. 16). These spectra indicate asymmetrical spectrum broadening in rhythm with the wing beat.

All behavioral experiments conducted to date demonstrate that echo modulations can certainly be used to discriminate fluttering from nonfluttering targets. Several neurophysiological approaches give additional support to this conclusion.

Schuller (1972) demonstrated, with evoked potential measurements, that horseshoe bats are very sensitive to small linear frequency modulations and Johnson, Henson and Goldman (1974) showed for Pteronotus parnellii that frequency modulations are encoded in cochlear microphonic potentials. Experiments in Rhinolophus ferrumequinum with sinusoidally amplitude modulated (Andreeva and Vasiliev, 1977; Schuller, 1979) and frequency modulated stimuli (Ostwald, 1978; Schuller, 1979) and with frequency modulations in Pteronotus parnellii (Suga and Jen, 1977) revealed a synchronization of modulation frequency and neuronal responses.

When stimulated with natural moth echoes reproduced with a laboratory computer, Ostwald (poster at this conference) demonstrated that neurons in the cf-processing area of Rhinolophus ferrumequinum respond in synchronization with the wing beat (fig. 17). This indicates that Doppler compensating bats are not only able to discriminate between echoes from fluttering insects and stationary targets but also to measure the wing beat frequency of their prey. This makes it possible that bats are capable of categorizing their prey according to differences in wing beat frequency.

The fine structure of moth echoes differs in various species but also depends on the angular orientation of the moth in the acoustical beam. If bats are capable of learning to discriminate the acoustical images of different species and of recognizing an individual species oriented at different angles, they would have additional cues for the classification or even the identification of their prey. This could be an explanation for the selective prey acquisition found in Pteronotus parnellii (Goldman and Henson, 1977). Further experiments are in progress to investigate this question.

Fig. 17. Neuronal responses recorded in the cf-processing area of the primary auditory cortex of <u>Rhinolophus ferrumequinum</u> when stimulated with a pure cf tone and an echo of a flying insect. The insect echo was reproduced with a laboratory computer using a carrier frequency equal to the frequency of the pure tone. Note the synchronization between neuronal responses and wing beat frequency (adapted from Ostwald, poster at this conference).

Feeding Behavior of Doppler Compensating Bats

Laboratory studies on the feeding behavior of <u>Rhinolophus ferrumequinum</u> have demonstrated that these bats only pursue moths which are flapping their wings, and pursuit occurs even when the insects fly in very confined spaces or sit on surfaces (Schnitzler and Hollerback, unpublished). Goldman and Henson (1977) observed similar behavior in <u>Pteronotus parnellii</u>.

Under natural conditions, rhinolophids usually forage near the ground or close to the branches of low trees, bushes and along walls (Eisentraut, 1950; Brosset, 1966; Wallin, 1969). It is even possible that they take insects from the foliage of trees (Griffin and Simmons, 1974) or from the ground (Blackmore, 1964). <u>Rhinolophus</u>

aethiops has also been observed hanging on a tree branch and hawking moths like a flycatcher by taking short flights and returning to the same vantage point (Shortridge, 1934).

Pteronotus parnellii have been observed foraging mainly within 3.5 m of the ground under a dense canopy of foliage (Bateman and Vaughan, 1974).

The foraging behavior of Hipposideros commersoni is described in an excellent paper by Vaughan (1977). These bats mainly forage in riverine vegetations using vantage points some 6 m above the ground. From these points they scan the environment with echolocation signals and hunt large beetles in the "flycatcher style", thereby demonstrating high precision in trajectory evaluation.

Since Hipposideros is very selective in its choice of prey, one can suggest that this species must have a highly developed target discrimination ability (Vaughan, 1977). The same can be concluded for Pteronotus, which also showed preferences in prey acquisition (Goldman and Henson, 1977). In rhinolophids, no observations concerning the selection of prey have been made to date.

These observations indicate that Doppler compensating bats often hunt in areas where there are many additional stationary targets (bushes, trees, ground). They therefore have to detect the insect echoes amongst heavy background clutter. The extreme specialization for the detection of echoes which are modulated by the insects' wing beats can be interpreted as adaptations to overcome this problem. Another point arising from these observations is that Pteronotus parnellii and Hipposideros commersoni are selective in their prey acquisition. This high degree of target discrimination can be explained by the ability of Doppler compensating bats to measure the wing beat frequency of their prey and maybe even to evaluate additional cues in the species specific fine structure of the modulated insect echoes.

The long cf-fm sounds of Doppler compensating bats are also of advantage in hunting insects in the "flycatcher style". The high duty cycle increases the chances of detecting an insect when scanning the environment and the high echo energy of the cf component allows the detection of insects over long ranges.

Since the feeding behavior of Doppler compensating bats from different families is similar, we suggest that echolocation systems with long cf-fm sounds, Doppler shift compensation and specialization of the hearing system for the evaluation of modulated echoes, can be interpreted as an evolutional adaptation allowing the capture of insects in a specialized ecological niche.

DETERMINATION OF OTHER TARGET PROPERTIES

Targets encountered by bats under natural conditions are characterized by many more features than range, angular direction, relative velocity and oscillatory movements. The additional target properties of size, shape, surface properties and resonance must also be encoded in the echo structure. This raises the question as to whether bats are able to evaluate these properties in order to classify or identify targets.

Discrimination experiments have been used to test the ability of some bats to discriminate between different target features. In some of these experiments an attempt was made to explain the discrimination performance on the basis of differences in echo structure.

Discrimination of Targets Differing in Size

In discrimination experiments with cylinders of different diameters a level of 75% correct choices was reached at echo intensity differences of 1 dB in Noctilio leporinus, of 2 dB in Myotis oxygnathus and of 4 - 5 dB in Rhinolophus ferrumequinum (Konstantinov, 1970; Airapetianz and Konstantinov, 1974). In Noctilio the discrimination threshold of 1 dB corresponds well to the intensity discrimination threshold of 1 - 2 dB found in an experiment where these bats had to discriminate one wire from two wires (Suthers, 1965). In further experiments with Rhinolophus, a discrimination threshold of 2.4 - 4.4 dB with spherical targets (Airapetianz and Konstantinov, 1974) and of 2.8 - 3.2 dB with circular planar targets (Fleissner, 1974) was measured. In Eptesicus fuscus an echo intensity difference of 1.5 - 3 dB was necessary to discriminate triangular targets of different size (Simmons and Vernon, 1971).

Discrimination of Targets Differing in Shape and Material

Dijkgraaf (1957) had previously demonstrated that Plecotus auritus can discriminate a disc from a cross, both of which had the same surface area. Airapetianz and Konstantinov (1965) successfully trained the same species to discriminate between a cube, pyramid and cylinder of equal volume. Myotis oxygnathus learned to discriminate triangles, squares and circles with equal surface areas (Konstantinov and Akmarova, 1968). Myotis oxygnathus and Rhinolophus ferrumequinum were able to discriminate a square with smooth edges from squares with serrated edges, and a square with holes in it from concave and convex squares (Airapetianz and Konstantinov, 1970). Vampyrum spectrum learned to discriminate a sphere from a prolate spheroid. The targets were designed so as to produce echoes of approximately equal overall amplitude (Bradbury, 1970). In this experiment, one bat appeared to be using differences in echo spectrum

and another differences in the overall intensity as discrimination
cues.

Konstantinov et al. (1973) tested the ability of Myotis
oxygnathus and Rhinolophus ferrumequinum to discriminate a regular
lattice of 7 metal rods (150 mm high and 3 mm in diameter) positioned along a 150 mm horizontally placed plank from a lattice with fewer
or more rods arranged on a plank of the same length. Myotis was
able to distinguish between the two targets with 75% correct choices
when the variable target contained only one more or one less rod.
Rhinolophus required a change of at least three rods for the same
discrimination. The discrimination performance in this experiment
was explained by the bats' ability to discriminate the different
interference patterns of the echoes returning from the different
targets (Malashin, 1970).

In discrimination experiments with similar targets of
different material it was found that Eptescus serotinus can distinguish between targets covered in the one case with glass and in
the other with velvet (Dijkgraaf, 1946). Myotis oxygnathus and
Rhinolophus ferrumequinum were able to discriminate an aluminum
plate from similar plates made from plexiglass, plywood and textolite
but not from brass and iron plates. The echo cues used for
discrimination in this experiment have not yet been identified.
The possibility of differing resonances causing differences in the
echo structure has been discussed by Airapetianz and Konstantinov
(1974).

Discrimination of Mealworms from Other Targets

Bats which have learned to catch mealworms projected into
their flight path are also able to discriminate these from other
targets (Webster, 1963a, 1963b, 1967a, 1967b; Webster and Durlach,
1963; Webster and Brazier, 1965, 1968; Griffin, Friend and Webster,
1965; Griffin, 1967). Several species of vespertilionids learned
to select and catch mealworms out of a cluster of discs of comparable
size (fig. 7). Experienced bats reached such a high discrimination
level that they caught more than 90% of mealworms and avoided
85 - 90% of the discs. The investigation of the target echo fine
structure led to the suggestion that the bats might be able to recognize spectral differences in the echoes of mealworms and discs
(Griffin, Friend and Webster, 1965; Griffin, 1967).

Discrimination of Targets with Different Depth Structure

In point targets, the returning echoes arrive as one wavefront
at the bat's ear. Three dimensional targets with several reflecting
surfaces return several echoes delayed temporally according to the
depth differences. Interferences between time delayed echoes create
a complex echo.

Simmons et al. (1974) designed a crucial experiment to determine whether bats are able to evaluate information contained in such
complex echoes and use it for the discrimination of targets differing in their depth structure. They trained Eptesicus fuscus to
discriminate a plate with holes of 8 mm depth from other plates
with hole depths between 6.5 - 8 mm. At depth differences below
0.6 - 0.9 mm, the bats were no longer able to discriminate between
the targets.

The analysis of echoes from different targets showed that the
interference between the echoes from the two reflecting surfaces
of each target caused differences in the echo spectra, these being
related to the hole depth (fig. 18). As the overall intensity of
the echoes differed only by 0.1 - 0.5 dB it was suggested that the
bats used the spectral differences and not the overall intensity
for discrimination. Similar experiments with Myotis myotis and
Megaderma lyra yielded comparable results (Habersetzer, Vogler and
Leimer, personal communication).

Many targets are structured in depth and therefore deliver
multiple echoes interfering with each other. The resulting spectral
cues are available for the information evaluation in the frequency
domain. In discrimination experiments, the observation that bats are
able to analyze and to learn to use this spectral information,
allows the conclusion that it is at least possible that they are
able to classify or even to identify targets on the basis of
spectral cues.

Under natural conditions, this identification mechanism may
frequently not be necessary. For instance, in bats hunting in the
open, it might be a good strategy to pursue each detected target
since the chance that the target is not an insect is remote. This
could explain why outdoor hunting Eptesicus pursue pebbles thrown
into their flight path when hunting in the open but learn to avoid
similar targets when they are offered together with mealworms in the
laboratory.

As a general rule it can be stated that fm signals of wide
bandwidth are useful for target discrimination and the echo
spectral cues may be a means of identification. Another result
of the experiments described above is that learning plays an
important role in target discrimination.

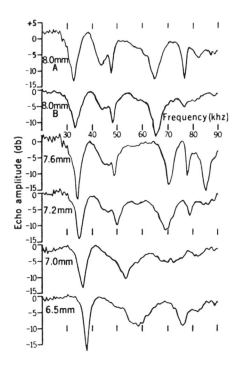

<u>Fig. 18.</u> Echo spectra of targets used in the hole-depth discrimi-
nation experiment. The targets were exposed to a simulated wideband signal. Amplitudes of the target spectra are expressed in
dB relative to those of a smooth target (adapted from Simmons
<u>et al.</u>, 1974).

BATS' RESISTANCE TO INTERFERENCE

    Under natural conditions, bats have the problem of separating
echoes from targets they are interested in from background echoes
(clutter), and of detecting their echoes in the presence of other
noise. This problem is especially difficult to solve when the
intensity of the useful echo is weak in comparison to the intensity
of the interference signals. In order to overcome such interference,
echolocation systems have evolved effective adaptations for the

gathering of biologically relevant information and for effectively
operating in the presence of interference encountered under
natural conditions.

Resistance to Noise Jamming

     In Plecotus townsendii, the ability to resist broad band noise
jamming was measured as an obstacle avoidance score reached at
different noise levels (Griffin and Grinnell, 1958; Griffin, McCue
and Grinnell, 1963); with low intensity noise, the bats performed
normally. With high intensity noise, the bats were reluctant to
fly. After dropping them into the air, the most skillful animals
still performed far above chance level. In these experiments,
the bats emitted signals similar to those produced without noise.
It was therefore possible to determine the detection range
and to calculate, from transmitted sound energy and wire diameter,
the echo energy at detection. The calculated data were then compared with predictions made under the assumption that the bats
might use an optimal filter receiver. This comparison indicated
that the bats performed even better than predicted.

     The explanation for this astonishing result was that the bats
reduced noise jamming by approaching the wire obliquely, so that
echoes and noise reached the ears from different directions. They
used the directionality of their hearing system to suppress the
unwanted interference.

     That hearing directionality is a very effective countermeasure
for the rejection of masking noise has also been demonstrated in
Rhinolophus ferrumequinum in experiments to measure behavioral
auditory directionality (Grinnell and Schnitzler, 1977) (fig. 6).
For example, a noise source masking a band of $\pm$ 3 kHz around the
signal frequency, positioned 30° laterally, required a 9 dB higher
SPL than one positioned directly ahead. An additional result of the
experiments with Plecotus is that the amplitude of the signals – at
least in hand-held bats – changed with the noise level.

     Russian scientists studied the ability of Myotis oxygnathus to
discriminate a cube from a cone or cylinder in the presence of noise.
With increasing noise levels, the signal durations were increased
from 1.5 – 2 msec (no noise) to 5 – 6 msec (Airapetianz and
Konstantinov, 1974).

     The ability of Myotis oxygnathus and Rhinolophus ferrumequinum
to detect a small sphere in the presence of wideband and bandlimited
noise has also been tested (Konstantinov, Movtchan and Makarov, 1973;
Makarov, 1973). Both species increased the intensity of their signals with increasing noise levels. The experiments with small-band

jamming noise in <u>Rhinolophus</u> revealed that the masking effect was
stronger when the fm-sweeps of the echoes were masked than when
the long cf components were jammed.

Simmons, Lavender and Lavender (1978) tested the range discrimination performance of <u>Eptesicus fuscus</u> under controlled noise
levels and analyzed the sounds used for discrimination. At noise
levels of -70 to -10dB (re-0.1 N/m^2) the bats showed a steady performance with about 80% correct responses in the discrimination of
a 2 cm target range difference. At 0 dB the discrimination performance declined to less than 60% correct responses and at +10 dB
the bats flew away.

Under quiet conditions and up to -50 dB, the target ranging
signals were 1 - 2 msec in duration and frequency modulated with a
strong second harmonic (fig. 19). With increasing noise level the
signal amplitude was increased as much as 6 to 10 dB and sound
durations up to 8 msec were obtained. Most of the increase in
sound duration was due to the addition of a nearly constant frequency tail, thus forming a fm short-cf signal (fig. 19).

When comparing the results of the different noise jamming
experiments the following tendencies are apparent: bats avoid
noise jamming by using the directionality of their echolocation
system so as to separate the useful echo from the unwanted interference. They also improve the signal to noise ratio when
increasing the amplitude and the duration and thus the energy
of the transmitted signal.

Clutter Rejection

Many bats can hunt insects in situations where there must be
many echoes from nearby objects. They even learn to identify and
selectively catch one mealworm out of a cluster of other ballistic
targets (fig. 7). For the explanation of this performance, we must
assume that they can recognize and separate useful echoes from
interfering clutter.

For Doppler-compensating bats we have already shown that the
utilized echoes from flying insects are characterized by typical
frequency and amplitude modulations caused by wing beats and
that a specialized hearing system is used for the selective
evaluation of this information.

In other bats, the separation of useful echoes from clutter
may be possible because of other typical echo features probably
contained in the echo spectrum.

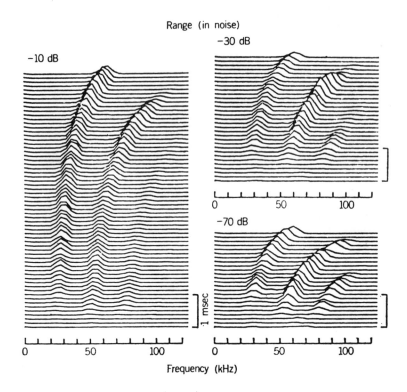

Fig. 19. Frequency structure of Eptesicus fuscus target-ranging signals used in different noise interferences. Noise in dB relative to a SPL of 0.1 N/m^2 (adapted from Simmons, 1978).

After a bat has made the decision that an echo should be utilized, it can use the echolocation system's high directional sensitivity to focus on the target in such a way that the echoes from cluttering targets are excluded. That bats actually use this strategy to separate a target out of clutter, has been demonstrated for Myotis lucifugus, Lasiurus borealis, Eptesicus fuscus and Rhinolophus ferrumequinum (Webster and Brazier, 1965, 1968).

EVOLUTIONARY ADAPTATION OF ECHOLOCATION TO FEEDING HABITS

Echolocation systems have evolved as effective adaptations for gathering behaviorally relevant information and suppressing unwanted interference. All bats have similar problems in finding their way and in avoiding obstacles, but owing to the wide variations in diet, there are very different problems in finding food. As a result, the differences found in their echolocation systems have very probably been developed under selective pressure as a means of finding the species specific nourishment.

The relationship between echolocation, feeding behavior and the acoustical environment in which bats seek their food has been discussed by many authors. The most recent publications in this field come from Novick (1977), Neuweiler (1977), Schnitzler (1978) and Simmons, Fenton and Farell (1979). In these publications, different species of bats have been mainly categorized according the types of echolocation signals present e.g.   long cf-fm, short cf-fm, fm-short cf and pure fm bats. This raises the question as to whether such a categorization according to sound structure alone reflects the different types of feeding behavior found in the different species.

Theoretically cf components allow the evaluation of Doppler shifts and improve the chances of detecting faint echoes, but for range measurement, evaluation of target fine structure and clutter rejection on the basis of spectral cues, they are not as effective as fm components. The use of different signal types in different species and also the change from one type to another within one species therefore reflects the use of different information gathering strategies.

On the basis of the data presented in this review, we can roughly categorize three different main types of echolocation systems, these corresponding to three different main types of feeding behavior.

Echolocation Systems of Insect Hunting Bats with Specializations for the Evaluation of Doppler Shifts

Bats with this type of echolocation system utilize loud cf-fm sounds; they compensate for Doppler shifts and have hearing systems specialized for the evaluation of frequency and amplitude modulations in echoes. This allows the detection and probably also the classification of fluttering insects even in dense clutter. The high signal energy of the cf component also allows the detection of insects over long ranges, thus making it possible to hunt for insects in the "flycatcher style". The basic frequency structure of

the echolocation signals remains similar in all orientation situations. Only the sound duration is changed. The terminal fm sweeps
are necessary for accurate range measurement, especially during
the approach and terminal phases. Ear movements are important for
the determination of angular direction.

All rhinolophids, probably all hipposiderids and Pteronotus
parnellii belong to this group. Further experiments must be conducted in order to determine whether some of the emballonurids
which produce cf-fm signals are also specialized for the evaluation of Doppler shifts.

Echolocation Systems of Insect Hunting Bats without Specializations
for the Evaluation of Doppler Shifts

This category includes bats which forage in open spaces and
species which occasionally or usually pursue insects in the vicinity
of obstacles. The echolocation signals of this group are of high
intensity. When searching for food, a wide variety of signal types
such as pure cf, short cf-fm, fm-short cf, shallow and steep fm
with and without harmonics have been recorded.

Theoretically, the cf components used in search flight by these
bats are as useful for the evaluation of Doppler shifts as those
in Doppler compensating bats. In most cases, however, the cf components are shorter in duration than those of the first group,
and the hearing systems of these bats are not specialized for the
evaluation of Doppler information. The ability to utilize Doppler
shifts is therefore almost certainly far less well developed here
than in Doppler compensating bats.

These bats, however, have the advantage that the high signal
energy of the cf components or of the shallow sweeps remains concentrated for the entire duration within the bandwidth of individual neuronal filters. The chance of detecting faint insect echoes
is therefore much improved (Grinnell and Hagiwara, 1972). The
use of such components of high signal energy thus reflects an
information gathering strategy oriented toward the detection of
faint echoes.

Pure fm signals, with and without harmonics, are usually emitted by these bats when they are close to targets and it has been
demonstrated through discrimination experiments that bats from this
group can learn to utilize spectral cues for target discrimination.
They also use the directionality of their echolocation system as
well as spectral cues to separate useful echoes from clutter.

The auditory sensitivity of the bats in this group is usually widely tuned over the whole frequency range of the echolocation sounds. The hearing systems have no specializations for the evaluation of Doppler shifts in the frequency range of the cf components or the shallow fm sweeps.

Some species use a wide variety of signal types. This demonstrates high adaptiveness to different information gathering problems. Species from different families belong to this group, varying from molossids hunting for insects in the open far away from obstacles to emballonurids which pursue insects in the forest. Further studies on echolocation behavior, feeding habits and hearing in different species may allow a further subdivision of this category.

Echolocation Systems of "Surface Bats"

Nycterids, megadermatids, Antrozous, Macrotus and Plecotus belong to a group specialized for preying on insects, arthropods or small vertebrates resting on surfaces (gleaning bats) and also bats which feed on fruit, pollen and nectar (phyllostomatids) or on blood (desmodontids). All these "surface bats" have the problem that the echoes from the targets in which they are interested are buried in dense clutter.

The echolocation signals of these bats are characterized by low intensities, short durations and shallow sweeps with many harmonics. These allow a good evaluation of target fine structure on the basis of spectral cues. This may aid in resolving the useful target echoes from the dense background clutter.

Several observations indicate that in "surface bats" nonecholocation cues may be used alone or as additional help to identify or localize food.

For instance, all "gleaning bats" have large ears which can be turned toward sound sources. This suggests a highly developed localization acuity due to passive hearing. In Megaderma lyra it has even been demonstrated (Fiedler, personal communication) that they switch off the emission of echolocation signals and hunt mice solely by passive hearing.

The musky odor of all flowers visited by nectar bats also suggests that smell plays an important role in the localization of these food sources. Another orientation aid may be the vision found in some species of this group.

There are probably many species of bats which do not fit completely into one of the proposed categories, e.g. all fm bats which hunt for flying insects but sometimes also catch insects on the ground. We hope that further field observations on echolocation and feeding behavior in more species of bats will allow us to gain a better understanding of the importance of echolocation as an evolutionary adaptation for information gathering problems in bats.

ACKNOWLEDGEMENTS

The research yielding new data published for the first time in this review has been supported by the Deutsche Forschungsgemeinschaft Grant No. Schn 138/7-9 and NIH Grant NS 12445.

We thank H.-H. Görs, U. Heilmann, W. Hollerbach, D. Menne, J. Ostwald, A. Rassa, S. Riediger, M.-L. Zander and M.M. Henson for critical discussions and technical assistance. Without their patience and valuable help this review would not have been finished on time.

REFERENCES

Airapetianz, E. Sh. and Konstantinov, A.I., 1970, Echolocation in Nature. Nauka, Leningrad (in Russian).

Airapetianz, E. Sh. and Konstantinov, A.I., 1970, Echolocation in Animals. Nauka, Leningrad. English Translation. Israel Program of Scientific Translations, Jerusalem, 1973.

Airapetianz, E. Sh. and Konstantinov, A.I., 1974, Echolocation in Nature. Nauka, Leningrad (in Russian).

Airapetianz, E. Sh. and Konstantinov, A.I., 1974, Echolocation in Nature. Nauka, Leningrad. English Translation. Joint Publications Research Service, No. 63328, 1000 North Glebe Road, Arlington, Virginia 22201.

Airapetianz, E. Sh. and Konstantinov, A.I., 1965, The problem of the role of echolocation in spatial analysis by bats. Bionika, p. 334, Nauka, Moscow (in Russian).

Airapetianz, E. Sh. and Vasiliev, A.G., 1970, The characteristics of the evoked responses in the auditory system of bats to ultrasonic stimuli of different fill frequency. Sechenov. Physiol. J. 56, 1721-1730 (in Russian).

Airapetianz, E. Sh. and Vasiliev, A.G., 1971, On neurophysiological mechanism of the echolocating apparatus in bats (frequency parameters). Internat. J. Neuroscience 1, 279-286.

Altes, R.A., 1973, Some invariance properties of the wide band ambiguity function. J. Acoust. Soc. Am. 53, 1154-1160.

Altes, R.A., 1975, Mechanism for aural pulse compression in mammals. J. Acoust. Soc. Am. 57, 513-515.

Altes, R.A., 1976, Sonar for generalized target description and its similarity to animal echolocation systems. J. Acoust. Soc. Am. 59, 97–105.

Altes, R.A. and Titlebaum, E.L., 1970, Bat signals as optimally tolerant waveforms. J. Acoust. Soc. Am. 48, 1014–1020.

Andreeva, N.G. and Vasiliev, A.G., 1977, Responses of the superior olivary neurons to amplitude modulated signals. Sechenov. Physiol. J. 63, 496–503 (in Russian).

Bateman, G.C. and Vaughan, T.A., 1974, Nightly activities of mormoopid bats. J. Mammal. 55, 45–65.

van Bergeijk, W.A., 1964, Sonic pulse compression in bats and people: a comment. J. Acoust. Soc. Am. 36, 594–597.

Beuter, K.J., 1977, Optimalempfängertheorie und Informationsverarbeitung im Echoortungssystem der Fledermäuse. In: Kybernetik 1977, G. Hauske and E. Butenandt, eds., pp. 106–125, München–Wien, Oldenbourg.

Blackmore, M., 1964, Order Chiroptera. In: The handbook of British mammals, N.H. Southern, ed., pp. 224–249, Blackwell Scientific Publication, Oxford.

Bradbury. J., 1970, Target discrimination by the echolocating bat, Vampyrum spectrum. J. Exp. Zool. 173, 23–46.

Brosset, A., 1966, La biologie des chiroptères. Masson et Cie., Paris.

Cahlander, D.A., 1963, Echolocation with wide-band waveforms. Thesis deposited in the library of MIT, Boston, Mass.

Cahlander, D.A., 1964, Echolocation with wide-band waveforms: Bat sonar signals. MIT Lincoln Lab. Techn. Rep. No. 271, AD 605322.

Cahlander, D.A., 1967, Discussion of Batteau's paper. In: Animal Sonar Systems, vol. II, R.–G. Busnel, ed., pp. 1052–1081. Laboratoire de Physiologie acoustique, Jouy–en–Josas.

Cahlander, D.A., McCue, J.J.G. and Webster, F.A., 1964, The determination of distance by echolocating bats. Nature 201, 544–546.

Curtis, W.E., 1952, Quantitative studies of echolocation in bats (Myotis l. lucifugus); Studies of vision in bats (Myotis l. lucifugus and Eptesicus f. fuscus) and quantitative studies of vision in owls (Tyto alba practincola). Thesis deposited in the library of Cornell Univ., Ithaca, N.Y.

Dijkgraaf, S., 1946, Die Sinneswelt der Fledermäuse. Experientia 2, 438–448.

Dijkgraaf, S., 1957, Sinnesphysiologische Beobachtungen an Fledermäusen. Acta Physiol. Pharmacol. Neerlandica 6, 675–684.

Eisentraut, M., 1950, Die Ernährung der Fledermäuse. Zoologische Jahrbücher 79, 115–177.

Flieger, E. and Schnitzler, H.–U., 1973, Ortungsleitungen der Fledermaus Rhinolophus ferrumequinum bei ein- und beidseitiger Ohrverstopfung. J. Comp. Physiol. 82, 93–102.

Fleissner, N., 1974, Intensitätsunterscheidung bei Hufeisennasen

(Rhinolophus ferrumequinum). Staatsexamensarbeit vorgelegt am
    Fachbereich Biologie der Universität Frankfurt/Main.
Glaser, W., 1971a, Eine systemtheoretische Interpretation der
    Fledermausortung. Studia Biophysica 27, 103-110.
Glaser, W., 1971b, Zur Fledermausortung aus dem Gesichtspunkt der
    Theorie gestörter Systeme. Zool. Jb. Physiol. 76 209-229.
Glaser, W., 1974, The hypothesis of optimum detection in bats
    echolocation. J. Comp. Physiol. 94, 227-248.
Goldman, L.J. and Henson, O.W. Jr., 1977, Prey recognition and
    selection by the constant frequency bat, Pteronotus p.
    parnellii. Behav. Ecol. Sociobiol. 2, 411-419.
Golubkov, A.G., Konstantinov, A.I. and Makarov, A.K., 1969,
    Effective range and sensitivity of the microchiropteran echo-
    location system. TR. LIAP 64, 117-127 (in Russian).
Gorlinsky, I.A. and Konstantinov, A.I., 1978, Auditory localization
    of ultrasonic source by Rhinolophus ferrumequinum. Kenya
    National Acad. for Advancement of Arts and Science 145-153.
Gould, E., 1955, The feeding efficiency of insectivorous bats.
    J. Mammal 36, 399-407.
Gould, E., 1959, Further studies of the feeding efficiency of bats.
    J. Mammal. 40, 149-150.
Griffin, D.R., 1944, Echolocation by blind men, bats and radar.
    Science 100, 589-590.
Griffin, D.R., 1953, Bat sounds under natural conditions, with
    evidence for the echolocation of insect prey. J. Exp. Zool.
    123, 435-466.
Griffin, D.R., 1958, Listening in the dark. Yale University Press,
    New Haven.
Griffin, D.R., 1962, Echoortung der Fledermause, insbesondere beim
    Fangen fliegender Insekten. Naturw. Rundsch. 15, 169-173.
Griffin, D.R., 1967, Discriminative echolocation in bats. In: Animal
    Sonar Systems, vol. I, R.-G. Busnel, ed., pp. 273-299. Labor-
    atoire de Physiologie acoustique, Jouy-en-Josas.
Griffin, D.R., 1971, The importance of atmospheric attenuation for
    the echolocation of bats (Chiroptera). Anim. Behav. 19, 55-61.
Griffin, D.R., Dunning, D.E., Cahlander, D.A. and Webster, F.A.,
    1962, Correlated orientation sounds and ear movements of horse-
    shoe bats. Nature 196, 1185-1186.
Griffin, D.R., Friend, J.H. and Webster, F.A., 1965, Target
    discrimination by the echolocation of bats. J. Exp. Zool.
    158, 155-168.
Griffin, D.R. and Galambos, R., 1941, The sensory basis of obstacle
    avoidance in flying bats. J. Exp. Zool. 86, 481-506.
Griffin, D.R. and Grinnell, A.D., 1958, Ability of bats to discrim-
    inate echoes from louder noise. Science 128, 145-147.
Griffin, D.R., and Hollander, P., 1973, Directional patterns of
    bats' orientation sounds. Period. Biol. 75, 3-6.
Griffin, D.R., McCue, J.J.G. and Grinnell, A.D., 1963, The resis-
    tance of bats to jamming. J. Exp. Zool. 152, 229-250.

Griffin, D.R., and Novick, A., 1955, Acoustic orientation of
    neotropical bats. J. Exp. Zool. 130, 251-300.
Griffin, D.R., and Simmons, J.A., 1974, Echolocation of insects by
    horseshoe bats. Nature 250, 731-732.
Griffin, D.R., Webster, F.A. and Michael, C.R., 1960, The
    echolocation of flying insects by bats. Anim. Behav.
    8, 141-154.
Grinnell, A.D., 1967, Mechanism of overcoming interference in
    echolocating animals. In: Animal Sonar Systems, Vol. I.
    R.-G. Busnel, ed., pp. 451-481. Laboratoire de Physiologie
    acoustique, Jouy-en-Josas.
Grinnell, A.D., 1970, Comparative auditory neurophysiology of
    neotropical bats employing different echolocation signals.
    Z. vergl. Physiol. 68, 117-153.
Grinnell, A.D. and Grinnell, V.S., 1965, Neural correlates of
    vertical localization in echolocating bats. J. Physiol.
    181, 830-851.
Grinnell, A.D. and Griffin, D.R., 1958, The sensitivity of
    echolocation in bats. Biol. Bull. 114, 10-22.
Grinnell, A.D. and Hagiwara, S., 1972, Adaptations of the auditory
    nervous system for echolocation. Studies of New Guinea bats.
    Z. vergl. Physiol. 76, 41-81.
Grinnell, A.D. and Schnitzler, H.-U., 1977, Directional sensitivity
    of echolocation in the horseshoe bat, Rhinolophus ferrumequinum.
    II. Behavioral directionality of hearing. J. Comp. Physiol.
    116, 63-76.
Gustafson, Y. and Schnitzler, H.-U., Echolocation and obstacle
    avoidance in the hipposiderid bat, Asellia tridens. J. Comp.
    Physiol., in press.
Hartridge, H., 1945a, Acoustic control in the flight of bats. Nature
    156, 490-494.
Hartridge, H., 1945b, Acoustic control in the flight of bats. Nature
    156, 692-693.
Johnson, R.A., Henson, O.W., Jr. and Goldman, L.R., 1974, Detection
    of insect wing beats by the bat, Pteronotus parnellii. J.
    Acoust. Soc. Am. 55, 53.
Kay, L., 1961a, The orientation of bats and men by ultrasonic echo-
    location. Brit. Commun. Electron. 8, 582-586.
Kay, L., 1961b, Perception of distance in animal echolocation.
    Nature 190, 361.
Kay, L., 1962, A plausible explanation of the bats echolocation
    acuity. Anim. Behav. 10, 34-41.
Konstantinov, A.I., 1969, The relationship of auditory perception
    and echolocation during detection and catching of
    insects by bats (Vespertilionidae). J. Evol. Biochem.
    Physiol. 5, 566-572 (in Russian).
Konstantinov, A.I., 1970, A description of the ultrasonic location
    system of piscivorous bats. Tez. Dokl. 23-y Nauchno-Tekhn.
    Konf. LIAP, Leningrad, 51-52.

Konstantinov, A.I. and Akhmarova, N.I., 1968, Discrimination
    (analysis) of target by echolocation in Myotis oxygnathus.
    J. Biol. Sci., Moscow Univ. 4, 22–28 (in Russian).
Konstantinov, A.I., Makarov, A.K. and Sokolov, B.V., 1978, Doppler-
    pulse sonar system in Rhinolophus ferrumequinum. Kenya National
    Acad. for Advancement of Arts and Science, 155–163.
Konstantinov, A.I., Makarov, A.K., Sokolov, B.V. and Sanotskaya,
    N.N., 1976, Physiological mechanism of the use of the Doppler
    effect in echolocation by the bat Rhinolophus ferrumequinum.
    J. Evol. Biochem. Physiol. 12, 413–418, (in English) and
    Zh. Evol. Biokhim. Fiziol. 12, 466–472 (1976) (in Russian).
Konstantinov, A.I., Malashin, V.I., Makarov, A.K., Kozlova, N.A.,
    Sokolova, N.N. and Chilingiris, V.I., 1973, The range and
    the principles of echolocation recognition in bats. Vopr.
    Sravn. Fiziol. Analizatorov No. 3 Izd-vo LGU (in Russian).
Konstantinov, A.I., Movtchan, V.N. and Makarov, A.K., 1973, Influ-
    ence of band limited noises on the efficiency of echolocation
    detection of target by Rhinolophus ferrumequinum. Period.
    Biol. 75, 7–11.
Konstantinov, A.I., Sokolov, B.V. and Stosman, J.M.A., 1967, Com-
    parative research on bat echolocation sensitivity. DAN SSSR
    175, 1418 (in Russian).
Konstantinov, A.I., Stosman, I.M. and Gorlinskiy, I.A., 1973,
    Characteristics of the directionality of reception and precision
    of localization of ultrasound by echolocating bats. Ref. Dokl.
    IV Vses. Konf.po.Bionike 4, 5156.
Long, G.R. and Schnitzler, H.U., 1975, Behavioral audiograms from
    the bat Rhinolophus ferrumequinum, J. Comp. Physiol. 100,
    211–220.
Makarov, A.K., 1973, Acoustic methods and apparatus for studying
    ultrasonic location in animals. Vopr. Sravn. Fiziol. Analiza-
    torov, No. 3, Izd-vo LGU, 77–86 (in Russian).
Malashin, V.I., 1970, Theoretical and experimental research on
    frequency-modulated signals reflected from objects being
    recognized by bats. Tez. Dokl. 23-y Nauchno-Teckn. Konf.
    LIAP, Leningrad, p. 57 (in Russian).
McCue, J.J.G., 1966, Aural pulse compression by bats and humans.
    J. Acoust. Soc. Am. 40, 545548.
McCue, J.J.G., 1969, Signal processing by the bat, Myotis lucifugus.
    J. Auditory Res. 9, 100107.
Möhres, F.P., 1963, Über die Ultraschallorientierung der Hufeisen-
    nasen (Chiroptera Rhinolophinae). Z. vergl. Physiol. 34,
    547–588.
Möhres, F.P. and Neuweiler, G., 1966, Die Ultraschallorientierung
    der Großblatt-Fledermäuse (Chiroptera-Megadermatidae). Z.
    vergl. Physiol. 53, 195–227.
Möller, J., Neuweiler, G. and Zöller, H., 1978, Response character-
    istics of inferior colliculus neurons of the awake cf-fm bat
    Rhinolophus ferrumequinum. J. Comp. Physiol. 125, 217–225.

Neuweiler, G., 1970, Neurophysiologische Untersuchungen zum Echo-
    ortungssystem der Großen Hufeisennase Rhinolophus ferrumequinum.
    Z. vergl. Physiol. 67, 273-306.
Neuweiler, G., 1977, Recognition mechanisms in echolocation of bats.
    In: Processing of complex acoustic signals T.H. Bullock, ed.,
    pp. 111-126. Dahlem Konferenzen, Berlin.
Neuweiler, G. and Möhres, F.P., 1967, Die Rolle des Ortungsgedächt-
    nisses bei der Orientierung der Grossblatt-Fledermaus Megderma
    lyra. Z. vergl. Physiol. 57, 147-171.
Neuweiler, G., Schuller, G. and Schnitzler, H.-U., 1971, On- and
    off- responses in the inferior colliculus of the greater horse-
    shoe bat to pure tones. Z. vergl. Physiol. 74, 57-63.
Neuweiler, G. and Vater, M., 1977, Response patterns to pure tones
    of cochlear nucleus units in the CF-FM bat Rhinolophus ferrum-
    equinum, J. Comp. Physiol. 115, 119-134.
Novick, A., 1963, Pulse duration in the echolocation of insects by
    the bat, Pteronotus. Ergebn. Biol. 26, 21-26.
Novick, A., 1965, Echolocation of flying insects by the bat
    Chilonycteris psilotis. Biol. Bull. 128, 297-314.
Novick, A., 1971, Echolocation in bats: some aspects of pulse
    design. Amer. Scientist 59, 198-209.
Novick, A., 1973, Echolocation in bats: a zoologist's view. J.
    Acoust. Soc. Am. 54, 139-146.
Novick, A., 1977, Acoustic orientation. In: Biology of bats,
    Vol. III, W.A. Wimsatt, ed., pp. 73-287.
Novick, A. and Vaisnys, J.R., 1964, Echolocation of flying insects
    by the bat, Chilonycteris parnellii. Biol. Bull. 127, 478-488.
Ostwald, J. 1978, Tonotopical organisation of the auditory cortex in
    the cf-fm bat Rhinolophus ferrumequinum. Verh. Dtsch. Zool.
    Ges., Gustav Fischer Verlag.
Peff, T.C. and Simmons, J.A., 1972, Horizontal-angle resolution
    by echolocating bats. J. Acoust. Soc. Am. 51, 2063-2065.
Pollak, G., Henson, O.W. and Novick, A., 1972, Cochlear microphonic
    audiograms in the "pure tone" bat, Chilonycteris parnellii
    parnellii. Science 176, 66-68.
Pye, J.D., 1960, A theory of echolocation by bats. J. Laryngol.
    Otol. 74, 718-729.
Pye, J.D., 1961a, Echolocation by bats. Endeavour 20, 101-111.
Pye, J.D., 1961b, Perception of distance in animal echolocation.
    Nature 190, 362-363.
Pye, J.D., 1963, Mechanism of echolocation. Ergebn. Biol. 26,
    12-20.
Pye, J.D., 1971, Bats and fog. Nature 229, 572-574.
Pye, J.D., Flinn, M. and Pye, A., 1962, Correlated orientation
    sounds and ear movements of horseshoe bats. Nature 196,
    1186-1188.
Pye, J.D. and Roberts, L.H., 1970, Ear movements in a hipposiderid
    bat. Nature 225, 285-286.

Roberts, L.H., 1972, Correlation of respiration and ultrasound production in rodents and bats? J. Zool. Lond. <u>168</u>, 439–449.

Schneider, H. and Möhres, F.P., 1960, Die Ohrbewegungen der Hufeisenfledermäuse (Chiroptera, Rhinolophidae) und der Mechanismus des Bildhörens. Z. vergl. Physiol. <u>44</u>, 1–40.

Schnitzler, H.-U., 1967, Discrimination of thin wires by flying horseshoe bats (Rhinolophidae). In: Animal Sonar Systems, Vol. I. R.G. Busnel, ed., pp. 69–87. Laboratoire de Physiologie acoustique, Jouy-en-Josas.

Schnitzler, H.-U., 1968, Die Ultraschall-Ortungslaute der Hufeisen Fledermäuse (Chiroptera-Rhinolophidae) in verschiedenen Orientierungs-situationen. Z. vergl. Physiol. <u>57</u>, 376–408.

Schnitzler, H.-U., 1970a, Echoortung bei der Fledermaus <u>Chilonycteris rubiginosa</u>. Z. vergl. Physiol. <u>68</u>, 25–39.

Schnitzler, H.-U., 1970b, Comparison of the echolocation behavior in <u>Rhinolophus ferrumequinum</u> and <u>Chilonycteris rubiginosa</u>. Bijdr. Dierk. <u>40</u>, 77–80.

Schnitzler, H.-U., 1971, Fledermäuse im Windkanal. Z. vergl. Physiol. <u>73</u>, 209–221.

Schnitzler, H.-U., 1973a, Control of Doppler shift compensation in the greater horseshoe bat, <u>Rhinolophus ferrumequinum</u>. J. Comp. Physiol. <u>82</u>, 79–92.

Schnitzler, H.-U., 1973b, Die Echoortung der Fledermäuse und ihre hörphysiologischen Grundlagen. Fortschr. Zool. <u>21</u>, 136–189.

Schnitzler, H.-U., 1978, Die Detektion von Bewegungen durch Echoortung bei Fledermäusen. Verh. Dtsch. Zool. Ges., p. 16–33, Gustav Fischer Verlag, Stuttgart.

Schnitzler, H.-U. and Grinnell, A.D., 1977, Directional sensivity of echolocation in the horseshoe bat, <u>Rhinolophus ferrumequinum</u>. I. Directionality of sound emission. J. Comp. Physiol. <u>116</u>, 51–61.

Schnitzler, H.-U., Schuller, G. and Neuweiler, G., 1971, Antworten des Colliculus inferior der Fledermaus <u>Rhinolophus euryale</u> auf tonale Reizung. Naturw. <u>58</u>, 617.

Schnitzler, H.-U., Suga, N. and Simmons, J.A., 1976, Peripheral auditory tuning for fine frequency analysis by the cf-fm bat, <u>Rhinolophus ferrumequinum</u>. III. Cochlear microphonics and N1-response. J. Comp. Physiol. <u>106</u>, 99–110.

Schuller, G., 1972, Echoortung bei <u>Rhinolophus ferrumequinum</u> mit frequenzmodulierten Lauten. J. Comp. Physiol. <u>77</u>, 306–331.

Schuller, G., 1979, Coding of small sinusoidal frequency and amplitude modulations in the inferior colliculus of "cf-fm" bat, <u>Rhinolophus ferrumequinum</u>. Exp. Brain Res. <u>34</u>, 117–132.

Schuller, G., Beuter, K. and Schnitzler, H.-U., 1974, Response to frequency shifted artificial echoes in the bat <u>Rhinolophus ferrumequinum</u>. J. Comp. Physiol. <u>89</u>, 275–286.

Schuller, G., Beuter, K. and Rübsamen, R., 1975, Dynamic properties of the compensation system for Doppler shifts in bat, Rhinolophus ferrumequinum. J. Comp. Physiol. 97, 113-125.

Shimozawa, T., Suga, N., Hendler, P. and Schuetze, S., 1974, Directional sensitivity of echolocation system in bats producing frequency-modulated signals. J. Exp. Biol. 60, 53-69.

Shortridge, G.C., 1934, The mammals of south west Africa. Heinemann, London.

Simmons, J.A., 1968, Depth perception by sonar in the bat Eptesicus fuscus. Ph.D. Dissertation, Princeton Univ., Princeton.

Simmons, J.A., 1969, Acoustic radiation patterns for the echolocating bats Chilonycteris rubiginosa and Eptesicus fuscus. J. Acoust. Soc. Am. 46, 1054-1056.

Simmons, J.A., 1971, Echolocation in bats: Signal processing of echoes for target range. Science 171, 925-928.

Simmons, J.A., 1973, The resolution of target range by echolocating bats. J. Acoust. Soc. Am. 54, 157-173.

Simmons, J.A., 1974, Response of a Doppler echolocation system in the bat, Rhinolophus ferrumequinum. J. Acoust. Soc. Am. 56, 672-682.

Simmons, J.A., Fenton, M.B. and O'Farrell, M.J., 1979, Echolocation and pursuit of prey by bats. Science 203, 16-21.

Simmons, J.A. and Lavender, W.A., 1976, Representation of target range in the sonar receivers of echolocating bats. J. Acoust. Soc. Am. 60 (suppl. 1), S5.

Simmons, J.A., Lavender, W.A. and Lavender, B.A., 1978, Adaptation of echolocation to environmental noise by the bat Eptesicus fuscus. Kenya National Academy for Advancement of Arts and Science 97-104.

Simmons, J.A., Lavender, W.A., Lavender, B.A., Doroshow, C.F., Kiefer, S.W., Livingston, R., Scallet, A.C., and Crowley, D.E., 1974, Target structure and echo spectral discrimination by echolocating bats. Science 186, 1130-1132.

Simmons, J.A. and Vernon, J.A., 1971, Echolocation: discrimination of targets by the bat Eptesicus fuscus. J. Exp. Zool. 176, 315-328.

Skolnik, M.I., 1962, Introduction to radar systems. McGraw-Hill, New York.

Skolnik, M.I., 1970, Radar handbook. McGraw-Hill, New York.

Sokolov, B.V., 1972, Interaction of auditory perception and echolocation in bats Rhinolophidae during insect catching. Vestn. Leningr. Univ. Ser.Biol. 27, 96-104 (in Russian).

Sokolov, B.V. and Makarov, A.K., 1971, Direction of the ultrasonic radiation and role of the nasal leaf in Rhinolophus ferrumequinum. J. Biol. Sci. Moscow Univ. 7, 37-44 (in Russian).

Sokolov, B.V. and Lipmanova, E.E., 1977, Echolocational estimation of the rate of motion of horseshoe bats. Vestn. Leningr. Univ. Biol. 3, 95-103 (in Russian).

Strother, G.K., 1961, Note on the possible use of ultrasonic pulse
    compression by bats. J. Acoust. Soc. Am. 33, 696-697.
Strother, G.K., 1967, Comments on aural pulse compression in bats
    and humans. J. Acoust. Soc. Am. 41, 529.
Suga, N., 1977, Amplitude spectrum representation in the Doppler-
    shifted-cf processing area of the auditory cortex of the
    mustache bat. Sicence 196, 64-67.
Suga, N., 1978, Specialization of the auditory system for reception
    and processing of species specific sounds. Fed. Proc. 37,
    2342-2354.
Suga, N. and Jen, P., 1976, Disproportionate tonotopic representation
    of processing cf-fm sonar signals in the mustache bat auditory
    cortex. Science 194, 542-544.
Suga, N. and Jen, P., 1977, Further studies on the peripheral auditory
    system of cf-fm bats specialized for fine frequency analysis
    of Doppler shifted echoes. J. Exp. Biol. 69, 207-232.
Suga, N., Neuweiler, G. and Moeller, J., 1976, Peripheral
    auditory tuning for fine frequency analysis by the cf-fm bat
    Rhinolophus-ferrumequinum. IV. Properties of peripheral
    auditory neurons. J. Comp. Physiol. 106, 111-125.
Suga, N. and O'Neill, W.E., 1978, Mechanism of echolocation in bats.
    TINS 1, 35-38.
Suga, N., O'Neill, W.E. and Manabe, T., 1978, Cortical neurons
    sensitivity to combinations of information-hearing elements of
    biosonar signals in the mustache bat. Science 200, 778-781.
Suga, N., Simmons, J.A. and Jen, PH.-S., 1975, Peripheral
    specialization for fine frequency analysis of Doppler-shifted
    echoes in the auditory system of the CF-FM bat Pteronotus
    parnellii. J. Exp. Biol. 63, 161-192.
Suga, N., Simmons, J.A. and Shimozawa, T., 1974, Neurophysiological
    studies on echolocation systems in awake bats producing cf-fm
    orientation sounds. J. Exp. Biol. 61, 379-399.
Suthers, R.A., 1965, Acoustic orientation by fish-catching bats.
    J. Exp. Zool. 158, 319-348.
Suthers, R.A., 1967, Comparative echolocation by fishing bats.
    J. Mammal. 48, 79-87.
Suthers, R.A., Thomas, S.P. and Suthers, B.J., 1972, Respiration,
    wing beat and ultrasonic pulse emission in an echolocating bat.
    J. Exp. Zool. 56, 37-48.
Vasiliev, A.G., 1971, Characteristics of electric responses of the
    cochlear nuclei in Vespertilionidae and Rhinolophidae to
    ultra-sonic stimuli with different fill frequency. Neuro-
    physiologica 4, 379-385 (in Russian).
Vasiliev, A.G., 1975, Characteristics of unit responses of the
    cochlear nuclei of bats Rhinolophidae to single and
    paired ultrasonic stimuli. Neurophysiology 7, 195-199
    (English translation).
Vasiliev, A.G., 1976, Characteristics of the responses of neurons in

the superior olive of bats in response to single and paired
    ultrasonic stimuli. Neirofiziologiya 8, 30–38 (in Russian).
Vasiliev, A.G. and Andreeva, N.G., 1971, Characteristics of the
    electric responses of medial geniculate body of Vespertilionidae
    and Rhinolophidae to ultrasonic stimuli with different fill
    frequency. Neirofiziologiya 3, 138–144 (in Russian).
Vasiliev, A.G. and Timoshenko, T.E., 1973, Characteristics of
    electric responses of superior olivary complex in
    Vespertilionidae and Rhinolophidae bats to ultrasonic
    stimuli with different fill frequency. Neirofiziologiya
    5, 33–39 (in Russian).
Vaughan, T.A., 1977, Foraging behavior of the giant leaf-nosed bat
    Hipposideros commersoni. East Afr. Wildl. J. 15, 237–250.
Wallin, L., 1969, The Japanese bat fauna. Zool. Bidr. Uppsala
    37, 408–413.
Webster, F.A., 1962, Mobility without vision by living creatures
    other than man (with special reference to the insectivorous
    bats). In: Proc. Mobil. Res. Conf. A.F.B., New York, J.W.
    Linser, ed., pp. 110–127.
Webster, F.A., 1963a, Active energy radiating systems: the bat and
    ultrasonic principles II, acoustical control of airborne inter-
    ceptions by bats. Proc. Int. Congr. Tech. and Blindness A.F.B.,
    New York 1, 49–135.
Webster, F.A., 1963b, Bat-type signals and some implications. In:
    Human factors in technology, Bennett, Degan and Spiegel, eds.,
    pp. 378–408, McGraw-Hill.
Webster, F.A., 1967a, Some acoustical differences between the bats
    and men. Proc. of the International Conference on Sensory
    devices for the blind. St. Dunstan's London, 63–87.
Webster, F.A., 1967b, Interception performance of echolocating bats
    in the presence of interference. In: Animal Sonar Systems,
    Vol. I, R.G. Busnel, ed., pp. 673–713. Laboratoire de
    Physiologie acoustique: Jouy-en-Josas.
Webster, F.A. and Brazier, O.B., 1965, Experimental studies on
    target detection, evaluation and interception by echolocating
    bats. Aerospace Medical Res. Lab., Wright-Patterson Air Force
    Base, Ohio, AD 628055.
Webster, F.A. and Brazier, O.G., 1968, Experimental studies on
    echolocation mechanisms in bats. Aerospace Medical Res. Lab.,
    Wright-Patterson Air Force Base, Ohio, AD 673373.
Webster, F.A. and Durlach, N.I., 1963, Echolocation systems of the
    bat. MIT Lincoln Lab. Rep. No. 41-G-3, Lexington, Mass.
Webster, F.A. and Griffin, D.R., 1962, The role of the flight mem-
    branes in insect capture in bats. Anim. Behav. 10, 332–340.

PERFORMANCE OF AIRBORNE BIOSONAR SYSTEMS:

II. VERTEBRATES OTHER THAN MICROCHIROPTERA

O. W. Henson, Jr. and H.-U. Schnitzler

University of North Carolina at Chapel Hill, and

Philipps-Universität, Lahnberge

INTRODUCTION

Biosonar is a means by which some vertebrates gain information about obstacles and prey or other aspects of their environment. It occurs in a variety of forms and it is accomplished by
the emission of brief signals and the subsequent perception and
analysis of echoes. It allows animals living under conditions of
limited visibility or with limited vision to obtain information
that may not be available or fully appreciated through other
sensory systems. Among terrestrial vertebrates the most sophisticated biosonar systems are encountered in the Microchiroptera
and indeed their lives are dependent upon their acoustic orientation capacities. By listening to echoes they can assess the
distance, direction, size, texture, shape and movement of obstacles and/or prey. The performance of the refined biosonar systems
in these animals is the subject of the second part of this review.
The first part will examine the echolocative capacities of terrestrial vertebrates whose lives are much less dependent on biosonar
and which have less sophisticated biosonar systems than
Microchiroptera.

Our knowledge of the use of biosonar by various animals is
based on behavioral experiments. In these experiments biosonar
signals are usually monitored in relation to the performance of
specific tasks and the performance of that task is quantified with
the system intact and experimentally altered. It is through such
experiments that biosonar has been convincingly demonstrated in
certain cave dwelling birds, cave dwelling Megachiroptera of
the genus Rousettus and in some of the Insectivora. In addition

it has been demonstrated that certain animals, such as rats and
man, may utilize biosonar in the absence of vision.

BIRDS

1.   Oilbirds (Steatornis caripensis)
     Echolocation by cave dwelling oilbirds was first described
by Griffin (1953) 25 years ago and since that time no additional
studies have been undertaken. In complete darkness the oilbirds
make loud clicking sounds which are clearly audible to man at a
considerable distance. Signal durations are so short that only a
few waves may be apparent on the display of an oscilloscope. The
signals have not been analyzed with modern signal processing
devices, but the major energy appears to be in the 6-10 kHz band.
Evidence strongly indicates that these birds use their biosonar
systems primarily, if not entirely, for the detection of the
rock walls within the caves where they roost. When placed in a
dark experimental chamber the birds immediately begin to emit
their echolocative signals and they can avoid direct collisions
with the walls of the chamber. If their ears are plugged, however, they typically collide with the first wall they encounter;
when the ears are subsequently unplugged they regain their ability
to avoid the walls with their previous skill level. Griffin noted
that the birds were unable to avoid an electric light cord which
was in their flight path, but no obstacle avoidance experiments
have been carried out with these birds.

2.   Cave swiftlets (Collocalia)
     Echolocation has been demonstrated in some but not all
representatives of the genus Collocalia. Those that echolocate
have been shown to produce click-like sonar signals with brief
durations and frequencies primarily in the 4-7.5 kHz band (Novick,
1959; Medway, 1959, 1967; Cranbrook and Medway, 1965; Griffin and
Suthers, 1970; Fenton, 1975).

     There have been three reports dealing with the ability of
Collocalia to avoid obstacles. Medway (1967) studied the ability
of C. esculenta and C. fuciphaga to avoid walls and a barrier of
vertically placed wooden rods 15 cm apart, each with a cross-
sectional area of 1 cm^2. On the basis of these studies he concluded that C. esculenta do not echolocate. On the other hand
C. fuciphaga was able to use biosonar with sufficient skill to
detect and avoid walls in total darkness; their ability to detect
the wooden rods was limited.

     Griffin and Suthers (1970) carried out experiments to test
the obstacle avoidance skills of C. vankorensis granti. Cylinders
were placed horizontally 40 cm apart and attached to a frame which
could be shifted in position. In addition, three rows of vertical

wires were hung from the ceiling of the chamber. The obstacles
used were 2 mm diameter plastic covered wires, 6.3 mm diameter iron
rods and 8 mm diameter plastic tubing. These experiments were the
first to seek information on the size range of objects which could
be detected and successfully avoided. The four birds tested avoided
the 6.3 mm diameter rods an average of 80% of the time (74%–92%
range); the 8 mm diameter plastic tubing 67% of the time (58%–80%
range); and the 2 mm diameter wires only 43% of the time (31%–50%
range). The reason the 6.3 mm rods were avoided better than the
8 mm tubing is not clear.

Fenton (1975) assessed the echolocative skills of C.
hirundinacea by constructing an obstacle course 10 M inside a mine
tunnel where the birds normally roosted. The obstacles were 10 mm
diameter wooden dowels, 1.5 M in length; 4 mm diameter pieces of
aluminum, 0.9 M in length; and 1.5 mm diameter copper wire strips,
0.9 M in length. Longer obstacles were not necessary since C.
hirundinacea invariably flew within 0.5 M of the tunnel ceiling.
Flight speed through the obstacle course was not known but as
the birds flew into the tunnel the average flight speed was about
10 M/sec. The data on avoidance were as follows:

obstacle diameter	percent misses	conditions
10 mm	91	flying out of mine
10 mm	73	flying into mine
4 mm	64	flying out of mine
4 mm	67	flying into mine
1.5 mm	62	flying out of mine
1.5 mm	55	flying into mine

From the published data it appears that some birds of a given
species are more skilled than others in avoiding obstacles; the
most skilled birds were usually able, at better than chance
levels, to avoid obstacles with diameters ranging from 4 mm to
10 mm.

MAMMALS

1.   Shrews
The utilization of echolocation by some species of shrews was
first demonstrated by Gould, Negus and Novick (1964). Shrews were
trained to locate and jump onto a small (12.7mm^2) platform 55–110 mm
below the rim of a circular disc. Experiments were carried out with
the masked shrew (Sorex cinereus), the Northern water shrew (Sorex
palustris), the wandering shrew (Sorex vagrans) and the short-tailed

shrew (Blarina brevicauda). It was shown that biosonar signal
emission was associated with searching for the platform. The sonar
signals were of low intensity and thus difficult to record. They
have been described as "pure" tones but they have not been carefully
scrutinized with sophisticated equipment. Under normal conditions
the shrews found the platform easily but when their ears were
plugged they were either unable to find it or found it with considerable difficulty.

    Buchler (1976) made more extensive studies on Sorex vagrans.
By testing the ability of these shrews to detect a 15 cm^2 barrier,
set at various distances along one limb of a Y-maze, he determined
that the greatest distance at which detection took place was about
65 cm. By varying the size of the barrier he found that the
smallest barrier that could be detected at a distance of 20 cm was
a 3.5 cm square (12.25 cm^2 area). Buchler was able to train the
shrews to discriminate between a solid barrier with a 4 cm square
hole and a similar barrier with no hole, when the barriers were
at distances ranging from 5 to 30 cm. On the basis of these and
other experiments it was concluded that: (1) the sonar system of
the wandering shrew is crude and operates only over a limited
range; (2) it is unlikely that it is used for prey identification
or detection; and (3) once an animal becomes familiar with an
area, spatial memory replaces echolocation as a means of orientation.

2.   Tenrecs
    Tenrecs (Hemicentetes semispinosus, Echinops telfairi and
Microgale dobsoni) have also been studied with a disc-platform
apparatus similar to that used for studying shrews (Gould, 1965).
The animals were trained to search along the rim of a 45.7 cm
diameter disc and locate a 10 cm^2 aluminum platform which was
11 cm below the disc and thus beyond the animals' reach. They were
rewarded when they successfully dropped onto the platform and
followed a ramp to the reward box. The platform was boiled in
potassium permanganate to minimize olfactory cues and its position
was changed before each trial. The experiments were carried out in
complete darkness; two animals were also blindfolded but this had
no effect on their performance. In six Hemicentetes short (10-12 mm)
tubes with 4 mm inside diameters were glued into the external acoustic meatus of each ear so that the echolocative skills could be
tested with both open and plugged tubes. When the tubes were open
all of the animals were able to locate and drop onto the platform,
but when the tubes were plugged with cotton, only two of six animals
were able to find the platform. How these two animals succeeded in
locating the platform is not known. Although the ear plugging
experiments were done only with Hemicentetes, one Echinops and one
Microgale were judged to utilize echolocation because they were able
to find the platform and because they, like Hemicentetes, made
clicking noises when searching for the platform. Centetes ecaudatus

were also observed to make clicking sounds and it is presumed that
they also echolocate. Of all the tenrecs studied by Gould,
Centetes made the loudest noises and these were said to be
audible to an observer at a distance of seven meters. According to
Gould, the sounds produced by Centetes were made by pressing the
tongue against the palate and then lowering it; the source of the
clicks produced by the other tenrecs was not clearly established.
The click rate was found to vary from animal to animal but in
general was slow, like that of shrews; the signal durations were
very short, 0.1-1.8 msec, and the main frequency components were
in the 5-17 kHz frequency band. Gould noted that Echinops
produce clicking noises while climbing in shrubbery and all of the
other tenrecs that he studied made clicking noises while exploring
strange places. His observations suggest that echolocation is used
under natural conditions for gaining general rather than very
specific information about the surroundings.

3.   Rousettus
       Three species of Rousettus (R. aegypticus, R. amplexicaudatus
and R. seminudus) have been studied to date and these are the only
Megachiroptera known to echolocate. Among Chiroptera they are
unique in their utilization of tongue clicks for echolocative
purposes (Kulzer, 1956, 1960; Möhres and Kulzer, 1956; Möhres, 1956;
Novick, 1958). The clicking sounds of these bats vary from about
2 to 4.5 msec in duration but only the initial 0.3-1.0 msec portion
of each click has a high amplitude. Different studies report that
the fundamental frequency is in the 10 to 17 kHz range with ultrasonic harmonic components extending up to 100 kHz. Certain harmonics
are emphasized, however, (ca 13, 26 and 38 kHz) and this has been
well demonstrated as being the result of resonance within the buccal
cavity (Roberts, 1975).

       When flying from tree to tree Rousettus make rapid and regular
successions of clicks which are clearly audible to the human ear.
The repetition rate and loudness of the clicks appear to increase
when the bats approach a branch or begin their take off. Our
knowledge of the ability of Rousettus to detect and avoid obstacles,
however, is limited to a single study on one bat (Griffin et al.,
1958). After considerable training this animal was able to avoid
1.0 mm cylindrical obstacles 68% of the time, 1.5 mm obstacles
77% of the time and 3 mm obstacles 85% of the time.

4.   Blinded Laboratory Rats
       Blinded laboratory rats appear to be able to utilize the
echoes of self-generated signals (sniffs, sneezes, clicks and
scratches) in order to gain some appreciation of their environment
(Rosenzweig et al., 1955). This was demonstrated by training
10 rats to choose between one of two elevated runways along which
they could walk toward a source of food; on one or the other runway

a randomly placed sheet of metal blocked the pathway. Seven of the rats learned to distinguish the open from the blocked runway at a 90% level (18 of 20 trials) when the metal barrier was perpendicular to the runway; when the barrier was turned at a 45 degree angle to reduce the echo intensity the rats lost this ability. They were also unable to choose the proper runway when their ears were plugged. The sounds produced by the rats were monitored and although rats are capable of producing ultrasonic signals they did not appear to use these when performing in the maze. In some trials the rats did not produce any sounds that could be detected by the observers, yet the rats made the correct decisions.

It has been suggested that other types of rodents might utilize echolocation under natural conditions, but the information presented to date is inconclusive (see Aryapet'yants et al., 1974).

5.   Blind Persons
    The amazing ability of some blind persons to detect and avoid obstacles has long been of interest (Diderot, 1749) and many remark-·able cases have been described and discussed (Hayes, 1935; Griffin, 1958; Rice, 1967, 1969). A number of experiments have been carried out on the obstacle detection skills of the blind; all have shown a high degree of variability and it is clear that some subjects are much more skilled than others. Some of the most extensive early experiments were carried out by Heller (1904) who reported that his blind students could detect a school chart (1 X 1.65 M) at distances of about 3 to 4 M by noting changes in the sounds of their own footsteps; he also found that they perceived a tactile sensation when they were 60-70 cm from the chart. That this tactile sensation was produced by acoustic signals reaching the ear and that echoperception was the basis of the tactile sensation known as facial vision was clearly demonstrated by Supa, Cotzin and Dallenbach (1944). Qualitative assessments of the capacity of blind persons to detect objects of different sizes and texture and at different distances and to localize an echo source have been made by Kellogg (1962), Rice, Feinstein and Schusterman (1965) and Rice (1967, 1969). In these experiments the subjects were allowed to make whatever natural sound they wished; this included talking, singing, whistling, hissing, finger snapping and tongue clicking.

Resolution

    Rice (1969) found one subject who had been blind since birth and who had superior echolocative skills; this individual could consistently, with 90% accuracy, detect the presence of a disc 50.8 mm in diameter and he could detect a disc 38 mm in diameter 75% of the time. The smallest target which could be detected was 27 mm in diameter. In these tests the targets were located 0.91 M in front of the subject. Other subjects, labeled "late blind", were judged

to be considerably less skilled than the above individual since
95-100% correct responses were obtained only when the diameter of
the disc was 104 mm or larger.

Distance perception

It is clear that acoustic delay between a signal and its
reflected echo can be appreciated by the human ear and an assessment
of this delay should provide a measure of target distance. Nevertheless, there have apparently been no attempts to test the ability
of blind persons to perceive the distance of targets; that is, on
the basis of echo information to state that an object is subjectively
at a specific range. Supa et al. (1944), however, did test for the
ability to detect barriers at unknown distances; subjects were asked
to approach the barrier as close as possible without touching it and
at this task the subjects were able to stop within several inches
of the barrier. Kellogg (1962) found that blind persons can also
resolve changes in the distance of targets. This was determined by
attaching a flat, 30.48 cm diameter, disc to a device that allowed
its position to be rapidly changed. Judgments were made in relation
to two successive target presentations. One presentation, the
"standard" target, was always positioned at a distance of 60.9 cm
and there were three fixed points where the targets could be positioned closer to the subject and three that were further away. In
these cases the subjects responded by stating whether a given target
was nearer or further away than the preceeding target in a pair of
presentations. Kellogg tested both blind and sighted (blindfolded)
persons; the blind persons were considerably more skilled than the
sighted ones. The data from Kellogg's graph No. 4 can be tabulated
as follows:

Distance of target from subject's face (in cm)	Percentage of time that the target was judged to be closer than the target at the standard distance	
	Blind subjects	Sighted (blindfolded) subjects
30.48	98.0%	56.0%
38.4	97.5	55.5
48.4	84.5	49.5
60.9 standard		
77.7	29.5	47.5
97.2	15.5	12.5
121.0	8.0	33.0

On the basis of these data Kellogg concluded that changes in distance of a little more than 10 cm could be appreciated by his skilled
blind subject and that this person could perceive differences in the

location of two targets better than a person using monocular vision.
This conclusion might be further tested by examining the ability of
subjects to detect two targets simultaneously, and to align the
targets side by side. One also must question whether these experiments were really testing distance perception or simply fine structural analysis of echoes.

Size discrimination

The ability of blind persons to discriminate among targets of
different sizes at set distances was also examined by Kellogg (1962).
In these experiments each target was set at the same distance and
seven different sized plywood discs (painted with a sand-texture
paint) were presented. The test subjects were asked whether the
target was larger or smaller than a 23.8 cm diameter standard. The
targets were always presented in pairs, but their order of presentation was random. In general it was found that the ability to
recognize targets that were smaller than the standard was good. One
subject could, with 100% accuracy, tell that a 14.7 cm target was
smaller than the 23.8 cm target at distances of 30.4 or 45.7 cm
from the face. In general the discrimination was not as good with
larger sized targets, especially at distances of 30.4 and
45.7 cm from the face. Although the subjects were undoubtedly
responding to differences in the physical characteristics of the
echoes it is not clear that there was an actual appreciation of
size differences.

Texture discrimination

Kellogg (1962) tested the ability of his subjects to discriminate among targets with different surface characteristics. The
surfaces of targets were covered with different materials (velvet,
denim, plain wood, painted wood, glass, metal) and placed 30.4 cm
in front of the subject's face. Each disc was paired with every
other disc by the method of paired comparisons. The results
indicate that the qualities of the echoes from hard surfaces can
be distinguished from those from soft surfaces. In these cases
it does not appear that there was a real appreciation of the surfaces
but rather that the echoes from the different targets sounded differently.

Shape discrimination

Rice (1966) presented some preliminary data on shape discrimination. Four blind subjects were presented with square, circular and
triangular shaped targets of equal area, and after training they were
able to discriminate among the three targets with 80% accuracy. One
subject was tested to determine whether an auditory concept of shape
had been formed but this did not seem to be the case. It is inter-

esting to note that the subject best able to discriminate shape used
hissing sounds of long duration rather than clicks.

Rice (1967) noted that some blind persons used scanning movements of their heads while attempting to detect a target. By comparing target detection capacities in subjects with mobile and
immobile heads, however, it was clear that head mobility was not
necessary for detection. It may have facilitated detection but made
little practical difference. Rice also tested performance of blind
persons with one ear plugged and compared binaural vs. monaural
echolocative capacities for targets at a distance of 30.4 cm. The
results indicated that binaural detection was superior to monaural
although the capacities were not dramatically different. Binaural
and monaural echolocative capacities were also compared when the
head was mobile and immobile.

The ability of blind subjects to localize the centers of targets
has also been tested (Rice, 1967); this was done by having each
subject point his nose at the middle of a target which was changed in
position from test to test. Localization ability was compared with
varying target size and distance, with and without the subjects' knowledge of results and with and without kinesthetic feedback. The
results showed that when a 16.2 cm target was presented at a distance of 0.91 M, the center of the target could be well localized
under all conditions; the accuracy of localization could be only
slightly improved with verbal or kinesthetic feedback. Since it is
possible that the ability to find the center of a target may vary
with targets of different diameter, Rice also used targets with
9.1 and 21.3 cm diameters. Target size, however, did not affect the
ability of the subject to localize the center nor did moving the
target from 0.91 to 1.52 M from the subject change the localization
acuity. To test localization acuity further, Rice (1967) asked the
subjects to find the left edge of a 16.2 cm target 0.91 M from the
subject's ear. On the basis of these experiments Rice concluded
that the ability to find the left edge was as good or better than
that of finding the center of the target. In these experiments it
was not clear whether the accuracy was due to kinesthetic adjustments
made by finding the center and moving left, or by actually listening
for an edge cue. Experiments were also conducted to evaluate the
extent of the arc over which targets could be localized from signals
directed straight ahead. Again, the subjects pointed their noses at
a 16.2 cm diameter target 0.91 M away after signals were emitted at
0° azimuth. The target was in one of 13 positions in a $\pm 90°$ arc, but
the accuracy was best between $\pm 30°$ of the midline. It is interesting
to note that the only subjects who could reliably detect targets
beyond 45° were those that were congenitally blind. These data
support prior studies indicating that blind persons are generally
superior to others in their ability to localize sound.

DISCUSSION

It is interesting to compare the echolocative capacities of
Rousettus with Collocalia since both of these animals have developed
biosonar systems which seem to be designed for general orientation
within the dark confines of caves and tunnels. As noted above,
experiments with one Rousettus revealed that 1 mm diameter cylindrical obstacles can be avoided 68% of the time, 1.5 mm diameter
obstacles 77% of the time and 3 mm diameter obstacles 85% of the
time. By contrast, comparable date on Collocalia shows 55-67%
avoidance for 1.5 mm diameter obstacles and 64-67% for 4 mm diameter
obstacles. It is clear from the studies of Roberts (1975) that sonar
 signals of Rousettus contain ultrasonic components of significant
strength and thus differ from the clicking noises of Collocalia in
being much broader in band width. It is also important to note that
the hearing capacities of birds at frequencies above 10 kHz, are much
inferior to those of Rousettus (Grinnell and Hagiwara, 1972). Thus,
it is to be expected that Rousettus should hear echoes from small
objects better than Collocalia. One must, however, be cautious in
comparing the performance of these animals because the tests performed encompass both echo perception and maneuverability; experimental
conditions under which the most critical tests were made were different (laboratory vs. natural roost, trained vs untrained, blindfolded vs. not blindfolded). In addition to comparative aspects of
maneuverability other information not known were the different speeds
at which the animals approached the obstacles, the distance at which
detection took place.

The levels of obstacle detection and avoidance skills in the
Microchiroptera are very clearly superior to those of Rousettus and
all of the other animals considered here. The biosonar signals of
Microchiroptera are much more structured and versatile and certain
aspects of the sense of hearing is clearly specialized for echo
perception. While Rousettus has difficulty detecting and avoiding
1 mm diameter wires, most Microchiroptera can avoid wires with diameters of 0.12 mm (Griffin, 1958; Konstantinov et al., 1966) and
Megaderma and Rhinolophus can avoid 0.06-.08 mm diameter wires
(Schnitzer, 1967, 1968).

There is no question but that the echolocative capacities of
shrews and tenrecs are crude in comparison with the Microchiroptera,
but they must be considered adequate for general orientation. One
of the more interesting aspects of biosonar signal emission by these
small terrestrial animals is the rate of emission. In
Microchiroptera rates of 10-150/sec are often cited while the rate
in shrews and tenrecs has typically been expressed in terms of 3
minute intervals. For tenrecs Gould (1965) gives values of 65 clicks
in some 3 minute intervals and for shrews 13-16. These slow rates
support the idea that the sonar systems of these animals are primarily used for general appreciation of the immediate environment.

It is tempting to compare the signal emission behavior of the shrews and tenrecs with blind persons. In working with blind persons, Rice (1969) noted that some of his subjects routinely made a few tongue click noises and from these they could obtain accurate information about the presence and location of objects over a wide range of positions and distances. He referred to this as an "auditory glance". It is curious that the biosonar capacities of the blind have not been pursued in more detail. There is much that might be learned and the principles applied to other animals.

ACKNOWLEDGEMENTS

We wish to thank Dr. M.M. Henson for discussions and technical assistance in the preparation of this manuscript, and we wish to acknowledge the support from the Deutsche Forschungsgemeinschraft (Grant No. Schn 138/7-9) and from the National Institutes of Health (Grant No. NS 12445).

ADDENDUM

A recent report on oilbirds by M. Konishi and E. I. Knudson (The oilbird: Hearing and Echolocation, Science, 204, 425-427, 1979) has clarified some of the questions concerning the frequency composition of the cries, the sense of hearing and echolocative skills of these animals. They found the sound energy of the click-like signals to be unevenly distributed in the 1-15 kHz range with most of the energy in the 1.5-2.5 kHz band. This band corresponds to the most sensitive range of the oilbirds' hearing as determined by cochlear microphonic and evoked neural potentials. The echolocative skills were tested by the ability of the birds to avoid disks with diameters of 5, 10, 20, 25, 30, 35 and 40 cm. The disks were spaced apart by 5 diameters in a square array consisting of rows and columns, and the number of disks in each array was adjusted to maintain a total disk surface of 1256 cm^2. Under these conditions the birds hit the 5 and 10 cm disks as if nothing existed in their paths. The first signs of avoidance appeared when 20 cm disks were presented, and all of the five birds tested were able to avoid the 40 cm disks. In test trials the obstacles were placed in a familiar passageway. On the basis of these studies the authors concluded that the role of echolocation in the oilbird is the detection of relatively large objects.

REFERENCES

Buchler, E. R., 1976, The use of echolocation by the wandering shrew (Sorex vagrans). Anim. Behav. 24, 858-873.

Cranbrook, Earl of, and Lord Medway, 1965, Lack of ultrasonic frequencies in the calls of swiftlets. Ibis 107, 259.

Diderot, D. Letter on the blind, 1749, in Early Philosophical Works, M. Jourdain, Trans. (Open Court Chicago, 1961), pp. 68–141 (cited from Rice, 1967).

Fenton, M. B., 1975, Acuity of echolocation in Collocalia hirundinacea (Aves: Apodidae), with comments on the distribution of echolocating swiftlets and molossid bats. Biotropica 7, 1–7.

Gould, E., 1965, Evidence for echolocation in the Tenrecidae of Madagascar. Proc. Amer. Phil. Soc. 109, 352–360.

Gould, E., N.C. Negus and A. Novick, 1964, Evidence for echolocation in shrews. J. Exp. Zool. 156, 19–38.

Griffin, D. R., 1953, Acoustic orientation in the oilbird, Steatornis. Proc. Nat. Acad. Sci. 39, 884–893.

Griffin, D. R. Listening in the Dark, Yale Univ. Press, New Haven, 1958.

Griffin, D. R. and R. A. Suthers, 1970, Sensitivity of echolocation in cave swiftlets. Biol. Bull. 139, 495–501.

Griffin, D. R., A. Novick and M. Kornfield, 1958, The sensitivity of echolocation in the fruit bat, Rousettus. Biol. Bull. 115, 107–113.

Grinnell, A. D. and S. Hagiwara, 1972, Studies on auditory physiology in nonecholocating bat, and adaptations for echolocation in one genus, Rousettus. Z. vergl. Physiol. 76, 82–96.

Hayes, S. P., 1935, Facial vision or the sense of obstacles. Perkins Publications, No. 12. Perkins Institution, Watertown Mass.

Heller, T. Studien zur Blindenpsychologie. Leipzig, W. Engelmann, 1904, 136 pp. (cited from Griffin, 1958).

Kellogg, W. N., 1962, Sonar system of the blind. Science 137, 399–404.

Konstantinov, A. I., B. V. Sokolov and I. M. Stosman, 1966, Comparative study of the sensitivity of echo-location in bats. Doklady Akad. Nauk SSSR 175, 1418–1421.

Kulzer, E., 1956, Flughunde erzeugen Orientierungslaute durch Zungenschlag. Naturwiss. 43, 117–118.

Kulzer, E., 1960, Physiologische und morphologische untersuchungen uber die erzeugung der orientierungslaute von slughunden der Gattung Rousettus. Z. vergl. Physiol. 43, 231–268.

Medway, Lord, 1959, Echolocation among Collocalia. Nature 184, 1352–1353.

Medway, Lord, 1967, The function of echonavigation among swiftlets. Anim. Behav. 15, 416–420.

Möhres, F. P., 1956, Uber die Orientierung der Flughunde (Chiroptera-Pleropiddae). Z. vergl. Physiol. 38, 1–29.

Möhres, F. P. and E. Kulzer, 1956, Uber die orientierung der flughunde (Chiroptera-Pteropididae). Z. vergl. Physiol. 38, 1–29.

Novick, A., 1958, Orientation in paleotropical bats. II. Megachi-
    roptera. J. Exp. Zool. 137, 443–462.
Novick, A., 1959, Acoustic orientation in the cave swiftlet. Biol.
    Bull. 117, 497–503.
Rice, C. E., 1967, Human echo perception. Science 155, 656–664.
Rice, C. E., 1969, Perceptual enhancement in the early blind?
    Psychol. Rec. 19, 1–14.
Rice, C. E. and S. H. Feinstein, 1965, The influence of target
    parameters on a human echo-detection task. Proc. Amer.
    Psychol. Assn., p. 65.
Rice, C. E., S. H. Feinstein. and R. J. Schusterman, 1965,
    Echo-detection ability of the blind: size and distance factors.
    J. Exp. Psychol., 70, 246–251.
Riley, D. A. and M. Rosenzweig, 1957, Echolocation in rats. J.
    Comp. Physiol. Psychol. 50, 323–328.
Roberts, L. H., 1975, Confirmation of the echolocation pulse pro-
    duction mechanism of Rousettus. J. Mammal. 56, 218–220.
Rosenzweig, M. R., D. A. Riley, and D. Krech, 1955, Evidence
    for echolocation in the rat. Science 121, 600.
Supa, M., M. Cotzen and K. M. Dallenbach, 1944, "facial vision"
    The perception of obstacles by the blind. Am. J. Psychol.
    57, 133–183.
Schnitzler, H.-U., 1967, Discrimination of thin wires by flying
    horseshoe bats (Rhinolophidae). In "Animal Sonar Systems"
    (R.-G. Busnel, ed.) pp. 69–87. Lab. Physiol. Acoust.,
    Jouy-en-Josas, France.
Schnitzler, H.-U., 1968, Die Ultraschall-Ortunglaute der Hufeisen-
    Fled-ermause (Chiroptera-Rhinolophidae) in verschiedenen
    Orientierungssituationen. Z. vergl. Physiol. 57, 376–408.

Chapter II
Echolocation Signals and Echoes

Chairman : D.R. Griffin

co-chairmen

W.W.L. Au   and   D. Pye

- <u>Underwater</u>
  Functional and descriptive anatomy of the bottlenosed
  dolphin nasolaryngeal system with special reference
  to the musculature associated with sound production.
      R.F. Green, S.H. Ridgway and W.E. Evans

  Electromyographic and pressure events in the nasolaryngeal system of dolphins during sound production.
      S.H. Ridgway, D.A. Carder, R.F. Green,
      S.L.L. Gaunt, A.S. Gaunt,and W.E. Evans

  Echolocation signals of the atlantic bottlenose dolphin
  (<u>Tursiops truncatus</u>) in open waters.
      W.W.L. Au

  Acoustics and the behavior of sperm whales.
      W.A. Watkins

  Click sounds from animals at sea.
      W.A. Watkins

  Signal characteristics for target localization and
  discrimination.
      K.J. Diercks

- <u>Airborne</u>
  Echolocation signals and echoes in air.
      J.D. Pye

  Echolocation ontogeny in bats.
      P.E. Brown and A.D. Grinnell

FUNCTIONAL AND DESCRIPTIVE ANATOMY OF THE BOTTLENOSED DOLPHIN NASOLARYNGEAL SYSTEM WITH SPECIAL REFERENCE TO THE MUSCULATURE ASSOCIATED WITH SOUND PRODUCTION

R. F. Green, S. H. Ridgway, and W. E. Evans[†]

Naval Ocean Systems Center
San Diego, Ca. 92132 U.S.A.
† Hubbs-Sea World Research Institute
San Diego, Ca. 92109 U.S.A.

INTRODUCTION

The nasolaryngeal system of the bottlenosed dolphin (Tursiops truncatus) is complex and highly specialized for respiration and sound production. The exact source and mechanism of sound production by dolphins is not understood. Electromyographic studies are essential for understanding the functional anatomy and mechanisms of sound production. Dissections and measurements have been made for the purpose of identifying external landmarks and other details necessary for inserting electrodes in the muscles of the naso-laryngeal area. Based upon these anatomic investigations, some functions of this musculature have been suggested, and the first electromyographic studies of dolphin acoustic mechanisms have been initiated.

Previous Studies

The biological community generally accepts the production of sound by the vast majority of living amniote vertebrates as beyond question. Many assume that our understanding of how this vast array of acoustic emissions are produced is at a high state of development. This is far from true. With the exception of humans, our knowledge of the mechanics of sound production in vertebrates, especially cetaceans, is based on indirect evidence and speculation (Kelemen, 1963; Gaunt and Wells, 1973; Kinne, 1975).

The mechanisms of sound production in most vertebrates are essentially modifications of the respiratory system and are in

reality an overlaid function. This factor more than any other has
contributed to compounding the problem of understanding the
mechanisms of sound production in marine mammals. We have not only
inherent complexities of the terrestrial mammalian nasolaryngeal
system but the added problems caused by modifications of the system
to permit a) breathing air while inhabiting an aquatic environment,
b) prolonged breath holding, c) diving to reasonably deep depth with
resulting pressure problems, and d) the ingestion of food without
taking water into the lungs. Despite years of discussions, special
seminars, experiments, morphological descriptions and several
studies of functional anatomy, uncertainty still exists, and the
details of production of the vast variety of sounds found in the
repertoire of cetaceans is still an unsolved mystery. Exact sites,
muscular or mechanical processes, aerodynamics (flow, pressure,etc)
and acoustical properties of the anatomic structures involved are
insufficiently known even in the bottlenosed dolphin (Tursiops
truncatus), the most intensively studied of all the species.

In birds, and more recently in studies of rodents and
chiropterans(Hersh, 1966; Roberts, 1973, 1975a, 1975b; Gaunt et al.,
1976) additional insight has been gained by using inhalation of
light gases (HeO$_2$), measurement of air flow and pressures, and
electrical potentials of muscles thought to be involved in sound
production in birds(Gaunt and Gaunt, 1977). We are now in the process
of applying these tools to the study of sound production mechanisms
in Tursiops. The data from these studies combined with the
information already available from analytical measurements near
presumptive sound production sites (Evans, 1973; Diercks et al.,
1971; Romanenko, 1974), x-ray cinematography (Gurevich, 1973; Norris
et al., 1971) and production of sound in light air (HeO$_2$) and
atmosphere (Evans and Rieger,1977, in preparation), should provide
the base for an accurate description for the total system.

There is no need for an additional detailed description of the
nasal sac system or the structure of the delphinid larynx; these
have been more than adequately provided by Lawrence and Schevill
(1956), Purves (1967), Mead (1972), and Schenkkan (1973). This
report is a documentation of the total nasolaryngeal system with
special reference to the musculature in relationship to external
features for use as a stereotaxic guide in electromyographic and
ultrasonic scanning studies.

Interest in how dolphins and other cetaceans produce sound goes
back more than 100 years. Murie(1871), interpreted " a double
raised smooth membranous fold" in the lower part of the larynx in
Risso's dolphin (Grampus griseus), to represent the vocal cords and
therefore to be the site of sound production. Dubois (1886), did
not find vocal cords in any of the cetaceans he examined. He did
agree, however, with Watson and Young (1879) who worked on beluga

whales, and Turner (1872), who studied fin whales. These workers assumed that the rear extensions of the arytenoid cartilages lie so close together that they would vibrate as air passed over them thus taking over the function of vocal cords. Lawrence and Schevill (1956) suggested the sounds were laryngeal in origin. As recently as 1967 Purves concluded that the mechanism of sound production in Phocoena phocoena is similar to sound production in man except that the aryepiglottal folds are called into play instead of the thyro-arytenoid (vocal) folds. Evans and Prescott (1962) suggested a dual sound source; the larynx for whistles and the nasal plugs and associated sacs for pulses. Norris (1964, 1969) favors the nasal sac system as the site of sound generation. Diercks et al. (1971) also suggested a sound production site in the region of the nasal plugs. Norris et al. (1971), Mead (1972), Schenkkan (1973), Evans and Maderson (1973), and Dormer (1974) further support the role of the nasal sacs and particularly the edges of the nasal plugs as the site of sound generation, especially for the pulses used in echo-location.

Some of the above conclusions concerning the site of sound production were derived by interpretation of anatomy while others were supported by a variety of experimental evidence. Roberts (1975a) has demonstrated that a comprehensive study of the anatomy of the rodent larynx, although informative, did not add much insight into a complete understanding of the function of this organ in the production of either audible or ultrasonic sounds. In further studies Roberts(1975b), using ablation of sections of the hypoglossal nerve, demonstrated the involvement of the intrinsic musculature of the larynx in the production of ultrasonic emissions. He also demonstrated that the muscles supplied by the hypoglossal nerves were not involved in the production of ultrasound. The use of this technique on delphinids, necessitating sacrifice of the animals, is impractical from an economic standpoint. Activity of specific muscles suspected of contributing to sound production can be measured using electromyographic (EMG) techniques. Studies of this nature also present some definite advantages over the nerve ablation method in that specific muscles can be studied. As far as can be determined from the literature no one has approached the problem by EMG studies. Since the muscles must ultimately produce most of the force to activate the mechanisms by which sounds are produced, we view it as essential to find out how they are functioning during the production of sounds.

Of several anatomic studies on the larynx and nasal sac systems, little has been reported that would be useful for EMG studies. The contents of this report developed out of a need for more detailed anatomic information with reference to external landmarks. Our intent is to supplement studies done by others so as to obtain the necessary information to do EMG's.

MATERIALS AND METHODS

The bottlenosed dolphin, <u>Tursiops truncatus</u> (Figure 1), has been extensively investigated, and its echolocation capability is well documented (Evans, 1973; Altes, 1974). Since this is the species on which we plan to do EMG studies, all of the dissections, measurements, and illustrations were of this species.

Dolphins are valuable animals, and specimens for anatomic dissection are often hard to come by. The specimens for this study all came from animals that died of natural causes in the wild, in ocenariums or in our own laboratory.

The laryngeal dissections used as many as seven different dolphins. This explains why the various views of the laryngeal cartilages, muscles, etc. are not of the same size. The larynx was taken at postmortem examination and preserved in 10% formalin. The transverse, frontal and oblique sections are freehand sections and vary in thickness as indicated.

We were able to obtain the head of a <u>Tursiops truncatus</u> that had been perfused with 10% formalin shortly after death. The head was sectioned into more than 30 transverse sections 1 cm thick.

Figure 1. Left lateral view of the bottlenosed dolphin, <u>Tursiops truncatus</u>.

The sections ranged from 8-11 mm in average thickness and varied
no more than 2 mm from side to side (Figure 30). The tissues
exterior to the cranial cavity were found to be well fixed, but the
brain·was in such poor condition that the brain sections were discarded and therefore not included in illustrations of the sections.
In some illustrations the details of cartilages, muscles and bone
have been outlined.

RESULTS AND DISCUSSION

The Larynx

     The larynx (Figure 2 and 3) is so specialized it is considered
to be an organ useful in characterizing the cetacea. It is intranarial, projecting upward from the floor of the pharynx with its
anterior end held firmly within the internal nares by a sphincter
composed of palato-pharyngeal muscle (Negus, 1931). In odontocete
cetaceans the opening of the external nares (blowhole) has migrated
to a position on top of the head so that when the animal surfaces
the opening into the nasal passage is the first part of the head to
be exposed. The elongated intranarial larynx functions to keep the
nasal air passages continuous with the glottis while the porpoise
dashes through the water open mouthed after its prey (Huxley, 1888).

     Laryngeal cartilages. With the exception of the so called
"aryteno-epiglottal beak", (Figure 2 and 3) the cetacean larynx is
similar in general structure to other mammalian larynges, being
composed of a cartilagenous framework held together by a number of
muscles.

     The thyroid cartilage (Figures 4,6,12,15,16,17,18,19,20,21,22,
25,26,27,and 28) is made up of right and left laminae fused at their
ventral margins without forming even the slightest laryngeal
prominence. Each lamina has two well developed processes; the
anteriorly directed cranial cornua and the posteriorly directed
caudal cornua, the latter being the better developed. The thyroid
cartilage forms the main body of the larynx. Most of the extrensic
muscles, which function to move the complete larynx, are attached
to this cartilage.

     The caudal cornua of the thyroid cartilage articulate with the
posterior-superior lateral margins of the cricoid cartilage by small
oval synovial joints. Anteriorly the thyroid cartilage also articulates with the epiglottal cartilage. This articulation is by a
fibrous joint that allows limited movement. Additional movement is
suggested, however, by a slight bending of the cartilage which is
especially thin just posterior to this joint.

Figure 2. Left dorso-lateral view of the laryngeal beak of a
bottlenosed dolphin.

Figure 3. A midsagittal section through the head of a bottlenosed
dolphin. A- the nasal area, B- the laryngeal area.

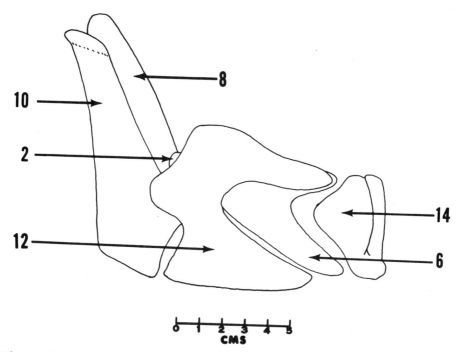

Figure 4. All laryngeal cartilages in approximate articulated position. 2- arytenoid cartilage, 6- cricoid cartilage, 8- cuneiform cartilage, 10- epiglottal cartilage, 12- thyroid cartilage, 14, 1st tracheal cartilage.

In Tursiops the cricoid cartilage (Figures 4,5,7,12,18,19,20, 21,22,24,25,26, and 27) is not a complete ring as is found in some cetaceans, but it is open ventrally, forming two rather long posteriorly projecting lateral cornua. These cornua almost fill the posterior lateral notch in each thyroid lamina. In addition to the joints formed between the cricoid and thyroid cartilages, there are much larger synovial joints along the anterior lateral margins of the body of the cricoid cartilage where the cricoid articulates with the arytenoid cartilages.

The delphinid epiglottal cartilage (Figures 4,5,8,12,13,14,15, 16,17,22,23,24,25,26, and 27) is much longer than the epiglottal cartilage in most mammals and has deepened medially to form a trough with thin lateral margins projecting posteriad to wrap around the anterior borders of the cuneiform and arytenoid cartilages, thereby forming the "laryngeal beak".

The arytenoid cartilages (Figures 4,5,9,12,17,18,19,22,24,25, and 26) articulate by well formed oval-shaped synovial joints with

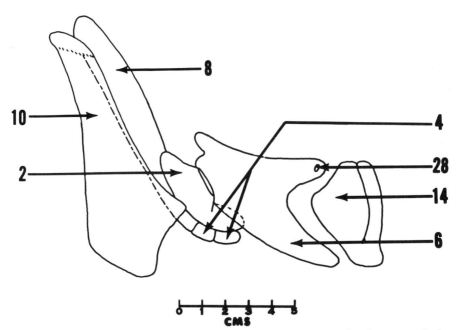

Figure 5. Laryngeal cartilages in approximate articulate position
except thyroid cartilage removed. 2- arytenoid cartilage,
4- corniculate cartilage, 6- cricoid cartilage, 8- cuneiform
cartilage, 10- epiglottal cartilage, 14- 1st tracheal cartilage,
28- articular facet to thyroid cartilage.

the anterior lateral margins of the cricoid cartilage. Each has a
well developed laterally projecting muscular process, but no vocal
processes are evident. The arytenoid cartilages also articulate
with the cuneiform cartilages.

The cuneiform cartilages (Figures 4,5,9,12,13,14,15,16,17,18,
19,20,22,23,24,25,26,and 27), form elongated blade-like processes
that face each other medially to lie in the trough formed by the
epiglottal cartilage. As indicated above, the cuneiform cartilages
articulate with the arytenoids. Two or more small cartilages,
probably representing the corniculate cartilages, attach to the
end of each cuneiform cartilage and also to the posterior two-fifths
of the ventral edge of the arytenoid.

The arytenoid, the cuneiform, and the corniculate cartilages
as described above have been referred to in various ways in the
literature on cetacean larynges. Benham (1901) called the structure
formed by these cartilages the arytenoid. D'Arcy-Thompson (1890)
called the blade-like cartilages the super-arytenoids. Howes (1879)

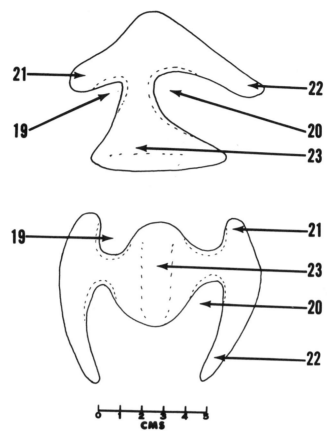

Figure 6. Lateral (top) and ventral (bottom) views of the thyroid cartilage. 19- anterior notch, 20- posterior notch, 21- cranial cornu, 22- caudal cornu, 23- lamina.

was the first to call the blades the cuneiform cartilages. Howes was supported by Cleland (1884), Purves (1967), and Dormer (1974).

Extrinsic laryngeal muscles. All the extrinsic muscles are paired except for the hyoepiglottal muscle which is single. To simplify description only the muscles on the left side are described.

The hyoepiglottal muscle (Figures 10,11,14,15,22,24,25,26, and 27) originates from the mid-dorsal basihyal segment of the hyoid bone, and inserts on the inferior one-half of the anterior edge of the epiglottal cartilage. This muscle probably functions to pull the epiglottal cartilage forward so as to enlarge the size of the anterior passageway in the laryngeal beak.

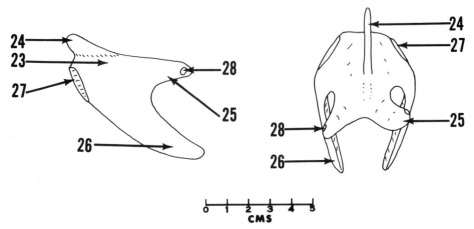

Figure 7. Left lateral (left) and dorsal (right) views of the cricoid cartilage. 23- lamina, 24- spine, 25- posterior dorsal cornu, 26- posterior ventral cornu, 27- articular surface to arytenoid cartilage, 28- articular facet to thyroid cartilage.

Figure 8. Lateral (left), anterior (middle), and posterior (right) views of the epiglottal cartilage. 29- lip, 30- trough.

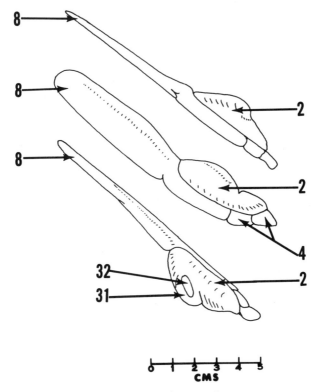

Figure 9. Ventral (top), lateral (middle), and dorsal (bottom) views of the arytenoid, cuneiform and corniculate cartilages. 2- arytenoid cartilage, 4- corniculate cartilage, 8- cuneiform cartilage, 31- muscular process, 32- articular surface to cricoid cartilage.

The sternothyroid muscle (Figure 28) originates on the anterior margin of the sternum, and inserts on the lateral surface of the thyroid cartilage just superior to and anterior to the insertion of the cricothyroid muscle. The sternothyroid probably functions to pull the complete laryngeal apparatus posteriad.

The thyrohyoid muscle (Figure 28) originates on the lateral surface of the thyroid cartilage just anterior and ventral to the insertion of the sternohyoid muscle. Its insertion is at the posterior margin of the basihyal and thyrohyal segments of the hyoid bone. It probably functions to pull the laryngeal apparatus anteriad, thus more firmly lodging the tip of the laryngeal beak in the internal nares.

The larynx is suspended from the base of the cranium by a highly complex series of muscles, the thyropalatine, the occipito-

Figure 10.Left lateral view of the larynx. 1- Cricoarytenoid-dorsal
muscle, 5- cricothyroid-oblique muscle, 6- cricoid cartilage,
7- cricothyroid-straight muscle, 9- hyoepiglottal muscle, 11- interarytenoid muscle, 14- 1st tracheal cartilage.

thyroid and the thyropharyngeal. These muscles attach to the dorsal
margin and anteromedial surface of the thyroid cartilage, and then
pass dorsad and anteriad to attach to the floor of the cranium. The
most anterior of these muscles is the thyropalatine,(a posterior
part of the palatopharyngeal muscle), which attaches to the medial
surface of the thyroid lamina and the thyroid ligament. The middle
muscle, the occipitothyroid, attaches to the dorsal margin of the
cranial cornu of the thyroid cartilage. The most posterior muscle,
the thyropharyngeal, attaches to the dorsal margin of the posterior
cornu of the thyroid cartilage. The anterior fibers of the palato-
pharyngeal muscle together with the pterygopharyngeal muscle form
the sphincter that partially fills the lower part of the internal
nares and holds the laryngeal beak in place.

    Intrinsic laryngeal Muscles. The cricothyroid muscle, (Figures
10,12,18,19,20,21,22, and 27) has its origin to the lateral posterior
surface of the lateral cornu of the cricoid cartilage.Its insertion
is to the thyroid cartilage along the inferior margin of the caudal
cornu, to the posterior notch and to the lateral surface of the
lamina just below the posterior notch. This muscle has two distinct

Figure 11. Lateral view of the larynx with thyroid cartilage and
cricothyroid muscles removed. 1- cricoarytenoid-dorsal muscle,
3- cricoarytenoid- lateral muscle, 6- cricoid cartilage, 9-hyo-
epiglottal muscle, 11- interarytenoid muscle, 14- 1st tracheal
cartilage, 15- thyroarytenoid muscle.

parts in the dolphin, a <u>dorsal</u> <u>oblique</u> <u>cricothyroid</u> and a <u>ventral</u>
<u>straight</u> <u>cricothyroid</u>.

    The <u>cricoarytenoid muscle</u>, (Figures 10,11,12,18,19,20,21,22,
24,25, and 26) has its origin over the dorsal surface of the cricoid
cartilage extending laterad to the medial surface of the caudal
cornu of the thyroid cartilage. The muscle is in two distinct parts,
the dorsal cricoarytenoid and the lateral cricoarytenoid. Because of
its attachment to the caudal cornu, Hosokawa (1950) calls this
muscle the ceratocricoarytenoideus. Both parts of the muscle insert
on the dorsal muscular process of the arytenoid cartilage. They
probably function to rotate the arytenoids lateral and posterio-
dorsad. This action increases inter-arytenoid space between the
edges of the arytenoids and thus the inter-cuneiform space between
the anterior edges of the cuneiform cartilages.

    The <u>thyroarytenoid muscle</u> (Figures 11,12,17,18,19,22,24,25,
26, and 27)lies on the medial surface of the thyroid cartilage and

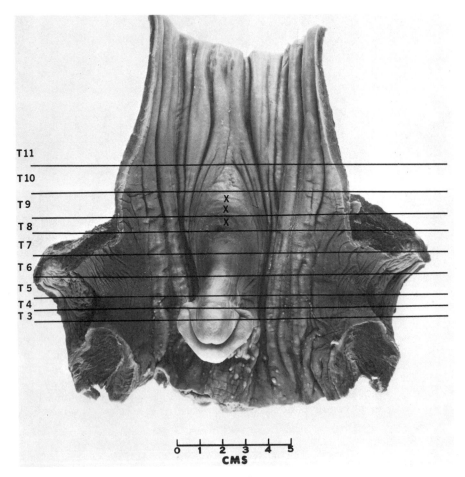

Figure 12. Dorsal view of laryngo-pharyngeal area showing location of transverse sections T3-T11. XXX: indicates where the cricoid crest can be palpated.

has its origin on the dorsal midline and medial surfaces. The fibers pass dorsad to insert on the ventral surface of the muscular process of the arytenoid cartilage. The anterior fibers pass laterad, over the posterior angle of the epiglottal cartilage, but do not appear to attach to the epiglottis in such a way as to act upon it. The thyroarytenoid probably functions to pull the arytenoid and cuneiform cartilages ventrad and mesad to decrease the angle between their medial surfaces. If the arytenoid and cuneiform cartilages are already approximated the thyroarytenoid muscles probably function to pull the arytenoids and cuneiforms deeper into the epiglottal trough.

Figure 13- Anterior section of section T3.   8- cuneiform cartilage,
9- hyoepiglottal muscle, 10- epiglottal cartilage.

Figure 14- Anterior surface of section T4.   8- cuneiform cartilage,
9- hyoepiglottal muscle, 10- epiglottal cartilage.

Figure 15. Anterior surface of section T5. 8- cuneiform cartilage,
9- hyoepiglottal muscle, 10- epiglottal cartilage, 11- thyroid
cartilage.

Figure 16- Anterior surface of section T6. 8- cuneiform cartilage,
9- epiglottal cartilage, 12- thyroid cartilage, 15- thyroarytenoid
muscle, 17- thyrohyoid muscle.

Figure 17. Anterior surface of section T7. 2- arytenoid cartilage, 8- cuneiform cartilage, 10- epiglottal cartilage, 11- interarytenoid muscle, 12- thyroid cartilage, 15- thyroarytenoid muscle, 17- thyrohyoid muscle.

Figure 18. Anterior surface of section T8. 1- cricoarytenoid- dorsal muscle, 2- arytenoid cartilage, 3- cricoarytenoid- lateral muscle, 5- cricothyroid- oblique muscle, 6- cricoid cartilage, 8- cuneiform cartilage, 11- interarytenoid muscle, 12- thyroid cartilage, 15- thyroarytenoid muscle.

Figure 19. Anterior surface of section T9. 1- cricoarytenoid-
dorsal muscle, 2- arytenoid cartilage, 3- cricoarytenoid- lateral
muscle, 4- corniculate cartilage, 5- cricothyroid-oblique muscle,
6- cricoid cartilage, 8- cuneiform cartilage, 12- thyroid cartilage,
15- thyroarytenoid muscle.

Figure 20. Anterior surface of section T10. 1- cricoarytenoid-
dorsal muscle, 3- cricoarytenoid- lateral muscle, 4- corniculate
cartilage, 5- cricothyroid-oblique muscle, 6- cricoid cartilage,
8- cuneiform cartilage, 12- thyroid cartilage.

Figure 21. Anterior surface of section T11. 1- cricoarytenoid-
dorsal muscle, 3- cricoarytenoid- lateral muscle, 5- cricothyroid-
oblique muscle, 6- cricoid cartilage, 7- cricothyroid- straight
muscle, 12- thyroid cartilage.

The interarytenoid muscle (Figures 10,11,12,17, and 18), joins
the dorsal margins of the cuneiform and to a lesser degree the
arytenoid cartilages. Some of the posterior fibers pass below and
join into the anterior edge of the dorsal cricoarytenoid muscle.
This muscle apparently functions to assist the cricoarytenoid
muscles in increasing the interarytenoid and intercuneiform spaces.

EMG Studies On Laryngeal Muscles

Since none of the laryngeal muscles can be recorded by the
use of surface electrodes because of the thick overlying blubber
and subcutaneous muscle, electrodes must be inserted directly into
the muscle to be studied. The bottlenosed dolphins available for
these studies were young adults 125-150 kg in weight and 2.3-2.5 m
in length. The measurements given below relate to animals in that
size range.

There are two approaches that can be used to insert an electrode
into the hyoepiglottal muscle. To begin the first approach (A in
Figure 29), palpate the posterior edge of the hyoid bone along the
midline. The electrode is inserted about 2.0 cm behind the posterior
edge of the hyoid bone and directed anteriorly at about $50-55^0$ so
as to pass between the hyoid and the epiglottal cartilage. The
muscle is between 4.5 and 5.0 cms deep and has a transverse width
of about 1.5 cm.

Figure 22- Left lateral view of the larynx indicating where transverse beak (TB) sections 1-7 and frontal (F) sections 8-11 were made.

The hyoepiglottal muscle can also be reached through the throat. The electrode would then be directed as illustrated at B in Figure 29, and the recording wires may be run out the subject's mouth.

To place an electrode into the sternothyroid muscle it was oriented using the baseline as illustrated in Figure 28, moving 3.5 cm to the left (animals left) of the midline to insert into the left sternothyroid muscle or 3.5 cm to the right of the mid-ventral line to insert into the right sternothyroid muscle. The electrode should be inserted at 90° to the body surface and placed 4.5 cm deep (A-Figure 28).

The palato-pharyngeal muscle, the muscle that helps to hold the larynx in place, and may also serve to differentially close off the internal nares, can be recorded by inserting an electrode as illustrated at C, Figure 29.

Figure 23. Dorsal surfaces of TB (transverse beak) sections 1-7.
8- cuneiform cartilage, 9- hyoepiglottal muscle, 10- epiglottal
cartilage.

Figure 24. Dorsal surface of section F8. 1- cricoarytenoid-dorsal muscle, 2- arytenoid cartilage, 3- cricoarytenoid-lateral muscle, 6- cricoid cartilage, 8- cuneiform cartilage, 9- hyoepiglottal muscle, 10- epiglottal cartilage, 15- thyroarytenoid muscle.

The electrode can be inserted into the cricothyroid muscle in much the same way as into the sternothyroid muscle, by using the line illustrated in Figure 28. Moving 2.5 cm lateral to the midventral line insertion can be made either to the left or right cricothyroid muscle. The electrode should be inserted at 90° to the body surface and placed 3.5-4 cm deep (C- Figure 28).

Figure 25. Dorsal surface of section F9. 1- cricoarytenoid-dorsal muscle, 2- arytenoid cartilage, 3- cricoarytenoid-lateral muscle, 6- cricoid cartilage, 8- cuneiform cartilage, 9- hyoepiglottal muscle, 10- epiglottal cartilage, 12- thyroid cartilage, 14- 1st tracheal cartilage.

To insert the electrode into the thyrohyoid muscle the posterior hyoid is palpated on the midline. Then move posteriad 2-2.5 cm along the midline and then from 2-2.5 cm laterad. The electrode should be inserted at 90° to the body surface. The

Figure 26. Dorsal surface of section F10. 1- cricoarytenoid-
dorsal muscle, 2- arytenoid cartilage, 3- cricoarytenoid-lateral
muscle, 6- cricoid cartilage, 8- cuneiform cartilage, 9- hyoepiglottal muscle, 10- epiglottal cartilage, 12- thyroid cartilage,
14- 1st tracheal cartilage, 15- thyroarytenoid muscle.

muscle is 3.5-4.0 cm deep and is backed up by the thyroid cartilage
and the base of the epiglottal cartilage (B- Figure 28).

The thyropalatine, the occipitothyroid and the thyropharyngeal
muscles lie lateral to the pharyngeal and can probably be

Figure 27. Dorsal surface of section F11. 4- corniculate cartilage, 5- cricothyroid-oblique muscle, 6- cricoid cartilage, 8- cuneiform cartilage, 9- hyoepiglottal muscle, 10- epiglottal cartilage, 12- thyroid cartilage, 14- 1st tracheal cartilage, 15- thyro-arytenoid muscle.

recorded by placing electrodes through the wall of the pharynx. If these muscles have an active role in sound production (other than suspending the larynx), we would speculate that they work together, therefore recordings from these muscles could be taken anywhere along the complete length of the larynx.

Figure 28. Ventral view of hyo-sternal area (the white line passes along the anterior edge of the flippers). 5- cricothyroid, 12- thyroid cartilage, 13- sternothyroid muscle, 17- thyrohyoid muscle, 18- panniculus carnosus muscle, 33- basihyal segment of hyoid bone, 34- thyrohyal segment of hyoid bone, 35-blubber, 36- thyroid gland. A- placement of sternothyroid muscle electrode, B- placement of thyrohyoid muscle electrode, C- placement of crico-thyroid muscle electrode.

     The cricoarytenoid muscles are located lateral and posterior to the cricoid crest and can be approached by palpating the anterior margin of the crest just posterior to the base of the laryngeal beak. Electrodes can be inserted into either the left or right cricoarytenoid muscle 2 cm posterior to and 1 cm left or right of the crest. There is a sheet of esophageal muscle about 0.5 cm thick which passes between the pharyngeal wall and the crico-arytenoid muscle. The electrode must be inserted through this muscle layer to reach the cricoarytenoid.

     The thyroarytenoid muscle is located approximately 2.3 cm lateral and ventral to the cricoid crest. The electrode should be

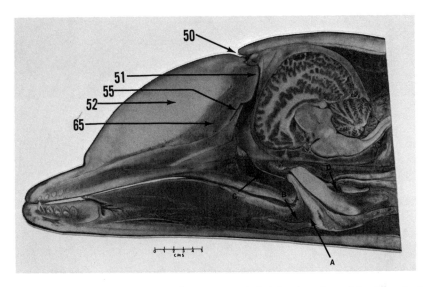

Figure 29. Mid-sagittal section of a dolphin head. 50- blowhole,
51- nasal plug, 52- melon, 55- premaxillary sacs, 65- nasal plug
muscle, A- placement of hyoepiglottal muscle electrode, B- alternate
placement of hyoepiglottal muscle electrode, C- placement of palato-
pharyngeal sphincter electrode, D- placement of interarytenoid
muscle elecrode.

inserted at a point 2 cm posterior to the posterior edge of the
epiglottal cartilage where the muscle is approximately 1 cm thick.

    The interarytenoid muscle passes just anterior to the cricoid
crest. The muscle extends approximately 1.5 cm to the left and
1.5 cm to the right of the midline and is approximately 2.0 wide
from front to back) and 1.0 cm thick along the midline. The
electrode is directed at an angle parallel to the dorsal edge of
the cuneiform cartilages for insertion into this muscle (D-Figure29).

                           The Nasal System

    The nasal system (Figure 3), includes the bony nasal passage;
the spiracular cavity; and the plugs, sacs, membranes and muscles
associated with the blowhole region, Due to their structure and
location, the internal nares with their plugs, sacs, and membranes
are considered to be the structures most likely involved in sound
production. Lawrence and Schevill (1956), Mead(1972), Schenkkan
(1973), and Dormer (1974) have given detailed descriptions and
illustrations of the nasal region in the bottlenosed dolphin.

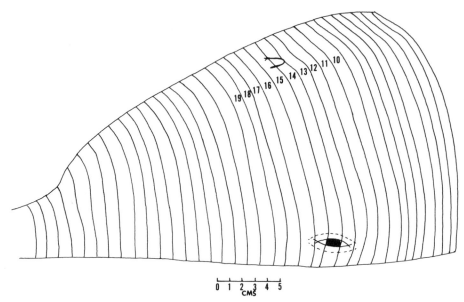

Figure 30. Left side of head showing the approximate location of
the cross sections through the nasal region. The numbers lateral
to the blowhole are section numbers referred to in the following
figures.

We were interested in the anatomy of this area specifically for
purposes of understanding dolphin sound production through ultrasonic scanning and EMG studies of function, therefore, we will
present only a general description here.

     The <u>bony</u> <u>nasal</u> <u>passage</u> passes upward through the skull in an
anteriorly curved, almost vertical plane (Figure 29). Within the
cranium the passage is separated by the bony and cartilagenous
septum into two sub-oval tubes. The upper part of the passageway
is a single chamber and is called the spiracular cavity (Von Baer,
 1926). It passes upward a short distance to a single opening,
the blowhole. The blowhole is more or less round when open, but is
"C" shaped when closed.

     The nasal sacs.The nasal sacs are extensively developed in
the bottlenosed dolphin. Four pairs of sacs communicate either
directly or indirectly with the spiracular cavity. The <u>premaxillary</u>
sacs (Figures 29, 33, and 34) are the largest and lie on top of the
premaxillary bones just anterior to the openings into the bony
nasal passage. They open into the spiracular cavity. The <u>vestibular</u>
<u>sacs</u> (Figures 32-37) are located just below the skin, projecting
posteriolaterad from the spiracular cavity into which they open.

The nasofrontal sacs (Murie,1871) are "U" shaped tubular sacs that pass around the spiracular cavity below the vestibular sacs.

The accessory sacs (Schenkkan, 1971) project laterally from the inferior vestibules; passages that connect the posterior naso-frontal sacs to the posterior spiracular cavity.

The nasal plugs (Lawrence and Schevill, 1956) protrude from the anterior spiracular wall just dorsal to the premaxillary sacs. They fit tightly against the posterior wall of the spiracular cavity to form a seal. The distal margins of the plugs form into lips that fit into the openings between the inferior vestibules and the posterior spiracular cavity.

·   Nasal musculature. The several layers of muscles associated with the sacs, plugs, and lips are also complexly developed. Huber (1934) was of the opinion that the rostral muscles were derived from the pars labialis of the maxillonasolabialis muscle and the layers of muscle around the blowhole to be derived from the pars nasalis of the maxillonasolabialis. Lawrence and Schevill (1956) and Mead (1972) concur with Huber's interpretation of these origins. Mead, however, suggests the division should be made between medial and lateral layers of muscle rather than between anterior and posterior muscles. This would result in the lateral, rostral, and superficial nasal muscles being derived from the pars labialis, and the medial rostral and deep nasal musculature being derived from the pars nasalis of the maxillonasolabialis muscle.

The posteroexternus (Figure 40) is the most superficial of the nasal sac muscles. Its origin extends along the posterior 1/2 of the supraorbital process of the maxilla bone, the temporal crest of the frontal bone, and nuchal crest (occipito-frontal suture). The muscle is generally quite thin and tendinous, with anterodorsad directed fibers passing over the vestibular sac to attach to the dense connective tissue in the posterior and lateral wall of the nasal passage. These anterior fibers are loosely attached over the dorsal surface of the vestibular sacs. The posterior fibers are directed anteriomediad to insert just above the anterior fibers.

This description of the posteroexternus is much the same as that given by Lawrence and Schevill (1956). It does not agree with Mead (1972), who describes the insertion of this muscle as being to the vertex and contralateral muscle. Lawrence and Schevill also illustrate this muscle to show its origin over most of the nuchal crest thus completely covering the anteroexternus muscle below it. Mead, however, illustrates the posteroexternus with its origin over the lower nuchal crest, with anteroexternus muscle exposed posteriorly. We have found the muscle to be as illustrated by Lawrence and Schevill about 80 percent of the time. About 20 percent

of the specimens that we dissected conform to Mead's description.
On occasion one side of the specimen will appear as described by
Mead while the other side will be structured like that described
by Lawrence and Schevill.

This muscle probably functions to pull back on the posterior
wall and   laterad on the lateral wall of the spiracular cavity. It
could also function as a compressor of underlying structures and to
restrict vertical expansion of the vestibular sac, as suggested by
Mead (1972).   Such a compressor function could also help in the
conservation of water and heat that would ordinarily be lost with
expired air (Coulombe et al., 1965).

The intermedius muscle originates under the posteroexternus
along the posterior 1/2 of the supraorbital process as well as from
the fascia over the anteroexternus and posteroexternus muscles.
It is more easily identified at its insertion end where the fibers
pass loosely into the dense connective tissue mass dorsal to the
nasal plugs. Mead (1972) also found the more dorsal fibers to be
associated with the anterior dorsal surface of the vestibular sac
This we found to be true in about 50 percent of the animals we
dissected. It is also the case on one side of the head, but not on
the other. This muscle is sometimes missing from both sides of the
head. This muscle possibly functions to pull the nasal plugs down
against the internal nares, but does not seem to be of great importance, as indicated by the generally weak insertion and variability.

The anteroexternus muscle (Figures 32 and 33) is a wide expan-
sive muscle which takes its origin from the anterior supraorbital
process posterior to the temporal and nuchal crests. The anterior
margin of the muscle folds back over itself giving the appearance
of a separate muscle. The posterior dorsal margin occupies the angle
between the anterior nuchal crest and the lateral margin of the
vertex. The anterior fibers insert into the dense connective tissue
beneath the anterior lip of the blowhole and into the lateral
anterior wall of the spiracular cavity. The posterior fibers insert
on the posterior and lateral wall of the spiracular cavity as well
as on the dorsal edge of the blowhole ligament. This muscle passes
both below and above the vestibular sac, completely surrounding it.

Anteriorly, the anteroexternus is difficult to separate from
the anterointernus muscle. The anterior fibers lie just posterior
to the posterior extension of the lateral rostral muscle. Posteriorly, the muscle can be separated more easily from the postero-
internus muscle, as the fibers run in different directions.

This description of the anteroexternus muscle is much like the
description of the same muscle as redefined by Mead. Lawrence and
Schevill refer only to the anterior fibers as the anteroexternus.

We found these anterior fibers to be distinctly separate in a few instances, but in most cases the anterior and posterior parts were not separable.

The anterior fibers of the muscle probably function to pull down on the nasal plug causing it to seal over the internal nares. These anterior fibers may, as suggested by Lawrence and Schevill, draw the vestibular sac forward and laterad. The posterior part of the muscle, as described by Mead, probably functions to draw the posterior lip backwards, opening the blowhole. Since the vestibular sacs are completely surrounded by this muscle it very likely functions to maintain the size and position of these sacs as well as compressing the sacs.

The posteriorinternus muscle (Figure 40) lies deep to the posterior three-quarters of the anteroexternus muscle as defined by Mead. Its origin is to the frontal bone, above the eye; to the dorsal medial margin of the supraorbital process, then dorsad to lie in the angle between the nuchal crest and the vertex. Its fibers pass anteriad and dorsad under the vestibular sacs to insert on the posterior and lateral wall of the nasal passage as well as to the lateral margins of the nasofrontal sacs. As Mead suggests, this muscle probably functions to pull the posterior wall forward, putting pressure on the blowhole ligament that in turn puts pressure on the lips of the nasal plugs causing the plugs to more effectively close the external bony passages.

The anterointernus muscle (Figures 32, 33, 35) is the deepest and most massive of the nasal muscles. The anterior part of this muscle as described by Mead includes what Lawrence and Schevill called the profundus muscle. In our dissections, we most often found the two parts of the muscle to be closely joined even though there was occasionally a distinct boundry between them. Anteriorly the fibers insert to the connective tissue mass anterior to the spiracular cavity and over the nasal plugs. The median and posterior fibers insert into the lateral wall of the nasal passage and to the nasofrontal sac.

This muscle is probably the most powerful of the nasal muscles and must therefore play a significant role in the function of the blowhole region. By drawing the anterior wall of the spiracular sac posteriad and by compressing the nasal plug mass it would play a significant role in closing the nasal passage. Purves (1967) suggested that the contraction of this muscle would compress the premaxillary sacs and function in the recycling of air.

The rostral muscle (Figure 31) consists of an elongated mass located on the lateral margins of the rostrum. Its origin is from the dorsal surface of the maxilla. The muscle is divisible into

lateral and medial parts with the lateral fibers inserting into the
lip and connective tissue lateral to the melon. The medial fibers
project dorsally and medially to insert into the melon and lateral
margin of the nasal plug. The medial fibers probably function to
alter the shape of the melon and may function in sound production
by action on the nasal plug. The muscle may also function in concert
with sound production if the various theories of acoustic beam forming by the dolphin melon are correct (Wood, 1964).

The nasal plug muscle is an elongated mass of fibers having
their origin on the premaxillary bones anterior to the premaxillary
sacs. The fibers insert into the dense connective tissue and fat of
the nasal plugs. There appear to be more muscle fibers in the right
side than in the left side of this muscle. This is probably due in
part to the fact that the right plug is as much as 2 to 2-1/2 times
as large as the left plug. This muscle most likely functions to pull
the nasal plugs forward and may be involved in sound production.

The intrinsic muscles of the nasofrontal sacs almost completely
surround the nasofrontal sacs. The muscles originate from the dense
connective tissue on the anterior wall of the vertex. Some fibers
pass laterad below the lateral arc of the nasofrontal sacs to merge
with the nasal plug muscle. Other fibers are associated with the
inferior vestibule passages. There are an almost unlimited number of
actions these fibers might have upon the sacs and passages they are
associated with.

Mead (1972) has described the diagonal membrane muscle,
originating on the anterolateral vertex deep to the intrinsic muscle
and inserting in the attached margin of the diagonal membrane. Mead
prefaced his description by stating that due to the complexity of this
this region the muscle has escaped the attention of earlier workers.
We have been very careful in searching for this muscle during our
last 4 dissections of the Tursiops nasal region. We have been
successful in locating fibers which fit Meads description but the
muscle is very diffuse and not well developed.

EMG studies on nasal musculature. To place electrodes in the
various layers of nasal musculature the sagittal section illustrated
in Figure 29, and the cross sections illustrated in Figures 31-39
can be scaled up to fit an individual experimental animal. This is
most easily done by measuring the width of the animal's head at the
posterior canthus of the eyes. Measurements can then be taken directly
from adjusted illustrations so as to know where and how deep to place
the electrodes. The anterior surface of section 13 is used as a
baseline since it can be easily referenced from external landmarks.

Through dissection and morphometric evaluation we have determined external landmarks and stereotaxic coordinates necessary for

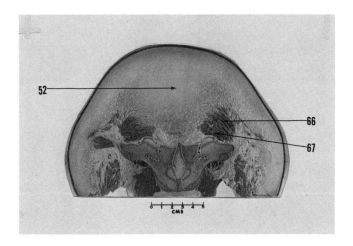

Figure 31. Anterior surface of section 19; 6 cm anterior to baseline (section 13). 52- melon, 66- lateral rostral muscle, 67- medial rostral muscle.

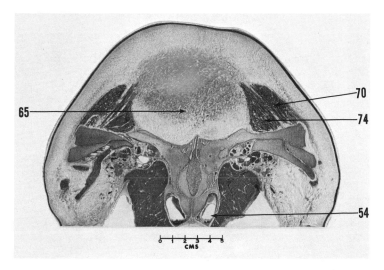

Figure 32. Anterior surface of section 16; 3 cm anterior to baseline (section 13). 54- pterygoid sinus, 65- nasal plug muscle, 70- medial rostral muscle, 74- anterior externus muscle.

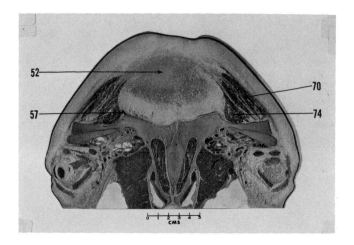

Figure 33. Anterior surface of section 15; 2 cm anterior to
baseline (section 13). 52- melon, 57- premaxillary sac,
70- anterior externus muscle, 74- anterior internus muscle.

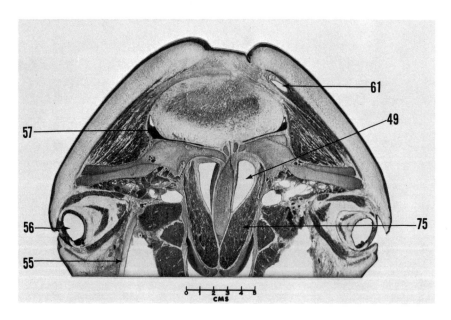

Figure 34. Anterior surface of section 14; 1 cm anterior to
baseline (section 13). 49- nasal passage, 55- mandible, 56- eye,
57- premaxillary sac, 61- vestibular sac, 75- palato-pharyngeal
sphincter.

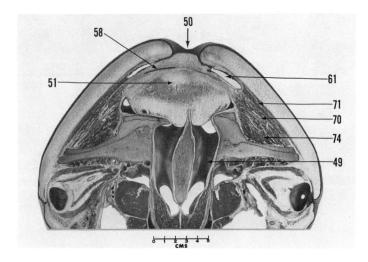

Figure 35. Anterior surface of baseline (section 13). This section passes 1/2 cm anterior to the back of the blowhole and 1/2 cm anterior to the posterior canthus of the eyes. 49- nasal passage, 50- blowhole, 51- nasal plug, 58- nasofrontal sac, 61- vestibular sac, 70- anterior externus muscle, 71- posterior externus muscle, 74- anterior internus muscle.

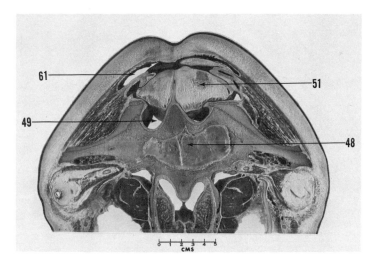

Figure 36- Posterior surface of section 13. This surface is 1 cm posterior to the anterior surface of section 13. 48- cranial cavity, 49- nasal passage, 51- nasal plug, 61- vestibular sac.

Figure 37. Anterior surface of section 12; 1 cm posterior to baseline (section 13). 48- cranial cavity, 51- nasal plug, 61- vestibular sac, 75- palato-pharyngeal sphincter.

Figure 38. Posterior surface of section 12; 2 cm posterior to the anterior surface of baseline (section 13). 48- cranial cavity, 51- nasal plug, 58- nasofrontal sac, 62-diagonal membrane.

Figure 39. Anterior surface of section 11; 2 cm posterior to baseline (section 13). 48- cranial cavity.

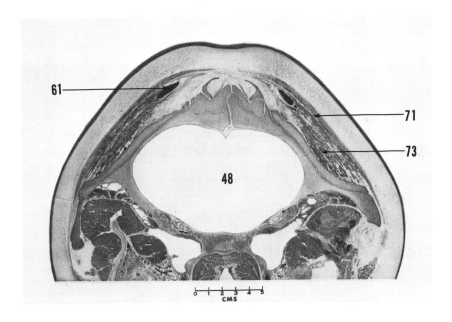

Figure 40. Anterior surface of section 10; 3 cm posterior to baseline (section 13). 48- cranial cavity, 61- vestibular sac, 71- posterior externus muscle, 73- posterior internus muscle.

placing electrodes in muscles suspected to be involved in sound production (see Ridgway, et al., this volume).

ACKNOWLEDGMENTS

We thank Dr. A. S. Gaunt for his many helpful suggestions and encouragement.

REFERENCES

Altes, R. A., 1974, Study of animal signals and neural processing with applications to advanced sonar systems, Rept. NUC TP 384, San Diego, Ca.

Benham, W. B., 1901, On the larynx of certain whales (Kogia, Balaenoptera and Ziphius, Proc. Zool. Soc., London, 286-300.

Cleland, P., 1884, On the viscera of the porpoise and white-beaked dolphin, J. Anat. Physiol., XVIII.

Coulombe, H. N., Ridgway, S. H., and Evans, W. E., Respiratory water exchange in two species of porpoise, Science, 149:86.

D'Arcy-Thompson, 1890, On the cetacean larynx, in: "Studies", Mus. Dundee.

Diercks, K. J., Trochta, R. T. Greenlaw, C. F., and Evans, W. E., 1971, Recording and analysis of dolphin echolocation signals, J. Acoust. Soc. Am., 49:1729.

Dormer, K. J., 1974, "The Mechanism of Sound Production and Measurement of Sound Processing in Delphinid Cetaceans", Ph. D. Diss., University of California, Los Angeles.

Dubois, E., 1886, Ueber den larynx (den cetacean), in: "Weber's Studien uber Saugetheire, 88-111.

Evans, W. E., and Prescott, J. H., 1962, Observations of the sound capabilities of the bottlenosed porpoise: a study of whistles and clicks, Zoologica, 47:121.

Evans, W. E., 1973, Echolocation by marine delphinids and one species of fresh-water dolphin, J. Acoust. Soc. Am., 54:191.

Evans, W. E., and Maderson, P. F. A., 1973, Mechanisms of sound production in delphinid cetaceans: a review and some anatomical considerations, Am. Zool., 13: 1205.

Evans, W. E., and Reiger, M. P., 1977, Production of sound in light air (HeO$_2$) atmosphere, in preparation.

Gaunt, A. S., and Wells, M. K., 1973, Models of syringeal mechanisms, Amer. Zool., 13:1227.

Gaunt, A. S., Gaunt S. L. L., and Hector, D. H., 1976, Mechanics of the syrinx in Gallus gallus. I. A comparison of pressure events in chickens to those in oscines, Condor, 78:208.

Gaunt, A. S., and Gaunt, S.L.L., 1977, Mechanics of the syrinx in Gallus gallus. II. Electromyographic studies of ad libitum vocalizations, J. Morphol., (in press).

Gurevich, V. S., 1973, A roentgenological study of the respiratory
    tract in Delphinus delphis, (Russ. Eng. Abstract), Zool. Zh.
    52:786.
Hersh, G., 1966, "Bird Voices and Resonant Tuning in Helium-Air
    Mixtures", Ph. D. Diss., University of California, Berkeley.
Hosakawa, H., 1950, On the cetacean larynx: with special remarks on
    the laryngeal sacs of the aryteno-epiglottal tube of the
    sperm whale, Whales Res. Inst., Science Report 3:23.
Howes, G. B., 1879, On some points in the anatomy of a porpoise,
    J. Anat. Physiol., 14:467.
Huber, E., 1934, Anatomical notes on Pinnipedia and Cetacea, Carn-
    egie Inst. Washington Publ. N°447:105.
Huxley, T. H., 1888, "A Manual of the Anatomy of Vertebrate Animals",
    D. Appleton and Co., 1,3,5 Bond St. New York.
Kelemen, G., 1963, Comparative anatomy and performance of the vocal
    organ in vertebrates, in: "Acoustic Behavior of Animals",
    R. G. Busnel, ed., Elsevier, Amsterdam.
Kinne, O., 1975, Orientation in space: animals-mammals, in: "Marine
    Ecology, Vol. 2", O. Kinne, ed., Wiley, London.
Lawrence, B., and Schevill, W. E., 1956, The functional anatomy of
    the delphinid nose, Mus. Comp. Zool. Bull., 114(4):103.
Mead, J. G., 1972, "On the Anatomy of the External Nasal Passages
    and Facial Complex in the Family Delphinidae of the Order
    Cetacea", Ph. D. Diss., University of Chicago.
Murie, J., 1871, On Risso's grampus, G. Rissoanus (Desm.), J. Anat.
    Physiol., 5:118.
Negus, V. E., 1931, "The Mechanism of the Larynx", C. V. Mosby Co.
Norris, K. S., 1964, Some problems of echolocation in cetacea ceta-
    ceans", in: "Marine Bioacoustics", W. N. Tavolga, ed., Perg-
    amon Press, New York.
Norris, K. S., 1969, The echolocation of marine mammals, in: "The
    Biology of Marine Mammals", H. T. Andersen, ed., Academic
    Press, New York.
Norris, K. S., Dormer, K. J., Pegg, J., and Liese, G. J., 1971,
    The mechanisms of sound production and air recycling in
    porpoises: a preliminary report," in:"Proc. 8th Annual
    Cong. Biol. Sonar and Diving Mammals".
Purves, P. E., 1967, Anatomical and experimental observations on the
    cetacean sonar system, in: "Animal Sonar Systems, Biology
    and Bionics", R. G. Busnel, ed., Imprimerie Louis Jean,
    GAP (Hautes Alpes), France.
Roberts, L. H., 1973, Cavity resonances in the production of orient-
    ation cries, Period. Biol., 75:27.
Roberts, L. H., 1975a, The functional anatomy of the rodent larynx
    in relation to audible and ultrasonic cry production, Zool.
    J. Linn. Soc., 56:255.
Roberts, L. H., 1975b, The rodent ultrasound production mechanism,
    in: "Ultrasonics".

Romanenko, Y. V., 1974, Physical fundamentals of bioacoustics, Fizi-
    cheskiye Osnovy Bioakustik, JPRS 63923, Moscow.
Schenkkan, E. J., 1971, The occurrence and position of the connecting
    sac in the nasal tract complex of small odontocetes, Mammalia
    (Cetacea), Beaufortia, 19:37.
Schenkkan, E. J., 1973, On the comparative anatomy and function of
    the nasal tract in odontocetes (Mammalia, Cetacea), Bijdr.
    Dierk., 43:127.
Turner, W., 1872, An account of the Great Finner Whale (Balaenoptera
    sibbaldii) stranded at Logneddery. Part 1. The soft parts,
    Trans. Roy. Soc. Edinburgh, 26:197.
Watson, M., and Young, A. H. 1879, The anatomy of the Northern Beluga
    (Beluga catadon Gray Gray: Delphinapterus leucas Pallas)
    compared with that of other whales, Trans. Roy. Soc. Edin-
    burgh, 29:393.
Wood, F. G., 1964, "Marine Bioacoustics", W. N. Tavolga, ed., Perga-
    mon Press, New York.

ELECTROMYOGRAPHIC AND PRESSURE EVENTS IN THE NASOLARYNGEAL SYSTEM

OF DOLPHINS DURING SOUND PRODUCTION

S. H. Ridgway, D. A. Carder, and R. F. Green

Naval Ocean Systems Center
San Diego, Ca. 92152 U.S.A.

A. S. Gaunt and S. L. L. Gaunt

Department of Zoology
Ohio State University
Columbus, Oh. 43210 U.S.A.

W. E. Evans

Hubbs-Sea World Research Institute
1720 South Shores Drive
San Diego, Ca. 92109 U.S.A.

INTRODUCTION

In electromyography fine wires are inserted into muscles to measure their activity periods. We have employed this technique to evaluate the activity of various muscles of the nasolaryngeal system of dolphins during the production of sounds. The anterior internus, the posterior internus, diagonal membrane muscle and the nasal plug muscle have been studied in greatest detail. These muscles fire just before and during the production of sounds whereas the hyoepiglottal muscle and the intercostal muscles do not. Simultaneous video recordings of blowhole movement and electromyogram (EMG) have been made to correlate such movement with EMG and sound production.

Catheters (5 to 8 Fr.) inserted into the nares, trachea and nasal sacs were used to measure pressure in various cavities during sound production. There were no pressure changes in the trachea.

This rules out the larynx as the primary sound source. Pressure increased in the nares and premaxillary sacs prior to each sound and dissipated after the sound. Catheters with inflated 15 ml balloons were placed in the vestibular sacs, premaxillary sacs and nares without preventing sound production. Pressure increased in both nares and both pre-maxillary sacs during sound production but not in the vestibular sacs.

Sounds are produced in the nasal system by action of the muscular nasal plug. Pressure in the nares increases, but we do not yet know exactly which membranes are vibrating or the direction of air flow in the system.

Previous Studies

Dolphins emit numerous diverse sounds that are generally described as whistles or pulses. A study of the mechanics of delphinid vocalizations is of interest from several points of view: 1) Delphinids appear to be remarkably effective in producing and broadcasting sound at minimum energetic cost both at the surface and to depths of at least a few hundred meters. 2) Some sounds (perhaps most) are used in a very effective echolocation system which can be better understood if the mechanism of sound generation is known. 3) If dolphins have an elaborate communications system based upon sounds, a better understanding of how the sounds are made will be useful in understanding their communications.

A number of anatomical studies have been done on the delphinid nasolaryngeal system (see Green et al this volume) and various theories have been proposed for how dolphins make sounds. Several authors, including Lawrence and Schevill, (1956), Purves, (1967) and Schenkken, (1973) have proposed that the sounds of cetaceans are made in and by the larynx while Diercks et al, (1971), Gurevich, (1973), Norris et al, (1971) and Dormer, (1979) have suggested sound production in the nasal system. Evans and Prescott, (1962), have suggested a combination system, whistles being produced by the larynx and pulses by the nasal system - "anatomical and behavioral evidence, as well as sound pressure measurements, indicate that echolocation clicks are produced in the nasal sac system of porpoises. Within the sac system the tubular sacs combined with the nasal plug nodes appear to be the site of sound production." Lilly and Miller, (1961), were the first to note that whistles and pulses could be produced simultaneously and Lilly (1963) reported that he "manually felt, over the skull, the vestibular sacs participating in the vocal process as these sacs filled and emptied during whistling. The mechanism of production of this whistle as well as other whistles is still under investigation. The number of the observed areas of production (inside head with tightly closed blowhole, inside head with open blowhole, and at blowhole slit) suggests that these animals can use each of several parts of the nose for these whistles, rather than just a single area" Lilly (1963).

Norris et al (1971), Hollien et al (1976) and Dormer (1979) have used X-rays to observe tissue changes in living, phonating dolphins. All of these investigations noted movements in the nasal system that correlated with sound production. There were no correlated movements in the larynx. Dormer (1979) says in summary that "the mechanism of sound production in porpoises consists of a series of muscular pumps, valves, and compliant sacs with the sound being generated at the nasal plug."

Although the X-ray studies have given many helpful insights such studies have not informed us as to how the sound is actually made. Therefore, we set out to study the mechanism of sound production in Tursiops and Delphinus using electromyography and pressure sensors so that the events of muscular contraction, pressure changes and sound production could be studied simultaneously.

MATERIALS AND METHODS

The experimental subjects were five Tursiops truncatus (three males and two females).

For the experiments the dolphins were placed in a special restraint device similar to the one figured elsewhere (Ridgway, this volume). Water covered about two-thirds of the body. Only the blowhole area and dorsum of the body was in air. An LC-10 hydrophone in a suction cup was attached to the melon for the purpose of recording the animals sonic output. A similar hydrophone was dangled in the water adjacent to the animal for a second channel of sound recording at 38 cm/sec on an FR-100 tape recorder. In addition, the output of the hydrophones were integrated and displayed on a polygraph along with all the physiological measures. In some experiments sound was recorded on video tape simultaneous with the dolphin's blowhole movements and polygraph tracings of EMG and pressure events.

Fine wires were placed into muscles (see Green et al, this volume) through hypodermic needles according to the method of Basmajian and Stecko (1962). The wires were connected by electrode clips, amplified (Grass model 7P511B or 7P3B) and displayed by the polygraph (Grass model 78D).

Catheters (5-8 Fr.) were inserted through the blowhole into the nares, vestibular sacs or premaxillary sacs. Some catheters had bulb tips that could be inflated with up to 15 ml of air or saline solution. In two experiments a catheter was inserted through a hypodermic needle into the trachea. The catheters were connected

to calibrated pressure transducers (Statham PM6TC + 5-350,
PM6TC + 10-350, and/or PM131TC + 5-350) and the analog pressure was
displayed on the polygraph.

RESULTS

EMG's from the following muscles have been studied during
sound production: Posterior Internus (PI), Anterior Internus (AI),
Nasal Plug Muscle (NPM), Hyoepiglottal Muscle (HM), Intercostal
Muscles (IC), Palatopharyngeal Muscle (PPM) and Diagional Membrane
(DM). PI, AI, NPM and DM were consistently associated with sound
production. PPM was often active during sounds, but its firing
was not as consistent as the former muscles during sound production.
IC and HM were active during only one type of sound -- the "Bronx
cheer" or "raspberry". This sound is made by the expulsion of large
amounts of air through the fluttering lips of the blowhole.

Figure 1. EMG, sound and pressure events in the nasal system of
Tursiops -- Right anteriorinternus muscle, anterior part (RAIA),
Right anteriorinternus muscle, posterior part (RAIP), Left anterior
internus muscle (LAI), Nasal plug muscle NPM, Whistles (W) and
Blows (B).

Muscles on both sides of the head were generally fired simultaneously during sounds and different parts of the same muscle had a similar EMG pattern (Figure 1). Occasionally less activity was seen in the right AI or PI than in the left and vice versa, but almost always there was some activity from both muscles of these two groups. In about ten percent of those sounds during which the DM was observed, it fired mainly on one side (Figures 2 and 3), but in most instances both sides were active (Figure 4).

Figure 2. EMG of the right and left diagonal membrane muscles during a short click train.

Figure 3. EMG of the right and left diagonal membrane muscles.

Sometimes the nasal plug (as viewed from above) moved toward the
right and sometimes toward the left. Although sound may be produced on one side or the other it seemed to require bilateral
muscle action in most cases.

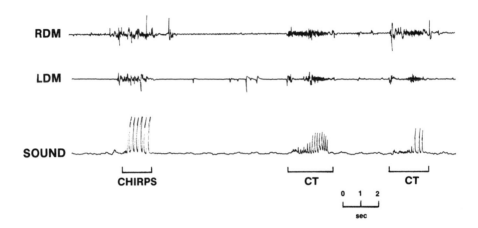

Figure 4. EMG of the right and left diagonal membrane muscles
during pulsed sounds described as chirps and click trains (CT).

      Catheters were placed in both nares, in vestibular sacs on both
sides, in premaxillary sacs on both sides and in the trachea.
Dolphins produced both whistles and pulses in varied form and
intensity with catheters in place. Pressure in both nares increased
immediately following muscular contractions (AI, NPM, PI). The
sound occurs as pressure reaches its peak. Pressure subsides just
after the end of the sound as muscle firing ceases (Figure 5).

      If we opened these intranarial catheters to atmospheric air,
sound production was greatly altered. If the catheter was small in
diameter, on the order of a millimeter or less, some short whistles
and short, very weak pulse trains were produced as air rushed from
the catheter. If the catheter was large in diameter, 3 or 4 mm,
the animal could not make sounds.

      We were not able to maintain catheters in the premaxillary sac
for periods of more than a few minutes at a time -- the animal kept
ejecting them. The dolphin was able to produce whistles and pulses
with catheters in the premaxillary sacs. Pressure in these sacs
was identical with that in the nares. Inflated balloons (15 ml) at
the end of catheters in the premaxillary sacs or in the nares did
not prevent sound production.

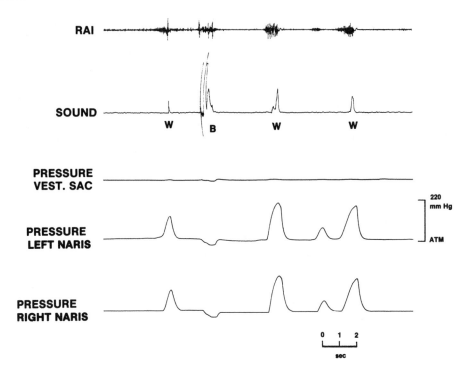

Figure 5. EMG and pressure events during three whistles (W) and a blow (B).

Figure 6. EMG and pressure events during several whistles (W) and a blow (B).

Pressure increases in the vestibular sacs often followed those in the nares, but were much smaller (Figure 6). Inflated balloon catheters placed in the vestibular sacs did not prevent sound production nor did the placement of open catheters, which allowed any pressure to escape, prevent sound production.

A positive pressure of about 25 mm Hg was maintained in the trachea between breaths. With each breath there was a negative pressure followed by a return to the positive pressure state. There were no changes in the intratracheal pressure during the production of whistles (Figure 7) or pulsed sounds.

Figure 7. Pressure events in the trachea (ITP) and Nares (INP) during four whistles and a blow.

We have not yet been successful in placing a catheter in the very small nasofrontal sacs nor in placing electrodes in the muscles surrounding it.

DISCUSSION

Our studies clearly show that _Tursiops_ produce sound in the nasal system and not in the larynx. Our most important evidence in

support of this can be seen in Figure 7. Pressure increased in the
nares before each whistle, but there was no change in intratracheal
pressure except during the course of blows. In humans there is
increased intratracheal pressure during phonation (Basmajian, 1978).
Further support for this conclusion came from the EMG data. If
vocalizations arose in the larynx, one would expect to see patterns
like those in man. During human phonations the IC muscles are
active. A preliminary burst of activity before each phrase is followed after an interval by increasing activity in the utterance of
the phrase (Basmajian, 1978). In the dolphin IC was inactive during
all of the common sounds. Only in the rarely produced "Bronx cheer"
was this muscle seen to fire in concert with sound. Bronx cheers
are apparently produced only when the dolphin's blowhole is out of
the water. In addition, the EMG from the HM of the dolphin showed
activity during the blow but not during sounds. In summary, the
EMG and pressure events that are related to phonation in the human
larynx were not observed during dolphin sounds, but EMG and pressure
events in the nasal system are highly correlated with sound production.

The vestibular sacs are not essential for sound production
since they can be isolated from the system without preventing sounds.
Some records show small pressure increases in these sacs during
sounds. We suspect that the vestibular sacs may be important as air
reservoirs. Air may be blown into the sacs during sounds then re-
circulated, thus saving air for phonation during long periods of
submergence. This function would be important at great depths where
air carried by the animal is compressed into a relatively small
volume.

Since dolphins must feed underwater, often at great depth, they
are continually subjected to changes in pressure. Air in the thorax
is compressed as the dolphin dives (Ridgway et al, 1969). If sound
were produced in the larynx the resonant characteristics of the
animal sounds would also be continuously changing. Since the volume
of the nasal system is small, air can be leaked from the lungs to
maintain volume as the lung is compressed. Because the air is
equilibrated at any depth muscles will be able to operate the sacs
and valves of the nasal system efficiently at depth and the acoustic
resonance need not change with depth.

Our data on pressure events demonstrates that the larynx must
be closed during sound production since large pressure changes
occur in the nares anterior to it, but none in the trachea just
posterior to and open to the larynx. Since sound is altered or
prevented by an open tube in the nares, the nares must be in direct
communication with structures that produced sound during the process.
The open tube in the nares simply prevents the nares from acting as
an efficient source of air pressure for the sound generating

apparatus. The presence of an inflated balloon in the nares does
not alter sound production since it does not prevent the space from
pressurizing. If sounds were being produced within the air column
of the nares the balloon would, however, be expected to alter the
sound.

Based on data collected using an array of transducers attached
to various locations on the gular, melon and rostral regions of a
trained Tursiops, Diercks et al (1971) identified the site of the
origin of sounds as an area 1.5 to 2.0cm beneath the blowhole
margin. This appears to correspond to the region where the internal
lip of the nasal plug abuts the diagonal membrane. Just lateral
to this the nasal plug nodes insert into the opening of the nasofrontal sacs. Glandular structures in this area (Evans and Maderson, 1973) might provide lubrication to tissue that might be
subjected to a high velocity airstream associated with sound
production. We do not agree with Evans and Maderson (1973) that
the nasofrontal sacs can summarily be written off as an important
component of the sound production mechanism. Their statement
referring to the great individual variability observed in the size
and form of these sacs that "the intra-specific variation which we
have commented upon has not yet been quantified, it seems most
unlikely that any significant variation would be permitted by
natural selection if the nasofrontal sacs played an active major
role in such a sophisticated function as delphinid phonation," is
not equivocal. Just such variation could provide individual acoustic differences which could be very adaptive for a highly social
species.

It appears to us that during phonation air is compressed in
the nares by depression of the nasal plug at the dorsal end and perhaps by action of the palatopharyngeal sphincter at the ventral end
of the nares. Air must pass from the nares, past the diagonal membrane at their upper end, through the spiracular cavity along the
posterior margin of the nasal plugs and perhaps past the nodes of
the nasal plug to the nasofrontal sacs. Although we have not yet
been able to measure pressure in the small nasofrontal sacs nor to
record EMG's of the intrinsic muscles that surround them, we suspect
that these sacs may be important in dolphin phonations. The sacs
could serve as resonators. Mucus in the sacs and muscles around
them could serve a damping function and help provide fine control
over frequency.

The gross and microanatomical nature of the nasal plug nodes,
diagonal membrane, and nasofrontal sacs, coupled with the acoustic
and now electromyographic and pressure measurements reported in
this paper strongly suggest that this total system is an excellent
candidate as the source and mechanism of sound production in most
delphinids.

REFERENCES

Basmajian, J. V. and Stecko, G. A., 1962, A new bipolar indwelling
    electrode for electromyography, J. Appl. Physiol., 17:849.
Basmajian, J. V., 1978, "Muscles Alive", The Williams and Wilkins
    Company, Baltimore.
Diercks, K. J., Trochta, R. T., Greenlaw, C. F., and Evans, W. E.,
    1971, Recording and analysis of dolphin echolocation signals,
    J. Acoust. Soc. Am., 49:1732.
Dormer, K. J., 1979, Mechanism of sound production and air recycling
    in delphinids: cineradiographic evidence, J. Acoust. Soc.
    Am., 65:229.
Evans, W. E., and Prescott, J. H., 1962, Observations of the sound
    capabilities of the bottlenosed porpoise: a study of whistles
    and clicks, Zoologica, 47:121.
Evans, W. E., and Maderson, P. F. A., 1973, Mechanisms of sound
    production in delphinid cetaceans: a review and some anatom-
    ical considerations, Am. Zool., 13:1205.
Green, R. F., Ridgway, S. H., and Evans, W. E., Functional and des-
    criptive anatomy of the bottlenosed dolphin nasolaryngeal
    system with special reference to the musculature associated
    with sound production, this volume.
Gurevich, V. S., 1973, A roentgenological study of the respiratory
    tract in Delphinus delphis, (Russ. Eng. abstract), Zool. Zh.,
    52:786.
Hollien, H., Hollien, P., Caldwell, D. K., and Caldwell, M. C., 1976,
    Sound production by the Atlantic bottlenosed dolphin, Tursiops
    truncatus, Cetology, 26:1.
Lawrence, B., and Schevill, W. E., 1956, The functional anatomy of
    the delphinid nose, Mus. Comp. Zool. Bull., 114(4):103.
Lilly, J. C., and Miller, A. M., 1961, Sounds emitted by the bottle-
    nosed dolphin, Science, 133:1689.
Lilly, J. C., 1963, Distress call of the bottlenosed dolphin: stimu-
    li and evoked behavioral responses, Science, 139:116.
Norris, K. S., Dormer, K. J., Pegg, J., and Liese, G. J., 1971, The
    mechanisms of sound production and air recycling in porpoises:
    a preliminary report, in: "Proc. 8th Annual Cong. Biol. Sonar
    and Diving Mammals".
Purves, P. E., 1967, Anatomical and experimental observations on the
    cetacean sonar system, in: "Animal Sonar Systems, Biology
    and Bionics", R; G. Busnel, ed., Imprimerie Louis Jean, GAP
    (Hautes Alpes), France.
Ridgway, S. H., 1979, Electrophysiological experiments on hearing
    in Odontocetes, this volume.
Ridgway, S. H., Scronce, B. L., and Kanwisher, J., 1969, Respiration
    and deep diving in the bottlenosed porpoise, Science, 166:1651.
Schenkkan, E. J., 1973, On the comparative anatomy and function of
    the nasal tract in Odontocetes (Mammalia, Cetacea), Bijdr.
    Dierk., 43:127.

ECHOLOCATION SIGNALS OF THE ATLANTIC BOTTLENOSE DOLPHIN (TURSIOPS TRUNCATUS) IN OPEN WATERS

Whitlow W.L. Au

Naval Ocean Systems Center

Kailua, Hawaii 96734

A wide variety of echolocation experiments performed with the Atlantic bottlenose dolphin (Tursiops truncatus) have indicated that these dolphins possess a highly sophisticated and adaptive sonar system. Results of discrimination experiments have shown that Tursiops can detect a 10% difference in the diameter of metallic spheres (Norris, Evans, Turner, 1967), material composition and thickness differences as small as 0.1 cm in metallic discs (Evans and Powell, 1967), differences between plates shaped as circles, triangles and squares independent of their cross sectional areas, as well as a 6% change in the diameter of the circles (Barta and Evans, 1970; Fish, Johnson, and Ljungblad, 1976), and a 0.8 dB difference in the target strength of corprene cylinders (Evans, 1973). Murchison and Penner (1975) have demonstrated detection ranges of 72.2 m for a 2.54-cm diameter solid steel sphere and 76.8 m for a 7.6-cm diameter water-filled sphere. Range resolution capabilities of 0.9, 1.5 and 2.8 cm for target ranges of 1,3, and 7 m have been reported by Murchison (1976). Nachtigall, Murchison and Au (1978) reported on the ease with which a Tursiops could discriminate between foam cubes and cylinders of different sizes. Hammer and Au (1978) found that an animal could recognize differences in the wall thickness of aluminum cylinders that were as small as 0.16 cm. They also found that the animal could detect differences between free-flooded aluminum, bronze and steel cylinders of equal dimensions.

The ability of Tursiops to perform these difficult target detection, recognition and discrimination tasks should be dependent on the kinds of echolocation signals that are being emitted, on the way the signals are beamed out and echoes

received by the animal, and on the signal processing schemes
being utilized. A considerable amount of information can be
obtained by an animal scanning across a target and examining the
target from different spatial aspects. Target information can
also be contained entirely within the echoes, independent of any
spatial clues. In this review, the echolocation signals used by
a number of T. truncatus in the open waters of Kaneohe Bay, Oahu,
Hawaii will be discussed. Since the natural habitat of T.
truncatus consists primarily of bays, harbors, inlets and other
littoral areas, having the capability of performing echolocation
experiments in the open waters of a bay is important from the
standpoint of working in an environment very similar to the
natural environment of these animals.

        Measurements of the echolocation signals emitted by two
Tursiops performing a target-detection experiment in Kaneohe Bay
led Au et al. (1974), to discover certain previously unknown
properties of the animals' sonar signals. They found that the
signals had peak energies at frequencies (peak frequencies)
between 120 and 130 kHz, which was over an octave higher than
previously reported peak frequencies between 30 and 60 kHz
(Evans, 1973). They also measured an average peak-to-peak click
source level of 220 dB re 1μPa at 1 yard, which represented a
level on the order of 30 dB higher than previously measured for
Tursiops. The use of high-frequency echolocation clicks was
attributed to the characteristics of the high ambient noise
environment of Kaneohe Bay. The effects of the ambient noise
on the animals will be discussed in Section II-B entitled
Spectrum Adaptation. The high source levels were attributed to
the absence of boundaries which could cause high reverberation
and to the long distances between the animals and the target.

        An example of a typical echolocation click train for an
animal performing a target detection task is shown in Fig. 1,
with the frequency spectrum plotted as a function of time on the
left and the individual click waveform displayed on the rights.
Five parameters are written beside each click waveform. The top
notations are the peak-to-peak source level in dB re 1μPa at 1
meter, followed by the peak frequency and the 3-dB bandwidth
which are separated by a slash (/). The parameters listed below
the signals are the cross-correlation coefficient between the
present signal and the previous one, followed by the time of
occurrence of the signal relative to the start of the click
train. The amount of information present in Fig. 1 is considerable and can be unwieldy especially if many tens of trials and
sessions are considered. In order to simplify the examination of
these signals, a process of averaging both the time domain and
frequency domain waveforms will be used. The averaging is
performed in software and is similar to the kinds of averaging

Figure 1. An echolocation click train for a _Tursiops truncatus_ performing a target detection task in Kaneohe Bay.

performed with real-time spectrum analyzers, and other digital
equipments. The average waveform and spectrum of the click
train displayed in Fig. 1 is shown in Fig. 2. Note the similarity between the average waveform and the individual signals of
Fig. 1. The higher level signals in most click trains are highly
correlated, whereas fluctuations in the waveshape are generally
associated with the lower level signals. Most of the waveforms
presented in this review will be in terms of their average waveforms. Some of the results that will be discussed have been
reported by Au et al. (1974); and Au, Floyd and Haun (1978),
although the bulk of the results have not yet been published.

I. ACOUSTIC FIELD OF AN ECHOLOCATING DOLPHIN

In order to facilitate the discussion of the acoustic field
of a dolphin, it is helpful to establish a frame of reference
with respect to the site at which the echolocation clicks are
produced. Although the exact location of click production and
the mechanisms involved are not yet known, indirect evidence
from anatomical investigations (Evans and Maderson, 1973),

Figure 2.   The averaged waveform and spectrum for the click
             train of Fig. 1.

acoustic measurements (Diercks, et al., 1971), ambient pressure measurements within the nares (Ridgway et al., 1974), and x-ray measurements (Norris et al., 1971; Hollien et al., 1976) all suggest that click signals are produced in the vicinity of the external nasal passage. Diercks et al., (1971), using multiple contact hydrophone data, places the source in the vicinity of the nasal plugs in the nares. Therefore, the region of the nasal passage approximately 3 cm below the blowhole will be assumed to be the site of click production in Tursiops.

A. Contact Hydrophone Measurements

The acoustic field in the immediate vicinity of a dolphin's head should be extremely complex because of the presence of air sacs, cranial bones and tissue having differing sound velocities, that could cause reflection and refraction of the acoustic pulses generated within the animal's head. The signal that is emitted into the water is formed by the superposition of the direct component of the generated pulse and the pulses that are modified by the various structures in the animal's head. Romenenko (1975) gave an example of an emitted pulse formed by the superposition of the direct pulse and a pulse reflected off the premaxillary sac. Dubrovskiy and Zaslavskiy (1975), using an optical technique with 5 mm diameter light bulbs inserted in the region of the plugs that shut the aditus to the tubular sacs, found the intermaxillary bone to be a principal contributor to the formation of the emission field. Bel'kovich and Dubrovskiy (1977), give an exellent summary of the Russian studies on the role of air sacs and cranial bones on the formation of the sound field of dolphins.

Simultaneous measurements of the signals in the near and far fields of a dolphin were made by Au et al., (1978) using a contact hydrophone placed either on the animal's melon or rostrum and a hydrophone located one meter from the animal's rostrum. Examples of the average signals of two echolocation click trains measured with the contact hydrophone placed on the melon and two click trains with the contact hydrophone placed on the animal's rostrum are shown in Fig. 3. Also included in Fig. 3 are the average signals measured by a hydrophone located one meter in front of the animal's rostrum. A comparison between the average signal measured at 1 m with the signal measured at the melon seems to indicate that there was a considerable amount of interference to the direct pulse by internally reflected and/or refracted pulses. Each cusp in the time-domain signal measured at the melon may indicate the arrival of a reflected or refracted component of the generated pulse. The peak frequencies for the signals measured at the melon were 8 and 6 kHz higher respectively, than the signals at 1 m. The 3-dB bandwidths were also smaller for the

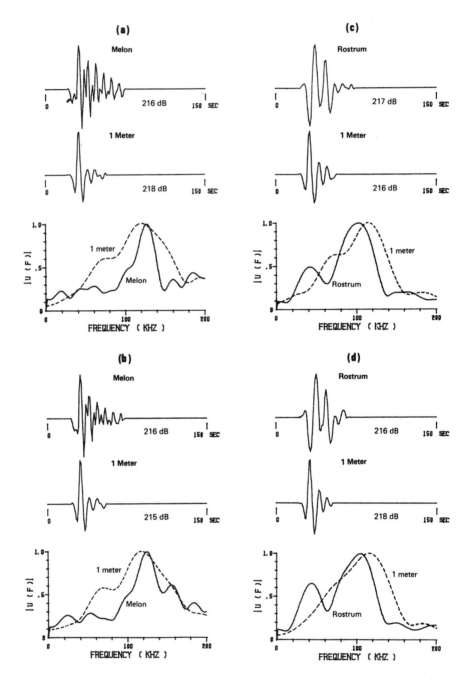

Figure 3.   Echolocation signals measured simultaneously at the melon or rostrum and at one meter.

signals at the melon. The higher peak frequencies and lower
bandwidths were probably due to the many cusps associated with
the signals measured at the melon.

The signals measured at the rostrum were not as complex as
those at the melon. The waveforms for the rostrum signals seem
to suggest that they were formed by the superposition of a direct
and a single internally reflected or refracted pulse. The peak
frequency of each rostrum signal was 11.7 kHz lower than the
signal at 1 m. This is not surprising since the signals that
were detected by the hydrophone on the rostrum traveled away from
the major axis of the projected beam.

Although the signals measured at 1 m in all cases were
similar in shape both in the time and frequency domain, the
corresponding signals measured at both the rostrum and melon were
considerably different. The melon signals seemed to be formed
by a large number of internally reflected and refracted pulses
that arrived very close together and soon after the arrival of
the direct pulse. This type of propagation is not surprising
since the melon region is close to the air sacs and areas of the
skull which could reflect acoustic energy into melon region.

The amplitude of the signals measured by contact hydrophones
placed on the melon and rostrum of an animal in relationship to
the signals measured in the far field was also studied by Au,
Floyd and Haun (1978). An animal was trained to station in a chin
cup that was supported by surgical rubber tubing. A vertical
array of three hydrophones was placed 1 m from the animal's
rostrum and a hydrophone was placed at 85 m which was 5 m beyond
the target. Two separate cases were considered, the first with
the contact hydrophone placed on the rostrum and the second
with the hydrophone on the melon. Since, the mean value of the
sound pressure level (SPL) measured at 1 m were within 0.9 dB in
both cases, the results of the two conditions were combined. The
results in terms of the peak-to-peak SPL are shown in Fig. 4.
with the SPL at 1 m obtained from the hydrophone in the vertical
array which detected the largest signal.

The dash curve represents the SPL at 1 m referenced to any
range r by considering the spherical spreading and absorption
losses. The absorption coefficient used was 0.44 dB/m which
corresponds to a typical Kaneohe Bay water temperature of 24°C
and salinity of 35 ppt and a signal frequency of 120 kHz. There
is good agreement between the SPLs at 1 and 85 m after accounting
for the transmission loss. The mean SPLs at 1 m and at the
rostrum and the melon were 217.7, 217.2, and 215.8 dB re 1 μPa,
respectively. Evans (1973), using an array of contact hydrophones,
found that the amplitude of a signal measured at the melon was
approximately equal to that measured at the rostrum. The 1.4 dB

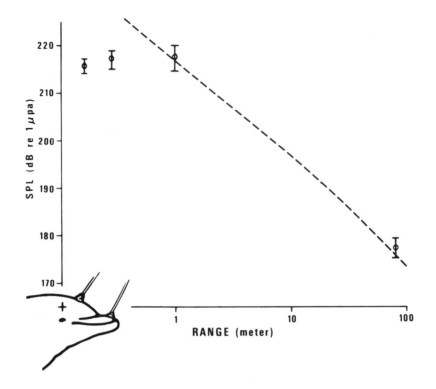

Figure 4.   Peak-to-Peak Sound Pressure level of echolocation
            signals as a function of range

difference between the melon and rostrum in fig. 4 is in general
agreement with the results of Evans (1973), considering the
variability in the data.   The reason for the amplitude of the
outgoing echolocation signal at the surface of the animal's head
being approximately the same as at a distance of 1 m may be
explained by the fact that the contact hydrophone was placed in
the near field of the animal's sound projection system.   Within
the near field, the SPL should vary considerably with distance
along the major axis of the beam, forming a number of peaks and
nulls.

B.   Near-to-Far Field Transition

A series of measurements was made by Au, Floyd, and Haun
(1978), to determine the distance from the tip of an animal's

rostrum where the amplitude of the acoustic pressure of the
echolocation signals began to satisfy the far-field condition by
decreasing as a function of $1/r$. An array of hydrophones was
placed at different distances from the tip of a rigid chin-cup
station aligned in the direction of the target to measure the
variation in the SPL as a function of distance away from the
animal. Initially all the hydrophones were submerged to the same
depth, with the acoustic center of each hydrophone at a depth
corresponding to the top of the animal's rostrum. It was
expected that this arrangement would measure the signal close to
the major axis of the vertical beam. However, the vertical
beam-pattern measurement to be discussed in Sec. IC showed that
the major axis was directed at an elevation angle of 20° above
the rostrum. Therefore a second series of measurements was made
with the hydrophones aligned along the major axis of the sound
beam.

       The results of both series of measurements are shown in
Fig. 5. Both sets of results agree fairly closely, with the
measured SPL difference departing from the $1/r$ function between
0.375 and 0.500 m in the first case and 0.500 and 0.625 m in the
second. Since the second series of measurements was performed
with the hydrophones aligned along the major axis of the vertical
beam, it should more accurately represent the propagation of the

Figure 5.   Relative sound pressure levels of transmitted signals
            across the transitional region between the acoustic
            near and far fields. The dash curve in each figure
            represents the spherical spreading loss with the
            reference point at 1 m.

signals close to the animal. Therefore, the region in which the
acoustic near field makes the transition to the far field exists
between 0.500 and 0.625 m from the tip of the animal's rostrum
for the high-frequency signals emitted by the animals in Kaneohe
Bay.· The location of the transition region for animals emitting
signals at peak frequencies different from the 110 to 130 kHz
typically used in Kaneohe Bay (see Section II-B) should vary
linearly with frequency.

C. Directional Pattern of Emitted Echolocation Signals

The echolocation signals of dolphins and other odontocetes
have been shown by a number of investigators to be directional.
Norris et al, (1961), observed that a blindfolded Tursiops could
not detect targets below its jaws and at elevation angles
greater than 90° above the rostrum. Evans, Sutherland, and Beil
(1964), used a Stenella longirostris cadaver and a Tursiops skull
in order to study the directional properties of echolocation
signals. With a cw sound source placed in the region of the
nasal sacs, they found that a definite beam was formed, and it
was directed 15° above the rostrum in the vertical plane and
forward in the horizontal plane. The beam was highly dependent
on frequency becoming narrower as the frequency increased.
Norris and Evans (1966), measured the directionality of Steno
bredanensis echolocation clicks by systematically moving a
single hydrophone placed at a fixed depth to different azimuths
as the animal echolocated on a target. The sound was found to be
directed forward in a narrow beam and was highly frequency
dependent. Schevill and Watkins (1966), studied the directionality of the clicks produced by an Orcinus orca and found that
the frequency content and amplitude of the clicks varied with
the orientation of the animal. As the whale turned away from
the hydrophone, the higher frequency components of the clicks
would progressively diminish in amplitude. Evans, (1973), used
an array of hydrophones spaced 5° apart in the horizontal plane
to measure the horizontal beam pattern of a restrained Tursiops.
Although several elevation angles were used, there is a distinct
possibility that none of Evan's, (1973), measurements were made
along the major axis of the beam in the vertical plane.

The vertical and horizontal beam patterns of a Tursiops
named Sven were measured by Au, Floyd and Haun (1978), with the
animal trained to station in a chin cup. The composite broadband
vertical beam pattern from nine trials is shown in Fig. 6a, and
the composite horizontal beam pattern from twelve trials shown
in Fig. 6b. These two plots show that the echolocations signals
were projected 2 to 3° to the left of center and at an elevation
angle of 20°. The 3-dB beamwidths were approximately 10° in the
vertical plane and 9.8° in the horizontal plane. The horizontal

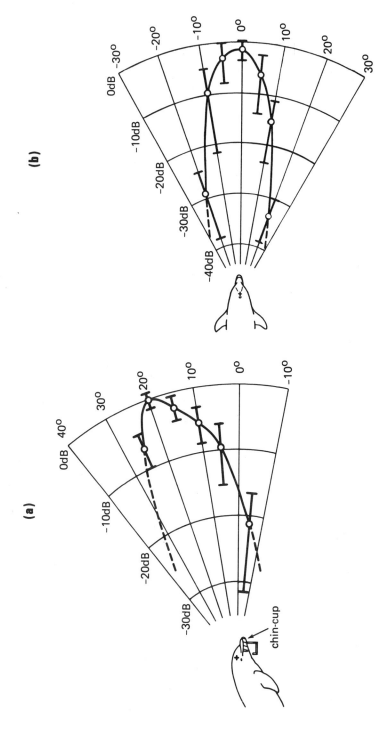

Figure 6. Composite broadband beam patterns for the echolocation signals projected by a Tursiops truncatus (a) in the vertical plane and (b) in the horizontal plane.

beam was measured along the axis of the vertical beam. The 20°
elevation of the vertical beam is in general agreement with the
15° angle measured by Evans, Sutherland, and Beil (1964), using a
Tursiops skull. The elevation of the beam axis would allow the
animal to direct its beam forward, since in a free swimming
posture, the rostrum of the animal frequently appears to be
pointed downwards at a 15° to 20° angle.

The responses of the LC-10 hydrophones used by Au, Floyd and
Haun (1978), to transient-like signals such as porpoise clicks
were not matched. The hydrophones indicated differences as much
as 10% in the peak frequency of the simulated porpoise clicks used
to calibrate the arrays. Therefore, a follow-on effort was made
using a BM-101 and four B&K-8103 hydrophones in order to obtain
better data on the waveform and frequency spectrum of the
projected signals as a function of angle in both the vertical
and horizontal planes. In response to simulated porpoise clicks,
each of the hydrophones indicated approximately the same peak
frequency, bandwidth and wave shape.

The follow-on series of array measurements was conducted
with an animal named Ekahi. Instead of a chin cup, a 40-cm rubber
hoop was used as the stationing device, which allowed the animal
to tilt its head downwards while echolocating. The target which
was located 6 meters from the hoop was submerged to the same
depth (76.2 cm) as the center of the hoop. The vertical array
was placed 1.5 m and the horizontal array 2.5 m from the hoop.
Ekahi was trained to enter the hoop and remain silent until an
audio tone was turned on cueing the animal to begin echolocating.
This technique of conditioned echolocation is discussed by
Schusterman, Kersting and Au (1979).

The composite broadband beam patterns for ten trials in each
plane are displayed in Fig. 7. The two sets of composite beam
patterns of Figs. 6 and 7 are similar, with Ekahi's beam being
slightly broader than Sven's. The 3-dB beamwidths were approximately 11.7° in the vertical plane and 10.7° in the horizontal
plane. Some of the differences between the two experiments may
have been due to different angular separations of the hydrophones
near the beam axis.

Examples of the average waveform and the corresponding
frequency spectrum as detected by the five hydrophones in the
vertical array of the second series of measurements are displayed
in Fig. 8. As the angle departed from the beam axis, the signals
in the time domain became progressively distorted, with respect
to the signal at 0°. In the frequency domain, the peak frequencies
decreased as the hydrophone angle departed from 0°. The presence
of multipaths is apparent for the signals measured above the

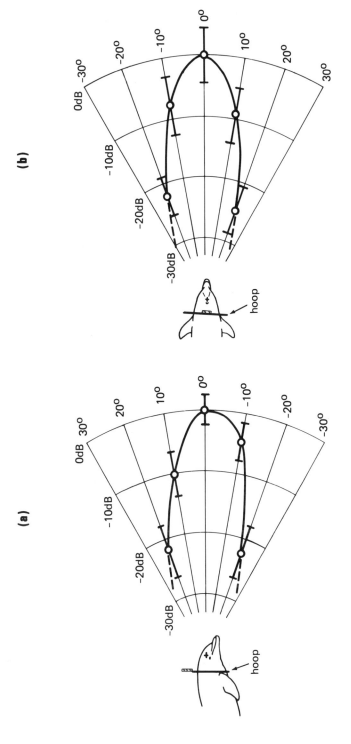

Figure 7. Composite broadband beam patterns obtained in the follow-on series of measurements using closely matched sets of hydrophones, (a) in the vertical plane and (b) in the horizontal plane.

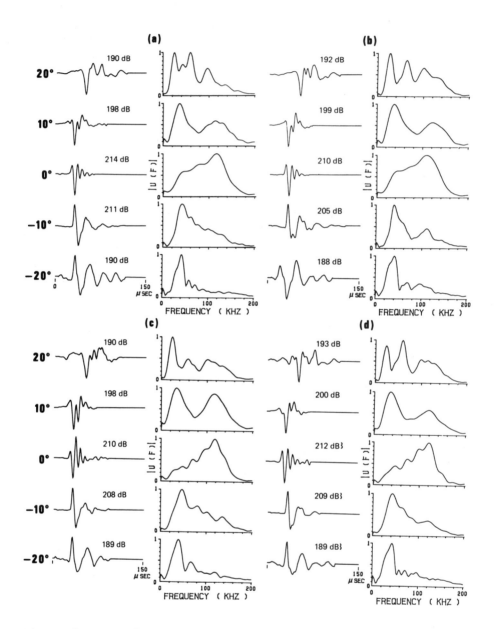

Figure 8.   Examples of the average waveforms and frequency spectra
             as a function of the elevation angle in the vertical
             plane.

beam axis at +10° and +20°. The multipaths may be due to internal reflections and refraction of the signals within the animal's
head, to the signals being radiated from different portions of a
finite source region, or to combinations of these possibilities.
For the signals measured above the major axis there are still
considerable amounts of energy at frequencies close to the peak
frequencies of the 0° signals, causing the spectra to have
multiple humps or peaks. This is not true for the signals
measured below the major axis. There seems to be a general shift
to lower peak frequencies with relatively little energy at the
frequencies close to the peak frequencies of the signals
measured along the major axis. The signals in the time domain
appear to be stretched with little multi-path effects.

The average waveforms and their corresponding frequency
spectra as detected by the five hydrophones in horizontal array
are displayed in Fig. 9 for three trials. As the angle departed
from the beam axis, the signals became progressively more distorted with respect to the signals measured at 0°. The signals
measured at -10° were the least distorted, having peak frequencies
close to those along the beam axis. As the angle departed from
the beam 0°, the peak frequencies shifted to lower values, with
the signals measured at -20° experiencing the greatest amount of
shift. The signals were not symmetrical about the beam axis
which is expected since the structure of the skull is not
symmetrical about the mid-line of the animal.

Besides the nasal sacs and the cranial bones, the melon may
also play a large role in the spatial directivity of the emitted
beam. In 1957, F. Wood (1964) suggested that the melon of a
porpoise might be a sound transducer helping to couple internally
generated sounds to sea water. He also speculated that the melon
may act as an acoustic lens to focus the sound into a narrow
beam. The beam patterns measured by Evans, Sutherland, and Beil
(1964), using a Tursiops skull were considerably broader than
the beam patterns of Figs. 6 and 7. This suggests that the melon
may play a role in focusing the outgoing acoustic energy, causing
a narrower beam to be radiated than would otherwise exist.
Romenenko (1975), studied the effects of a dolphin's skull, with
and without soft tissue present, on the horizontal beam pattern.
He found that for frequencies of 80 kHz and higher, the beams
measured with the whole head were approximately half as wide as
the beams obtained with the skull alone. Norris and Harvey (1974)
examined the sound velocity of the lipids in the melon and found
a definite structure that could cause the focusing of the outgoing
acoustic energy. By slicing the melon of a dead animal in a regular fashion and measuring the sound velocity of various sections
within each slice, they found a low-velocity core extending from
just below the anterior surface towards the right nasal plug and

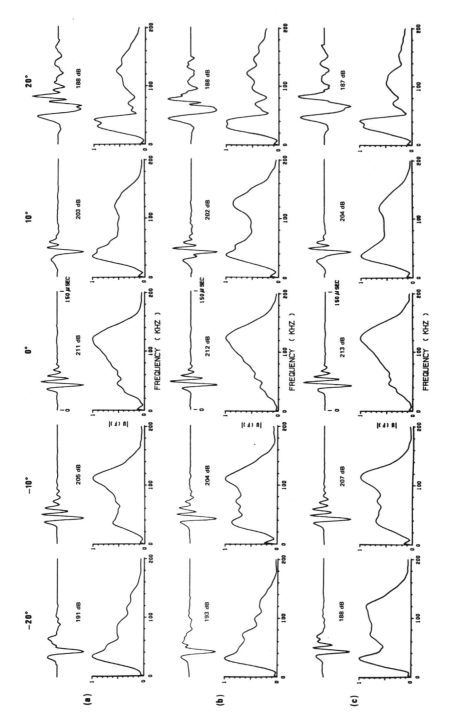

Figure 9. Examples of the average waveforms and frequency spectra as a function of angle in the horizontal plane.

a graded outer shell of higher velocity tissue. Such a velocity gradient would cause the signal to be focused both in the vertical and horizontal planes. The amount of refraction and the subsequent effect on the overall beam patterns have not been determined yet.

II. CHARACTERISTICS OF ECHOLOCATION SIGNALS

A. Click Intervals

The click intervals associated with two animals echolocating on a target located at ranges between 54.9 and 73.2 m was examined by Au et al. (1974). Two examples showing the variations in the click interval on a pulse-to-pulse basis, one with the target present and the other with the target absent, are shown in Fig. 10. One interesting feature shown in Fig. 10 is the variation in the interval from pulse to pulse. The variation seems to be fairly regular, with the click interval increasing, then decreasing, and once again increasing, forming a cyclic pattern. The variations depicted in Fig. 10 were typical of all the click trains examined. The click intervals when the target was present were all greater than the two-way transit time for an acoustic signal to travel to the target and back, indicating that the animals emitted each successive pulse after the arrival of the echo signal from the preceding pulse. This fact was also observed by Morozov et al. (1972), in their study. It is interesting to note that the click intervals for the target-absent case were also greater than the two-way transit time had a target been present. This implies that the animals had a general expectation of hearing the target echo at a specific time and may have been performing a "mental time-gating" of the echoes.

Accurate measurements of the time difference between the click intervals and the two-way transit time can only be made if the animal's position with respect to the target is well known. Several experiments have been conducted at the Naval Ocean Systems Center (NOSC) with an animal either stationing in a hoop or in a chin cup so that the distance between animal and target was fixed and could be accurately measured to within several cm. The average click interval per trial for a target distance of 6 m and 1 m is displayed in Fig. 11. In the 1 m case, the animal wore a blindfold. It is clear from the figure that the click intervals were longer than the two-way transit times. The average lag time between the reception of the echo and the transmission of the next pulse was 21.6 ± 10.6 msec at 6 m and 18.6 ± 8.3 msec at 1 m. Morozov, et al., (1972), reported a mean difference of 20 msec, which was fairly constant over a distance of 40 to 4 m as the animal closed in on a target. The click intervals for Tursiops, measured by Evans and Powell (1967) with a simultaneous

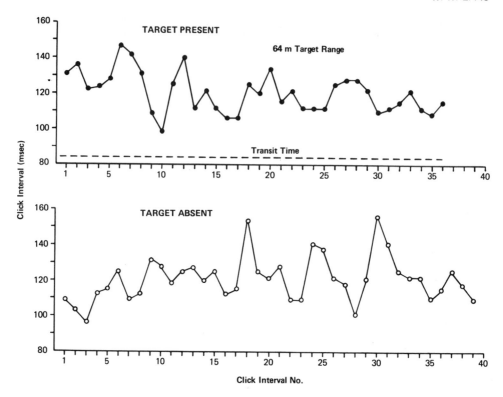

Figure 10.   Examples of the variations in the click interval
             during two echolocation trials.

audio-video monitor system, showed an almost constant mean lag
time of 15.4 msec for target ranges from 1.4 m to 0.4 m. As the
animal closed within 0.4 to 0.03 m the mean lag time decreased
to a minimum of 2.5 msec. Therefore, the mean lag time for
target ranges greater than 0.4 m is approximately 15-22 msec.
This implies that the maximum mean repetition rate should be
between 45 and 67 pulses per second, for targets at ranges
greater than 0.4 m, and that it may take an animal between 15 and
22 msec to process an echo. At the very close ranges (less than
0.2 m), the animal may possibly process several echoes at a time
instead of single echoes, as seems to be the case involving
targets at ranges greater than 0.4 m.

B. Peak Frequency and Bandwidth

     Histograms of the peak frequencies and the 3-dB bandwidths
of the signals used by four animals performing different echolocation tasks in Kaneohe Bay are presented in Fig. 12.

Figure 11.   Average click intervals per trial for two ranges.

The average peak frequency and bandwidth along with the time
bandwidth product of the signals used to determine the histograms
are shown in Table 1. The dolphins, Ehiku and Heptuna, were
involved in a target detection task at ranges greater than 50 m
with a 7.6 cm diameter water-filled sphere as the target. The
signals used by Ehiku and Heptuna were measured with a hydrophone located at 73 m. Ekahi was involved in a shape discrimination study at a range of 6 m with foam spheres and cylinders
as targets. The cylinders had lengths that were less than 5 cm
and diameters that were less than 3.81 cm. Sven was involved in
a target recognition and discrimination task at a range of 6 m
with different diameter and material composition cylinders that
were 17.8 cm in length. The same hydrophone was used in the
measurements of the signals emitted by these four animals.

The frequency histograms indicate that with the exception of
Heptuna, at least 70% of the clicks had peak frequencies
between 110 and 130 kHz. Only 60% of Heptuna's clicks had peak
frequencies between 110 and 130 kHz. The peak frequencies most
favored by Ehiku, Ekahi and Heptuna fell in the interval
between 110 and 120 kHz. The average peak frequencies for these
three animals in Table 1 are similar, with values close to 115
kHz. Sven showed a preference for the frequency interval between
120 and 130 kHz, although its average value of 121.1 kHz is close
to the lower limit of the 120-130 kHz interval.

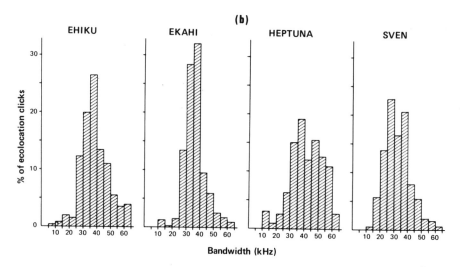

Figure 12. Histograms of peak frequency and bandwidth for four
            Tursiops truncatus performing echolocation tasks in
            Kaneohe Bay.

     The half-power bandwidth of the animal's echolocation signal
is a parameter that has not received much attention in the past.
However, this parameter is an important one since the information
content of a signal is directly related to its bandwidth
(Berkowitz, 1965). The bandwidth histograms indicate that at
least 75% of the signals had bandwidths greater than 25 kHz with
the frequency interval between 30 and 40 kHz being the most common.

Signals that are representative of the average signal used by each animal are shown in Fig. 13.   These signals have peak frequencies and bandwidths that were closest to the average peak frequency and bandwidth displayed in Table 1.   Included in Fig. 13 is the envelope of the autocorrelation function $\rho(\tau)$ of the signals.   The time resolution constant $\Delta\tau$ and the time-bandwidth product $\beta\tau$ included in Fig. 13 and Table 1, are signal parameters commonly used in radar and sonar signal analysis.   The time resolution constant is directly related to the autocorrelation function of the signal and was originally defined by Woodward (1953) as

$$\Delta\tau = \frac{\int\limits_{-\infty}^{\infty} |\rho(\tau)|^2 \, d\tau}{|\rho(0)|^2} = \frac{\int\limits_{-\infty}^{\infty} |U(f)|^4 \, df}{[\int\limits_{-\infty}^{\infty} |U(f)|^2 \, df\,]^2} \tag{1}$$

The rms bandwidth β and duration τ about the energy centroid of the signal are generally used to compute the time-bandwidth product. The parameters β and τ are defined as (Burdic, 1968)

$$\beta^2 = \frac{\int\limits_{-\infty}^{\infty}(f - f_o)^2 \, |U(f)|^2 \, df}{\int\limits_{-\infty}^{\infty} |U(f)|^2 \, df} \qquad \tau^2 = \frac{\int\limits_{-\infty}^{\infty}(t - t_o)^2 \, |\mu(t)|^2 \, dt}{\int\limits_{-\infty}^{\infty} |\mu(t)|^2 \, dt} \tag{2}$$

where the centroids f_o and t_o are defined as

$$f_o = <f> = \frac{\int\limits_{-\infty}^{\infty} f \, |U(f)|^2 \, df}{\int\limits_{-\infty}^{\infty} |U(f)|^2 \, df} \qquad t_o = <t> = \frac{\int\limits_{-\infty}^{\infty} t \, |\mu(t)|^2 \, dt}{\int\limits_{-\infty}^{\infty} |\mu(t)|^2 \, dt} \tag{3}$$

$U(f)$ is the transform of the signal envelope $\mu(t)$.   The time-bandwidth product using the expressions for β and τ given in equation 2 will have a lower limit of $1/(4\pi)$.

Table 1.   The average signal parameters for 4 Tursiops truncatus

animal	f_p (kHz)	BW (kHz)	$\beta\tau$	$\Delta\tau$ (μsec)
Ekahi	114.9±5.0	41.5±12.4	0.12	15.3
Ehiku	115.0±3.9	40.8± 3.6	0.12	11.8
Heptuna	116.2±5.4	45.6± 6.9	0.15	14.5
Sven	121.1±3.2	38.0±12.2	0.12	13.6

Figure 13. Representative signals having approximately the same
peak frequency and bandwidth as the average values of Table 1.

C.   Spectral Adaptations

Spectral adaptation of the echolocation signals used by the
animals in Kaneohe Bay due to the ambient noise condition has been
reported by Au, Floyd and Haun (1978). The peak frequencies in
Table 1 indicate the animals typically used signals with peak
frequencies between 100 and 130 kHz. These peak frequencies are
over an octave higher than previously reported values between 30
and 65 kHz, (Evans, 1973), for Tursiops. Au, et. al., (1974),
noted that the ambient noise in Kaneohe Bay, is dominated by the
presence of snapping shrimp, and has a typical spectrum that is
relatively flat between 40 and 100 kHz, dropping off rapidly for
frequencies above 100 kHz. They estimated that the animals should

be masked by the noise up to a frequency of approximately 137 kHz, so that in the reception of the target echoes, the received signal-to-noise ratio becomes the most critical parameter. Substantial improvement to the signal-to-noise (S/N) ratio could be achieved by operating at peak frequencies above 100 kHz. The average peak frequencies listed in Table 1 are very similar, yet the tasks the animals were involved in were completely different. This suggests the dominant role of the ambient noise on the selection of the proper peak frequency by the animals. It will be shown in Section IID that Ekahi and Sven enjoyed at least a 20 dB greater signal-to-noise ratio than Heptuna and Ehiku, and therefore did not need to operate at frequencies beyond 100 kHz. However, less acoustic energy was required to achieve a given or desired S/N ratio by emitting clicks with frequencies greater than 100 kHz.

The question of whether a dolphin adaptively changes the spectrum and shape of its echolocation signals when examining different targets has generated a considerable amount of speculation and a number of hypotheses with very little data from meticulously conducted experiments. The mean peak frequencies and the standard deviations of the echolocation signals shown in Fig. 14 can be used to examine this question. Ekahi was initially involved in an experiment to determine his ability to discriminate between foam spheres and cylinders. In any given trial, one target out of a set of three foam spheres and five foam cylinders would be presented six meters from the animal stationed in a hoop. The six targets were designed to have similar target strengths to eliminate target strength differences as clues. Ekahi was later used in a discrimination experiment with metallic cylinders as targets. The metallic cylinders were at least 3.5 times longer than the foam cylinders of the initial experiment. Sven was involved in an experiment to test his ability to recognize two specific aluminum cylinders from cylinders made of the different materials listed in Fig. 14. Although the metallic cylinders used with Ekahi were completely different from the foam spheres and cylinders, there was no significant shift in the peak frequency of the signals used by Ekahi. The same could be said for the data collected with Sven. It is important to realize that the data of Thompson and Herman (1975), indicate that for frequencies between 100 and 130 kHz, a bottlenose dolphin cannot perceive pitch differences of less than 8 kHz. Therefore, the small frequency differences shown in Fig. 14 should not be perceived by the animals. The data of Fig. 14 seems to indicate that for the conditions existing in Kaneohe Bay, the animals do not purposefully adjust the peak frequency of the emitted signals to "match" the targets being investigated. Bel'kovich and Dubrovskiy (1977), arrived at a similar conclusion for dolphins kept in tanks at their facility.

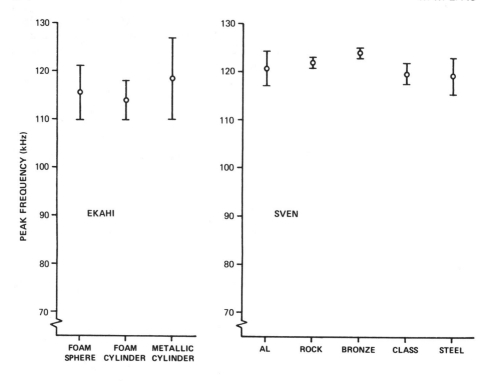

Figure 14. Average peak frequencies for the signals used by Ekahi and Sven while performing in different discrimination experiments.

Changes in the spectral composition of the signals in a pulse train do occur. However, Bel'kovich and Dubrovskiy (1977), argued that any attachment of significance to these changes are most likely due to incorrect interpretation of the results, or are based on insufficient statistics. To illustrate the need to examine dolphin signals from a statistical basis, consider the spectral-time plots for six echolocation trials by Ekahi on the same target in Fig. 15. The signals were measured with Ekahi performing in the shape discrimination experiment, and the spectral-time plots pertain to trials in which a 10.2 cm foam sphere was presented to the animal. In cases (a) through (c), there are obvious changes in the spectral composition of the signals, and observations based on a single or a few displays may lead to the conclusion that the animal was adaptively adjusting his pulses to maximize the information available in the echoes. However, when

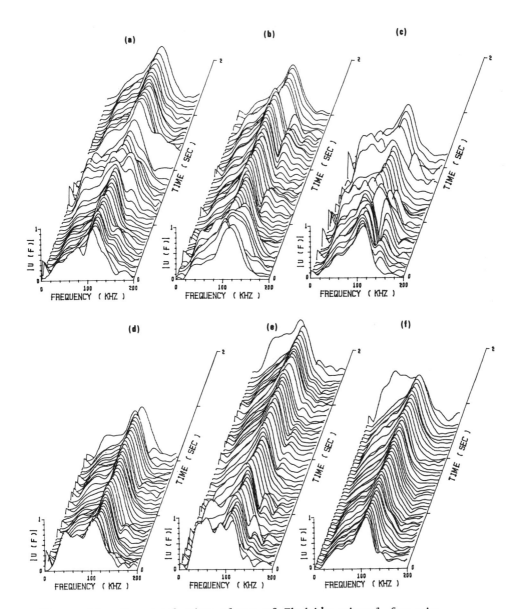

Figure 15. Spectral-time plots of Ekahi's signal for six different trials with the same target present.

comparing plots a through c, there does not seem to be any consistent pattern in which the signals were being altered.
Furthermore, the animal also emitted signals that were fairly
consistent and steady, as can be seen in plots d through f. Yet
all of these spectral-time plots were for the same target. The
spectral changes may be due to the animal turning momentarily
away from the target, internal steering of the beam even when
its rostrum was directed to the target, fluctuations due to its
click mechanism, or lapses in the attention given by the animal
during a trial.

It is possible that spectral adaptations were not observed
for these animals because of the dominant influence of the
ambient noise condition. Since the animals preferred to use
signals with peak frequencies between 110 and 130 kHz, any changes
or frequency adjustment within this range may have insignificant
effects on the echoes. In an environment where an animal is not
masked by the noise, and can achieve approximately the same S/N
ratio per unit of effort at any frequency between 30 and 130 kHz,
spectral adaptations may possibly be present.

D. Click Source Levels

The source level of the echolocation signals used by the
animals should depend on the S/N ratio of the received echoes
that the animals require or desire to have in order to perform a
given task. The SPL of the echoes from a target can be expressed
by the equation

$$EL(dB) = SL(dB) - 40 \log r - 2\alpha(f_p)r + TS(dB) \qquad (2)$$

Where: EL =   echo level in dB

SL =   source level in dB

r =   target range

$\alpha(f_p)$ =   absorption coefficient at the
peak frequency of the signal

TS =   target strength of the target
in dB

Equation 2 simply states that the amplitude of the return echo is
equal to the amplitude of the outgoing signal minus the propagation and reflection losses. The peak-to-peak source level as
a function of the total loss due to propagation and reflection
for five different animals under different circumstances is
depicted in Fig. 16. The data show that there is a general

Figure 16. Peak-to-peak source level as a function of the total
acoustic energy loss for five different animals
performing different echolocation tasks.

increase in the source level as the total losses increase. However, the source level does not increase at the same rate as the
total loss. This indicates that the animals seem to prefer to
operate at a high S/N ratio. The decrease in the loss in moving
from case i to case a is over 62 dB, and the corresponding
decrease in the source level was only 12 dB.

The results of Fig. 16 also seem to indicate that range
has a more powerful influence on the source level than target
strength. In cases a and e, with both targets at 6 m, there is
a 26 dB difference in the target strength but only a 4 dB
difference in the source level. In cases c and e involving the
same animal, the difference in target strength is 10.1 dB,
whereas the difference in the source level was only 1 dB. These
small changes in source level with target strength suggest that
the animals operated at a high S/N ratio when the targets were
at 6 m and changes in the target strength were not large enough
to cause appreciable changes in the source level. There is
hardly any difference in the source level for cases a through e
with the targets at 6 m. However, when the target range increased causing large changes in the total loss, the source level
also increased considerably.

The maximum average peak-to-peak source level in a trial was
recorded at 227.6 dB for one of the trials in case h of Fig. 16.
The largest single click measured was 230 dB, emitted by Heptuna
in case h. These source levels may seem inordinately high,
especially when comparing with conventional man-made sonars, and
considering the amount of energy required to project high level
signals into the water. However, one must realize that peak-to-
peak levels are being considered for the animals, the signals are
very short, and the beam is relatively narrow. The rms pressure
for these dolphin signals is typically 15 to 20 dB lower than the
peak-to-peak amplitudes. For comparing transient signals similar
to dolphin signals, it is more meaningful to refer to the energy
flux density of the acoustic wave than to the sound pressure
level (Urick, 1967). The energy flux density is defined as the
time integral of the instantaneous intensity. For signals
reported in case h of fig. 16, the average energy flux density was
167.4 ± 1.3 dB re $(1 \; \mu Pa)^2$s. The rms pressure of a cw pulse tone
that would have an equivalent energy flux density as the average
echolocation signal of case h can be expressed as

$$SPL(dB) = 167.4 - 10 \log (t) \qquad\qquad (3)$$

For a 10 msec long pulse, an rms source level of 187.4 dB re 1 µPa
would have the same energy flux density as the dolphin signals of
case h of Fig. 16.

III. DISCUSSION

There are many difficulties inherent in the area of measurement and analysis of echolocation signals used by dolphins, and

therefore there are large gaps in our understanding of these
signals and how they are used. Most of the results presented
in this review are averaged kinds of results. Within the range
of average characteristics, there are often many variations that
are not explainable or easily characterized. One of the major
problems in studying dolphin signals involves the sheer mass of
data that is collected in just one recording session. We typically record approximately 60 to 70 trials during a recording
session, with each trial resulting in anywhere from tens of
clicks to several hundred. If these numbers are compounded by
several days of recording, the total amount of data recorded per
animal per experiment can be overwhelming. It generally requires
several orders of magnitude more time and effort to digitize the
analog tapes, analyze the digital information, and finally display the results. Imagine having 60 to 70 spectral history and
waveshape plots like those shown in Fig. 1, for each recording
session. Added to these plots could be the frequency distribution
plots and graphs of the variations of certain parameters associated with the signals on a click-to-click basis.

The approach taken in the measurement and analysis of
dolphin signals is one in which the animals were allowed to
achieve a solution or solutions to a specific sonar problem before
the signals were recorded. Diercks (1972), suggested that a more
meaningful approach would be to vary the animal's sonar problem
and monitor the development of its solution, which would enable
us to understand the signal design utilized by the animal in
order to achieve a solution. This approach is universally
recognized as having considerable merit, but has not been used in
the United States because of the cost that would be involved. It
is simpler to record a short sequence after stabilization in the
animal's performance is achieved. The alternative would be to
perform many recordings over an indeterminate period during the
development of the solution. A feature that would be desirable
in the second approach is to have the capability of real-time or
near real-time analysis of the results. However, before the
second approach is even considered, effective ways of handling the
mass of data that would be collected and require analysis must be
devised.

For the present, we are confined to results that at best can
only approximate the acoustic field of these animals. A dolphin
is not like a standard transducer, which can be positioned precisely in a given location and excited in a desired manner by an
electrical signal. Instead, we are dealing with a relatively
large aquatic animal often weighing in excess of 130 kg, and
operating in an environment which is not conducive for tightly
controlled acoustic experiments. Even if good physical control
could be exercised over a dolphin, other factors such as the
possibility of internal beam steering, behavioral control,

motivation, attention spans, and other psychological influences must be considered. If an animal has the freedom of exciting its source region in a spatially non-uniform and irregular manner, it could cause considerable variations in the signal within a click train on a click-to-click basis. The attachment of significance to these variations or any variations within any click train is difficult since it is not possible to know if fluctuations exist because the animal decided to momentarily "look" at something else besides the target, or vary the degree of attention paid to each echo. Fluctuations in the signals may also be present due to changes in the degree of effort expended by the animal to produce the pulses in the train. Our experience has shown that these animals tend to optimize their behavior to accomplish a task using the least amount of energy as possible. In an echolocation experiment involving noise masking, and using two separate animals, when the masking level is high making the task difficult or impossible, the animals would often go through the motions of echolocating and responding without emitting a single click, or emit only a small fraction (less than 20%) of the number of clicks they would otherwise use. All of these difficulties are inherent when dealing with a complex creature such as a dolphin, and can be overcome only by continual refinements in methodology, behavioral control, and acoustic measurements and analysis.

REFERENCES

Au, W. W. L., Floyd, R. W., Penner, R. H., and Murchison, A. E., 1974, Measurement of echolocation signals of the Atlantic bottlenose dolphin, Tursiops truncatus Montagu, in open waters, J. Acoust. Soc. Am., 56:1280.

Au, W. W. L., Floyd, R. W., and Haun, J. E., 1978, Propagation of Atlantic bottlenose dolphin echolocation signals, J. Acoust. Soc. Am., 64:411.

Barta, R. E., and Evans, W. E., 1970, Private communications in Fish, Johnson, and Ljungblad (1976).

Bel'kovich, V. M., and Dubrovskiy, N. A., 1977, Sensory bases of Cetacean orientation, U.S. Joint Publication Research Service JPRSL/7157, May 27.

Burdic, W. S., 1968, "Radar Signal Analysis", Prentice-Hall, Inc., Englewood Cliff, N. J.

Diercks, K. J., Trochta, R. T., Greenlaw, C. F., Evans, W. E., 1971, Recording and analysis of dolphin echolocation signals, J. Acoust. Soc. Am., 49:1729.

Diercks, K. J., 1972, Biological sonar systems: a bionics survey, Applied Research Laboratories Technical Report N° 72-34 (ARL-TR-72-34), 190.

Dubrovskiy, N. A., and Zaslavskiy, G. L., 1975, Skull bone, dolphin
    sounding pulses,"Dolphin Echolocation" (Joint U.S. publica-
    tion Research Service, JPRS 65777, Sept. 79).
Evans, W. E., Sutherland, W. W., and Beil, R. G., 1964, The direct-
    ional characteristics of delphinid sounds, in: "Marine Bio-
    Acoustics,"W. N. Tavolge, ed., Pergamon Press, New York.
Evans, W. E., and Powell, B. A., 1967, Discrimination of different
    metallic plates by an echolocating delphinid, in: "Proc.
    Symp. Animal Sonar Systems: Biology and Bionics", R. G. Bus-
    nel, ed., Laboratoire de Physiologie Acoustique, Jouy-en-Jo-
    sas, France.
Evans, W. E., 1973, Echolocation by marine delphinids and one species
    of fresh-water dolphin, J. Acoust. Soc. Am., 54:191.
Evans, W. E., and Maderson, P. F. A., 1973, Mechanisms of sound
    production in delphinid cetaceans: a review and some anatom-
    ical considerations, Am. Zool., 13:1205.
Fish, J. F., Johnson, C. S., and Ljungblad, K. K., 1976, Sonar tar-
    get discrimination by instrumented human divers, J. Acoust.
    Soc. Am., 59:602.
Hammer, C., and Au, W. W. L;, 1978, Target recognition via echo-
    location by an Atlantic bottlenose porpoise (Tursiops trun-
    catus), J. Acoust. Soc. Am., 64, Suppl. N° 1:587.
Hollien, H., Hollien, P., Caldwell, D. K., and Caldwell, M. C.,
    1976, Sound production by the Atlantic bottlenose dolphin
    Tursiops truncatus, Cetology, 26:1.
Morozov, B. P., Akapiam, A. E., Burdin, B. I., Zaitseva, K. A.,
    Sokovykh, Y. A., 1972, Tracking frequency of the location
    signals of dolphins as a function of distance to the target,
    Biofizika, 17:139.
Murchison, A. E., and Penner, R. H., 1975,Open water echolocation
    in the bottlenose dolphin (Tursiops truncatus): metallic
    sphere detection threshold as a function of distance, in:
    "Proceedings of Conference on the Biology and Conservation
    of Marine Mammals," University of California, Santa Cruz, Ca.
Murchison, A. E., 1976, Range resolution by an echolocating dolphin
    (Tursiops truncatus), J. Acoust. Soc. Am., 60:S5.
Nachtigall, P. E., Murchison, A. E., and Au, W. W. L., 1978, Dis-
    crimination of solid cylinders and cubes by a blindfolded
    echolocating bottlenose dolphin (Tursiops truncatus), J.
    Acoust. Soc. Am., 64, Suppl. 1:587.
Norris, K. S., Prescott, J. H., Asa-Dorian, P. V., and Perkin, P.,
    1961, Experimental demonstration of echolocation behavior
    in the porpoise Tursiops truncatus (Montagu), Biol. Bull.,
    120:163.
Norris, K. S., and Evans, W. E., 1966, Directionality of echolocation
    clicks in the rough-tooth porpoise, Steno bredanensis (Lesson)
    in: "Marine Bioacoustics", W. N. Tavolga, ed., Pergamon
    Press, New York.
Norris, K. S., Evans, W. E., and Turner, R. N., 1967, Echolocation
    in an Atlantic bottlenose porpoise during discrimination,

in: "Proc. Symp. Animal Sonar Systems: Biology and Bionics",
R. G. Busnel, ed., Laboratoire de Physiologie Acoustique,
Jouy-en-Josas, France.

Norris, K. S., Dormer, K. J., Pegg, J., and Liese, G. J., 1971, The
mechanisms of sound production and air recycling in porpoises:
a preliminary report, in: "Proc. 8th Annual Cong. Biol. Sonar
and Diving Mammals".

Norris, K. S., and Harvey, G. W., 1974, Sound transmission in the
porpoise head, J. Acoust. Soc. Am., 56:659.

Romanenko, Y. V., 1973, Investigating generation of echolocation
pulses in dolphins, Zool. Zh., 11:1698, JPRS 61553.

Romanenko, Y. V., 1974, Physical fundamentals of bioacoustics,
Fizicheskiye Osnovy Bioakustik, JPRS 63923, Moscow.

Schevill, W. E., and Watkins, W. A., 1966, Sound structure and directionality in Orcinus (Killer Whale), Zoologica, 51:71.

Schusterman, R. J., Kersting, D. A., and Au, W. W. L., 1979, Stimulus control of echolocation pulses in Tursiops truncatus,
this volume.

Thompson, R. K. R., and Herman, L. M., 1975, Underwater frequency
discrimination in the bottlenose dolphin (1-140 kHz) and
human (1-8 kHz), J. Acoust. Soc. Am., 57:943.

Urick, R., 1967, "Principle of Underwater Sound", McGraw-Hill, New
York.

Wood, F. G., 1964, "Marine Bioacoustics", W. N. Tavolga, ed., Pergamon Press, New York.

Woodward, P. M., 1953, "Probability and Information Theory with
Applications to Radar", Pergamon Press, New York.

ACOUSTICS AND THE BEHAVIOR OF SPERM WHALES

William A. Watkins

Woods Hole Oceanographic Institution

Woods Hole, Massachusetts 02543

The first good recordings of underwater sperm whale sounds that we have were made 27 years ago (22 April 1952, R/V Caryn at 38°29'N, 69°29'W) and five years later the sounds were identified definitely and described (Worthington and Schevill 1957). At every opportunity since then, we have stopped with these whales and listened underwater (Watkins 1977). Repeated observations have gradually sorted out many of the sounds. The overlapping clatter of click series that are usually heard in the presence of these whales have given way with time to acoustic sequences that apparently have characteristics identifiable with individuals. The unorganized welter of sound can be simplified to series of pulses that are related and that can sometimes be traced for hours. Whales underwater have been located and tracked by their own sounds. Yet with all this, the sounds are still not well enough correlated with the whales' actions to provide anything but glimpses into the role that acoustics plays in the behavior of sperm whales.

The literature dealing with the underwater sounds from sperm whales shows the sequence of our perception of that behavior: The sounds were described first by Worthington and Schevill (1957). A sample recording and analysis of sperm whale sounds was published by Schevill and Watkins (1962). A short description was given by Schevill, Backus, and Hersey (1962), and Backus and Schevill (1966) presented a more detailed analysis of these sounds. A variety of sounds were reported in the presence of sperm whales by Perkins, Fish, and Mowbray (1966). A description of sounds produced by these whales during a whale hunt was given by Busnel and Dziedzic (1967). Sonobuoy measurements of the sound characteristics were made by Dunn (1969) and Levenson (1974).

Norris and Harvey (1972) suggested mechanisms for sound production. Arrays of hydrophones (Watkins and Schevill 1972,
Watkins 1976) were used with sperm whales to study acoustic responses and behaviors including their reaction to pinger sounds
and sub-surface swimming and diving (Watkins and Schevill 1975),
spatial distribution underwater (Watkins and Schevill 1977a),
and exchanges of stereotyped pulse sequences (Watkins and
Schevill 1977b). Diving behavior of whales being chased during
marking operations was assessed by means of sonar equipment and
reported by Lockyer (1977). The study of acoustic behavior of
sperm whales was reviewed by Watkins (1977).

   The characteristics of sperm whale sounds, the comparisons
with equivalent sounds of other odontocetes and the behavioral
associations that are discussed here summarize our understanding
to date, both from published and as yet unpublished observations
at sea.

METHODS

   Whenever we have located sperm whales at sea, either acoustically or by sight, we have tried to move upwind, stop engines,
and listen underwater. Several hundred hours of sperm whale
sounds have been recorded, from both single hydrophones and
three-dimensional arrays of hydrophones. We have attempted to
correlate the sounds with any behaviors that were observed,
usually short glimpses of parts of whales occasionally seen at
the surface.

   The acoustic recordings have been analyzed for spectra of
their sounds and for relationship to other sounds. Sound sequences from groups of whales and from individuals have been
traced over several hours of listening. Array data has been
analyzed for location of nearby sound sources and for direction
to more distant sources. Intensities and frequency relationships
have been compared. A wide spectrum of listening gear has been
utilized, including narrowband sonar systems, medium bandwidth
(30 kHz) gear, and calibrated wide bandwidth (to 200 kHz) instrumentation equipment. Analysis systems were designed to permit detailed scrutiny of particular sounds or sequences of sound,
and included display by means of oscillographs for time series,
sound spectrographs for frequency comparisons, and computer
analyzers for location analyses. The published studies have
specified equipment and techniques used.

CHARACTER OF SPERM WHALE SOUNDS

   The sounds we have heard from sperm whales have always been
impulsive—clicks at a variety of repetition rates. These were
sharp onset, broadband pulses with energy as high as 30 kHz or

more, when the whales were within 20 m. From such nearby whales,
major emphases in frequency were often found in the 10 to 16
kHz range. With distance, of course, the higher frequencies
were attenuated by the environment so that at 2 km most of the
audible energy was in frequencies below 5 or 6 kHz, with apparent emphases at 2 to 4 kHz. After listening for a time to a
particular sperm whale, it was possible to judge its relative
distance from the hydrophone by an assessment of the amount of
high-frequency energy audible in the click sounds. The lower
frequency content of sperm whale clicks was variable but usually
extended well below 100 Hz. Secondary pulse structure in the
clicks often provided strong low-frequency energy. We have not
found much directionality in the clicks of sperm whales and we
have found no differences that could be attributed to depth.

The level of sperm whale sounds was found to be highly variable. As much as 26 dB difference was observed between sounds
a few seconds apart in the same series, with approximately
equivalent spectra from apparently the same whale. Calculations
of maximum sound levels for sperm whale clicks also were variable on different occasions from 65 to about 80 dB, relative 1
dyne/cm^2. Thus, with substantial energy below 1 kHz, and
favorable local oceanographic conditions, sperm whales could be
heard at distances that often were found to be greater than 10
km. Under the same conditions, it was possible also to barely
hear whales clicking at distances of only a few hundred meters.
These differences were estimated to be as much as 40 dB. The
whales varied the level of their sounds apparently at will.

The duration of individual clicks from sperm whales varied
from 2 to 30 msec or more, depending on the structure of secondary pulses present in many of the clicks (see Backus and
Schevill 1966). The longer sequence of sound in individual
clicks enhanced their low frequency content so that the longer
clicks sometimes blended into the background reverberation.
Multiple sound paths and surface reflections often made detailed
analyses of the sounds difficult to sort out. Clicks from individual whales (equivalent high frequency spectra, same direction to source, and without change in repetition sequence)
sometimes switched from longer reverberant pulses to short clicks
with little internal structure.

The sperm whale clicks were produced occasionally in very
slow sequence, almost singly, one in 5 or 10 sec for sequences
of 2 to 10 or more clicks. Usually, however, the clicks were
heard at repetition rates of 1.5 to 3 per sec, and characteristically repeated at very regular intervals (see Backus and
Schevill 1966). Click series without interruption or much change
in rate have been measured for sequences lasting 20 min or more.

Click series of only 1 to 10 sec were also commonly heard. Longer series were usually produced at a slower and more regular rate,
shorter series were sometimes at fast rates, to 60 or more per
sec. Sometimes the terminal portion of long series and more
often the end of short series were variable and became irregular
in click rate (Watkins and Schevill 1977b).

Click Patterns

The irregular click patterns that sometimes formed the
terminal portions of some of the click series were occasionally
repeated at the end of other click series by the same whales.
These end patterns or codas were repeated sometimes also as
separate stereotyped click sequences. The codas were formed of
3 to 40 or more clicks in about one second and produced in
specific temporal patterns of click repetition. Each coda pattern appeared to be unique for specific whales within acoustic
range over at least a few hours. Though codas were not often
heard, sometimes they were repeated almost exactly, up to 66
times in a 15 min period. Codas were produced in particular
situations and were occasionally heard as exchanges between whales
(Watkins and Schevill 1977b).

Because sperm whales tended to maintain relatively regular
clicking rates during their longer click sequences, it was often
possible to follow the sounds of individuals. During these
sequences, it was possible also to note patterns of reverberation
and reflection as well as relative spectra and click structure
that served to help identify succeeding sound sequences from the
same whales. In addition, analyses of multiple hydrophone array
recordings often provided direction (in the vertical as well as
the horizontal plane) to the sound source. And occasionally,
this could be refined to give a location for the source. Thus,
in spite of irregularity in clicking patterns and silent periods,
we were able sometimes to trace much of the acoustic output of
individual whales over as much as four hours. This is also the
maximum that we have been sure we were still with the same sperm
whales.

Individual whales did not produce sounds continuously.
They often were silent at the surface, then as a dive began,
they would click for a short time after submergence. Of a
group of whales, there was usually one or two that were dominant
acoustically, and clicked more often, more continuously than the
others. But even these whales were silent underwater for periods
of a few seconds to 10 min at a time. Whales sometimes apparently took turns clicking. Sometimes they all clicked at
once, producing as many overlapping click series as there were
whales in the group. And sometimes all were silent for short

periods. Usually, at least one whale in a group (the acoustically dominant whale) could be heard to start clicking again
within one to two minutes. Some whales produced only one or
two identifiable click sequences within a dive cycle of 20 min
or more. Usually, however, when several whales were together
or within a few km of each other, at least a few click series
were audible from each whale.

ACOUSTIC BEHAVIOR

Lone sperm whales out of acoustic range of other whales
seldom produced any audible sounds. But, if sounds could be
heard underwater from distant sperm whales, the whale that was
apparently by itself produced occasional click sequences. A
lone whale, that joined other whales became as loquacious as
the others increasing its acoustic activity as it approached
the group. As whales separated, less sounds were produced.

At the surface, whales in a group together were relatively
silent, and they usually dove within a short time of each other.
They often began their dives at steep vertical angles with flukes
raised above the water as they started down. From our four-
hydrophone array data, we found that the whales often began
clicking at 5 to 25 m depth, their vertical dive angles were now
only 10° to 15°, and their swimming speeds were 2 to 4 km per
hour. Underwater, the whales invariably spread out so that received angles to the sound sources indicated a widening distribution. The whales separated as they dove and apparently occupied a large volume of water, perhaps foraging as they maintained
separations of several hundred meters (Watkins and Schevill
1977a). Maximum dive depths were estimated at 500 to 800 m.
Whales from each of the different source directions periodically
clicked for a few minutes usually at slow (2 to 3 per second),
regular rates, often overlapping with the click sequences of
others. Though widely separated underwater, as the whales came
toward the surface, the sound source directions also came together and merged. Then the whales reappeared at the surface
within a short distance of each other, usually less than 50 m.

Sperm whales that passed close to our drifting ship or near
the buoyed hydrophone cables have often produced long series of
relatively rapid clicks (25 to 50 per sec) usually at a very
regular rate. The endings of such series have sometimes changed
to irregular patterns of click repetition with variable rates and
missed beats. The last few clicks have sometimes been repeated
at the same repetition patterns as separate codas or as the same
coda at the end of another click sequence. During these close
approaches by sperm whales, click repetition rates usually have
not varied relative to distances from obstacles or other whales.

One whale came toward the ship and passed a few meters below the keel on its way down for a dive. Its click series was rapid and very regular with little obvious variation, starting at the surface and continuing till the whale was well away from the ship and our cables.

Why Do Sperm Whales Click?

Our own bias has assumed echolocation for sperm whales to conform with the demonstrated capabilities of many other odontocetes, and it would seem convenient for them if they did. The spectra and general form of sperm whale clicks has seemed to support this notion. The frequency range is suited for echolocation of obstacles, and prey. Other features of their sounds, however, do not so easily fit echolocation: (1) Sperm whale clicks do not appear to be highly directional. (2) Click repetition rates are generally very regular and have not varied with changing distances to an approaching obstacle. (3) Sperm whales are silent for long periods, especially when they are alone. (4) The level of their clicks appears to be generally greater than that required for echolocating prey or obstacles. The sounds may be heard over such long ranges that environmental distortion of the sound paths doubtless would be limiting. (5) Individual clicks are usually too long for convenient echolocation (a 30 msec click would obscure echo information from the first 22 m), and the duration of clicks usually remains constant throughout the click series.

Other odontocetes encountered at sea and monitored with the same equipment have shown different characteristics from sperm whales. These have included *Stenella plagiodon*, *Stenella coeruleoalba*, *Delphinus delphis*, *Lagenorhynchus acutus*, *Lagenorhynchus albirostris*, *Grampus griseus*, *Globicephala melaena*, etc. All of these are considered to be echolocators (though most have not yet been tested). (1) Their clicks have been highly directional, so much so that hydrophones at less than 100 m to the side or rear of the animal have often failed to pick up the sounds. (2) Click repetition rates have varied with the distance of the animal to the target. (3) Clicks are produced nearly continuously with only short silent periods. (4) The levels of their clicks seem to match the requirements of the echolocation tasks, and the energies are efficiently distributed. (5) Individual clicks are usually short and become even shorter at higher repetition rates.

Though their sounds are not arranged in ways that would seem convenient for echolocation, nevertheless, sperm whales may utilize the echo information. Echoes from prey and obstacles at some ranges may be heard between steady outgoing pulses, and the rate of clicking may be modified enough to accommodate some changing distances. Increasing click repetition rate with shortening distance to a target (echolocation run) is commonly heard from other echolocating odontocetes, but we have only heard a similar sequence once from sperm whales--a recording by Leanne Hinton and Kenneth Bloom in 1968 off northern Chile (Norris and Harvey 1972). Other echo information such as distances from the surface or bottom probably could also be noted, as well as general details of sound scattering layers nearby.

The acoustic behavior of sperm whales, though not apparently convenient for echolocation, does seem to fit a context of communication, perhaps to keep in contact with other sperm whales underwater. A number of our observations make this plausible: the exchange of codas, the relative silence of whales at the surface together, the increase in clicking underwater from widely separated whales, their return to the surface together, the silence of lone whales, the usual composition of the sounds that makes click sequences audible and distinctive even at a distance.

Probably, both echo information and communication are a part of their acoustic behavior. Our observations seem to relate the sounds more clearly to social interaction, but we may also be dealing with a different bio-sonar system for sperm whales.

Acknowledgements

This work has been supported by the Oceanic Biology Programs of the Office of Naval Research, N00014-74-C0262 NR083-004. The analyses and observations of sperm whales at sea have been shared by many, particularly by William E. Schevill, whose work in 1949 began the scientific investigation of whale sounds. He participated directly in most of the sperm whale observations and has reviewed this manuscript. I am grateful for his helpful encouragement. The manuscript has been prepared by Karen E. Moore and Shirley Waskilewicz. This is contribution No. 4252 from the Woods Hole Oceanographic Institution.

REFERENCES

Backus, R. H., and Schevill, W. E., 1966, Pyseter clicks, in:
    "Whales, Dolphins, and Porpoises", K. S. Norris, ed., Univ.
    of California Press, Berkeley.
Busnel, R. G., and Dziedzic, A., 1967, Observations sur le comporte-
    ment et les émissions acoustiques du cachalot lors de la
    chasse, Bocagiana, 14:1.
Dunn, J. L., 1969, Airborne measurements of the acoustic character-
    istics of a sperm whale, J. Acoust. Soc. Am., 46:1052.
Levenson, C., 1974, Source level and bistatic target strength of
    the sperm whale (Physeter catodon) measured from an oceano-
    graphic aircraft, J. Acoust. Soc. Am., 55:1100.
Lockyer, C., 1977, Observations on diving behaviour of the sperm
    whale Physeter catodon, in: "A Voyage of Discovery",
    M. Angel, ed., Pergamon Press, Oxford.
Norris, K. S., and Harvey, G. W., 1972, A theory for the function of
    the spermaceti organ of the sperm whale (Physeter catodon L.)
    NASA Special Publication 262:397.
Perkins, P. J., Fish, M. P., and Mowbray, W. H., 1966, Underwater
    communication sounds of the sperm whale, Physeter catodon,
    Norsk Hvalfangst-Tidende, 55(12):225.
Schevill, W. E., Backus, R. H., and Hersey, J. B., 1962, Sound pro-
    duction by marine animals, in: "The Sea, Vol. 1", M. N. Hill,
    ed., J. Wiley and Sons, New York.
Schevill, W. E., and Watkins, W. A., 1962, "Whale and Porpoises
    Voices. A Phonograph Record", Woods Hole Oceanographic
    Institution, Woods Hole, Mass., 24 pages and phonograph
    record.
Watkins, W. A., and Schevill, W. E., 1972, Sound source location by
    arrival-times on a nonrigid three-dimensional hydrophone
    array, Deep-Sea Research, 19:691.
Watkins, W. A., and Schevill, W. E., 1975, Sperm whales (Physeter
    catodon) react to pingers, Deep-Sea Research, 22:123.
Watkins, W. A., 1976, Biological sound-source locations by computer
    analysis of underwater array data, Deep-Sea Research, 23:175.
Watkins, W. A., 1977, Acoustic behavior of sperm whales, Oceanus,
    20:50.
Watkins, W. A., and Schevill, 1977a, Spatial distribution of Physeter
    catodon (sperm whales) underwater, Deep-Sea Research, 24:693.
Watkins, W. A., and Schevill, W. E., 1977b, Sperm whale codas,
    J. Acoust. Soc. Am., 62:1485.
Worthington, L. V., and W. E. Schevill, 1957, Underwater sounds
    heard from sperm whales, Nature, 180:291.

CLICK SOUNDS FROM ANIMALS AT SEA

William A. Watkins

Woods Hole Oceanographic Institution

Woods Hole, Massachusetts   02543

Echolocation signals from passing cetaceans (*Delphinidae, Phocoenidae*, etc.) at sea generally exhibit somewhat different characteristics from those of the same species in captivity that are involved in discrimination experiments in echolocation. Presumably, these differences are partly because the echolocation task at sea would have been primarily for orientation (as suggested by Norris, Evans, and Turner 1967; Norris 1969). Generally, the clicks from groups of cetaceans passing within a few meters of a hydrophone at sea are (1) relatively low frequency (rarely energy above 50 kHz from any species), (2) mostly low level, and (3) often at slow, variable repetition rates. Under these circumstances, the clicks may be used primarily in social contexts.

Such click sequences from individual animals usually are very difficult to follow for more than a few seconds. It is seldom possible to know which animal produced the sounds that were heard, how deep it was, how far away it was, or what its orientation to the hydrophone might have been. Overlapping sounds from several sources often occur as other animals are apparently stimulated to sound production by the sounds they hear. Overriding louder squeals or bursts of clicks from nearby animals often overload and distort the recording of the lower level clicks.

The highly directional character of echolocation clicks, so noticeable for captive animals (Norris et al. 1961; Schevill and Watkins 1966; Norris and Evans 1967; and Au, Floyd, and Haun 1978) also is very evident in listening to animals at sea. As groups of delphinids pass and individuals approach a hydrophone within 20 m or less, the click sounds often are rich in frequencies to 30 or

50 kHz and the levels increase predictably with shortened distances.
Then, as an animal changes orientation relative to the hydrophone,
the high frequencies decline sharply and the levels of even the
lower frequency click components may drop off rapidly even though
distances to the animal may continue to shorten as it comes abreast.
When these cetaceans investigate hydrophones and cables at close
range, their echolocation clicks dramatically increase in apparent
intensity and bandwidth and often overload linear recording
circuits. Frequencies to 150 kHz may be heard at these times.

These differences in sound level and frequency spectra that
depend on the orientation of the animals also make it difficult to
locate or to track animals by their own sounds. With multiple
hydrophones separated by only 30 m, it is the lower frequency
click components only that are picked up by the hydrophones to the
side or below or to the rear of the animal. These transient lower
frequency components of clicks are easily masked by ambient noise,
and their level often is not high enough to be heard well on all
hydrophones in an array.

It is seldom possible, therefore, to describe accurately the
echolocation clicks of cetaceans at sea. Their general characteristics and similarities are noticeable, but specific sound
structures are described best from more controlled situations. For
example, we were aware of the click sounds of *Phocoena phocoena*
(harbor porpoise) from encounters with them at sea, but it was not
until a sequence of four increasingly more controlled experiments
with captive animals that additional details of their click sounds
were available (Busnel, Dziedzic, and Andersen 1963; Schevill,
Watkins, and Ray 1969; Dubrovskii, Krasnov, and Titov 1970; Møhl
and Andersen 1973). The older acoustic measurements with limited
bandwidth, lower frequency equipment did not detect high frequency
emphases and consequently assessments of source level were often
too low, though lower frequency components were accurately represented (Watkins 1974). Some of the differences in the characteristics of echolocation clicks that have been reported for the same
species of cetaceans can be attributed to differences in equipment bandwidth, as well as differences in behavior, and in
echolocation tasks.

Comparisons of cetacean sounds at sea also must account for
their voluntary control of sound production. Frequency, spectral
emphasis, pulse rate and sound level can be varied by the animal
at will (demonstrated particularly through hydrophone array
analyses, as in Watkins and Schevill, 1974). Depending on behavioral
activity and level of interest, the click sounds from the same group
of animals can vary widely. The longer duration tonal (squeal,
whistle) and pulsed (squawk) sounds from cetaceans often are
species-specific and appear to be louder at sea than their click

sounds. These longer sounds are probably not used for echolocation, and they are not considered here.

The assessments of click characteristics from animals at sea
outlined below are mostly from our unpublished observations, based
in large part on wide bandwidth (200 kHz) recording using three-
dimensional arrays with 30-m or more separation between hydrophones (Watkins and Schevill 1972). Typically, our ship is
stopped and quieted upwind or in the path of the animals so that
they may pass close to the hydrophones. In most instances the
animals are in groups of three to several hundred, depending on
species and behavior, they are apparently undisturbed, and usually
they ignore the ship and its hydrophone array. The animal sounds
appear to be representative of behaviors that do not require much
echolocation effort; therefore, these click sounds may be primarily
for social contact. Some of the distinctions in click structure
stand out and appear to be species specific, while other click
variations are more readily explained by circumstantial or behavioral differences. Care in obtaining undistorted recordings (particularly difficult in high frequency and transient recording) and
in utilizing equivalent listening systems for each encounter vali-
dates such comparisons.

Clicks heard at sea from many of the smaller cetaceans are
not easily separable by species, particularly true for most delphinids (4.5 m or less), even though their squeals (whistles) and
pulsed sounds often are species-specific. *Delphinus delphis*,
Grampus griseus, *Lagenorhynchus acutus*, *L. albirostris*, *L. obscurus*,
Stenella coeruleoalba, *S. longirostris*, *S. plagiodon*, *Tursiops*
aduncus, and *T. truncatus* that have passed by our hydrophones at
sea have produced clicks that were superficially very much alike.
Differences usually were attributable to variable orientations of
the animals. Anomalous click spectra (such as those from *Lageno-*
rhynchus australis in an exceptionally quiet background, Schevill
and Watkins 1971) may represent responses to behavioral or environmental situations. The clicks heard at sea from delphinids passing
by our hydrophones were short, broadband, rapid rise-time pulses
with maximum frequencies at 30 to 150 kHz, and variable frequency
emphases. Click terminations were often lost in reverberation and
surface reflections, and though the higher frequency components
were produced in less than 0.5 msec, there usually was a longer
total pulse envelope with considerable energy in frequencies well
below 500 Hz. Levels varied widely with orientation and behavior,
but maximum source levels under these circumstances were about
60 dB at 1 m (re 1 dyne/cm^2), about half the dB level measured in
discrimination echolocation by Au, Floyd, and Haun (1978). The
differences between echolocating captives and clicks from the same
species at sea have been noted also by other investigators (Busnel
and Dziedzic 1966, for example), and for other species (*Steno*

bredanensis, Norris and Evans, 1966).

Under the same "at sea" conditions as for the smaller species listed above, delphinid species of 5 m or larger produced clicks that often had more low frequency energy and less obvious high frequency emphasis, slower average click rates, and somewhat highter maximum click levels at the lower frequencies. The stronger lower frequency content of the clicks (relative to clicks of the smaller delphinids) was particularly noticeable when two species of two sizes (*Globicephala melaena* and *Stenella coeruleoalba*, for example) passed by the hydrophone within a few seconds of each other. Apparently, a species size gradient (*Tursiops*, *Pseudorca*, *Globicephala*, *Orcinus*) roughly matches the increasing low frequency emphasis or perhaps decreasing high frequency emphasis in the clicks of animals recorded at sea. Differences in the clicks of echolocating captives and those of *Globicephala* at sea were noted by Evans (1973). Smaller animals of larger species also fit the general pattern for their species (Schevill and Watkins, 1966). In at least the largest of these, *Orcinus*, there also is a matching reduction in the high frequency hearing capability (Hall and Johnson, 1972). With the animals' behavior and orientation to the hydrophone, clicks from these larger species also varied widely in repetition rate, in level, and in frequency content. Sometimes the clicks had emphases at particular frequencies, and sometimes they had broader bandwidth characteristics that may have reflected individual preference.

Phocoena phocoena is very difficult to hear at sea. Their click sounds are relatively low level (maximum about 30 dB at 1 m re: 1 dyne/cm^2 when facing a hydrophone) and they have characteristic low frequency emphases often near 2 kHz (Schevill, Watkins, and Ray, 1969); Møhl and Andersen, 1973; Watkins, 1974). High frequency emphases above 100 kHz (Dubrovskii, Krasnov, and Titov, 1970; Møhl and Andersen, 1973) and also noted in *Phocoena* clicks during echolocation experiments. These higher frequency components have not been recorded at sea perhaps because of rapid attenuation and their low level, their high directionality, or like the delphinids, *Phocoena* may not emphasize the high frequencies unless they need to do so. We have recorded somewhat similar sounds from *Phocoenoides dalli* (see also Ridgway, 1966), and from four species of *Cephalorhynchus* (Watkins, Schevill, and Best, 1977). These species have many of the same characteristics in the low frequencies at sea, suggesting that if echolocation were tested, other higher frequency click components also would be found.

Clicks with tonal emphases at specific frequencies from 2 to 20 kHz are produced by *Monodon monoceros*, narwhal (Watkins, Schevill and Ray, 1971; Ford and Fisher, 1978) and in two of the three observations reported, the animals appeared to be feeding and perhaps echolocating. If so, as in *Phocoena*, the clicks probably also have higher frequency emphases (not recorded by the 24 kHz equipment). In contrast, another species of the family Monodontidae, *Delphinapterus Leucas*, the beluga, produced short clicks with broadband characteristics. Our recordings of beluga clicks at sea were with equipment bandwidths of 4 kHz, 30 kHz, and 200 kHz, and all indicate short (often less than 0.1 msec), highly directional clicks with little low frequency energy. The click sounds from passing beluga at sea were not readily detected unless the animal was close to a hydrophone or turned toward a hydrophone (often suddenly overloading the equipment as it turned).

The differences between the click sounds heard from cetaceans passing in the open ocean and the clicks produced by the same species during echolocation experiments appear to be related to differences in (1) behavior - social vs. task motivation, (2) types of echolocation - relative position vs. target discrimination, (3) orientation - seldom facing hydrophone vs. careful positioning, and (4) competing ambient sound. The sounds heard from social groups of cetaceans at sea, therefore, do not demonstrate their acoustic capabilities but are indicative of their usual patterns of sound production. Similarly, analyses of clicks produced during echolocation experiments show the animal's careful adjustment of click formation to suit difficult tasks and meet training requirements, but may not give the acoustic parameters that can be expected during other behaviors.

Acknowledgements

Support for these studies has been from the Oceanic Biology Program of the Office of Naval Research, N00014-74-C0262 NR083-004. Wm. E. Schevill shared in these observations and read the manuscript. We thank F. G. Wood for his review. Karen E. Moore and Elaine M. Ellis have prepared the manuscript. This is contribution N° 4300 from the Woods Hole Oceanographic Institution.

REFERENCES

Au, W. W. L., Floyd, R. W., and Haun, J. E., 1978, Propagation of
    Atlantic bottlenose dolphin echolocation signals, J. Acoust.
    Soc. Am., 64:411.
Busnel, R. G., Dziedzic, A., and Andersen, S., 1963, Acoustique
    physiologie - sur certaines caracteristiques des signaux
    acoustiques du marsouin Phocoena phocoena, L. C. R. Acad.
    Sc. Paris, 257:2545.
Busnel, R. G., Dziedzic, A., 1966, Acoustic signals of the pilot
    whale Globicephala melaena and of the porpoise Delphinus
    delphis and Phocoena phocoena, in: "Whales, Dolphins, and
    Porpoises", K. S. Norris, ed., Univ. California Press, Berk-
    eley.
Dubrovskiy, N. A., Krasnov, P. S., and Titov, A. A., 1971, On the
    emission of echolocation signals by the Azov sea harbor
    porpoise, Akust. Zh., 16:521; Sov. Phys. Acoust., 16:444.
Evans, W. E., 1973, Echolocation by marine delphinids and one species
    of fresh-water dolphin, J. Acoust. Soc. Am., 54:191.
Ford, J. K. B., and Fisher, H. D., 1978, Underwater acoustic signals
    of the narwhal (Monodon monoceros), Can. J. Zool., 56:552.
Hall, J. D., and Johnson, C. S., 1972, Auditory thresholds of a
    killer whale Orcinus orca Linnaeus, J. Acoust. Soc. Am., 51:
    515.
Møhl, B., and Andersen, S., 1973, Echolocation: high-frequency
    component in the click of the harbour porpoise Phocoena ph. L.,
    J. Acoust. Soc. Am., 54:1368.
Norris, K. S., Prescott, J. H., Asa-Dorian, P. V., and Perkin, P.,
    1961, Experimental demonstration of echolocation behavior in
    the porpoise Tursiops truncatus (Montagu), Biol. Bull., 120:
    163.
Norris, K. S., and Evans, W. E., 1966, Directionality of echoloca-
    tion clicks in the rough-tooth porpoise, Steno bredanensis,
    (Lesson), in: "Marine Bioacoustics", W. N. Tavolga, ed.,
    Pergamon Press, New York.
Norris, K. S., Evans, W. E., and Turner, R. N., 1967, Echolocation
    in an Atlantic bottlenose porpoise during discrimination,
    in:"Proc. Symp. Animal Sonar Systems: Biology and Bionics,"
    R. G. Busnel, ed., Laboratoire de Physiologie Acoustique,
    Jouy-en-Josas, France.
Norris, K. S., 1969, The echolocation of marine mammals, in: "The
    Biology of Marine Mammals", H. T. Anderson, ed., Academic
    Press.
Ridgway, S. H., 1966, Dall porpoise, Phocoenoides dalli (True):
    observations in captivity and at sea, Norsk Hvalfangst-
    Tidende, 5:97
Schevill, W. E., and Watkins, W. A., 1966, Sound structure and
    directionality in Orcinus (Killer Whale), Zoologica, 51:71.
Schevill, W. E., and Watkins, W. A., 1971, Pulsed sounds of the
    porpoise Lagenorhynchus australis, Breviora, 366.

Schevill, W. E., Watkins, W. E., and Ray, C., 1969, Click struc-
     ture in the porpoise, Phocoena phocoena, J. Mamm., 50:721.
Watkins, W. A., Schevill, W. E., and Ray, C., 1971, Underwater
     sounds of Monodon (narwhal), J. Acoust. Soc. Am., 49:595.
Watkins, W. A., and Schevill, W. E., 1972, Sound source location
     by arrival-times on a nonrigid three-dimensional hydrophone
     array, Deep-Sea Research, 19:691.
Watkins, W. A., 1974, Bandwidth limitations and analysis of ceta-
     cean sounds, with comments on delphinid sonar measurements
     and analysis, (K. J. Diercks, R. T., Trochta, and W. E.
     Evans, J. Acoust. Soc. Am., 54:200 (1973)). J. Acoust. Soc.
     Am., 55:849.
Watkins, W. A., and Schevill, W. E., 1974, Listening to Hawaiian
     spinner porpoises, Stenella cf. longirostris, with a three-
     dimensional hydrophone array, J. Mamm., 55:319.
Watkins, W. A., Schevill, W.E., and Best, P. B., 1977,Underwater
     sounds of Cephalorhynchus heavisidii (Mammalia: Cetacea),
     J. Mamm., 58:316.

SIGNAL CHARACTERISTICS FOR TARGET LOCALIZATION AND DISCRIMINATION

K. Jerome Diercks

Applied Research Laboratories

University of Texas, Austin, Texas 78712

Included in this review are hypotheses regarding exploitable clue structure in target echoes, analyses of echolocation signal forms and how the characteristics of signals might influence echolocation behavior, and explicit, although heuristic, models of specific echolocation phenomena.

EXPLOITABLE ECHO CLUE STRUCTURE

A target in water may present a complex idiosyncratic response to broadband insonification, a phenomenon of the geometrical and physical (elastic) properties of the target. Targets differing in one or more properties will present different responses (echoes) to the same insonification (Hickling, 1962, 1964; Metsaveer, 1971; Ryan, 1978). A target that is insonified by a signal that is long in duration (relative to the time extent of the target) and swept in frequency will yield an echo whose amplitude fluctuates as the frequency changes (describing the "frequency response" of the target over the swept band), a result of interference between waves reflected from different boundaries or traversing different paths within the scatterer. Locally intense amplitude peaks spanning narrow frequency bands within the swept band of the echo may occur, and it is speculated that these may be detected and exploited by the animal to effect discrimination between differing targets. Also, the amplitude fluctuations within the echo may be periodic with one or more periods idiosyncratic of the target, and differences in periodicities or period patterns may be exploited for target differentiation.

In a purely theoretical work Koshovyy and Mykhaylovs'kyy
(1972) assess the utility of both the local peak and periodicity
clues for differentiating simple geometric forms (cylinders) that
differ in both size and material. They calculate the frequency
response for a specific metallic cylinder (Duralumin) over the
nondimensional frequency range $0<ka<25$ ($ka = 2\pi a/\lambda$, where a =
target radius and λ = wavelength in the surrounding medium).
This response displays a large amplitude peak localized around ka
= 2, with lesser peaks (or well delineated nulls) occurring at
intervals of $\Delta ka \simeq 3$. It is assumed by the authors that cylinders
of other materials will exhibit analogous peaks (at other frequencies) in their responses over the same frequency band--or, by
implication, the band that is relevant to the echolocator--which
differentiate the materials. The occurrence of such peaks may or
may not occur (Hickling, 1962, 1964; Ryan, 1978) rendering this
clue unreliable.

However, these authors further show a linear relationship
between the velocity of a compressional wave in the target
material and the principal (reviewer's emphasis) periodicities of
the target's responses, which does provide for a reliable (and
psychophysically manageable) discriminatory clue. Targets that
differ slightly in the periodicities of their frequency responses
would be more difficult to discriminate than those with larger
differences, a prediction borne out by experiment (Babkin et al,
1971; Dubrovskii and Titov, 1975). Dubrovskii and Titov, with a
bottlenose dolphin T. truncatus, showed a nominal 0.5 ka difference between frequency response periodicities was required for
reliable discrimination between two sets of different metal
spheres.

Diercks et al. (1968, 1975), using recorded swept frequency
(FM) echoes from different metallic spheres, demonstrated the
utilization of these clues--local frequency peaks, and response
periodicities--by human listeners. Their results also indicated
a 0.4-0.5 ka difference in response periodicities required for
reliable discrimination. They further showed that although
discrimination could be effected by the amplitude variations of
the echo envelope (AM-only echo), the task was noticeably easier
when each amplitude peak was uniquely associated with a different
localized frequency band (FM echo).

While this "model" of echo (target) discrimination has
received some experimental support, its validity hinges on the
assumed signal form, a relatively long duration swept frequency
pulse, which is not characteristic of delphinid echolocators (at
least in captivity). It is generally accepted that the delphinid
echolocation signal form is impulsive in nature; specific parameter
values are species dependent (Diercks, 1972; Evans, 1973).

A target that is insonified by an impulsive signal will return a train of image, or at least similar pulses to the receiver, depending upon the structural and acoustical complexity of the scatterer (Diercks and Hickling, 1967; Shirley and Diercks, 1970; Ryan, 1978). This echo form is akin to the impulse response of the target, and the Fourier spectrum of the echo is the frequency response of the target (over the bandwidth of the signal).

It is possible, although unlikely, that the animal performs a time-amplitude analysis of echoes (impulse trains) from differing targets, detects differences in the times of occurrence and relative amplitudes of the pulses making up the echoes from each, and discriminates on this clue structure. It is more likely that the animal performs a time-frequency transformation of the echo yielding a neural equivalent of the target's frequency response which is then assessed in a manner analogous to that postulated above for swept frequency signals, i.e., detection of differences in (transformed) echo "periodicities or period patterns" and of pitch differences between localized amplitude peaks.

Thus, the model advanced by Koshovyy and Mykhaylovs'kyy is applicable to both signal forms, depending on the assumptions made about the transmitter or receiver.

ANALYSES OF SIGNAL FORMS

While the Russian literature on delphinid echolocation abundantly displays diverse signal forms recorded during various echolocation tasks (Ayrapet'yants et al., 1969; Dubrovskiy et al., 1971; Zaslavskii, 1972; Romanenko, 1973, 1974; Ayrapet'yants and Konstantinov, 1974; Golubkov et al., 1975), analyses and characterizations of these have appeared simplistic or do not exist. In general, their results have impacted speculation on the anatomical nature and location of the signal generator. Other results, discussed in context below, have caused or influenced the development of conceptual models of specific echolocation phenomena.

Dziedzic, Escudié, and their colleagues in France have examined signal forms of various species of echolocating dolphins, including the low frequency emissions of Phoecena phoecena (Escudié et al., 1971; Dziedzic et al., 1974; Dziedzic et al., 1975; Dziedzic and Alcuri, 1977; Dziedzic et al., 1977). Using the ambiguity function

$$\chi_s(\tau,\eta) = \int_T S(t)S(\eta t-\tau)dt \quad ,$$

where   S(t) = echolocation signal,

          τ = echo signal delay, and

          η = Doppler coefficient,

these investigators have shown that the generic forms of signals employed by echolocating dolphins--impulsive "clicks" and short duration, frequency modulated "chirps"--are highly tolerant to relative motion between target and emitter, i.e., they are Doppler tolerant. For example, for the impulsive signals of Delphinus delphis, for relative velocities spanning the range -75 to 75 m/sec, much larger than is likely ever to occur in reality, the Doppler coefficient η varies from 0.95 to 1.05, and over this range $\chi_s(\tau,\eta) \simeq \chi_s(0,1)$. Similar results were obtained for a short duration chirp recorded from this species, however, with increased ambiguity in range resolution (Dziedzic et al., 1974).

Analysis of low frequency chirp and click signals from Phoecena phoecena revealed similar Doppler tolerances (Dziedzic et al., 1977). These signals displayed increasing frequency modulation, i.e., signal period decreasing with duration, in contrast to comparably Doppler tolerant signals with decreasing modulation examined by other investigators (Altes, 1974; Altes and Skinner, 1977).

The import of these results is that with a broad bandwidth, e.g., 3 to 4 octaves, phase perturbations of the signal, regardless of its form are ineffectual.

From the study of the low frequency signals of Ph. ph. it was concluded that the FM chirp, in addition to its Doppler tolerance, is characteristically like certain theoretically optimal signals for use in a noisy and/or reverberant environment, implying its use in these conditions, with employment of a relatively much higher frequency band emission when required for increased resolution (Dziedzic et al., 1977).

An acoustical simulation by Dziedzic et al. (1975) of a dolphin-like echolocation click showed effects of spatiotemporal filtering on the signal parameters: the spectrum was narrowed and displaced toward a lower frequency band. The ambiguity surfaces recorded for the signal recorded on the axis of transmission and at an angle of 12° off this axis both revealed a strong tolerance to Doppler, i.e., insensitivity to relative velocities vectored along off-transmission axes.

A study of form recognition by Tursiops truncatus revealed pronounced changes in signal characteristics occurring at a "threshold" distance from the targets (4 m distance from 20 cm diam polygonal plates) (Dziedzic and Alcuri, 1977). A typical signal for animal-target distances greater than 4 m was like a damped sinusoid, or an impulse followed by ringing, with a spectral peak at 50-55 kHz and range resolution (first zero of autocorrelation) of nominal 55 cm. At distances shorter than 4 m the typical signal was like a single sine wave, at much lower frequency, 9-30 kHz, with 3 to 5 times less range resolution, 130-260 cm.

The conclusion reached in this study was that target differentiation occurred during the initial phase of approach, at distances greater than 4 m, using the higher frequency broader bandwidth signal. That the animal significantly degraded its resolution during the final 4 m of approach to the targets is difficult to rationalize. It is perhaps worth noting that the "character" of these results supports the observation by Au et al., (1974) of use of a significantly higher frequency emission by Tursiops at long ranges in open water than that reported by others for this species at relatively shorter distances in pools (Diercks et al., 1971, 1973; however, see also, Golubkov et al., 1975).

MODELS OF ECHOLOCATION PHENOMENA

Livshits (1975) postulates a pattern recognition model for target detection, identification (or differentiation), and ranging that employs a dual channel correlator with a pattern memory in one channel. Echoes from targets at different ranges are crosscorrelated with a delayed stored replica of the emitted signal. Spatiotemporal filtering, Doppler, etc. will distort the echo waveform, leading to correlations less than maximal. At the same time these same echoes are being compared, by crosscorrelation, with stored (remembered) echo patterns in the "identification" channel of the processor. The stored pattern that correlates best with each signal is then crosscorrelated with the delayed stored replica of the emission in the detection channel.

Only echoes whose crosscorrelations exceeded an initial threshold value during the first correlation are strobed for the second correlation, which, if this exceeds a second, higher threshold, constitutes target recognition (through memory). It is possible that the initial correlation of an echo signal with the stored emission could exceed the higher (second) threshold; memory scanning and re-correlation would still need be performed to accomplish recognition.

A corollary procedure in this model would be for the animal to emit one or more of the stored signal patterns identified during the first memory search in hopes of achieving an enhanced,

suprathreshold crosscorrelation of the echo following this second
emission, i.e., a process of "signal design." If the latter
correlation exceeded the higher threshold value, target recognition would again have been accomplished. However, it would not
be unlikely that the correlation with the modified stored
emission might be degraded relative to the first time, leaving
the echolocator uncertain of how to proceed.

Echolocation signal modifications that are apparently target
related have been reported, but the data are equivocal. Although
signal design is an appealing notion, incontestable evidence of
its occurrence in echolocating animals is lacking.

It is possible that this model could be tested by accumu-
lating and examining large numbers of signals obtained using
acoustically known targets to (1) yield a reliable evaluation of
signal variation, and (2) to compare reliable signal variations
with target acoustical characteristics to yield an assessment of
the apparent occurrence and nature of signal "design."

Livshits (1974) has also postulated a model for rapid "beam
scanning" by dolphins based on the assumptions of a swept frequency, or frequency modulated transmission and a dispersive
medium between the signal generator and the radiating surface of
the emitter. Although anatomy is not specifically addressed, it
is implicit that the signal generator is thought to lie near the
nasal plugs or sacs and that the dispersive medium is the melon.

In its simplest expression, the model assumes that the transmission path lengths from the generator to the radiating surface
are unequal, as shown below.

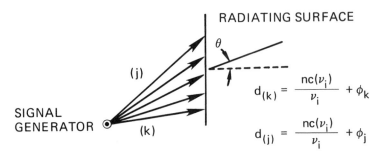

Fig. 1.

This causes at frequency ν_i a "phase gradient" across the aperture of the radiator leading to propagation along an axis oblique to the radiating surface. A change to frequency $\nu_i+\Delta\nu$ will cause a different phase gradient or pattern across the aperture leading to propagation along an axis directed at a slightly different angle from the surface. It is seemingly unnecessary to invoke a frequency dependence of the transmission velocity in the medium (melon), i.e., dispersion, except to bound the phase gradient across the aperture, e.g., $-\pi<\phi\leq\pi$.

Assuming that a narrow beam is formed at each discrete frequency ν_i and radiated in a direction Θ_i, then sweeping frequency from ν_i to ν_k will cause the radiated beam to scan from Θ_i to Θ_k. The scan rate will depend on the rate of frequency change. It follows that only a fractional part of the duration of the swept frequency signal is radiated in any direction Θ--akin to a rotating directed transmission (RDT)--thereby increasing range resolution.

An echo, which is a fractional part of the generated waveform, is crosscorrelated with a delayed stored replica of the of the latter to yield a narrow correlation peak at the locus of ν corresponding to Θ thereby providing a target bearing measure (assuming synchronous transmission), and through pulse compression by correlation, further enhanced range resolution.

While there seems to be evidence for the occurrence of rapid scanning (Bel'kovich and Reznikov, 1971) and of lateral beam steering (Ayrapet'yants and Konstantinov, 1974), the many shortcomings of the model should be evident. The formation of a narrow beam at ν_i directed along Θ_i assumes a steady state across the radiating aperture. The nonsteady state inherent in a swept frequency transmission does not preclude beam sweeping but necessitates an elaborate frequency dependent velocity function for the transmission medium to provide the well behaved beam forming and scanning process described (which very well may exist). The occurrence and characteristics of radiated sidelobes and the ambiguities they might cause are circumvented by assuming that they do not exist (and if they do, phase shading may be employed to suppress them).

The processing scheme seems particularly naive. Crosscorrelation of the expected short duration echo--perhaps only one or two cycles of the swept frequency signal--with the swept frequency signal would result in correlation sidelobes of amplitude nearly equal to the correlation "peak" and, thus, would cause appreciable ambiguity in target bearing resolution. Also, such an echo would have a small time-bandwidth product $\simeq 1$, providing little processing gain and negligible enhancement of range resolution.

If it is accepted that a narrow beam is formed and rapidly scanned to present effectively only a short duration transmission, like a transient, that differs in frequency with direction, which result has been observed in animal studi-s (Diercks, et al., 1971; Romanenko, 1973) and in syntheses (Diercks, et al., 1973; Dziedzic et al., 1975), then the specification of a long duration FM signal is not damaging. Indeed, the model could readily manage either the stability (Diercks et al., 1971) or instability (Romanenko, 1973) of the signals recorded from animals.

The appearance of direction dependent frequency differences is not sufficient to "demonstrate" the model; these would also appear with a broad bandwidth radiation directed along the axis of the radiator (i.e., rostrally in the animal). The model provides impetus for careful observation of apparent and obvious scanning during animal echolocation. Reliable observation of apparent scanning not correlated with related head or body motions would indicate that the transmission model described could be operating and would thus impact many extant concepts and hypotheses regarding the anatomical structures used for and the mechanics of signal generation.

REFERENCES

Altes, R. A., 1974, Study of animal signals and neural processing with applications to advanced sonar systems, Rept. NUC TP 384, San Diego, Ca. 92132.

Altes, R. A., and Skinner, D. P., 1977, Sonar velocity resolution with a linear-period-modulated pulse, J. Acoust. Soc. Am., 61:1019.

Au, W. W. L., Floyd, R. W., Penner, R. H., and Murchison, A. E., 1974, Measurement of echolocation signals of the Atlantic bottlenose dolphin, Tursiops truncatus, Montagu, in open waters, J. Acoust. Soc. Am., 56;1280.

Ayrapet'yants, E. S., Golubkov, A. G., Ershova, I. V., Zhezherin, A. R., Zvorykin, V. N., Korolev, V. I., 1969, Echolocational differentiation and description of radiated pulsations in dolphins, Doklady Akad. Nauk SSSR, 188:1197.

Ayrapet'yants, E. S., and Konstantinov, A. E., 1974, "Ekholokatsiya V Prirode", JPRS 63328-1, 2 , Leningrad.

Babkin, V. P., Dubrovskiy, N. A., Krasnov, P. S., Titov, A. A., 1971, Discrimination of the materials of spherical targets by the bottlenose dolphin, Tr. Akust. Inst., 17:80.

Bel'kovich, V. M., and Reznikov, A. Y., 1971, What's new in dolphin sonar?, Priroda, 11:84.

Diercks, K. J., 1972, "Biological Sonar Systems: a Bionics Survey," Applied Research Laboratories Tech. Rpt. N° 72-34 (ARL-TR-72-34), 190 (5 September, 1972).

Diercks, K. J., 1975, Listener discrimination of broadband FM echoes
    from simple elastic targets, J. Acoust. Soc. Am., 58, Suppl.
    1, S103.
Diercks, K. J., and Hickling, R., 1967, Echoes from hollow aluminum
    spheres in water, J. Acoust. Soc. Am., 41:380.
Diercks, K. J., Ryan, W. W., Mikeska, E. E., and Weisser, F. L.,
    1968, (see Diercks, 1975), Applied Research Laboratories
    Tech. Memo., N° 68-22 (ARL-TM-68-22), 46, (25 November, 1968).
Diercks, K. J., Trochta, R. T., and Evans, W. E., 1973, Delphinid
    sonar: measurement and analysis, J. Acoust. Soc. Am., 54:
    200.
Diercks, K. J., Trochta, R. T., Greenlaw, C. F., and Evans, W. E.,
    1971, Recording and analysis of dolphin echolocation signals,
    J. Acoust. Soc. Am., 49:1729.
Dubrovskiy, N. A., Krasnov, P. S., and Titov, A. A., 1971, On the
    emission of echolocation signals by the Azov Sea harbor
    porpoise, Akust. Zh., 16:521; Sov. Phys. Acoust., 16:444.
Dziedzic, A., Escudié, B., Guillard, P., Héllion, A., 1974, Evi-
    dence of tolerance to the effect of Doppler in the sonar
    emissions of Delphinus delphis, C. R. Acad. Sci. Paris,
    279(Ser. D):1313.
Dziedzic, A., Escudié, B., and Héllion, A., 1975, Realistic methods
    for analysis of biological sonar signals, Ann. Telecomm.,
    30,270.
Dziedzic, A., and Alcuri, G., 1977, Acoustic recognition of forms
    and characteristics of the sonar signals of Tursiops trun-
    catus, C. R. Acad. Sci. Paris, 285(Ser. D):981.
Dziedzic, A., Chiollaz, M., Escudié, B., and Héllion, A., 1977,
    Some properties of low frequency sonar signals of the dol-
    phin Phocoena phocoena, Acustica, 37:258.
Escudié, B., Héllion, A., and Dziedzic, A., 1971, Results from
    studies of air and marine sonars through signal processing
    and spectral analysis, in: "Proc. Troisième Colloque sur le
    Traitement du Signal et ses Applications", Nice, 1-5 June,
    1971:533.
Evans, W. E., 1973, Echolocation by marine delphinids and one species
    of fresh-water dolphin, J. Acoust. Soc. Am., 54:191.
Golubkov, A. G., Korolev, V. I., Antonov, V. A., and Ignat'eva, E.
    A., 1975, Comparison of dolphin echolocation signals with
    results of optimal signal calculations, Doklady Akad. Nauk
    SSSR, 223:1251.
Hickling, R., 1962, Analysis of echoes from a solid elastic sphere
    in water, J. Acoust. Soc. Am., 34:1582.
Hickling, R., 1964, Analysis of echoes from a hollow metallic sphere
    in water, J. Acoust. Soc. Am., 36:1124.
Koshovyy, V. V., and Mykhaylovs'kyy, V. M., 1972, The operating
    principle of the echolocating system of marine animals,
    Dopovidi Akad. Nauk Ukrayins'kayi RSR, Ser. A, Fizyko-techhn.
    ta matemat. Nauky, Ukranian, 12:1097, JPRS 58344.

Livshits, M. S., 1974, Some properties of dolphin hydrolocator
     from the viewpoint of correlation hypothesis, Biofizika, 19:
     916, JPRS 64329 (17 March, 1975).
Livshits, M. S., 1975, Correlation model of the recognition of ob-
     jects by echolocating animals, Biofizika, 20:920.
Metsaveer, J., 1971, Algorithm for calculation of echo-pulses from
     elastic spherical shells in fluid by summation of wave groups,
     Inst. of Cybernetics, Acad. Sci. Eston. SSR Tallinn, Pre-
     print 3, 40.
Romanenko, Y. V., 1973, Investigating generation of echolocation
     pulses in dolphins, Zool. Zh., Moscow, 11:1698, JPRS 61553.
Romanenko, Y. V., 1974, Physical fundamentals of bioacoustics,
     Fizicheskiye Osnovy Bioakustik, Moscow, JPRS 63923.
Ryan, W. W., Jr., 1978, Acoustical reflections from aluminum cylin-
     drical shells immersed in water, J. Acoust. Soc. Am., 64:
     1159.
Shirley, D. J. and Diercks, K. J., 1970, Analysis of the frequency
     response of simple geometric targets, J. Acoust. Soc. Am.,
     48:1275.
Zaslavskii, G. L., 1972, Investigation of the location signals of
     the bottlenosed dolphin using a two-channel system of re-
     cording, Biofizika, 4:717, Biophysics, 17:753.

ECHOLOCATION SIGNALS AND ECHOES IN AIR

J. David Pye

Department of Zoology and Comparative Physiology,

Queen Mary College, Mile End Road, London E1 4NS

1. INTRODUCTION : A PERSONAL VIEW

Since the Frascati meeting in 1966, there has been considerable expansion in our understanding of the acoustic structures
used by bats and birds for echolocation in air. In part this has
been due to the development of instrumental techniques which were
already undergoing something of a revolution at the time.   High-
speed tape-recorders suitable for bat recording had only become
available a few years previously and their use in conjunction with
Sonagraph sound spectrum analysers was then becoming widespread.
This combination was rapidly replacing the analysis of waveforms
by photographic sampling in real time from an oscilloscope screen
which, due to complex phase changes, is open to ambiguous interpretation of harmonics or frequency changes.

The sonagraph has sometimes been accused of generating
spurious harmonics, but if used with oscilloscope level monitors
which can cope with the peak levels of brief signals in a way that
V-U meters cannot, it is capable of displaying the harmonic
structure with considerable clarity and fidelity. It is true that
phase information is lost and the resolution of rapid transient
structures is poor, but these features can be depicted in
oscilloscope waveform displays which then become complementary
to the spectrum x time display of the sonagram.

Now the fashion is again changing and analyses are increasingly
being undertaken by digital computers instead of the preprogrammed
analogue sonagraph. In the case of spectrum x time displays, the
advantages are somewhat debatable since the 'mountain range' display

may conceal smaller 'foothills' lying 'behind', whereas ,the Z-modulation of a sonagram reveals all, especially if used with contouring.   But the advantage of the digital computer lies in its flexibility of programming and in the variety of other derived data that it can present.   Thus the analysis of echolocation signals is again moving into a new era of greater subtlety, where description is giving way to statements of real functional significance.

This development was foreshadowed 13 years ago at Frascati by the pioneering work of D. A. Cahlander. He not only presented the first application of radar theory to the echolocation signals of bats but also derived formulations of the theory which are applicable to the 'wideband' nature of many such signals. Understanding of the need for such an approach, and of the need to revise the assumptions that are made with relatively narrowband radar signals, is now rapidly increasing.

Other developments during the period under review include the discovery of some new kinds of pulse structure and a much wider taxonomic sample of species studied: it is now possible to say something about pulses detected from about 200 species from 16 families of bats.   That is nearly one third of all species, the major taxonomic omissions being three tiny families comprising only four species in all.

On the other hand, the miniaturisation of recording equipment, so that it can easily be used from battery power in the field, has made it increasingly clear that many species are flexible in their operation and cannot be specified by a single 'definitive' pulse structure.   Several species that are highly stereotyped in their behaviour indoors are found to vary from apparently pure Doppler operation to high-resolution, Doppler-tolerant range measuring when observed under varying and more natural conditions.   This should increase our caution when speaking of 'the pulses' of a given species, since these may only be used in the situation observed and really represent part of a larger repertoire.   Moreover the data available at present do contain some taxonomic surprises and raise puzzling problems about the evolution of echolocation signals in bats.

Finally there has been some progress in understanding the generation of the signals, their propagation in the atmosphere and the properties of certain important targets as echo reflectors. An increasing number of nocturnal insects are now known to detect bats and to take evasive action; three more families have been added in the last 13 years and the details of this behaviour are now better understood.

These are the matters that form the content of this review and they will now be set out formally and as objectively as possible. It is hoped that the above introduction will help to put them into context historically, functionally and with regard to the reviewer's brief for this meeting.

2. FREQUENCY PATTERNS

The signals emitted by, and assumed to be used by, different bats show a great variety of structure which makes them difficult to classify in a single, systematic continuum. To these must be added the sounds emitted by other animals, especially the birds. Frequency, frequency pattern, intensity, duration and harmonic content all show a range of characteristics, many of them assorted independently in the total population. Simmons (in press) has discussed the development of these signals mainly by reference to frequency patterns, duration and harmonic content, while Novick (1971) discussed pulse design almost exclusively by reference to duration.

It is, however, now clear from radar/sonar theory that frequency pattern plays a major role in determining the accuracy and resolution of different signals. Frequency pattern also correlates closely with the characteristics of the auditory system and the general behaviour of each species. It is frequency pattern, therefore, that probably forms the best basis for classification and comparison, although other factors must not be overlooked. The sounds that have been recorded from bats then fall into three general categories: broadband impulse sounds, frequency modulated (f.m.) sweep sounds and constant frequency (c.f.) sounds. Each of these will be discussed separately but, since the distinctions are not always sharp, a more general treatment must be given first.

Broadband impulse sounds are those in which there is no clearly defined, coherent 'carrier' frequency, no evidence of frequency modulation and an amplitude pattern that is rapid and transient - ie. 'clicks'. A number of species come close to this description, notably Rousettus and the birds. It is debatable, however, whether the 15-20 ms multi-click bursts of Aerodramus maximus and Steatornis show a coherent wave structure or merely consist of independent transient impulses. Even the very brief clicks of Aerodramus vanikorensis and other swiftlets can be considered as simply a packet of a very few cycles (Figure 3).

F.m. sweep and c.f. signals would seem to be self-explanatory terms and radar theory suggests that they are used to measure target range and radial velocity respectively. However, purely c.f. pulses are rare and much more commonly a long c.f. portion

is associated with an initial sweep (upwards or downwards) and/or
a terminal downsweep.   The variety of such intermediate patterns
is shown diagrammatically in Figure 1, while some actual analyses
are given in Figure 2.

Furthermore, some pulses have such brief periods of constant
frequency that it is doubtful whether they confer any appreciable
degree of velocity sensitivity.   Simmons et al (1975) proposed
two categories of c.f./f.m. pulses: 'short' c.f./f.m. with
durations less than 5 ms, and 'long' c.f./f.m. with durations
greater than 6 ms.   This scheme has recently been challenged by
Gustafson and Schnitzler (in press) on the grounds that some
species emit pulses in each category (it is not clear whether the
authors wish in this way to categorise individual pulses or the
species that emit them;   the subject of flexibility of acoustic
behaviour by a single species is discussed in Section 7).

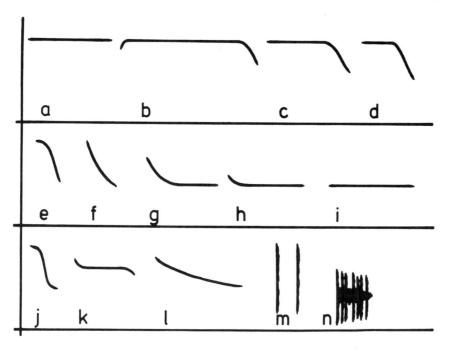

Figure 1.   Some frequency patterns of echolocation sounds used in
air; no harmonic structure is indicated. (a) to (i) form a
sequence from high c.f. through c.f./f.m., pure f.m. and f.m./c.f.
to low c.f.   The lowest row shows two forms of 'sigmoid' pulse
(j & k), shallow f.m. (l), a double broadband impulse sound (m)
and a multi-click burst (n).

Pye (1978) has calculated the velocity sensitivity of a range of c.f. signals by comparison with the value of 0.1 m.s^{-1} measured in <u>Rhinolophus ferrumequinum</u> by Simmons (1974); the theoretical limits are better than this by a factor of just over two.   Even c.f. durations of only 5 ms should permit a sensitivity better than 1 m.s^{-1} at 35 kHz, reducing to 0.23 m.s^{-1} at 150 kHz. Although Simmons et al (1975) appear to disagree, Pye argued that a sensitivity little better than flight velocity, say 5-10 m.s^{-1}, would be useful at least for target detection. On the basis of this argument, and for convenience, no distinction will be made here between different durations of constant frequency; all signals containing any perceptible c.f. component will be classed as c.f. pulses, whether or not this is associated with f.m. sweeps.

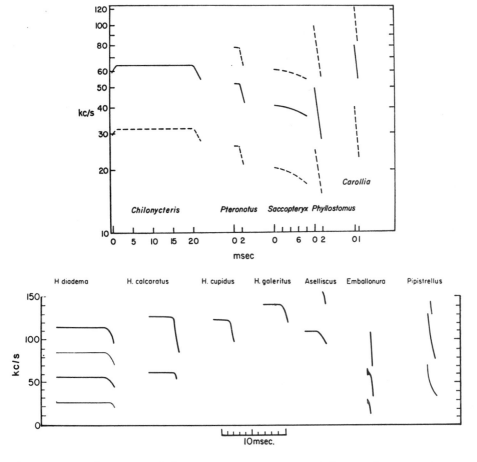

Figure 2.   Diagrams of frequency patterns and harmonic structures in pulses of bats from the New World (Grinnell, 1970) and the Old World (Grinnell and Hagiwara, 1972).

Figure 1 also shows a 'sigmoid' f.m. pattern (j) and the
pattern which Griffin called a 'pothook' (e). The frequency is
not quite constant in either of these cases and they are perhaps
best classed as f.m. pulses for want of information about their
functional significance.

Descriptions of bat signals owe much to the pioneering
comparative studies of Griffin, Novick, Möhres and Kulzer in 1955-63
although more recent techniques have led to a revision of some
details. In the following three sections material from papers
published since Frascati will be reviewed together with a general
picture of our present understanding. Not all papers will be cited
individually since the overall account owes much to previous work.
Further details have been culled from: Bradbury (1970), Griffin
and Simmons (1974), Grinnell (1970), Grinnell and Hagiwara
(1972a, b), Konstantinov and Sokolov (1967), Möhres and Neuweiler
(1966), Peff and Simmons (1971), Pye (1968a, 1972, 1973, 1978), Pye
and Roberts (1970), Roberts (1972a, b; 1973), Schnitzler (1968,
1970a, b; 1971, 1973), Simmons (1969, 1970, 1971, 1973), Simmons
et al (1979), Suthers (1967), Suthers et al (1972) and Suthers and
Fattu (1973). Larger reviews for the period include: Airapet'yants
and Konstantinov (1970/1973), Sales and Pye (1974) and Novick (1977).

3.   BROADBAND IMPULSE SOUNDS

Sounds of this general category are used in air by the two
known groups of echolocating birds: the South American oil bird,
Steatornis caripensis, and some of the South-East Asian cave
swiftlets, Aerodramus (formerly Collocalia). They are also emitted
by megachiropteran fruit bats of the genus Rousettus and probably
by the closely related genus Stenonycteris.

All these sounds share one feature: they are composed of
brief, click-like waveforms that are not coherent, or only barely
so, within each pulse; the rapid, transient-like nature of their
amplitude patterns leads to considerable band-spreading. There is
no clear carrier frequency or indication of frequency modulation.
The sounds fall into two groups as recently pointed out by Medway
and Pye (1977): double clicks with an internal interval of 15-40 ms
as in Rousettus and some Aerodramus spp, and 'single' clicks as in
Steatornis and Aerodramus maximus. The 'single' clicks, however,
are individually complex, each consisting of a rapid burst of sound
impulses lasting for up to 25 ms. The spectral range is largely
ultrasonic in Rousettus (though also clearly audible to man) while
in all the birds there is little or no energy above 10-15 kHz.

No new work on Steatornis is known to this reviewer since
1966. However, Sales and Pye (1974) and Medway and Pye (1977)
have analysed an original recording (made in Trinidad in 1963) and

Figure 3. Oscillograms of some broadband impulse sounds.
(a) Aerodramus fuciphagus: a train of 5 double clicks and 2
clicks in detail; (b) Aerodramus vanikorensis: a train of 6
double clicks and 2 clicks in detail; (c) Aerodramus maximus:
a train of 5 multi-click bursts and 2 bursts in detail; (d) a
double click of Rousettus aegyptiacus (left) and two multi-click
bursts of Steatornis caripensis in detail. All time markers =
10 ms. From Medway and Pye (1977).

have generally confirmed the earlier (1954) descriptions by
Griffin.   Energy was found from 1 kHz to 12 kHz with the peak at
2-4 kHz and a lesser one at 5-7 kHz (Griffin reported 7-8 kHz).
Each click consisted of up to seven impulses lasting up to 20 ms
in all (Figure 3d); click rate varied from 4-7 s^{-1}, when flying
round a large chamber of the cave, to 11-12 s^{-1} when landing on
the nest.   Observation of birds feeding at a nut tree suggested
that no signals are emitted outside the cave although the night
was dark。

Medway and Pye (1977) have reviewed the recent literature on
Aerodramus, including papers by Fenton (1975), Griffin and Suthers
(1970), Medway (1967, 1969), Medway and Wells (1969), Pecotich

(1974) and Penny (1975). There is now evidence that 11 of the 15 species can echolocate by emitting clicks, while 2 appear not to and the status of 2 is unknown. Four species have been studied in sufficient detail for descriptions of their sounds to be compared:

Table 1. Echolocation sounds in *Aerodramus* spp.

Species	Click	kHz	Source
A.h.hirundinaceus	Double	2-15+	Fenton (1975)
A.vanikorensis granti	Double	2-16	Griffin & Suthers (1970)
A.v.salangana	Double	2-9+	Medway & Pye (1977)
A.v.natunae	Double	2-10	Medway & Pye (1977)
A.f.fuciphagus	Double	2-9+	Medway & Pye (1977)
A.maximum lowi	'Single'	1.5-7+	Medway (1959, Medway & Pye (1977).

Double clicks have a constant internal interval of 15-16 ms, but Griffin and Suthers (1970) reported that in A.v.granti the clicks were occasionally treble. Although the spectrum is much lower and the first click is of lower amplitude than the second, the double clicks strongly resemble those of Rousettus (Figure 3, a, b, d). The frequency ranges given above are revised from earlier recordings. This conflicts with the argument of Sales and Pye (1974) that a difference in wavelength could account for the superior obstacle avoidance ability of A.vanikorensis (avoids 6 mm wires with 4.5-7.5 kHz; Griffin and Suthers, 1970) when compared with A.fuciphagus (hits wooden rods of 1 cm using 1.5-4.5 kHz; Medway, 1967, 1969). It seems that further work with consistent techniques is necessary, but Fenton (1975) found that both sounds and the ability to avoid thin rods were similar in A.hirundinaceus to those reported for A.vanikorensis by Griffin and Suthers (1970).

Aerodramus maximus clicks appear to be complex and 'messy' when recorded indoors, as if superimposed on multiple echoes. In a few of the clearest cases, however, it is apparent that each click consists of a burst of about six impulses lasting for a little over 20 ms (Figure 3c) and is thus very similar to the clicks of Steatornis (Medway and Pye, 1977). The clicks of A.unicolor (then Collocalia brevirostris unicolor), described briefly by Novick (1959), may also be of this kind.

There has been little new work on the echolocation sounds of rousettine fruit bats. Sales and Pye (1974) examined new recordings of Rousettus aegyptiacus from Uganda and of R.amplexicaudatus from Malaysia. No differences could be distinguished in the click structures or spectra (Figure 4) although the interclick interval was about 50% greater in R.amplexicaudatus. A computer interval

histogram run on a recording of hundreds of clicks from several
R.aegyptiacus flying in a large outdoor aviary, showed a peak at
20 ms and negligible scatter outside 18-22 ms. This is considerably
less than observed by Kulzer (1958, 1960) in this species from
Egypt (20-44 ms) and may represent a geographical variation in
behaviour.   Grinnell and Hagiwara (1972b) found that clicks of
R.amplexicaudatus from New Guinea had a spectrum of at least
10-90 kHz (peak 20-60 kHz) and an interclick interval that was
consistently about 30 ms. (Woolf, 1974, briefly reported pairing
of f.m. pulses in young Eptesicus fuscus, with an interpulse
interval of 20-40 ms).

     Finally, Kingdon (1974) has reported that the closely related
Stenonycteris lanosus lives in dark caves like Rousettus and is
able to echolocate by emitting clearly audible clicks.   No
recordings or analyses were made but the sounds are said to be
"chinks, that seem .... quieter and more 'musical' than the clicks
of Rousettus". If there were "only the faintest glimmer" of
light, they flew silently.   This is in contrast to the less closely
related Lissonycteris angolensis which Lawrence and Novick (1963)
showed to have no echolocation.

     Echolocation by shrews was first reported for three species
of Sorex and Blarina brevicauda by Gould et al in 1964.   The
sounds they recorded showed a rapid series of closely spaced
clicks of wide bandwidth.   Grunwald (1969) was unable to confirm
echolocation in two species of Crocidura although ultrasounds were
recorded while sniffing. Sales and Pye (1974) reported clicks with
energy from 10 kHz to 70-110 kHz from C.russula (a species used by
Grunwald) as they investigated a microphone; these clicks were
emitted singly or in rapid bursts of 8-25 with interclick
intervals as short as 1 ms. Buchler (1976) has produced fresh

Figure 4.   Sonagrams and oscillograms of double clicks from
Rousettus aegyptiacus (left) and R.amplexicaudatus (right).
From Sales and Pye (1974).

evidence of echolocation in <u>Sorex vagrans</u>.   Recordings from this
species showed double clicks with a spectrum of 18-60 kHz (peak at
20-40 kHz) and an interclick interval averaging 8.2 ms (range
5.3-11.7 ms).   The first click was slightly "shorter" than the
second.   The resemblance between these sounds and those of birds
and fruit bats is striking.

Sounds produced by other terrestrial mammals, including tenrecs
and rats are reviewed in posters at this conference by Buchler and
Mitz and by Chase respectively.

4. F.M. SWEEP PULSES

The short, harmonically rather pure, deep frequency sweep
pulse was originally discovered and intensively investigated by
Griffin and his coworkers (see eg. Griffin, 1958) in <u>Myotis
lucifugus</u>.   It was for some time thought to be typical of all
Vespertilionidae and still seems to be so for many genera under all
conditions observed.   In cruising flight the duration is 1-4 ms
during which the frequency falls by about an octave.   Amplitude
increases to a maximum about the middle of the pulse and then
decays again so that precise duration and frequency limits of the
sweep are difficult to define - values are extended by better
signal-to-noise ratios.

The swept carrier wave is generally the fundamental, for a
weak second harmonic component often appears towards the end of
the pulse.   The sweep has often been referred to as linear f.m.
and sometimes as a logarithmic sweep or a hyperbolic sweep.
However, Cahlander (1967) at Frascati presented analyses made with
an 'uncorrected' period meter (one giving period instead of its
reciprocal, instantaneous frequency) and showed that the period
$(1/f)$ is linear.   Whether this holds more generally for f.m.
sweeps is not yet clear.   But Altes and Titlebaum (1970) showed
theoretically that, because of the broadband nature (ie. high
bandwidth to centre-frequency ratio, often approaching unity), a
linear period sweep is optimally Doppler tolerant (see Section 8).
It is thus ideal for range measurement even if the range is
changing, although a Doppler-coupled ranging error is then introduced.   This kind of pulse thus appears to lie at one end of the
velocity/range measuring continuum and to be highly specialised.

During the interception of prey, negotiation of an obstacle
or landing, the pulse repetition rate increases from 10-15 s^{-1} to
100-200 s^{-1}, the pulse duration falls to 0.25 ms or so, the centre
frequency and sweep width are decreased;   there is often a
pronounced increase in the second harmonic intensity though the
total intensity falls due to weakening of the fundamental.   This
pattern is shown in Figure 5.   The high repetition rate produces

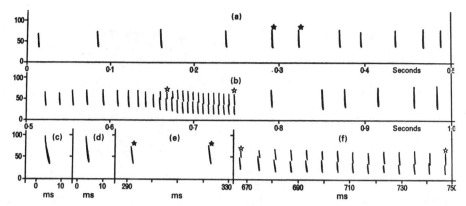

Figure 5.   Sonagram analysis of <u>Myotis lucifugus</u> flying indoors.
(a) and (b) are a continuous record including an avoidance 'buzz';
(c) and (d) are loud pulses from another recording;   (e) and (f)
show the starred pulses from (a) and (b) in detail.   From Sales
and Pye (1974).

a brief buzzing note from an ultrasound detector in real time and
the term 'interception buzz' has come to be applied to the
sequence itself.   Cahlander (1967) showed at Frascati that range
discrimination is improved in interception pulses and Doppler
coupling is virtually eliminated from the ambiguity diagram of a
single pulse.

The pure f.m. sweep is not restricted to the Vespertilionidae,
but some other families and even many vespertilionids show a much
greater harmonic content.   Here the depth of the fundamental sweep
is reduced but bandwidth is maintained by harmonics up to the fourth
or fifth.   A strong second harmonic is seen for instance in the
cruising pulses of <u>Plecotus</u>, <u>Nyctalus</u> and <u>Pipistrellus</u> (Figures 6
and 12).   Such pulses are also found in some Emballonuridae,
Natalidae and Molossidae.   Higher order harmonics are seen in
orally emitted pulses of some Emballonuridae, Rhinopomatidae,
Mormoopidae and Mystacinidae, and in nasally emitted pulses of
Nycteridae, Megadermatidae, Phyllostomidae and Desmodontidae.

The early work of Griffin and others was achieved by photographic sampling of oscilloscope waveforms.   Since some of the
waveforms are complex in short, multiple harmonic sweeps, they are
easily misinterpreted unless supported by spectral analysis as
is now provided by tape recordings.   Thus the relative intensities
and order of harmonics may be obscured and it is now known that,
especially in nasally emitting bats, the fundamental is often weak
or indetectable.   Also frequency modulation is difficult to detect

Figure 6. Sonagrams and the corresponding oscillograms of six f.m. pulses with different harmonic complexity. From Sales and Pye (1974).

in short pulses and this is exacerbated by fixed formant filtering
(see Section 9) which accentuates different harmonics in turn at
the same frequency. Thus pulses of Megadermatidae and Nycteridae
were initially described (Novick, 1958a; Möhres and Neuweiler,
1966) as c.f. pulses with no frequency sweep. Novick (1977) and
others persist in this interpretation although it can be clearly
shown (Sales and Pye, 1974, and elsewhere) that a sweep is present
as in most pulses of the Phyllostomidae (Figure 6). The reviewer
can now confirm this general picture for Lavia frons, Cardioderma
cor, Megaderma lyra, M.spasma, Nycteris thebaica, N.hispida, N.arge,
N.macrotis and N.aethiopica, although the waveforms closely
resemble those published by other authors.

Another variant of the f.m. sweep pulse is shown by Molossidae
and some Vespertilionidae such as Nyctalus (when hunting in the
open). This is a long, shallow sweep (Figure 1, 1). Here there
appears to be little harmonic content although the propagation
distance may suppress higher harmonics. The bandwidth is therefore often quite small and the sweep may become so shallow that
the pulses become effectively constant frequency. Frequency
itself is generally low in these pulses, ranging from 15 kHz to
35 kHz or so. Occasionally, hunting molossids (which show a
great variety of signal structures) produce pulses of a deeper
'sigmoid f.m.' shape as in Figure 1, j.

Other variations of f.m. are associated with periods of
constant frequency and with flexible behaviour as discussed in
Sections 5 and 7.

5. CONSTANT FREQUENCY PULSES

The first c.f. pulses were discovered by Möhres in 1953 in
Rhinolophus ferrumequinum. Because of their extreme length (c.f.
commonly over 70 ms) the signals of Rhinolophidae might be regarded
as the extreme form of velocity sensing systems. But such signals
always contain a downward frequency sweep at the end and commonly
a smaller upward one at the beginning (Figure 1, b). Pure c.f.
signals with no sweeps are uncommon but are found for instance in
Noctilio or in the outdoor cruising pulses of molossids and some
vespertilionids. These signals may be of 'high' frequency (ie. at
the top of sweeps seen in other pulses) as in Noctilio (Figure 1, a)
or 'low' frequency (at the bottom) as in molossids and vespertilionids (Figure 1, i).

Since Frascati, Schnitzler and his associates have clearly
shown (1968-in press) that Rhinolophus detects upward Doppler
shifts in c.f. echoes and adjusts the emitted frequency so as to
keep wanted echoes at a fixed frequency. This Doppler compensation thus results in a fine variation of emitted c.f. values from

a 'resting' frequency to as much as 8 kHz below this (in
R.ferrumequinum). This behaviour is correlated with striking
auditory adaptations shown by electrophysiology (Neuweiler and his
associates), by behavioural tests (Long) and by aural anatomy
(A. Pye and Bruns). Simmons (1974) has measured velocity sensitivity in R.ferrumequinum by behavioural methods, showing a
response to 0.1 m.s^{-1}, corresponding to the theoretical limit for
a pulse of this frequency and a duration of 20 ms.

Schnitzler has also demonstrated Doppler compensation in
Pteronotus parnellii of the Mormoopidae (Schnitzler 1970a,b, 1971)
and in Asellia tridens of the Hipposideridae (Gustafson and
Schnitzler, in press), both of which emit pronounced c.f. portions
with terminal sweeps. Actual durations of c.f. vary between
families and with behaviour: typical values are 8-70 ms in
Rhinolophidae, 5-25 ms in Hipposideridae and 5-27 ms in P.parnellii.
In all cases the longest pulses are emitted singly at low repetition
rates in cruising flight but in interception or when negotiating an
obstacle the c.f. part shortens and repetition rates rise as high as
80-100 s^{-1} (Figure 7).

C.f. pulses with various degrees of terminal sweep have also
been found in some Emballonuridae such as Rhynchonycteris and
there is circumstantial evidence that in crowded caves they are
emitted by some Phyllostomidae such as Lonchophylla, Carollia or
Glossophaga (see Section 7). Thus c.f. pulses may occur in all
four Superfamilies of the Microchiroptera; it remains to be seen
whether they are always associated with velocity sensing as theory
would suggest.

Figure 7. Sonagram analysis of Pteronotus parnellii flying
indoors. Long c.f./f.m. pulses, emitted singly or in groups in
cruising flight, give way to shorter c.f. in the extended avoidance sequence. Predominant second harmonic. From Schnitzler
(1970a).

It is a striking feature of the 'obligate' c.f. signals that the second harmonic is generally the strongest component. This was first noticed by Novick (1958a) in some African rhinolophids and now seems to be the general rule. Even though the fundamental and third harmonic may be of negligibly low intensity when the bat is in good condition, they can appear quite strongly in bats that are ailing or if the vocal tract is off-tuned by light bases (see Section 9). This is especially true of the nasally emitting bats (Rhinolophidae and Hipposideridae); for the orally emitting c.f. species (Emballonuridae and Mormoopidae) there are appreciable levels of fundamental, third harmonic and even traces of the fourth harmonic. In Noctilio and the c.f. signals of Molossidae and Vespertilionidae ('facultative' c.f. forms) the fundamental is predominant.

Very short periods of c.f. before an f.m. sweep (Figure 1, d, e) have been discussed in Section 2. They have been recorded from all the Mormoopidae except Pteronotus parnellii. This poses an evolutionary problem, for P.parnellii shows extreme specialisations that are largely convergent on the Rhinolophidae and Hipposideridae; yet P.personata especially is very similar in form but acoustic-ally very different. This also shows in the auditory systems of these bats as shown physiologically (Grinnell, 1970) and anatomi-cally (A. Pye, poster, this conference).

A further feature to emerge since Frascati is the use of two constant frequencies, or a bimodal spectrum. Some pulses have a higher c.f. frequency than others, with the intermediate band occupied only by down-sweeps from the higher pulses. This can take two forms: bimodal populations and bimodal emission by single individuals. The first was seen in Hipposideros commersoni (Pye, 1972) from Kenya (Figure 8). It seems likely that this was a mixed roost of H.c.gigas and H.c.marungensis, but Triaenops persicus (=afer) and to a lesser extent Hipposideros caffer from the same cave showed a tendency to be bimodal in the presence of conspecifics though not when isolated. A functional interpretation of the 'silent band' in terms of Doppler spreading was tentatively proposed.

Individual bimodality has been found in three species of Emballonuridae and in all of them the pulse pattern tends to be sigmoidal c.f. (Figure 1, k). In both Saccopteryx leptura and S.bilineata the individual bats emit pulses of higher and lower frequency alternately (Pye, 1973, 1978). This occurs whether the bats are flying singly or in company, indoors or hunting freely. The frequency interval between the two kinds of pulses is 2-2.5 kHz, corresponding to the Doppler shift produced by the apparent flight speed. In outdoor recordings the lower frequency of an approaching bat is commonly the same as the upper frequency when it recedes

Figure 8. Histograms of c.f. frequency in <u>Hipposideros</u> <u>commersoni</u>
from Kenya. (a) 50 pulses recorded in the <u>roost cave</u>; (b) 10
pulses from each of 5 bats recorded separately. From Pye (1972).

(Figure 9). But considerable variation in the frequency interval
is found at times so that one or both frequencies appear to change
at the source. Interception uses only the lower frequency and
non-flying bats use only the upper frequency. Again an interpretation was attemped on the hypothesis that the lower frequency
is pre-adapted for Doppler compensation.

In the third form, <u>Coleura</u> <u>afra</u>, two frequencies are only
found in bats hunting freely (mentioned briefly by Pye, 1978).
When flying indoors or even when emerging from the roost cave, only
the upper frequency was detected. The possibility of Doppler
compensation in these species and in other bats that use c.f. only
outdoors (such as <u>Nyctalus</u> and <u>P.pipistrellus</u>) may be resolved by
the use of Doppler radar observation (Pye, 1978; Halls, 1978).

6. AMPLITUDE AND AMPLITUDE PATTERNS

A number of factors make it difficult to obtain reliable and
comparable measurements of the amplitude or intensity of ultrasonic
signals. Suitable calibrated microphones and other equipment
certainly exist and it is easy to produce measures of given pulses
as detected. The difficulty with defining absolute values lies
with the animal, since bats are often quite directional in their
emissions (see Section 10) and actively turn their heads unless
intensively trained. Also there appear to be considerable pulse-
to-pulse variations and even hand-held or perching bats are
difficult to keep at the correct distance from the microphone,

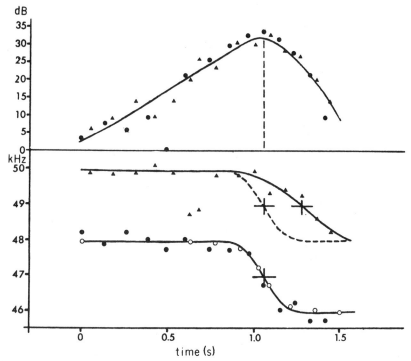

Figure 9. Analysis of amplitude (above) and frequency (below) of
Saccopteryx bilineata flying past the microphone in the open. Open
circles represent an imaginary source of 47 kHz moving at 7 m.s^{-1}
within 77 cm of the microphone.   From Pye (1978).

which, at the often used value of 10 cm, is quite critical.
Really useful values would have to be obtained from flying bats and
this is clearly impossible at present. Detailed statistical analysis might be useful but does not appear to have been attempted.

      Most statements about the 'loudness' of bat signals are
therefore rather subjective and relative, relying on the observer's
experience of how much amplifier gain is necessary in making
recordings.   Novick has recently (1977) presented similar conclusions and has withdrawn the maximum values of over 1000 μbar
(100 Pa) at 10 cm reported earlier. It seems that the loudest bats
may reach 100-200 μbar at 10 cm while the quieter, 'whispering'
bats only give 2-5 μbar - a total range of about 40 dB.

      Most purely insectivorous bats, or at least those that hunt
for flying insects, are loud, almost independently of frequency
pattern.   The most intense include Molossidae, Vespertilionidae,
Emballonuridae, Noctilionidae, Rhinolophidae, Hipposideridae and
Mormoopidae.   Insectivorous bats that feed on perching insects,

such as Nycteridae and Plecotus of the Vespertilionidae, tend to be
soft; nevertheless, with a strategically placed microphone, this
reviewer has recorded Nycteris arge hawking for insects in the open
with very faint pulses. Megadermatidae, Phyllostomidae and
Desmodontidae are also very soft although again some phyllostomids
and some megadermatids may sometimes hawk for insects.
Rhinopomatidae and Mystacinidae are also of fairly low intensity.

It seems that despite the lack of hard data, intensity is the
one characteristic for which a correlation with feeding habits can
be attempted although the above exceptions suggest caution. It is
also the only measure which is reasonably consistent within each
family. Neither of these points is at all true for frequency
patterns since short f.m., multiple harmonic sweeps are used for
every type of feeding (except fishing?), while insects are caught
by pulses of every type of frequency pattern. Taxonomic contrasts
of frequency pattern are numerous: eg. Pteronotus parnellii versus
the other Mormoopidae, or Rhynchonycteris and Saccopteryx versus
Taphozous and Peropteryx in the Emballonuridae.

Amplitude patterns within each pulse are very interesting and
give useful clues to pulse generation (see Section 9). Early
observations by Griffin and Novick showed that short f.m., multiple
harmonic sweeps of some phyllostomids may have an envelope with two
peaks. Pye (1967, 1968a) suggested that this is due to a fixed
formant filter accentuating first one harmonic and later the one
above during the f.m. sweep. Thus both amplitude peaks have the
same frequency as seen in the periodicity of the waveform.

The same argument applies to the terminal sweep of c.f./f.m.
pulses of Rhinolophidae, Hipposideridae and Pteronotus parnellii.
The amplitude of the sweep may decay rapidly from that of the
intense c.f. part (sweep-decay pulses) or it may rise to a new peak
from a weaker c.f. part before decaying again (sweep-peak pulses).
This variation has been confirmed in Rhinolophus by Schnitzler
(1967) who showed that sweep-decay is used by a resting or
cruising bat while sweep-peaks are used at higher repetition
rates when landing (see Section 9).

One of the most striking cases is that of Asellia tridens in
which the c.f. amplitude may fall so low that it disappears into
the baseline on the oscilloscope (-20 dB = 1/10 amplitude). This
led Möhres and Kulzer (1955) to suppose that the c.f. part can be
eliminated and for practical purposes perhaps it is (as intensity
if proportional to amplitude squared). However, Pye and Roberts
(1970) showed that the duration of the c.f. part is unaffected and
a sonagraph display with a 40 dB dynamic range (100:1 in amplitude)
shows much the same frequency pattern (Figure 10).

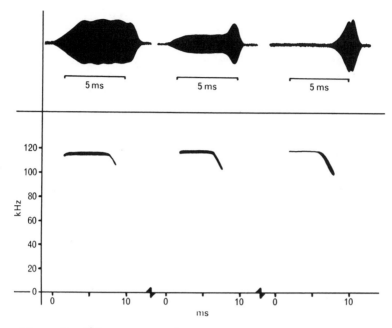

Figure 10.   Oscillogram envelopes and sonagrams of three pulses
from Asellia tridens.   Left: a sweep-decay pulse;   centre: a
sweep-peak pulse;   right: a pulse in which the c.f. part is
present but is so attenuated that it is masked by noise in the
linear-amplitude oscilloscope picture and may be effectively f.m.
only.   From Sales and Pye (1974).

7.   FLEXIBILITY OF PULSE STRUCTURE

At one time it seems to have been tacitly assumed that each
species used a characteristic pulse structure with only limited
changes such as the cruising and interception pulses.   The advent
of field recording since Frascati has considerably changed this
picture.   Although it is often difficult to identify bats in the
wild, this can be done often enough to show that some species
behave very differently under different conditions (reviewed by
Pye, 1973, 1978;   Sales and Pye, 1974).   This is not always the
case, for Griffin and Simmons (1974) found the outdoor hunting
signals of Rhinolophus ferrumequinum to be just like those of bats
avoiding obstacles indoors.   They found interception 'buzzes',
although Airapet'yants and Konstantinov (1970/73) found that
Rhinolophus falls silent just before interception, when recorded
in a flight cage.

But striking changes are seen in some vespertilionids when
recorded in the wild.   Scotophilus nigrita then introduces a
low-c.f. component while cruising (Figure 11) while Pipistrellus

pipistrellus and Nyctalus noctula commonly eliminate the f.m.
sweep altogether (Figure 12) as first reported by Pye (1967).
There appears to be a close correlation here with the altitude of
the bat.   Nyctalus commonly flies high and eliminates f.m.;
P.pipistrellus only eliminates f.m. when flying above about 5 m.
At lower altitudes, cruising pulses are like those of Scotophilus,
while close to the ground or indoors the c.f. part is rare.   Even
Myotis, which previously seemed to be quite stereotyped, may also
have flexible species: Fenton and Bell (in press) have recently
described a variety of cruising pulses from species in the western
U.S.A., including f.m./c.f. with a low c.f. of up to 5 ms in
M.volans.

     Oscillograph records by Griffin (1958) of Eptesicus fuscus
flying in the open appeared to show similar behaviour with very
long pulses.   Simmons, Lavender and Lavender (1978) recently
rexamined these records and confirmed this conclusion.   Furthermore
they were able to reproduce the behaviour in the laboratory by
subjecting trained E.fuscus to broadband random noise.   At high
noise levels the bats lengthened their pulses to include a terminal
c.f. part and also reduced the relative levels of harmonics.   It
seems that a longer pulse (of greater energy) and a narrower

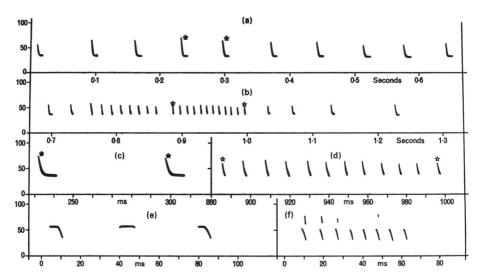

Figure 11.   (a–d) Sonagram analyses of Scotophilus nigrita hunting
in the open.   (a) and (b) form a continuous sequence;   (c) shows
the starred cruising pulses and (d) the starred buzz sequence in
more detail.   (e) and (f) show a cruising triplet and a short buzz
from Noctilio leporinus flying in a large outdoor cage.
From Sales and Pye (1974).

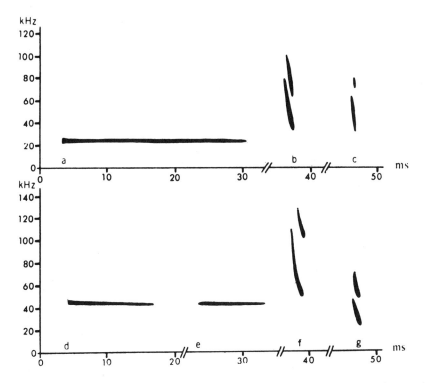

Figure 12.   Sonagrams of pulses from Nyctalus noctula (above) and
Pipistrellus pipistrellus (below).   (a, d, e):   outdoor cruising
pulses;   (b, f):   indoor cruising pulses;   (c, g):   interception
buzz pulses.   From Pye (1978).

bandwidth would be ideal adaptations to echo detection in noise.
But since the precision of range measurement is thereby reduced,
the bats may have then to rely on more velocity (Doppler) information.   A. Pye (1978) has shown that an African serotine, E.rendalli,
has a cochlear structure with many features of c.f. bats and has
suggested that this species may use c.f. extensively although it did
not when recorded in a quiet laboratory.

The most extreme example in the Vespertilionidae has been
described by Simmons and O'Farrell (1977) in Plecotus phyllotis.
When hunting outdoors, this bat emits a pure c.f. of 27 kHz for
20-200 ms followed after a few ms silence by a sweep from 24 kHz
to 12 kHz with some harmonics.   The second harmonic sweep of 40-22
kHz shows an amplitude reduction at 28-25 kHz, the c.f. frequency.
Sweeps can also be emitted alone, without the c.f. part.   Indoors
the bats emit short sweeps with two components, equal to the second
and fourth harmonics of the outdoor pulses, or with one component

equal to the fifth harmonic.   This repertoire offers formidable
problems to a formant interpretation.

Further changes have been reported for unidentified, high-
flying molossids (probably Tadarida spp in Africa and Molossus spp
in Panama) by Sales and Pye (1974).   They found considerable
variability in duration and frequency pattern.   Simmons, Lavender,
Lavender et al (1978) have described similar diversity in Tadarida
brasiliensis and T.macrotis which were highly adaptive to acoustic
conditions: from pure c.f. through f.m./c.f. to pure f.m. sweeps.
Noise interference indoors produced an increase in intensity and
duration and (surprisingly) there was some increase in bandwidth
in T.brasiliensis.

Another kind of variation was discussed in Section 6:   the
pulses of Asellia, described by Möhres and Kulzer (1955) as
'kombinierter-typ' pulses.   Not only do they combine c.f. and f.m.
but either can be accentuated to the virtual exclusion of the
other (Figure 10).

Variability has also been described in both species of Noctilio
by Suthers (1965, 1967;   Suthers and Fattu, 1973), whose findings
were confirmed for N.leporinus by Sales and Pye (1974).   Cruising
pulses are c.f./f.m. but when fishing two such pulses commonly occur
together with a c.f. pulse (lasting up to 11.6 ms with little or no
perceptible sweep) to form a cruising triplet (Figure 11, e).
Suthers has also described c.f./f.m. pulses with a slow sweep lasting up to 10 ms or so.   Bats approaching each other on collision
courses emit 'honks' consisting of a brief c.f. part followed by an
extended deep sweep of as much as two octaves.   Interception buzzes
eliminate the c.f. part while the f.m. sweeps shorten and fall in
frequency as in Myotis (Figure 11, f).

Bradbury (1970) has reported striking flexibility in the
phyllostomid, Vampyrum spectrum.   Indoor discrimination flights
used short pulses with a double sweep, shown to be the third harmonic plus a subsequent but overlapping fourth harmonic.   Duration
shortened from 1.5 ms to 0.4 ms as the target was approached.
This pattern suggests a very sharply tuned formant filter.   In a
large outdoor flight cage there were also long pulses up to 15 ms
in duration with a very shallow sweep.   These had a predominant
fundamental and decreasing harmonics up to the seventh and thus
showed little sign of formant filtering.   Figure 13 shows one
pulse of each type.

Pye (1973) produced circumstantial evidence that in a cave in
Panama other phyllostomids, principally Lonchophylla robusta but
also Carollia perspicillata and/or Glossophaga soricina, may emit
long c.f./f.m. pulses similar to those of Pteronotus parnellii.
When recorded in the laboratory, all these gave only short f.m.,

Figure 13. Two consecutive pulses (with some echoes) of a
<u>Vampyrum</u> <u>spectrum</u> flying in a large outdoor cage.   The shorter
pulse is typical of pulses used in discrimination flights indoors.
From Bradbury (1970).

multiple harmonic pulses 'typical' of their family.

From the above it is now clear that, if observations are made
under a variety of conditions, cases of pronounced flexibility can
be found in all four Superfamilies of the Microchiroptera. Further
species are sure to be found as observations and techniques are
extended.   It must also be assumed that the changes seen are
adaptive. Simmons et al (1975; Simmons, in press) discussed the
concept of various strategies in the detection of echoes from
different kinds of targets under varying conditions such as clutter
and ambient noise;   it seems that at least some bats can vary
their stratagems in order to optimise their performance.

It is also clear that if pulse structures cannot be 'assigned'
to each species as characteristic, they cannot be used with any
confidence as a taxonomic criterion. Recently Simmons (in press)
has attempted a discussion of the 'phylogeny and evolution' of
echolocation in bats but without implying a relationship to the
phylogeny of the bats themselves.   Despite the stimulatory nature
of this synthesis, three fixed points seem fairly safe.   The
primitive type of pulse is probably the multiple harmonic type with
shallow or no f.m.;   the pure fundamental, deep f.m. sweep
(apparently used almost invariably by <u>Myotis</u>) is the most advanced
form of range-measuring signal;   and the very long c.f./f.m. signal
(apparently used invariably by <u>Rhinolophus</u>) is the most advanced

form of velocity-measuring signal.

8. RADAR SIGNAL THEORY

At Frascati Cahlander (1967) presented a pioneering study of
the application of radar theory to the echolocation signals of
bats. After a theoretical consideration of wideband signals, he
computed ambiguity diagrams for six pulses of <u>Myotis lucifugus</u>
and an outdoor cruising pulse of <u>Lasiurus borealis</u>.

The need for reworking the theory arises from the invalidity
of certain assumptions made in radar theory. In radar the bandwidth
is so low compared with the centre frequency that Doppler shifts
may be regarded (eg. Bird, 1974) as an incremental offset, thus
simplifying the calculations without loss of accuracy. But with
wideband signals, which in bats may exceed an octave, the true
Doppler shift must be treated as a factor, compressing or expanding
the time scale, which Cahlander called $\propto$ (see also Kelly and
Wishner, 1965).

Similar treatments have been applied to wideband sonar signals,
for example by Russo and Bartberger (1964) and by Kramer (1967).
Further differences due to propagation in air and to the nature of
bats' targets are outlined by Pye, J.D. (this conference, poster).

The ambiguity diagram may be prepared either by assembling a
series of cross-correlations between the signal and Doppler-shifted
versions of the signal or by assembling Fourier spectra of products
between the signal and time-shifted versions of the signal (Halls,
this conference, poster).   Bird (1974) has shown that these two
methods correspond to two different approaches to receiver design:
the correlation or matched filter receiver and the Fourier transform
or beatnote receiver. Either system may correspond to an 'ideal'
receiver and both have been suggested as models of echo processing
in bats.

It is often assumed that a sharply peaked ambiguity surface
represents both high accuracy and good resolution. But Bird (1974)
has shown that these are related but distinct concepts. A sharp
peak gives high accuracy but its 'foothills' may mask weaker echoes
from other targets.   Improved resolution may therefore be gained
by degrading the accuracy, ie. by deliberately mismatching the
filter in correlation receivers.

It is clear that ambiguity diagrams offer detailed information
about the potential capability of bat signals, either as single
pulses or as groups or trains of pulses. It seems reasonable to
assume that the animals will exploit such capability due to

Figure 14. Range discrimination in three species of bats. Theoretical performance is compared with actual achievement in distinguishing the nearer of two targets. Top left: Eptesicus fuscus; top right: Phyllostomus hastatus; lower left: Tadarida brasiliensis; lower right: the autocorrelation function of a T.brasiliensis pulse. From Simmons (1970); Simmons, Lavender, Lavender et al (1978).

adaptiveness under survival pressure. This has actually been demonstrated for some species by Simmons (1970, 1971, 1973; Simmons, Lavender, Lavender et al, 1974, 1978). He prepared autocorrelation functions (without Doppler shifts) of the pulses of three species and from these calculated the theoretical limits of range discrimination. The performance of trained, non-flying bats in distinguishing between two targets at different ranges was then measured and the combined results are shown in Figure 14.

    Altes and Titlebaum (1970) developed theoretical aspects of Cahlander's work and showed that for wide-band f.m., optimum Doppler tolerance is provided by a linear period sweep. The close

approximation of many bats' sweeps to this form was described in
Section 4. Glaser (1971, 1974) has also tackled the application of
theory to bat signals; he assumes linear f.m. sweeps and some
other postulates that are frankly puzzling. Escudié et al (1976
and poster, this conference) have also been active in the theoretical consideration of bat signals in relation to echolocation
performance; they assume hyperbolic f.m. in Myotis and other
vespertilionids.

Despite the insights provided by theoretical work, this must
be supported by computation of ambiguity diagrams for actual signals.
This second aspect of Cahlander's work has now been taken up by
three groups: Beuter (in press and this conference), Pye and Halls
(this conference, posters) and Tupinier et al (in press). All have
chosen to effect the computation by cross-correlation and this has
been achieved digitally (Beuter; Escudié; Tupinier et al, in
press), by analogue methods (Cahlander) or by a 'hybrid' system
using digitally operated CCD delay lines and analogue correlation
(Halls). At first sight analogue computation by the Fourier
spectra method looks attractive but it requires single sideband
mixers of exceptional bandwidth.

The results of Beuter form a separate review of this
conference.

9. SOUND GENERATION

Before Frascati the main work on sound production in bats was
that of Kulzer (1958, 1960) on Rousettus and Novick and Griffin
(1961) on five families of the Microchiroptera. Since then the
literature has increased considerably.

The click sounds of Rousettus are, as Kulzer showed, produced
by clicking the tongue. Pye (1967) suggested that resonance
of the buccal cavity would then influence both the spectrum and the
envelope of emitted pulses. Roberts (1975) tested this by recording R.aegyptiacus and R.amplexicaudatus in air and in 80% He:
20% O_2 gas mixtures (He-O_2). Velocity of sound was monitored by
the resonant frequency of a spark-excited tube in order to allow for
air leakage. The spectra were complex but all identifiable points
were raised in He-O_2: in one case between 1.56 and 1.58 times for
a measured increase in sound velocity of 1.59 times.

Little is known about the production of the somewhat similar
sounds by echolocating birds. The slender evidence available for
Aerodramus has been discussed by Medway and Pye (1977) who favoured
some form of oral emission rather than a direct linkage from wing
movements in flight as postulated by Harrison (1966).

In the Microchiroptera it appears that ultrasound is vocally
generated by the vibration of laryngeal membranes driven by respiratory air-flow (Novick and Griffin, 1961). Several aspects of this
mechanism have been confirmed and developed by more recent work.
Laryngeal structure has been examined in 13 species of 7 families by
Wassif and Madkour (1968/9). They described skeletal structure,
drawing attention to the pronounced tracheal and ventricular pouches
that may act as resonators in Rhinolophus, Asellia and Nycteris.

    Correlation of pulse production with respiratory air-flow has
been achieved with a hot-wire anemometer by Schnitzler (1968) in
Myotis myotis, Rhinolophus ferrumequinum and R.euryale, and by
Roberts (1972b) in R.luctus, Eptesicus serotinus and Plecotus
auritus. Suthers et al (1972; Suthers and Fattu, 1973) used a
thermistor anemometer on Phyllostomus hastatus and Eptesicus fuscus.
All authors agree that pulse production occurs primarily during
exhalation although it can occur during inhalation when interrup-
tions of air-flow(mini-exhalations) could be seen. Schnitzler
(1971) also correlated pulse production with wing-beat in Myotis
lucifugus and Carollia perspicillata, while Suthers et al (1972)
observed respiratory air-flow, wing-beat and pulse production
simultaneously in flying Phyllostomus hastatus. Inhalation occurs
on the down-beat of the wing, exhalation and the majority of
pulses on the up-beat.

    Laryngeal action has been further studied in Eptesicus fuscus
by Suthers and Fattu (1973) who measured sub-glottic pressures.
The values during phonation were high, up to 60-70 cm H_2O. They
suggested that the vocal folds form a powerful glottal stop while
the thin, paired vocal and ventricular membranes are ultrasonic
generators during the pulse. This is consistent with their own
histological and myophysiological observations and with the
denervation experiments of Novick and Griffin (1961). Tidal volume
as measured with a body plethysmograph varied from 0.3 ml for
single pulses to 0.6 ml for rapid pulse trains.

    Schuller and Suga (1976) denervated laryngeal muscles,
recorded from active cricothyroid muscle and electrically stimulated the cricothyroid in Rhinolophus. Unilateral denervation of
the superior laryngeal nerve reduced the c.f. frequency by 4-6
kHz while bilateral denervation reduced it by as much as 30 kHz
with the introduction of prominent harmonics. Both spike recording
and stimulation suggested that the cricothyroid muscle is
responsible for tuning the c.f. and is thus the mediator of Doppler
compensation behaviour (See Section 5).

    At Frascati Pye (1967) suggested that vocal tract resonance
should act as a formant in filtering the sounds actually emitted,
thus accounting for the relative amplitudes of harmonics,

progressive phase changes during f.m. and some aspects of envelope
shape.   An electrical model of such a system was described later
(Pye, 1968a).   Light gases should not affect the glottal fundamental or harmonic frequencies but should strongly influence their
relative intensities as in human vowel sounds (this argument
appears to have been misunderstood by Novick, 1977).   Preliminary
results with hydrogen-air mixtures were described for Artibeus
jamaicensis (1967) and Rhinolophus hipposideros (1968a).   In
Rhinolophus even slight mis-tuning caused the fundamental to appear
strongly, suggesting that this is normally suppressed by a tuned-
reject resonance.   It is therefore striking that Schuller and
Suga (1976) found that harmonics emitted after bilateral superior
laryngeal nerve section in R.ferrumequinum were "much weaker" if
they fell close to the normal fundamental c.f. frequency.

More careful and sophisticated experiments have now been
performed using He-O2 by Schnitzler (1970a, b, 1973) on Rhinolophus
ferrumequinum and Pteronotus parnellii, and by Roberts (1973) on
R.ferrumequinum, R.luctus, Hipposideros diadema, Desmodus rotundus,
Eptesicus serotinus, Pipistrellus pipistrellus and Plecotus auritus.
All c.f. bats behaved as predicted, with no change in frequency but
a progressive shift of energy from the second to the third harmonic
and strengthening of the fundamental.   Roberts also found this last
effect in Desmodus and suggested that the nasal cavity suppresses
the fundamental in other nose-leaf bats.   He was, however, unable
to find any influence of He-O2 in vespertilionids and concluded
that here the purity and amplitude pattern of the fundamental sweep
are determined at the laryngeal level.

Roberts (1972a) also investigated the suggestion that sweep-
peaks and sweep-decay patterns in c.f. bats could be selected by a
change in glottal frequency.   High-resolution analysis by Lissajou's
figures of c.f. frequency in pulses of each type from six species
of Rhinolophidae, five species of Hipposideridae and in Pteronotus
parnellii failed to show any correlation (such a correlation would
of course be inconsistent with present knowledge of Doppler
compensation).   He concluded that the changes may be brought
about by retuning the formant frequency or by 'source generated'
amplitude changes.

Recently Griffiths (1978) has examined the cricothyroid muscle
in Mormoopidae and suggested that its posterior branch may be
responsible for varying the vocal tract resonance and thus the
harmonic content of the pulse.   Suthers and Durrant (poster, this
conference) recorded from and denervated both branches of this
muscle in Pteronotus parnellii.   Results show that both the anterior
and posterior branches contribute to the frequency of the c.f.
pattern of the pulse - the former branch being particularly
important at the start of the pulse, the latter affecting the

terminal f.m. sweep.   Of special interest is the fact that the
posterior, but not the anterior branch also plays an important role
in controlling the harmonic content of the emitted sound.   During
vocalization in 20% oxygen and 80% helium, the harmonic emphasis
is switched from the second to the first and third harmonics.

An important development in studies of sound production was
the demonstration by Suga and Schlegel (1972, 1973; Suga, Schlegel
et al 1973) that normal orientation pulses can be elicited from
awake bats by electrical stimulation of the midbrain.   This was
followed up by Suga, Schlegel et al (1973) in seven species of
three families (Noctilio leporinus, Pteronotus parnellii,
P.suapurensis, Eptesicus fuscus and three species of Myotis),
representing a wide range of pulse types.   The emitted sounds were
typical of those used in echolocation and various pulses of each
known repertoire could be elicited by exciting different regions of
the brain.   Pulses produced by this technique have been used in
the study of laryngeal mechanisms (Suthers and Fattu, 1973) and of
directional radiation patterns (Shimozawa et al, 1974; Schnitzler
and Grinnell, 1977; see Section 10).

Jen and Suga (1976) have also shown that motor control of
laryngeal and middle-ear muscles is coordinated, potentials appearing in the former about 3 ms before the latter.   Both sets of
muscles also respond to acoustic stimuli, with the laryngeal
muscles contracting about 3 ms after the middle-ear muscles.   Jen
and Ostwald (1977) used this cricothyroid muscle response as a
test of hearing: response threholds with downward sweeping f.m.
stimuli were lower than with either upward sweeps or tone pulses at
the best frequency.

10. SOUND RADIATION PATTERNS

Griffin and Hollander (1973) discussed the difficulties of
measuring the sound field around active bats, but considerable
success has been achieved with animals under various degrees of
constraint.

Simmons (1969) measured directionality in Pteronotus parnellii
at 56 kHz, its c.f. frequency, and in Eptesicus fuscus at its
greatest amplitude (about 30 kHz depending on direction).   He used
four microphone positions in the horizontal plane: 0° (straight
ahead), 22.5^{o}, 45° and 90° to one side.   Both species showed half
the pressure amplitude (-6 dB) at 22.5^{o}; at greater angles
Pteronotus was more directional (-20 dB at 45°, -26 dB at 90°)
than Eptesicus (-9 dB at 45°, -14 dB at 90°).   These results were
compared with Möhres figures (1967) for Rhinolophus ferrumequinum
(quoted as -6 dB at 22.5^{o}) and Megaderma lyra (-6 dB at 35°)

although Möhres (1967) and Möhres and Neuweiler (1966) actually
showed "intensity" plots with -6 dB at 55° for Megaderma.

Schnitzler (1968) referred to "beamwidth" in Rhinolophus
ferrumequinum and R.euryale as 20° both horizontally and vertically.
Thus it appears that, at least near the axis, a mouth-horn can be
as directional as a nose-leaf. Strother and Mogus (1970) calculated
theoretical patterns for a number of species, predicting -6 dB at
28° for Pteronotus parnellii and at 43° for R.ferrumequinum.

Griffin and Hollander (1973) compared 0° and 90° amplitudes in
Myotis lucifugus and Eptesicus fuscus. Higher frequencies at the
top of the fundamental sweep were more directional (25 dB) than
lower ones at the bottom (10 dB). In E.fuscus the second and third
harmonics were also examined and despite their higher frequency
gave only about 10 dB for 0°/90°. But the end of the second
harmonic sweep (frequency equal to the start of the fundamental
sweep) rose to a value of 23 dB. This suggests that the polar
diagram has side-lobes at harmonic frequencies.

Shimozawa et al (1974) measured detailed patterns around
Myotis grisescens and M.lucifugus. The bats' heads were fixed
and stereotyped pulses were elicited by brain stimulation.
Readings were taken every 10° at 55 kHz, 75 kHz and 95 kHz, and
two detailed plots are shown in Figure 15. In the horizontal plane
-6 dB points are at 40° for 55 kHz, 33° for 75 kHz and 30° for
95 kHz; in the median plane the main beam axis was 5-10° below
the horizontal.

A similar examination of the c.f. pattern in Rhinolophus
ferrumequinum was made by Schnitzler and Grinnell (1977) and the
results are shown in Figure 15; -6 dB points are at 25° laterally
and above, but at 70° below (similar values were given by
Airapet'yants and Konstantinov, 1970/73). Interference with the
upper part of the nose-leaf produces distortion of the pattern,
especially in the vertical plane (Figure 15), while blocking one
nostril disrupts it severely. These results and those of Shimozawa
et al have been combined with directional hearing patterns to
indicate the total pattern of echolocation sensitivity (Grinnell
and Schnitzler, 1977; Shimozawa et al, 1974).

Peff and Simmons (1971) measured horizontal angular discrimination by behavioural methods. E.fuscus could distinguish 6-8°
and Phyllostomus hastatus (-6 dB beamwidth 40-50°) gave 4-6°.
Clearly directionality is much sharpened by the receiver, and transmitter directionality serves mainly to enhance signal energy at the
target.

Figure 15. Directional emission patterns for two bats.
Left: patterns at 2 dB intervals for Myotis grisescens at two
frequencies; from Shimozawa et al (1974); right: patterns at
3 dB intervals for Rhinolophus ferrumequinum, above, the normal
bat, below after covering the upper part of the nose-leaf with
vaseline; from Schnitzler and Grinnell (1977).

11. PROPAGATION

At Frascati Pye (1967) expressed some concern about the
possibility of velocity dispersion in the propagation of ultrasound
in air.   It now seems that this is unlikely to be a significant
effect over the biological spectrum, at least in reasonably normal
concentrations of carbon dioxide.   No progressive phase shift can
be seen between a fundamental and its harmonics for a range of
frequencies over at least several metres.   So it seems likely
that Pierce's original observations (1925) were perturbed by 'tube

effects' in his interferometer. Further confirmation of this
point is still required, however.

Another propagation effect, dispersive absorption in air, has
been reviewed in detail by Griffin (1971). The effect may be
profound at higher frequencies and is strongly dependent on the
water vapour content of the air: 150 kHz at 3% water vapour
suffers a loss of about 15 dB.m^{-1}. Water vapour content can be
derived from temperature and humidity. For bats, most conditions
will be bounded by 3% water vapour (say 100% relative humidity at
24°C) and 0.5% water vapour (50% R.H. at 8°C or 25% R.H. at 20°C).
Griffin presented the absorption data as a function of % water
vapour, with frequency as the parameter, but for some purposes it
may be more convenient to read against frequency, with % water
vapour as the parameter. Some of Griffin's data have therefore
been replotted in this way in Figure 16. Recent figures from an
unpublished Boeing lab report (Evans, L.B., 1974) and a National
Physical Laboratory report (Bazley, 1976) suggest that all these
values should be approximately halved. Evans also states that
temperature effects are small for a given humidity. Delany (1977)
has discussed the historical appreciation of this factor and has
reviewed other aspects of ultrasound propagation in air.

Figure 16. Atmospheric attenuation as a function of frequency at
three values of water vapour content. The dashed lines for underwater attenuation are in dB.km^{-1}. Atmospheric data in dB.m^{-1}
replotted from Griffin (1971); (Pye, in press).

Discussing the strong attenuation, Griffin (1971) pointed out
that because of its 'low-pass filter' properties, air is a "very
murky medium" especially if it is damp and the frequency high.
Thus bats needing to detect echoes from long distances would be
obliged to use low frequencies and indeed high flying species
(molossids and some vespertilionids such as Nyctalus) use lower
frequencies than others.   It may be that a ground echo is necessary
for orientation and maximum altitude could thus be frequency
dependent.

MÜhres (1953) suggested that vectorial addition of sounds from
the two nostrils of Rhinolophus, half a wavelength apart, could
partially overcome the inverse square law in the axial direction.
He plotted actual measurements of sound pressure amplitude against
distance.   But it is intensity that decreases as the inverse
square, so that pressure (proportional to the square root of
intensity) decreases as the inverse distance.   MÜhres' graphs are,
indeed, a very good fit for an inverse distance relation.   His
vectoral argument, when calculated quantitatively, shows that,
because the nostrils are so close together, there is negligible
interaction on the axial direction except within a few mm of the
nose-leaf.   For 80 kHz and nostril spacing ($\lambda/2$) 2.1 mm, the
vector sum is 99.9% effective (less than 0.01 dB deficit) at only
2.37 cm.   At all distances of importance, the 'far field', the
axial propagation behaves as if there were a single source.

A further influence on propagation could be the presence of
water droplets in the air (Pye, 1971).   Lord Rayleigh (1896) calculated that spherical liquid drops in air show a conformational

Figure 17.   Resonant frequencies for spherical water droplets in
air from a formula given by Rayleigh.   From Pye (1971).

vibration at a frequency related to their diameter. Extrapolation
suggests that droplets of 100 μm resonate at 11 kHz and 18 μm at
150 kHz (Figure 17).   Stable fogs contain droplets from 100 μm to
1 μm and such suspended droplets should be very nearly spherical,
giving sharp tuning (high Q).   Thus fog should absorb biological
ultrasound and reradiate it slowly over a considerable time.   To
echolocation, fog would appear 'inky black' with perhaps a faint
after-glow. Artificial fogs in the laboratory certainly produce
extremely high attenuation at all ultrasonic frequencies and field
observations suggest that bats avoid discrete fog banks; indeed
their sounds cannot be detected by a bat detector from inside the
fog wall.

12.   ECHOES

The echo return from natural objects with broadband signals is
complex.   Much radar theory assumes a point target so that the
echo is identical to the transmitted signal except for a delay due
to round-trip propagation, a frequency shift due to radial velocity
and an attenuation due to propagation losses and target cross-
section.   But real targets present a more complicated problem and
many, such as insect prey, will have dispersive cross-sections as
well as being aspect sensitive and temporally variable.   The
reflective properties of many targets, even simple geometrical ones,
are too complex for theoretical analysis and are probably best
tested empirically as in Simmons' experiments with trained bats (eg.
Simmons and Vernon, 1971).

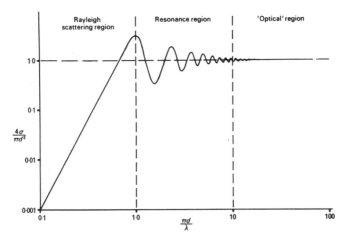

Figure 18.   Echo cross-section (σ) of a sphere as a function of
wavelength (λ).

The simplest, aspect insensitive, case is a sphere whose back-scattering cross-section as a function of wavelength is shown in Figure 18. At long wavelengths (greater than the target circumference, πd) the echo intensity is proportional to wavelength to the minus four (classical Rayleigh scattering). Only when wavelength is less than one tenth of the circumference is echo strength effectively independent of wavelength. Thus small targets are high-pass filters in respect of the echoes they produce. Small insects are neither spherical nor aspect-insensitive and their beating wings must also lead to temporal fluctuations; nevertheless they too must act as high-pass reflectors.

In some insects, such as moths, the wings are large enough to produce strong, specular echoes at normal incidence. The reflective properties of actual moths in tethered flight have been measured by Roeder (1963, 1967c); he showed very clearly that the echo strength is very aspect-sensitive and increases enormously when the wings are at certain critical positions. Johnson et al (1974) briefly drew attention to the Doppler shifts to be expected from moving insect wings. Recently Schnitzler (1978; and in press) has shown that flight movements of insect wings produce an asymmetrical echo spectrum due to interaction between amplitude modulation and Doppler shift; he has discussed in some detail the implications of this complex echo for the echolocation of c.f. bats.

Many insect prey species are not merely passive reflectors but detect the ultrasound emitted by bats up to 30-40 m away. Responses range from turning away from distant bats to erratic, evasive manoeuvres or power diving to the ground from nearer ones. Members of the Arctiidae and Ctenuchidae may reply by emitting a train of ultrasonic clicks, sometimes at very high repetition rates (recent references Fenton and Roeder, 1974; Fullard and Fenton, 1977; Fullard, 1979). These moth sounds may mask the echo of the bat's cry or warn of distastefulness or even mimic the warning signals of other distasteful species (Dunning, 1968).

Much has been learned of these various echolocation counter-measures in the last 13 years (partly reviewed by Sales and Pye, 1974) and their study may give useful insight to the operation of echolocation systems. Noctuid moths have been studied in great detail by Roeder (1966 a-d, 1967 a-c, 1969 a-c, 1970, 1971, 1972, 1974 a,b, 1976; Roeder and Fenton, 1973; Roeder and Payne, 1966) and by Adams (1971, 1972), Agee (1967, 1969 a-c, 1971), Ghiradella (1971), Goldman and Henson (1977), Lechtenberg (1971), Paul (1973, 1974), and Payne, Roeder and Wallman (1966). The abdominal tympanic organ of geometrids has been examined by Roeder (1974a) and a strikingly different, 'non-tympanic' ear has been discovered in the mouthparts of certain sphingids (Roeder, Treat and Vandeberg, 1968, 1970; Roeder and Treat, 1970; Roeder, 1972,

1976).   An ultrasonic ear that can initiate evasive behaviour
has also been discovered on the forewing of certain neuroptera
(Miller and MacLeod, 1966) and Miller has studied this system and
its associated behaviour in some detail (1970, 1971, 1974, 1975,
in press and this conference).

Thus no less than five different ultrasonic hearing organs
are now known to mediate the evasion of bats in at least seven
distinct groups of nocturnal flying insects. The evolutionary
significance of this variety, in view of the relatively recent
predation pressure of insectivorous bats, has been indicated by
Pye (1968b).

Larger, distributed targets such as foliage and rough ground
must give complex reflections although flat surfaces, such as some
single leaves, act as specular reflectors. Smooth ground or water
surfaces would ideally give an 'image' of the bat at a range equal
to twice the bat's altitude.   Ripples on water, however, give
strong reflections over wide angles.   A charming and thought-
provoking picture of the world as seen (or not seen) by a bat
using only ultrasonic echoes has been given by Lewin (1978).

References

Adams, W.B., 1971. Intensity characteristics of the noctuid acoustic
    receptor, J.gen.Physiol, 58: 562-579.
Adams, W.B., 1972. Mechanical tuning of the acoustic receptor of
    Prodenia eridania (Cramer) (Noctuidae), J.exp.Biol., 57: 297-308
Agee, H.R., 1967. Response of acoustic sense cell of the bollworm
    and tobacco budworm to ultrasound, J.econ.Entomol., 60:
    366-369.
Agee, H.R., 1969a, Response of flying bollworm moths and other
    tympanate moths to pulsed ultrasound, Anns.ent.Soc.Amer.,
    62: 801-807.
Agee, H.R., 1969b, Response of Heliothis spp. (Lepidoptera:
    Noctuidae) to ultrasound when resting, feeding, courting,
    mating or ovipositing, Anns.ent.Soc.Amer., 62: 1122-1128.
Agee, H.R., 1969c, Acoustic sensitivity of the European corn
    borer moth, Ostrinia nubilalis,Anns.ent.Soc.Amer., 62:
    1364-1367.
Agee, H.R., 1971, Ultrasound produced by wings of adults of
    Heliothis zea, J.Insect Physiol., 17: 1267-1273.
Airapet'yants, E.Sh. and Konstantinov, A.I., 1970/73,
    "Echolocation in Animals", Nauka, Leningrad, published 1970
    (English translation 1973, Israel Program of Scientific
    Translations).
Altes, R.A. and Titlebaum, E.L., 1970, Bat signals as optimally
    doppler tolerant waveforms, J.acoust.Soc.Amer., 48: 1014-1020.
Bazley, E.N., 1976, Sound absorption in air at frequencies up to
    100 kHz, National Physical Laboratory Acoustics Report No.
    Ac 74.
Bird, G.J.A., 1974, Radar Precision and Resolution, Pentech Press,
    London.
Bradbury, J.W., 1970, Target discrimination by the echolocating
    bat, Vampyrum spectrum, J.exp. Zool, 173: 23-46.
Buchler, E.R., 1976, The use of echolocation by the wandering
    shrew (Sorex vagrans), Anim.Behav., 24: 858-873.
Cahlander, D.A., 1967, in "Animal Sonar Systems: biology and
    Bionics", ed. Busnel, R.-G., Jouy-en-Josas, pp. 1052-1081.
Delany, M.E., 1977, Sound propagation in the atmosphere: a
    historical review, Acustica, 38: 201-223.
Dunning, D.C., 1968, Warning sounds of moths, Z.Tierpsychol., 25:
    129-138.
Escudié, B., Hellion, A., Munier, J. and Simmons, J.A., 1976,
    Etude theoretique des performances des signaux sonar
    diversité de certaines chauves-souris à l'aide du traitement
    du signal, Rev.d'Acoust., 38: 216-229.
Evans, L.B., 1974, Atmospheric absorption of sound: temperature
    dependence, Report No. D3-9190 (code no. 81205)., Boeing
    Co., Wichita, Kansas.

Fenton, M.B., 1975, Acuity of echolocation in Collocalia hirundinacea
    (Aves: Apodidae), with comments on the distributions of
    echolocating swiftlets and molossid bats, Biotropica, 7: 1-7.
Fenton, M.B. and Bell, G.P. (in press), Echolocation and feeding
    behaviour in four species of Myotis (Chiroptera), Can.J.Zool.
Fenton, M.B. and Roeder, K.D., 1974, The microtymbals of some
    Arctiidae, J.Lepid.Soc., 28: 205-211.
Fullard, J.H., 1979, Behavioral analyses of auditory sensitivity
    in Cycnia tenera Hübner (Lepidoptera: Arctiidae), J.comp.
    Physiol., 129: 79-83.
Fullard, J.H. and Fenton, M.B., 1977, Acoustic and behavioural
    analyses of the sounds produced by some species of Nearctic
    arctiidae (Lepidoptera), Can.J.Zool., 55: 1213-1224.
Ghiradella, H., 1971, Fine structure of the noctuid moth ear.
    I. The transducer area and connections to the tympanic
    membrane in Feltia subgothica, Haworth, J.Morph., 134: 21-46.
Glaser, W., 1971, Zur Fledermausortung aus dem Gesichtspunkt der
    Theorie gestörter Systeme, Zool.Jb.Physiol., 76: 209-229.
Glaser, W., 1974, Zur Hypothese des Optimalempfangs bei der
    Fledermausortung, J.comp.Physiol., 94: 227-248.
Goldman, L.J. and Henson, O.W., 1977, Prey recognition and selection
    by the constant frequency bat, Pteronotus parnellii, Behav.
    Ecol.Sociobiol., 2: 411-419.
Gould, E., Negus, N.C. and Novick, A., 1964, Evidence for
    echolocation in shrews, J.exp.Zool., 156: 19-38.
Griffin, D.R., 1954, Acoustic orientation in the oil bird,
    Steatornis, Proc.nat.Acad.Sci., 39: 884-893.
Griffin, D.R., 1958, "Listening in the Dark", Yale Univ. Press.
Griffin, D.R., 1962, Comparative studies of the orientation sounds
    of bats, Symp.zool.Soc.Lond., 7: 61-72.
Griffin, D.R., 1971, The importance of atmospheric attenuation
    for the echolocation of bats (Chiroptera), Anim.Behav., 19:
    55-61.
Griffin, D.R. and Hollander, P., 1973, Directional patterns of
    bats' orientation sounds, Period.Biol., 75: 3-6.
Griffin and Novick, 1975, Acoustic orientation of neotropical
    bats, J.exp.Zool., 130: 251-300,
Griffin, D.R. and Simmons, J.A., 1974, Echolocation of insects by
    horseshoe bats, Nature, Lond, 250: 731-732.
Griffin, D.R. and Suthers, R.A., 1970, Sensitivity of echo-
    location in cave swiftlets, Biol.Bull., 139: 495-501.
Griffiths, T.A., 1978, Modification of M.Cricothyroideus and the
    larynx in the Mormoopidae, with reference to amplification
    of high-frequency pulses, J.Mammal., 59: 724-730.
Grinnell, A.D., 1970, Comparative auditory neurophysiology of
    neotropical bats employing different echolocation signals,
    Z.vergl.Physiol., 68: 117-153.
Grinnell, A.D. and Hagiwara, S., 1972a, Adaptations of the auditory
    nervous system for echolocation: studies of New Guinea bats,
    Z.vergl.Physiol., 76: 41-81.

Grinnell, A.D. and Hagiwara, S., 1972b, Studies of auditory neurophysiology in non-echolocating bats and adaptations for echolocation in one genus, Rousettus, Z.vergl.Physiol., 76: 82-96.

Grinnell, A.D. and Schnitzler, H.U., 1977, Directional sensitivity of echolocation in the horseshoe bat, Rhinolophus ferrumequinum; II Behavioral directionality of hearing, J.comp.Physiol., 116: 63-76.

Grünwald, A., 1969, Untersuchungen zur Orientierung der Weisszahnspitzmäuse (Soricidae-Crocidurinae), Z.vergl.Physiol., 65: 191-217.

Gustafson, Y. and Schnitzler, H.U., in press, Echolocation and obstacle avoidance in the hipposiderid bat, Asellia tridens, J.comp.Physiol.

Halls, J.A.T., 1978, Radar studies of bat sonar, Proc.Fourth Int. Bat Res.Conf., eds. Olembo, R.J., Castelino, J.B. and Mutere, F.A., Kenya Lit. Bureau, Nairobi, 137-143.

Harrison, T., 1966, Onset of echo-location clicking in Collocalia swiftlets, Nature, Lond., 212: 530-531.

Jen, P.H.S. and Ostwald, J., 1977, Response of cricothyroid muscles to frequency-modulated sounds in FM bats, Myotis lucifugus, Nature, Lond., 265: 77-78.

Jen, P.H.S. and Suga, N., 1976, Coordinated activities of middle-ear and laryngeal muscles in echolocating bats, Science, 191: 950-952.

Johnson, R.A., Henson, O.W. and Goldman, L., 1974, Detection of insect wingbeats by the bat, Pteronotus parnellii, J.acoust. Soc.Amer., 55: S53 (Abstract).

Kelly, E.J. and Wishner, R.P., 1965, Matched-filter theory for high-velocity accelerating targets, I.E.E.E. Trans.mil., Electr., Mil-9: 56-69.

Kingdon, J., 1974, East African Mammals: an Atlas of Evolution in Africa, Vol. IIA, Academic Press, London & New York.

Konstantinov, A.I. and Sokolov, B.V., 1967, Use of active and passive location by bats in catching insects, 3rd Conf.Ecol. Physiol.Biochem.Morphol., Novosibirsk (in Russian).

Kramer, S.A., 1967, Doppler and Acceleration Tolerances of High-gain Wideband Linear F.M. Correlation Sonars, Proc. I.E.E.E., 55: 627-636.

Kulzer, E., 1958, Untersuchungen über die Biologie von Flughunden der Gattung Rousettus Gray, Z.Morph.und Ökol der Tiere, 47: 374-402.

Kulzer, E., 1960, Physiologische und Morphologische Untersuchungen über die Erzeugung der Orientierungslaute von Flughunden der Gattung Rousettus, Z.vergl.Physiol., 43: 231-268.

Lawrence, B. and Novick, A., 1963, Behavior as a taxonomic clue: relationships of Lissonycteris (Chiroptera), Breviora, 184: 1-16.

Lechtenberg, R., 1971, Acoustic response of the B cell in noctuid moths, J.Insect Physiol., 17: 2395-2408.

Lewin, R., 1978, Bats-eye view, Trends in Neurosciences, 1: 38.

Medway, Lord, 1959, Echo-location among Collocalia, Nature, Lond., 184: 1352-1353.

Medway, Lord, 1967, The function of echo-navigation among swiftlets, Anim.Behav., 15: 416-420,

Medway, Lord, 1969, Studies of the biology of the edible-nest swiftlets of South East Asia, Malayan Nature Jour., 22: 57-63.

Medway, Lord and Pye, J.D., 1977, Echolocation and the systematics of swiftlets, in: Evolutionary Ecology, eds. Stonehouse, B. and Perrins, C., Macmillan, London, 225-238.

Medway, Lord and Wells, D.R., 1969, Dark orientation by the giant swiftlet, Collocalia gigas, Ibis, 111: 609-611.

Miller, L.A., 1970, Structure of the green lacewing tympanal organ (Chrysopa carnea, Neuroptera), J.Morph., 131: 359-382.

Miller, L.A., 1971, Physiological responses of green lacewings (Chrysopa, Neuroptera) to ultrasound, J.Insect Physiol., 17: 491-506.

Miller, L.A., 1974, (Abstract) The behavioral response of green lacewings to ultrasound, Amer.Zool., 13: 1258.

Miller, L.A., 1975, The behavior of flying green lacewings, Chrysopa carnea, in the presence of ultrasound. J.Insect Physiol., 21: 205-219.

Miller, L.A., in press, Interactions between bats and green lacewings (Chrysopa, Insecta) in free flight, Proc. 5th Int. Bat Res.Conf., Texas Tech.Univ.Press.

Miller, L.A. and MacLeod, E.G., 1966, Ultrasonic sensitivity: a tympanal receptor in the green lacewing, Chrysopa carnea, Science, 154: 891-893.

Möhres, F.P., 1953, Über die Ultraschallorientierung der Hufeisennasen (Chiroptera: Rhinolophinae), Z.vergl.Physiol., 34: 547-588.

Möhres, F.P., 1967, Ultrasonic orientation in megadermatid bats, Animal Sonar Systems: biology and bionics, ed. Busnel, R.-G., Jouy-en-Josas, 115-127.

Möhres, F.P. and Kulzer, E., 1955, Ein neuer, kombinierter Typ der Ultraschallorientierung bei Fledermäusen, Naturwissenschaften, 42: 131-132.

Möhres, F.P. and Kulzer, E., 1956a, Über die Orientierung der Flughunde (Chiroptera, Pteropodidae), Z.vergl.Physiol., 38: 1-29.

Möhres, F.P. and Kulzer, E., 1956b, Untersuchungen über die Ultraschallorientierung von vier afrikanischen Fledermaus-familien, Verh.der Deutsch.zool.Ges. in Erlangen, 1955, 59-65.

Möhres, F.P. and Neuweiler, G., 1966, Die Ultraschallorientierung der Grossblat-Fledermäuse (Chiroptera, Megadermatidae), Z.vergl.Physiol., 53: 195-227.

Novick, A., 1958a, Orientation in paleotropical bats: I. Microchiroptera, J.exp.Zool, 138: 81-154.

Novick, A., 1958b, Orientation in paleotropical bats: II.
    Megachiroptera, J.exp.Zool., 137: 443-462.
Novick, A., 1959, Acoustic orientation in the cave swiftlet,
    .Biol.Bull., 117: 497-503.
Novick, A., 1962, Orientation in neotropical bats: I. Natalidae
    and Emballonuridae, J.Mammal., 43: 449-455.
Novick, A., 1963, Orientation in neotropical bats: II.
    Phyllostomidae and Desmodontidae, J.Mammal., 44: 44-56.
Novick, A., 1971, Echolocation in bats: some aspects of pulse
    design, Amer.Scient., 59: 198-209.
Novick, A., 1977, Acoustic orientation, in "Biology of Bats",
    ed. Wimsatt, W.A., 3: 73-287.
Novick, A. and Griffin, D.R., 1961, Laryngeal mechanisms in bats
    for the production of orientation sounds, J.exp.Zool., 148:
    125-146.
Paul, D.H., 1973, Central projections of tympanic fibres in
    noctuid moths, J.Insect Physiol., 19: 1785-1792.
Paul, D.H., 1974, Responses to acoustic stimulation of thoracic
    interneurons in noctuid moths, J.Insect Physiol., 20: 2205-2218.
Payne, R.S., Roeder, K.D. and Wallman, J., 1966, Directional
    sensitivity of the ears of noctuid moths, J.exp.Biol., 44:
    17-31.
Pecotich, L., 1974, Grey swiftlets in the Tully River Gorge and
    Chillagoe caves, Sunbird, 5: 16-21.
Peff, T.C. and Simmons, J.A., 1971, Horizontal-angle resolution by
    echolocating bats, J.acoust.Soc.Amer., 51: 2063-2065.
Penny, M., 1975, The Birds of Seychelles and the Outlying Islands,
    Collins, London.
Pierce, G.W., 1925, Piezoelectric crystal oscillators applied to
    the precision measurement of the velocity of sound in air and
    carbon dioxide at high frequencies, Proc.Amer.Acad.Arts Sci.,
    60: 271-302.
Pye, A., 1978, Aspects of cochlear structure and function in bats,
    Proc.Fourth Int.Bat Res.Conf., eds. Olembo, R.J., Castelino,
    J.B. and Mutere, F.A., Kenya Lit.Bureau, Nairobi, 73-83.
Pye, J.D., 1967, Synthesizing the waveforms of bats' pulses,
    Animal Sonar Systems: biology and bionics, ed. Busnel, R.-G.,
    Jouy-en-Josas, 43-65.
Pye, J.D., 1968a, Animal sonar in air, Ultrasonics, 6: 32-38.
Pye, J.D., 1968b, How insects hear, Nature, Lond., 218: 797.
Pye, J.D., 1971, Bats and fog, Nature, Lond., 229: 572-574.
Pye, J.D., 1972, Bimodal distribution of constant frequencies in
    some hipposiderid bats (Mammalia: Hipposideridae), J.Zool.,
    166: 323-335.
Pye, J.D., 1973, Echolocation by constant frequency in bats,
    Period.biol., 75: 21-26.
Pye, J.D., 1978, Some preliminary observations on flexible echo-
    location systems, Proc.Fourth Int.Bat Res.Conf., eds. Olembo,
    R.J., Castelino, J.B. and Mutere, F.A., Kenya Lit.Bureau,
    Nairobi, 127-136.

Pye, J.D., in press, Why ultrasound?, Endeavour.

Pye, J.D. and Roberts, L.H., 1970, Ear movements in a hipposiderid bat, Nature, Lond., 225: 285-286.

Rayleigh, Lord, 1896, The Theory of Sound (2nd edn), Macmillan, London.

Roberts, L.H., 1972a, Variable resonance in constant frequency bats, J.Zool., 166: 337-348.

Roberts, L.H., 1972b, Correlation of respiration and ultrasound production in rodents and bats, J.Zool., 168: 439-449.

Roberts, L.H., 1973, Cavity resonances in the production of orientation cries, Period.biol., 75: 27-32.

Roberts, L.H., 1975, Confirmation of the echolocation pulse production mechanism of Rousettus, J.Mammal., 56: 218-220.

Roeder, K.D., 1963, Echoes of ultrasonic pulses from flying moths, Biol.Bull., 124: 200-210.

Roeder, K.D., 1966a, Auditory system of noctuid moths, Science, 154: 1515-1521.

Roeder, K.D., 1966b, Acoustic sensitivity of the noctuid tympanic organ and its range for the cries of bats, J.Insect Physiol., 12: 843-859.

Roeder, K.D., 1966c, A differential anemometer for measuring the turning tendency of insects in stationary flight, Science, 153: 1634-1636.

Roeder, K.D., 1966d, Interneurones of the thoracic nerve cord activated by tympanic nerve fibres in noctuid moths, J.Insect Physiol., 12: 1227-1244.

Roeder, K.D., 1967a, Prey and predator, Bull.ent.Soc.Amer., 13: 6-9.

Roeder, K.D., 1967b, Turning tendency of moths exposed to ultrasound while in stationary flight, J. Insect Physiol., 13: 873-888.

Roeder, K.D., 1967c, Nerve cells and insect behaviour, Harvard Univ. Press, Cambridge, Mass., revd. edn.

Roeder, K.D., 1969a, Acoustic interneurons in the brain of noctuid moths, J.Insect Physiol., 15: 825-838.

Roeder, K.D., 1969b, Brain interneurons in noctuid moths: differential suppression by high sound intensities, J.Insect Physiol., 15: 1713-1718.

Roeder, K.D., 1969c, Acoustic interneurons in the brain of noctuid moths, J.exp.Zool., 134: 127-157.

Roeder, K.D., 1970, Episodes in insect brains, Amer.Scient., 58: 378-389.

Roeder, K.D., 1971, Acoustic alerting mechanisms in insects, in "Orientation: Sensory Basis", ed. Adler, H.E., Ann.N.Y. Acad.Sci., 188: 63-79.

Roeder, K.D., 1972, Acoustic and mechanical sensitivity of the distal lobe of the pilifer in choerocampine hawkmoths, J.Insect Physiol., 18: 1249-1264.

Roeder, K.D., 1974a, Responses of the less sensitive acoustic sense cells in the tympanic organs of some noctuid and geometrid moths, J.Insect Physiol., 20: 55-66.

Roeder, K.D., 1974b, Acoustic sensory responses and possible bat
    evasion tactics of certain moths, Can.Soc.Zool.ann.Meet.Proc.,
    Univ.New Brunswick, 71-78.
Roeder, K.D., 1976, Joys and Frustrations of doing research,
    Perspectives in Biol.Med., winter, 231-245.
Roeder, K.D. and Fenton, M.B., 1973, Acoustic responsiveness of
    Scoliopteryx libatrix L. (Lepidoptera: Noctuidae), a moth that
    shares hibernacula with some insectivorous bats, Can.J.Zool.,
    51: 681-685.
Roeder, K.D. and Payne, R.S., 1966, Acoustic orientation of a moth
    in flight by means of two sense cells, Symp.Soc.exp.Biol., 20:
    251-272.
Roeder, K.D. and Treat, A.E., 1970, An acoustic sense in some
    hawkmoths (Choerocampinae), J. Insect Physiol., 16: 1069-1086.
Roeder, K.D., Treat, A.E. and Vandeberg, J.S., 1968, Auditory
    sense in certain sphingid moths, Science, 159: 331-333.
Roeder, K.D., Treat, A.E. and Vandeberg, J.S., 1970, Distal lobe
    of the pilifer: an ultrasonic receptor in choerocampine
    hawkmoths, Science, 170: 1098-1099.
Russo, D.M. and Bartberger, C.L., 1964, Ambiguity diagram for
    linear F.M. sonar, J.acoust.Soc.Amer., 38: 183-190.
Sales, G.D. and Pye, J.D., 1974, "Ultrasonic Communication by
    Animals", Chapman and Hall, London.
Schnitzler, H.U., 1967, Discrimination of thin wires by flying
    horseshoe bats (Rhinolophidae), Animal Sonar Systems: biology
    and bionics, ed. Busnel, R.-G., Jouy-en-Josas, 69-87.
Schnitzler, H.U., 1968, Die Ultraschall-Ortungslaute der Hufeisen-
    Fledermäuse (Chiroptera-Rhinolophidae) in verschiedenen-
    Orientierungssituationen, Z.vergl.Physiol., 57: 376-408.
Schnitzler, H.U., 1970a, Echoortung bei der Fledermaus, Chilonycteris
    rubiginosa, Z.vergl.Physiol., 68: 25-38.
Schnitzler, H.U., 1970b, Comparison of the echolocation behaviour
    in Rhinolophus ferrumequinum and Chilonycteris rubiginosa,
    Bijd.tot de Dierk, 40: 77-80.
Schnitzler, H.U., 1971, Fledermäuse im Windkanal, Z.vergl.Physiol.,
    73: 209-221.
Schnitzler, H.U., 1973, Die Echoortung der Fledermäuse und ihre
    hörphysiologischen Grundlagen, Fortschritte der Zool., 21:
    136-189.
Schnitzler, H.U., 1978, Die Detektion von Bewegungen durch
    Echoortung bei Fledermäusen, Vehr.Dtsch.Zool.Ges., 1978: 16-33.
Schnitzler, H.U., in press, Detection of the fluttering movements
    of insects by constant frequency bats, Proc. 5th Int.Bat Res.
    Conf., Texas Tech.Univ.Press.
Schnitzler, H.U. and Grinnell, A.D., 1977, Directional sensitivity
    of echolocation in the horseshoe bat, Rhinolophus ferrumequinum;
    I. Directionality of sound emission, J.comp.Physiol., 116:
    51-61.

Schuller, G. and Suga, N.,   1976, Laryngeal mechanisms for the
    emission of CF-FM sounds in the Doppler-shift compensating
    bat, Rhinolophus ferrumequinum, J.comp.Physiol., 107: 253-262.
Shimozawa, T., Suga, N., Hendler, P. and Schuetze, S., 1974,
    Directional sensitivity of echolocation system in bats
    producing frequency-modulated signals, J.exp.Biol., 60: 53-69.
Simmons, J.A., 1969, Acoustic radiation patterns for the echo-
    locating bats, Chilonycteris rubiginosa and Eptesicus fuscus,
    J.acoust.Soc.Amer., 46: 1054-1056.
Simmons, J.A., 1970, Distance perception by echolocation: the
    nature of echo signal-processing in the bat, Bijd.tot de Dierk,
    40: 87-90.
Simmons, J.A., 1971, Echolocation in bats: signal processing of
    echoes for target range, Science, 171: 925-928.
Simmons, J.A., 1973, The resolution of target range by echolocating
    bats, J.acoust.Soc.Amer., 54: 157-173.
Simmons, J.A., 1974, Response of the Doppler echolocation system in
    the bat, Rhinolophus ferrumequinum, J.acoust.Soc.Amer., 56:
    672-682.
Simmons, J.A., in press, Phylogenetic adaptations and the evolution
    of echolocation in bats (Chiroptera), Proc.5th Int.Bat Res.
    Conf., Texas Tech.Univ.Press.
Simmons, J.A., Fenton, M.B. and O'Farrell, M.J., 1979, Echolocation
    and pursuit of prey by bats, Science, 203: 16-21.
Simmons, J.A., Howell, D.J. and Suga, N., 1975, Information content
    of bat sonar echoes, Amer.Scient., 63: 204-215.
Simmons, J.A., Lavender, W.A. and Lavender, B.A., 1978, Adaptation
    of echolocation to environmental noise by the bat, Eptesicus
    fuscus, Proc.Fourth Int.Bat Res.Conf., eds. Olembo, R.J.
    Castelino, J.B. and Mutere, F.A., Kenya Lit.Bureau, Nairobi,
    97-104.
Simmons, J.A., Lavender, W.A., Lavender, B.A., Childs, J.E.,
    Hulebak, K., Rigden, M.R., Sherman, J., Woolman, B and
    O'Farrell, M.J., 1978, Echolocation by free-tailed bats
    (Tadarida), J.comp.Physiol., 125: 291-299.
Simmons, J.A., Lavender, W.A., Lavender, B.A., Doroshow, C.A.,
    Kiefer, S.W., Livingston, R., Scallet, A.C. and Crawley, D.E.,
    1974, Target structure and echo spectral discrimination by
    echolocating bats, Science, 186: 1130-1132.
Simmons, J.A. and O'Farrell, M.J., 1977, Echolocation by the
    long-eared bat, Plecotus phyllotis, J.comp.Physiol., 122:
    201-214.
Simmons, J.A. and Verton, J.A., 1971, Echolocation: discrimination
    of targets by the bat, Eptesicus fuscus, J.exp.Zool., 176:
    315-328.
Strother, G.K. and Mogus, M., 1970, Acoustical beam patterns for
    bats: some theoretical considerations, J.acoust.Soc.Amer.,
    48: 1430-1432.

Suga, N. and Schlegel, P., 1972, Neural attenuation of responses
     to emitted sounds in echolocating bats, Science, 177: 82-84.
Suga, N. and Schlegel, P., 1973, Coding and processing in the
     auditory system of FM-signal-producing bats, J.acoust.Soc.
     Amer., 54: 174-190.
Suga, N., Schlegel, P., Shimozawa, T. and Simmons, J., 1973,
     Orientation sounds evoked from echolocating bats by electrical
     stimulation of the brain, J.acoust.Soc.Amer., 54: 793-797.
Suthers, R.A., 1965, Acoustic orientation by fish-catching bats,
     J.Zool., 158: 319-348.
Suthers, R.A., 1967, Comparative echolocation by fishing bats,
     J.Mammal., 48, 79-87.
Suthers, R.A. and Fattu, J.M., 1973, Fishing behaviour and acoustic
     orientation by the bat (Noctilio labialis), Anim.Behav., 21:
     61-66.
Suthers, R.A., Thomas, S.P. and Suthers, B.J., 1972, Respiration,
     wing-beat and ultrasonic pulse emission in an echolocating
     bat, J.exp.Biol., 56: 37-48.
Tupinier, Y., Biraud, Y., Chiollaz, M. and Escudie, B., in press,
     Analysis of vespertilionid sonar signals during cruise,
     pursuit and prey capture, Proc.Fifth Int.Bat Res.Conf., Texas
     Tech.Univ.Press.
Wassif, K. and Madkour, G., 1968-9, The structure of the hyoid bone,
     larynx and upper part of trachea in some Egyptian bats,
     Bull.Zool.Soc.Egypt, 22: 15-26.
Woolf, N.K., 1974, Ontogeny of bat sonar: paired FM signalling,
     J.acoust.Soc.Amer., 55: S 53 (abstract).

ECHOLOCATION ONTOGENY IN BATS

Patricia E. Brown and Alan D. Grinnell

Departments of Biology and Physiology, and Ahmanson
Laboratory of Neurobiology, Brain Research Institute;
University of California, Los Angeles; Los Angeles,
California 90024

ECHOLOCATION ONTOGENY IN BATS

The ontogeny of echolocation in bats has received relatively
little attention. Some bats do not hear at birth and are incapable
of emitting the high frequency sounds necessary for accurate echolocation. No bat can fly at birth. Even bats born in a relatively
advanced state, with the ability to hear and emit ultrasounds, still
must learn to fly and correlate their outgoing pulses and returning
echoes in a three-dimensional world before they are weaned. A study
of echolocation ontogeny must integrate physiology, morphology, and
behavior. The emission of short, high frequency sounds depends on
the development of laryngeal musculature, but these ultrasonic pulses
are of little value if the auditory system cannot process them or if
the animal cannot maintain the high body temperatures necessary for
ultrasonic hearing. All these systems develop simultaneously, and
we have yet to assess the role of learning versus maturation in echolocation ontogeny. Our understanding of echolocation in adult bats
can be enhanced through knowledge of how this complex sensory system
develops.

Development of Vocalizations

All bats are born in a relatively helpless state and are totally
dependent on their mothers for a period of weeks or months postnatally
(Orr, 1970). Gould (1975) has classified baby bats of eight genera
as either precocial (eyes and ears open at birth, body covered with
fur, emitting more than one type of vocalization), altricial (eyes
and ears closed, skin naked, emitting only one type of vocalization),

or as intermediate between these two extremes. The degree of morphological development at birth is not necessarily related to phylogeny.
An example would be the vespertilionids Antrozous pallidus (Brown,
1973, 1976; Gould, 1975) and Myotis (see refs. below). Antrozous
is truly altricial, with eyes and ears sealed for at least four days
postnatally (Fig. 1), while Myotis velifer (Brown, unpublished data)
and M. lucifugus (Gould, 1971, 1975) are alert and mobile at birth.
Smaller bats generally produce relatively large young compared to
the adult (Orr, 1970). Bats of the family Phyllostomatidae are typically born in an advanced state (body covered with fur and eyes open)
and develop quite rapidly.

FM Bats

        In most instances, there is a correlation between the degree of
morphological development at birth and the types of sounds emitted
by the infant. For example, newborn Antrozous emit only continuous
loud isolation calls when separated from their mothers (Brown, 1973,
1976). These calls which are longer, louder, and lower in frequency
than echolocation pulses, serve to promote contact between the mother
and young. Isolation calls appear to be ubiquitous among baby bats,
and whereas they differ somewhat in structure between species, they
all serve a common function (Gould, 1970, 1971, 1975a, 1975b, 1977,
1979; Gould, Woolf, and Turner, 1973; Brown, 1973, 1976; Matsumura,
1979; Schmidt, 1972; Porter, 1977; Orr, 1954; Davis et al., 1968; Jones,
1967; Douglas, 1967; Pearson et al., 1952; Kulzer, 1962; Kay and
Pickvance, 1963; Kunz, 1973). A double-note (a pair of long-short
or short-long pulses) is also present at birth in some species, but
appears at a later age in others (Gould et al., 1973; Gould, 1975a;
Woolf, 1974). Short FM sounds, the apparent precursors of later FM
echolocation pulses, are emitted at birth by some precocial species
(i.e. Macrotus and Carollia) but generally do not appear for several
days (Eptesicus) to a week (Antrozous) postnatally. Woolf (1974) rec-
ognizes isolation calls, double notes, and FM pulses as an ontogenetic
sequence, and has grouped them into a "sonar family". Whereas these
categories of sounds may be structurally similar, isolation calls are
not necessarily the precursors of adult echolocation pulses, although
in some species this may be the case. In only four genera of bats
have the sound emissions in known-age bats been analyzed throughout
their postnatal development. These include Myotis (M. lucifugus,
Gould, 1971, 1975a; M. velifer, Brown, unpublished observations; M.
oxygnathus, Konstantinov, 1973), Eptesicus fuscus (Gould, 1971, 1975b;
Woolf, 1974), Antrozous pallidus (Brown, 1973, 1976), and Rhinolophus
ferrumequinum (Konstantinov, 1973; Matsumura, 1979; Glenis
Long, personal communication). At present it is not possible to postulate a general plan for bat sonar ontogeny, and each species must
be examined individually. The vespertilionids Antrozous, Eptesicus,
and Myotis all differ in the time course of their vocal development.
In Antrozous, it appears that echolocation pulses and communication
sounds have different origins. As previously stated, Antrozous emits
only FM isolation calls at birth (Fig. 2). As the bat matures, the
isolation call shortens and increases in frequency, until by 20 days

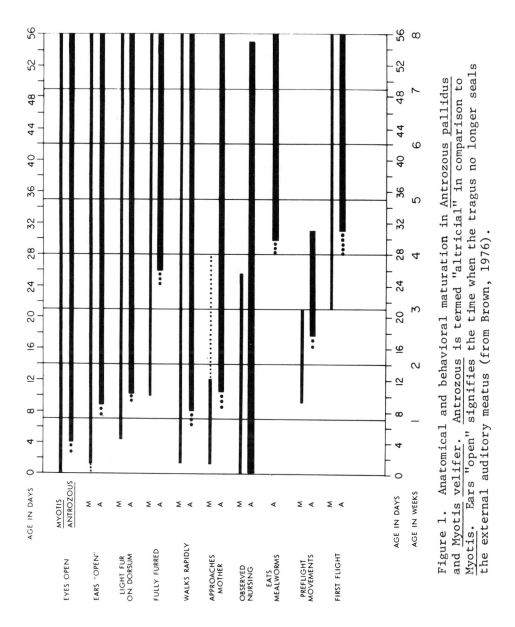

Figure 1. Anatomical and behavioral maturation in Antrozous pallidus and Myotis velifer. Antrozous is termed "altricial" in comparison to Myotis. Ears "open" signifies the time when the tragus no longer seals the external auditory meatus (from Brown, 1976).

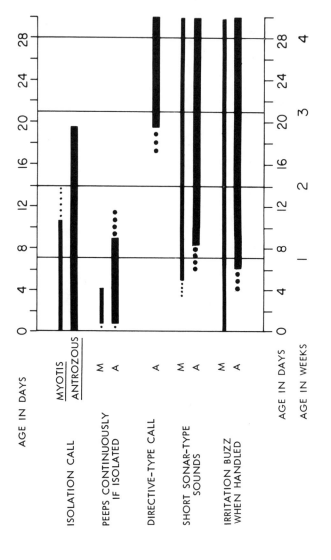

Figure 2. Development of vocalizations in Antrozous pallidus and Myotis velifer. The dots represent the earliest or latest age at which a vocalization was recorded, while the dark lines show the average age. There is no known analogue in Myotis to the adult directive of Antrozous (from Brown, 1976).

postnatally it resembles the adult "directive," a communication sound.
Double note isolation calls are emitted by some baby Antrozous and
not others. These may function as a vocal signature for individual
recognition (Brown, 1976). On about the fifth day postnatally, a
shorter 10-20 msec FM pulse is emitted in addition to the longer (60
msec) isolation call. This is the apparent precursor of the short
orientation sound first emitted between seven and nine days after
birth (Fig. 3a). This short ultrasonic pulse is emitted as the now-
mobile baby crawls around exploring its environment. It gradually
shortens and increases in frequency and repetition rate (Table 1),
evolving into the adult orientation pulse before the first flight at
four to five weeks (Fig. 3b,c). The early sonar pulses are rich in
harmonics. In some instances, it appears that the fundamental is
dropped during ontogeny, with subsequent sound energy being funneled
into the second harmonic. In the majority, however, there is a gradual increase in fundamental frequency and a subsequent loss of harmonics.

Myotis velifer (Brown, unpublished data) follows the same sequence as Antrozous except that the time course is compressed (Fig.
1,2), and this more advanced bat crawls actively at two days after
birth and emits short FM pulses in addition to longer isolation calls
and double notes. Since all of these categories of sounds are present
simultaneously, it does not appear that the isolation call evolves
into the double note and sonar pulse, nor that they represent a continuum. It is possible that even though isolation calls are the
acoustic precursors of echolocation pulses (i.e. they preceed them
developmentally), the isolation calls still remain to serve a different function. No analogue to the adult "directive" of Antrozous
exists for Myotis. The isolation call is emitted less frequently as
the bat matures, disappearing entirely by fourteen days postnatally.
Infant Myotis lucifugus were found to emit isolation calls, double
notes, and short FM pulses (Gould, 1971, 1975a). These babies interspersed sonar signals and isolation calls depending on their state
of excitation. If a day old Myotis lucifugus was held by the nape
of the neck and moved up and down, it emitted sonar calls. In contrast, Konstantinov (1973) reported that the larger European Myotis
oxygnathus emitted only loud, audible isolation calls for the first
ten days after birth. These isolation calls apparently evolved into
adult-like echolocation signals by 25 days postnatally, although the
higher frequencies utilized by adults were not recorded until 45 days,
an age when insect pursuit is normally initiated in this species.
This observation supports the hypothesis proposed by Gould (1970,
1971, 1977) and Woolf (1974) that sonar calls are derived from isolation calls.

Eptesicus is intermediate between Myotis and Antrozous both in
its degree of morphological development and its vocalizations (Gould,
1971). For the first three days postnatally, only an isolation call
is emitted in response to stressful situations. Although infant

Figure 3. A–C: Tracings of sound spectrographs of orientation pulses
                 for <u>Antrozous pallidus</u>: A – 9 days; B – 16 days;
                 C – adult.

          D–F: <u>Rhinolophus ferrumequinum</u>: D – newborn isolation
                 call; E – 8 days double note; F – 14 days orientation
                 pulse. The CF portion of adult pulses is 3 kHz
                 higher. These tracings are from recordings made by
                 Long of a wild colony.

          H,I: <u>Pteronotus parnellii</u>: H – newborn isolation call;
                 I – adult orientation pulse.

          J,K: <u>Pteronotus suapurensis</u>: J – newborn isolation call;
                 K – adult orientation pulse.

Table 1. Average characteristics of <u>Antrozous pallidus</u> echolocation
pulses during ontogeny. Repetition rate and interpulse interval vary
with activity. Before 28 days when they can fly, measurements of
repetition rates are from crawling bats, while rates for the 46-day
and adult are from bats flying in a room without obstacles.

AGE days	FREQUENCY KHZ	PULSE DURATION msec.	INTERPULSE INTERVAL msec.	REPETITION RATE pulses/sec.
6	30→10	20	158	8
17	40→17	4	30	24
26	60→23	3	22	28
46	70→20	2.5	17	35
82 (adult)	80→27	2	12	45

<u>Eptesicus</u> continue to emit isolation calls under certain situations;
they are replaced at three to five days by double notes, either in
long-short or short-long combinations (Gould <u>et al</u>., 1973; Gould,
1975b; Woolf, 1974). At approximately five days of age, short (3
msec) sonar sounds are produced often as paired signals. They differ
from double notes in being of equal duration. Double notes continue
to function in communication, while the short FM sounds evolve into
echolocation pulses by day 21. <u>Eptesicus</u> begins to fly at approximately 25 days postnatally. Woolf (1974) contends that double notes
are the precursors of sonar sounds.

    Woolf (1974) demonstrated that the development of the FM pulses
was not affected by deafening due to bilateral malleus removal in
infant <u>Eptesicus</u>. In order to alter the normal acoustic environment,
Gould (1975b) hand-raised some infant <u>Eptesicus</u> while others were
raised by mothers whose superior laryngeal nerves had been cauterized.
These infants emitted normal sonar pulses when recorded as adults,
suggesting a strong genetic component in the ontogeny of the sonar
signal.

CF Bats

The three genera discussed in detail up to this point all utilize FM pulses for echolocation as adults, and it is not surprising
that the earliest sounds emitted by these bats are long FM isolation
calls that, though longer and lower in frequency, are recognizable
as members of the same "sonar family" (Woolf, 1974). It is therefore
not surprising that neonatal vocalizations of bats employing CF or
CF/FM pulses as adults should contain primarily constant frequencies.
Several recent studies have documented constant frequency vocalizations in juvenile bats of the Old World families Hipposideridae
(Gould, 1979; Grinnell and Hagiwara, 1972) and, Rhinolophidae (Kay and
Pickvance, 1963; Matsumura, 1979; and Long, personal
communication). Whereas the adult bats emit all sonar pulses via
the nostrils, neonates appear to emit high intensity isolation calls,
double notes, and sonar calls primarily via the mouth. This same
pattern has also been observed in phyllostomatids (Gould, 1977),
where adults emit relatively low intensity sounds nasally, but neonates produce higher intensities orally. The nature of this ontogenetic shift from oral to nasal vocalizations in Rhinolophus was
first reported by Konstantinov (1973), and has been subsequently documented by Sumiko Matsumura (1979 and personal communication) under
laboratory conditions, and by Glenis Long (personal communication)
in a natural roost. Bats less than one week old emit predominantly
"noisy" or broad band isolation calls through the mouth (Fig. 3d),
although some pure tone pulses are emitted nasally. Often intermediate combinations of nasal and oral sounds occur in the same pulse.
These infant pure tone pulses differ from those of the adults. Infant
pulses contain slight FM descending components both at the beginning
and end of the pulse (Fig. 3e), while the adults begin each pulse
with a long CF portion and end with an FM tail (Fig. 3f) (Schnitzler,
1968). The infant sounds are rich in harmonics, these being supressed
and filtered in the adult sonar pulses with most sound energy present
in the second harmonic (Novick, 1958; Pye, 1968). Halls (in prep.) has
found that within roosts, adults do emit "modified" shortened CF
pulses, with more FM present and harmonics, presumably for communication. As Rhinolophus matures, nasal signals replace oral ones,
so that at three weeks the oral mode is found only in the first syllable of the "attractive" (= isolation) call. By four weeks when
this bat begins to fly, all oral calls have disappeared with the
exception of communicative distress calls. In addition to the shift
from oral to nasal, and "noisy" to pure tones, the character of the
nasal tones changes with age. The fundamental frequency increases
from 12 kHz in neonates to 36 kHz characteristic of the adults. During development, the fundamental and third harmonic are suppressed,
so that only the second harmonic remains. The anatomical basis for
the shift from oral to nasal sound emission arises from the closing
of the laryngo-nasal junction which is incomplete in the week-old
bat. This allows air flow either through the nose or mouth, and
accounts for the high degree of variability in infant vocalizations
(Matsumura, 1979).

Long (personal communication) found a similar acoustic sequence in the European horseshoe bat (Rhinolophus ferrumequinum ferrumequinum) while recording vocalizations in a natural maternity roost in the attic of a house. However the time frame was compressed as this subspecies begins to fly at 14-16 days postnatally. The CF portion of the adult echolocation pulse is 83 kHz compared to 72 kHz for R. ferrumequinum nippon. In both cases this is the second harmonic. Long found that in infant vocalizations the greatest sound energy was present in the fundamental and third harmonic. She also noted that, in contrast to FM bats, the duration of the pulse (CF component) increased with age and showed less variability than in infant bats.

Matsumura (in preparation) has made an observation in the context of mother-young communication that might have important implications in the development of echolocation behavior in young Rhinolophus. During reunions, before the infant can fly, antiphonal calling by mother and infants ensues. In the final stages of this "duetting", the infant appears to copy the pure tone emission pattern of the mother (albeit 3 kHz lower). At the beginning of a vocal exchange the mother seems to actively synchronize her "lead signals" (pure tone pulses rich in harmonics analogous to "directives" in Antrozous) with the infant's two to three syllable pure tone pulse (nasal sound). Later the infant begins to emit single syllable pulses, and actively synchronizesthese so that they regularly overlap with the mother's emissions, in effect copying her pattern of vocalizations.

Adult Hipposideros emit echolocation pulses containing an initial CF component followed by an FM tail. The duration and frequency of the CF portion varies with the species, as does the sweep of the FM tail. Among five species of Hipposideros recorded by Gould (1979) in Malaysia and four by Grinnell and Hagiwara (1972) in New Guinea, H. diadema emits a long CF/short FM pulse, structurally similar to Rhinolophus ferrumequinum but shorter in duration, while H. ridleyi produces a short CF/long FM. The sonar calls of the other species are intermediate between these. Due to problems of maintaining healthy nursing bats in captivity, Gould was unable to trace sonar development in any individual, but rather obtained cross-sectional data by recording sounds from different size/age classes. Infant Hipposideros, like Rhinolophus, emit via the mouth high intensity isolation calls, double notes, and CF pulses that are rich in harmonics. In addition to isolation calls, one day old H. diadema produced a CF/FM signal similar to that of the adult with predominant energy in the second harmonic ($\sim$40 kHz) whereas only the third harmonic is presented in the adult sonar pulse ($\sim$60 kHz). In H. diadema from New Guinea (Grinnell and Hagiwara, 1972) four harmonics are present in the adult pulse with the second (58 kHz) and fourth (116 kHz) being dominant. The long low CF isolation calls recorded in infant H. bicolor and cineraceus are similar to those emitted by newborn Rhinolophus (Fig. 3d) and Pteronotus (Fig. 3h,j). In Hippo-

sideros the double notes and precursors of sonar pulses contain ascending and descending FM components (Gould, 1979; Grinnell and Hagiwara, 1972) which also resemble those of week-old Rhinolophus (Fig.
3e).

In Panama, an attempt was made to trace the ontogeny of CF/FM
sonar pulses in Pteronotus parnellii and suapurensis, members of the
New World family Mormoopidae (Brown and Grinnell, unpublished data).
Mother bats refused to care for their babies in captivity, and hand-
rearing was not possible. Recordings were made of newborn vocalizations which were limited to long, low frequency CF pulses rich in
harmonics (Fig. 3h,j), similar to those recorded in Rhinolophus (Fig.
3d) and Hipposideros. Although the adult sonar calls are quite
different in P. parnellii and suapurensis, the infant isolation calls
are structurally quite similar. Unfortunately it was not possible
to determine whether these early vocalizations were precursors of
the adult orientation pulse. Just as Woolf (1974) has postulated an
FM "sonar family", a similar "family" grouping appears to exist for
CF bats of diverse phylogenies.

                        The Development of Hearing

When new-born pallid bats (Antrozous) failed to behaviorally
respond to their mothers' directives (communication sounds), one
logical conclusion was that they could not hear. This hypothesis was
confirmed by neurophysiological recordings of single units and evoked
potentials from the posterior colliculus in anesthetized bats in
response to auditory stimuli (Brown, Grinnell, and Harrison, 1978).
In fact there was no auditory response in bats until six days postnatally, and then only to loud, low frequency sounds such as contained in the mother's directives. By this age, the infant's ears
are open and erect, the tragus no longer sealing the external meatus.
Absolute sensitivity and frequency range increase rapidly until by
24 days after birth, the evoked potential audiogram resembles that
of the adult (Fig. 4). Short ultrasonic pulses are not emitted until
day nine, and parallel the ability to hear frequencies in that range.
This corresponds to the age at which the baby bat is able to crawl
rapidly and explore its environment. Except for the first week of
its life, the infant bat emits only sounds which it can actually
hear (Fig. 5).

Besides the high frequency sensitivity lacking in infant Antrozous, another important adaptation for echolocation is a high degree
of temporal resolution or fast auditory recovery time. Forty msec
or longer are required for the auditory system to respond to the
second of two identical stimuli (i.e. the outgoing pulse and returning
echo) compared with less than four msec in the adult (Fig. 6). This
also develops rapidly in the juvenile bat, so that fast temporal
resolution has developed by the time of first flight at one month.

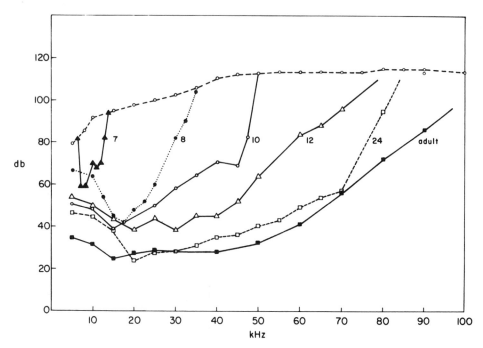

Figure 4. Audiograms of <u>Antrozous</u> <u>pallidus</u> as determined by evoked
potentials recorded from the posterior colliculus. The age in days
is to the right of each curve. No evoked responses were observed
in bats younger than 6 days. By 24 days, both frequency range and
sensitivity resembled that of adults. Top dotted line represents
maximum speaker output in this and subsequent graphs of evoked
potentials. (From Brown <u>et al</u>., 1978.)

        Even after they begin to answer their mothers' vocalizations,
infant <u>Antrozous</u> appear to have difficulty in localizing the sound
source. This behavioral observation is substantiated by collicular
responses to sounds presented at different angles. While evoked potentials show little directionality at any age, single unit recordings
in bats older than three weeks (but not younger than this) show a
high degree of directionality suggesting that binaural interactions
develop relatively late.

        <u>Myotis</u> <u>velifer</u> are born in a more advanced state than <u>Antrozous</u>
(Fig. 1), and at two days after birth are capable of emitting short
sonar sounds while crawling. Evoked potentials were first recorded
in two-day old bats (Brown and Grinnell, unpublished data). Increased
sensitivity to high frequency sounds developed more rapidly than in
<u>Antrozous</u>, with the evoked potential audiogram in a two-week old

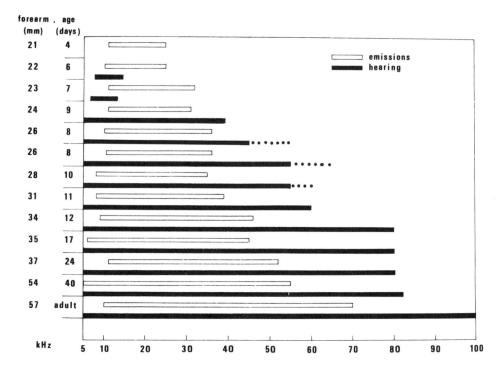

Figure 5. Range of frequencies emitted (first harmonic) compared to frequencies actually heard by Antrozous pallidus at different ages. Bats are arranged according to forearm length, which is a better index of maturation than absolute age. Range of hearing was determined by evoked potentials recorded with gross tungsten electrodes. Dotted lines indicate responses to higher frequencies in single units. Except during the first week postnatally, pallid bats emit sounds which they can hear. (From Brown et al., 1978.)

Myotis resembling that of an adult (Fig. 7). As in Antrozous the ability to hear high frequency sounds parallels the bat's ability to emit them (Fig. 8).

In contrast to Myotis velifer, young Myotis oxygnathus develop more slowly, with no behavioral or electrophysiological responses to sound before ten days of age, with adult sensitivity attained by 25 days. During this period temporal resolution improved from a partial recovery time of 20 msec in ten day old M. oxygnathus to 4 msec in adults. There was also a dramatic improvement in the ability of the auditory system to follow a train of stimuli (Konstantinov, 1973; Konstantinov and Stosman, 1972).

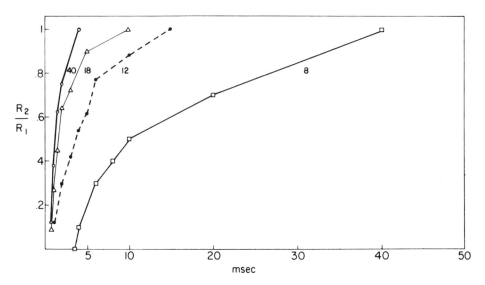

Figure 6.   Recovery of responsiveness of evoked potentials in
Antrozous pallidus of different ages (age in days next to each
curve). R_2/R_1 represents the ratio of amplitudes of the second
response to the first when two identical stimuli are presented.
A value of 1 signifies full response recovery.   The abscissa shows
the interval in msec separating the two stimuli.   (From Brown et al.,
1978.)

        Development of the auditory system is more rapid in the CF/FM
bat Rhinolophus ferrumequinum.   Stosman and Konstantinov (1972) first
recorded evoked potentials to auditory stimuli in week old bats, but
both frequency range and sensitivity rapidly increased until by 18-
21 days the audiograms resembled those of the adult.   However the
extremely sharp decrease in auditory threshold in the region ($\sim$70 kHz)
corresponding to the CF component of the adult echolocation signal
was not noted until 25 days postnatally.   Recovery of full responsiveness to the second of two identical stimuli within 10 msec was
first recorded in 12-16 day old bats.   Adult levels of temporal resolution were attained by 18-20 days, just before Rhinolophus begins
to fly.

        Grinnell and Hagiwara (1972) studied the electrophysiological
responses of a juvenile Hipposideros galeritus and an immature, non-
volant H. calcaratus.   Although the exact age of these bats was not
known, the evoked potentials were tuned to lower frequencies than in
the adults, corresponding to the lower emitted pulse frequencies of
the juvenile bats.   They postulated that "the range of frequencies
emitted may be regulated to correspond to the gradual changes in

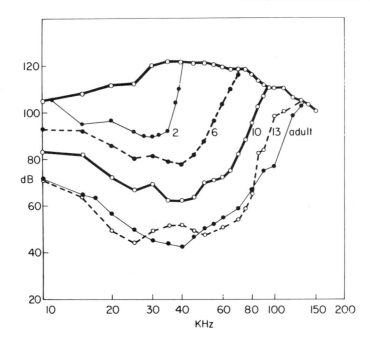

Figure 7. Audiograms of <u>Myotis velifer</u> as determined by evoked
potentials recorded from the posterior colliculus. The age in days
is to the right of each curve. Responses were first recorded in
two-day old bats, and by 13 days the audiogram resembled that of
an adult.

auditory capabilities as the inner ear matures" (Grinnell and Hagiwara, 1972).

    Baby <u>Pteronotus parnellii</u> and <u>suapurensis</u> are morphologically
immature and helpless at birth, and relatively immobile with their
eyes closed and skin naked. They emit a continuous isolation call,
which unfortunately is ignored by their mothers in captivity, resulting in surplus of one-day old subjects and none of later ages.
They would certainly be classified as altricial (Gould, 1975a), yet
their auditory responsiveness is the most developed of any bat studied
to date. Day old <u>P. suapurensis</u> have a tuning curve which closely
parallels that of the adult although slightly less sensitive (Fig. 9).
Day-old <u>Pteronotus parnellii</u> exhibit almost the full range of frequency response found in the adult, but lack the sharp tuning present
near the frequency of the CF portion of the sonar pulse (Brown and
Grinnell, unpublished data). They do exhibit the greatest sensitivity
at 20 kHz, a frequency represented in neonatal sound emissions (Fig.
10).

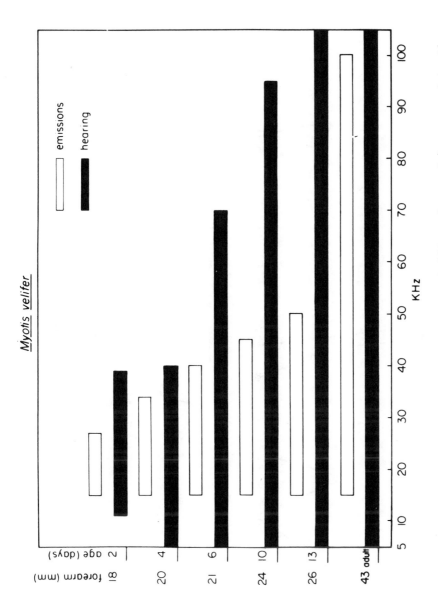

Figure 8. The range of frequencies emitted (first harmonic) compared to those heard by Myotis velifer at different ages and forearm lengths. After the age of two days, this species emits sounds containing frequencies that they can hear.

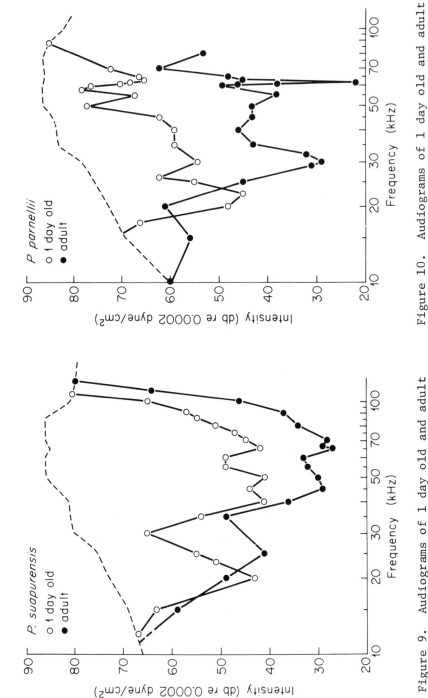

Figure 9. Audiograms of 1 day old and adult Pteronotus suapurensis as determined by evoked potential recordings from the posterior colliculus.

Figure 10. Audiograms of 1 day old and adult Pteronotus parnellii as determined by evoked potential recordings from the posterior colliculus.

Unfortunately, the development of the auditory system has not been followed in most bats, even those in which the adult capabilities have been well-documented. In those species that have been studied in detail, we find dramatic changes in the auditory system, observable from day to day, throughout ontogeny. An increase in absolute sensitivity is coupled to a shift in tuning toward higher frequencies correlated with an increase in the frequencies contained in vocalizations. There is a rapid acceleration in the rate of recovery of responsiveness, following initial exposure to a sound, probably one of the most impressive adaptations of the auditory system for echolocation (Grinnell, 1963b; Grinnell and Hagiwara, 1972b). Unfortunately the accompanying anatomical changes necessary to elucidate the observed neurophysiological maturation have not been described in baby bats, but probably reflect postnatal development both in the cochlea and neural circuitry (Brown, Grinnell and Harrison, 1978). Studies on other altricial mammals have shown that the cochlea and interneuronal connections in the brain continue to develop after birth. For example, the mouse inner ear is not fully differentiated until ten days postnatally (Weibel, 1957; Kikuchi and Hilding, 1965; Mikaelian and Ruben, 1965; Sher, 1971). In rats, major developments. of neural circuits in the forebrain continues for a month after birth (Caley, 1971), while brain stem connections proliferate for at least three weeks postnatally (Endroczi and Hartmann, 1968). It is conceivable that the long latencies and lack of synchrony observed in the first auditory evoked potentials recorded in baby bats arise from neural information being carried by a few, small, non-myelinated fibers with non-uniform diameters. The change in frequency range and sensitivity in bats can probably be attributed to the maturation of the cochlea. Studying tonotopic organization of best frequencies of units of Antrozous has shown that in an eight-day old bat they are uniformly distributed throughout the colliculus with best response at 15-20 kHz (Brown, Grinnell and Harrison, 1978). With an increase in age the surface units remain unchanged in frequency preference, but the tuning in deeper units is to higher frequencies (20-40 kHz in 10-day old bats, and 50-80 kHz in adults). Since even the earliest units detected appear to have complex response areas with narrowly tuned and closed curves, suggesting convergence and lateral inhibition, it is postulated that hair cells and primary afferents change their tuning to account for the frequency shifts. All of these hypotheses await actual confirmation by detailed ontogenetic neuroanatomical studies of the bat's auditory system.

Development of Echolocation Behavior

It is difficult to assess the roles of learning versus maturation in echolocation ontogeny since perfection of flight and echolocation skills must both occur simultaneously. Must bats learn to associate objects with returning echoes? How important is early experience to later mastery of echolocation skill? What is innate and what is learned?

A few observations have been made on the early flight behavior
of bats, but no comprehensive studies have been published on early
echolocation behavior either in natural or laboratory situations.
From our developmental studies of hearing and vocalizations, we are
now aware that these systems mature before the age of first flight.
However, juvenile bats may continue to nurse for as long as a month
after they actually begin to fly. Schmidt and Manske (1973) reported
nine-month old vampire bats nursing in a captive colony. This overlap between first flight and weaning allows time for perfection of
echolocation behavior before a bat has to forage on its own. Probably
a certain period is necessary in all species.Brown (unpublished data)
has noted a low survival rate (as judged by band recoveries) among
captive-born Antrozous, later released as adults into a wild colony.
Their chance of survival increased if they were released with their
mothers, but was never as great as if they were born and reared in
the wild colony. Captive-born bats may be still searching for the
dish of mealworms, never having developed an accurate auditory image
of their natural food (i.e. ground-dwelling arthropods for Antrozous).
The basic problem with field studies of early echolocation behavior,
is the inability to record from known-age bats. While this is rectified in captive-born bats, the natural context and relevant stimuli
may be absent. We have studied echolocation ontogeny in Antrozous
in captivity for ten years and recognize the natural sequence of
sonar pulse development. This past spring and summer (1978), Brown
monitored a natural maternity colony of pallid bats on Santa Cruz
Island, California. The problem of individual identification was
overcome by collaring mothers and pre-volant young (age determined
by forearm length)with FM radio transmitters tuned to narrow frequency bands. This enabled her to follow the movements and tape record the vocalizations of individual bats on their first flights. On
Santa Cruz Island, pallid bats roost in two-story brick barns, utilizing spaces within the thick brick walls as diurnal retreats.
Adults night roost from the attic rafters. Infant bats spend the
first weeks of life between the bricks, where their opportunity for
acoustic feedback is limited. Between three and four weeks postnatally they crawl around on top of the bricks after the mother de-
parts for nightly foraging periods. Within the next week they begin
to make short flights between rafters in the barn. The FM orientation
pulses emitted at this time contain lower frequencies than those of
the adults and are interspersed with communication sounds (directives). The repetition rate is relatively high (20-30/sec) and fairly
constant, with no apparent increase before take-off or landing.
Around five weeks postnatally the young bats venture out of the roost
for the first time, often accompanying their mothers. These initial
flights are usually directly to another building within a 200 meter
radius. Sometimes they spend most of the night flying back and forth
between buildings. During this time the juveniles' echolocation
pulses are distinguishable from those of the adults by their slightly
lower frequency content. Between five and six weeks the young bat

will spend periods of five to ten minutes away from the roost, and
during this time they are often detected circling low over adjacent
fields. The echolocation pulses of adults and juveniles are now
indistinguishable. By seven weeks, the bats' nocturnal activity
pattern is similar to the adults. They are away from the roost for
approximately two hours after dark, gather with others in the night
roost for most of the night, and often engage in a predawn hunting
foray. In the night roost, juveniles less than eight weeks old were
observed nursing. Therefore, even if unsuccessful in their early
hunting attempts, they were assured of a meal. Since our recording
equipment is not portable enough to allow us to follow a flying bat
in the field, we are now developing a crystal-controlled transmitter
with a condenser microphone to remotely monitor sonar signals.

Buchler (in preparation) has also conducted field observations
on early echolocation behavior in Myotis lucifugus. His colony
roosted under the shingles of a building and did not have access to
the interior of the building in which to practice initial flights.
Although Myotis develops more rapidly, the basic pattern was the
same as in Antrozous. He attached radioactive tantalum tags to six
infant bats in order to follow their movements. On some bats capable
of flight he attached temporary chemoluminescent spheres. At about
fifteen days of age, young bats were observed to leave the roost after
the adults had departed. They would fly slowly near the roost, landing on large stationary objects such as walls and tree trunks. During
this time they emitted pairs of echolocation pulses, a phenomenon not
observed in older foraging bats or adults. Buchler notes that paired
ultrasonic pulses are also produced by wandering shrews, tenrecs and
swiftlets, animals that use their pulses presumably for obstacle
avoidance. They may also serve this function for naive bats. Adult
M. lucifugus do not emit the double note while hunting. At this early
age the juvenile bat may be forming an "auditory" picture of its immediate environment. Within a few nights the young bats venture into
the surrounding woods and attempt their first insect captures. Some
bats employ a "flycathcatcher" technique, flying out from a tree trunk
to ambush an insect, then returning to it. At this intermediate stage,
the juvenile bat remains stationary and the target moves, while in
the final and most difficult step, a moving bat is confronted with
moving prey. Older bats gradually increase their foraging range,
flight duration and stamina, but do not yet perform the elaborate
aerial pursuit maneuvers observed in adults. They do begin to emerge
at the same time as the adults and fly with them along traditional
routes to feeding areas. It is difficult to assess how much information is transferred from adult to juvenile bats about foraging
areas and techniques.

In the laboratory we can study echolocation development under controlled conditions although it is possible that some aspects of behavi-
or may not develop at natural rates. Wire avoidance tests have been a
standard method for assessing bat echolocation ability. Normally, fly-

ing bats will increase the repetition rate of echolocation pulses as they approach and detect an object (Griffin, 1958). By measuring the point at which this increase occurs, the distance of initial detection can be determined.

The initial flights of five week old pallid bats through vertical strings were recorded in our laboratory. Often only a slight increase in repetition rate or none at all occurred on string approach, and more likely than not a collision resulted. By the second or third pass an increase in repetition rate was noted prior to negotiating the obstacle. Many flights were usually necessary before their accuracy at avoidance equalled that of adults. Many did not possess the flight skills to accomplish this maneuver. In some instances, a week of practice in a large flight cage free of obstacles was necessary before they could avoid the strings. This may be analogous to the practice period for young wild bats in the barn attic or between buildings.

Pallid bats were raised by their mothers under a variety of conditions in an attempt to alter their early acoustic environment and hence auditory experience. The control situation was a large flight room. The most natural of these altered environments was raising the bats in small containers approximating a crevice. When removed from their homes at five weeks of age, these bats had no difficulty in echolocating or flying, yet did not possess the stamina of those raised in the large flight room. In an attempt to prevent them from hearing their own echoes, some bats were reared in foam-lined, sound-proof, "anechoic" boxes, and others in chambers in which they were constantly bombarded with loud "white noise". Neither condition affected their future ability to echolocate. Deficits were apparent, however, in bats that had been fitted with monaural or binaural ear plugs. If the plugs remained intact between seven and twenty-one days postnatally, echolocation ability as an adult was impaired. Although there were a few exceptions, the majority of these bats were unable to avoid string obstacles less than 50% of the time compared with 70% for controls. All bats of any group that could not fly well were eliminated from the testing procedure. The monaural group showed consistently poorer performance than the binaurals. This deficit still exists more than two years later. We could not block the hearing of self-emitted sounds, but possibly the perception of echoes is important during a critical developmental period.

Gould (in preparation) also attempted to alter the early acoustic environment of _Eptesicus fuscus_ by raising them in a helium-oxygen atmosphere, where sound travels approximately 1.5 times faster than in air. The performance of these bats on initial flights in air through wire obstacles was tested and compared to that of control, captive-raised young, and wold bats with the same forearm length. Surprisingly, no difference was observed in bats raised in the He-O_2 chamber. All bats on their initial flights increased their pulse repetition rate upon obstacle approach. For young bats the distance at which this change in repetition rate occurred was half that of adults. Even on initial flights, many naive bats could avoid the obstacles. Pulse patterns changed with age and experience. There was an increase in pulse interval during the search phase and a decrease in pulse interval during the terminal phase of target approach, coupled with a decrease in pulse duration.

These preliminary observations suggest that some aspects of echolocation ontogeny such as the production of sonar pulses and auditory tuning are genetically predetermined and refractory to experimental manipulation. Others, such as the development of echolocation behavior, may be influenced by experience and environment. Current research is now aimed at clarifying these relationships.

Acknowledgements

This research was supported by National Science Foundation grants. We would like to thank Drs. Jim Simmons, Brock Fenton, and Jean Harrison for the use of their equipment; Mr. R. Roverud, Ms. C. Brown and Dr. Timothy Brown for assistance in the field; and Ms. N. Meyers for the development of the bat radio transmitter. Special appreciation is due Dr. T. Brown for comments on clarity and style of this manuscript.

References

Brown, P.E., Vocal communication and the development of hearing in the pallid bat, Antrozous pallidus. Ph.D. Thesis, UCLA, 144pp., 1973.

Brown, P.E., Vocal communication in the pallid bat, Antrozous pallidus. Zeit Tier Psychol., 41, 34-54, 1976.

Brown, P.E., Grinnell, A.D. and Harrison, J.E., The development of hearing in the pallid bat, Antrozous pallidus. J. Comp. Physiol., 126, 169-182, 1978.

Caley, D.W., Differentiation of the neural elements of the cerebral cortex in the rat. In: Cellular aspects of neural growth and differentiation. Ed., D. Pease, University of California Press, Los Angeles, pp. 73-102, 1971.

Davis, W.H., Barbour, R.W. and Hassell, M.D., Colony behavior of Eptesicus fuscus. J. Mammal, 50, 729-736, 1968.

Douglas, A.M., The natural history of the ghost bat, Macroderma gigas in Western Australia. W. Austral. Nat., 10, 125-137, 1967.

Gould, E., Echolocation and communication in bats, In: About Bats. Eds. B. Slaughter and D. Walton, Southern Methodist Univ. Press, Dallas, pp. 144-161, 1970.

Gould, E., Studies of maternal-infant communication and development
    of vocalizations in the bats Myotis and Eptesicus, Communications
    in Behavioral Biology. Part A, 5, No. 5, pp. 263-313, March,
    1971.
Gould, E., Neonatal vocalizations in bats of eight genera. J. of
    Mammalogy, 56, no. 1, 15-29, February, 1975a.
Gould, E., Experimental studies of the ontogeny of ultrasonic vocal-
    izations in bats. Dev. Psychobiol., 8, 333-346, 1975b.
Gould, E., "Echolocation and Communication" in Biology of Bats of
    the New World Family Phyllostomatidae, Part II, Ed. Baker, R.J.,
    J.K. Jones, Jr., and D.C. Carter, Special Publications the
    Museum Texas Tech. University, No. 13, pp. 247-279, 1977.
Gould, E., Infant calls of 10 species of Malaysian bats (Micro and
    Megachiroptera). Poster presented at Animal Sonar Systems
    Jersey Symposium, 1979.
Gould, E., Woolf, N.K. and Turner, D.C., Double-note communication
    calls in bats: occurrence in three families, J. of Mammal.,
    Vol. 54, No. 4, 998-1001, 1973.
Grinnell, A.D. and Hagiwara, S., Adaptations of the auditory nervous
    system for echolocation. Studies of New Guinea bats, Z. vergl.
    physiologie, 76, 41-81, 1972a.
Grinnell, A.D. and Hagiwara, S., Studies of auditory neurophysiology
    in nonecholocating bats, and adaptations for echolocation in one
    genus, Rousettus. Zeit. Vergl. Physiol., 76, 82-96, 1972b.
Grinnell, A.D., The neurophysiology of audition in bats: Temporal
    parameters. J. Physiol.(Lond)., 167, 67-96, 1963.
Jones, C., Growth, development, and wing loading in the evening bat
    Nycticeius humeralis (Rafinesque). J. Mammal, 48, 1-19, 1967.
Kay, L. and Pickvance, T.J., Ultrasonic emissions of the lesser
    horseshoe bat, Rhinolophus hippersideros (Bech), Proceed. Zool.
    Soc., Lond., 141, 163-171, 1963.
Kikuchi, K. and Hilding, D.A., The development of the organ of Corti
    in the mouse. Acta Oto-Laryngal, 60, 207-222, 1965.
Konstantinov, A.I., Development of echolocation in bats in postnatal
    ontogenesis. Period Biol., 75, 13-19, 1973.
Konstantinov, A.I. and Stosman, I.M., Electrical activity in the
    inferior colliculus under the effect of ultrasound stimuli in
    ontogeny in bats of the genus Myotis oxygnathus.Zh. Evol. Bio-
    khim. Fiziol., 8, 182-188, 1972.
Kulzer, E., Über die jugendentwicklung der Angola-Bulldog fleder
    maus Tadarida (Mops) condylura, (A. Smith, 1933) (Molossidae).
    Säugeteirkundl. Mett., 10, 116-124, 1962.
Kunz, T.H., Population studies of the cave bat (Myotis velifer):
    reproduction, growth, and development. Occ. papers, Mus. Nat.
    Hist., Univ. of Kansas, 15, 1-43, 1973.
Matsumura, S., Mother-infant communication in a horseshoe bat (Rhino-
    lophus ferrumequinum nippon), I. Development of vocalization.
    Submitted to J. Mamm., 1979.
Mikaelian, D. and Ruben, R.J., Development of hearing in the normal
    CBA-J mouse. Acta Oto-Laryngal, 59, 451-461, 1965.

Orr, R.T., Natural history of the pallid bat, Antrozous pallidus.
    Proc. Calif. Acad. Sci., 28, 165-248, 1954.
Orr, Robert T., "Development: Prenatal and Postnatal" in Biology
    of Bats. Vol. 1, Chapter 6, Ed. by W.A. Wimsatt, Academic Press,
    New York, pp. 217-231, 1970.
Pearson, O.P., Koford, M.R. and Pearson, A.K., Reproduction of the
    lump-nosed bat (Corynorhinus rafinesquii) in California. J.
    Mammal., 33, 273-320, 1952.
Porter, F.L., Social behavior and acoustic communication in the bat,
    Carollia perspicillata. Ph.D. dissertation, Washington University,
    169pp., 1977.
Schmidt, V., Social calls of juvenile vampire bats (Desmodus rotundus)
    and their mothers. Bonn. Zool. Beitr., 4, 310-316, 1972.
Schmidt, V. and Manske, V., Die jugendentwicklung der vampirfleder-
    mäuse (Desmodus rotundus). Z. Saugetierkunde, 38, 14-33, 1973.
Schnitzler, H.U., Die ultraschull-ortungslaute der Hufeisen-Fleder-
    mäuse (Chiroptera-Rhinolophidae) in verschiedenen Direntierung-
    ssituationen. Z. Vergl. Physiol., 89, 275-286, 1968.
Sher, A.E., The embryonic and postnatal development of the inner ear
    of the mouse. Acta Oto-Laryngal (Suppl), 285, 1-77, 1971.
Stosman, I.M. and Konstantinov, A., Characteristics of evoked poten-
    tials of inferior colliculus of the bat Rhinolophus ferrum-
    equinum. Zh. Evol. Biokhim i Fiziol., 8, 612-616, 1972.
Weibel, E.R., Zur kenntnis der differenzieurungs-vörgange im epithel
    des ductus cochlearis. Acta Anatomica, 29, 53-90, 1957.
Woolf, N., The ontogeny of bat sonar sounds: with special emphasis
    on sensory deprivation. Unpublished doctoral thesis, Johns
    Hopkins University, Baltimore, Maryland, 140pp., 1974.

Chapter III
Adaptiveness of Echolocation

Chairman : G. Neuweiler

co-chairmen

F.G. Wood **and** M.B. Fenton

- <u>Underwater</u>
  Adaptiveness and ecology of echolocation in toothed whales.
  F.G. Wood and W.E. Evans

- <u>Airborne</u>
  Adaptiveness and ecology of echolocation in terrestrial (aerial) systems.
  M.B. Fenton

ADAPTIVENESS AND ECOLOGY OF ECHOLOCATION IN TOOTHED WHALES

F. G. Wood* and W. E. Evans**

*Naval Ocean Systems Center, San Diego, CA 92152

**Hubbs-Sea World Research Institute, San Diego, CA 92109

Of all aquatic animals only odontocete cetaceans are known to have a well developed sonar system. While certain other marine vertebrates will be discussed briefly in the latter part of this review, we are primarily concerned with the toothed whales.

EARLY BEHAVIORAL OBSERVATIONS

Before discussing the ecological and adaptive aspects of echolocation in odontocetes, it may be of value to briefly review the 30 year history of observations of this phenomenon.

In 1947, Arthur F. McBride, curator of Marine Studios (later Marineland of Florida), recorded some observations of dolphin (*Tursiops truncatus*) behavior that suggested to him the existence of "some highly specialized mechanism enabling the porpoise to learn a great deal about his environment through sound." The observations were made in the course of capture operations undertaken in very turbid water and usually at night. McBride noted that the animals: (a) would avoid a fine-mesh net but readily charge a large (10-inch square) mesh net, and (b) would immediately go over the net at the point where the cork line was momentarily pulled below the surface by others striking the net from below. The possibility of bioluminescent clues was considered, but no bioluminescence was visible to a human eye above the surface. McBride was reminded of the "sonic sending and receiving apparatus which enables the bat to avoid obstacles in the dark, "and, further, saw possible significance in the enormous development of the dolphin's cerebral cortex and the obvious importance of the acoustic

sense to these animals. This was the first direct inference
backed by good evidence that dolphins use sound for navigation.

Following McBride's untimely death, his notes remained in
the files of the Marineland of Florida research laboratory until
one of us (F.G.W.) happened on them in the early 1950's. During a
visit to Marineland by W. E. Schevill in 1955, the notes were
called to his attention. By then there was evidence that the
bottlenose dolphin could both emit and hear sounds in the ultrasonic frequency range (> 20 kHz). This led Winthrop Kellogg to
speculate increasingly on the existence of dolphin sonar (Kellogg
and Kohler, 1952; Kellogg, Kohler and Morris, 1953; Kellogg, 1953).
Schevill thought that McBride's priority in this matter should be
recognized, and offered to submit the notes for publication. Thus,
McBride's observations, with an introductory note by Schevill,
became a historically important item in the literature (McBride,
1956).

Well before McBride's observations became known, a number of
workers had harbored the suspicion that cetaceans could echolocate.
The first rather crude experiments to determine the hearing range
of the Atlantic bottlenose dolphin revealed that these animals
would respond to ultrasonic frequencies (Kellogg and Kohler, 1952;
Kellogg, Kohler and Morris, 1953; Schevill and Lawrence, 1953).
The turbid coastal waters inhabited by *Tursiops truncatus* suggested
that this species, and logically others, might well use echolocation for food finding and navigation. However, Schevill and
Lawrence (1953), in their auditory response experiment with a
dolphin in the near-opaque waters of a fenced off lagoon at
Florida's Marineland, listened carefully for echolocation signals
but heard none. To them the question remained open.

When one of us (F.G.W.) came to the Florida Marineland as
curator in 1951, he found that one of his predecessors had
acquired a U.S. Navy underwater sound projector (type CFF 78187).
Using this as a hydrophone, in conjunction with an amplifier, tape
recorder, and speaker or headset, he began to record and listen to
the sounds produced by the dolphins, *Tursiops* and, usually, one or
two spotted dolphins, *Stenella plagiodon*, that were maintained in
a large oceanarium tank. The hydrophone was suspended in front of
a porthole so that it was possible, while listening, to also record
notes on the behavior of the animals. From this work (Wood, 1952,
1953) two observations are relevant here. Whenever the transducer
was hung in the tank, the dolphins exhibited their normal curiosity
of any strange object or new introduction. As they swam by a few
feet from the transducer, they would peer at it and at the same
time emit rasping and grating sounds. This suggested that they
were "echo-investigating" the hydrophone. The same response could
be elicited by other objects, such as a bucket or length of pipe.

In retrospect, it seems apparent that although the animals could see the object clearly, they were also relying on the different kind of information derived from the echoes of their sound emissions.

The other observation of interest in the present context had to do with the animals during feeding. When a diver descended into the tank and began his circuits of the bottom, handing out fish, a constant subdued combination of rasping and mewing sounds were heard. When the last fish was handed out, the diver would raise the empty basket above his head, and the dolphins, which had been clustering about him, would immediately leave and display no further interest. At this time the undercurrent of sound would cease or diminish quite abruptly. Again, in light of later knowledge it seems clear that the mewing and rasping sounds were echolocation pulse trains which sounded subdued because the animals were echolocating directly on the fish being handed out some distance from the hydrophone. When they echolocated on the hydrophone they often overloaded the equipment. It could be inferred from this that echolocation is very closely tied to feeding activity, even occurring when the food is clearly visible.

The later experiments of Schevill and Lawrence (1956) and Kellogg (1958) effectively removed any doubt that the Atlantic bottlenose dolphin does indeed have a sonar capability, although in none of this work was vision completely excluded. Incontrovertible evidence was provided by Norris et al. (1961) who successfully blindfolded a dolphin by covering its eyes with latex suction cups. Their trained Atlantic bottlenose swam normally when blindfolded and avoided obstacles, including pipes suspended vertically to form a maze. The dolphin oriented to fragments of fish tossed in the water and took them as they drifted downward. Norris noted that when a bit of food drifted below the level of the melon, the animal did not respond to it. From this he speculated that the sounds were directional and being projected from the melon.

During the same period in the latter half of 1959 when Norris and his colleagues were attempting to devise a blindfold for dolphins, one of us (F.G.W.) was similarly engaged, unsuccessfully, at the Florida Marineland. Hearing of Norris' success using suction cups to occlude a dolphin's vision, Wood used the same technique to blindfold a recently caught *Tursiops*. The animal was untrained and had to be caught in order for the suction cups to be applied. Once they were in place, however, the dolphin swam about the tank with no indication that it lacked vision, and went directly to fish tossed in the water, taking them as they drifted down.

This Atlantic *Tursiops* had been taken in turbid inshore waters. Norris (1969) described three later observations of dolphins taken from clear oceanic waters which became disoriented when unable to use vision. A *Delphinus delphis*, which emits typical click trains at sea, when introduced into a tank filled with dirty water "may hit the walls frantically, force its head out of water, and sometimes may drown in its frenzy." A trained Pacific whitesided dolphin, *Lagenorhynchus obliquidens*, when blindfolded promptly ran into the side of its tank, and only after a period of days learned to navigate without vision. In the third instance a *Tursiops* captured in Hawaiian waters was taught to push a paddle on command. After being blindfolded the animal failed utterly to find the paddle despite the fact that it was 20 cm in diameter and covered with acoustically reflective foam Neoprene.

From these observations Norris concluded that there is very likely a large learned component in the echolocation behavior of marine mammals, and that animals living in clear seas may use their systems in quite different ways from those that inhabit turbid inshore waters.

The early speculation regarding a sonar capability in dolphins was based on behavioral observations. It was reinforced by the assumption that such a capability would have ecological adaptiveness for animals that navigate and catch their actively swimming prey in turbid water or in darkness, as at night or at depths.

THE INVESTIGATION OF ODONTOCETE ECHORANGING AND CONSIDERATION OF ITS ADAPTIVENESS

Observations and experiments in the 1950's confirmed the existence of odontocete echolocation and provided some insight into its adaptive features. The work that followed in the 60's concentrated on refining knowledge of the effectiveness and complexity of this remarkable sensory specialization.

Norris et al. (1967) and Evans and Powell (1967) experimentally demonstrated that blindfolded Atlantic bottlenose dolphins not only detect, avoid, or find objects, but can also discriminate finite differences in "targets" that affect their acoustic reflectivity, e.g., size, shape, and mass (density). During this same time period Busnel and colleagues in France and Denmark had demonstrated that the harbor porpoise, *Phocoena phocoena*, is also an echolocator (Busnel and Dziedzic, 1967).

The results of these studies were presented, along with correlative data on the hearing of the species concerned, at the international conference on biosonar held in Frascati, Italy, in 1966. The work received comment and critique from biologists who had been studying the sonar system of bats and from sonar engineers who also participated in the conference (Busnel, 1967). This critical look at echolocation in dolphins and porpoises raised significant questions which were to shape some of the research that followed. The demonstrated capabilities were impressive, but how important was this sophisticated sensory capability in navigation or food finding? What acoustic information was being used by the animals studied? The bottlenose dolphin and harbor porpoise inhabit inshore waters and even rivers where the water is murky and sometimes full of suspended debris that could represent false targets and navigational hazards. Did open-sea species such as the common dolphin and pilot whales echolocate? Although short-duration clicks had been heard from all odontocetes recorded, evidence of production of click trains was not evidence of an echolocation capability, although many authors referred to these acoustic emissions as echolocation clicks.

In 1965 Soviet scientists began to replicate some of the work of Western investigators using those species they had available in the Black Sea: the harbor porpoise, *Phocoena phocoena*, the common dolphin, *Delphinus delphis*, and a subspecies of *Tursiops*. By 1973 the literature had increased immensely. The number of species for which an echolocation capability had been demonstrated experimentally was growing, the source levels and frequency content of the pulses used were being measured with increasing accuracy, and signal processing techniques were being employed to gain a greater understanding of the total delphinid sonar system (Evans, 1973).

In 1967 the odontocete species that were known to use echolocation were *Phocoena phocoena* and *Tursiops*. By 1973 several additional species had been added to this list. Penner and Murchison's work (1970) on *Inia* indicated that this species could discriminate the difference in diameter of thin wires presented as pairs behind a visually opaque, acoustically transparent screen. The common dolphin, *Delphinus delphis*, could detect differences in complex geometric figures as well as *Tursiops* could (Belkovich et al., 1969). A preliminary study was done with a Pacific pilot whale, *Globicephala*, trained to wear latex eye cups and retrieve rings thrown into the water (W. E. Evans and L. L. Clarke, 1972, unpub.). As mentioned earlier, a Pacific whitesided dolphin, *Lagenorhynchus obliquidens*, learned to avoid obstacles without the aid of vision (Norris, 1969). This same species was tested with the task used by Evans and Powell (1967) to demonstrate echolocation discrimination in *Tursiops* and performed as well as the latter species (Evans, 1973). During a study designed to determine hearing thresholds, a captive killer whale, *Orcinus orca*, was also trained to retrieve a ring while blindfolded (Hall and Johnson, 1971).

The instrumentation now being used was considerably more sophisticated than that used during the 1950-1960's. The total recording system of hydrophones, associated amplifiers, and tape recorders had much broader responses — 100 Hz - 120 kHz typically, compared to 50 Hz - 20 kHz. Information on frequency and waveform characteristics of the pulse increased significantly, but the complexity of interpreting the data also increased.

Most of the experiments being done under laboratory conditions in the 1970's were still asking questions as to just how good the echolocation system is, without much reference to what this all means to the animal in the real world. In contrast, much of the information on bats was based on field observations and recordings of sounds produced by various species in free flight and during feeding.

Questions as to the ecological and adaptive nature of echoranging in various species of small toothed whales were raised in 1969 by Norris, and were again briefly addressed by Evans (1973). In Evans' discussion of the echolocation pulse wave-forms used by various species, he divided the animals by type of habitat. The small odontocetes inhabit very diverse aquatic environments. Some species are found exclusively in rivers and lakes (e.g., *Inia*, *Platanista*). Some prefer tidal flats, muddy bays, and estuaries (e.g., *Phocoena*, *Tursiops*). Some species occur almost exclusively in the open sea (e.g., *Delphinus*), while others are generally found in the open sea but venture from time to time into the shallower waters around islands (e.g., *Globicephala* and *Lagenorhynchus*). A few primarily inhabit arctic waters, close to the ice pack or up rivers and fjords (e.g., *Delphinapterus* and *Monodon*). Two odontocetes, *Orcinus* and *Tursiops*, appear to have a greater zoogeographic range and occupy a wider variety of aquatic environments than any others. The echo-location pulses produced by various species were placed by Evans in specific categories: 1) narrow-band, low-frequency sonic pulses, 400 Hz to 20 kHz, with peak energy at 2 to 4 kHz; 2) broad-band sonic to low-frequency ultrasonic, 100 Hz to 30 kHz, with peak energy at 16 to 20 kHz; 3) broadband, high-frequency ultrasonic, 16 to 150 kHz, with peak energy at 60 to 80 kHz; and 4) broadband sonic to ultrasonic, 200 Hz to 150 kHz, with peak energy occurring at 30 to 60 kHz.

The first of these categories (narrow-band, low-frequency sonic) was originally believed to be descriptive of *Phocoena*, *Inia*, and *Delphinapterus*, but improvements in instrumentation and measur-ing techniques revealed that assignment to this category was more a function of limitations in the recording systems than it was the nature of the signals used by the animals. However, this category still has some validity, since later studies have shown that these species may operate in more than one mode.

An additional error can result from the directional characteristics of echolocation signals, the frequency content of recorded
pulses depending on the angle of the animal to the hydrophone
(Norris and Evans, 1967; Au et al., 1978; Watkins, this volume).
The high frequency content of the pulse is directed essentially to
the front of the animal. At 10^0 to 15^0 to the right or left the
frequency decreases rapidly (Norris and Evans, 1967).

Although the harbor porpoise, Amazon River dolphin, and
beluga were originally thought to produce pulses with predominantly
or only low-frequency content, this was not consistent with the
observed echolocation detection and discrimination capabilities,
nor, in the case of *Phocoena*, with the finding that its hearing
extended beyond 120 kHz (Mohl and Andersen, 1970). The later
studies using broadband instrumentation demonstrated that while
all of these species produced pulses with low-frequency components,
when echolocating under conditions of poor or no visibility their
pulses contained considerable energy in the ultrasonic range
(Dubrovsky et al., 1971; Herald et al., 1969; Penner and Murchison,
1970; Gurevich and Evans, 1977; Mohl and Andersen, 1973).

It is of interest to note that these species, as well as the
Indus River dolphin, *Platanista indi*, which also falls in this
category (Herald et al., 1969), share certain environmental and
ecological similarities. They inhabit turbid waters, feed on very
small prey, and must frequently have to distinguish their prey from
drifting debris.

While the only other member of the Monodontidae, the narwhal,
differs from the beluga ecologically in some respects, being a deep
water species (Sergeant, 1978), teuthophagous, and apparently
diving to considerable depths when feeding (Tomilin, 1957), its
pulses, like those of the beluga, are relatively narrow band and
quite structured. The narwhal has a low band from .5 to 5 kHz
and a high band from 12 to 24 kHz (Ford and Fisher, 1978; see also
Watkins and Schevill, 1971).

In contrast to the species discussed above, all of which
occupy rather narrow ecological niches, *Tursiops* emits broadband
pulses. As mentioned earlier, populations of *Tursiops* occur in a
wide range of habitats, from the open sea to bays and rivers.
From measurements made in tanks, most of the energy in *Tursiops*
pulses was found to peak between 35 and 60 kHz (Evans, 1973). However, in a target detection experiment conducted with two Atlantic
bottlenose dolphins in the open waters of a Hawaiian bay, Au et al.,
(1974) recorded pulses with peak energies between 120 and 130 kHz.
The conditions at the site--shallow water with high ambient noise
conditions caused primarily by snapping shrimp--were very similar
to environments in which *T. truncatus* is often found. By shifting

the peak frequencies of their pulses to well above 100 kHz the
animals were presumably optimizing the signal-to-noise ratio, and
in so doing revealed an adaptive capability not previously observed
under the less realistic conditions of a laboratory tank.

In Figure 1 representative pulse waveforms for several of the
species discussed are shown. Detailed acoustic and ecological data
for *Tursiops*, *Delphinus*, *Orcinus*, *Inia*, *Phocoena*, and *Delphinapterus*
are presented in the Appendix.

In April 1969, in the course of conducting a series of echolocation discrimination tests (Diercks et al., 1971) with a female
Atlantic bottlenose dolphin named Scylla, it was decided to conduct
an experiment that had been suggested many times: testing the
efficiency of a blindfolded *Tursiops* in catching a live free-
swimming fish. The dolphin, as in the discrimination tests, was
instrumented with seven broadband hydrophones fabricated at the
Applied Research Laboratories, University of Texas. Five of the
hydrophones were directed inward to monitor the transmitted echolocation signals, and two, attached to the blindfolding eyecups,
were directed forward to detect target echo signals (Fig. 2). The
seven hydrophones were connected through an umbilical to poolside
preamplifiers and magnetic tape recorders. In addition, two
Atlantic Research (now Celesco) Type LC-10 broadband hydrophones
were positioned in front of the two targets and a third hydrophone,
USRD/NRL Type BM101A, was located behind the target plane, all
three connected via amplifiers to tape recorders. The movements
of the dolphin were monitored by two video cameras, one above the
pool and one underwater. The configuration of the test tank and
equipment is shown in Fig. 3.

Echolocation signals were detected by the monitoring hydrophones during all discrimination trials prior to the live fish
experiment. In the last trial before the fish was introduced the
targets were a metal sphere and a dead fish suspended by its tail.
Following this, a metal sphere was suspended in the left target
location to maintain semblence of another routine echolocation
trial. As Scylla approached the target area, emitting echolocation
clicks, a live fish (a 20 cm sargo, *Anisotremus davidsoni*) something she had not encountered during her several years in captivity,
was dropped into the water at the right hand target location.

The dolphin, without emitting any detectable echolocation
signals, positioned herself adjacent to the fish as indicated in
Fig. 4. She maintained this position with respect to the fish as
it swam around the perimeter of the pool. No echolocation signals
were detected by any of the ten hydrophones during this circuit.
When the fish reached the nylon net suspended in the passage
connecting the two pools it darted through. Scylla rolled and

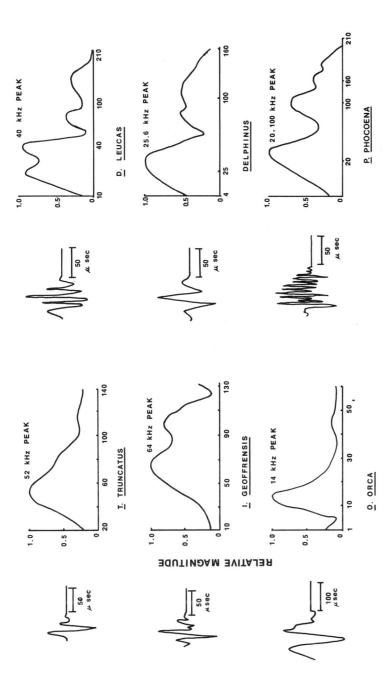

FREQUENCY, kHz

FIGURE 1. Representative echolocation signal, pulse waveforms, and associated spectra for six Odontocete cetaceans.

FIGURE 2.   Transducer locations of the head of the bottlenose dolphin
used in live fish experiment.

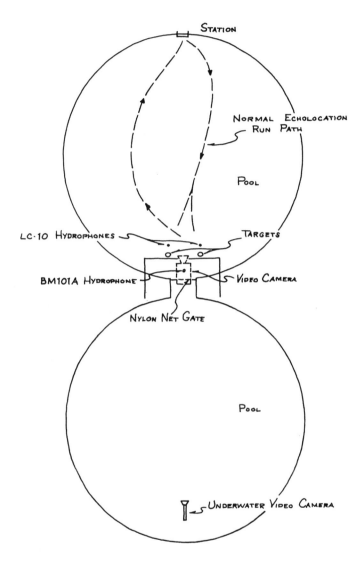

FIGURE 3.   Schematic of experimental pools showing path of animal
during 'normal' echolocation run on non-moving targets.

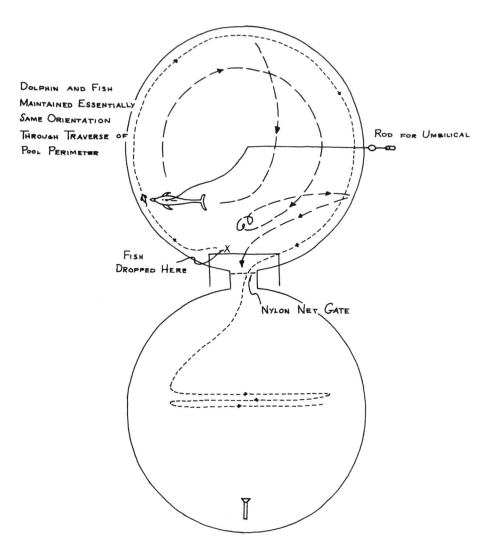

FIGURE 4.   Path of instrumented bottlenose dolphin and fish during
            the first trial.

became entangled in the hydrophone umbilical. She was then held at the side of the pool while all of the attached hydrophones and the two eyecups were removed. When released she returned to the nylon net where she could now see the fish in the adjacent pool. At this time she began to emit clicks at a rate that rose as high as 1500 per second. The repetition rate reached a level that she was apparently unable to maintain; her head would jerk, she would stop emitting, then begin again to the next head jerk, and so on. Subsequently, the fish was netted and tossed back into the test pool where Scylla, again blindfolded, now caught it and returned it to her trainer, all without emitting a detectable sound. This was repeated four more times, with the fish returned dead after the fifth retrieval. It was then hand-fed to her, and she readily ate it.

This experiment is described in some detail because in its various circumstances it is probably unique, and it appears to be significant. For the first part of the test we can only surmise that passively received acoustic information enabled the blind-folded dolphin to track the fish. Since she made no attempt to catch it during a complete circuit of the tank, and later retrieved it five times, we can also surmise that she didn't con-sider the live fish a food object. Scylla's performance indicated that these animals are able to detect, track, and catch fish with-out using echolocation or vision. This mode of acoustic behavior may be more commonly used in nature than we tend to suppose, and perhaps is attributable to the "mixed blessings" of echolocation discussed by Fenton (this volume).

Scylla's vocal behavior and head jerks when she could see the fish in the other pool but not get at it because of the net is probably best explained as an emotional outburst. Long-term obser-vations of several *Delphinus*, recorded on magnetic and video tape at the Hubbs-Sea World Research Institute, strongly suggest that click train emissions often occur in social contexts and represent emotional and perhaps communicative behavior.

We can study the odontocete echolocation capability as a relatively well-defined system with three primary functional compo-nents: 1) sound production, 2) echoreception, and 3) acoustic echo processing. In the context of ecological adaptations, however, as this last experiment demonstrated, echolocation cannot be viewed as distinct from another of the animals' acoustic attributes: that having to do with reception and interpretation of sounds occurring in their environment. This latter capability poses difficult prob-lems for the investigator. Our perception of the world is inherently limited by our particular ensemble of sensory equipment. We cannot know, or even imagine, the perceptual world of other animals that differ markedly from us in sensory capabilities. We can experimentally "ask questions" of them using behavioral or electrophysiological techniques, and obtain certain information regarding their sensory threshold ranges and apparent limitations. But we cannot know the complete gestalt of their perception and how that perception is adapted to the environment in which they live.

Still other aspects of sound production in odontocetes that apparently cannot be viewed as completely distinct from echoloca- tion are those having to do with emotional state (as seemed to be demonstrated in the live fish experiment) and communication. It has commonly been assumed that clicks are echolocation signals, while whistles and other phonations are the ones that have communi- cative or emotional significance (see, for example, McBride and Kritzler, 1951; Wood, 1953; Schevill and Lawrence, 1956; Kellogg, 1961). However, there is growing evidence that clicks may serve other purposes. Lilly and Miller (1961) assigned a communicative function to slow click trains of *Tursiops truncatus*. Schevill and Watkins (1962) and Watkins (this volume), while not precluding sperm whale clicks for echo information, provide evidence that clicks are used as communication signals. In a well controlled experiment designed to test whether one bottlenose dolphin could tell another, separated by a visual barrier, to push one or the other of two paddles, depending on whether the first animal saw a flashing or steady light, the two performed with a very high percentage of correct responses. Although both whistles and clicks were emitted by the first dolphin, analysis of the tapes revealed that performance success depended directly on her emission of click trains (Bastian, 1967; see also Evans and Bastian, 1969, p. 433, for the conclusion that the animals had, in effect, trained them- selves during the long period of preparation for the tests, and that intentional communication was not demonstrated).

One other possible function of clicks has been suggested. In consideration of the problem sperm whales have in catching squid in the darkness of great depths, the fact that otherwise healthy sperm whales with badly deformed lower jaws have been taken by whalers, and other factors, Soviet investigators have speculated that a sperm whale may search the water in front of it with lower- frequency echolocation. When a squid or fish is discovered and approached, the whale alters its click structure to higher frequen- cies, which narrows and concentrates the beam so that the prey is stunned and can be readily seized (Berzin, 1971).

ODONTOCETE ACOUSTICS AND ANATOMICAL MODIFICATIONS

The importance of echolocation to toothed whales is indicated by modifications of the skull, ear bones, and soft structures of

the head. Some of these features are apparent in early fossil
remains from the Lower Miocene. We consider it possible that echolocation was at one time more important to the success of these
animals than it is now, and that other acoustic attributes and
capabilities have been acquired (perhaps because of the "mixed
blessings" of echolocation; see Fenton, this volume) that reduce
reliance on echolocation without, however, loss of the potential
for making such fine discriminations as have been demonstrated
under experimental conditions.

The odontocete skull is unique among mammals, including
mysticetes, in its asymmetry, the result of enlargement of dorsal
elements on the right side. No reversal of this asymmetry has
been reported. The degree of asymmetry varies, reaching its most
extreme form in the sperm whale, *Physeter*, and pygmy sperm whales,
Kogia (Raven and Gregory, 1933).

Some years ago it was suggested that the puzzling asymmetry
of the odontocete skull was an adaptation related to the development of a sound-producing organ and represented a modification of
the skull having to do with echolocation (Wood, 1964, pp. 395-396).
The idea stemmed, in part, from dissection of a *Kogia* head, reveal-
ing in the right nasal passage a pair of tightly opposed horny lips
which appeared to constitute an admirable structure for pneumatic
sound production. This *museau de singe* (monkey's mouth) is presumed to be homologous with the *museau de singe* of the sperm whale,
a similar structure in the right nasal passage but located forward
of the spermaceti organ, not behind it as in *Kogia*. The right
nasal system of *Physeter* is also more specialized in other respects
(Raven and Gregory, 1933).

Schenkkan and Purves (1973) identify the *museau* as the
homologue of the external lips of the right naris in other
odontocetes, but on the basis of its "generally low level of
muscular control" and their presumption that as a sound generator
it would be very expensive in its use of air, they reject an
acoustic function for the *museau*, assigning instead a valving role
and attributing sound production to the larynx, not only in *museau*-
bearing forms but all odontocetes. We will return to this point
further on.

From detailed examination of the right nasal complex of the
sperm whale, Norris and Harvey (1972) argued that the *museau* is
indeed a sound generator and that the spermaceti organ acts as a
reverberation chamber in the production of the burst-pulse signals
of *Physeter*. Mackay (this volume) agrees that the spermaceti organ
of *Physeter* is involved in sound emissions but postulates a different mode of response to sounds produced by the *museau*. The work
of Karol et al. (1978) on a pygmy sperm whale also supports the

view that both the *museau* and spermaceti organ are components of the sound production system in *Kogia*, although, considering the topological relationship of the two structures, they propose a wave guide function for the spermaceti organ.

It has been suggested (Norris and Harvey, 1972) that the deeply basined forehead of *Physeter* which is covered over most of its area by the posterior wall of the frontal sac (a diverticulum of the right nasal passage), is indicative of a sound reflective structure. This might also apply to the enlarged right side of the bipartite basining of the *Kogia* skull in which the ethmoid septum is deflected to the left (see Raven and Gregory, 1933).

It may be noted that the left nasal passage in physeterids is considerably larger and follows a more direct course to the left boney naris, which is correspondingly larger than the right. There seems no reason to doubt that the left passage, as suggested by Raven and Gregory, is for fast inhalation and exhalation, while (contrary to Raven and Gregory) the right passage and associated structures have a phonatory function.

The skulls of bottlenose whales, Ziphiidae, also exhibit marked asymmetry, and the left boney naris is conspicuously larger than the right, as shown in True (1910). In addition, a *museau* and what is apparently a spermaceti organ have been reported in the ziphiids *Ziphius cavirostris* (Norris and Harvey, 1972), *Mesoplodon bidens* (Schenkkan, 1973), and *Hyperoodon ampullatus* (*museau*: Schenkkan, 1973; spermaceti organ: Tomilin, 1957). The ziphiids are also characterized by a maxillary crest which in all but one species forms a basin vertical to the long axis of the rostrum (Moore, 1968). According to Fleischer (1978), both physeterids and those ziphiids that have been examined lack a tympanic membrane, having instead a thin, ceramic-like boney plate.

Along with the above likenesses it may be noted that the sperm whale and some, at least, of the ziphiids are apparently the deepest and longest divers of all cetaceans (Caldwell et al., 1966; Rice, 1978) and that the predominant food of both families is cephalopods (Tomilin, 1957). We think it likely that the similar acoustic modifications of the skull and ecological characteristics of the two families are more than coincidental, especially since they apparently are attributable to parallel evolution, not close phylogenetic relationship (Miller, 1923).

Except for sounds made by the sperm whale (see Watkins, this volume) almost nothing is known about the phonations of other members of these families. Faint click trains have been recorded from a *Kogia* by using a contact microphone in air (Caldwell et al., 1966) and pulsed sounds have been recorded from a *Mesoplodon* at sea

(D. K. Ljungblad, pers. comm). Although Watkins presents evidence that *Physeter* click emissions he has recorded generally had more the characteristics of communication signals than echolocation pulses, it seems inherently unlikely, considering the complexity of the right narial passage and associated structures, along with the mode of feeding at great depths, that sperm whales lack a well developed sonar system. Further observations may clarify what appears to be an improbable limitation on the function of acoustic emissions by these animals. As Watkins suggests, we may be dealing with a different type of system.

The earliest known fossil delphinids, like the physeterids and ziphiids, have come from Lower Miocene strata. Although asymmetry of the skull and narial structures is present in dolphins, it is less extreme than in the aforementioned families. Sound production is still incompletely understood. The majority of workers have placed sound generation in the complex nasal system, although one early hypothesis was that clicks are produced in the nasal system and whistles by the larynx. The literature is briefly reviewed by Ridgway et al. (this volume) who then present compelling evidence that (in *Tursiops*) clicks and whistles originate in the nasal system and not, as maintained by Schenkkan and Purves (1973) in the larynx. The work of Ridgway et al. also clarifies the ability of *Tursiops* to simultaneously emit whistles and clicks, as previously reported by Lilly and Miller (1961) and Evans and Prescott (1962).

Schenkkan and Purves state that the *museau* of physeterids and ziphiids is homologous with the external lips of the right naris in other odontocetes. Norris (1969) has reported that in the common dolphin, *Delphinus delphis*, the lips of the nasal plugs (presumably what Schenkkan and Purves referred to as the external lips) are asymmetrical, with the right hand lip approximately twice the size of the other lip. This form of asymmetry has also been seen in a variety of other delphinids examined by one of us (W.E.E.).

Anatomically, the highly asymmetrical skull and other head structures of the physeterids especially, but also the ziphiids, appear to represent a more extreme departure from a primitive and more generalized condition than does the delphinid head. If the *museau* is a sound generator, as seems probable, it indicates specialization of one sound system. On the other hand, *Tursiops* has a bilateral, though still asymmetrical, system which is highly versatile in sound production and, in conjunction with neural processing, known to be capable of permitting very fine discriminations. A basically bilateral but asymmetrical arrangement of narial structures is present in other delphinids (see Schenkkan, 1973).

We suggest that the narial system and associated structures of the sperm and bottlenose whales are specialized for sonic

emissions adaptive for the ecology of these two families. The
Tursiops acoustic emission system may be more of a general purpose
nature, although variations among the delphinids (and other
odontocetes) may reflect differences in their ways of life.

THE ODONTOCETE BRAIN AND AUDITION

Basic to the echolocation capability of odontocetes is their
neuroprocessing of such sounds--the integration, analysis, and
interpretation of acoustic information contained in the echoes.
Neural processing requires neural tissue. It is well known that
all odontocetes, and some in particular, have very impressive
brains (see, for example, Morgane and Jacobs, 1972). Although
there is much variation in the size of odontocete brains, both
absolutely and by whatever index of cephalization one chooses to
use, in general the brains of these animals are characterized, in
gross morphology, by large size, with greater width and height than
length; great development of the cerebral hemispheres; and intricate convolutions of the cortex. Morgane and Jacobs (1972)
identify three major arcuate lobar formations in each cerebral
hemisphere, the limbic, paralimbic, and supralimbic lobes, in their
terminology. The paralimbic lobe, bounded principally by the
limbic cleft (sulcus splenialis) and paralimbic cleft (sulcus
entolateralis) is, according to Morgane and Jacobs, unique to
odontocetes and its homology unknown. The supralimbic lobe,
bounded mediad by the paralimbic cleft and laterad by the ectosylvan fissure, is the largest of the three divisions and is
subdivided into three arcuate lobules. Morgane and Jacobs state
that these lobules and defining sulci exhibit great variation in
position and configuration, not only in different species but
within the same species and even between right and left hemispheres
of the same individual, a condition characteristic of the so-called
"association" areas of the cerebrum. These are most extensive in
man but also relatively large in other higher primates and in
elephants (Kruger, 1966). Association areas are generally thought
to be concerned with the more complex mental processes.

More recently, Ladygina and Supin (1977) studied the location
of sensory projection zones of the cerebral cortex by electrophysiological mapping. According to them, the areas termed by Morgane
and Jacobs the paralimbic lobe and the superior (i.e., mediad)
lobule of the supralimbic lobe constitute the visual zone, while
most of the large intermediate lobule of the supralimbic lobe is
auditory area. Ladygina and Supin note that the location of projection areas in the bottlenose dolphin and harbor porpoise--and
probably, therefore, in other odontocetes--differs appreciably
from their location in terrestrial mammals studied. The visual
area is shifted rostrally, and the auditory area, which is the

largest, occupies most of what is usually designated as the supra-
sylvian gyrus and not the ectosylvan gyrus, or temporal lobe, as
in terrestrial mammals. In short, the greater part of what Morgane
and Jacobs consider to be probable "associational" areas are,
according to the work of Ladygina and Supin, visual and auditory
projection zones. Considering both the anatomical and electrophysiological evidence, it is perhaps permissable to wonder if the
importance of audition to odontocetes, especially in echolocation,
may have brought about a cortical organization analagous to that
represented by the association areas of other large brained but
less acoustically oriented mammals.

If we accept Jerison's (1973) "principal of proper mass,"
which states that "the mass of neural tissue controlling a particular function is appropriate to the amount of information processing involved in performing the function," we might hypothesize
that the high degree of encephalization of odontocetes reflects,
to some considerable extent at least, their known and presumed
acoustic attributes and capabilities. The only other hypothesis is
that these animals--at least those with the biggest brains--possess
an intelligence comparable to, if not exceeding, that of man
(Lilly, 1961, 1963, 1967). But even Lilly has acknowledged that
"Convincing scientific evidence of cetacean intelligence remains
to be established, and arguments comparing levels of human and
dolphin are philosophical not scientific" (Lilly, 1978).

In the remainder of this dissertation we will examine the
evidence and arguments dealing with large odontocete brains and
with odontocete bioacoustics and other factors that may be
associated with the high degree of encephalization of these animals.
This would seem to have a bearing on the ecological adaptiveness of
echolocation in this particular group.

Paleontological and Anatomical Evidence

As pointed out by Jerison (1973), the history of the brain in
whales is a history of early enlargement; "dramatically notable by
the Miocene in the ancestors of living porpoises...The evolution
of the whale was associated with the evolution of sensorimotor
adaptations for maneuvering in water and, even more dramatically,
the evolution of the auditory and sound producing systems for echolocation and for the perception of objects by their echoes" (pp.
346 and 351). Jerison notes, in part from paleontological evidence,
the enlargement of the acoustic components of the eighth nerve, the
inferior coliculi of the midbrain, the entire cerebral cortex, and,
in association with the cortex, the cerebellum, all as parts of
these adaptations.

In Oligocene odontocetes there was no sign of the asymmetry that now characterizes the skulls of toothed whales, or of the basining of the facial region that in some odontocetes is associated with the presence of a spermaceti organ that may play an important part in echolocation. Both of these features are apparent "in muted form" in early Miocene fossils (Whitmore and Sanders, 1976). Fleischer (1976) has presented evidence that the adaptation of the cochlea for high-frequency sound reception, and therefore echolocation, occurred in the Squalodontoidea during the late Oligocene. At the same time, these animals developed the multilayer acoustic shield at the forehead, in a process that is part of the telescoping of the odontocete skull described by Miller (1923). Fleischer considers this shield as particularly focussing the ultrasonic portion of the outgoing sound and insulating the hearing organs from the emitted pulses. In the Miocene, according to Fleischer, the hearing organ of the odontocetes was fully adapted for a sonar system.

There appears, then, to be at least a rough chronological correlation between the development of echolocation and the development of a large brain.

The Mysticeti, it may be noted, have symmetrical skulls and lack a spermaceti organ or melon (Harrison and King, 1965), as well as, apparently, anything like the narial asymmetries and specializations in odontocetes. Fleischer (1976), from a study of fossil material, concluded that "There is good evidence that neither the Archeoceti nor the Mysticeti ever possessed a sonar system. In contrast to the Odontoceti, the Mysticeti adapted their hearing organ to the perception of sound of very low frequencies." Mysticete sounds are typically very low frequency moans (Schevill, 1964); the complex "song" of *Megaptera* (Payne and McVay, 1961) appears to be a special case.

The ears of terrestrial mammals are inefficient sound receivers under water, and the earliest adaptations in the evolution of odontocete acoustic mechanisms must have involved alterations of the ancestral ears, designed for airborne sounds, to more efficient receivers of waterborne sounds. An aqueous medium, however, provides the potential for much better reception of sound, since the density differences between water and the bone and soft tissue of a submerged animal are a very small fraction of the differences between air and the bone and soft tissue of a terrestrial animal. Norris (1968) considers this the most important factor in the evolution of odontocete acoustic mechanisms, and cites Hickling (1962) who pointed out that we would expect environmental sounds to induce little vibration in terrestrial animals except at very localized points, while much vibration would be induced in aquatic organisms by environmental sounds, and at a

large variety of loci. Norris documents the extensive modification of the odontocete skull and head structures that he considers
most probably attributable ultimately to the much reduced density
difference encountered when the ancestral forms of the odontocetes
entered the water.

We will speculate that increase in brain size was a development that accompanied these modifications. An increment in hearing
sensitivity through an incremental adaption of the receiving system
for waterborne sounds may have resulted in an adaptive increase in
neural tissue for the integration, analysis, and interpretation of
the acoustic information now available. This, in turn, might permit the utilization of further adaptive refinements of the aural
structures, with, then, a further development of neural tissue for
processing the increase in available acoustic information, and so
on. At some point, perhaps quite early, a rudimentary echolocation
system appeared, and because of its adaptive value could have
accelerated the process, with the introduction now of neural components for a more and more refined control of sound emissions, along
with an increased capacity for processing the echo information as
well as sounds received from the environment. The development of
communication signals may have contributed to neural control and
processing capabilities, but the meager evidence for vocalizations
having apparent communication significance in delphinids (for
example, the repetitive whistle of a separated calf (McBride,
1940); the "mating call" of a bull *Tursiops* (Tavolga and Essapian,
1957); the "distress call" (Lilly, 1963; Wood 1973), and the
"signature whistle" concept (Caldwell and Caldwell, 1965, 1971,
1972)) suggests that they are of an "intuitive" nature, rather than
having the characteristics of formal language (the difference
between "Ouch" and "Fire is hot"). But this issue must remain open.

The Brains of Odontocetes (and Bats)

"Brains are metabolically expensive and don't get bigger
(phylogenetically) unless in some fashion they are more than paying for their upkeep" (Hockett, 1978). There is evidence that
odontocete brains as large as those in some modern delphinids
(taking into consideration body size) had evolved by early Miocene
times (Jerison, 1973). This suggests that the major period of
brain development occurred long ago and that any later modifications have been of a less radical nature.

Hockett (1973, p. 416) speaking of man, states that after
steady growth of the hominid brain for two million years, its size
seems to have leveled off about 50,000 years ago. Why did it stop?
Hockett suggests that mutations produced some crucial reorganiza-
tion of the workings of the brain whereby greater efficiency was

effected without continued metabolically expensive increase in
size. Today, all our brains are fully human, and though differing
in size the differences are irrelevant. Why, then, do sizes differ?
Because, Hockett holds, they differed 50,000 years ago when size
was important, and because since then there has been no selective
reason anywhere for the size to change. The range in size of
odontocetes, both as to body and brain, exceeds that of any other
suborder of mammals, and the application of Hockett's statements
to cetaceans is tenuous, but what he says may provide a valid
insight.

It is a common tendency to think of large brains as indicating
"intelligence," although there is no universally agreed upon
definition of the term. Jerison (1973) avoids the usual difficulties by suggesting that "Biological intelligence may be nothing
more (or less) than the capacity to construct a perceptual world."
For us, this is the "real" world of which we are conscious. Other
animals may have "real" worlds, too, but these worlds will differ
according to how the animal's brain does the work of integrating
sensory and motor events.

Jerison sees two important dimensions of biological intelligence. One is the extent to which sensory capabilities have been
integrated with one another, and the second is the extent to which
behavior in response to sensory information is flexible and adjustable to inconsistencies in that information.

It is easier to appreciate now, Jerison continues, why gross
measures of the brain should be most closely related to biological
intelligence. "The number of neurons and the complexity of their
interconnections should reflect the degree to which sensory systems have been elaborated and interconnected; it might make little
difference from the point of view of intelligence which systems
have actually been emphasized in particular species."

After discussing other indices of encephalization and some of
the difficulties encountered with them, Jerison describes a measure
he developed that he calls the "encephalization quotient" or EQ.
The EQ is related to "average" brain: body relations of the mammals
as a whole. It is the ratio of actual brain size to expected brain
size and is defined by the equation $EQ = \frac{E}{.12P^{2/3}}$, in which E is
brain weight in grams, and P is body weight in grams. (For a
detailed explanation of how this equation was developed the reader
is referred to Jerison, 1973).

By this equation, the EQ of an "average" living mammal is
unity. A higher EQ denotes a brain size larger than that expected
for the size of the animal. The value of EQ for *Homo sapiens*, if
E=1350 g and P=70,000 g, is 6.6.

It is interesting to apply Jerison's formula to odontocetes, more for the purpose of intercomparisons than for ascertaining the degree of departure from "expected" brain size. Table 1 gives brain and body weights and corresponding EQs for animals of a number of different species. We present these figures with a few qualifications. In some instances we have brain-body weights for only a single individual. Data for clearly immature animals could not be used since young animals have relatively larger brains for their body size than adults. In this regard *Inia* data are probably borderline; however, since the animals were females, and females apparently do not grow as large as males (see Layne, 1958), the EQ figures may not be far off.

Table 1. Brain weight (E), body weight (P) and
encephalization quotient (EQ) of certain
odontocetes. All weights in grams.

Species	E	P	EQ($\bar{x}$)
Tursiops truncatus (N=3)[1]	1511	99,000	5.9
Lagenorhynchus obliquidens (N=2)[1]	1211	105,500	4.5
Grampus griseus (N=1)[2]	2551	400,000	4.0
Delphinus delphis (N=9)[2]	802	69,000	4.0
Globicephala macrorhynchus (N=1)[2]	2798	522,000	3.5
Inia geoffrensis (N=2)[3]	618	62,400	3.0
Phocoenoides dalli (N=1)[1]	793	110,000	2.9
Orcinus orca (N=1)[2]	6138	2,409,000	2.9
Platanista gangetica (N=4)[4]	295	59,750	1.6

[1]Ridgway et al., 1966; [2]Ridgway and Brownson, 1979;
[3]Gruenberger, 1970; [4]Pilleri and Gihr, 1970

If a correction for blubber weight (not an "average" mammal constituent) were incorporated, the EQs would be higher. Slijper (1962) states that blubber constitutes 30-35 percent of body weight in dolphins. On the other hand, cetacean data constituted only a small fraction of the mammalian brain-body weights used by Jerison in formulating his equation; it is likely that this in some degree counteracts the blubber weight factor and reduces any deviation from more accurately computed EQs. A different kind of error can result when dealing with very large animals; there is

considered to be an upper limit on the allometric relationship
between brain and body weight beyond which brain size does not
keep pace with body size.   Jerison thinks it probable that when
selection pressures toward enlargement of the body result in body
size above about 1000 kg (in land mammals, at least) they can act
more or less independently of those toward enlargement of the
brain.   The body weight of the killer whale in Table 1 is such
that the computed EQ is undoubtedly too low; we consider this in
our discussion of the table.

These general comments aside, what are we to make of the
range of EQ values shown?   First, it is apparent that, with the
exception of *Platanista*, all are considerably higher than
Jerison's "average" mammal, with *Tursiops* having an EQ in the
range of that of *Homo sapiens*.   A second observation is that there
is not a clear correlation with what is known about the echolocation capabilities of the different species.   However, *Tursiops*,
which has an excellent echolocation capability utilizing very
broadband pulses, has a very broad geographic range, is found in
varied types of aquatic habitats, and subsists on a wide variety
of food organisms.   It also adapts readily to captivity.   For
Lagenorhynchus, considering all of these factors, an EQ of around
4.5 appears appropriate.   The other species, although demonstrating
excellent echolocation capabilities, are more specialized with
respect to their echolocation pulses, geographic range, habitats,
food items, or a combination of these.   Within the freshwater
family Platanistidae, the higher EQ of *Inia*, even allowing for the
possibly too high values in Table 1, may reflect the fact that it
has functional eyesight, while *Platanista* does not.

The EQ of *Orcinus* is almost certainly too low on the basis of
allometric considerations.   We have noted above factors other than
echolocation that seem to relate to EQ values.   The killer whale
has probably the greatest geographic range of any odontocete (from
the Arctic to Antarctic), is found from shallow, restricted waters
to the open sea, has a widely varied diet, and adapts very well to
conditions of captivity.   On the other hand, *Orcinus* preys on
large fishes and also on other marine mammals, including whales,
which are active noise sources.   Locating food would appear to be
less of a problem for killer whales than for most other odontocetes.
Also, the computational capacity necessary to handle information
in the bandwidth from 20 Hz to 30 kHz is certainly less than that
needed for a 100 kHz bandwidth.   However, we suspect that if an
accurate EQ could be computed for *Orcinus* it would be well up the
scale, possibly in the range of *Tursiops*.   (See Appendix for relevant information on *Orcinus* and other species alluded to above.)

Eisenberg and Wilson (1978), using regression analysis,
plotted cranial volume:body weight ratios for 225 species of bats
of 14 families.   Their data indicated that the aerial insectivores
that rely almost completely on echolocation have the lowest brain

size to body weight ratios, while the frugivores of the suborder
Megachiroptera and those members of the microchiropteran family
Phyllostomatidae which have secondarily become fruit eaters, have
the largest. Insectivorous bats that discretely sample microhabitats or utilize a complex foraging strategy involving vision
and sonar to exploit micro-niches have intermediate ratios.
Eisenberg and Wilson modified Jerison's EQ formula by incorporating their extensive bat data (Jerison's sample included few
bats). The computed EQs correlated extremely well with trophic
roles, i.e., nectarivore and frugivore, carnivore, piscivore, and
so on in descending order of mean EQ values down to the aerial
insectivores. It was even possible to suggest that a species
suspected of being an aerial insectivore is instead, on the basis
of its EQ, a foliage gleaner.

With regard to the largest brains, Eisenberg and Wilson submit that "a foraging strategy based on locating relatively large
pockets of energy-rich food that are unpredictable in temporal and
spacial distribution necessitates the use of a complicated information storage and retrieval system involving input from several
sense organs."

The situation with respect to toothed whales is by no means
as clear cut as it seems to be with bats. We have presented
evidence for at least a rough chronological correlation between
the initial development of an echolocation system (quite different
from that of insectivorous bats, as discussed further on) and the
acquisition of a large brain. We suggest that other factors have
contributed to the variation in relative brain size seen in recent
toothed whales. In terms of Jerison's encephalization quotient,
higher values appear to reflect greater adaptability, as seen in
broad geographic range, penetration into a variety of ecologic
niches, a wide spectrum of food items, and adaptability to conditions of captivity. All of this, along with a very broadband
sonar, is descriptive of *Tursiops* which, so far as known, has the
highest odontocete EQ.

Integration of sensory inputs also represents a measure of
adaptability. We infer sensory integration in an animal that has,
for example, both good eyesight and a well developed sonar system;
the simultaneous use of vision and hearing seemed indicated by the
observations described early in this paper of the dolphins that
peered at, but also echolocated on, objects hung in an oceanarium
tank, and both looked at and echolocated on the fish being handed
out. In the discussion of Table 1, we suggested that the EQ of
Platanista, lower than that of the confamilial genus *Inia*, is at
least in part attributable to its lack of functional vision.

All of these factors, ecological and sensory, appear to bear
some relation to brain size in odontocetes. They indicate the
extent of variability of situations or circumstances the animal

may experience. Put another way, they determine the potential
scope of the animal's (in the species sense) perceptual world —
Jerison's concept of "biological intelligence."

While echolocation would seem to have a high adaptive value
for an aquatic vertebrate, there is hardly any evidence that it
has evolved in animals other than toothed whales. The sea catfish,
Arius felis, has been shown by Tavolga (1977) to have a rudimentary
acoustic orienting capability. Echolocation in pinnipeds and
penguins has been suggested by Poulter (1963, 1967), but attempts
to experimentally demonstrate such an ability in pinnipeds have
been unsuccessful (Evans and Haugen, 1963; Schusterman, 1966;
Scronce and Ridgway, this volume), and penguins have not been
studied in this respect. It remains possible that an echolocation
capability will be found in some of the Pinnipedia, such as those
species that are ice dwelling in the Arctic and Antarctic and have
a need to feed in total darkness and find breathing holes. At
present, however, there is no indication of the existence of a
complex aquatic biosonar system other than in odontocetes. Other
aquatic vertebrates seem to have taken the path of other sensory
systems or combinations of systems.

Even those dolphin species that have demonstrated a highly
developed sonar capability may use it less than we tend to suppose,
as pointed out with respect to the live fish experiment. Every
year a large number of dolphins and porpoises, including species
known to have an excellent sonar system, become entrapped in gill
nets, purse seines, and nets set off South African beaches to
protect bathers from sharks (Working Party on Marine Mammals, 1977,
pp. 100-107). The target strength of some, at least, of these nets
should make them readily detectable. While such entrapments may
result from inattention or concentrating on the pursuit of prey,
the fact remains that we have no idea how much of the remarkable
detection and discrimination capability of these animals is used
in the real world and how much of it is an untapped capacity
brought out by an artificial selection process called "training."

We cannot avoid further consideration of bats, which with
brains weighing only a fraction of a gram are skilled echolocators.
We have argued that, other factors notwithstanding, the large size
of the odontocete brain probably reflects in some considerable
degree the hypertrophy of neural tissue for processing acoustic
information, especially that contained in echoes, and for very
precise control of transient broadband sound emissions.* The idea

---

*The weakly electrogenic Mormyridae, freshwater fishes of Africa,
may, in the enormous development of their cerebellum, represent an
analogous example. Very brief polyphasic electric pulses, emitted
at a variable rate from a few to 130 per second (unless the fish
is "silent") provide an electrical field, any interruption of

has been expressed by others. Tomilin (1968), for example, hazarded the guess that "The use of echolocation and the echolocation apparatus may have played as important a role in the formation of the dolphin brain as did the hands, work, and articulate speech in the development of the human brain." The minute brain of insect-catching bats does not negate our hypothesis, but we need to consider the bat brain and some of the differences between the microchiropteran and odontocete echolocation systems.

One significant difference lies in the nature of the sound emissions, the bats using FM sweeps (ignoring here the finer distinctions among different species) and the odontocetes using transient broadband clicks. We would expect that processing of echoes from extremely brief broadband clicks emitted at rates up to 600 per second or more would require a much greater neural processing capability. Also, the sensorimotor control of such pulses, including the ability to shift peak energy to different parts of their frequency spectrum, would seem to be a much more complex task (Au et al., 1974).

Although sound attenuation is greater in air than in water, the air-target density match is such that echolocation and signal processing should be much simpler than in an aquatic environment where objects are generally more closely density-matched to water.

The bat system is highly specialized for a specific task (and apparently has been for 50 or 60 million years); the odontocete system appears to be more versatile and flexible. Also, the interpretation of acoustic information from the environment may have contributed to the neural processing capacity of cetaceans.

It is interesting that of the terrestrial/aerial vertebrates known or suspected to use echolocation only the insectivorous bats use time-frequency calls. All of the others, including the frugivorous bats of the genus *Rousettus*, produce broadband pulses (Fenton, this volume). In none of these is there any indication of a highly developed sonar system; the broadband pulses, so far as known, serve only for detection of obstacles in the path of the

which can be detected. By their electrosensory system mormyrids can avoid obstacles and detect prey at a distance. The pulses also appear to have communication significance (Moller and Baur, 1973). In contrast, the weakly electric (except for the electric eel, which also has a powerful discharge) Gymnotidae produce monophasic electric pulses at, in most species, a constant frequency. Although able to detect obstacles and prey by their electric field, gymnotids lack a large cerebellum. The "gigantocerebellum" of mormyrids, which produce complex pulses under greater control, is generally considered to be predominantly an association center for the electrosensory system (Lissmann, 1958; Bennett, 1971).

echolocator. Fenton suggests that the use of broadband clicks for
echolocation (in terrestrial settings) does not confer a strong
adaptive advantage. He notes, however, that the facultative use
of a broadband pulse form of echolocation by a wide variety of
animals indicates that it does provide some advantages.

Phyletically, it would seem that echolocation by broadband
clicks represents the "conventional" mode of biosonar in animals.
The bats with their FM sweeps constitute a unique exception
(unless the frequency-modulated chirps of some odontocetes have a
significant sonar function; see Diercks' review, this volume).

There has evidently been little or no selective pressure to
develop further the low-order broadband system that some terres-
trial/aerial vertebrates possess. This is obviously not true of
the odontocetes. The adaptive value of a sonar system for
swift-swimming, deep-diving, fish or squid eaters seems apparent,
especially given the sound-conducting properties of water. Considering the prevalence of a broadband form of echolocation, however
rudimentary, in terrestrial/aerial vertebrates, it is not surprising that odontocete sonar is of the same type. Broadband
clicks provide, in fact, distinct advantages in an aquatic environment because they are better for coping with reverberation —
the reflection of sound off surface, bottom, and particulates in
the water column.

We can assume that at some point in the history of odontocetes
an incipient echolocation capability involving broadband pulse
emissions appeared. There followed, then, evolutionary selection
for sound generation, reception, and processing systems that would
make the most of broadband emissions. Given the potential to
develop a broadband system, including very fine sensorimotor control of their sound emissions and a highly developed neuroprocessing component, the odontocetes followed the evolutionary
course that was most adaptive for them.

ACKNOWLEDGEMENTS

K. J. Diercks and R. T. Trochta kindly provided information
and sketches of the live fish experiment. We are obliged to V. S.
Gurevich for making us aware of and providing translations of most
of the Soviet literature cited. The figure showing pulse wave
forms and the drawings and layouts of the Appendix material were
the work of illustrator Ann Dawes. Finally, we must express our
appreciation to Jan Letourneau, Karie Wright, and Yvonne Moreno
for their skill and patience in typing early drafts, and to Jan,
especially, for the final, much revised, copy.

APPENDIX

Natural History and Echolocation Data
on Six Species of Odontocetes

ATLANTIC BOTTLENOSE DOLPHIN *(Tursiops truncatus)*

Natural History Data

Size: to 3.5 meters

Distribution: Found world wide in temperate and tropical waters. Most common in bays, rivers, estuaries, and mud flats. Also seen far at sea.

Food preference: Catholic feeders, predominantly fish but squid, octopus and a variety of large crustaceans also taken.

Swimming and diving: Will reach burst speeds of 18-20 knots for short periods, can sustain 4-5 knots for extended periods. Can travel 100 nautical miles/day. Dives to depths of over 350 meters have been recorded.

General information: Best studied of all marine delphinids. Adapts well to captivity. Has been bred and raised in captivity and some individuals have survived and thrived for over 20 years. Most of our available data on sound production in general and echolocation specifically is based on observations of this species. Good data are available on behavior physiology, nutritional requirements, also only species for which good data on visual acuity are available.

Echolocation Data

Signal type: pulse   Duration: 10-200 μsec

Peak frequency (kHz): 15-130

Measured source levels: Peak levels 155-204 dB re: 1 μPa (Ref. Evans, 1973; Babkin and Dubrovsky, 1971; Alexeyeva et al., 1971; Golubkov, 1972).

Detection: ability to detect spheres made of various materials, diameters from 3.3 cm to 7.6 cm at ranges of 6 meters to 73 meters. (Ref. Titov, 1972; Au et al., 1974).

Discrimination: Capability to discriminate objects of same size, different materials excellent, also shape discrimination. Can discriminate difference in steel spheres which differ in diameter by 4 mm (Ref. Evans, 1973; Dubrovsky et al., 1970; Norris et al., 1967; Fadeeva, 1973; Golubkov, Ivanenko, 1970; Dziedzic, Nachtigale, Murchison this vol; Evans and Powell, 1967).

Hearing: Range 75 Hz to 150 kHz, Maximum sensitivity 20-80 kHz (Ref. Johnson, 1966; Morozov et al., 1972).

COMMON DOLPHIN *(Delphinus delphis)*

Natural History Data

Size: to 2.6 meters

Distribution: World wide in temperate and tropical seas (water temperature range 10° to 28°C). Usually found in the deep open ocean outside depths of 200 meters, except frequently found in vicinity of banks and sea mounts in deep water. Probably the most numerous of all the cetaceans.

Food preference: Small schooling fishes and squid. Known to feed on anchovy, herring, a variety of mid-water fishes including myctophids and deep-sea smelt. In Eastern Pacific primarily a nocturnal feeder, taking advantage of photophobic organisms that migrate to the surface at night.

Swimming and diving: Very fast burst speeds to at least 24 knots. Can sustain speeds of 6 knots for long periods. Radio tagged individuals moved more than 110 nautical miles in 24 hours diving at night to depths in excess of 220 meters.

General information: This is one of the three species most heavily impacted by mortality associated with the tropical Eastern Pacific Yellowfin Tuna purse seine fishery. Has been maintained successfully in captivity for over 8 years and used extensively as an experimental animal, especially in USSR (Black Sea).

Echolocation Data

Signal type: pulse

Duration: 35-350 μsec

Peak frequency (kHz): 20-100 (Ref. Evans, 1973; Titov, 1972; Bel'kovich and Reznikov, 1971)

Measured source level (at 1 meter): peak level 140 dB re 1 μPa

Echolocation performance
Detection: 13 cm sculpin at 2.6 meters (Titov, 1972)

Discrimination: Several experimental demonstrations of ability to detect differences in shape, size, and material, e.g. spheres, cylinders, pyramids, 13 cm Herring from 13 cm metal "fish". (Ref. Titov, 1972; Bel'kovich et al., 1969; Bel'kovich and Borisov, 1971; Demitrieva et al., 1972; Bel'kovich, 1970; Rezvov et al., 1973; Gurevich, 1968).

Hearing: Range measured 100 Hz to 280 kHz, maximum sensitivity at 60 to 100 kHz (Ref. Bel'kovich and Solntseva, 1970).

KILLER WHALE *(Orcinus orca)*

Natural History Data

Size: to 9 meters

Distribution: World wide from northern to southern polar waters. Throughout its range seems to prefer coastal areas and often enters bays, estuaries and river mouths in search of food. Sightings far at sea are not rare.

Food preferences: Known to feed on squid, large fish (tuna, salmon), sea turtles, seabirds, marine mammals, including large baleen whales.

Swimming and diving: Very fast swimmer, may be fastest of all the cetaceans. Sprint speeds in excess of 25 knots have been estimated. Can cruise for extended periods at speeds of 6-8 knots. Dives to depths of 250 meters have been observed under controlled conditions.

General information: Killer whales have now been successfully maintained under controlled conditions for several years. They are possibly most adaptable cetacean in captivity. They are very tractable and have been trained to participate in a variety of psycho-physical studies.

Echolocation Data

Signal type: pulse

Duration: 0.5-1.5 msec

Peak frequency (kHz): 14, range energy 100 Hz to 30 kHz
Measured source level (at 1 meter): 178 dB re 1 μPa (Diercks *et al.*, 1971)

Echolocation performance
Detection: Can detect a 5 cm diameter air-filled plastic cylinder at 6 meters.

Discrimination: At present there are no data available on the echolocation discrimination capability of this species (Evans, 1973).

Hearing: Range of hearing measured 500 Hz to 31 kHz, maximum sensitivity at 15 kHz (Ref. Hall and Johnson, 1972).

HARBOR PORPOISE *(Phocoena phocoena)*

Natural History Data

Size: to 1.8 meters

Distribution: Common in the North Atlantic with a northern limit of Iceland, White Sea, and the Davis Straights. Southern limit is Cape of Good Hope. Southern form may be different sub-species. Also in Eastern North Pacific near the United States and Japan. Inhabits the Black Sea, Sea of Azov and the Mediterranean Sea. Rarely in tropical waters. Usually in shallow water bays, rivers and estuaries.

Food preference: Main food items herring, small cod, sole; they also eat squid and crustaceans.

Swimming and diving: Does not stay submerged for more than three to four minutes. Seldom found in water deeper than 150 meters, so probably not a deep diver. Fast burst speeds when feeding to 15 knots or more. Cruising speed much slower, 2-4 knots.

General information: Has been successfully maintained at aquaria in United States and Europe (USSR). Majority of research on this species done in USSR and Europe (Denmark).

Echolocation Data

Signal type: pulse

Duration: 40-200 μsec

Peak frequency (kHz): 20-150 (Ref. Dubrovsky *et al.,* 1970; Möhl and Andersen, 1975)

Measured source level (at 1 meter): peak level 112 dB re 1 μPa

Echolocation performance
Detection: 0.2 diameter steel wire and nylon thread diameters of 0.9 mm. Also detection of 75 mm diameter cylinders of various materials at 8-11 meters (Ref. Busnel *et al.,* 1967; Zaslavsky *et al.,* 1969).

Discrimination: Can detect difference of 1.75 cm in diameter of steel cylinders at 2 meters (Ref. Titov, 1972).

Hearing: Range measured, 1 kHz-150 kHz, maximum sensitivity at 8, 32, 64 kHz (Ref. Andersen, 1970; Sukhoruchenko, 1973).

WHITE WHALE, BELUGA *(Delphinapterus leucas)*

Natural History Data

Size: to 4.3 meters

Distribution: Arctic waters of Atlantic, Pacific, and White seas. Seen as far south as 50°N, but this is rare. Frequents shallow bays and river mouths. Occasionally ascends rivers several hundred miles to feed and bear young. Also found offshore in southern part of range.

Food preference: Feeds on a variety of small fishes, including caplin salmon smolt, and small cod; also squid and a variety of benthic crustaceans.

Swimming and diving: Burst speeds to 15 knots probably faster. Slow cruising, 2-4 knots. In bays and rivers spends much time lolling at surface. Diving ability not known. Can easily breath-hold for more than 8 minutes. Feeds on organisms that are benthic at more than 300 meters.

General information: A very vocal cetacean and one of the first recorded. Possibly first species successfully kept in captivity. Has reproduced in captivity but has not successfully raised young. Several specimens have and are being maintained in United States, Canada and Europe. Echolocation ability and hearing only recently studied.

Echolocation Data

Signal type: pulse

Duration: 20-100 μsec

Peak frequency (kHz): 40, 80, 120
Measured source level (at 1 meter): Peak levels of 160-180 dB re 1μPa possibly capable of higher.

Echolocation performance
Detection: Detection of 10 cm^2 plastic square plate at 8 meters. Also can detect 5 mm ceramic cylinder at 5 meters.

Discrimination: Can discriminate flat square, triangle from three dimensional target (pyramid). Can discriminate a 3 cm x 3 cm square from 2 cm x 2 cm square. (Ref. Gurevich and Evans, 1976).

Hearing: Range measured 1 kHz to 123 kHz. Maximum sensitivity 60-65 kHz (Ref. White *et al.*, 1978).

BOUTU, AMAZON RIVER DOLPHIN *(Inia geoffrensis)*

Natural History Data

Size: to 2.5 meters

Distribution: Found in the Amazon and Orinoco rivers and their tributaries. Will occasionally venture in the brackish bays at the mouth of the rivers.

Food preference: Eats mainly bottom dwelling fish, especially a species or armoured catfish. Also will eat freshwater crustacea.

Swimming and diving: In its native habitat speeds of 15 knots have been observed. Usually animal cruises at 2-3 knots. Experimental tests of diving ability have not been conducted, however can easily hold breath for 4 minutes.

General information: Has been successfully maintained for periods in excess of 10 years at several aquaria in the U.S., Europe, and Japan. Trains well, and is an excellent experimental animal.

Echolocation Data

Signal type: pulse

Duration: 15-100 μsec (Ref. Diercks *et al.,* 1971; Evans, 1973)

Peak frequency (kHz): 60-65
Measured source level (at 1 meter): peak level 146 dB re: 1 μPa (Ref. Evans, 1973)

Echolocation performance
Detection: 1.4 mm diameter wire at approximately 1 meter (Penner and Murchison, 1970).

Discrimination: Detection of 2.4 mm difference in diameter of cork/neoprene cylinders of same length (Ref. Diercks *et al.,* 1971; Evans, 1973).

Hearing: Range measured 1 kHz-105 kHz, Maximum sensitivity 30-50 kHz (Jacobs and Hall, 1972).

REFERENCES

Abramov, A. P., E. Sh. Ayrapet'yantz, V. I. Burdin, A. G. Golubkov, I. V. Ershova, A. R. Zhezherin, V. I. Korolev, Yu. A. Malyshev, G. K. Ul'yanov, and V. B. Fradkin 1971 Investigations of the dolphin's capabilities to differentiate volumetric targets according to their linear dimensions and material. Proceedings of the 8th All-Union Acoustical Conference, p. 3. (In Russian)

Alexeyeva, T. V., A. G. Golubkov, and I. V. Ershova 1971 On the problem of the active width of the spectrum of dolphins' echolocation signals. Trudy Akusticheskogo Instituta 17:99-103. (In Russian)

Andersen, S. 1970 Auditory-sensitivity of the harbour porpoise, *Phocoena phocoena*. In: Investigations of Cetacea, G. Pileri, Ed. Vol. 2, pp. 255-258. University of Berne, Switzerland

Au, W. W. L., R. W. Floyd, R. H. Penner, and A. E. Murchison 1974 Measurement of echolocation signals in the Atlantic bottlenose dolphin, *Tursiops truncatus* Montagu, in open waters. Jour. Acoust. Soc. Am. 56(4):1280-1290

Au, W. W. L., R. W. Floyd, and J. E. Haun 1978 Propagation of Atlantic bottlenose dolphin echolocation signals. Jour. Acoust. Soc. Am. 64:411-422

Ayrapet'yantz, E. Sh., A. G. Golubkov, I. V. Ershova, A. R. Zhezherin, V. N. Zvorykin, and V. I. Korolev 1969 Echolocation differentiation and properties of the emitted sounds of dolphins. Doklady Academii Nauka 188(5):1197-1199. (In Russian)

Babkin, V. P., and N. A. Dubrovsky 1971 Range and noise resistance of the bottlenosed dolphin echolocator during detection of various targets. Trudy Akusticheskogo Instituta 17:29-42. (In Russian)

Babkin, V. P., N. A. Dubrovsky, P. S. Krasnov, and A. A. Titov 1971 Material discrimination in spherical targets by the bottlenosed dolphin. Proceedings of the 8th All-Union Acoustical Conference, p. 5. (In Russian)

Bastian, J. R. 1967 The transmission of arbitrary environmental information between bottlenose dolphins. In: Animal Sonar Systems: Biology and Bionics, R.-G. Busnel, Ed. Vol. II, pp. 803-873. Laboratoire de Physiologie Acoustique, Jouy-en-Josas, France.

Bel'kovich, V. M.   1970   Mechanism of sound generation in Cetacea. In: Ekholocatsiya v Prirode, Nauka, Leningrad, pp. 297-310. (In Russian)

Bel'kovich, V. M., V. I. Borisov, V. S. Gurevich, and N. L. Krushinskaya 1969 The ability of echolocation in *Delphinus delphis*. Zoologicheskii Zhurnal 48(6):876-884.   JPRS 48780.

Bel'kovich, V. M., V. I. Borisov, V. S. Gurevich, and N. L. Krushinskaya 1969 Echolocational capabilities of the common dolphin (*D. delphis*). Zoologicheskii Zhurnal 48(6):876-883. (In Russian)

Bel'kovich, V. M., and V. I. Borisov 1971 Echolocational discrimination of the complex target by dolphins. Trudy Akusticheskogo Instituta 17:19-23. (In Russian)

Bel'kovich, V. M., V. I. Borisov, V. S. Gurevich, N. L. Krushinskaya, and I. L. Dmitrieva 1972 Echolocation discrimination of distance between the targets (angular resolution) by dolphins. Proceedings of the Abstracts of the 5th All-Union Conference on Marine Mammals, Makhachkala, Part 2:30-33. (In Russian)

Bel'kovich, V. M. and A. E. Resnikov 1971 New in dolphin echolocation. Priroda 7:71-75. (In Russian)

Bennett, M. V. O.   1971   Sensory systems and electric organs. In: Fish Physiology, Vol. 5, W. S. Hoar and D. J. Randall, Eds. Academic Press.

Berzin, A. A. 1971 The Sperm Whale. Pacific Scientific Research Institute of Fisheries and Oceanography (USSR). Trans. by Israel Program for Scientific Translations, publ. by U. S. National Technical Information Service, Springfield, Va. 393 pp.

Busnel, R.-G., A. Dziedzic, and S. Andersen 1965 Seuils de perception du systeme sonar du Marsouin, *Phocoena phocoena* en fonction du diametre d'un obstacle filiforme. Comptes Rendu des Seances de l'Academie des Sciences, Paris 260(1):295-297.

Busnel, R.-G., Ed. 1967 Animal Sonar Systems: Biology and Bionics. Two volumes. Laboratoire de Physiologie Acoustique, Jouy-en-Josas-78, France.

Busnel, R.-G. and A. Dziedzic 1967 Resultats metrologiques experimentaux de l'echolocation chez le *Phocoena phocoena*, et leur comparaison avec ceux de certaines chauves-souris. In: Animal Sonar Systems: Biology and Bionics, R.-G. Busnel, Ed. Vol. I, pp. 307-335. Laboratoire de Physiologie Acoustique, Jouy-en-Josas-78, France.

Caldwell, D. K. and M. C. Caldwell 1965 Individualized whistle contours in bottlenose dolphins. Nature 207:434-435.

Caldwell, D. K., M. C. Caldwell, and D. W. Rice 1966 Behavior of the sperm whale, *Physeter catodon* L. In: Whales, Dolphins, and Porpoises, K. N. Norris, Ed. Univ. of California Press

Caldwell, D. K., J. H. Prescott, and M. C. Caldwell 1966 Production of pulsed sounds by the pygmy sperm whale, *Kogia breviceps*. Bull. So. Calif. Acad. Sci. 65(4):246-248.

Caldwell, M. C., N. R. Hall, and D. K. Caldwell 1971 Ability of an Atlantic bottlenose dolphin to discriminate between, and potentially identify to individual, the whistles of another species, the spotted dolphin. Cetology No. 6, 6 pp.

Caldwell, D. K. and M. C. Caldwell 1972 Senses and communication. In: Mammals of the Sea: Biology and Medicine, S. H. Ridgway, Ed. pp. 466-502. Chas. C. Thomas, Springfield, Illinois.

Diercks, K. J., R. T. Trochta, C. F. Greenlaw, and W. E. Evans 1971 Recording and analysis of dolphin echolocation signals. Jour. Acoust. Soc. Am. 49(6, pt. 1):1729-1732.

Diercks, K. J., R. T. Trochta, W. E. Evans 1973 Delphinid sonar; measurement and analysis. Jour. Acoust. Soc. Am. 54:200-204.

Dmitrieva, I. L., N. L. Krushinskaya, V. M. Bel'kovich, and A. V. Shurkhal 1972 Echolocational discrimination of the shape of different subjects by dolphins. Proceedings of the Abstracts of the 5th All-Union Conference on Marine Mammals, Makhachkala, Part 2:76-80. (In Russian)

Dubrovsky, N. A., P. S. Krasnov, and A. A. Titov 1970 On the problem of emitting ultrasonic echolocational signals by harbor porpoise (*Ph. phocoena*). Akusticheskii Zhurnal 4:521-525. (In Russian)

Dubrovsky, N. A. and P. S. Krasnov 1971 Discrimination of the solid spheres according to their materials and dimensions by the bottlenosed dolphin (*T. truncatus*). Trudy Akusticheskogo Instituta 17:9-18. (In Russian)

Dubrovsky, N. A., P. S. Krasnov, A. A. Titov 1971 On the emission of ultrasound echolocation signals by the Azov Sea harbor porpoise. Akist. Zh. 16:521-525; Sov. Phys.-Acoust. 16:444-447.

Dubrovsky, N. A. and G. L. Zaslavsky 1973 Temporal structure and directivity of sound emission by the bottlenosed dolphin. Proceedings of the 8th All-Union Acoustical Conference pp. 56-59. (In Russian)

Eisenberg, J. F. and D. E. Wilson 1978 Relative brain size and feeding strategies in the Chiroptera. Evolution 32:740-751.

Evans, W. E. 1973 Echolocation by marine delphinids and one species of fresh-water dolphin. Jour. Acoust. Soc. Am. 54(1):191-199.

Evans, W. E. and J. H. Prescott 1962 Observations of the sound production capabilities of the bottlenose porpoise: A study of whistles and clicks. Zoologica 47:121-128.

Evans, W. E. and Ruth Haugen 1963 An experimental study of the echolocation ability of the California sea lion *Zalophus californianus* (Lesson). Bull. So. Calif. Acad. Sci. 62:165-175.

Evans, W. E. and B. A. Powell 1967 Discrimination of different metallic plates by an echolocating delphinid. In: Animal Sonar Systems: Biology and Bionics, R.-G. Busnel, Ed. Vol I, pp. 363-383. Laboratoire de Physiologie Acoustique, Jouy-en-Josas-78, France.

Evans, W. E. and J. Bastian 1969 Marine mammal communication: social and ecological factors. In: The Biology of Marine Mammals, H. T. Andersen, Ed. pp. 425-475.

Fadeeva, L. M. 1973 Discrimination of spheres with various echosignal structure by dolphins. Proceedings of the 8th All-Union Conference on Marine Mammals Part 1, p. 134. (In Russian)

Fleischer, Gerald 1976 Hearing in extinct cetaceans as determined by cochlear structure. Jour. Paleon. 50(1):133-152.

Fleischer, Gerald 1978 Evolutionary principles of the mammalian middle ear. Advances in Anatomy, Embryology, and Cell Biology, Vol. 55, Fasc. 5, 70 pp.

Ford, J. K. B. and H. D. Fisher 1978 Underwater acoustic signals of the narwhal (*Monodon monoceras*). Canadian Jour. Zool. 56:552-560.

Golubkov, A. G. and Yu. V. Ivanenko 1970 Study of the range resolution ability of the dolphin's echolocator. Proceedings of the 23rd scientific-technical conference of LIAP p. 65. (In Russian)

Golubkov, A. G., I. V. Ershova, I. V. Korolev, and Yu. A. Malyshev 1972 On energetical parameters of the Black Sea bottlenosed dolphin echolocational apparatus. Trudy LIAP 76:9-12. (In Russian)

Gruenberger, H. B. 1970 On the cerebral anatomy of the Amazon dolphin, *Inia geoffrensis*. In: Investigations on Cetacea V.II G. Pilleri, Ed. University of Berne, Switzerland.

Gurevich, V. S. 1969 On ability of the common dolphin (*D. delphinus*) to discriminate geometrical targets by means of echolocation. Vestnik Moskovskogo Universita 3:109-112. (In Russian)

Gurevich, V. S. and W. E. Evans 1976 Echolocation discrimination of complex planar targets by the beluga whale (*Delphinapterus leucas*). Jour. Acoust. Soc. Am. 60(Suppl. 1)S5-S6.

Hall, J. D. and C. S. Johnson 1971 Auditory thresholds of a killer whale *Orcinus orca* Linnaeus. Jour. Acoust. Soc. Am. 51(2):515-517.

Harrison, R. J. and J. E. King 1965 Marine Mammals. Hutchinson University Library, London.

Herald, E. S., R. L. Brownell, Jr., F. L. Frye, E. J. Morris, W. E. Evans and A. B. Scott 1969 Blind river dolphin: First side-swimming cetacean. Science 166:1408-1410.

Hickling, R. 1962 Analysis of echoes from a solid elastic sphere in water. Jour. Acoust. Soc. Am. 34(10):1582-1592.

Hockett, C. F. 1973 Man's Place in Nature. McGraw-Hill.

Hockett, C. F. 1978 In search of Jove's brow. American Speech 53(4):243-313.

Irvine, L. 1939 Respiration in diving mammals. Physiol. Rev. 19:112-134.

Jerison, H. J. 1973 Evolution of the Brain and Intelligence. Academic Press.

Karol, R., C. Litchfield, D. K. Caldwell, and M. C. Caldwell 1978 Compositional topography of the melon and spermaceti organ lipids in the pygmy sperm whale *Kogia breviceps*: Implications for echolocation. Mar. Biol. 47:115-123.

Kellogg, Remington 1928 The history of whales. Quart. Rev. Biol. 3:29-76, 174-208.

Kellogg, W. N. 1953 Ultrasonic hearing in the porpoise, *Tursiops truncatus*. Jour. Comp. and Physiol. Psych. 46(6):446-450.

Kellogg, W. N. 1958 Echo ranging in the porpoise. Science 128:982-988.

Kellogg, W. N. 1961 Porpoises and Sonar. Univ. Chicago Press.

Kellogg, W. N. and R. Kohler 1952 Reactions of the porpoise to ultrasonic frequencies. Science 116(3018):250-252.

Kellogg, W. N., R. Kohler, and H. N. Morris 1953 Porpoise sounds as sonar signals. Science 117:239-243.

Korolev, L. D., N. V. Lipatov, R. N. Resvov, M. A. Savel'ev, and A. B. Flenov 1973 Investigation of the spatial directivity diagram of the dolphin sound emitter. Proceedings of the 8th All-Union Acoustical Conference 1:33. (In Russian)

Kruger, Lawrence 1966 Specialized features of the cetacean brain. In: Whales, Dolphins, and Porpoises, K. S. Norris, Ed., pp. 232-252. Univ. of California Press.

Ladygina, T. F. and A. Ya. Supin 1977 Localization of sensory projection zones in the cerebral cortex of the bottlenosed dolphin, *Tursiops truncatus*. Zurnal Evolyutsionnoy Biokhimii i Fiziologii No. 6, pp. 712-718. (In Russian)

Layne, J. N. 1958 Observations on freshwater dolphins in the Upper Amazon. Jour. Mamm. 39(1):1-22.

Lilly, J. C. 1961 Man and Dolphin. Doubleday and Co.

Lilly, J. C. 1963 Critical brain size and language. Perspectives in Biology and Medicine 6:246-255.

Lilly, J. C. 1967 The Mind of the Dolphin. Doubleday and Co.

Lilly, J. C. 1978 Dolphin and the law. Ltr. to the Honolulu (Hawaii) Advertiser, 31 January

Lilly, J. C. and A. M. Miller 1961a Vocal exchanges between dolphins. Science 134(3493):1873-1876.

Lilly, J. C. and A. M. Miller 1961b Sounds emitted by the bottlenose dolphin. Science 133(3465):1689-1693.

Lissmann, H. W. 1958 On the function and evolution of electric organs in fish. Jour. Exp. Biol. 35:156-191.

McBride, A. F. 1956 Evidence for echolocation in cetaceans. Deep Sea Research 3:153-154.

McBride, A. F. and H. Kritzler 1951 Observations on pregnancy, parturition, and postnatal behavior in the bottlenose dolphin. Jour. Mamm. 32:251-266.

Miller, G. S., Jr. 1923 The telescoping of the cetacean skull. Smithsonian Misc. Collections 76(5):1-71.

Møhl, B. and S. Andersen 1973 Echolocation: high frequency component in the click of the harbour porpoise. Jour. Acoust. Soc. Am. 54(5):1368-1372.

Moller, P. and R. Bauer 1973 "Communication" in weakly electric fish, *Gnathonemus petersii* (Mormyridae). Interaction of electric organ discharge activities of two fish. Animal Behavior 21(3):501-512.

Moore, J. C. 1968 Relationships among the living genera of beaked whales. Fieldiana:Zoology 53(4):209-298.

Morgane, P. J. and N. S. Jacobs 1972 Comparative anatomy of the cetacean nervous system. In: Functional Anatomy of Marine Mammals, R. J. Harrison, Ed., Vol. 1, pp. 117-244. Academic Press.

Norris, K. S. 1968 The evolution of acoustic mechanisms in odontocete cetaceans. In: Evolution and Environment, E. T. Drake, Ed., pp. 297-324.

Norris, K. S. 1969 The echolocation of marine mammals. In: The Biology of Marine Mammals, S. Andersen, Ed., pp. 391-423. Academic Press.

Norris, K. S., J. H. Prescott, P. V. Asa-Dorian 1961 An experimental demonstration of echolocation behavior in the porpoise, *Tursiops truncatus* (Montagu). Biol. Bull. 20(2):163-176.

Norris, K. S. and G. W. Harvey 1972 A theory for the function of
    the spermaceti organ of the sperm whale. In: Animal Orienta-
    tion and Navigation, S. R. Galler et al., Eds., National
    Aeronautics and Space Administration, Washington, D.C.

Norris, K. S. and W. E. Evans 1967 Directionality of echoloca-
    tion clicks in the rough-tooth porpoise, *Steno bredanensis*
    (Lesson). In: Marine Bioacoustics, Vol. 2, W. N. Tavolga,
    Ed., pp. 305-314. Pergamon Press.

Norris, K. S., W. E. Evans, and R. N. Turner 1967 Echolocation
    in an Atlantic bottlenose porpoise during discrimination.
    In: Animal Sonar Systems: Biology and Bionics, R.-G. Busnel,
    Ed., Vol. I, pp. 409-437. Laboratoire de Physiologie
    Acoustique, Jouy-en-Josas-78, France.

Payne, R. S. and S. McVay 1971 Songs of humpback whales.
    Science 173(3997):585-597.

Penner, R. H. and A. E. Murchison 1970 Experimentally demon-
    strated echolocation in the Amazon River porpoise, *Inia
    geoffrensis* (Blainville). Naval Undersea Center (San Diego,
    Calif.) Technical Publication 187, Rev. 1, 26 pp.

Pilleri, G. and M. Gihr 1970 Brain-body weight ratio of
    Platanista gangetica. In: Investigations of Cetacea, Vol.
    II, G. Pilleri, Ed., University of Berne, Switzerland.

Poulter, T. C. 1963 Sonar signals of the sea lion. Science
    139:753-755.

Poulter, T. C. 1967 Systems of echolocation. In: Animal Sonar
    Systems: Biology and Bionics, R.-G. Busnel, Ed., Vol. I,
    pp. 157-183. Laboratoire de Physiologie Acoustique,
    Jouy-en-Josas-78, France.

Raven, H. C. and W. K. Gregory 1933 The spermaceti organ and
    nasal passages of the sperm whale (*Physeter catodon*) and
    other odontocetes. Am. Mus. Novitates No. 677, 18 pp.

Resvov, R. N., Ya. A. Savel'ev, and A. B. Flenov 1973 The study
    of dolphin's echolocator ability in detection of different
    targets, difference in range, and angle on vertical and
    horizontal planes. Proceedings of the 8th All-Union Acousti-
    cal Conference Part 1:134-135. (In Russian)

Rice, D. W. 1978 Beaked whales. In: Marine Mammals, Delphine
    Haley, Ed., pp. 89-95. Pacific Search Press.

Ridgway, S. H., N. J. Flanagan, and J. G. McCormick 1976
Brain-spinal cord ratios in porpoises: possible correlation
with intelligence and ecology. Psychon. Sci. 6:491-492.

Ridgway, S. H. and R. W. Brownson 1979 Relative brain sizes and
cortical surface areas in odontocetes. In preparation.

Schenkkan, E. J. 1973 On the comparative anatomy and function of
the nasal tract in odontocetes (Mammalia, Cetacea).
Bijdragen Tot de Dierkunde 43(3):127-159.

Schenkkan, E. J. and P. E. Purves 1973 The comparative anatomy
of the nasal tract and the function of the spermaceti organ
in the Physeteridae (Mammalia, Odontoceti). Bijdragen Tot de
Dierkunde 43(1):93-112.

Schevill, W. E. 1964 Underwater sounds of cetaceans. In:
Marine Bio-Acoustics, W. N. Tavolga, Ed., pp. 307-316.

Schevill, W. E. and B. Lawrence 1953 Auditory responses of a
bottlenose porpoise *Tursiops truncatus* to frequencies above
100 kHz. Jour. Exp. Zool. 124(1):147-165.

Schevill, W. E., W. A. Watkins, and C. Raye 1963 Underwater
sounds of pinnipeds. Science 141:50-53.

Schevill, W. E. and B. Lawrence 1965 Food-finding by a captive
porpoise, *Tursiops truncatus*, Breviora 53:1-15.

Schevill, W. E., W. A. Watkins, and C. Ray 1969 Click structure
in the porpoise, *Phocoena phocoena*. Jour. Mamm. 50(4):721-
728.

Schusterman, R. J. 1966 Underwater click vocalization by a
California sea lion: Effects of visibility. Psych. Rec.
16:129-136.

Sergeant, D. E. 1978 Ecological isolation in some cetacea. In:
Recent Advances in the Study of Whales and Seals, A. N.
Severtsov, Ed., Nauka, Moskow.

Slijper, E. J. 1962 Whales. Basic Books, Inc., New York.

Tavolga, W. N. 1977 Mechanisms for directional hearing in the
sea catfish (*Arius felis*). Jour. Exp. Biol. 67:97-115.

Tavolga, M. C. and F. S. Essapian 1957 The behavior of the
bottlenosed dolphin (*Tursiops truncatus*): Mating, pregnancy,
parturition, and mother-infant behavior. Zoologica 42, Part
I, pp. 11-34.

Titov, A. A.   1972   Investigation of the sound activities and
     peculiarities of the Black Sea dolphin's echolocator.   Ph.D.
     Thesis, Karadag.   (In Russian)

Tomilin, A. G.   1957   Mammals of the U.S.S.R. and Adjacent
     Countries.   Vol. IX Cetacea.   Izdatel'sto Akademi Nauk,
     Moscow, U.S.S.R.

Tomilin, A. G.   1968   Factors promoting powerful development of
     the brain in Odontoceti: Trudy Vsesoyuznogo
     Sel'skokhozyaystvennogo Instituta Zaochnogo Obrasovaniya,
     No. 31, pp. 191-200.   In Russian.   JPRS No. 49777.

True, F. W.   1910   An account of the beaked whales of the family
     Ziphiidae in the collection of the United States National
     Museum, with remarks on some specimens in other American
     museums.   U. S. National Museum Bulletin 73, 89 pp., 42
     plates.

Watkins, W. A. and W. E. Schevill   1971   Underwater sounds of
     Monodon (Narwhal).   Jour. Acoust. Soc. Am. 49(2), Pt. 2:
     595-599.

Webster, F. A.   1963   Active energy radiating systems: The bat
     and ultrasonic principles II; acoustical control of airborne
     interceptions by bats.   In: Proceedings of the International
     Congress on Technology and Blindness, Vol. 1, pp. 49-135.

Whitmore, F. C., Jr., and A. E. Sanders   1976   Review of the
     Oligocene Cetacea.   Syst. Zool. 25(4):304-320.

Wood, F. G., Jr.   1952   Porpoise sounds.   A phonograph record of
     underwater sounds made by *Tursiops truncatus* and *Stenella
     plagiodon*.   Publ. by Marineland Research Laboratory, Marine-
     land, Florida.

Wood, F. G., Jr.   1953   Underwater sound production and concurrent
     behavior of captive porpoises, *Tursiops truncatus* and
     Stenella plagiodon.   Bull. Mar. Sci. Gulf and Caribbean
     3:120-133.

Wood, F. G.   1964   General Discussion.   In: Marine Bio-Acoustics,
     W. N. Tavolga, Ed., pp. 395-396.   Pergamon Press.

Working Party on Marine Mammals   1977   In: Mammals in the Seas,
     Vol. 1, Report of the FAO Advisory Committee on Marine
     Resources Research.   Food and Agriculture Organization of
     the United Nations, Rome.

Yablokov, A. V. 1962 The key to a biological riddle - the sperm whale at a depth of 2,000 meters. Priroda 4:95-98. (In Russian)

Yablokov, A. V., V. M. Bel'kovich, and V. I. Borisov 1972 Whales and Dolphins, Parts I and II. Kity i Del'fini, Izd-vo Nauka, Moscow. Translation JPRS 62150-1. 528 pp.

Zaslavsky, G. L. 1971 On directivity of sound emission in the Black Sea bottlenosed dolphin. Trudy Akusticheskogo Instituta 17:60-70. (In Russian)

Zaslavsky, G. L. 1972 Study of echolocational Signals of dolphins by means of two-channel registration system. Biofisika 4:717-720. (In Russian)

Zaslavsky, G. L. 1974 Experimental investigations of the spatial-temporal structure of the echolocational signals of dolphins. Ph.D. Thesis. (In Russian)

Zaslavsky, G. L., A. A. Titov, and V. M. Lekomtsev 1969 Investigations of the echolocational abilities of the harbor porpoise (*Ph. phocoena*). Trudy Akusticheskogo Instituta 8:134-138. (In Russian)

ADAPTIVENESS AND ECOLOGY OF ECHOLOCATION IN TERRESTRIAL (AERIAL) SYSTEMS

M. Brock Fenton

Department of Biology, Carleton University, Ottawa,

Canada, K1S 5B6

There are two clearly defined approaches to echolocation by terrestrial (aerial) animals based on the types of orientation sounds they use, namely broadband clicks or structured (time – frequency) calls that may be either broadband or narrowband. The broadband clicks are usually produced in pairs (Buchler and Mitz, this volume) and are associated with the location of obstacles in the paths of the echolocators, while the structured calls are used to identify specific features about the target and its position. To date, the only terrestrial animals that are known to use structured calls are bats in the suborder Microchiroptera; all other terrestrial echolocators, including the Megachiroptera, use broadband clicks (Table 1).

Echolocation is one method of obtaining information about one's surroundings under conditions of poor lighting and it shares many features with electrolocation, a-system of orientation also effective in darkness (Hopkins 1976; 1977). Unfortunately, although it is clear that echolocation is polyphyletic in the animal kingdom, we have no clear idea of how it arose. Observations of bears, aardvarks and ratels using a loud snort and its associated echo to identify underground cavities (Kingdon 1977), suggests that the use of echoes to obtain information about the environment is not restricted to the 'conventional echolocators' (Table 1). Ewer (1968) proposed that echolocation calls are derived from communication signals, but there is no proof for this hypothesis. Data from the ontogeny of bat vocalizations (e.g. Gould 1975; this volume; Brown 1976; Brown and Grinnell, this volume) show no conclusive evidence of a transition from communication to echolocation calls in all of the species examined to date, although some workers have suggested otherwise (Gould 1971; Konstantinov 1973). Echolocating animals produce an

Table 1. Terrestrial (aerial) animals known or suspected to use echolocation (from a variety of sources including: Griffin 1958; Airapet'yants and Konstantinov 1973; Sales and Pye 1974; Buchler 1976; Novick 1977, and specific papers cited below).

animals	broadband calls	structured calls
Amphibia, Apoda, Caeciliidae		
Dermophis cf. septentrionalis[1]	X	
Aves, Caprimulgiformes, Steatornithidae		
Steatornis caripensis	X	
Apodiformes, Apodidae		
Collocalia vanikorensis[2]	X	
Collocalia hirundinacea[3]	X	
Collocalia brevirostris	X	
Collocalia maxima	X	
Collocalia fuciphaga	X	
Mammalia, Insectivora, Soricidae		
Sorex vagrans	X	
Sorex cinereus	X	
Sorex palustris	X	
Blarina brevicauda	X	
Cryptotis parva	X	
Tenrecidae		
Hemicentetes semispinosus	X	
Echinops telfairi	X	
Microgale dobsoni	X	
Chiroptera, Pteropodidae		
Rousettus spp.	X	
Microchiroptera, all families		
all species		X
Rodentia, Gliridae		
Glis glis	X	
Muridae		
Clethrionomys gapperi	X	
Mesocricetus auratus	X	
Rattus norvegicus[4]	X	
Primates, Hominidae		
Homo sapiens	X	

[1]Thurow and Gould 1977; [2]Griffin and Suthers 1970; n.b. the classification of swiftlets is a topic of some debate, and there are workers who classify these species of swiftlets in the genus Aerodramus, a name associated with a subgenus by other workers − see Pye, this volume; [3]Fenton 1975; [4]Chase, this volume.

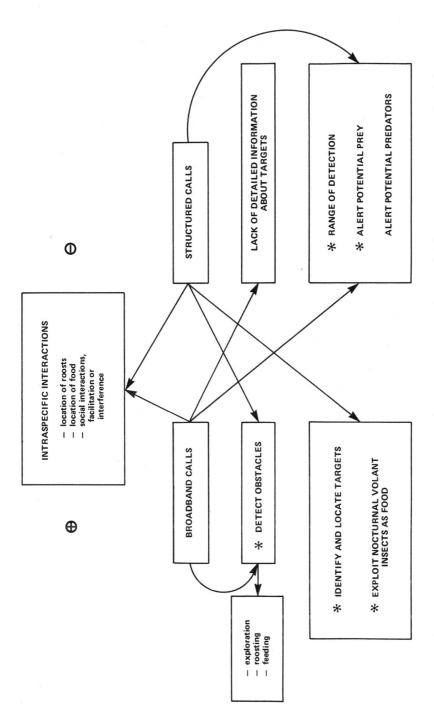

Figure 1. An outline of the advantages, disadvantages and mixed blessings associated with echolocation by terrestrial (aerial) animals. The asterisks indicate advantages or disadvantages documented by experimental evidence.

array of vocalizations which serve a variety of functions from
mother-young interactions (e.g. Brown 1976; Barclay 1978; Gould,
this volume) to mating (Barclay and Thomas 1979).

In short, we have no working hypothesis about how echolocation
evolved. One way of clarifying our knowledge about this subject is
to consider the advantages and disadvantages associated with
echolocation. What follows here is a review of the sparse evidence
on the adaptiveness of echolocation in terrestrial (aerial) animals,
considering the advantages, disadvantages and mixed blessings
associated with it (Fig. 1).

ADVANTAGES OF ECHOLOCATION

Detection of Obstacles

Broadband clicks or highly structured calls can be used by
animals to detect the presence of obstacles in their paths. The size
of the obstacles detected is partly a function of the wavelengths of
the sounds produced (in broadband clicks, presumably the portion of
the call containing the most energy) and the size and proximity of
the target. Detection of obstacles independent of light conditions
entails several important advantages to the animal. These benefits
are clearer for flying echolocators such as bats and birds, than
they are for terrestrial or fossorial species, partly because there
are many more data for the fliers, especially the insectivorous bats.

Perhaps the principal advantage to all echolocators using
broadband clicks, and a side benefit, albeit an important one for
species using structured calls, is to make activity patterns
independent of light conditions, allowing extensive activity at
night or in poorly illuminated habitats during the day. For shrews
echolocation allows poor resolution of habitat details but
facilitates exploration activity (Buchler 1976). Buchler (1976)
showed that Sorex vagrans tended to echolocate more when exploring
unfamiliar territory than when using familiar areas. A direct
benefit to shrews familiar with a large array of microhabitats could
be increased access to food, an important point for an animal on a
tight energy budget (Randolf 1973). Although I found no comparable
data for other small terrestrial echolocators, similar constraints
and advantages probably exist for their echolocation behaviour.

Activity independent of light conditions has made caves
available to flying echolocators, providing benefits associated with
feeding, breeding and predators. Most birds do not use fruit as
food for nestlings because it does not provide adequate protein for
rapid growth, so that nestlings reared on fruit grow more slowly and
are in the nest longer, increasing the time they are exposed to
predators. By nesting inside caves and reducing the threat of

predation on nestlings, oilbirds (Steatornis caripensis) can exploit
fruit, an abundant, low protein food, for their nestlings and
themselves (Morton 1973).

Caves provide nesting or roosting sites that are relatively
protected against many predators, although some arthropods prey on
the eggs and young of some swiftlets (Collocalia spp.; Medway 1962).
However, large populations of bats and/or swiftlets using the
entrances to caves are an excellent food source for many predators,
ranging from snakes to small Carnivora and a variety of birds of
prey (Gillette and Kimbrough 1970). Thus exploitation of caves by
large numbers of echolocators has inherent disadvantages typical of
refuging populations (Hamilton and Watt 1970).

The large surface areas of the interiors of caves present an
abundance of nest and roost sites, particularly for species such
as swiftlets that can build their nests on relatively smooth walls,
or animals such as Rousettus spp. that do not build nests. Since
populations of some species of swifts appear to be limited by the
availability of nest spaces (Lack 1956), the ability to echolocate
and nest in caves may have fostered an increase in populations of
echolocating swiftlets (Fenton 1975). Echolocation allows birds to
come and go from their roosts at night, permitting Collocalia spp.
access to the night skies where they could use vision to track insect
prey (Fenton 1975).

Detection of obstacles by echolocation has permitted animals
to sever the close ties of their activity patterns with ambient
light levels, allowing nocturnal activity, increased opportunity for
exploration and exploitation of caves as nesting and roosting sites.
Caves may protect some animals, particularly young, from predators,
allowing exploitation of abundant low protein food (Steatornis) and
may have contributed to an expansion of populations by increasing
the number of reproducing individuals in the population (Collocalia).

Locate and Identify Targets

Orientation sounds with specific time-frequency structures
allow species in the Microchiroptera to gather specific information
about targets in their paths. Specifically these calls permit
location and tracking of flying insects (Griffin, Webster and
Michael 1960; Simmons, Fenton and O'Farrell 1979) and determination
of the general identity of the potential target (Goldman and
Henson 1977; Schnitzler 1978). Using structured calls some bats can
detect fish or other prey on or near the surface of water with
enough precision to permit capture (Suthers 1965). Other species
use echolocation to locate insects on vegetation or other surfaces,
permitting capture by gleaning (Fenton and Bell 1979). Although the
location and specific identification of targets is closely related
to the ability to exploit populations of nocturnal insects as food

(see below), it also allows accurate evaluation of details of roosts
and other environmental factors.

The versatility of roosting and social behaviour known from
the Microchiroptera, in addition to their diverse feeding habits,
probably reflects the advantage conferred by an echolocation system
that allows an animal to obtain specific, detailed information about
targets, regardless of the light conditions.

Access to Volant Nocturnal Insects as Food

Insectivorous bats use structured calls to exploit nocturnal
insects, particularly smaller ones, as food. These are a resource
largely unavailable to animals that orient by vision, except under
conditions where a predator can track prey against a constant,
light-coloured background. Birds, including many species of owls
and most Caprimulgiformes, regularly feed on nocturnal volant
insects which they locate and track visually using the night sky
as a background. Echolocating insectivorous bats and a variety of
spiders exploit nocturnal flying insects as food, but their prey-
catching prowess is not constrained by light.

Small nocturnal insects are an important food source, access
to which provides predators with a rich, if not superabundant, food
supply. A number of studies has shown that temperate insectivorous
bats feed heavily on insects in proportion to their availability
(Black 1974; Belwood and Fenton 1976; Bradbury and Vehrencamp 1976;
Anthony and Kunz 1977) and none has provided evidence to support
the idea that food is limiting for populations of insectivorous bats.
In areas where 10 or 12, or even 25 species of insectivorous bats
feed in the same place at the same time, there is no evidence of
competition between species or individuals for food (Bell 1979
Fenton and Thomas 1979). The population levels of insectivorous
bats are very high in some areas, suggesting that nocturnal insects
are not in short supply in spite of astounding levels of predation
associated with insectivorous bats (Davis, Herreid and Short 1962).
These data all clearly demonstrate that nocturnal insects are an
abundant food supply.

A reflection of the effect of this food resource is the fact
that the majority of insectivorous bats are small (less than 40 grams
as adults; Fig. 2), while all of the nocturnal insectivorous birds
are much larger (over 40 grams as adults; Fenton 1974; Fenton and
Fleming 1976). The preponderance of small insectivorous bats
suggests that bats are very effective as small, nocturnal
insectivores and much less effective as larger, nocturnal
insectivores.

One of the principal advantages of echolocation is access to
the food resource represented by small nocturnal insects and in

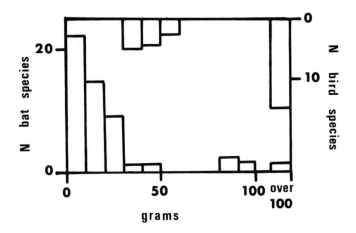

Figure 2. Size distribution (by weight) of the species of animalivorous and insectivorous bats and nocturnal birds which occur in Guyana (adapted from Fenton and Fleming 1976).

this context, I consider echolocation and flight to be the keys to the success of the Microchiroptera.

DISADVANTAGES OF ECHOLOCATION
Lack of Detailed Information about Targets

This disadvantage applies only to species using broadband clicks and is directly comparable to the lack of the ability for flapping flight in flying squirrels. The relative lack of resolution associated with broadband clicks is probably more a function of information processing by the animal than a limitation inherent in the signals themselves.

Range

One of the principal disadvantages of echolocation is its limited effective range in air, a situation potentially more hazardous for a flying than for a walking or running animal. Griffin (1971) discussed the role of attenuation in echolocation by bats and his data emphasized that bats using higher frequency calls (which provide better resolution of target information) suffered more from interference by attenuation. Pye (1971) demonstrated how rain and fog could hamper echolocation by increasing attenuation and providing additional clutter and phantom targets. It is clear that atmospheric conditions can strongly influence the effectiveness, especially the range of echolocation.

It is difficult to determine the effective range of echolocation. Buchler (1976) suggested that for shrews distance perception using echolocation was less than 1 m and Novick (1977), in his summary of data on acoustic orientation by bats, provided figures of 2 or 3 m for the effective range of echolocation of insects and large obstacles by many Microchiroptera. Data on the range of echolocation are usually associated with laboratory studies of bats orienting towards preferred roosts (Novick 1977), although Griffin, Webster and Michael (1960) provided data for hunting Myotis lucifugus in the lab that showed that these bats fixed on their insect prey at a range of about 0.8 m. Unfortunately, the distance over which a bat or other echolocator fixes on its target is the minimum effective range, not the maximum effective range.

Vaughan's description of feeding behaviour of Hipposideros commersoni (1977), the largest insectivorous bat, suggests an effective range of less than 10 m in the field. Field observations of Myotis volans feeding in Arizona suggest an effective range of 5 - 10 m, based on the distances at which individual bats fixed on their insect prey (Fenton and Bell 1979). In the field, feeding Myotis lucifugus and M. californicus appeared to locate prey at much shorter distances (less than 1 m) and were able to make repeated attempts to capture insects in a short distance, a behaviour not observed in M. volans (Fenton and Bell 1979).

It is clear that different species of bats, sometimes in the same genus, show different depths of perception through echolocation. Brosset (1966) had suggested this when he proposed two basic feeding strategies for insectivorous bats, long range and short range. His observations on long range species were based mainly on field observations of Taphozous peli, which appeared to locate insect targets at 15 m, while his suggestions about short range operators were based largely on the report of Griffin, Webster and Michael (1960).

Intensity of echolocation calls probably influences the range at which an animal can detect and identify a target. Griffin (1971) estimated that the maximum effective range for Hipposideros galeritus was no more than 2 m. This species produced a relatively low intensity call, stronger than those typical of Nycteris spp., but much weaker than the high intensity calls associated with Myotis spp., Eptesicus spp., or Rhinolophus spp.

When the effective range of echolocation is considered in terms of body lengths of the animal involved, a problem of scaling is obvious, further influenced by the speed at which the echolocator is travelling. Larger, faster species will be adversely affected by a short effective range (relative to their size) simply because they may have reached a target before they have had time to

process information in the returning echoes. Slower, smaller
species will be less drastically affected in this context.
Considerations of food availability notwithstanding, the relative
dearth of larger insectivorous bats (Fig. 2) may be a reflection
of the limited range effectiveness of echolocation.

   Larger insectivorous bats appear to have at least two strategies
for reducing the constraints associated with short range operation.
Hipposideros commersoni often hunts from a perch, flying out only
to capture vulnerable insects (Vaughan 1977), alleviating in part
the restrictions of range, while saving on energy spent in flight.
Production of loud orientation calls with fundamental frequencies
in lower ranges (15 - 20 kHz; Fig. 3) could also reduce the effects
of atmospheric attenuation, increase the effective range and
provide adequate details about larger insect targets appropriate
for larger bats.

Figure 3.   Fast Fourier Transform analysis of calls used for
echolocation by Collocalia hirundinacea (5.2 kHz), an unidentified
molossid in Cote d'Ivoire (17.0 kHz), Eptesicus fuscus (29.2 kHz),
Myotis lucifugus (44.2 kHz), Hipposideros cyclops (58.9 kHz),
Pipistrellus nanus (82.6 kHz), Rhinolophus landeri (100.6 kHz) and
Hipposideros ruber (139.4 kHz).   In each case, at least 30 calls
were averaged by a Princeton Applied Research Model 4513 FFT to
produce these spectra.

Alerting Prey

   The presence of ears that are sensitive to the echolocating
cries of bats in many moths (Roeder 1975) and some lacewings
(Miller 1975) demonstrates an important disadvantage for animals
which use echolocation for hunting, specifically, alerting potential

to the presence and approach of a predator. Roeder (1967) and Miller (pers. comm.) have determined that moths and lacewings that can hear bats have a 40% less chance of being captured by echolocating bats than species lacking ears(see also Miller, this volume).

In the Lepidoptera ears are polyphyletic, appearing in different places in different families (Table 2). The ears of moths are, like those of other insects, tuned by their physical characteristics including the diameter of, and the tension on, the tympanum (Michelson 1971; Adams 1972). It is possible for an echolocating bat to adjust its orientation calls to make them less conspicuous to a moth, and thus to reduce the advantage to the moth (Fig. 4; see also Fullard and Fenton, this volume). Crucial adjustments for a bat in the context of foiling a moth include flight speed and the intensity and frequency of orientation calls, modifications to reduce the warning time available to the insect (Fenton and Fullard 1979).

Table 2. The locations of tympana in different families of moths known to include tympanate species. Data from Roeder (1974; 1975), Sales and Pye (1974), and James H. Fullard (pers. comm.)

family	thorax	first or second abdominal segment	maxillary palp
		locations of ears	
Thaumetopoeidae	X		
Agaristidae	X		
Lymantriidae	X		
Noctuidae	X		
Pyralidae		X	
Arctiidae	X		
Notodontidae	X		
Geometridae		X	
Drepanidae		X	
Thyatiridae		X	
Thyretidae	X		
Sphingidae[1]			X

[1] only in the Choerocampinae

Some details of moth defense and possible bat counters have been reviewed in the light of tuning characteristics of moths ears (Fullard and Fenton, this volume). These data suggest that the hearing ability of insects could exert an important selection pressure on the echolocation behaviour of bats. It is important to

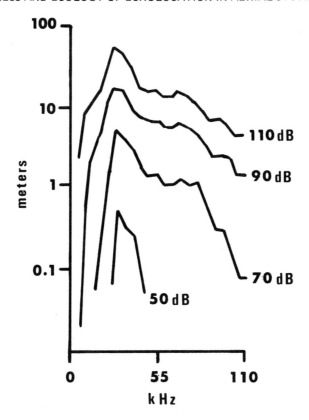

Figure 4. A representation of the maximum distances at which
a tympanate moth (Tmetopteryx bisecta ; Notodontidae) can detect
the echolocation cries of bats based on auditory threshold values.
These figures are derived by transforming audiogram values for the
moth with the data of Griffin (1971) on attenuation, and they
illustrate the effects of initial intensities of bat echolocation
calls (dB re 2 x 10^{-5} N/m^2; 10 cm), at 20% relative humidity.

note, however, that at least 65% of the bat species in any area,
and 100% in some areas, use FM echolocation calls that sweep through
the range of maximum sensitivity of tympanate moths (Fig. 5). Bats
which have countered moth defenses by adjusting their echolocation
calls appear to be consistently in a minority - one reason that
their counters are effective. The ultrasonic sounds produced by
some moths, usually arctiids, are an extension of insect hearing
defense against bats. These sounds cause some bats to reject
otherwise palatable food (Dunning and Roeder 1965), and are produced
just as the bat closes with the moth. The acoustic characteristics
of the clicks (power spectra and frequency-time structure) closely

match those of FM orientation calls produced by many bats as they
close with their insect prey.   The clicks of the moths may be
initially processed as echoes by the bats' auditory systems,
resulting in disruption of information processing and producing a
Protean effect (Fullard, Fenton and Simmons 1979).   A bat-
deterrent effect has also been suggested for the clicks of Peacock
Butterflies (Møhl and Miller 1976).

Figure 5. The audiogram of the African geometrid _Gorua apicata_
superimposed on an FFT analysis of 400 bat echolocation calls
recorded in the field in Cote d'Ivoire (dots).   The moth is most
sensitive in the frequency range with the greatest energy of
echolocation calls.

        In summary, echolocation calls may adversely affect the
hunting effectiveness of insectivorous bats by alerting potential
prey and providing them with time for escape maneouvers. The tuning
characteristics of moth ears allow bats to reduce the advantage
gained by the moth by adjusting the frequency and intensity of their
calls.   Countering insect defensive strategies may represent an
important selective force on bat echolocation.

                            Alerting Predators

        Trains of pulses emitted by an animal as it moves through an
area not only produce information about targets, but also provide
potential predators with specific details about the location and

course of the echolocator. Thus a shrew or bat producing
orientation calls could reveal its presence and course to a
predator and significantly reduce its chances of survival (assuming
that the predator can hear the calls).

There are relatively few predators which specialize on bats or
other echolocators, and the inability of birds to hear sounds above
20 kHz (Schwartzkopff 1973) precludes their exploitation of
ultrasonic echolocation cries to locate prey. Hawks and owls do
not appear to specialize on shrews as food, although these small
mammals comprise a consistently small portion of the diets of many
raptors and owls (e.g. Craighead and Craighead 1956), and they are
also taken by many mammalian predators. The Bat Hawk (Macheiramphus
alcinus) feeds heavily on bats (Fenton, Cumming and Oxley 1977)
which it locates using vision. Bat Hawks feed in groups of bats or
swiftlets that are leaving or returning to caves or other
roosts.

Carnivorous bats may be the predators best suited to exploit
the information available in the orientation calls of bats of
other animals. However, in spite of the suggestions that species
such as Vampyrum spectrum feed heavily on bats (e.g. Howell and
Burch 1974), one recent study of the diet of V. spectrum in Costa
Rica showed that bats did not constitute a large portion of its
diet (Vehrencamp, Stiles and Bradbury 1977).

There appear to be no data supporting the possibility that
the echolocation calls of other animals are used by predators as
aids to locating food. However, it may be significant that some
broadband echolocators cease production of their clicks when not
orienting in darkness (e.g. Collocalia hirundinacea; Fenton 1975),
while others continue to produce their cries, even when flying in
bright moonlight (e.g. Rousettus aegyptiacus; Thomas and Fenton 1978).

MIXED BLESSINGS

Information about the presence and course of an echolocator
provided by its calls is available to conspecifics. For example,
bats that increase their pulse repetition rates as they close with
insect targets reveal to eavesdroppers the presence of food. In
some settings listening for feeding buzzes could permit members of
the same colony to converge upon and exploit a rich patch of food,
but in other instances (e.g. territorial species), it may facilitate
interference with feeding behaviour of the individual that located
the food source.

In Arizona Fenton and Morris (1976) observed that some bats
occasionally responded to the playback of feeding buzzes by
increased activity around the speaker, suggesting that some

insectivorous bats exploit feeding calls of others as a means of
locating food. Some colleagues and I have conducted a number of
similar experiments involving the broadcast of feeding buzzes
produced by Myotis lucifugus in areas where conspecifics were flying.
The results mirror those of the earlier study (Fenton and Morris 1976)
since in most instances the playbacks elicited no response, but
on several occasions bat activity in the vicinity of the speaker
increased drammatically when the feeding buzzes were presented. In
southeastern Ontario M. lucifugus commonly forages in groups,
feeding for the most part on hatches of aquatic insects (Belwood
and Fenton 1976), and in this species in this setting it is
adaptive for individuals to be able to locate and exploit swarms of
insects.

The patchy distribution of many populations of nocturnal
insects may be best exploited by species that effectively communicate
information about the location of rich patches of food. In this
context the inherent communication properties of echolocation calls
may maximize use of nocturnal insects as food. However, territorial
insectivorous bats (e.g. Wallin 1961; Tuttle, Stevenson and
Rabinowitz 1978) can monitor intrusions by other bats into their
territories and distinguish between feeding and cruising trespassers.

In many instances insectivorous bats quickly locate and exploit
new roost opportunities (Brosset 1966; Fenton 1970), and roost
availability may limit populations in, and community structures of,
some bat faunas (Humphrey 1975). The echolocation calls of
individuals near or moving towards a roost may be used by others to
locate groups of conspecifics or roosts. Playback experiments we
have conducted in the vicinity of hibernation sites during swarming
behaviour have clearly demonstrated that M. lucifugus respond to the
calls of conspecifics and form large groups of flying bats in the
vicinity of the source of the calls. In this case we played recorded
echolocation calls of M. lucfugus about 50 m from the entrance to
an abandoned mine which is used each winter as a hibernation site
by about 5,000 bats. The echolocation calls (typical FM cries)
were presented to the M. lucifugus as they passed through the area
and within several minutes we consistently accumulated a column of
20 to 50 bats flying around in the vicinity of the speaker. This
response was not elicited by 40 kHz pure tones or by tape noise.

In this situation, bats were obviously using the echolocation
calls of others to locate groups of conspecifics. Vaughan and
O'Shea (1977) described the rallying behaviour typical of Antrozous
pallidus returning to a day roost in the morning. In that case
the bats produced distinctive 'directive' calls, and it is not clear
if the bat echolocation calls were also important in the responses
of individuals to the aggregation of conspecifics. There is no
evidence of a 'directive' call in the repertoire of M. lucifugus
(Barclay 1978).

Exploitation of orientation calls of conspecifics may be important in a number of bat social interactions. Again the calls may facilitate some interactions and increase the chances of interference in others. In some cases modifications to echolocation calls may enhance their effectiveness in communication. For example, the honking of one Noctilio leporinus at another was described by Suthers (1965) and we have observed this behavior in M. lucifugus, M. volans, and Eptesicus fuscus. In other cases calls produced by bats during social interactions may be drastically different from echolocation calls, probably to enhance their effectiveness (e.g. the copulation call of M. lucifugus; Barclay and Thomas 1979).

Mixed blessings in terms of the use of echolocation calls by conspecifics probably extend to other social interactions among insectivorous bats. Territorial behaviour has been reported from several species (e.g. Wallin 1961; Bradbury and Emmons 1974; Tuttle, Stevenson and Rabinowitz 1978) and individuals may be able to recognize their neighbours by their echolocation calls.

While working at the Rockefeller University Field Station near Millbrook, New York in July 1976, I occasionally observed chases involving pairs of Eptesicus fuscus. The chases sometimes involved physical contact, usually when the chasing bat struck the leader, and often vocalizations which were audible to me. Over a grassy clearing surrounded on three sides by deciduous woods a number of bats fed each evening, including M. lucifugus as they passed through en route to other areas, and up to 5 E. fuscus that remained in the vicinity of the clearing for at least 90 minutes after dark. One of the E. fuscus set up a patrol flight pattern in the clearing by flying a large four-leafed-clover pattern about 100 m across, over the other bats feeding in the area. Occasionally this bat left its flight path and pursued another E. fuscus, chasing it out of the immediate area before returning to its patrol path. The patrolling bat made no effort to chase away all of the other bats (M. lucifugus or E. fuscus) feeding below it. On the evening of 19 July 1976 I played the echolocation calls of an E. fuscus recorded 5 km away to the patrolling bat, causing it to leave its flight patrol and to chase the nearest E. fuscus. In 30 minutes I was able to elicit 7 such altercations, where the average number of altercations on 10 previous and 5 subsequent nights was 2 (range 0 to 3). Until the chases were initiated, all of the bats involved in the area appeared to have been producing 'typical' FM echolocation calls.

Since I cannot show that the patrolling bat was the same individual from night to night, and because I am not certain if the conspecifics feeding in the area were the same from night to night, it is not possible to provide a detailed interpretation of these observations. However, for the purpose of this discussion, the results strongly suggest that the patrolling bat responded to the echolocation calls of the intruder which it could distinguish from the familiar calls of other conspecifics feeding below it. Vision seems unlikely as an important factor in this situation as the patrolling bat was above the other bats and swooped down to chase intruders.

As we gather more information about the behaviour of echolocating animals in the field, it should become increasingly apparent that the orientation calls are used in social contexts and may be pirated. Interactions with conspecifics should demonstrate that echolocation offers a mixed blessing, an asset in some settings and a liability in others.

ADAPTIVENESS OF ECHOLOCATION

The relative lack of echolocation behaviour in terrestrial settings outside the Microchiroptera suggests that the use of broadband clicks for orientation does not confer a strong adaptive advantage to animals that practise it. On the other hand, the facultative use of echolocation by a variety of animals (e.g. man, and perhaps bears, aardvarks and ratels) demonstrates some advantages when lighting is poor or non-existant. Echolocation involving the use of structured calls, whether broadband or narrowband, represents a special adaptation (sensu Van Valen 1971) that has resulted in the occupation of a new adaptive zone, principally the exploitation of nocturnal insects as food.

STILL TO DO

There are a number of important questions to be addressed in the context of the adaptiveness of echolocation. How frequently do insectivorous bats turn off their echolocation and rely on other sensory input to locate food (e.g. Fiedler, Habersetzer and Vogler, this volume)? What is the importance of the sounds and smells of potential prey as cues for insectivorous bats? Do bats capable of producing structured calls occasionally use broadband clicks? How do individuals adjust the intensities of their calls in specific situations where a quiet or a loud call might be most effective?

Fortunately the number of researchers going into the field to observe the natural behaviour of bats and other echolocators is increasing along with the number of workers studying different aspects of echolocation in the lab. Furthermore, the availability of portable and relatively inexpensive apparatus for studying and monitoring the calls of echolocators has increased (Simmons, Fenton, Ferguson, Jutting and Palin 1979) allowing field workers to put themselves into the acoustic world of the animals they study.

Researchers wishing to study the behaviour of bats in the field can often exploit their feeding habits. Specifically, species in many families will feed at concentrations of insects around lights, while other species predictably visit feeding sites. In either situation, biologists can anticipate the response of the animal and study its feeding behaviour. However, it is important to remember that observing an animal producing a pattern of calls which we associate with echolocation does not prove that the animal was echolocating and not using other sensory cues to find its prey.

ACKNOWLEDGEMENTS

I am grateful to Robert Barclay, Gary Bell, Patricia Brown, Edward Buchler, James Fullard, Alan Grinnell and James Simmons for reading this manuscript and making helpful suggestions about it. L. Van Every kindly prepared Figure 1. My research on bats has been supported by National Research Council of Canada Operating and Equipment Grants.

REFERENCES

Adams, W.B. 1972. Mechanical tuning of the acoustic receptor of Prodenia eridania (Cramer) (Noctuidae). J. exp. Biol. 57: 297-304.

Airapet'yants, E. Sh., and A.I. Konstantinov. 1973. Echolocation in animals (translated from Russian). Israel Prog. Sci. Transl. Jerusalem: Keter Press.

Anthony, E.L.P. and T.H. Kunz. 1977. Feeding strategies of the little brown bat, Myotis lucifugus, in New Hampshire. Ecology 58: 775-786.

Barclay, R.M.R. 1978. Vocal communication and social behaviour of the little brown bat, Myotis lucifugus, (Chiroptera : Vespertilionidae). M.Sc. Thesis, Department of Biology, Carleton University, Ottawa, Canada.

Barclay, R.M.R. and D.W. Thomas. 1979. Copulation call of the bat, Myotis lucifugus; a discrete, situation specific communication signal. J. Mamm. 60: in press.

Belwood, J.J. and M.B. Fenton. 1976. Variation in the diet of Myotis lucifugus (Chiroptera : Vesperitilionidae). Can. J. Zool. 54: 1674-1678.

Bell, G.P. 1979. Summer habitat use and response to food patches by insectivorous bats in a desert community. M.Sc. Thesis, Department of Biology, Carleton University, Ottawa, Canada.

Black, H.L. 1974. A north temperate bat community: structure and prey populations. J. Mamm. 55: 138-157.

Bradbury, J.W. and L.H. Emmons. 1974. Social organization of some
    Trinidad bats. I Emballonuridae. Z. Tierpsychol. 36: 137-183.
Bradbury, J.W. and S.L. Vehrencamp. 1976. Social organization and
    foraging in emballonurid bats. I Field studies. Behav. Ecol.
    and Sociobiol. 1: 337-381.
Brosset, A. 1966. La biologie des chiroptères. Masson et Cie, Paris.
Brown, P.E. 1976. Vocal communication in the pallid bat, Antrozous
    pallidus. Z. Tierpsychol. 41: 34-54.
Buchler, E.R. 1976. The use of echolocation by the wandering shrew
    (Sorex vagrans). Anim. Behav. 24: 858-873.
Craighead, J.J. and F.C. Craighead. 1956. Hawks, owls and wildlife.
    Wildlife Management Institute.
Davis, R.B., C.F. Herreid II and H.L. Short. 1962. Mexican free-
    tailed bats in Texas. Ecol. Monogr. 32: 311-346.
Dunning, D.C. and K.D. Roeder. 1965. Moth sounds and the insect-
    catching behavior of bats. Science 147: 173-174.
Ewer, R.F. 1968. Ethology of mammals. Paul Elek Scientific Books,
    London.
Fenton, M.B. 1970. Population studies of Myotis lucifugus
    (Chiroptera : Vespertilionidae) in Ontario. Life Sci. Contr.,
    R. Ont. Mus. no. 77: 1-34.
    -          1974. The role of echolocation in the evolution of
    bats. Am. Nat. 108: 386-388.
    -          1975. Acuity of echolocation in Collocalia hirundinacea
    (Aves : Apodidae), with comments on the distributions of
    echolocating swiftlets and molossid bats. Biotropica 7: 1-7.
Fenton, M.B. and G.P. Bell. 1979. Echolocation and feeding behaviour
    in four species of Myotis (Chiroptera). Can. J. Zool. 57:
    in press.
Fenton, M.B., D.H.M. Cumming and D.J. Oxley. 1977. Prey of Bat
    Hawks and availability of bats. Condor 79: 495-497.
Fenton, M.B. and T.H. Fleming. 1976. Ecological interactions
    between bats and nocturnal birds. Biotropica 8: 104-110.
Fenton, M.B. and J.H. Fullard. 1979. The influence of moth hearing
    on bat echolocation strategies. J. comp. Physiol. in press.
Fenton, M.B. and G.K. Morris. 1976. Opportunistic feeding by
    desert bats (Myotis spp.). Can. J. Zool. 54: 526-530.
Fenton, M.B. and D.W. Thomas. 1979. Dry season overlap in activity
    patterns, habitat use and prey selection by sympatric
    African insectivorous bats. Biotropica in press.
Fullard, J.H., M.B. Fenton and J.A. Simmons. 1979. Jamming bat
    echolocation: the clicks of archtiid moths. Can. J. Zool. 57:
    647-649.
Gillette, D.D. and J.D. Kimbrough. 1970. Chiropteran mortality.
    IN About Bats, B.H. Slaughter and D.W. Walton (eds), Southern
    Methodist Univ. Press, Dallas, pp. 262-283.
Goldman, L.J. and O.W. Henson, Jr. 1977. Prey recognition and
    selection by the contrast frequency bat, Pteronotus parnellii
    parnellii. Behav. Ecol. Sociobiol. 2: 411-420.

Gould, E. 1971. Studies of maternal-infant communication and development of vocalizations in the bats Myotis and Eptesicus. Comm. Behav. Biol. Part A 5(5): 263-313.

- 1975. Neonatal vocalizations in bats of eight genera. J. Mamm. 56: 15-29.

Griffin, D.R. 1958. Listening in the dark. Yale University Press. New Haven.

- 1971. The importance of atmospheric attenuation for the echolocation of bats (Chiroptera). Anim. Behav. 19: 55-61.

Griffin, D.R. and R.D. Suthers. 1970. Sensitivity of echolocation in cave swiftlets. Biol. Bull. 139: 495-501.

Griffin, D.R., F.A. Webster and C.R. Michael. 1960. The echolocation of flying insects by bats. Anim. Behav. 8: 141-154.

Hamilton, W.J. III and K.E.F. Watt. 1970. Refuging. Ann. Rev. Ecol. Syst. 1: 263-286.

Hopkins, C.D. 1976. Stimulus filtering and electroreception: tuberous electroreceptors in three species of gymnotid fish. J. comp. Physiol. 111: 171-207.

- 1977. Electric communication. IN How animals communicate. T. Sebeok (ed.), Indiana University Press, Bloomington, Indiana.

Howell, D.J. and D. Burch. 1974. Food habits of some Costa Rican bats. Revta. Biol. trop. 21: 281-294.

Humphrey, S.R. 1975. Nursery roosts and community diversity of Nearctic bats. J. Mamm. 56: 321-346.

Kingdon, J. 1977. East African mammals, an atlas of evolution in Africa, Volume IIIA, Academic Press, London.

Konstantinov, A.I. 1973. Development of echolocation in bats in postnatal ontogenesis. Period. Biol. 75: 13-19.

Lack, D. 1956. Swifts in a tower. Methuen, London.

Medway, Lord. 1962. The swiftlets (Collocalia) of Niah Cave, Sarawak. Ibis 104: 45-66.

Michelson, A. 1971. The physiology of the locust ear. II Frequency discrimination based upon resonances in the tympanum. Z. vergl. Physiol. 71: 63-101.

Miller, L.A. 1975. The behaviour of flying green lacewings, Crysopa carnea in the presence of ultrasound. J. Insect Physiol. 21: 205-219.

Møhl, B. and L.A. Miller. 1976. Ultrasonic clicks produced by the peacock butterfly: a possible bat-repellent mechanism. J. exp. Biol. 64: 639-644.

Morton, E.S. 1973. On the evolutionary advantages and disadvantages of fruit-eating in tropical birds. Am. Nat. 107: 8-22.

Novick, A. 1977. Acoustic orientation. IN Biology of Bats, Volume III, W.A. Wimsatt (ed.), Academic Press, New York, pp. 74-287.

Pye, J.D. 1971. Bats and fog. Nature 229: 572-574.

Randolph, J.C. 1973. Ecological energetics of a homeothermic predator, the short-tailed shrew. Ecology 54: 1166-1187.

Roeder, K.D. 1967. Nerve cells and insect behavior, revised edition. Harvard Univ. Press, Cambridge.

—   1974. Acoustic sensory responses and possible bat
    evasion tactics of certain moths. Can. Soc. Zool. Ann. Meeting
    Proc., Univ. New Brunswick, Fredericton, pp. 71-78.

—   1975. Neural factors and evitability of insect
    behaviour. J. exp. Zool. 194: 75-88.

Sales, G. and J.D. Pye. 1974. Ultrasonic communication by animals.
    Chapman and Hill, London.

Schwartzkopff, J. 1973. Mechanoreception. IN Avian Biology, D.S.
    Farner and J.R. King (eds.), Volume 3, Academic Press,
    New York.

Schnitzler, H.-U. 1978. Die Detektion von Bewegungen durch
    Echoortung bei Fledermäusen. Verh. Dtsch. Zool. Ges. 1978:
    16-33.

Simmons, J.A., M.B. Fenton and M.J. O'Farrell. 1979. Echolocation
    and pursuit of prey by bats. Science 203: 16-21.

Simmons, J.A., M.B. Fenton, W.R. Ferguson, M. Jutting and J. Palin.
    1979. Apparatus for research on animal ultrasonic signals.
    Life Sci. Misc. Pub., R. Ont. Mus.

Suthers, R.A. 1965. Acoustic orientation by fish-catching bats.
    J. Zool. (London) 158: 319-348.

Thomas, D.W. and M.B. Fenton. 1978. Notes on the dry season
    roosting and foraging behaviour of Epomophorus gambianus and
    Rousettus aegyptiacus (Chiroptera : Pteropodidae). J. Zool.
    London 186: 403-406.

Thurrow, G.R. and H.J. Gould. 1977. Sound production in a caecilian.
    Herpetologica 33: 234-237.

Tuttle, M.D., D.E. Stevenson and A.R. Rabinowitz. 1978. Foraging
    site selection and territoriality of Myotis grisescens. Paper
    presented at the Fifth International Symposium on Bat Research,
    Albuquerque, New Mexico.

Van Valen, L. 1971. Adaptive zones and the orders of mammals.
    Evolution 25: 420-428.

Vaughan, T.A. 1977. Foraging behaviour of the giant leaf-nosed bat
    (Hipposideros commersoni). E. Afr. Wildl. J. 15: 237-249.

Vaughan, T.A. and T.J. O'Shea. 1976. Roosting ecology of the pallid
    bat, Antrozous pallidus. J. Mamm. 57: 19-42.

Vehrencamp, S.L., F.G. Stiles and J.W. Bradbury. 1977. Observations
    on the foraging behavior and avian prey of the Neotropical
    carnivorous bat, Vampyrum spectrum. J. Mamm. 58: 469-478.

Wallin, L. 1961. Territorialism on the hunting ground of Myotis
    daubentoni. Säugetierk. Mitt. 9: 156-159.

Chapter IV
Auditory Processing of Echoes

Chairman : L. Kay

co-chairmen

C. Scott Johnson and G. Neuweiler

- <u>Underwater</u>

    Sound reception in the porpoise as it relates to echo-
    location.
      J.G. McCormick, E.G. Wever, S.H. Ridgway and
      J. Palin.

    Behavioral measures of odontocete hearing.
      A.N. Popper

    Electrophysiological experiments on hearing in
    odontocetes.
      S.H. Ridgway

    Peripheral sound processing in odontocetes.
      K.S. Norris

    Cetacean brain research : Need for new directions.
      T.H. Bullock

    Important areas for future cetacean auditory study.
      C. Scott Johnson

- <u>Airborne</u>

    Auditory processing of echoes : peripheral processing.
      G. Neuweiler

    Organizational and encoding features of single neurons
    in the inferior colliculus of bats.
      G.D. Pollack

    Auditory processing of echoes : representation of acoustic
    information from the environment in the bat cerebral cortex.
      N. Suga and W.E. O'Neill

SOUND RECEPTION IN THE PORPOISE AS IT RELATES TO ECHOLOCATION

James G. McCormick
Department of Otolaryngology
Bowman Gray School of Medicine
Wake Forest University
Winston-Salem, North Carolina   27103

E. G. Wever
Auditory Research Laboratories
Princeton University
Princeton, New Jersey   08540

S. H. Ridgway
Naval Ocean Systems Center
San Diego, California   92152

J. Palin
Auditory Research Laboratories
Princeton University
Princeton, New Jersey 08540

INTRODUCTION

Previous to the work of our group which was published in 1970 (McCormick, et al., 1970), all theories of hearing in Cetacea were based on dissections and experiments with dead specimens. Such experiments continue to be published in the literature to this day, and just as the earlier studies of dead material did, they only serve to confound the many students and investigators of Cetacean hearing, especially those who have little formal training in the science of physiological acoustics.

Our study was conducted with careful procedures and controls for artifacts, utilizing the electrophysiological method best suited for the study of middle and external ear mechanics of the

live animal--the A.C. cochlear potentials, which were discovered
and elucidated by our colleague Professor E. G. Wever.

As is true with most scientific works, the details of our
original paper of 1970 were read by few people, and retained by
even fewer people. Word of mouth--investigator to investigator--
has changed the context of our work so that many times we cannot
recognize our own work when we hear the "word of mouth" report
of our study at meetings. And so, many of our colleagues and
close friends have lost the essence of our publication.

On the other hand, we ourselves are all too late with the
new unpublished material which we present in this publication.
Furthermore, we have done our friends and colleagues an injustice
by not defining more specifically what we mean when we say the
porpoise ear receives sound by bone conduction. In a summation
statement, Dr. C. Scott Johnson, one of the outstanding pioneers
of the study of porpoise hearing, aptly pointed out the fact that
there are many theories and mechanisms of bone conduction in the
literature, thus presenting a semantic barrier to the understanding
of our work. In part, our new material presented here is directed
at rectifying this problem.

A second important point to be made here is that our work
supports and is in agreement with the heroic electrophysiological
research on the auditory system of the porpoise carried out by
Professors Theodore Bullock, Allen Grinnell, N. Suga, and their
colleagues in Japan (1968). Their study, although primarily
concerned with the central nervous system processing of auditory
signals by the porpoise, came before our work, and it was an
inspiration to some of our studies. We shared our new techniques
for anesthesia and major surgery on the porpoise (Ridgway and
McCormick, 1967) with Professor Bullock's group, and they in turn
shared their first electrophysiological experiences on the auditory
system of the porpoise with us.

Thirdly, this work is directed at clarifying the fact that we
find our experimental results in agreement with Professor Kenneth
Norris's theory of porpoise echolocation (1964). We feel that our
efforts, along with the work of Bullock and colleagues, support
the echo reception half of Norris's concept; and the classic work
of Bill Evans and his colleagues on the projection areas for sonar
in the porpoise supports the other half of Norris's hypothesis
(c.f. Diercks, et al., 1971).

Finally, recent work by Sam Ridgway and his colleagues has
added new credence to our speculation that the maintenance of air
over the round window membrane in the porpoise is important for
reception of sound. Ridgway's laboratory has produced evidence

for a mechanism in the porpoise which could easily carry out such
a requirement--the air pulse pressure system in the nares area of
the porpoise is directly connected to the porpoise middle ear
cavity via the eustacean tube (see Ridgway, Carder, Green, Gaunt,
Gaunt, and Evans this volume).

BONE CONDUCTION HEARING IN THE PORPOISE

   Although it is generally accepted that the porpoise evolved
from a land mammal with an "air" ear (Romer, 1966), the results
reported by McCormick, et al. (1970) support the idea that the
meatus and ear drum of the porpoise no longer function in the
hearing process. Since the cochlear potential resulting from
point source vibration of the lateral meatus was not altered by
dissection and separation of the medial meatus pathway, one must
conclude that the meatus of the porpoise is no more important for
sound conduction than the surrounding tissue of the head. The
fact that relatively high pressure applications at the ear drum
failed to damp the cochlear potential argues strongly for the dis-
involvement of the ear drum with the hearing process. This disin-
volvement is also confirmed by the fine anatomical study of the
inner ear which revealed that the ear drum is not directly
connected to the malleus.

   Since McCormick, et al., found that the Cetacean ear has no
direct functional or anatomical coupling between its degenerated
aerial system and the middle ear ossicular chain, they proposed
that it seemed appropriate to analyze this system in terms of
direct vibration modes of stimulation such as bone conduction and
tissue conduction. We believe that the evolution of the porpoise
ear has taken advantage of such a direct vibrational system, and
we should like to lead into a development of this thesis with a
presentation of bone stimulation theories. Wever and Lawrence
(1954) have formulated an excellent critique of bone stimulation
theories. As a springboard into an understanding of the operation
of the Cetacean ear, we take advantage of their discussion and
figures.

   Two questions strike us as primary to our evolutionary theory
of the porpoise ear. Since we are proposing that the ear of the
land mammal lost its aerial function with adaptation to an aquatic
environment, how are the sensory hair cells of the cochlea stimulated in an aerial sound system, and is the stimulation phenomenon
the same for bone conduction?

   The state of affairs for cochlear hair cell stimulation with
aerial sound hearing is well explained in the following quote from
Lawrence (1950, p. 1020):

"In order for the hair cells to function properly
they must be immersed in fluid. If these highly
specialized cells were like those of other tissues
in the body they could get their energy supply
from capillaries, but the very nature of these
cells activity precludes the presence of any
structure that might dilate, contract, or pulse.
The function of the hair cell is not only to be
stimulated mechanically but to be stimulated by
forces related only to the form of external
physical energy called sound. So of necessity
the hair cell is immersed in fluid from which
it can absorb nourishment without any extraneous,
disturbing influences to make it give off false
potentials. In this way the hair cell is very
happily situated; it can conveniently carry out
its metabolic processes without disturbance, and
it sits between two structures, the basilar
membrane and the tectorial membrane, so that if
the former can be made to move or vibrate a
squeezing of the hair cell will result."

Good experimental evidence has been marshalled in support
of the idea that the mode of the excitation of the cochlear endorgan in bone conduction is the same as that for aerial. The
critical studies are outlined by Wever and Lawrence (1954) and by
Hood (1962). Bekesy (1932) was the first to conclude that cochlear
excitation is the same for bone and aerial stimulation. He found
that the perception of a bone conducted tone of 400 cycles per
second could be cancelled out by aerial sound of the same frequency
but 180° out of phase. Lowy (1942) applied Bekesy's technique of
phase cancellation while measuring the cochlear potential of cats
and guinea pigs with stimulation in the frequency range of 250 to
3000 cycles per second. Cancellation obliterated the cochlear
potential, and moving the recording electrode to different positions along the cochlear boundary did not alter the cancellation
phenomenon. In a more comprehensive study, Lowy's findings were
extended and confirmed by Wever and Lawrence (1954). They found
total vectorial cancelling of air- and bone-conducted stimuli in
the frequency range of 100 to 15,000 cps.--providing that all the
stimuli were below the level of overloading. The development of
our theory leads us to the consideration of two different theories
of bone conduction--one theory for the low frequencies and another
theory for the high frequencies. Therefore it is significant to
note here that Wever and Lawrence (1954) find the same mode of
excitation of the organ of Corti for low frequency bone stimulation
as for high frequency bone stimulation.

The foregoing arguments make it clear that an evolutionary

change of the porpoise ear to a direct vibrational system does not
in itself call for a modification of the nature of hair cell stimulation found in the land mammal ear. However, I believe the case
was not so simple for evolution with regard to the mode of bone
stimulation to be utilized by the porpoise ear. There are two
different theories of bone conduction for the air mammal ear--
translational and compressional. There are pros and cons with
these two different systems which have definite significance for
the development of the function of the porpoise ear as we know it
today. To convey our point we must first explain the two theories.

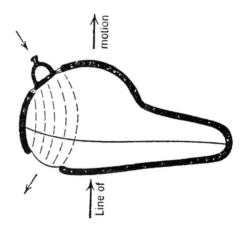

Figure 1. The translatory mode of bone conduction. When the
petrous bone is moved along the line indicated by the heavy arrows
the cochlear fluid is displaced toward the round window and the
basilar membrane is depressed. From Wever and Lawrence (1954).

     The principles of translational bone conduction are depicted
in Figure 1. This mode calls for a relative movement between the
footplate of the stapes and the otic capsule. Vibration of the
skull is believed to set the otic capsule in motion relative to
the footplate of the stapes; thus the cochlear fluid is displaced
toward the round window and the basilar membrane is depressed.
This type of bone conduction exploits the same stapes-oval window
"piston-cylinder" system which functions in aerial hearing.

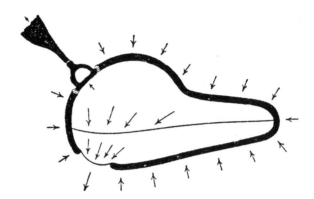

Figure 2. The compressional mode of bone conduction, as conceived
by Herzog. When the cochlear capsule is compressed from all sides
its fluid contents are pressed out, mainly taking the path of
least resistance through the round window. From Wever and Lawrence
(1954).

Compressional bone stimulation, proposed by Herzog (1930),
does not call for a functioning of the stapes "piston-cylinder"
system. Instead it is postulated that the skull when stimulated
with a bone conduction receiver sets up compressional waves which
tend to force the fluid in the cochlear capsule out the round
window (see Figure 2). Compression of the cochlear capsule is
thought to result in fluid being forced out the round window for
two reasons. First, the mass of the stapes in the oval window on
the vestibular side of the basilar membrane functions as a load
which favors the release of the compressional pressure at the
round window where there is no loading mass. Secondly, the 5/3
vestibular to tympanic fluid ratio is believed to favor a depression of the basilar membrane towards the round window during otic
capsule compression.

What empirical evidence supports the translational and compressional theories of bone conduction? The first contribution
to an understanding of bone conduction theory was made by Bekesy
in 1932--later expanded in 1948. As illustrated in Figure 3
Bekesy discovered that when the head is driven with a bone stimulator at low frequencies it vibrates as a unit mass in a translational manner. However, as the frequency of the bone stimulation
is raised through 1000 cycles per second the skull begins to
vibrate in a compressional mode--with forehead stimulation, the
forehead and occiput began to get out of phase with one another.
At 1500 cycles per second the front and back regions of the head
were 180° out of phase.

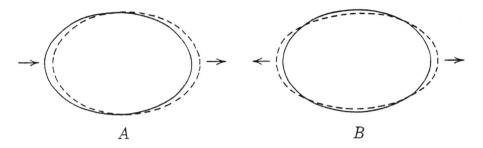

Figure 3.   From Wever and Lawrence (1954): The form of motion of
the head in bone conduction according to Bekesy. For low tones
the motion is as represented at A, with the head moving as a
whole, and for high tones it is as represented at B, with opposite
surfaces moving in contrary directions.

       Bekesy's experimental results call for the operation of a
translational system up to 1000 cycles per second, a combination
translational-compressional system to 1500 cycles per second, and
compressional mode alone for the high frequencies above 1500
cycles per second. In accordance with translational and compressional theories these findings predict an operational stapes
"piston-cylinder" system for low frequency bone stimulation, and
inoperation of the stapes for high frequency bone stimulation. A
few ingenious experiments demonstrate this point while further
elucidating the nature of bone stimulation:

       Stacy Guild (1936), in concert with a low frequency transla-
tional theory, found 512 cycles per second bone stimulation most
effective when applied to an osseous pathway in the same direction
as the stapes oval window system. Guild determined this through
thousands of post-mortem histological examinations of temporal
bones. The temporal bones were obtained from patients in the
Johns Hopkins Hospital that had had audiometric bone and aerial
hearing tests before death.

       In a consideration of the "nerve deafness" bone stimulator
tests in which a bone stimulator is applied to the side of the
head, Guild found that when all osseous pathways to the otic capsule were intact except for the sub-aditus trabeculae there was a
marked degree of impairment of hearing by bone conduction. The
sub-aditus trabeculae run between the medial part of the posterior
wall of the external auditory canal and the inferolateral aspect
of the prominence of the horizontal semicircular canal. Patients
with unmolested trabeculae did not show a deficit in bone stimulation hearing.

Most remarkable of all, Guild found a patient with chronic
otitis media who heard better by bone conduction than by air
conduction--this patient's sub-aditus trabeculae were not fractured.

In direct support of the operation of the stapes in low
frequency bone stimulation Guild (1936, p. 747) notes the following:
"The combination of anatomic arrangements is such that sound waves
transmitted to the otic capsule via the subaditus trabeculae
approach the organ of Corti from a direction and in a manner more
nearly like sound waves through the air conduction mechanism than
do those that come to the otic capsule via the other osseous
pathways."

Using 435 cycles per second bone stimulation, Barany (1938)
corroborated Guild's hypothesis with an entirely different experimental approach. As portrayed in Figure 4, Barany masked one
ear of human subjects with white noise while he applied a point
source bone stimulator to different locations of the head. By
varying the phase and loudness of an aerial sound to the unmasked
ear until cancellation of the bone stimulation perception
occurred, Barany was able to determine the effective loudness of
bone stimulation to different loci on the head. As Guild's work
predicts, the most effective bone stimulation was obtained with

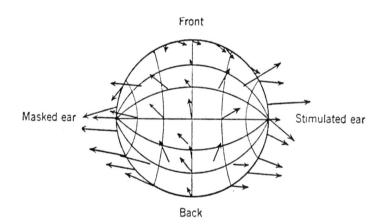

Figure 4. Effect of the point of application of a bone-conduction
stimulus to the surface of the head. The head is seen from above,
with the forehead at the top of the drawing. The left ear was
masked and the effective loudness of the sound was measured in the
right ear. The origins of the arrows represent the points of stimulation, their lengths represent the relative loudness, and their
directions represent the phase relations. From Barany's work
referenced in Wever and Lawrence (1954).

applications in the same direction as that in which the stapes
operates.

Happily for both theories of bone stimulation, Smith (1943)
has demonstrated the degree of involvement of the stapes in bone
conduction from 100 cycles per second to 10,000 cycles per second.
He recorded the cochlear potentials of cats, first with the stapes
in its natural condition, and secondly with the stapes immobilized
by a string around it connected to a pin in the bulla. The string
could be released to return the stapes to its normal condition.
"Translational" low frequency reduction of the cochlear potential
occurred with stapes immobilization, and high frequency "compressional" cochlear potential reduction did not occur during stapes
immobilization.

Smith has included in Figure 5 the changes of cochlear
responses for air conduction as well as bone conduction. Since
air conduction utilizes the same stapes piston system as translational bone conduction it is also reduced--adding to the evidence
for stapes involvement in translational bone conduction.

Returning to the work of McCormick, et al.(1970), what now
do these theories of bone conduction say for the operation of the
porpoise ear? Although, as we have seen, the introduction of a
land ear into an aquatic environment does not call for a new mode
of stimulation of the organ of Corti, it does restrict the auditory
endorgan to what we believe is a relatively inefficient method of
stimulation for the high frequencies, i.e., a compressional
system.

Figure 5. Changes of cochlear responses for air and bone conduction caused by a mechanical fixation of the stapes. From observations in the cat by Smith (1943).

The most effective machinery developed by the land ear for transmission of sound energy to the inner ear is the stapes–piston cylinder system. Even at the level of the fish it becomes apparent that the development in evolution of a "piston–cylinder system" is a distinct advantage for the stimulation of an auditory endorgan. Vallancien (1963, p. 540) notes that: "Fish possessing a Weberian apparatus respond to higher frequencies and lower intensities than fish not so equipped."

The porpoise makes considerable use of high frequency hearing. In order to perform this and also to exploit the piston–cylinder system of the land mammal ear, we propose that the evolutionary development of the porpoise ear extended the operation of translational system over the entire frequency sensitivity range of the animal.

With regard to a compressional mode of bone stimulation, Wever and Lawrence (1954, p. 226) have noted that "this mode of vibration can be realized in a solid only if it is contained in a denser solid whose flexions compress it equally from all sides." Although these requirements are partially met in the human ear and other land mammal ears, it is apparent to us from a consideration of the gross anatomy of the porpoise ear that evolution has built a system which will not respond to compressional movements of the skull. The porpoise has a bulla which is completely separated from the skull. The bulla is connected to the skull by cartilage, connective tissue, and fat--thus anatomically isolating it from the skull proper.

Less obvious, but very important for a translational system, the bulla of the porpoise ear is much denser than the skull, and the walls of the cochlear capsule are quite thick. Boenninghaus (1904) is the first we know of to note that the Odontocete bulla is much more dense than that of the land mammal.

The principles set down by Wever and Lawrence (1954, p. 266) and the anatomical relationships just described for the porpoise suggest acoustic isolation of the ear from the skull. The body mapping study confirmed the acoustic isolation of the ear from the skull: Cochlear potential readings taken with bone stimulation of the upper jaw and points over the dorsal portions of the skull were consistently, from animal to animal, much lower than signals elicited from sensitive areas on the lower jaw. The vector stimulation of the teeth of the upper jaw as opposed to that on the lower jaw also corroborated the idea of acoustic isolation of the bulla.

In our 1970 publication we only showed a summary figure of our auditory body mapping studies (see Figure 6). Some of the

Figure 6. An outline of the body of a dolphin (<u>Tursiops truncatus</u>) showing by the three dotted areas the regions of greatest sensitivity to sounds. (After McCormick, et al., 1970)

detailed data of our experiment are presented for the first time in Figures 7A, 7B, 8A, 8B, and 9.

The fact that the axis of the stapes of the porpoise was noted to project on an imaginary line through the aspect of the lower jaw which was found to be a sensitive reception area in the body mapping study suggests the operation of Barany's phenomena for the porpoise. As shown in Figures 10A and 10B the bone stimulation of the lower jaw of the porpoise was most effective when directed back toward the bulla.

Such a high frequency reproduction of Barany's results (the body mapping studies were all done with stimulation above 20kHz) would not be expected in the human since a high frequency "compressional system" would not be selective for areas of the body linked to the functioning of the stapes piston-cylinder system. The porpoise CNS auditory "body mapping" at 30 kHz of Professor Bullock's group (1968) agrees very well with our cochlear potential mapping of the porpose at 20K Hz (McCormick, et al., 1970). See Figures 11A, 11B, and Table One. We have no higher frequency data to compare with the Bullock study's 65 kHz data.

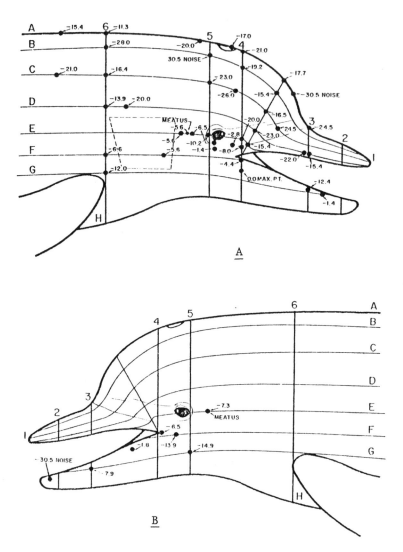

Figure 7. Cochlear potentials for 23,277 Hz point source bone stimulation plotted in db relative to the highest reading for porpoise number one (Tursiops truncatus). (A) is ipsilateral and (B) is contralateral to the round window recording electrode.

Figure 8. Cochlear potentials for 23,420 Hz point source bone stimulation plotted in db relative to the highest reading for porpoise number two (<u>Tursiops</u> <u>truncatus</u>). (A) is ipsilateral and (B) is contralateral to the round window recording electrode.

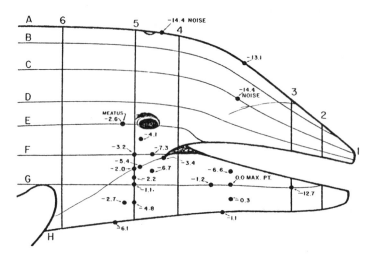

Figure 9. Cochlear potentials for 19,762 Hz point source bone
stimulation plotted in db relative to the highest reading for
porpoise number three (<u>Lagenorhynchus obliquidens</u>).

Table 1. Averages of lowest collicular evoked potential thresholds
found in Stenella to sounds presented via hydrophone pressed
against different parts of the head. Thresholds are shown in dB
relative to the most sensitive spot on the side of the contralateral
lower jaw

	High frequencies (60 kHz/s up)	Low frequencies (17-30 kHz/s)
Under tongue	+ 9 dB	+ 21 dB
Ventral surface under lower jaws	+ 9	+ 5
Contralateral lower jaw	0	0
Ipsilateral lower jaw	− 5	+ 1
Contralateral melon	−16	− 19
Ipsilateral melon	−11	− 17
Midline of melon	−14	− 19
Contralateral external meatus	−23	− 8
Ipsilateral external meatus	−22	− 14
Blowhole	−22	− 16
Rostrum	−42	− 27
		(no sensitivity)

After Bullock, et al. (1968).

A

B

Figures 10A and 10B. Two photographs of a <u>Tursiops truncatus</u> skull are used to illustrate the cochlear potentials resulting from stimulating the teeth of the lower jaw of a live <u>Tursiops truncatus</u>. The arrows represent the angle at which a bone stimulator was applied to the teeth. The numbers in the white ovals are the cochlear potentials (in microvolts) obtained during the particular stimulation trial. The stimulation frequency was 20,305 Hz.

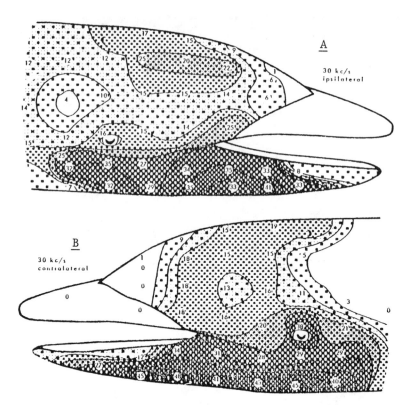

Figure 11A and 11B. Pattern of sensitivity of an individual
Stenella to sound produced by a hydrophone pressed against the
head surface at the points shown. The numerical values represent
attenuation at threshold; therefore, the largest numbers represent
greatest sensitivity. Contour lines are drawn at intervals of 5
db in sensitivity. Recording was from the inferior colliculus.
Note that sensitivity was greatest along the side of the contra-0
lateral mandible (except for under the tongue), and on the ipsilateral melon. After Bullock, et al., (1968).

     The 18 dB drop in the cochlear potential observed with damp-
ing of the ossicular chain during high frequency stimulation
certainly argues further for an operational ossicular chain at
high frequency, and thus a high frequency translational system in
the porpoise. It will be recalled that in Smith's experiment a
restriction of the ossicular chain of the cat had no effect on the
high frequency cochlear potential--as would be predicted for a
high frequency "compressional" land mammal ear.

The drops of the cochlear potential observed with damping of the bulla suggest that movement of the bulla in response to sound energy is critical for the hearing process. Since the malleus is fused to the bulla, the ossicular chain must function as a rigid spring mounted inside the tympanic cavity. When sound energy passing through the body tissues sets the bulla in motion, the inertia and stiffness of the ossicular chain probably cause the stapes to move relative to the oval window of the petrous bone-- the petrous bone being part of the moving bulla. Disconnecting the malleus of the porpoise from the bulla as we did in 1970 should have little effect on the sound reception of the porpoise since the middle ear chain would still function as an inertial mass lagging the motion of the bulla and petrous bone.

REFERENCES

Barany, E. A., 1938, Contribution to the physiology of bone con-
      duction, Acta Oto-laryngology, Supp. 26:223.
von Békésy, G., 1932, Zur theorie des hörens bei der schallaufnahme
      durch knochenleitung, Ann. Physik., 13:111.
von Békésy, G., 1948, Vibration of the head in a sound field and
      its role in hearing by bone conduction, J. Acoust. Soc. Amer.,
      20:749.
von Békésy, G., 1960, "Experiments in Hearing", McGraw-Hill, New
      York.
Bullock, T. H., Grinnell, A. D., Ikezono, E., Kameda, K., Katsuki,
      Y., Nomoto, M., Sato, O., Suga, N., and Yanagisawa, K., 1968,
      Electrophysiological studies of central auditory mechanisms
      in Cetaceans, Zeitschrift für vergleichende Physiologie, 59:
      117.
Boenninghaus, G., 1904, Das ohr des zahnwales, zugleich ein beitrag
      zur theorie der schalleitung, Zoologische Jahrbucher Abthei-
      lung für Anatomie und Ontogene der Tiere, 19:189. (English
      translation by E. G., Wever).
Diercks, K. J., Trochta, R. T., Greenlaw, C. F., and Evans, W. E.,
      1971, Recording and analysis of dolphin echolocation signals,
      J. Acoust. Soc. Amer., 49:1729.
Guild, S. R., 1936, Hearing by bone conduction: the pathways of
      transmission of sound, Annals of Otology, Rhinology, and
      Laryngology, 17:207.
Herzog, H., 1926, Das knochenleitungsproblem; theoretische erwäg-
      ungen, Zeits. f. Hals-Nasen-Ohrenheilk, 15:300.
Herzog, H., 1930, Die mechanik der knochenleitung im modellversuch,
      Zeits, f. Hals-Nasen-Ohrenheilk, 27:402.
Hood, J. D., 1962, Bone conduction: a review of the present posi-
      tion with a special reference to the contributions of Dr.
      George von Békésy, J. Acoust. Soc. Amer., 34:1325.

Lawrence, M., 1950, Recent investigations of sound conduction, part I. The normal ear, Annals of Otology, Rhinology, and Laryngology, 59:1020.

Lowy, K., 1942, Cancellation of the electrical cochlear response with air and bone conducted sound, J. Acoust. Soc. Amer., 14:156.

McCormick, J. G., Wever, E. G., Palin, J., and Ridgway, S. H., 1970, Sound conduction in the dolphin ear, J. Acoust. Soc. Amer., 48:1418.

Norris, K. S. 1964, Some problems of echolocation in Cetaceans, in: "Marine Bio-acoustics, Vol. 1", W. N. Tavolga, ed., Pergamon Press, New York.

Norris, K. S., 1969, The echolocation of marine mammals, in: "The Biology of Marine Mammals", H. T. Andersen, Academic Press, New York.

Ridgway, S. H., 1972, Homeostasis in the aquatic environment, in: "Mammals of the Sea: Biology and Medicine", S. H. Ridgway, ed., Charles C. Thomas, Illinois.

Ridgway, S. H., and McCormick, J. G., 1967, Anesthetization of porpoises for major surgery, Science, 158:510.

Ridgway, S. H., McCormick, J. G., and Wever, E. G., 1974, Surgical approach to the dolphin's ear, J. Exp. Zool., 188, N° 3:265.

Ridgway, S. H., Carder, D. A., Green, R. F., Gaunt, A. S., Gaunt, S. L. L., and Evans, W. E., 1979, Electropyographic and pressure events in the nasolaryngeal system of dolphins during sound production, this volume.

Romer, A. S., 1966, "Vertebrate Paleontology", 3rd Edition, The University of Chicago Press, Chicago.

Smith, K. R., 1943, Bone conduction during experimental fixation of the stapes, J. Exp. Psychol., 33:96.

Tavologa, W. N., 1965, Review of marine bio-acoustics, state of the art: 1964, Technical report: NAVTRADEVCEN, U. S. Naval Training Device Center, Port Washington, N.Y.

Vallancien, B., 1963, Comparative anatomy and physiology of the auditory organ in vertebrates, in: "Acoustic Behavior of Animals", R. G. Busnel, ed., Elsevier Press, New York.

Wever, E. G., 1959, The cochlear potentials and their relation to hearing, Annals of Otology, Rhinology and Laryngology, 68:975.

Wever, E. G., and Lawrence, M., 1954, "Physiological Acoustics", Princeton University Press, New Jersey.

Wever, E. G., McCormick, J. G., Palin, J., and Ridgway, S. H., 1971, The cochlea of the dolphin, Tursiops truncatus, Proc. National Acad. Sciences, 68:2381.

Wever, E. G., McCormick, J. G., Palin, J., and Ridgway, S. H., 1971, The cochlea of the dolphin, Tursiops truncatus. I. General morphology, Proc. National Acad. Sciences, 68:2381.

Wever, E. G., McCormick, J. G., Palin, J., and Ridgway, S. H., 1971, The cochlea of the dolphin, Tursiops truncatus. II. The basilar membrane, Proc. National Acad. Sciences, 68:2708.

Wever, E. G., McCormick, J. G., Palin, J., and Ridgway, S. H., 1971,
    The cochlea of the dolphin, Tursiops truncatus. III. Hair
    cells and ganglion cells, Proc. National Acad. Sciences,
    68:2908.
Wever, E. G., McCormick, J. G., Palin, J., and Ridgway, S. H., 1972,
    Cochlear structure in the dolphin, Lagenorhychus obliquidens,
    Proc. National Acad. Sciences, 69:657.

BEHAVIORAL MEASURES OF ODONTOCETE HEARING

Arthur N. Popper

Department of Anatomy, Georgetown University

Washington, D.C. 20007

Much of what is now known about hearing mechanisms in vertebrates has come from studies using behavioral and psychophysical
techniques to ask questions regarding auditory detection and processing. These studies have included determination of detection
capabilities for a variety of signals presented in a quiet environment and in the presence of other signals. Such studies are
valuable in determining the resolving and analysis capabilities
of the auditory system. Other investigations have been directed
at determining an animal's ability to discriminate between signals
that differ in a variety of parameters, including intensity, frequency, spectral components, position in space, or temporal relationships. Such studies provide insight into detection capabilities
of an animal, as well as provide information on mechanisms of
acoustic processing at different levels of the auditory system.

While psychophysical studies have proven particularly fruitful
in enhancing our understanding of the auditory system in terrestrial
mammals, it has not been until the past 10 to 15 years that carefully controlled investigations have been performed on odontocetes.
Certainly the technical difficulties associated with any behavioral
study of a large mammal, especially involving sound in water, have
hampered such investigations. However, information about the
auditory system of these animals is of considerable interest since
it has become clear that many odontocete species are exceptionally
well adapted for analysis of their environment using acoustic signals. The studies discussed below, in spite of very limited data
on intra- and inter-specific differences, clearly demonstrate that
the auditory system in representatives of several odontocete families is extraordinarily well adapted for the use of sound in intra-
and inter-animal communications.

AUDITORY SENSITIVITY

Measures of pure tone auditory sensitivity provide insight into
the sound detection capabilities of a species, most significantly
in terms of the range of frequencies that can be heard and the
minimum sound level detectable (or threshold) at each of the frequencies.

Behavioral and physiological studies of auditory sensitivity
have been done in a number of odontocete species, starting with
investigations by Kellogg (1953) and Schevill and Lawrence (1953a,
1953b) who first demonstrated that the bottle-nosed dolphin, Tursiops truncatus, could detect sounds about 100 kHz. Subsequent
to these studies, and particularly since Johnson's study in 1967,
there have been a number of investigations using careful behavioral,
psychophysical and acoustical techniques to measure hearing sensitivity in five odontocete species (Fig. 1). Additional studies,
using electrophysiological techniques to record evoked potentials
from the inferior colliculus (Bullock et al., 1968), have provided
data on range of sensitivity in several additional species (broken
lines in Fig. 1). Care must be taken, however, in comparing the
behaviorally and physiologically determined thresholds since data
determined physiologically are relative to an arbitrary threshold
rather than 1 microbar (dyne/cm^2). Consequently, while it is appropriate to compare data determined with the two approaches in
terms of the frequency range over which sounds can be detected,
comparisons of absolute sensitivity cannot be made.

The species for which behaviorally determined data are available (Fig. 1) include the bottle-nose dolphin, Tursiops truncatus
(Johnson, 1966, 1967), the Amazon River dolphin, Inia geoffrensis
(Jacobs & Hall, 1972), Orcinus orca (Hall & Johnson, 1972), the
common dolphin, Phocoena phocoena (Andersen, 1971), and the beluga
whale, Delphinapterus leucas (White et al., 1978). Auditory sensitivity for each of these species except Phocoena were measured
using similar behavioral techniques. In each case, thresholds were
determined using an operant conditioning paradigm where the animal
had to make a discrete and quantifiable response when it detected
a sound. In each study, attempts were made to assess thresholds in
a relatively quiet environment. However, it must be pointed out
that Johnson (personal communication) now suggests that thresholds
determined behaviorally at frequencies above 50 kHz may actually
not be absolute values, but instead, may have been limited by
masking thermal noise.

Although there are several trends that can be discussed regarding the data in Figure 1, it is apparent that there is substantial
agreement among all species regarding the shape of the sensitivity
curves. The one exception to these data is reported in a study by
Supin and Sukhoruchenko (cited in Ayrapet'yants and Konstantinov,

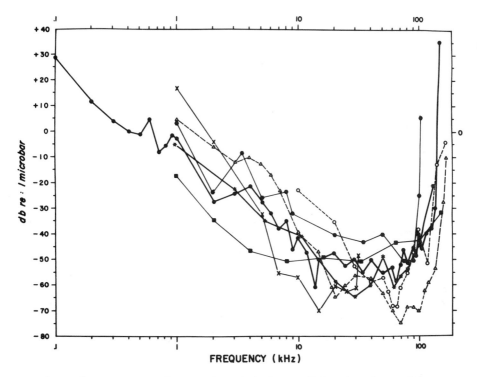

Fig. 1. Odontocete Auditory Sensitivity. Behavioral studies are
indicated by a solid line and electrophysiological studies by a
broken line.   Symbols: * -Delphinapterus leucas (composite of two
animals, White et al., 1978); ✳ Inia geoffrensi (Jacobs & Hall,
1972); × -Orcinus orca (Hall & Johnson, 1972); ■ -Phocoena phocoena
(Andersen, 1971); o -Stenella caeruleoalba (Bullock, et al., 1968);
Δ-Tursiops gilli (Bullock et al., 1968); ●-Tursiops truncatus
(Johnson, 1967).

1974). These workers, using a "galvanic skin response" to measure
sensitivity to tones, reported that Phocoena was able to detect
sounds from 4 to 180 kHz with a maximum sensitivity of about -52dB
(re: 1 μbar) at 128 Hz. This is in contrast to Andersen's findings
(Fig. 1) of best sensitivity between 8 and 30 kHz for Phocoena, and
a sensitivity of -35 dB at 128 kHz. The significant differences in
the data obtained by Andersen and the Russian workers may be methodological, but this cannot be said for certain since little information is available regarding the procedures used by Supin and
Sukhorunchenko.

Several points may be made regarding data on pure tone sensitivity for odontocetes. The range of frequencies detectable by all the
species studied (including species investigated electrophysiologi-

cally) with the exception of Orcinus, is wider than that of non-echolocating terrestrial mammals. There are considerable inter-specific differences in auditory sensitivity among the odontocetes studied behaviorally, with Tursiops, Delphinapterus, Stenella and Phocoena having a wider range of sensitivity than Inia or Orcinus. Best sensitivity, however, is seen in Orcinus, which was able to detect signals as low as -72 dB (re: 1 µbar) at 16 kHz (Hall & Johnson, 1972). Since all studies, with the exception of one on Delphinapterus by White et al. (1978), were with a single specimen, we do not know for certain whether the inter-specific variation seen in the behavioral data reflects true species differences in sensitivity, or whether the data would be less variable were studies done with more than one specimen. Sensitivity for two specimens of Delphinapterus, one male and one female, did show some substantial differences, particularly at higher frequencies where the male could detect lower intensity sounds than the female. However in the region of best sensitivity for both animals (30-90 kHz), the female was about 4 dB more sensitive than the male (White et al., 1978). It is significant that the variability in hearing between the two specimens of Delphinapterus is substantially less than inter-specific variability between the five odontocetes shown in Figure 1. If we can extrapolate to the other species, we might tentatively suggest that intra-specific variability among odontocetes will be substantially less than the inter-specific variability seen in Fig. 1. Thus, the variation most likely represents real species differences in the ability to detect pure tones.

AUDITORY DISCRIMINATION

Behavioral studies of pure tone sensitivity provide information about the general range of sounds detectable by an animal. However, such data tell little about the actual functioning of the auditory system or about the capabilities of an animal for the detection of biological signals used in communications. Communicative signals in many vertebrates, as in dolphins, are complex and may include considerable variation in frequency, intensity, pulse patterns and duration. Only with studies of an animal's ability to handle these functions is it possible to accurately assess its ability to detect and analyze signals that hold relevance to inter- or intra-specific communications.

Frequency Discrimination

Odontocete sounds cover a broad frequency spectrum (e.g. Evans, 1973). The energy in the echolocation clicks may cover a 20-50 kHz band, with the specific spectrum varying with the species and the echolocating situation (Au et al., 1973; Evans, 1973). Communicative signals generally contain energy below 20 kHz and consist of pure tone whistles which are often frequency modulated. It has been suggested that analysis of target information during echolocation involves spectral analysis, and the differences in frequency spectra between emitted and returned signals would provide the animal with information about the nature of the target (e.g. Norris, 1969). Consequently, the ability to analyze frequency would hold considerable importance to a dolphin in echolocation, as it would in social communication where whistles with different frequency components may be of communicative significance.

Data from several experiments have demonstrated that Tursiops can discriminate small changes in frequency. These data are also supported by physiological and morphological studies showing that the peripheral and central regions of the auditory system are well adapted for frequency analysis. Studies by Bullock et al. (1968) have demonstrated that the inferior colliculus in anesthetized animals was very responsible to small changes in frequency. Investigations of the morphology and innervation of the cochlea in several different delphinid species by Wever et al. (1971, 1972) show that the cochlea is anatomically suited for good frequency analysis. Wever and his colleagues suggest that the very large number of ganglion cells innervating the cochlea in several delphinid species, particularly as compared with man (although there are about the same number of sensory hair cells in both species), is possibly indicative of a good frequency analysis system.

Direct studies of discrimination capabilities in Tursiops have been performed using behavioral measures of the discrimination limens (DL's), or the minimum difference between two frequencies that an animal can discriminate successfully at a particular frequency. The first study (Jacobs, 1972) found that the relative DL's (DL/F where F is frequency in Hertz) ranged from 0.28% to 1.4% for frequencies from 1 to 90 kHz. The best relative DL's occurred between 5 and 20 kHz where the animal could discriminate signals that were 0.3% to 0.4% apart. This work was replicated and extended by Herman and Arbeit (1972) and by Thompson and Herman (1975),who found that the best discrimination for Tursiops was for frequencies from 2 to 53 kHz where the relative discrimination limens were from 0.2% to 0.4%. Frequency discrimination was consistently better than 0.89% at all other frequencies studied except at 1 and 140 kHz where the relative DL's rose to 1.4%. Significantly, the data for the studies by Jacobs and by Herman and his colleagues on two different animals are similar to one another, with the only major difference being the range of best discrimination capabilities. This difference may possibly be explained, in part, by the poor health of Jacob's animals, which died before the completion of his study.

While not directly comparable to the data for Tursiops due to methodological differences, data on frequency discrimination have also been reported by Sukhoruchenko (1973) for Phocoena phocoena.

Sukhoruchenko reported relative discrimination limens of 0.1% to
0.29% from 3 to 225 kHz. It is striking that while these data are
similar to the data for Tursiops below about 53 kHz, the DL's for
Phocoena are essentially flat over the whole range of hearing.
This finding is contradictory to all other studies of mammalian
frequency discrimination, including those for Tursiops, where the
relative DL's function rises steeply towards the end of the animal's
range of auditory sensitivity (Fay, 1974).

Intensity Discrimination

The ability to discriminate sound intensity may be of considerable significance to odontocetes if amplitude variation of spectral components of echolocation clicks is important in target detection and discrimination. Few data are available regarding the
intensity discrimination capabilities of odontocetes, although some
data lead to the suggestion that several species may be able to
discriminate signals that differ by as little as 1 dB (Bullock
et al, 1968; Johnson, 1967), an ability which is on the same order
as that in humans (Reisz, 1928). The only study of intensity discrimination was done by Burdin et al. (1973), who determined, in
behavioral experiments, that Tursiops could discriminate between
two broad-band (50-100 kHz) white noise signals that differed in
amplitude by 4% to 7%. Experiments only indirectly related to intensity discrimination in odontocetes indicate a discrimination
ability of about 1 dB (Evans, 1973), which is comparable to data
from preliminary behavioral studies (Johnson, 1967) and physiological recordings from the inferior colliculus (Bullock et al., 1968).

Temporal Analysis

Several investigations have been performed on the ability of
Tursiops to compare sound quality and to determine the temporal
resolving ability for sounds of different durations. Johnson
(1968) measured auditory sensitivity for pure tones of different
durations and found that sensitivity in Tursiops, as in man, becomes
poorer as the duration of the pulsed pure tones is decreased below
100 to 200 msec. The data for Tursiops closely resembles that for
humans. In both species, resolution of very short signals improves
substantially when a series of short-duration signals come in rapid
succession, permitting temporal summation of the signal by the
auditory system.

Discrimination of signals of different durations has been investigated by Yunker and Herman (1974), who found that Tursiops
could discriminate between two tones that differed in duration by
as little as 8% (e.g., between 300 and 324 msec signals). While
Yunker and Herman obtained similar results for a variety of different tonal signals, they did not use any signals shorter than 300
msec in their study. Thus, it remains to be seen whether temporal

resolution would stay the same or change for signals more closely
approximating the duration of echolocation-like pulses found in
delphinids.

Critical Bands and Masking

Most "normal" behavioral situations involve detecting and
discriminating signals in noisy environments rather than detecting
signals in relatively quiet environments such as a laboratory tank
(which may also contain a good deal of thermal noise about 50 kHz,
resulting in masking at these frequencies) (Johnson, personal communication). Consequently, it is of considerable importance that
any animal be able to detect signals in the presence of noise. In
the case of odontocetes, it is also necessary that the animals be
able to discriminate their own signals from those of nearby sound
producing conspecifics.

The mechanism involved in discrimination of signals in noise,
and for spectral analysis of signals in mammals, is generally considered to center around the concept of the critical band. This
concept suggests that the mammalian auditory system is functionally
analogous to a series of narrow band-pass filters (Green and Swets,
1966), each of which passes signals only within its own frequency
range. This effectively eliminates noise that is not within a
particular band of signals, thereby enhancing the signal-to-noise
ratio around the signal. Studies of the critical band attempt to
determine how finely tuned the filters in the ear are through
measurement of the ability of an animal to detect the presence of
a signal that is imbedded in noise. Investigations with humans
have demonstrated the presence of 24 non-overlapping critical bands
within the approximate frequency range from 50 kHz to 16 kHz
(Scharf, 1970). These critical bands may represent distinct loci
on the basilar membrane and correspond to constant distances along
the membrane (Greenwood, 1961).

Johnson (1968) measured critical ratios (a function related
to the critical band) (Reisz, 1928) for Tursiops by determining
differences in threshold levels for selected pure tone frequencies
as a function of spectrum level of the masking noise at the same
frequency. He found evidence for up to 40 critical bands in
Tursiops. The critical ratios for Tursiops are similar to those
for the ringed seal (Pusa hispida) (Terhune and Ronald, 1975), humans
(Hawkins and Stevens, 1959) and the domestic cat (Watson, 1963).
However, Tursiops appears to have narrower critical ratios than
these other species at frequencies where corresponding data are
available.

SOUND LOCALIZATION

A fundamental auditory task is the determination of the position in space of a sound source (Erulkar, 1972; Mills, 1972) and,
in fact, it has been suggested that adaptation for sound localization has had broad significance in the evolution of the mammalian
auditory system (Masterton et al., 1969). The process of sound
localization is of particular importance for odontocetes since it
would be used in the determination of the position of other animals
or, in part, in the location of a target during echolocation.
While investigations of odontocete sound localization have been
limited, recent experiments have demonstrated an excellent ability
to do localization (McDonald-Renaud, 1974; Renaud and Popper, 1975).

Renaud and Popper (1975) studied localization in an open body
of water with an animal that was trained not to move its head during
the behavioral localization task (head movement enhances detection
abilities). Renaud and Popper measured the ability of Tursiops to
discriminate between targets that were coming from the animal's midline, as opposed to targets located off of the midline. This determination of minimum audible angle was done with 3 msec duration
pure tone pulses for frequencies from 6 to 100 kHz. The minimum
audible angle from 20 to 90 kHz was between 2° and 3°, indicating
that an animal could successfully discriminate between sounds that
were on its midline from sounds coming from 2-3° off of the midline.
At 6, 10, and 100 kHz, the minimum audible angle was between 2.8°
and 4.0°.

Since 3-msec duration pure tone pulses are not very similar
to echolocation sounds, Renaud and Popper (1975) measured minimum
audible angles using trains of echolocation-like clicks, each of
which was a 35 μsec transient with energy centered at 64.35 kHz.
The minimum audible angle for this signal was 0.9°, leading to the
suggestion that the broad energy spectrum and/or rapid onsets in
this signal enhanced the localization ability of the animal.

Sounds are often not located along an animal's midline, but
instead come from anywhere around the body. Consequently, Renaud
and Popper (1975) determined the minimum audible angle using sounds
centered at different points on the animal's horizontal plane.
When the center point for the signals were changed to 15° and 30°
to the side of the animal the minimum audible angle for the pure tone
pulses was 1.3 to 1.5° for the 15° azimuth, while the MAA dropped
to 5° when the azimuth was 30°, suggesting a slight improvement in
localization ability when sounds are centered away from the animal's
horizontal midline as compared to sounds on this midline.

The ability to localize sounds above and below an animal is
of particular importance for aquatic and flying animals where
behaviorally significant sounds may come from anywhere in three

dimensions. Consequently, Renaud and Popper (1975) studied the
vertical localization ability of Tursiops and found data indicating
an ability that is comparable to discrimination thresholds on the
horizontal plane. However, Renaud and Popper (1975) pointed out
that the mechanism used for localization on the vertical plane
may be different from that used for horizontal localization by
Tursiops.

It is difficult to compare the localization for Tursiops determined by Renaud and Popper (1975) with other odontocetes due to
significant methodological differences between experiments. Di-
yachenko and colleagues (cited in Ayrapet'yants and Konstantinov,
1974) reported that Tursiops could localize correctly in 90% of
trials when sound came from speakers 1.5° apart. While these were
long duration sounds, and experimental details were lacking, the
data compare reasonably well with those of Renaud and Popper.
Dudok Van Heel (1959, 1962) studied Phocoena phocoena and found
discrimination ability of 11° for a 3.5 kHz pure tone and 8° for
a 6 kHz pure tone. Andersen (1971), also using Phocoena, found a
discrimination ability of 3° for a 2 kHz pure tone. However, both
experiments with Phocoena used unrestrained animals, and stimulus
control was weak. Further, the frequencies used were far below the
range of best auditory sensitivity of Phocoena. Thus, it was
possible for these animals to "scan" the environment with their
heads during the position discrimination task, thus improving
directional discrimination.

DISCUSSIONS AND CONCLUSIONS

It is apparent, in spite of limited ata, that several species
of odontocete have exceptionally good hearing capabilities that is,
as far as is now known, rivaled by few other mammals. Since less
data are available for bats, it is still not possible to state
whether the auditory capabilities in bats and odontocetes run parallel courses. However, we can compare behavioral auditory data
between odontocetes and another group of marine mammals, the pinnipeds. Data for a number of pinniped species demonstrates excellent hearing capabilities, although no way on a par with that
of most odontocetes. For example, in measures of pure tone sensitivity, the California sea lion (Zalophus californianus) was found
to only be able to detect sounds to about 50 kHz (Schusterman et
al., 1972), and sensitivity at most frequencies was substantially
poorer than for Tursiops. While the harbor seal (Phoca vitulina)
is capable of detecting signals over almost as wide a range as
Tursiops (Møhl, 1968), sensitivity for this species, and that for
several other pinnipeds with similar hearing ranges, is substantially poorer than for Tursiops above 30 kHz. Frequency discrimination capabilities compared between Tursiops and other animals
in-water show that this species has as good, if not better, dis-

crimination ability than man below 2000 Hz (Thompson & Herman, 1975) and exceeds that of Phoca vitulina (Møhl, 1967), and Zalophus (Moore & Schusterman, 1978) by about 2½ times. Similarly, the ability to discriminate intensity appears to be somewhat better in Tursiops than in Zalophus (Moore & Schusterman, 1976). While data are scanty for the pinnipeds, it appears that the critical ratio for Tursiops is slightly narrower than for Pusa hispida (Terhune & Ronald, 1975). However, the differences in critical ratios between Tursiops, Pusa, and several other mammalian species may not be significant at frequencies for which there are overlapping data (Gurevich, 1970).

Finally, the ability to discriminate the position of a sound source in space appears to be considerably better in Tursiops than in other aquatic mammals including Zalophus californianus (Moore & Au, 1975; Moore, 1975; Gentry, 1967) and Phoca vitulina (Møhl, 1964; Terhune, 1974).

While it is not now reasonable to make any sort of conclusive statement, it does appear that the excellent auditory capabilities seen in many odontocetes may be related to the use of high frequency sounds for echolocation and determination of information about the nature of a target using acoustic reflection. It is, however, imperative that additional studies be undertaken with odontocetes, making use of behavioral techniques, to extend our knowledge of hearing abilities of these species. While replicated data on frequency discrimination in Tursiops and on hearing sensitivity in Delphinapterus leucas indicate that the general trend in sensitivity and discrimination shown by the data for all species is real, it will still be of considerable value to have additional replicated data on hearing capabilities of a single species, such as Tursiops, as well as an extension of the data to additional species. It would be of particular interest to obtain psychophysical data for a species of delphinid, such as Orcinus, in which hearing and/or the range of the frequency components of the echolocation click, are different from that in Tursiops.

Clearly, the excellent ability to detect, and presumably analyze sounds, as demonstrated by the behavioral data currently available, correlates well with the sophisticated echolocation ability of these species. Without the refined ability to detect and analyze sounds it is highly likely that such an echolocation ability would not be possible in these species.

ACKNOWLEDGMENTS

Portions of the work presented here were supported by the Office of Naval Research, Contract N00123-74-C-0660. Preparation of the manuscript was supported in part by Research Career Development Award NS-00312 from the National Institute of Neurological and Communicative Disorders and Stroke.

REFERENCES

Andersen, S. Auditory sensitivity of the harbour porpoise Phocoena phocoena. In: Investigations on Cetacea. (Ed. by G. Pilleri) Benteli Ag. Vol. 3. pp. 255-259 (1971a).

Andersen, S. Directional hearing in the harbour porpoise Phocoena phocoena. In: Investigations on Cetacea. (Ed. by G. Pilleri) Benteli Ag. Vol. 3. pp. 260-264 (1971b).

Au, W.W.L., Floyd, R.W., Penner, R.H. & Murchison, A.E. Measurement of echolocation signals of the Atlantic bottlenose dolphin, Tursiops truncatus Montagu, in open water. J. Acoust. Soc. Amer., 56: 1280-1290 (1974).

Ayrapet'yants, E.S. & Konstantinov, A.I. Echolocation in Nature. Joint Publication Research Service. #JPS 63326-1-2 (1974).

Bullock, T.H., Grinnel, A.D., Ikezono, E., Kameda, K., Katsuki, Y., Nomoto, M., Sato, O., Suga, N. & Yanagisawa, K. Electrophysiological studies of the central auditory mechanisms in cetaceans. Z. vergl. Physiol., 59: 117-156 (1968).

Burdin, V.I., Markov, V.I., Reznik, A.M., Skoriyakov, V.M. & Chupakov, A.G. Determination of the differential intensity threshold for white noise in the bottlenose dolphin (Tursiops truncatus Barabasch). In: Morphology and Ecology of Marine Mammals (Ed. by K.K. Chapskii & V.E. Sokolov) Wiley, NY p. 112 (1973).

Dudok van Heel, W.H. Audio-direction findings in the porpoise Phocoena phocoena. Nature 183: 1063 (1959).

Dudok van Heel, W.H. Sound and cetacea. Netherlands J. Sea Res. 1, 407-507 (1962).

Erulkar, S.D. Comparative aspects of spatial localization of sound. Physiol. Rev., 52: 237-260 (1972).

Evans, W.E. Ecolocation by marine delphinids and one species of freshwater dolphin. J. Acoust. Soc. Amer., 54: 191-199 (1973).

Fay, R.R. Auditory frequency discrimination in vertebrates. J. Acoust. Soc. Amer., 56: 206-209 (1974).

Gentry, R.L. Underwater auditory localization in the California sea lion (Zalophus californianus). J. Aud. Res., 7: 187-193 (1967).

Gourevitch, G. Detectability of tones in quiet and in noise by rats and monkeys. In: Animal Psychophysics (Ed. by W.C. Stebbins) Appleton-Century-Crofts, NY pp. 67-98 (1970).

Green, D.M. & Swets, J.A. Signal Detection Theory and Psychophysics. Wiley, NY (1966).

Greenwood, D.N. Critical bandwidths and the frequency coordinates of the basilar membrane. J. Acoust. Soc. Amer., 33: 1344-1356 (1961).

Hall, J.D. & Johnson, C.S. Auditory thresholds of a killer
  whale Orcinus orca Linnaeus. J. Acoust. Soc. Amer., 51:
  515-517 (1972).

Hawkins, J.E. & Steven, S.S. The masking of pure tones by white
  J. Acoust. Soc. Amer., 22: 6-13 (1950).

Herman, L.M. & Arbeit, W.R. Frequency discrimination limens in
  the bottle-nose dolphin: 1070 KC/S. J. Aud. Res., 12: 109-120
  (1972).

Jacobs, D.W. Auditory frequency discrimination in the Atlantic
  bottlenose dolphin, Tursiops truncatus Montagu: A preliminary
  report. J. Acoust. Soc. Amer., 52: 696-698 (1972).

Jacobs, D.W. & Hall, J.D. Auditory thresholds of a freshwater
  dolphin, Inia geoffrensis Blainville. J. Acoust. Soc. Amer.,
  51: 530-533 (1972).

Johnson, C.S. Auditory thresholds of the bottlenose dolphin
  (Tursiops truncatus Montagu). U.S. Naval Ordinance Test
  Station NOTSTP 4178, 25 pp. (1966).

Johnson, C.S. Sound detection thresholds in marine mammals.
  In: Marine Bio-Acoustics II. (Ed. by W.N. Tavolga) Pergamon
  Press, NY pp. 247-260 (1967).

Johnson, C.S. Masked tonal thresholds in the bottlenosed porpoise.
  J. Acoust. Soc. Amer., 44: 965-967 (1968).

Kellogg, W.N. Ultrasonic hearing in the porpoise, Tursiops
  truncatus (Mont.). Aquatic mammals, 4: 1-9 (1953).

Masterton, B., Heffner, H., & Ravizza, R. The evolution of
  human hearing. J. Acoust. Soc. Amer. 45: 966-985 (1969).

McDonald-Renaud, D.L. Sound localization in the bottlenose
  porpoise, Tursiops truncatus (Montagu). Ph.D. Thesis, University
  of Hawaii (1974).

Mills, A.W. Auditory localization. In: Foundations of Modern
  Auditory Theory (Ed. by J.V. Tobias) Academic Press, NY Vol.
  II. pp. 303-348 (1972).

Möhl, B. Preliminary studies on hearing in seals. Vidensk.
  Medd. fra Dansk Naturh. Foren., 127: 283-294 (1964).

Möhl, B. Frequency discrimination in the common seal and a
  discussion of the concept of upper hearing limit. In: Underwater Acoustics. (Ed. by V.M. Alberts) Plenum, NY Vol. 2.
  pp. 43-54 (1967).

Möhl, B. Auditory sensitivity of the common seal in air and
  water. J. Aud. Res., 8: 27-35 (1968).

Moore, P. Underwater localization of click and pulsed pure tone
  signals by the California sea lion (Zalophus californianus).
  J. Acoust. Soc. Amer., 57: 406-410 (1975).

Moore, P. & Au, W. Underwater localization of pulsed pure tones
  by the California sea lion (Zalophus californianus). J. Acoust.
  Soc. Amer., 58: 721-727 (1975).

Moore, P. & Schusterman, R.J. Discrimination of pure tone intensities by the California sea lion. J. Acoust. Soc. Amer.,
  60: 1405-1407 (1976).

Reisz, R.R. Differential sensitivity of the ear for pure tones.
    Phys. Rev., 31: 867-875 (1928).
Renaud, D.L. & Popper, A.N. Sound localization by the bottlenose
    Porpoise, Tursiops truncatus. J. Exp. Biol., 63: 569-585 (1975).
Scharf, B. Critical bands. In: Foundations of Modern Auditory
    Theory (ed. by J.V. Tobias) Academic Press, NY Vol. I. pp.
    157-202 (1970).
Schevill, W.E. & Lawrence, B. Auditory response of a bottlenosed
    porpoise, Tursiops truncatus, to frequencies above 100 KC.
    J. Exp. Zool., 124: 147-165 (1953a).
Schevill, W.E. & Lawrence, B. High-frequency auditory responses
    of a bottlenosed porpoise, Tursiops truncatus (Montagu).
    J. Acoust. Soc. Amer., 25: 1016-1017 (1953b).
Schusterman, R.J., Balliet, R.F. & Nixon, J. Underwater audiogram
    of the California sea lion by the conditional vocalization
    technique. J. Exp. Anal. Behav., 17: 339-350 (1972).
Schusterman, R. J. & Moore, P. A. The upper limit of underwater
    auditory frequency discrimination in the California sea lion.
    J. Acoust. Soc. Amer., 63: 1591-1595 (1978).
Sukhoruchenko, M.N. Frequency discrimination in dolphins (Phocoena
    phocoena). Fiziol. AH. SSR. IM. M. Sechenova, 59: 1205-1210
    (1973).
Terhune, J.M. Directional hearing of a harbor seal in air and
    water. J. Acoust. Soc. Amer., 56: 1862-1865 (1974).
Terhune, J.M. & Ronald, K. Masked hearing thresholds of ringed
    seals. JH. Acoust. Soc. Amer., 58: 515-516 (1975).
Thompson, R.K.R. & Herman, L.M. Underwater frequency discrimin-
    ation in the bottlenosed dolphin (1-140 kHz) and human (1-
    8 kHz). J. Acoust. Soc. Amer., 57: 943-948 (1975).
Watson, C. Masking of tones by noise for the cat. J. Acoust.
    Soc. Amer., 35: 167-172 (1963).
Wever, E.G., McCormick, J.G., Palin, J. & Ridgway, S.H. The
    cochlea of the dolphin Tursiops truncatus: General morphology.
    Proc. National Acad. Sci., 68: 2381-2385 (1971).
Wever, E.G., McCormick, J.G., Palin, J. & Ridgway, S.H. Cochlear
    structure in the dolphin, Lagenorhynchus obliquidens. Proc.
    National Acad. Sci., 69: 657-661 (1972).
Yunker, M.P. & Herman, L.M. Discrimination of auditory temporal
    differences in the bottlenose dolphin and by the human.
    J. Acoust. Soc. Amer., 56: 1870-1875 (1974).

ELECTROPHYSIOLOGICAL EXPERIMENTS ON

HEARING IN ODONTOCETES

Sam H. Ridgway

Naval Ocean Systems Center

San Diego, Ca.   92152   U.S.A.

INTRODUCTION

   Although electrophysiological techniques have not been extensively employed in the study of cetacean hearing, there are more
than a dozen significant studies that will be briefly reviewed in
this paper. These techniques have been useful for measuring hearing thresholds by evoked responses and classical conditioning
(several animals each of Phocoena and Tursiops). The auditory
cortex has been located at the dorsum of the cerebral hemisphere
only 1.5 to 3.0 cm lateral to the midline and adjacent to visual
and somatosensory cortex. The midbrain auditory structures of
dolphins are specialized for ultrabrief, ultrasonic, fast-rising,
closely spaced sounds like echolocation clicks, and temporal resolution is rapid. Frequency-modulated tones produce maximum responses
both at midbrain locations and at cortical recording sites, suggesting the importance of such signals to the animals. Methods and
equipment are now available for broader application of electrophysiological techniques without expenditure of valuable animals.
These techniques can contribute further insight concerning how
odontocetes process sound and what characteristics of the acoustic
information package are most important to the animal.

Previous Studies

   Among cetaceans only the small toothed whales, commonly called
dolphins or porpoises, have been studied by electrophysiological
techniques. The advent of safe and effective anesthesia techniques
(Ridgway and McCormick, 1967) promised to open the way for a
thorough physiological assessment of the unusually large brains of

483

these aquatic mammals. However, because of several factors, including protective legislation and economics, many types of investigations, such as single-unit recording and mapping of the cerebral cortex, have not been possible except in the Soviet Union.

Since much of the Soviet literature is not generally available in the West, Bullock and Gurevich (1979) have recently completed a review. Ladygina and Supin (1970) published the first Soviet investigations in this field. They studied the auditory evoked potential (AEP) with electrodes placed in small holes drilled in the skull of unanesthetized dolphins. Since 1970, the Soviet investigators have produced an impressive list of reports, primarily from the laboratory of A. Ya. Supin. Much of this is apparently reviewed and updated in a recent book by Supin, Mukhametov, and Ladygina (1978) that is not yet available in English translation. This paper will mention a few of the Soviet studies that appear to be of major significance and will also discuss some Western studies that predate most of the Soviet work.

LITERATURE REVIEW AND COMMENTS

Bullock et al (1968) were the first to employ electrophysiological techniques in the study of cetacean hearing. They studied the AEP from midbrain auditory structures, and among other things they provided the first physiological evidence for the lower jaw sound path postulated by Norris (1964). This is a landmark paper that deserves careful reading. Conclusions of major importance are as follows: 1) Temporal resolution of successive sounds was found to be extremely rapid. 2) There was no evidence of facilitation of response to the second of two identical stimuli as had been observed in echolocating bats. 3) Dramatic changes in AEP waveform, but not amplitude, were produced by distortion of the sound field by holding a sheet of paper between the sound source and the lower jaw. 4) Gentle stroking of the lower jaw or splashing water on the jaw masked response to sounds, but electrical stimulation of the skin of the jaw did not produce responses or mask AEPs. 5) Small changes in stimulus frequency resulted in changes of AEP amplitude and waveform. 6) Frequency-modulated (FM) tones were more likely to produce a large AEP than were constant frequency tones. In one case, a 100-kHz tone produced a weak AEP, while a 135-kHz tone produced none. An FM pulse starting at 100 kHz and sweeping upward to 135 kHz caused a large AEP. The effect was also present for tones sweeping down in frequency. 7) This investigation showed that the AEP audiogram was similar to the behavioral curve reported by Johnson (1966) for Tursiops, but with a slightly lower upper limit of about 135 kHz. 8) An AEP to a given tone pip could be masked only by a relatively narrow band of frequencies surrounding that stimulus (Bullock et al, 1968).

Bullock and Ridgway (1971 and 1972) recorded and telemetered electrical activity from the inferior colliculus or nucleus of the lateral lemniscus and from a few locations in the cerebral cortex. The recording electrodes were placed stereotaxically with the animal under anesthesia. The experimental animals were trained to emit on command a series of clicks at a rate of 15 to 300 per second. (These clicks are similar to those used in echolocation.) The voluntary clicks resulted in AEPs. However, the clicks of highest intensity often evoked quite modest AEPs while clicks of much weaker intensity gave maximal AEPs, suggesting important differences in click composition. Artificial echoes showed recovery to be more rapid (0.5 ms including the duration of the click) following the animal's own click than following an artificial conditioning tone burst of equal evoking power.

The typical midbrain AEP was found to be specialized for ultrasonic, ultrabrief, fast-rising, closely spaced sounds like the echolocation clicks. No potential was evoked if the frequency was below 5 kHz or if the rise time was above 5 ms. At several cerebral locations (mainly in the posterior lateral temporal cortex), however, long latency, long duration, slowly recovering potentials were evoked by frequencies below 5 kHz, whether fast or slowly rising acoustic envelopes were used (Bullock and Ridgway, 1972).

Another electrophysiological technique, cochlear potential recording, was employed by McCormick et al (1970) in studies of sound conduction in the dolphin ear. Since a study employing cochlear potentials will be discussed by McCormick in this volume, the subject will not be further mentioned.

Lende and Akdikmen (1968) and Lende and Welker (1972) did some limited investigations on cortical mapping in Tursiops. A small somatic motor area was found in the frontal pole (Lende and Akdikmen, 1968). These investigators, however, found it necessary to remove most of the melon and much of the blowhole and nasal sac system for access to frontal cortex. Since these ablated structures probably require fine motor control to produce the varied complex sounds of odontocetes (see Ridgway et al this volume), the nasal system will probably occupy a large area of motor cortex. Thus, its removal precludes studies of its representation.

Ladygina and Supin (1970 and 1977) used AEPs to locate sensory projection zones in the cerebral cortex of Phocaena and Tursiops. They reported that visual and acoustic regions in both species are shifted to the dorsal surface of the hemisphere and that all projection zones are adjacent to each other (cortical projection areas forming a single complex are also found in Monotremata and Edentata). The auditory cortex was found approximately 1.5 to 3.0 cm lateral to the sagittal suture of the skull. Thus, as compared to other mammals, there is an apparent shifting of the auditory area from

the temporal to the parietal lobe. Zanin et al (1978) duplicated
some of these findings. It should be mentioned, however, that a
tonotopic map of the cochlea, based on sound frequency areas on the
cortex, has not been done. Additional areas of auditory sensitivity
may be found in temporal cortex which is less accessible in cetaceans (Cf. Bullock and Ridgway, 1972).

Popov and Supin (1976 and 1978) in studies of Phocaena and
Tursiops concentrated on changes in AEPs from the auditory cortex
produced by varying the characteristics of sound stimulation. They
found distinct changes in the cortical AEPs in response to switching
the sound on and off compared to changes in the duration of the
signal and to changes in the frequency of the signal. The threshold
for AEPs in the auditory cortex of Tursiops was found to be close
to the behavioral hearing threshold.

Voronov and Stosman (1977) studied the frequency-threshold
characteristics of subcortical elements of the central auditory
system of Phocaena; they used 200-μs clicks and 5-ms tones. Both
"on" and "off" responses were measured for the tone stimuli. In
all structures examined, maximum sensitivity was found in the 100-
to 120-kHz range (with preception extending to 200 kHz), the same
as that found by Sukhoruchenko (1971) using classical conditioning
of the galvanic skin response (GSR). The frequency bandwidth at a
level 10 dB over the lowest thresholds was 40 to 50 kHz in the
cochlear nuclei, 70 to 90 kHz in trapezoid body, 70 to 80 kHz in
Superior olive, 45 to 55 kHz in lateral lemniscus, and 35 to 45 kHz
in inferior colliculus and medial geniculate. There was an additional maximum of sensitivity from 20 to 80 kHz in the lateral
lemniscus (bandwidth 30 to 40 kHz), inferior colliculus (bandwidth
40 to 50 kHz), and medial geniculate (bandwidth 40 to 50 kHz).
"Off" responses had a lower threshold up to about 150 kHz. From
150 to 200 kHz both "on" and "off" responses had the same threshold,
which showed an elevation of 10 to 20 dB per 10 kHz in frequency
increase.

Supin and Sukhoruchenko (1970 and 1974) and Sukhoruchenko
(1971 and 1973) used a classical conditioning method to measure
hearing thresholds. After restraining the experimental animal with
straps in a small rubber-lined tank, the investigators recorded
respiration, heart action, and the GSR. They paired a tone (the
conditioned stimulus) with an electric shock (the unconditioned
stimulus). Ten to fifteen pairings of the tone and shock were
usually sufficient to develop a conditioned reflex (respiratory
pause or GSR). Every tenth to fifteenth tone was reinforced, bringing about a stable response that did not extinguish. By attenuating
the tone and noting the disappearance of the response, a complete
audiogram was done in a single experiment or within two or three
experiments. This made it possible to test a reasonable number of
experimental animals.

The technique was extended (Supin and Sukhoruchenko, 1974) to measure differential frequency thresholds. An FM tone at an intensity approximately 40 dB higher than the threshold of audition in the frequency band under test was used as the conditioned stimulus. After a conditioned reflex was developed to FM tones as distinct from unmodulated tones, the range of the frequency modulation was lowered until the reflex disappeared (threshold). In Phocaena, the differential frequency threshold was found to be 10 Hz at 3 kHz increasing to 300 Hz at 190 kHz. Above 22 kHz, the differential thresholds were approximately constant at 0.1 to 0.2% at any given frequency. Thompson and Herman (1975) using behavioral techniques found values as low as 0.2 to 0.3% between 6 and 20 kHz, but at 140 kHz their value (for Tursiops) was 1.4% (seven times coarser than the Soviet findings) and no responses were obtained above 150 kHz. These findings are especially noteworthy since they provide strong evidence for frequency analysis over the entire range of hearing in Phocaena and Tursiops.

Recent discoveries have shown that electroencephalographic (EEG) recordings made in the far field (i.e., scalp in humans, vertex of the skull, mastoid process) can be used to obtain AEPs that reflect activity even at the brain stem level. Thus, a great deal of information about different levels of the auditory system can be collected without placing electrodes in the brain (Galambos and Hecox, 1977).

Seeley et al (1976) used such recordings in a technique for rapidly assessing the hearing of Tursiops. A single monopolar electrode was placed into the vertex of the skull 5.8 cm posterior

Figure 1. Tursiops positioned in restraint apparatus to assess hearing using the AEP (Seeley et al, 1976).

to the caudal margin of the blowhole. A reference electrode was
built into a suction cup (7.5 cm in diameter) and attached with
active vacuum to the skin surface of the head rostral to the first
electrode. The experimental animal was restrained with straps in
a foam, plastic-padded box. The box was partially filled with
water that just covered the dorsal surface of the snout (figure 1).
The animal was kept wet by frequent misting from a garden sprayer
during recording sessions of 4 to 7 hr.

Because of the proximity of the massive nasal muscles to the
brain (see Green et al this volume) and to the recording electrodes,
each breath at average intervals of approximately 20 s resulted in
artifact several times larger than the EEG. In addition, other
gross body movements produced large artifact. Seeley et al (1976)
electronically removed this movement artifact by using an artifact-
inhibiting system (figure 2). The system provided a 13.5-s delay
between data recording and computer analysis of average transients.
In this time period the stimulus epochs containing artifact were
eliminated, making it possible to obtain an AEP with as few as 16
stimuli; however, 48 stimuli were more commonly used (figure 3).

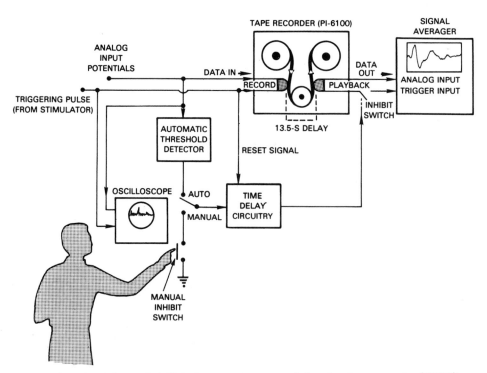

Figure 2. Artifact-inhibiting system used by Seeley et al (1976).

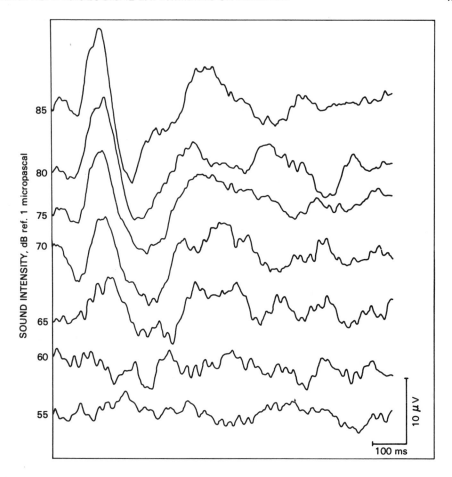

Figure 3. Effect of change in stimulus intensity on auditory
evoked potentials in an unanesthetized Tursiops. Each potential
represents an average of 48 sound stimuli at 50 kHz. Upward
deflection indicates negativity of the vertex relative to the
reference electrode (Seeley et al., 1976).

        This technique was used to obtain audiograms from seven
Tursiops of both sexes and various ages. Most of these audiograms
were similar to the behavioral curve of Johnson (1966), but one
31-year-old female was virtually deaf, showing a 50 to 70 dB higher
threshold over most of the frequency range.

        Ridgway and Seeley (1979) used implanted telemetry with elec-
trodes positioned over the dura at the vertex and 9.0 cm lateral
over temporal cortex to record an EEG continuously from an unrestrained dolphin. Four different sound stimuli were used for AEP
studies: 1) a 500 ms tone with slow (10 ms) rise and fall,

2) a pulse train with the same envelope as the tone, i.e., 10-ms
rise and fall, containing 100 20-μs pulses 5 ms apart, 3) a single
20-μs pulse, and 4) an FM tone sweeping up in frequency from 20 to
130 kHz (slow rise and fall). Each stimulus was presented at
intervals of 5 s, and 64 responses were averaged for each plot.

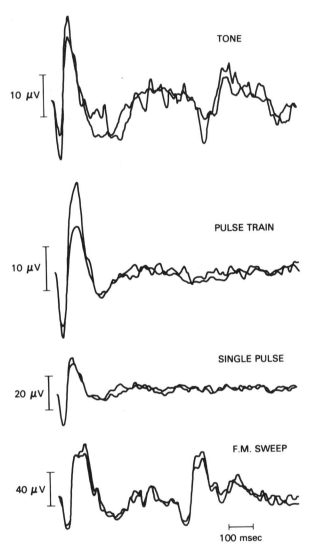

Figure 4. Each line consists of 64 1-s averages of a _Tursiops_ EEG
in response to a different sound stimulus. Upward deflection
indicates negativity of the vertex relative to an electrode over
the right temporal cortex (Ridgway and Seeley, 1979).

All of the responses began with a positive component peaking at about 30 ms followed by a negative component at about 100 ms. The tone burst and FM stimuli showed a strong "off" response after the stimuli at 500 to 600 ms. It is interesting to note that the pulse train of the same duration did not show this response. The AEPs produced by FM sweeps were much larger than with the other stimuli employed. These recordings from the cortex are consistent with the earlier findings of Bullock et al. (1968) at the subcortical levels. Many dolphin whistles are long FM sweeps, and the consistent findings that this type of signal produces large responses indicates that the FM signals are probably very important to the animals. Thus, they perhaps deserve intensive studies equivalent to the attention given the brief clicks known to be important in echolocation.

FUTURE STUDIES

Future studies at NOSC will employ improvements in the techniques reported by Seeley et al. (1976) for recording AEPs at various levels of the central auditory system of odontocetes. Auditory processing at the brain stem reflected by the averaged brainstem response, sensory processing at the thalamus and cortex reflected by longer latency AEPs, and cognitive processing that may be reflected in endogenous potentials such as the P300 or "Ah Ha" response are three major areas of study that could yield considerable insight concerning how the odontocetes process sound and what characteristics of the acoustic information package are important to the animal.

REFERENCES

Bullock, T. H., Grinnell, A. D., Ikezono, E., Kamedo, K., Katsuki, Y., Nomoto, M., Sato, O., Suga, N., and Yanagisawa, K., 1968, Electrophysiological studies of central auditory mechanism in cetaceans, Zeitschrift für vergleichende Physiologie, 59:117.

Bullock, T. H., and Gurevich, V. S., 1979, Soviet literature on the nervous system and psychobiology of cetaceans, International Review of Neurobiology, (in press).

Bullock, T. H., and Ridgway, S. H., 1971, Evoked potentials in the auditory system of alert porpoises (Cetacea) and sea lions (Pinnipedia) to their own and to artificial sounds, Proc. Int. Un. Physiol., IX:89.

Bullock, T. H., and Ridgway, S. H., 1972a, Neurophysiological findings relevant to echolocation in marine mammals, in: "Animal Orientation and Navigation", S. R. Galler, K. Schmidt-Koenig, G. J. Jacobs, R. E. Belleville, eds., NASA, U.S. Govn't Printing Off., Washington, D.C.

Bullock, T. H. and Ridgway, S. H., 1972b, Evoked potentials in the
    central auditory system of alert porpoises to their own and
    artificial sounds, J. Neurobiol., 3:79.
Galambos, R. and Hecox, K, 1977, Clinical applications of the brain
    stem auditory evoked potentials in man,"Psychopharmacology
    Correlates of Evoked Potentials", Vol. 2, J. E. Desmedt, ed.,
    Krager, Basel.
Johnson, C. S., 1966, Auditory thresholds of the bottlenosed
    porpoise (Tursiops truncatus Montagu), N.O.T.S. TP 4178.
Ladygina, T. F., and Supin, A. Y., 1970, The acoustic projection in
    the dolphin cerebral cortex, Fiziol. Zh. SSSR im. I. M.
    Sechenova, 56:1554.
Ladygina, T. F., and Supin, A. Y., 1977, Localization of sensory
    projection zones in the cerebral cortex of the bottlenosed
    dolphin, Tursiops truncatus, Zh. Evolyutsionnoy Biokhimii
    I Fiziol.
Lende, R. A., and Akdikmen, S., 1968, Motor field in the cerebral
    cortex of the bottlenosed dolphin, J. Neurosurgery, 29:495.
Lende, R. A., and W. I. Welker, 1972, An unusual sensory area in the
    cerebral neocortex of the bottlenose dolphin, Tursiops
    truncatus, Brain Research, 45:555.
McCormick, J. G., Wever, E. G., Palin, J., and Ridgway, S. H., 1970,
    Sound conduction in the dolphin ear, J. Acoust. Soc. Amer.,
    48:1418.
Norris, K. S., 1964, Some problems of echolocation in cetaceans,
    in: "Marine Bio-Acoustics", W. N. Tavolga, ed., Pergamon
    Press, New York.
Popov, V. V., and Supin, A. Y., 1976, Determination of the hearing
    characteristics of dolphins by measuring induced potentials,
    Fiziol. Zh. SSSR im. I. M. Sechenova, 62:550.
Popov, V. V., and Supin, A. Y., 1976, Responses of the dolphin
    auditory cortex to complex acoustic stimuli, Fiziol. Zh.
    SSSR im. I. M. Sechenova, 62:1780.
Popov, V. V., and Supin, A. Y,, 1978, Electrophysiological studies
    of the auditory system of the Tursiops truncatus, in:
    "Morskiye Mlekopitayushchiye. Resul'taty i Metody Issledovan-
    iya",Izdatel'stvo Nauka, Moscow.
Ridgway, S. H., Carder, D. A., Green, R. F., Gaunt, A. S., Gaunt,
    S. L. L., and Evans, W. E., 1979, Electromyographic and
    pressure events in the nasolaryngeal system of dolphins
    during sound production, this volume.
Ridgway, S. H., and McCormick, J. G., 1967, Anesthetization of
    porpoises for major surgery, Science, 158:510.
Ridgway, S. H., and Seeley, R. L., 1979, Auditory evoked responses
    of the dolphin cortex to four different sound stimuli, in
    preparation.
Seeley, R. L., Flanigan, W. F., and Ridgway, S. H., 1976, A tech-
    nique for rapidly assessing the hearing of the bottlenosed
    porpoise Tursiops truncatus, Naval Undersea Center TP 552,
    San Diego, California.

Sukhoruchenko, M. N., 1971, Upper limit of hearing of dolphins
    with reference to frequency, Tr. Akust. Inst. Moscow, 17:54.
Sukhoruchenko, M. N. 1973, Frequency discrimination of dolphin
    (Phocoena phocoena), Fiziol. Zh. SSSR im. I. M. Sechenova,
    59:1205.
Supin, A. Y. and Sukhoruchenko, M. N., 1970, The determination of
    auditory thresholds in Phocoena phocoena by the method of
    skin galvanic reaction, Tr. Akust. Inst. Moscow, 12:194.
Supin, A. Y., and Sukhoruchenko, M. N., 1974, Characteristics of
    acoustic analyzer of the harbor porpoise Phocoena phocoena,
    in: "Morfologiya, Fiziologiya i Akustika Morshkikh Mlekopita-
    yushchikh", V. Y. Sokolov, ed., Izdatel'stvo Nauka, Moscow.
Supin, A. Y., Mukhametov, L. M., Ladygina, R. F., Popov, V. V.,
    Mass, A. M., and Polyakova, I. G., 1978,"Electrophysiological
    Studies of the Dolphin's Brain", V. E., Sokolov, ed.,
    Izdatel'stvo Nauka, Moscow.
Thompson, R. K. R., and Herman, L. M., 1975, Underwater frequency
    discrimination in the bottlenose dolphin (1-140 kHz) and in
    human (1-8 kHz), J. Acoust. Soc. Amer., 57:943.
Voronov, V. A., and Stosman, I. M., 1977, Frequency-threshold
    characteristics of subcortical elements of the auditory
    analyzer of the Phocoena phocoena porpoise, Zh. Evolyutsionnoy
    Biokhimii I Fiziol., 6:719.
Zanin, A. V., Bibikov, N. G., Vodyanaya, E. G., 1978, Evoked pot-
    entials of the Tursiops under the stimulation with short
    sound signals, in: "VII-aya Vsesoyuznayz Konferentsiyz po
    Morskim Mlekopitayushchim", Simpheropol'.

PERIPHERAL SOUND PROCESSING IN ODONTOCETES

Kenneth S. Norris

Coastal Marine Laboratory
University of California - Environmental Studies
Santa Cruz, California 95064, U.S.A.

A decade ago two divergent theories describing peripheral sound processing in odontocetes had been formulated. Their differences were unresolved. One view held that sounds were generated in the larynx, radiated through the soft tissue of the throat or transmitted up through the skull and rostrum. Reception was thought restricted to the region of the external auditory meatus, a pinhole in most odontocetes. Sounds gathered at this region were thought to be transmitted to the middle ear via the narrow and sometimes occluded external auditory canal and the tympanic ligament (Fraser and Purves, 1959 ; Purves, 1966). The second view suggested quite radical modification of both sound reception and transmission paths. Sounds were envisioned as being produced by extra-laryngeal structures adjacent to the nasal passages in the forehead, and transmitted and transduced into the water via the fatty melon of the odontocete forehead, or through the mesorostral canal of the odontocete snout. The external ear canal was viewed as non-functional, at least for high frequency sounds, and instead sound was postulated to enter the lower jaw through a thinned posterior portion of the mandible (the acoustic window, or pan bone) and to be transmitted via the fat filled mandibular canal to the tympan-operiotic complex of the middle ear. (Evans and Prescott, 1962 ; Norris, 1964 ; Norris, 1968). These modifications were presumed to have answered the multiple problems attendant upon the re-entry of a terrestrial mammal into the aquatic environment. Major selective pressures involved in such change resulted from partial loss of sensitivity, partial loss of binaurally based directionality, and partial loss of discrimination. These problems sprang from impedance missmatches between the terrestrial ear and water, complexity of the aquatic acoustic environment, and velocity related problems (sound travels

495

approximately 5 times as fast in water as air). Also involved was
the new found need to recycle air for phonation in an aquatic
animal living much of its life below the surface, in deep water.

Much new evidence now suggests that the second view is generally correct. Major evolutionary changes of the sorts envisioned
seem to have occurred. Only the broad outlines are known, however,
and many simple functions are poorly understood or remain to be
investigated.

THE MELON

The porpoise forehead, anterior to the blowhole, is, in most
odontocete species, formed of a convex fatty tissue mass lying on
the dorsal surface of the bony rostrum (Fig. 1). Smooth fibrous
blubber covers its exterior surface and grades inward to an oil-

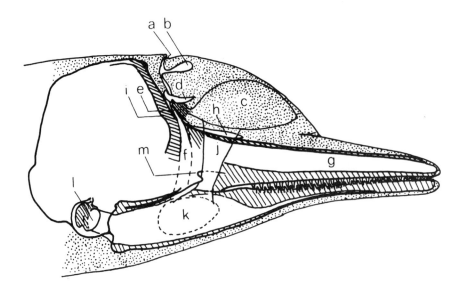

Fig. 1. Structures in the delphinid odontocete head : a, blowhole ;
         b, vestibular sac ; c, melon ; d, tubular sac ; e, nasal
         valve with lip entering tubular sac ; f, internal nares ;
         g, mesorostral (or ethmoid) cartilages ; h, premaxillary
         nasal sac ; i, cribriform plate ; j, proposed wave guide
         from melon to outer tissue of mandible, passing anterior to
         antorbital notches ; k, acoustic window in mandible, and
         the site of overlying fatty window ; l, attachment point of
         mandibular wave guide onto bulla ; m, antorbital notch.

rich fatty core that lies over the center of the rostrum and the
dorsal groove of the mesorostral canal. Embracing the melon is the
rostral muscle that inserts on the rostrum and terminates in the
fibrous tissue on either side of the melon. In the mid-sagittal
plane the melon is an oblate spheroidal mass lying on the rostrum
posterior to the beak, and in Delphinus delphis ends in the tissue
of the right nasal plug lip. Behind it lies the complexity of the
sacs of the dorsal narial canals, and the blowhole. To some extent
the vestibular and pre-maxillary sacs embrace the posterior melon,
and are thought to contribute to the shaping of emitted sound
fields, and to aid in isolation of the middle ears from emitted
sounds (Norris, 1968 ; Giro and Dubrovsky, 1974).

     The melon has been proposed as a sound focussing lens that
may serve to shape emitted sounds prior to their entry into water.
Reflection from air sacs and bone, both impedance missmatched to
the melon tissue, has been proposed as a focussing means (Norris,
1964), as has a differential transmission velocity topography within the melon (Norris and Harvey, 1974). The melon is probably
deformable through the action of the embracing muscles and this is
clearly the case in the Beluga Delphinapterus leucas, whose melon
seems to be deformed foreward and thinned when the animal ensoni-
fies objects of interest with click trains (G. Carlton Ray,
personal communication).

CHEMICAL COMPOSITION OF MELON AND JAW FAT

     A transverse section of the melon reveals regular and obvious
topographic variations in the fatty tissue from the surface to the
core. The surface blubber is very tough, with fat droplets tightly
held in a connective tissue matrix in most odontocete species.
Medially this changes until at the core the translucent fat is so
oil-rich and lightly held that slight pressure will express abundant
clear oil from the tissue. Through my sponsor at the time, the
Office of Naval Research, I was able to interest Dr. Donald Malins,
a lipid biochemist, in the problem. His work, and that of his
colleague Usha Varanasi quickly revealed unique biochemical topography in the odontocete melon and jaw lipids. Interest among bioche-
mists in this unique system spread and several laboratories in
North America and Europe have contributed studies. This work has
linked biochemical structure, molecular branching and chain length
as determinants of acoustic transmission velocity. Lipids in
"acoustic tissues" are different from those in surrounding tissues.
They contain triacylglycerols and wax esters rich in short and
medium chain (C_5-C_{12}) acids, while blubber is composed primarily of
long chain triacylglycerols (C_{14}-C_{22}). In delphinids and monodontids,
isovaleric and longer iso acids are present in large amounts in
areas implicated in acoustic transmission (Varanasi and Malins,
1971 ; Wedmed et al., 1973 ; Litchfield and Greenberg, 1974 ;

Varanasi, Feldman and Malins, 1975 ; Malins and Varanasi, 1977 ;
Blomberg, 1978). Certain areas, such as the acoustic window and
inner melon, contain almost exclusively 1, 3 diisovaleroyl-2-
acylglycerols and isovaleroyl wax esters. This topography has been
associated with a sound transmission velocity topography in the
melon ; the central melon being of lower transmission velocity than
the outer areas and encasing blubber (Norris and Harvey, 1974).

A sharp biochemical difference has been found between the
acoustic fatty tissues of delphinids and the same tissues in
physeterids, ziphiids and platanistids (Litchfield, Greenberg and
Mead, 1976). This difference accounts for the different odors
emanating from fatty tissues of the two assemblages, and seems to
me to be phylogenetically related. Ziphiids and physeterids are
typified by waxy fats of little odor, while delphinid fats are
typically pungent with the clinging odor of isovalerate derivatives.

Interestingly, the unique fat composition and distribution
variations mentioned above closely define tissues that have been
implicated in acoustic processing in both the lower jaw and melon,
and are sharply different from body fat (Litchfield, Greenberg and
Mead, 1976). Thus it seems valid to speak of "acoustic tissues"
as I and others have done.

EXPERIMENTAL STUDIES

Contact hydrophone studies, in which small receiving hydrophones can be pressed or attached by suction to various loci on the
porpoise head have shown that impulsive or click-type sounds are
emitted in structured fashion from the anterior melon and the tip
of the upper jaw (Diercks et al., 1971 ; Evans, 1973 ; Norris and
Harvey, 1974). High frequencies and intensities are focussed foreward so that for Tursiops truncatus, all sound above 100 kHz is
typically restricted in a beam to a few degrees of arc, centering
directly ahead of the rostrum, and upward (Evans, 1973). A similar
field has been described for the roughtooth porpoise, Steno bredanensis (Norris and Evans, 1967), and a field emphasizing lower
frequencies with a small high frequency component has been outlined
for the killer whale, Orcinus orca (Schevill and Watkins, 1966).
Sound fields emanating from the side of the delphinid head are
restricted to lower frequencies, and the entire emitted beam of
all of these species drops both in frequency and intensity nearly
to zero at 90° to the longitudinal exis. This corresponds closely
to the common observation of workers recording in the field, that
when delphinids pass a recording station, significant sound levels
are only recorded when animals face or are passing the hydrophone.

These fields result from sources that have been localized
behind the melon (see Ridgway et al., this volume). On anatomical

grounds, Evans and Prescott (1962), Norris (1966) and Evans and
Maderson (1973) postulated a nasal plug source as the most likely
locus. In 1968 I suggested that from the stand-point of impedance
matching the most likely mechanism was a lip vibrating in an air
column, with the generated sound propagating inside the lip itself
passing into the melon and hence to the water, thus avoiding the
major expected energy loss if vibration of the air column itself
was the major source. Evans (1973) has proposed a friction model
in which a lip rubs over another surface, releasing energy as the
rubbed tissue folds and restitutes elastically.

   Evans and Prescott (1962) were able to produce click trains
in an excised porpoise head by forcing air through the bony nares
from the basicranial space. The sequence of events in chirp production in Stenella longirostris was visualized by Norris, Dormer,
Pegg and Liese (1971) using x-ray cinematography, using a living
animal trained to phonate while in air. Phonation was precisely
correlated with the passage of small blebs of air upward past the
nasal plugs. Such air passage inflated the vestibular sac. Periodic
emptying of this sac with the return of air into the basicranial
space without opening of the blowhole was postulated as the air
cycling system for porpoises phonating below the surface. Air sac
inflation was noted to provide adequate sound mirrors for sound
projection. This work has recently been expanded and extended by
Dormer (1974). The elegant electromyographic work of Ridgway and
his colleagues in this volume is consonant with these x-ray studies
and extends them with precise timing of the muscular events in
phonation, and clearly shows that in those studies at least, no
sounds were being produced by the larynx since its muscles were
inactive during phonation. Even though Purves (1966) was able to
produce sound from the porpoise larynx by insertion of a Galton
whistle into the tip of the laryngeal spout and application of air
under pressure, I do not feel there is any evidence that clearly
implicates the larynx in porpoise sound production, though future
studies may so implicate it. Thus far, then, electromyographic,
x-ray, and sound field measurements all implicate a source at the
level of the nasal plugs. Of course, some sounds are produced by
odontocetes at the blowhole valve by expressing air past it into
the water, and other sounds are made by jaw clapping and aerial
behavior, but these are not of concern to us here.

   Only one direct measurement, taken for other purposes, shows
sound focussing in the melon. Norris and Harvey (1974) attempting
to find a pathway by which received sound might enter the melon and
reach the cochlea, ensonified an excised bottlenose porpoise
(Tursiops truncatus) head with 25 sec. clicks from a click generator
at various angular positions around the preparation in the horizontal plane. With a small receiving hydrophone implanted approximately
in the center of the melon sound, levels above calibration

level for an all water path were produced when the source was 20°
to either side of the midline and horizontal to the axis of the
melon. When the sound source was rotated ahead of the preparation,
sending sound directly into the apex of the melon, a reduction in
received sound was obtained.

That the melon shows sound transmission velocity topography
sufficient to account for beam production was shown by Norris and
Harvey (1974) who measured velocity at many points on tissue slices
comprising in composite, the entire melon. A low velocity core
along the longitudinal axis of the melon was found to grade into
a higher velocity tissue closer to the melon surface.

THE SPERMACETI ORGAN

The spermaceti organ is found in varying form in the sperm
whale Physeter catodon, the pygmy sperm whales, Kogia sp, and in
the various Ziphiid whales. Its function has long been a subject of
speculation. Suggestions have ranged from a "force pump" (Raven and
Gregory, 1933), a hydrostatic organ (Clarke, 1978), to a sound
generation system (Norris and Harvey, 1972 ; Mackay, this volume).

In the sperm whale it is an enormous elongate spermaceti oil-
filled structures that may comprise 1/4 the length of a large adult
whale. Bounded on both ends by air sacs that are diverticula of the
right nasal passage, and underlain by a channellike passage of the
same nasal passage, it may overhang the lower jaw by as much as
10 % of an adult animal's length. Anteriorly, the organ grades into
the tissue of the upper lip of a broad horizontal pair of lips that
block the narial passage, the museau du singe, or Monkey's Muzzle.
This strange structure is thought by some to be a sound generator
(Norris and Harvey, 1972 ; Mackay, this volume).

Recently four of the various views about its function have
been published, and their differences remain unresolved. Norris
and Harvey (1972) noting the peculiar reverberent pulses often
emitted by the whale, in which an initial strong pulse decays
through a series of 7 or more pulses in a diminuendo through about
25 m sec., postulated that the organ is a reverberation chamber in
which an initial sound is produced at the museau producing a reverberent signal between the two end air sacs which act as sound
mirrors. A model using design criteria determined from dissections,
produced signals closely similar to those of the whale. They were
able to use the structure of clicks to check the model, using a
click sequence of a small whale that surfaced alongside a vessel of
known length. Clicks were used to predict whale length and checked
against the actuality of the sighted animal. The correspondence was
close. Nonetheless, as Mackay indicated (this volume) compensating
errors seem to have been involved in this calculation. The velocity

of spermaceti used in the Norris and Harvey model postulated a
closed organ pipe model for reverberation. The spermaceti organ
is more probably likened to an open pipe because its rear wall is
bounded by the compliant proximal sac. A halving of wavelength
approximately compensates for the velocity error. Mackay suggests
that air bubbles in the spermaceti sample may have given the peculiar velocity measure, which was made by Dr. Diercks of the University of Texas. This author suggests that this may be correct and
that the source of such bubbles may have been the harpoon explosion
that killed the animal. Mackay (this volume) proposed a bugle model,
noting that sperm whale sounds are not notably directional as
compared to those of other odontocetes. It seems to me that lack
of strong directionality is to be expected of a signal that peaks
between 6-12 kHz, and that, like the killer whale Orcinus orca,
(Schevill and Watkins, 1966) there may be high frequency components
to the sound that have yet to be recorded, and that may be closely
focussed foreward, and may be of much lower intensity than the
remainder of the signal. Mackay reports that when a fathometer was
playing into the spermaceti organ of his specimen a reverberent
pulse resulted. Only further studies using living animals seem likely
to resolve these different views. In my view, other recent suggestions seem unconvincing for various reasons. The overriding reason
for skepticism is that in each case the model put foreward has not
been directed by any observed behavior by the animal itself that
would suggest these adaptive systems. For example, it seems gratui-
tous to me to construct a model for the spermaceti organ as a
hydrostatic structure where there is no evidence that sperm whales
habitually hover in nature, as Clarke (1979) suggests they may do.

Kozak (1974) postulates a "video acoustic system" in which
the spermaceti organ is a sound receptor, using the serious fluid-
filled blisters on the posterior wall of the proximal sac as a
receptor field. The idea is that clicks are generated by the museau
and return, as echos to a receptor field on the sac wall, to provide
an acoustic image of objects in the environment. Nervous innervation,
ending as free nerve endings, as shown by staining the tissues of
the blister-covered wall is presented as supportive evidence. No
transduction system is demonstrated that would convert received
sound into nerve impulses and sensation and the innervation seems
scattered and slight. The geometry of the spermaceti organ seems
better suited to sending than receiving in that it narrows anteriorly
and enters the upper lip of the museau, and is surrounded laterally
and above by a thick meshwork of ligamentous cables. Nonetheless,
Bullock et al. (1968) demonstrated high sound sensitivity for the
porpoise melon, and no transduction system seems to exist there,
in a region primarily innervated by the trigeminal nerve, which has
not previously been involved in sound processing. The sperm whale,
like other odontocetes shows a good separation of emitting and
receiving structures, isolation being achieved primarily by air sac

barriers, and the sperm whale shows a well developed lower jaw reception system that seems clear evidence of reception at that locus. Nonetheless, I am not yet ready to dismiss Kozak's ideas entirely, but await further data. Clarke (1979) postulates that the sperm whale uses the spermaceti organ to establish neutral buoyancy during dives, by changing the temperature of the contained spermateci which concurrently undergoes slight density changes. My basic doubts about the idea are several. First, the spermaceti organ seems ill designed to be a heat exchange organ. Blood flow within the spermaceti is very modest indeed, the spermaceti case being so clear of all obstructions that whalers could bail its contained spermaceti oil out with buckets. As Schenkkan and Purves (1973) point out, vascular heat exchange requires cooling of blood and hence either cooling of the entire animal, or a heat loss mechanism that has yet to be demonstrated.

To bolster the argument for the organ as a heat exchange structure, Clarke postulates that the whale draws water into its right nostril, cooling the spermaceti from below. Once again, this seems a very inefficient design for heat exchange, in which the approximately 300 liters of waxy oil (Schenkkan and Purves, 1973) lie on top this flattened air passage, in the nearly cylindrical organ whose crossectional diameter is a meter or more. Such inhalation of water, if it can be done past the museau du singe, which seems to be a valve whose closure is directed against such entry, seems a cumbersome and peculiar adaptation for an animal that may phonate with the same structure.

The time course of density change under even the most extreme rates of heat exchange is slow enough that significant portions of many dives would find density change in process when thermoclines are reached. The occurence, or even existence of such thermoclines, is variable in the world oceans, and finally, there is no evidence for whales hovering, but instead for a fanning out of schools underwater, and continued foreward movement (Watkins, 1977). Watkins (this volume) found that not all animals phonate but some may be silent for fairly long periods during dives, but that some phonation in sperm whale schools is the norm. If the silent animals are hydrostatically equilibrating, what are the phonating animals in the same school doing ?

Another view is contained in an anatomical work by Schenkkan and Purves (1973). Their studies lead them to propose that the spermaceti organ assists in evacuation of the lungs prior to a deep dive and that it serves as an air reservoir for phonation at extreme depths. Perhaps because their preparation for the sperm whale was a fetal animal they were unable to conceive that the right nasal passage of the animal can express air into the left passage through a connecting vestibule, and thus allow air recycling at depth, and also phonation during dives that on occasion last more than an hour. They postulate phonation in the sperm whale larynx, a view as we note from the work of Ridgway (this volume) not substantiated for other odontocetes. I should note however, that the sperm whale is very different anatomically from the delphinids used by Ridgway. Schenkkan and Purves' contention that the spermaceti organ assists in lung evacuation seems without support. First, there is an enormous developmental investment by the animal in the spermaceti organ, which extends a quarter the length of the animal, and which involves many very intricate structures. It is hard for me to beleive that it was developed as a means for lung evacuation when other diving whales do not have it, and evacuate their lungs into the trachea and air sacs of the head by simple provision of muscular means of lung compression and of structures that provide varying degrees of resistance to pressure. During diving the collapsible rib cages of odontocetes direct air into the somewhat more pressure resistent trachea, and the air sacs of the skull. This process of evacuation probably involves muscular activity at alveolar sphincters, intercostal muscles and elsewhere. But it seems unreasonable to invoke the modest buoyancy of the spermaceti organ as a major feature of such evacuation. Changes in animal orientation underwater would reduce or reverse such an affect in any case.

Suffice it to say, there exists no definitive evidence of how sperm whales use their spermaceti organs. I feel direct experiments will be needed involving living animals, for us to learn its function. To me, not surprisingly, postulates for a phonation and sound processing function seem the most likely, and the best supported by knowledge of the animal in nature.

THE ODONTOCETE SKULL

Various attempts have been made to correlate skull contours with emitted sound fields (Evans, Sutherland and Beil, 1964 ; Dubrovsky and Zaslavski, 1970). Most of these studies have sprung from observation of the marked asymmetry found in most Delphinid, Physeterid and Ziphiid skulls, and assume that the bone-tissue interface serves as a sound reflector. This view may go back to my own early work (Norris, 1964) which pictured such a relationship. I now feel that the effects are mostly secondary ones, with the bony tissues supporting soft tissue structures that are directly involved in sound processing. In this view the bones give shape and rigidity to soft structure processing systems, involving muscles, fatty sound processing conduits, air film reflectors, and soft tissue sound generation structures. The mesorostral canal and the mandible have sound processing functions. The canal seems to provide a channel for sound conducted in its contained cartilage and may be involved in shaping the content and form of emitted signals from the rostral tip.

At its medial end the mesorostral cartilage enters the isthmus between the two nasal plugs, near the site of sound production in Delphinids, and at its distal end, its cartilage actually extends beyond the ivory-like encasing premaxillary bone to end in a rounded tip just beneath the skin. As noted earlier, this structure does carry high frequency sounds (Evans, 1973 ; Purves, 1966) as determined by sound recordings made near or on it.

THE MANDIBLE

It seems clear now that environmental sounds enter the odontocete mandible by passing through overlying blubber through what I have termed the "acoustic window", penetrating the very thin "pan bone" of the rear mandible, enter the mandibular fat body and are transmitted to the middle ear. Considerable controversy exists about the route by which such sounds enter the cochlea (see McCormick et al., this volume ; Fleischer, this volume).

The odontocete lower jaw and its adjacent structures have been greatly modified for this sound transmission function. First, the zygomatic arch which overlies the posterior portion of the mandible has been reduced from a very stout bony strut in many terrestrial mammals, to a thin flexible bony strand in most odontocetes (not platanistids). It is presumed that such reduction interference by the bone in received sound fields (Norris, 1968). The oval fatty tissue of the acoustic window is either completely free of muscle strands, or nearly so, and it directly overlies the thinnest part of the mandible, or pan bone (Fig. 2). Its tissue has been found to be composed of "acoustic fat" like that of the melon and mandibular canal, and not found elsewhere in the odontocete body (Varanasi and Malins, 1971). Sound apparently may also enter the mouth of odontocetes, presumably during the last stages of feeding or biting, since high sensitivity to sound has been found for the base of the porpoise tongue (Bullock et al., 1968). Here there is no bony barrier since the mandible is excavated and the mandibular fat body directly underlies the tissue at the inner angle of the jaw and mouth cavity. Sounds, hitting the animal outside the mouth however, must traverse the pan bone to reach the mandibular fat body, and this bone, which is very thin (to 0.2 mm) has been found to vary widely between species in both thickness and thickness topography. A sound processing function has been postulated for this topography and the physical relations that allow partial sound analysis by movements of this bone across complex sound fields have been presented (Norris, 1968). In effect, the pan bone of most odontocetes flares at a small angle to the mid-sagittal axis of the animal, and sounds returning from directly foreward hit the bone in many odontocetes at an angle that can be expected to attenuate them somewhat. Ten degrees or more to either side, transmission into the fat body, which will vary according to wavelength, can be expected

Fig. 2.   Delphinid lower jaw, showing the tissue overlying the jaw
          with its fatty "acoustic window", and the angle of
          incidence (θ) of sound coming from directly ahead of the
          animal.

to increase as normality is approached. An optimum angle, less
than normal, is expected because sounds hitting the jaw at right
angles can be expected to transmit poorly down the jaw because
reflections would be involved. Optimum transmission between 15-30°
of incidence is expected because the path to the bulla is direct.

    As the animal moves its head during scanning (Kellogg, 1960 ;
Norris et al., 1961) the animal will scan across sound fields in
which both transmitted intensity and frequency composition change
can be expected. Once inside the mandibular fat body, which is a
well defined subcylindrical mass of fat flaring posteriorly, sounds
may be transmitted directly to the tympanic bulla, to which the fat
body attaches. The transmission route is direct and relatively
short in most odontocetes, though some, such as the Platanistids,
show complexity that is not understood. The fat body, it has been
postulated (Norris and Harvey, 1974), may function as a tuned
receiver, related to frequencies of most interest, as does the
human external ear canal.

    Some experimental evidence exists that supports these conten-
tions. Studies of cochlear microphonics or evoked midbrain potentials in relation to projected sound have been performed (McCormick
et al., 1970 ; Bullock et al., 1972). Both studies show sensitivity to brief clicks or pips of sound to be highest over the lower
jaw, and as much as 6 times as high there as for sound projected at

the external auditory meatus. Bullock et al. (1968) found further that the tongue was highly sensitive, and anomalously, that the porpoise melon was 5.5 times as sensitive as the external auditory canal to clicks. They also found that jaw sensitivity to these sounds was highest in the contralateral colliculus, indicating a normal path through the cochlea, with a crossover before reaching the collicular level. On the other hand, reception in the melon proved best in the ipsilateral colliculus, suggesting either a crossover prior to reaching the cochlea, as for example if sound passed through the melon and down the subcutaneous fat over the antorbital notches to the opposite mandibular fat body, or deeper in the acoustic system of the brain. Such superficial passage over the antorbital notches was tested by Norris and Harvey (1974) by implantation of small hydrophones in the proposed sound path and ensonification of the opposite side the melon. Such transmission was not found to occur. Also possible is some form of reception in the melon itself, perhaps mediated by the trigeminal nerve. The observation remains a mystery.

Angular changes in received intensity have been demonstrated by hydrophone implantation inside the mandibular fat body and subsequent ensonification, using freshly dead porpoises. A partial null for sounds projected at O directly ahead of the animal, was found using hydrophone implants in the mandibular fat body, as was predicted by structure. Highest received levels were found between 20–25° to the side of the animal's longitudinal axis, also as predicted. Two fold increases in intensity were found for implants in the fat body just anterior to its insertion on the bulla. This indicates a structural intensification of received sound, like that in a hearing trumpet, or the external ear canal of man (Norris and Harvey, 1974).

Finally, hydrophone implants inside the peribullary air sinus showed no reception at all in an ensonified animal, indicating that this diverticulum of the eustacian canal indeed isolates the middle ear eomplex to sounds, except those reaching the ear by tissue pathways (Norris and Harvey, 1974).

Taken in its entirety, the odontote head seems adapted to sound emission from the forehead and upper jaw, with the melon and air sacs providing a means for internal sound processing (Giro and Dubrovsky, 1974) and impedance matching prior to emission into the water. Air sacs especially, seem positioned to isolate emission sources from reception locations. Reception for clicks at least, seems isolated in the lower jaw and tongue (except for the anomalous observation of such reception in the melon, by Bullock et al., 1968). In both emission and reception, sounds are processed by structures composed of special acoustic fats not found elsewhere. They seem to function by differential transmission velocity topography, and by means of tissue interfaces of various sorts. The entire system seems to represent an adaptive response to the problems faced by a terrestrial animal entering water, including extreme impedance missmatching, and radically different environmental transmission velocities, which all but nullify normal aerial directional hearing.

REFERENCES

Blomberg, J., 1978, Functional aspects of odontocete head oil lipids with special reference to pilot whale head oil, Prog. Chem. Fats and other Lipids, 16:257.

Bullock, T.H., Grinnell, A. D., Ikezono, E., Katsuki, Y., Nomoto, M., Sato, O., Suga, N., and Yanigasawa, K., 1968, Electrophysiological studies of the central auditory mechanisms in cetaceans, Z. vergl. Physiol., 59:117.

Bullock, T. H., and Ridgway, S. H., 1972, Evoked potentials in the central auditory system of alert porpoises to their own and artificial sounds, J. Neurobiol., 3:79.

Clarke, M. R., 1978, Buoyancy control as a function of the spermaceti organ in the sperm whale, J. Mar. Biol. Assoc., United Kingdom, 58:27.

Clarke, M. R., 1979, The head of the sperm whale, Scientific Amer., 1:128.

Diercks, K. J., Trochta, R. T., Greenlaw, R. L., and Evans, W. E., 1971, Recordings and analysis of dolphin echolocation signals, J. Acoust. Soc. Amer., 49:1729.

Dormer, K. J., 1974, The mechanism of sound production and measurement of sound processing in delphinid cetaceans, Ph. D. Diss., University of California, Los Angeles.

Dubrovsky, N. A., and Zaslavski, G. L., 1975, Role of the skull bones in the space-time development of the dolphin echolocation signal, Sov. Phys. Acoust., 21 (Trans. Amer. Inst. Physics.).

Evans, W. E., 1973, Echolocation by marine delphinids and one species of fresh water dolphin, J. Acoust. Soc. Amer., 54 (1):191.

Evans, W. E., and Prescott, J. H., 1962, Observations of the sound production capabilities of the bottlenose porpoise: a study of whistles and clicks, Zoologica, 47:121.

Evans, W. E., Sutherland, W. W., and Beil, R. G., 1964, The directional characteristics of delphinid sounds, in: "Marine Bio-Acoustics", W. N. Tavolga, ed.

Evans, W. E., and Maderson, P. F. A., 1973, Mechanisms of sound production in delphinid cetaceans: a review and some anatomical considerations, Amer. Zool., 13:1205.

Fraser, F. C., and Purves, P. E., 1959, Hearing in whales, Endeavour, 18:93.

Giro, L. R., and Dubrovsky, N. A., 1974, The possible role of supracranial air sacs in the formation of echo ranging signals, A. Kurstichevkiy Zhurnal, 20:706.

Kellogg, W. N., 1960, Auditory scanning in the dolphin, Psychol. Rec., 10:25.

Kozak, V. A., 1974, The "video acoustical system" of the sperm whale, Trans.: Zhur. Evolyutsionnoi Biokhimii i Fiziologii, 10(3): 276.

Litchfield, C., and Greenberg, A. J., 1974, Comparative lipid patterns in the melon fats of dolphins, porpoises, and toothed whales, Comp. Biochem. Physiol., 478:401.

Malins, D. C., and Varanasi, U., 1977, Acoustic pathways in the cetacean head: assessment of sound properties through the use of a new microtechnique, (abst.), Proc. 2nd Conf. on Biol. of Marine Mammals, 36.

McCormick, J. G., Wever, E. G., Palin, J., Ridgway, S. H., 1970, Sound conduction in the dolphin ear, J. Acoust. Soc. Amer., 48:1418.

Norris, K. S., 1964, Some problems of echolocation in cetaceans, in: "Marine Bioacoustics", Pergamon Press, New York.

Norris, K. S., 1968, The evolution of acoustic mechanisms in odontocete cetaceans, in: "Evolution and Environment", E. T. Drake, ed., Yale University Press, New Haven.

Norris, K. S., Presscott, J. H., Asa-dorian, P. V. and Perkins, P., 1961, An experimental demonstration of echolocation behavior in the porpoise, Tursiops truncatus (Montagu), Biol. Bull., 120:163.

Norris, K. S., Evans, W. E., and Turner, R. N., 1966, Echolocation in an Atlantic bottlenose porpoise during discrimination, in: "Les Systèmes Sonars Animaux, Biologie et Bionique, Tome I.", R. G., Busnel, ed., Laboratoire de Physiologie Acoustique, INRA-CNRZ, Jouy-en-Josas.

Norris, K. S., and Evans, W. E., 1967, Directionality of echolocation clicks in the rough-tooth porpoise, Steno bredanensis (Lesson), in: "Marine Bio-Acoustics, Vol. 2", W. N. Tavolga, ed., Pergamon Press, New York.

Norris, K. S., Dormer, K. J., Pegg, J., and Liese, G. J., 1971, The mechanisms of sound production and air recycling in porpoises: a preliminary report, in: "Proceedings of the 8th Annual Conference Biol. Sonar and Diving Mammals".

Norris, K. S., and Harvey, G. W., 1972, A theory for the function of the spermaceti organ of the sperm whale (Physeter catodon L.), National Atmos. and Space Admin., Special Publication, 262:397.

Norris, K. S., and Harvey, G. W., 1974, Sound transmission in the porpoise head, J. Acoust. Soc. Amer., 56:659.

Purves, P. E., 1966, Anatomical and experimental observations on the cetacean sonar system, in: "Les Systèmes Sonars Animaux, Biologie et Bionique, Tome I", R. G. Busnel, ed., Laboratoire de Physiologie Acoustique, INRA-CNRZ, Jouy-en-Josas.

Raven, H. S., and Gregory, W. K., 1933, The spermaceti organ and nasal passages of the sperm whale (Physeter catodon) and other odontocetes, Amer. Mus. Novitates, 677:1.

Schenkkan, E. J. and Purves, P. E., 1973, The comparative anatomy
        of the nasal tract and the function of the spermaceti organ
        in the Physeteridae (Mammalia, Odontoceti), Bijdragen tot de
        Dierkunde, 43:93.
Schevill, W. E., and Watkins, W. A., 1966, Sound structure and dir-
        ectionality in Orcinus (Killer Whale), Zoologica, 51:71.
Varanasi, U., and Malins, D. G., 1971, Unique lipids of the porpoise
        (Tursiops gilli): Differences in triacyl glycerols and wax-
        esters of acoustic (mandibular canal and melon) and blubber
        tissues, Biochem. and Biophys. Acta, 231:415.
Varanasi, U., Feldman, H. R., and Malins, D. C., 1975, Molecular
        basis for formation of lipid sound lens in echolocating cet-
        aceans, Nature, 255:340.
Watkins, W. A., 1977, Acoustic behavior of sperm whales, Oceanus,
        50-58.
Wedmid, Y., Litchfield, C.,Ackman, R. G., Sipos, J. C., Eaton, C. A.,
        and Mitchell, E., 1973, Heterogeneity of lipid composition
        within the cephalic melon tissue of the pilot whale (Globi-
        cephala melaena), Biochem. and Biophys. Acta, 326:439.

CETACEAN BRAIN RESEARCH: NEED FOR NEW DIRECTIONS

Theodore Holmes Bullock

Neurobiology Unit, Scripps Institution of Oceanography,
and Department of Neurosciences, School of Medicine,
University of California, San Diego, La Jolla, CA 92093

Whereas neuroscience is a rapidly moving field, characterized by the application of many new techniques, progress in understanding the greatest brains has slowed down in the west and is only forging ahead in a few Soviet laboratories in particular directions. The virtual absence of work on the cetacean brain in the U.S.A. is attributable to a combination of cost, the Marine Mammal Protection Act, and public sentiment against exploiting these animals. Two things need to be said here, in this historic international forum. One is that I support the Marine Mammal Protection Act and believe most of my colleagues do; I am working to extend similar protection to primates. The other is that new research is needed and justified especially on the brain and behavior, albeit with a high threshold of justification, that is, for questions of particularly high significance and to which answers are likely.

Doomed animals such as those that beach themselves to die of natural causes, those that are harvested in Japan and other countries for food and those that are accidentally trapped in commercial fishermen's nets, should not be wasted when humane studies could contribute to our meager knowledge of the unique brain. Hopefully, there will be far fewer such animals within a short time, as commercial fishing is stopped. In the meantime, a few of the hundreds still landed would permit the first application of modern methods of brain anatomy and some important new physiology and pharmacology.

The Soviet work referred to is not well known elsewhere. Actually several hundred active scientists in the USSR are publishing currently on the cetacean brain and behavior. Bullock and Gurevich (1979) have reviewed the literature since 1960; they conclude that there is a need for Western workers to study selected items in

511

Russian and for Soviet authors to publish full papers, with methods
and data, in the international journals more often than they do.
Otherwise the animals they have used are not able to contribute as
much as they could to the accumulation of knowledge about them.
Bullock and Gurevich also point out the concentration of Soviet
attention to certain scientific areas, leaving other problems relatively undeveloped. Even in the areas of maximum attention some of
the work bears confirmation or re-examination with newer methods.

In the remaining paragraphs, I will touch upon a few of the
areas where new work is especially needed.

ANATOMY OF THE BRAIN

Comparative neurology is now in a renaissance with major new
findings and reinterpretations, due largely to a battery of recently
discovered techniques that permit tracing connections between
distant structures, thus establishing their real place in the system,
and allowing comparisons on a more rational, informed basis.

Almost no modern methods have been applied to the study of
cetacean brain anatomy, especially the powerful axon transport
techniques. Probably very few workers have even had well fixed,
i.e. perfused brains for classical methods. A special feature of
the new axon transport techniques for tracing pathways is that hours
following injection of the innocuous marker substance (e.g. an amino
acid or a peroxidase) suffice for its distribution along axons, so
that the methods are useable on doomed animals that will survive
only for a day or so. Injection under anesthesia would employ
routine technique used in human stereotaxic surgery to place a long
needle at a chosen site in the brain.

The reason that major new insights into the evolution of the
mammalian brain are to be expected is that cetaceans have evolved
separately so far that structures such as the olives, the cerebellum,
and the cerebralneocortex are not now satisfactorily compared to the
structures of the same names in other mammalian orders. Some structures are unaccountably small such as the auditory dorsal cochlear
nucleus and major parts of the limbic system, including the hippo-
campus. Others are of prodigious size and very likely have hitherto
unrevealed specializations such as the inferior colliculus which is
12 times as big, in a dolphin brain of 780 gms., as the homologous
structure in the human.

PHYSIOLOGY OF THE BRAIN

I will mention only one technique still hardly used on cetaceans, although Seeley, Flanigan and Ridgway (1976) demonstrated its

feasibility. This is the averaged evoked potential recorded through
the skull, without entering the brain. An enormous literature shows
its usefulness in studying the human brain. It can be a quick and
objective way of plotting response against frequency of sound for
example, or the effect of interval in pairs of clicks or the best
parameters, rate and span of FM. In recent experiments with Ridgway
and colleagues we have found that, as in the human, responses of
several brain stem centers early in the auditory pathway can be
studied, as well as later responses of higher centers. Evoked
potentials might be able to show that two stimuli are discriminable
by the brain, or that a continuum of slightly different stimuli are
categorized into two distinct classes, like phonemes, or that one
hemisphere does this more than the other. There are special waves
associated in the human with recognition and cognitive processes;
these have not yet been identified in any non-human species.
Clearly this powerful tool should be used to assess many aspects.

By extending the technique to intracranial electrodes, such as
are in standard use in human stereotaxic recording, a large additional domain of insights based on more localized and sensitive
recording, can be obtained, as demonstrated first by Bullock and
Ridgway (1971, 1972a,b) and further by a Soviet group (reviewed in
Supin et al., 1978), and on localized electrical stimulation as
demonstrated by Lilly (1958, and see Lilly and Miller, 1962). In
the absence of mortality data from the Soviet experience the method
must be considered to have appreciably more risk in dolphins than
in humans although this may in fact be much lower now than some
years ago, especially if the electrodes are only left in place for
some days. It is well suited to the class of animals mentioned in
Section II which are expected to survive only a day or so. A long
list of deep as well as localized surface structures become accessible and many of their functional relations can be revealed by
electrical recording. Given the special interest and value of
understanding the functioning cetacean brain it will probably be
worthwhile to employ multichannel recording, current source density
analysis, focal and diffuse electric stimulation, cryoprobes and
controlled focal release of drugs to bring out the comparisons and
contrasts with brains in other mammals.

This is not the place to give background and protocol for
specific experiments. The point of this note is to say that, with
all due caution, and appropriate due process, modern techniques for
brain research should be applied to cetaceans because the appreciation and understanding of their unique status in respect to
nervous system achievement amply justify the humane imposition of
low risk procedures.

REFERENCES

Bullock, T. H., and Gurevich, V. S., 1979, Soviet literature on the
    nervous system and psychobiology of cetaceans, International
    Review of Neurobiology, (in press).
Bullock, T. H., and Ridgway, S. H., 1971, Evoked potentials in the
    auditory system of alert porpoises (Cetacea) and sea lions
    (Pinnipedia) to their own and to artificial sounds, Proc.
    Int. Un. Physiol. Sci., IX:89.
Bullock, T. H., and Ridgway, S. H., 1972a, Neurophysiological find-
    ings relevant to echolocation in marine mammals, in: "Animal
    Orientation and Navigation", S. R. Galler, K. Schmidt-Koenig,
    G. J. Jacobs, R. E. Belleville, eds., NASA, U.S. Govn't
    Printing Off., Washington, D.C.
Bullock, T. H., and Ridgway, S. H., 1972b, Evoked potentials in the
    central auditory system of alert porpoises to their own and
    artificial sounds, J. Neurobiol., 3:79.
Lilly, J. C., 1958, Electrode and cannulae implantation in the
    brain by a simple percutaneous method, Science, 127:1181.
Lilly, J. C., and Miller, A. M., 1962, Operant conditioning of the
    bottlenose dolphin with electrical stimulation of the brain,
    J. Comp. Physiol. Psychol., 55:73.
Seeley, R. L., Flanigan, W. F., and Ridgway, S. H., 1976, A tech-
    nique for rapidly assessing the hearing of the bottlenosed
    porpoise Tursiops truncatus, Naval Undersea Center TP 552,
    San Diego, California.
Supin, A. Y., Mukhametov, L. M., Ladygina, T. F., Popov, V. V.,
    Mass, A. M., and Polyakova, I. G., 1978, "Electrophysiologi-
    cal Studies of the Dolphin's Brain", V. E. Sokolov, ed.,
    Izdatel'stvo Nauka, Moscow.

IMPORTANT AREAS FOR FUTURE CETACEAN AUDITORY STUDY

C. Scott Johnson

Naval Ocean Systems Center

San Diego, California 92152

INTRODUCTION

We have now accumulated enough information on cetacean hearing to discover some anomalies, that is, experimental results that we do not understand as well as we would like to. Three of these anomalies will be discussed. In each case more data is necessary in order to clarify the question. Until this is accomplished, speculation as to the basis of the problems is of limited value.

AUDITORY THRESHOLDS AND CRITICAL RATIOS

In the frequency region above about 50 kHz, cetacean hearing appears to be limited by thermal noise (C. S. Johnson, 1979). This is an important result because it means that at least for the frequencies above 50 kHz, and quite possibly for those below as well, we need not perform the laborious duty of determining absolute thresholds. Since the thresholds are always thermal noise limited in healthy animals, the only relevant numbers are the steepness of the cut-off at the upper limit of hearing and the critical ratios.

The critical ratios referred to here are as defined by Fletcher (1940) and are estimated by measuring auditory thresholds of pure tones superimposed on a background of continuous broad-band noise. Unfortunately, critical ratios have only been determined for one, T. truncatus (C. S. Johnson, 1968), and those values are in some question (c. s. Johnson, 1979). The reason for this is that the threshold values predicted using the thermal noise spectrum and the measured critical ratios differ from those measured on several cetacean species by about an order of magnitude, i.e., 10 dB. This means that either the critical ratio values are too large by a

515

factor of ten, which is quite a large discrepancy, or these animals
can hear 10 dB below the noise in a critical band. This is obviously an anomalous situation that can only be resolved by doing more
masked threshold experiments.

FREQUENCY MODULATED SIGNALS

The following account relates one instance of unexplained
behavior related to the use of FM signals.

This author worked closely with Jacobs (1972) on the reported
frequency discrimination experiment. The original procedure was
to train the animal to respond by pushing on one lever when it
heard an intermittently pulsed tone of constant frequency, and to
a second lever when tones of two different frequencies were pulsed
alternately. The duration of the individual pulses was a few
tenths of a second and the series of pulses was played for a few
seconds. In the one case the animal was presented with a da da da
da etc. signal and the other a da dee da dee etc. signal. Once
this task was learned the separation in frequency between the
alternating tones could be reduced until the discrimination failed
and the just noticeable frequency determined.

This procedure was tried for several weeks using two different
animals and no matter how different we made the frequencies of the
two tones, neither animal performed above the chance level in
making the discrimination. Many different pulse lengths and overall
stimulus durations were tried to no avail. Finally, in desperation,
the signals were changed from pulsed tones of equal or different
frequencies to a constant frequency tone versus a sine wave modulated FM signal. By the end of the first day following the change
in the signals, the animal was performing the discrimination with
near perfection! Unfortunately at the time it was not possible to
explore this problem further and it has not been investigated since.

It is obvious from this experience that the type of modulation,
in this case pulsed versus sine wave, is of great importance to the
animal and emphasizes the importance of investigating the significance of LPM and other types of modulation.

CRITICAL TIME INTERVAL

Soviet scientists, Belkovich and Dubrovsky (1976), p. 91, and
Velmin and Dubrovsky (1976), have reported some very interesting
results indicating that something anomalous may be happening when
pulse echo separations are in the range of 200 to 300 μsec.

The first reference above describes an experiment in which two
T. truncatus were trained to discriminate two click trains; one of
single pulses and one of pairs of pulses. The second pulse of the

pair train was varied in both time separation from the first pulse
and in amplitude relative to the first pulse. Their results show
that the animals' performance was good in making the discrimination
as the separation of the paired pulses decreased from 2,000 μsec
until a separation of about 500 μsec. The scores of both animals
decreased rapidly with decreasing pulse separation to about 90 μsec.
At this point the score of animal number 1 began improving rapidly
as pulse separation decreased while that of animal number 2 continued to decline. The data imply that as the pulse separation
decreased the acoustic stimuli perceived by the animals changed,
e.g., from pulse pairs to a time separation pitch signal or other.
Animal number 1 caught on to the change while animal number 2 did
not. Similar behavioral effects have been reported before (C. S.
Johnson, 1971).

In the experiment reported by Velmin and Dubrovsky (1976), a
single T. truncatus was required to discriminate between two trains
of paired pulses. One of the pulse trains always had a pulse
interval of 100 μsec while the interval in the second train was
varied from 150 μsec to 2,000 μsec. The results show the animal's
performance scores at 100 percent until the pulse separation reached
200 μsec. Performance dropped with increasing separation above to
200 μsec to 0 percent at about 320 μsec and all larger separations.
As in the first experiment it appears that the stimulus changed
between 200 and 300 μsec and the animal didn't catch on to the
change.

What is happening here? The distance from the blow hole of a
T. truncatus, approximate location of the sound source, to the tip
of the nose and back to the ear bones is about 75 cm. This is
presumably the minimum round trip distance encountered by the animal
in echolocating. A 75 cm round trip distance gives a pulse-echo
time interval of 500 μsec as the minimum pulse-echo interval of
value to the animal while echolocating. If what is happening is a
transition from perception of individual echo pulses to the time
separation pitch for range determination since such short (300 μsec)
pulse-echo intervals are never encountered while echolocating. It
would be very useful to both verify these results and to extend
them to intervals of greater than 2,000 μsec (a corresponding
target range of 140 cm).

Another important question is: under what conditions can we
compare data from passive listening multiple pulse experiments with
that obtained from active echolocation experiments?

REFERENCES

Bel'kovich, V. M., and Dubrovskiy, N. A., 1976, "Sensory Bases of
    Cetacean Orientation",Nauka, Leningrad (English translation
    JPRS L/7157).
Fletcher, H., 1940, Auditory patterns, Rev. Mod. Phys., 12:47.
Jacobs, D., 1972, Auditory frequency discrimination in the Atlantic
    bottlenosed dolphin, Tursiops truncatus, Montagu: a prelim-
    inary report, J. Acoust. Soc. Amer., 52:696.
Johnson, C. S., 1971, Auditory masking of one pure tone by another
    in the bottlenosed porpoise, J. Acoust. Soc. Amer., 49:1317.
Johnson, C. S., 1968, Masked pure tone thresholds in the bottle-
    nosed porpoise, J. Acoust. Soc. Amer., 44:965.
Johnson, C. S., 1979, Thermal noise limit in cetacean hearing,
    submitted for publication in J. Acoust. Soc. Amer.
Velmin, V. A., and Dubrovskiy, N. A., 1976, On the auditory analysis
    of pulsed sounds by dolphins, Doklady Akademii Nauka SSSR,
    225:470.

AUDITORY PROCESSING OF ECHOES: PERIPHERAL PROCESSING

Gerhard Neuweiler

J.W.Goethe-Universität, Zoologisches Institut

Frankfurt, Germany

COCHLEA – THE FREQUENCY ANALYZER OF THE AUDITORY SYSTEM

In mammals sounds hitting the ear release a travelling wave on the Basilar Membrane (BM). The BM acts as a bank of mechanical low pass filters where high sound frequencies produce a maximal wave amplitude in the basal part of the BM and low ones in the apical region. The place of maximal vibration is considered the site of the BM where that frequency is represented. From base to apex the width of the BM gradually increases and the thickness decreases resulting in a continuous decline of the stiffness of the BM which correlates to the frequency representation along the course of the BM. From base to apex gradually lower frequencies are represented, each octave covering an equal distance on the BM.

This uniform pattern of logarithmic frequency representation in the mammalian cochlea has been recently challenged by physiological and anatomical studies in bats. Already in 1964, 1966 and 1967 A.Pye found that in some bat species, notably Rhinolophus, thickness and width of BM are regionally differentiated and depart from a smooth gradual change. In search of the 83 kHz filter in horseshoe bats Bruns (1976) described several structural specializations of the basal and outer (pectinata) part of BM (e.g. thickenings of the BM and secondary spiral lamina, minimal width of the BM) which abruptly disappear at the 4.3 mm site (from base) of the BM. By frequency mapping he showed that the filter frequency of about 83 kHz is represented exactly at that site where the sudden discontinuity of the stiffness gradient occurs (Fig. 1).

Furthermore these studies (Bruns 1976a, b) disclosed that the narrow frequency band from 82 – 86 kHz containing most relevant

Fig. 1.

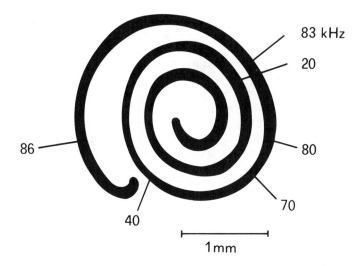

Fig. 2. Disproportional representation of frequencies on the basilar membrane in the greater horseshoe bat (Rhinolophus ferrumequinum). Basilar membrane is projected into a transmodiolar plane.
The frequency range of the constant frequency part of echoes from
82 to 86 kHz is vastly expanded. (From Bruns and Schmieszek, in
press).

Fig. 1. (Opposite page) The filter tuned to the constant frequency part of the echoes in the cochlea of the greater horseshoe bat.
Upper graph: Thickness of basilar membrane from base to apex (upper abscissa). Lower abscissa: Frequency representation on the basilar membrane as determined by frequency mapping. CF: frequency
range of the constant frequency part of the echoes. FM and shaded
area: frequency range of the final frequency modulated part of the
echoes. Lower graph: Tuning curves of cochlear nucleus units, inferior collicular and behavioral audiogram in horseshoe bats. Ordinate: Thresholds in relative dB-units. Note: Precise correlation
of sharp tuning in the 83-86 kHz range with its representation on
the thickened part of basilar membrane. (Bruns and Schmieszek,i.p.).

echo informations is represented on the BM in a vastly expanded fashion covering a length of the BM otherwise used for a complete octave, e.g. 80 - 40 kHz. This basal region of the BM in horseshoe
bats may be called a frequency magnifier or an acoustic fovea (see
Fig. 2, Schuller and Pollak, in press).

An analysis of the basilar membrane fine structure (Bruns, in
prep.) disclosed that within the basal, specialized "foveal" region
the inner and the outer part of the BM form rigid plates only weakly
coupled mechanically by a thin sulcus. Capacity of independent vibration of the outer parter is further facilitated by a loose connection of the outer BM to the cochlear bony wall. As a result for
stimulus frequencies above 80 kHz (and not for lower frequencies!)
the inner and outer part of the BM vibrate out of phase (Wilson
1977, Wilson and Bruns, in prep.). These antiphasic movements may
create radial fluid motions in the organ of Corti partially cancelling out longitudinal travelling wave vibrations. By this interaction of longitudinal and transverse motions deflection peaks of the
BM caused by a travelling wave may become narrowly localized. Thus
antiphasic movements of the two uncoupled BM-parts may assist in a
precise frequency analysis (second mechanical filter, guard-ring
principle of Wilson 1977).

All these structural specializations of the BM and its anchoring system together with additional specializations of receptor
arrangements and spacings (Bruns, in press) along with the antiphasic vibration mode for the filter frequency clearly demonstrate that
the horseshoe bat has a mechanically specialized part of the BM
resulting in a narrow band pass filter matched to the constant frequency component of the echoes and also possesses a frequency magnifier or acoustic fovea built into the cochlea for the filter frequencies from 82 to 86 kHz. Bruns (in prep.) incorporated all these data into a model of cochlear function based on Steele's vibration model for the mammalian cochlea. This specialized and sophisticated case of mechanical adaptation to hearing in a narrow frequency band and acute frequency discrimination strongly supports
the notion that in the mammalian auditory system the frequency spectrum is mechanically analyzed by the cochlea without additional
neuronal filtering. The frequency resolution of an auditory system
is that of its basilar membrane.

These structural data fairly well corroberate cochlear microphonic (CM) recordings. In all bat species commonly investigated,
the CM-audiograms match the frequency bands of the specific echolocation sounds. In "long CF/FM" - bats the CM-audiogram includes
a narrow filter for the CF - frequency (Pollak et al., 1972). Suga
et al. (1975) elaborated on the CM of Pteronotus parnellii, a
"long CF/FM" - bat. As demonstrated by slow decay times of cochlear
microphonics they described ringing of a mechanical resonator tuned
to 61.1 kHz, i.e. 0.5 to 1.0 kHz below the best frequency of sum-

mated auditory nerve responses. A similar resonator has been described in horseshoe bats (Schnitzler et al., 1976). However, in
these bats only a small dip of no more than 14 dB was recorded in
the cochlear microphonic audiogram 0.5 to 1.0 kHz below the frequency of the constant frequency component in the heard echoes.
Since part of the experiments were performed in anaesthetized bats
and since the site of the recording electrode greatly influences
the results, cochlear microphonic recordings have to be interpreted
cautiously.

All microphonic studies in Pteronotus including detuning experiments of the 62 kHz - filter by overstimulation (Pollak et al.,
in press) are in favor of a resonance system whereas the vibration
measurements of the basilar membrane in Rhinolophus indicate other
possible mechanisms which, however, also may show ringing qualities.

The horseshoe bat's basilar membrane is certainly a specialized
but not an exceptional case. Morphological differentiations of the
basilar membrane are recorded for other bat species and other mammals from different parts of the BM and a detailed inspection with
precise frequency mapping methods may disclose specialized regions
and differentiated frequency representation in frequency ranges specifically relevant for a species' behavioral repertoire. We should
depart from the notion that in all mammalian basilar membranes the
frequency range heard by a species is represented in a uniform logarithmic array.

An interesting comparative case is Pteronotus parnellii, a
"long CF/FM" - bat with a CF-frequency of 62 kHz and a correspondingly tuned narrow filter in its cochlear microphonic audiogram
(Pollak et al., 1972). However, no morphological specializations
focusing on a certain site of the basilar membrane have been recorded so far (M. Henson 1978). Since no frequency mapping is available for the cochlea of Pteronotus a detailed comparison with that
of Rhinolophus is not yet possible, but it is conceivable that these two unrelated bat species achieved the same filtering precision
by different cochlear mechanical adaptations.

NEUROANATOMY
A) INNERVATION DENSITIES OF THE COCHLEA AND TONOTOPY

Apart from dolphins, bats have the largest number of spiral
ganglion cells. The organ of Corti is highly innervated; in Myotis
lucifugus the ratio of ganglion to hair cells is 55300/3500 (700
inner hair cells and 2800 outer hair cells) or 15:1 (Ramprashad et
al., 1978). The density distribution along the basilar membrane is
not uniform and reaches a maximum of 22:1 at the end of the lower
basal turn. In the "long CF/FM" - bat Pternotus parnellii the in-

nervation appears particularly dense at the beginning and at the
end of the basal turn with sparse fibers inbetween (M.Henson 1978)

In horseshoe bats the total number of spiral ganglion cells
only amounts to 16000 as in the case of Plecotus austriacus (Firbas 1972). The innervation density in the basal region is low
(500 fibers/mm length of BM) and suddenly increases to a maximum
of 1750 fibers/mm at the transition from the first to the second
half turn where the frequencies of the final FM-sweep of the echolocation sounds are represented (Bruns and Schmieszek, in press).
80 to 90 % of all afferent fibers run to the inner hair cells.
About 8 to 16 fibers converge onto one inner hair cell in the
"acoustic fovea" - region, 24 fibers in the region where the frequencies of the FM-sweep are represented and each inner hair cell
in the region representing lower frequencies is innervated by
about 15 fibers. Although the innervation of the "acoustic fovea"-
region seems to be low it is actually high related to the frequency

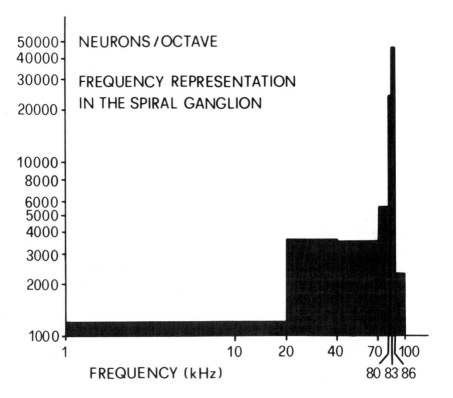

Fig. 3. Frequency representation of spiral ganglion neurons in
greater horseshoe bats. Ordinate: number of neurons/octave; abscissa: best frequency of neuron as determined by its afferent origin
on the basilar membrane. (From Bruns and Schmieszek, in press).

range covered. Since the frequency range from 82 to 86 kHz spreads over most of the lower basal turn 3360 spiral ganglion cells or 21% of the total cell number innervate the basilar membrane representing this narrow frequency range (Fig. 3.). This example illustrates that functional conclusions based on cochlear hair cell and innervation densities can only be made when the frequency representation on the basilar membrane has been mapped.

In all auditory nuclei of bats and other mammals so far studied neurophysiologically and/or neuroanatomically a tonotopic arrangement of the neurons' best frequencies has been reported. In our lab for instance we have anatomical and/or physiological data in the horseshoe bat's spiral ganglion, cochlear nuclei, superior olivary complex, inferior colliculus, medial geniculate body and the auditory cortex (also see Ostwald 1978) unequivocally demonstrating tonotopic organization. Within this tonotopic order the frequency band of 82 to 86 kHz always is overrepresented. As already mentioned one fifth of the afferent fibers come from that region of the basilar membrane where 82 to 86 kHz are represented. Thus the peripheral "acoustic fovea" is faithfully reflected in the afferent fiber population (Fig. 3.), wired through and quantitatively maintained throughout all nuclei of the central auditory system up to the auditory cortex. This is a striking analogy to the foveal representation in the visual central nervous system (Schuller and Pollak, in press). Thus in the horseshoe bat, central nervous overrepresentation of a narrow frequency band always is the result of a corresponding blow-up of that frequency range on the basilar membrane.

It would be most interesting to investigate whether the central overrepresentation of behaviorally relevant frequency bands and its tracing back to a corresponding overrepresentation and ordering in the peripheral auditory system could be applied not only to horseshoe bats but to other bat species and perhaps to all mammals also. The first crucial test for this hypothesis would be the frequency map of the basilar membrane and a quantitative count of innervation densities in Pteronotus parnellii.

If the hypothesis of the peripheral origin of frequency overrepresentation could be generalized, overrepresentation would become a mere special case of the more general fact that all auditory nuclei are organized in a tonotopic fashion, always originating in in the corresponding orderly representation of the frequencies on the basilar membrane.

B) ASCENDING AUDITORY PATHWAY

Neuroanatomical studies of the Chiropteran auditory system have long been neglected. Schweizer (1978) has studied the cyto-

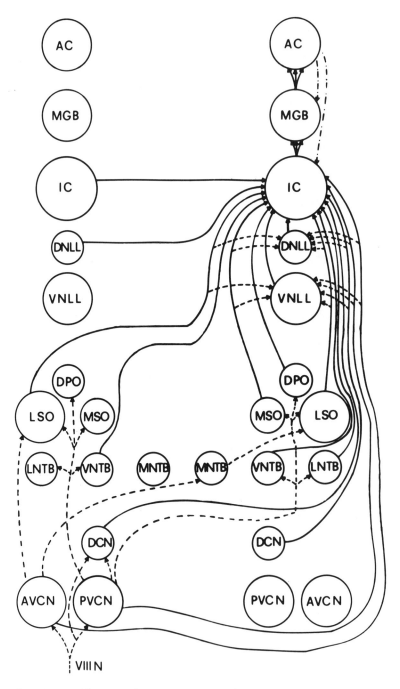

Fig. 4. Ascending auditory pathway in bats (Legend opposite page).

architecture and projections to and from the inferior colliculus
(IC) in horseshoe bats, Rhinolophus ferrumequinum, and Zook et al.
(in press) have made a similar study in Pteronotus parnellii and
Artibeus jamaicensis. Both authors used horseradishperoxidase (HRP)-
tracing methods.

In all three bat species the auditory pathways do not diverge
from the principle mammalian scheme. Except for the medial nucleus
of the trapezoid body (MNTB, a relay nucleus for the projection
from the contralateral anteroventral cochlear nucleus AVCN to the
lateral superior olivary nucleus LSO) all peripheral nuclei project
to the inferior colliculus resulting in 15 different inputs from
the auditory periphery to the inferior colliculus. These inputs come
from: contralateral anteroventral, posteroventral and contra- and
ipsilateral dorsal cochlear nuclei, from contra- and ipsilateral
ventral and ipsilateral lateral nuclei of the trapezoid body, from
ipsi- and contralateral lateral and ipsilateral medial superior
olivary nucleus, from ipsilateral dorsal periolivary nucleus, from
contra- and ipsilateral dorsal and ipsilateral ventral nuclei of
the lateral lemniscus and finally from contralateral inferior colliculus (Fig. 4.).

In contrast the higher auditory centers receive only input
from one auditory nucleus each, the medial geniculate body (MGB)
receives ascending input from the inferior colliculus and the auditory cortex (AC) from the medial geniculate body. The auditory cortex projects back into the medial geniculate body and into the inferior colliculus but not further peripherally.

In all bats the auditory nuclei are largely hyperthrophied
including auditory cortex and medial geniculate body, contrary to
earlier findings (Poljak 1926). The most hyperthrophied nuclei in
echolocating bats are the anteroventral cochlear nucleus and the
medial nucleus of the trapezoid body pointing to the functional importance of the lateral superior olivary nucleus in bats.

Fig. 4.   Scheme of the ascending auditory pathway in Rhinolophus
ferrumequinum. Left ipsi- and right column contralateral nuclei.
Below IC not all interconnections between the nuclei are shown.
Dashed lines: pathways not directly projecting to IC. Broken lines:
Descending projections from auditory cortex. AC auditory cortex;
AVCN anteroventral cochlear nucleus; DCN dorsal cochlear nucleus;
DNLL dorsal nucleus of lateral lemniscus; DPO dorsal periolivary
nucleus; IC inferior colliculus; LNTB lateral nucleus of trapezoid
body; LSO lateral superior olivary nucleus; MGB medial geniculate
body; MNTB medial nucleus of trapezoid body; MSO medial superior
olivary nucleus; PVCN posteroventral cochlear nucleus; VNLL ventral
nucleus of lateral lemniscus; VNTB ventral nucleus of trapezoid body;
VIII N auditory nerve. (Adapted from Schweizer 1978).

The wiring diagram of the three species studied reveals a remarkable divergence of the afferent auditory information into 15 peripheral nuclei and a subsequent convergence of all peripheral outputs upon one single nervous center, the inferior colliculus (Fig. 4.). Auditory information is already frequency analyzed in the cochlea and the frequency information is allotted to specific auditory nerve fibers. The complete set of frequency analyzed information is projected upon both ventral cochlear nuclei and upon the dorsal cochlear nucleus directly or via the posteroventral cochlear nucleus. As will be shown in the chapter on specialized peripheral neurons complex neuronal processing starts already in the cochlear nuclei and continues on the next stages, the trapezoid body , superior olivary complex and the lemniscal nuclei. About the specific nature of processing performed in the different nuclei little is currently known.

A good example of our scant knowledge can be illustrated by considering the superior olivary complex, the most peripheral nuclei where binaural interaction may occur. According to a theory of Irving and Harrison (1967) the presence of a large lateral superior olivary nucleus and the absence of the medial one is related to small head size and the use of binaural intensity differences for high frequency sound localization. On the other hand a large medial superior olivary nucleus is related to a large head size and the use of binaural time differences for sound localization and subsequent tracking of the sound direction by eye movements. Accordingly they described echolocating bats as not possessing a medial superior olivary nucleus. But Rhinolophus ferrumequinum, Molossus ater and Molossus molossus as well as Pteronotus parnellii (Schweizer 1978, Zook et al., in press) do have a well developed medial superior olivary nucleus. From his HRD-labeling data Schweizer (1978) concludes that the prominence of either the medial or lateral nucleus is related to the frequency range heard by a mammal since in the lateral nucleus high frequencies are preferentially represented whereas the medial superior olivary nucleus shows retrograte labeled cells from inferior collicular regions where low frequencies are represented.

Both Zook et al. and Schweizer obtained remarkably similar results with some minor exceptions: Zook et al. describe an intermediate lemniscal nucleus in Pteronotus which may be homologous to the dorsal part of the ventral nucleus of lateral lemniscus described in Rhinolophus. In Pteronotus the anteroventral cochlear nucleus is reported to project bilaterally to the next nuclei whereas in Rhinolophus it seems to project only contralaterally. Since in Artibeus labeled cells were found bilaterally throughout the olivary complex Zook et al. concluded that this bat does not have a medial superior olivary nucleus that is considered to project only to the ipsilateral inferior colliculus. This issue requires more detailed comparative studies.

Both studies contain rich informations on the cytoarchitecture of the nuclei so that neurophysiologists now can aim at describing functionally the local subdivisions and morphologically identified neuron types.

The convergence of all peripheral outputs upon the inferior colliculus assigns a central rank in neuronal processing to this midbrain nucleus. The direct and the parallel multistage projections from the cochlear nuclei to the colliculus reminds one of Licklider's delay-line theory describing an analyzing mechanism in the time domain (Tobias, 1970).

In an echolocating bat, auditory information has to be transformed in appropriate motor commands governing the flight musculature. The linkage between auditory and motor system in horseshoe bats is represented by a large projection from the inferior colliculus and from the auditory cortex (Radtke, 1979) to the lateral pontine nuclei. The latter project to the enormously hyperthrophied paraflocculus and to the vermis of the cerebellum.

Collicular projections are also sent to the formatio reticularis and to the central grey matter where vocalization of echolocation sounds can be electrically elicited (Suga et al., 1975). The auditory cortex is also connected to the grey matter and to other nuclei related to vocalization, e.g. the area cinguli, the amygdala, nucleus caudatus and an area in the frontal cortex (Radtke, 1979). The interaction of auditory and vocalization system will be dealt with in the last chapter.

BINAURAL ANALYSIS OF ECHOES

It is widely believed that bats rely on binaural intensity differences for determining horizontal angular positions of a sound source. There exist many evoked potential and single unit studies from superior olivary complex to higher nuclei demonstrating that the discharge rate may encode azimuth differences of a loudspeaker position as small as 0.5 to $2.5°$ (Grinnell and Hagiwara, 1972, Jen and Suga, 1974; Shimozawa et al. 1974; Jen, 1978).

Schlegel (1977 and in prep.) experimentally tested the binaural interaction of stimulus intensities with the use of earphones in collicular, lemniscal and superior olivary neurons of horseshoe bats. In all neurons recorded, the tuning curves were the same for ipsi- and contralateral stimulation. 70% of all units recorded were excited by contralateral and inhibited by ipsilateral stimulation (E/I-types of neurons). This result confirms to the classical concept of binaural intensity difference coding. However the discharge of the E/I-neurons not only depends on the binaural intensity differences but also on the absolute intensity level rendering the discharge information ambiguous. This equivocality has to be eliminated

at a higher interacting level. Assuming that a 5% change of discharge rate is detected by the brain these neurons could encode
azimuth differences of about 1°.

By periodically amplitude modulating the pure tone intensities
diametrically in the ipsi- and contralateral earphone, Schlegel
(in prep.) simulated a sound source moving horizontally around the
head of the horseshoe bat at different speeds (4°/sec to 1000°/
sec). He tested the reaction of superior olivary, lemniscal and collicular neurons to movement of direction and to its speed. Most
neurons faithfully encode the azimuth angle of the sound source
irrespective of movement direction and speed (proportionalists in
the classification of Schlegel). Some neurons responded only to a
narrow angular sector with a marked "best angle". These neurons
also encode the speed of the sound source movement by increasing
discharge rates for faster movements ("speed-encoders"). They do
not discriminate the direction of the movement. However, there are
other speed encoders which are vigorously excited by an angular
movement from the contra- to the ipsilateral side and are immediately inhibited by movement in the opposite direction (movement direction encoder, see Fig. 5.).

It has to be emphasized that most of these speed and direction
sensitive neurons also encode the direction of a stationary sound
source, sometimes only in a limited horizontal angular range. Yet
there are neurons which do not respond to any stationary sound
source at all but only discharge when the sound source is moving
in the horizontal plane (movement detectors, Fig. 5.). This detailed study shows that directional information and directional movement of a sound source are intricately analyzed and processed at
peripheral levels of the auditory system.

Although most units respond to several parameters there exist
many specialists responding preferentially or even exclusively to
a single parameter such as movement or a certain direction of movement, demonstrating once more that specialized (feature) detectors do exist at peripheral levels.

Similar studies on binaural time differences are much in demand. In this context the few data already available in our lab
suggest that the generally accepted statement of bats using only
binaural intensity differences for directionality encoding may have
been a hasty conclusion.

COMPARATIVE ASPECTS
A) NEURAL PROCESSING IN FM – AND "LONG CF/FM"- BATS

The numerous bat species use a wide range of different echolocation
sounds. Although there exists a continuum of structural features

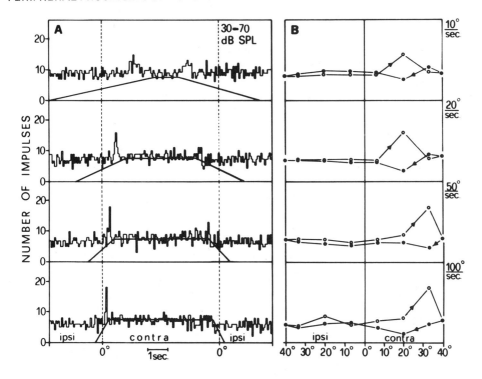

Fig. 5. Inferior collicular neuron selectively sensitive to direction of angular movement in Rhinolophus ferrumequinum. A) PST-histograms to tone pips at the unit's best frequency. Sound source is
apparently moved from 40° ipsi- to 40° contralateral and reverse by
diametrically changing the intensity from 30 to 70 dB SPL between
both earphones until complete reversal (70 to 30 dB) during a time
course as indicated by the slopes of the trapezoid. As steeper the
slopes as faster the directional movement (right hand ordinate).
Broken vertical line marks midline position (i.e. equal intensities
in both earphones) of sound source. B) Discharge rate of the unit
as in A but plotted against angular position in the horizontal plane. Open circles: movement from ipsi- to contralateral, black dots:
movement from contra- to ipsilateral side. Note: unit preferentially responds during movement and unit reacts by strong excitation to
sound source movement from ipsi- to contralateral and a shallow inhibition by movement in the reverse direction. (From Schlegel, in
prep.).

one may classify the bat species as FM - bats using broadband downward frequency modulated signals (often with harmonics), "short
CF/FM"-bats emitting an FM-sweep preceeded by a pure tone (CF) of
up to 8 msec duration, and "long CF/FM"-bats emitting a narrow FM-
sweep preceded by a pure tone of 10 to 100 msec duration (Grinnell,
1973).

Even at the level of the cochlear nucleus remarkable differences of response patterns to pure tones appear between FM- and "long CF/FM"-bats (Fig. 6.). In horseshoe bats many cochlear neurons have very complex response patterns displaying simultaneous excitatory and inhibitory interactions of various durations and elicited by a single pure tone stimulation. Frequently longlasting inhibition of spontaneous activity is elicited by the onset of a pure tone. Thus inhibition is a dominating part of neural activity in the cochlear nucleus of horseshoe bats (Neuweiler and Vater, 1977) as in many other mammals.Half of the neuron sample recorded were phasic on.

Fig. 6. Differences of neuronal response patterns to pure tones between FM- and "long CF/FM"- bats. In FM-bats (left column) cochlear nucleus units mostly respond tonically whereas in "long CF/FM"-bats (right column) they display a variety of response patterns, a a large fraction responds phasic on. More than 90% of inferior collicular neurons in FM-bats respond phasic on whereas in "long CF/FM"-bats most of them respond in complex patterns dependent on stimulus frequency. (Adapted from Suga, 1964 and Möller et al., 1978).

In contrast no such phasic on - neurons were found in the cochlear nucleus of the FM-bat Myotis lucifugus and the units recorded
uniformly responded by a tonic excitatory pattern enduring as long
as the stimulus lasted (Suga, 1964). Specialized neurons responding
only to certain stimulus parameters have not been described in the
cochlear nucleus of FM-bats whereas in horseshoe bats different specialized cochlear neurons have been recorder (Neuweiler and Vater,
1977).

In the inf.colliculus a similar striking difference of neural
processing between the two groups of bats reappears. In the horseshoe bats collicular neurons show response patterns similar to neurons in the cochlear nucleus with a trend to even greater complexity
and longlasting inhibitory effects, a feature of neuronal processing
seemingly neglected by auditory physiologists. The percentage of
phasic on - neurons drops to about 30 to 40% of some 600 neurons recorded in Rhinolophus ferrumequinum. However, in the FM - bats
Mollossus ater and Molossus molossus, 95% of 329 collicular neurons
recorded were not spontaneously active and responded only with phasic on to pure tones (Vater et al.,1979).

Again in complete contrast to the "long CF/FM" - bats in FM-
bats a large majority of collicular neurons so far recorded are phasic on - neurons with no spontaneous activity. One is impressed by
the uniformity of response patterns in FM - bats and the wide diversity of neuronal responses in "long CF/FM" - bats (Fig. 6.).

The same is true for inferior collicular responses in the FM-
bats Tadarida brasiliensis (78% phasic on-responders; Pollak et al.,
1978) and Myotis lucifugus (95% phasic on-responders; Suga, 1969).
Thus in neuronal peripheral processing a dichotomal trend appears
between FM - bats on the one hand and the "long CF/FM"- bats on the
other hand. Whereas the neuronal activities of cochlear nucleus and
inferior colliculus of the latter group agrees with that in other
mammals the neural processing in the FM - bat group is more uniform
and specialized. The prevalence of phasic on - neurons and the lack
of spontaneous activity above the cochlear nucleus level is the result of strong dominant inhibition and may be related to the possible timing capacities of such neurons (see also next review by
G.Pollak).

The benefit of emitting a short CF-component in echolocation
sounds is still a puzzling question. In the auditory nuclei of
"short CF/FM"-bats the frequencies of the CF - part are never overrepresented as they are so extravagantly in "long CF/FM"-bats. Grinnell and Hagiwara (1972) provide neurophysiological evidence that
the short CF - part facilitates the response to the terminal FM -
component of the echo. However, in some other bats sometimes also
emitting a short CF-part no such facilitation was observed in single unit recordings (Vater and Schlegel 1979).

B) OPTIMAL ADAPTATIONS

Since the introduction of optimal filter theory into experimental echolocation research by Simmons (1971) and Glaser (1974) the search of neuronal correlates of optimal adaptations to specific echolocation sounds became topical. As already mentioned, audiograms reach minimal thresholds in frequency ranges of the species' echolocation sounds. Even in closely related species, e.g. Molossus molossus and Molossus ater the minimal thresholds of audiograms are 5 kHz apart related to the main frequency range of the sounds emitted at 35-25 kHz in M. ater and 45-30 kHz in M. molossus (Vater et al., 1979).

In a detailed study in the above mentioned species Vater and Schlegel (1979) showed that most collicular neurons respond equally sensitive to CF- and FM-signals of the same duration. However, about one third of the units recorded responded with up to 6o dB lower thresholds to a signal mimicking the FM-component of the echolocation sound. FM-signals with longer sweeps than those emitted were less effective. These results are in agreement with those found in Tadarida brasiliensis by Pollak et al. (1978).

Thus there exists an adaptation, though suboptimal, of parts of the auditory system to the structure of the echolocation sound. Possible neuronal mechanisms for this adaptation to FM - echolocation sounds are discussed in Vater and Schlegel (1979).

Optimal receivers detect signals they are matched to much better in noise than other receiving systems. In fact this improvement in signal to noise - detectability is the major asset of a matched filter. The signal to noise ratio of its output should improve by a factor of 2 TB (T effective duration, B effective bandwidth of signal; Glaser, 1974). In our lab signal to noise ratios of threshold responses in collicular neurons were measured with pure tone signals in Rhinolophus ferrumequinum, a "long CF/FM"-bat, and with frequency downward modulated signals in Molossus ater and M. molossus, so called FM - bats. Both signals mimick the species-specific echolocation signals.

Threshold signal to noise ratios of these units range between +4o to - 20 and rarely reach values predicted by matched filter theory. However, in Molossid bats FM-sweeps are better detected in noise than pure tones by about 10 dB, thus indicating an adaptation of the receiver system to the echo signals, though again suboptimal (Fig. 7.).

Additionally in Molossus, collicular neurons detect a masked signal of long duration much easier than a short one which nicely relates to Simmons' finding that Tadarida, another FM - bat, lengthens its echolocation sounds in the presence of noise. The better detectabi-

Fig. 7. Comparison of threshold - signal/noise - ratios (Threshold-SNR) to downward sweeping frequency modulated (FM) and constant frequency (CF) signals of equal duration (2 msec) for inferior collicular neurons in two FM-bats, Molossus molossus (open circles) and Molossus ater (black dots). Broken line represents equal values for both CF- and FM- signal. CF-signals were at best frequency, all FM-signals (modulation heights 40, 20 or 10 kHz) were centered at best frequency. Note: a majority of units has lower threshold - SNR for FM- than for CF-signals. (From Vater and Schlegel, 1979).

lity of long sweeps is readily explained by the higher energy content of the long sweep.

SPECIALIZED PERIPHERAL NEURONS

In auditory research the dispute on the concept of specialized (feature) detectors continues and has been recently revived by the spectacular findings in the auditory cortex of the mustache bat (Suga and his associates, see also review by Suga and O'Neill in this book). All detector concepts imply a hierarchical neuronal processing with unspecific neurons in the periphery and more and more selective neurons in the next stages converging onto the highly specific "detector" neuron in the cortex.

As a matter of fact there is ample evidence that specialized neurons do exist in peripheral nuclei and that their number probably increases at higher levels. There are specialized neurons in the superior olivary complex only responding when a stimulus is moving in the horizontal plane from ipsi- to contralateral side; it will not respond to a stationary or oppositely moving sound source (Schlegel, in prep.; see also Fig. 5.). There are neurons in the cochlear nucleus of horseshoe bats only responding to a pure tone component of the echolocation sound, they will not react to any noise band as narrow as 120 Hz (Neuweiler and Vater 1977). There are neurons in the inferior colliculus which do not respond to a pure tone but vigorously fire when the carrier frequency is modulated (Schuller, 1979). Apparently specialization (or selectivity or special rejection mode) is achieved parallelly within peripheral and central stages.

The auditory system deals with continua of bioacoustic dimensions such as frequencies, frequency modulations, narrow noise bands, amplitude modulations and so on (see also Scheich, 1977). Specialized neurons respond exclusively or preferentially to certain sectors or combinations of these continual dimensions and reject other signals.

The main issue is the problem if these specialized neurons are matched to behaviorally significant signals only or if specialized neurons exist along all bioacoustic dimensions irrespective of the known existance of correspondant signals in the acoustic world of a species. To be sure, there is ample evidence in mammals and in birds that specialized neurons matched to behaviorally relevant or species-specific signals do exist. However, since we hardly offer any other complex stimuli other than behaviorally relevant ones to an auditory system we can not expect to find specializations other than those that we asked for.

In FM- as well as in "long CF/FM"- bats, specialized neurons have been recorded not matching the structure of the specific echolocation sounds (Suga, 1973; Neuweiler and Vater, 1977; Vater et al., 1979). Recently it was shown that some bat species may vary the time structure of their echolocation sounds to a greater extent than breviously found and one might argue that these behaviorally unspecified specialized neurons may match to echolocation sounds or to communication calls hitherto undiscovered. In no case has the specifity of these neurons been tested with real signal structures emitted by the bat. Thus these neurons may be still considered as evidence of specialization in the auditory nervous system not implying immediate behavioral relevance. Apparently the brain has the capacity of discriminating complex stimuli in a much larger matrix surpassing the limited range of what we call behaviorally relevant.

Moreover recent research in awake squirrel monkey indicate
that detector neurons in the auditory cortex change their selective
(or rejective) qualities in time to an unexpected extent (Manley
et al., 1978). Apparently those detector neurons have much more dynamic properties than we like them to have for our theories. Unfortunately due to convenience of the currently adopted neurophysiological data collecting and processing methods longlasting response
variability of neurons is hardly recorded and induces neurophysiologists to look at neuronal properties statically. In this context
the few studies at hand already indicate that auditory neurons may
dramatically change their response characteristics related to stimulus history. We should design dynamic auditory detector hypothe-
sises by not only playing around in the threedimensional space of
a brain, i.e. in the frequency-domain, but by expanding our thinking into the time dimension. Any scheme of the auditory system or
a subsystem not integrating its functions and functional states in
time is unrealistic.

All data presented here indicate that specialized neurons
exist parallel at peripheral and central levels of the neural auditory system and along all bioacoustic dimensions. Accordingly we
expect a matrix of specialized neurons far beyond those behaviorally relevant. Within the large range of specialized neurons those
may become overrepresented by experience or by heritage which match
to frequently encountered auditory signals. The outlines presented
here are compatible with the concept of a "focal property selective
system" forwarded by Scheich (1977).

THE MOVING PREY DETECTING SYSTEM OF THE HORSESHOE BAT
- A CASE OF MULTIPLE AND CONVERGING SPECIALIZATIONS

The long constant frequency (CF) component of 10 to 100 msec
duration emitted by horseshoe bats is an excellent signal for detecting frequency (FM) and amplitude (AM) modulations caused by
sound reflections from moving wing beating prey. The emitted frequency of 81 to 85 kHz (interindividual variation) is simply a carrier frequency for those complex modulated echo informations.

The ear of the horseshoe bat has a built-in mechanical filter
narrowly tuned to the individual carrier frequency of the echoes
(Fig. 1 and 2, Neuweiler, 1970; Bruns 1976a, b; Schuller and Pollak, in press; see also chapter 1 of this review). The narrow frequency band of about 82 to 86 kHz containing the biologically relevant signals from moving prey is represented on the basilar membrane in a magnified, vastly expanded fashion as a kind of an acoustic fovea (Fig. 2., Bruns, 1976b; Schuller and Pollak, in press).

This foveal arrangement results in an extremely sharp tuning
with Q-values up to 500 in the afferent influx for the frequency

band of 82 to 86 kHz. This sharp tuning is preserved throughout the
auditory system with Q_{10dB}-values of single units frequently as high
as 400 to 500 (highest Q-values in other mammals and in FM-bats:
20 to 60). In horseshoe bats precise tuning has been found in the
auditory nerve (Neuweiler and Vater, 1977), in the cochlear nucleus
(Suga et al. 1976), in the inferior colliculus (Neuweiler, 1970;
Möller et al. 1978) and in the auditory cortex (Ostwald, 1978).
However, it has to be emphasized that the extremely sharp tuning in
the filter band is only due to mechanical specializations of the
horseshoe bat's cochlea. There occurs no additional neuronal tuning,
as expressed by Q-values of the units recorded. As in the eye the
fovea and its functional consequences are a result of peripheral
receptor organization.

In an evoked potential study, Schuller (1972) discovered that
the inferior colliculus responds to frequency modulation depths as
small as 10 Hz as long as the carrier frequency lies between 82 and
86 kHz. Schuller postulated that the horseshoe bat takes advantage
of the fine FM- and AM-sensitivity for detecting wing beating prey.
In a recent behavioral study Goldmann and Henson (1977) proved that
this hypothesis is correct in another "long CF/FM"-bat, Pteronotus
parnellii.

Collicular single units faithfully encode frequency modulations
of the carrier frequency with modulation rates of up to 500 Hz and
modulation depths as low as 10 Hz. These neurons resolve frequency
changes periodically repeated every 3 - 5 msec (Fig.12 in the next
review by Pollak; Schuller, 1979). Schuller also described a great
sensitivity to amplitude modulations down to modulation indeces
of 3-5%. Whereas sensitivity to amplitude modulations is a general
feature of collicular neurons, sensitivity to low frequency modulation depths is restricted to the neurons with best frequencies in
the filter band possessing extremely narrow response areas. The FM-
sensitivity is the result of this narrow tuning originating in the
frequency expanded part of the cochlea, i.e. the acoustic fovea
(Schuller, 1979).

Consequently high sensitivity to FM-signals in neurons with
best frequencies between 82 and 86 kHz is already encountered in
the cochlear nucleus (Vater, in prep.). Responsiveness to frequency modulation rates is even better in the anteroventral cochlear
nucleus reaching a maximum of 900 Hz compared to maximal 500 Hz in
the inferior colliculus. Interestingly FM-signals are differentially processed in the different subdivisions of the cochlear nucleus.
Tonically (primary like) responding neurons, preferentially occurring in the anteroventral cochlear nucleus most faithfully encode
timestructures of a sinusoidally frequency modulated signal, whereas phasically responding , built-up and complex neurons,mostly ocur-
ring in the dorsal cochlear nucleus, are less suitable or uncapable
of following periodic modulations (Vater, in prep.).

Specialized neurons were also found in the inferior colliculus as well as in the cochlear nucleus. Many collicular neurons preferentially or exclusively encoded FM - signals only at low or medium sound intensity levels (10 to 60 dB SPL) and rejected FM - signals of higher intensities. This feature was not related to upper threshold tuning curves and may be interpreted as an adaptation to echo listening (Pollak and Schuller, in prep.). Other neurons in both cochlear and collicular nuclei only responded to FM-stimuli but failed to fire to the carrier frequency at any intensity or to any amplitude modulation.

These detailed recent studies amply prove that the capacity to respond to minute frequency and amplitude modulations again is due to peripheral adaptations and may be fully explained by the narrow tuning areas originating in the cochlea, at least as far as FM-encoding is concerned. Selective responsiveness to modulations of the carrier frequency caused by potential wing beating prey also starts at the level of the cochlear nucleus.

There are more general adaptations of the periphery facilitating hearing of the pure tone echo component in horseshoe bats. Under natural conditions the bat never hears a single echo but has to listen to at least two overlapping signals, the emitted sound and the echo (or a time smeared sequence of echoes).

In collicular neurons of horseshoe bats as of other mammals the response to a second signal is easily inhibited by a preceeding one. However there is an interesting exception to this rule: in neurons with best frequencies within the filter region (82-86 kHz) this inhibition of the response to the second tone (i.e. echo) is completely missing when the frequency of the preceeding signal (i. e. the emitted sound) lies in the frequency band of 78 to 82 kHz which actually corresponds to the frequencies emitted by a hunting horseshoe bat. Yet the frequencies of the emitted echolocation sounds not only prevent inhibition but in addition enhance the response to the subsequent echo by increasing firing rates and decreasing thresholds up to 20 dB (Fig. 8.; Möller, 1978). It has to be emphasized that lack of inhibition and facilitation is only elicited by the narrow frequency band of 78 to 82 kHz, actually emitted by a flying horseshoe bat.

We have always suspected that vocalization may neuronally condition auditory neurons to echo - listening. Recently Schuller (in press) has shown that, indeed, some collicular neurons only respond to a pure tone or only encode FM-signals (echoes from wing beating prey) when the horseshoe bat is vocalizing. Acoustic stimulation with the same stimuli but without vocalization does not elicit any response in these neurons (Fig. 10). This fascinating interaction of vocalization and auditory system shall be dealt with in the last chapter.

Fig. 8. Lack of inhibition and enhancement of response by two tone stimulation in inferior colliculus neurons of Rhinolophus ferrumequinum. Left figure: response to a fixed tone (echo) is inhibited by a continuous tone (shaded area) except to a narrow frequency range of 78-82 kHz, corresponding to the frequencies actually emitted by a flying horseshoebat. Solid line and black dots: tuning curve of the unit; BEF Best Excitatory Frequency; BIF Best Inhibitory Frequency. Right Figure: Frequency range of a first tone (hatched column) causing facilitation of response to a second fixed tone as shown by evoked potential amplitudes (solid line, black dots). Fixed tone : 60 dB SPL, frequency at Reference Frequency (RF), i.e. frequency of the echo, preceded by a first tone of variable frequencies (abscissa) and 90 dB SPL. Delay time between both stimuli 10 msec, stimulus duration 30 msec each. Horizontal left bar marks e.p.amplitude for fixed tone alone. Right hand ordinate: thresholds of e.p. to single tone stimulation. Broken line and open circles: Threshold curve for on-responses. Dotted line and black squares: Threshold curve for off-responses. Note: Frequency range causing facilitation coincides with that causing lack of inhibition (left figure). (From Möller, 1978).

Finally the flying horseshoe bat has to eliminate its own travelling speed from the movement detecting system, since the own flight speed shifts the carrier frequency of the echoes to higher values and out of the center frequency of the cochlear filter. This uncoupling of the movement detector from the bat's own movements is achieved by Doppler effect compensation (Schnitzler, 1968). The bat, after listening to the echo frequency , decreases the frequency of the next emitted sound in such a way that the carrier frequency of the echoes is kept close to the center frequency of the cochlear filter at 83 kHz. The bat maintains the echo frequency with

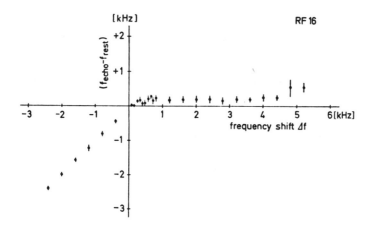

Fig. 9. Doppler-shift compensation in horseshoe bats (Rhinolophus
ferrumequinum). Ordinate: Difference between heard echo frequency
(f_{echo}) and frequency emitted (f_{rest}). Abscissa: experimentally induced frequency shifts of the echo (f). For frequency shift up
to 4 kHz above f-rest the bat keeps the echofrequency close to f-
rest by lowering the emitted frequency (horizontal sequence of dots
in the upper right quadrant). Frequency shifts below f-rest are not
compensated for (lower, left quadrant). (From Schuller et al.,1974).

an accuracy of $\pm$ 50 Hz (Schuller et al., 1974; Fig. 9.). This sophisticated feed-back system for locking the echo carrier frequency to
the center frequency of the receiver (f_{rest} in Fig. 9.) only functions when emitted sound and returning echo overlap by at least
15 msec (Schuller, 1977). The capacities of the feed-back control
system have been elegantly described by Schuller et al. (1975).
Ingeniously the feed back system reacts so sluggishly that it only
compensates frequency shifts of the complete echoes and never reacts
to short shifts occurring within the pure tone part which actually
are the signals the bat needs to hear.

All these many specializations of mechanical and neuronal nature enumerated in this chapter occur between cochlea and inferior
colliculus and converge into a highly sophisticated echolocaction
system specialized in detecting moving prey. The adaptations to moving prey detection so far described are: 1) sharply tuned mechanical filter in the cochlea matched to the 83 kHz carrier frequency
of the echo; 2)an acoustical fovea in the cochlea matched to the
filter frequency range and subsequent overrepresentation of the
filterfrequencies throughout the auditory nervous system; 3) high
sensitivity to minute frequency modulations of the echo carrier frequency; 4) lack of inhibition and facilitation of neuronal respon-

ses to echoes by frequencies of preceeding emitted sound (Fig. 8.);
5) conditioning of auditory neurons to FM-stimuli by vocalizing an
echolocation sound; 6) Dopplershift compensation (Fig. 9.).

A final question has to be answered. Many insectivorous bats
hunt insects on the wings. Why do only a few of them (e.g. mustache bat and horseshoe bat) emit long pure tones exploiting the encoding capacities of this signal for movements?
For a bat hunting in open air any echo received may indicate
a potential prey (or predator!) since obviously no nonflying object
will be on air. Thus for such a bat there is no need of differentiating living prey against stationary objects.
In contrast, a bat searching for flying insects close to foliage or any other dense background will receive a sequence of time
smeared echoes destroying the signal structure (e.g. FM-sweeps)
and deteriorating the detectability of the prey. However, time
smeared echoes of a pure tone signal retain the signal structure,
namely a pure tone, as long as the background is stationary whereas
any moving object in front of the dense target will "pop up" in the
pure tone as a complex small modulation. Even when the background
is randomly moving (e.g. foliage) wing beating prey may be easily
discriminated against the random modulations of the echo (caused by
background movements) by the periodicity of modulations coming from
beating wings. The latter argument may be also the reason why these
bats use very long pure tones in search flights.

Thus it appears that the use of a long pure tone component in echolocation is an adaptation for hunting flying insects close to dense
backgrounds (see also Glaser,1974 and Brosset, 1966). Of course
this movement detecting system only works when the receiver is adapted to the pure tone as is the case in horseshoe bats and in mustache bats (Fig. 1.).

INTERACTION OF VOCALIZATION AND AUDITORY SYSTEM

Vocalization centers may control auditory input in two ways:
a)peripherally by eliciting contraction of middle ear muscles just
prior to or during vocalization or
b) directly by exerting some neuronal influence onto auditory neurons.

Both ways have been shown to exist: the peripheral one by Henson (1965) and Suga and Jen (1975) and the central one by Suga and
Schlegel (1972) and Suga and Shimozawa (1974). In Myotis grisescens
the response of lateral lemniscal neurons to vocalized echolocation
sounds was attenuated up to 25 dB compared to responses to an identical play-back. Since this attenuation was never observed in summating potential recordings of the auditory nerve (N_1) this neuronally conveyed damping influence elicited by vocalization commands

Fig. 10. Vocalization conditions responsiveness of collicular neurons in Rhinolophus ferrumequinum to frequency modulated signals.
PST-histograms to a vocalized echolocation sound alone (VOC); to
an artificial pure tone at the same frequency and intensity as the
vocalized pure tone part of the echolocation sound (AS); to a
simultaneous presentation of vocalization and the pure tone (VOC+
AS) ; to a sinusoidally frequency modulated tone with carrier frequency equal to that of the vocalized tone and a modulation depth
of ± 500 Hz (AS(FM)); to a combination of vocalization and frequency modulated signal (echo) (VOC+AS(FM)); and to a combination of
a pure tone at the frequency of the vocalized pure tone part and
the frequency modulated signal (AS+AS(FM)). Note: Frequency modulation is strongly encoded during an ongoing vocalization. The
same neuron fails to respond at all when stimulated with the same
signals but without vocalization (graphs in the lowest row).
(From Schuller, in press.).

has to occur somewhere after the auditory nerve and before the inferior colliculus. This neural attenuation serves a general protection of the ear from selfstimulation by vocalization.

    Recently Schuller (in press) reported fascinating specific and
facilitatory influences of vocalization commands on auditory neurons. In the inferior colliculus many units responded differently
to a vocalized and to an identical nonvocalized signal. This vocal
effect was not due to middle ear muscle activities. Among those neurons influenced by vocalization commands were some specialized to

respond to an FM-signal (mimicking an echo returning from a wing
beating insect) only when the bat vocalized (Fig. 10.). If the same
neuron was stimulated with the same stimulus set in the nonvocali-
zing bat it did not respond at all. In fact the vocalized signal
could not be replaced by any other identical or slightly varied
signal in order to evoke the FM-encoding capacity of the neuron.
To avoid any misinterpretation it has to be emphasized that only
few neurons showed this specific behavior.

Apparently these neurons are sensitized to FM-stimuli by activities of the vocalization system. This enhancing effect is lost
when the FM-stimulus is presented 40 to 60 msec after onset of vocalization or after the emitted sound is terminated. Since the duration of the vocalized echolocation sound could not be manipulated
it is not clear if the sensitizing effect is exerted as long as vocalization endures or if it operates during a fixed time window of
40 to 60 msec triggered by onset of vocalization.

These new results open up an entirely new aspect of auditory
physiology. The interaction of vocalization and auditory system
has to be studied neurophysiologically and neuroanatomically in
great detail. These linking pathways are not only of interest for
a Doppler-shift compensating or any other echolocating bat but also
for each acoustically communicating mammal and especially for humans. The few results already available should us make careful in
transforming responses of auditory neurons into behavioral relevance.

ACKNOWLEDGEMENTS

This research was supported by DFG Br 593/2, Ne 146/9 and A.v.Hum-
boldt - Foundation.
I thank Rosa Veltlin and Beatus Habkern for valuable assistance in
preparing this review.

REFERENCES

Brosset, A., 1966, La biologie des Chiropteres, Masson, Paris.
Bruns, V., 1976a, Peripheral auditory tuning for fine frequency
      analysis by the CF-FM bat, Rhinolophus ferrumequinum. I.Mecha-
      nical specializations of the cochlea. J.comp.Physiol.A 106,
      77-86.
Bruns, V., 1976b, Peripheral auditory tuning for fine frequency
      analysis by the CF-FM bat, Rhinolophus ferrumequinum. II.Fre-
      quency mapping in the cochlea. J.comp.Physiol.A 106, 87-97.
Bruns, V., in prep., Adaptations for frequency sharpening in the
      cochlea of the Greater Horseshoe Bat. II. Functional morpho-
      logy of the basilar membrane and its anchoring system.

Bruns, V., and Goldbach, M., in prep., Adaptations for frequency
    sharpening in the cochlea of the Greater Horseshoe Bat. I.Re-
    ceptors and tectorial membrane.
Bruns, V., and Schmieszek, E., in prep., Regional variation in the
    innervation pattern in the cochlea of the Greater Horseshoe
    Bat.
Firbas, W., and Einzinger, H., 1972, Über das Ganglion spirale der
    Chiroptera. Z.Säugetierk. 37, 321–326.
Glaser, W., 1974, Zur Hypothese des Optimalempfangs bei der Fleder-
    mausortung. J.comp.Physiol. 94, 227–248.
Goldman, L.J., and Henson, O.W., 1977, Prey recognition and selec-
    tion by the constant frequency bat, Pteronotus p. parnellii.
    Behav.Ecol.Sociobiol. 2, 411–420.
Grinnell, A.D., 1973, Neural processing mechanisms in echolocating
    bats, correlated with differences in emitted sounds. J.Acoust.
    Soc.Am., 54, 147–156.
Grinnell, A.D., and Hagiwara, S., 1972, Adaptation of the auditory
    system for echolocation: studies of New Guinea bats. Z.vergl.
    Physiol. 76, 41–81.
Henson, M.M., 1978, The basilar membrane of the bat, Pteronotus p.
    parnellii. Amer.J.Anat. 153, 143–158.
Henson, O.W., 1965, The activity and the function of the middle-ear
    muscles in echolocating bats. J.Physiol. 180, 871–887.
Irving, R., and Harrison, J.M., 1967, The superior olivary complex
    and audition: a comparative study. J.comp.Neurol., 130, 77–86.
Jen, P.H.-S., 1978, Electrophysiological properties of auditory
    neurons in the superior olivary complex of echolocating bats.
    J.comp.Physiol., 128, 47–56.
Jen, P.H.-S., and Suga, N., 1974, Coding of directional information
    by neurons in the superior olivary complex of bats. J.Acoust.
    Soc.Amer., 55, S52.
Manley, J.A., and Müller-Preuß, P., 1978, Response variability of
    auditory cortex cells in the squirrel monkey to constant acou-
    stic stimuli. Exp.Brain Res., 32, 171–180.
Möller, J., 1978, Response characteristics of inferior colliculus
    neurons of the awake CF-FM bat, Rhinolophus ferrumequinum.
    II.Two-tone stimulation. J.comp.Physiol. A, 125, 227–236.
Möller, J., Neuweiler, G., and Zöller, H., 1978, Response characte-
    ristics of inferior colliculus neurons of the awake CF-FM bat,
    Rhinolophus ferrumequinum. I. Single-tone stimulation. J.comp.
    Physiol. A, 125, 217–225.
Neuweiler, G., 1970, Neurophysiologische Untersuchungen zum Echo-
    ortungssystem der Großen Hufeisennase Rhinolophus ferrum equi-
    num. Z.vergl.Physiol., 67, 273–306.
Neuweiler, G., and Vater, M., 1977, Response patterns to pure tones
    of cochlear nucleus units in the CF-FM bat, Rhinolophus fer-
    rumequinum. J.comp.Physiol. A., 115, 119–133.
Ostwald, J., 1978, Tonotope Organisation des Hörcortex der CF-FM-
    Fledermaus Rhinolophus ferrumequinum. Verh.Dtsch.Zool.Ges.,
    1978, p.198.

546 G. NEUWEILER

Poljak, S., 1926, Untersuchungen am Octavussystem der Säugetiere und an dem mit diesem koordinierten motorischen Apparaten des Hirnstammes. J.Psychol.Neurol. 32., 170-231.

Pollak, G., Henson, O.W., and Johnson, R., in press, Multiple specializations in the peripheral auditory system of the CF-FM bat, Pteronotus parnellii. (J.comp.Physiol.).

Pollak, G., Henson, O.W., and Novick, A., 1972, Cochlear microphonic audiograms in the pure tone bat Chilonycteris parnellii parnellii. Science, 176, 66-68.

Pollak, G., Marsh, D., Bodenhamer, R., and Souther, A., 1978, A single unit analysis of inferior colliculus in unanesthetized bats: response patterns and spike count functions generated by CF and FM sounds. J.Neurophysiol., 41, 677-691.

Pollak, G., and Schuller, G., in prep., Encoding properties of collicular neurons for sinusoidal frequency modulations at different intensities in Rhinolophus ferrumequinum.

Pye, A., 1964, A comparative anatomical study of the auditory apparatus of selected members from the orders Chiroptera and Rodentia. Ph.D. Thesis, London Univ.

Pye, A., 1966, The structure of the cochlea in Chiroptera. I.Microchiroptera: Emballoruroidea and Rhinolophoidea. J.Morph. 118, 495-510.

Pye, A., 1967, The structure of the cochlea in Chiroptera. III.Microchiroptera: Phyllostomatidae. J.Morph. 121, 241-254.

Radtke, S., 1979, Struktur und Verschaltung des Hörcortex der Grossen Hufeisennase (Rhinolophus ferrumequinum). Staatsex.Thesis, Univ. Frankfurt.

Ramprashad, F., Money, K.E., Landolt, J.P., and Laufer, J., 1978, A neuro-anatomical study of the cochlea of the little brown bat, Myotis lucifugus. J.Comp.Neurol. 178, 347-364.

Scheich, H., 1977, Central processing of complex sounds and feature analysis, in "Recognition of complex acoustic signals" Th.H. Bullock ed., Dahlem Konferenzen, Berlin.

Schlegel, P., 1977, Directional coding by binaural brainstem units of the CF-FM bat, Rhinolophus ferrumequinum. J.comp.Physiol. 118, 327-352.

Schlegel, P., in prep., Coding of moving sounds by binaural brainstem units in the Greater Horseshoe Bat, Rhinolophus ferr.

Schnitzler, H.-U., 1968, Die Ultraschall-Ortungslaute der Hufeisen-Fledermäuse in verschiedenen Orientierungssituationen. Z.vergl. Physiol. 57, 376-408.

Schnitzler, H.-U., Suga, N., and Simmons, J.A., 1976, Peripheral auditory tuning for fine frequency analysis by the CF-FM bat, Rhinolophus ferrumequinum. III. Cochlear microphonics and auditory nerve responses. J.comp.Physiol. A, 106, 99-110.

Schuller, G., 1972, Echoortung bei Rhinolophus ferrumequinum mit frequenzmodulierten Lauten. Evoked potentials im Colliculus inferior. J.comp.Physiol. 77, 306-331.

Schuller, G., 1974, The role of overlap of echo with outgoing echolocation sound in the bat Rhinolophus ferrumequinum. Naturw.

61, 171-172.

Schuller, G., 1979, Coding of small sinusoidal frequency and amplitude modulations in the inferior colliculus of "CF-FM" bat, Rhinolophus ferrumequinum. Exp. Brain Res. 34, 117-132.

Schuller, G., in press, Vocalization influences auditory processing in collicular neurons of the CF-FM bat, Rhinolophus ferrumequinum. (J. Comp. Physiol.).

Schuller, G., Beuter, K., and Rübsamen, R., 1975, Dynamic properties of the compensation system of Doppler shifts in the bat Rhinolophus ferrumequinum. J. Comp. Physiol. 97, 113-125.

Schuller, G., Beuter, K., and Schnitzler, H.-U. 1974, Response to frequency shifted artificial echoes in the bat Rhinolophus ferrumequinum. J. Comp. Physiol. 89, 275-286.

Schuller, G., and Pollak, G., in press, Disproportionate frequency representation in the inferior colliculus of horseshoe bats: evidence for an acoustic fovea. (J. Comp. Physiol.)

Schweizer, H., 1978, Struktur und Verschaltung des Colliculus inferior der Großen Hufeisennase (Rhinolophus ferrumequinum). Diss.-Thesis, University of Frankfurt, Germany.

Shimozawa, T., Suga, N., Hendler, P., and Schuetze, S., 1974, Directional sensitivity of echolocation system in bats producing frequency modulated signals. J. Expl. Biol. 60, 53-69.

Simmons, J. A., 1971, Echolocation in bats: signal processing of echoes for target range. Science 171, 925-928.

Suga, N., 1964, Single unit activity in the cochlear nucleus and inferior colliculus of echolocating bats. J. Physiol. 172, 449-474.

Suga, N., 1969, Classification of inferior collicular neurons of bats in terms of responses to pure tones, FM sounds, and noise bursts. J. Physiol. 200, 555-574.

Suga, N., 1973, Feature extraction in the auditory system of bats, in: "Basic Mechanisms in Hearing", A. R. Møller, ed., pp. 675-744. Academic Press, New York and London.

Suga, N., and Jen, P.H.-S., 1976, Disproportionate tonotopic representation for processing CF-FM sonar signals in the mustache bat auditory cortex. Science, 194, 542-544.

Suga, N., and Jen, P.H.-S., 1975, Peripheral control of acoustic signals in the auditory system of echolocating bats. J. Expl. Biol. 62, 277-311.

Suga, N., Neuweiler, G., and Möller, J., 1976, Peripheral auditory tuning for fine frequency analysis by the CF-FM bat, Rhinolophus ferrumequinum. IV. Properties of peripheral auditory neurons. J. Comp. Physiol. A, 106, 111-125.

Suga, N., and Schlegel, P., 1972, Neural attenuation of responses to emitted sounds in echolocating bats. Science, 177, 82-84.

Suga, N., and Shimozawa, T., 1974, Site of neural attenuation of responses to self-vocalized sounds in echolocating bats. Science, 183, 1211-1213.

Tobias, J. V., Ed., 1970, Foundations of modern auditory theory. Academic Press, New York and London.

Vater, M., in prep., Responses of cochlear nucleus neurons to comp-
    lex stimuli in the CF-FM bat, Rhinolophus ferrumequinum.
Vater, M., and Schlegel, P., 1979, Comparative auditory neurophysio-
    logy of the inferior colliculus of two Molossid bats, Molossus
    ater and Molossus molossus. II. Single unit responses to fre-
    quency-modulated signals and signal and noise combinations.
    J. Comp. Physiol. A, 131, 147-160.
Vater, M., Schlegel, P., and Zöller, H., 1979, Comparative auditory
    neurophysiology of the inferior colliculus of two Molossid bats,
    Molossus ater and Molossus molossus. I. Gross evoked potentials
    and single unit responses to pure tones, J. Comp. Physiol. A,
    131, 137-146.
Wilson, J. P., 1977, Towards a model for cochlear frequency analysis,
    in: "Psychophysics and Physiology of Hearing", E. F. Evans and
    J. P. Wilson, Eds., Academic Press, New York and London.
Zook, J. M., and Casseday, J. H., in press, Identification of audi-
    tory centers in lower brain stem of two species of echolocating
    bats: evidence from injection of horseradish peroxidase into
    inferior colliculus.
Zook, J. M., Siegel, B. M., and Casseday, J. H., in prep., Projec-
    tions to inferior colliculus in the mustache bat, Pteronotus
    parnellii parnellii as seen by retrograde transport of horse-
    radish peroxidase.

ORGANIZATIONAL AND ENCODING FEATURES OF SINGLE NEURONS IN THE

INFERIOR COLLICULUS OF BATS

George D. Pollak

Department of Zoology, University of Texas

Austin, Texas 78712 U.S.A.

INTRODUCTION

The past five years have witnessed a remarkable increase in the number of studies concerned with the neural basis of echolocation. These studies have dealt with features of the peripheral auditory system (Bruns 1976a,b; Henson 1978; Pollak et al. 1979; Schnitzler et al. 1976; Suga et al. 1975), the lower central parts of the auditory pathway, such as the cochlear nucleus (Neuweiler and Vater 1977) and superior olivary complex (Jen 1978), as well as with higher regions such as the inferior colliculus (Möller et al. 1978a,b; Schuller, 1979a; Schuller and Pollak 1979) and cortex (reviewed by Suga). But beyond the sheer proliferation, the reports have provided much more sophisticated and detailed insights into the principles that underly both the structural and functional organization of the Chiropteran auditory system than were known before. In the previous section, Dr. Neuweiler reviewed the data from studies of the cochlea and lower brain stem centers of the horseshoe bat whereas in the following section Drs. O'Neill and Suga have summarized their findings from the cortex of the long CF/FM mustache bat. In this chapter, I shall attempt to bridge the intervening gap by reviewing the results obtained from experiments on the inferior colliculus.

Recent neurophysiological studies have focused on two types of echolocating bats. One type is characterized by bisonar cries which are composed primarily of loud but brief downward sweeping frequency modulated components. I shall refer to this group as the FM bats. Among these are bats of the genus _Myotis_ and _Tadarida_, the two genera which have been most intensively studied.

The second type are the long CF/FM bats so called because
their echolocation calls are composed of a long constant frequency
(CF) component followed by a brief downward sweeping FM portion.
The best known of the long CF/FM bats are the greater horseshoe
bat, Rhinolophus ferrumequinum, and the mustache bat, Pteronotus
parnellii.   Since the behavior as well as the neurophysiology differ in a number of significant ways between the FM as compared to
the long CF/FM bats, I shall treat them in separate sections.   In
the section below I shall first review the results from neurophysiological studies conducted on FM bats and in the subsequent section
results from recent studies concerned with the processing of the
CF component in long CF/FM bats will be presented.

FM BATS
Tuning Curves

One of the fundamental characteristics of all auditory neurons
is the tuning curve. The tuning curve defines the range of frequencies capable of evoking discharges in a given neuron and thus
is a reflection of the range of supra-threshold excitatory synaptic
inputs, originating in the cochlea and auditory nerve, that are
mapped upon an individual neuron. The majority of tuning curves
obtained from collicular neurons are more or less triangular in
shape with the frequency at the apex of the triangle requiring the
lowest intensity to evoke discharges. (Grinnell 1963a; Suga 1964b;
Pollak et al. 1978). This frequency will hereafter be called the
best frequency (BF) of the neuron.

The tuning curves of the population of collicular neurons
differ with regard to their widths (Suga 1964b; Pollak et al. 1978).
Some are quite wide and receive excitatory inputs from a large
region of the cochlea while others are narrower reflecting a more
restricted excitatory input from the cochlea. The sharpness of a
tuning curve is typically expressed as a Q value, defined as the
BF divided by the bandwidth of the tuning curve at a level 10 db
above the BF. Hence, a high Q value means that the tuning curve
is very sharp while a small Q value means that the tuning curve
is broad.

In FM bats, such as Myotis and Tadarida, the majority of
collicular neurons have relatively broad tuning curves (Suga
1964b; Pollak et al. 1978). The Q values typically range from
5 to 10 or 15 with only a very small number of neurons having Q
values above 20, i.e., only a small segment of the population is
sharply tuned.

Auditory neurophysiologists commonly interpret the sharpness of tuning as an indication of how well the system can discriminate small frequency differences. A neuron having a small

Q value is thought to be poorly suited for distinguishing or recognizing small frequency differences because the cell responds to a wide range of frequencies. Thus, an organism whose nervous system is composed only of broadly tuned neurons is considered to have a lesser ability to discriminate small frequency differences than is an animal having a large population of very sharply tuned neurons. Although this interpretation may be true for some neurons, I shall present evidence in a later section which demonstrates that in some neurons at least, the tuning curve has no predictive value for ascertaining a neuron's frequency selectivity.

Tonotopic Organization

Whereas the tuning curve is a measure of the extent to which the cochlea maps upon an individual neuron, the manner in which BFs vary as a function of position within the IC provides an index of how the cochlea maps upon the entire expanse of this nuclear region. Studies of FM bats indicate that the IC is tonotopically organized where the mapping of frequency is expressed as a systematic shift in the BF. In both Myotis (Grinnell 1963a) and Tadarida (Pollak et al. 1978) low frequencies are represented in the dorsal portions of the IC while higher frequencies are represented in progressively more ventral parts of the colliculus. Frequency, then, is mapped upon the IC in a systematic fashion which, in principle, is typically mammalian and bears a striking resemblance to the tonotopic organization of the colliculus of other mammals that have been studied (Aitkin 1976).

Discharge Patterns Evoked by Tone Bursts

When tone bursts at the frequency of the neuron's BF are presented, a wide variety of discharge patterns can be evoked among the population of collicular neurons (Suga 1973; Pollak et al. 1978; Vater et al. 1979; Vater & Schlegel 1979). These patterns however fall into two general categories: 1) phasic patterns and 2) tonic or sustatined patterns. Within the category of phasic responders, three basic types are recognized (Pollak et al. 1978): phasic on neurons, phasic on-off neurons, and phasic burst neurons. Several varieties of sustained patterns can also be distinguished. The two basic types are found in roughly the following proportions: phasic 78%, sustained 22%.

I shall not discuss further the discharge features evoked by tone bursts. The reason is that very different results obtain when FM bursts, simulating the natural orientation cries, are presented (Pollak et al. 1977c, 1978). These differences are reflected in two general ways. First, there is a change in the proportion of units having phasic as compared to tonic discharge patterns. Whereas about 20% of the neurons have tonic patterns when driven with tone bursts, this pattern never occurs with FM signals. Second, a unit having a phasic pattern to both FM and tone bursts may exhibit radical differences in latency, synchrony and firing rate changes with intensity depending upon which signal is presented. In short, the encoding features of a neuron may be, and usually are, considerably different with FM signals as compared to its discharge characteristics observed with tone bursts. The important point is that since loud FM bats use, for the most part, brief FM signals for echolocation the response features evoked by tone bursts cannot be invoked in explanations of how neurons encode information extracted from echoes.

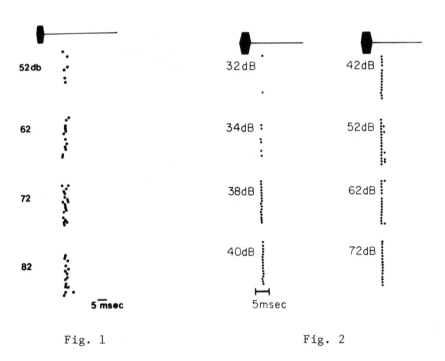

Fig. 1   Fig. 2

Fig. 1. pELR pattern evoked by 4.0 msec FM bursts that swept downward from 50–25 kHz. Threshold was about 47 dB. Latencies were 52 dB, 14.9 ± .98 msec; 62 dB, 13.87 ± 1.08 msec; 72 dB, 13.23 ± .73 msec; 82 dB, 13.06 ± .65 msec.

Fig. 2. An example of a pCLR pattern. Signals were 4.0 msec FM bursts that swept downward from 50–25kHz. Threshold was 32dB. Latencies were 34 dB, 7.83 msec ± 245 µsec; 38 dB, 7.20 msec ± 150 µsec; 72 dB, 6.90 msec ± 117 µsec. Note overall latency change determined 2dB above threshold (34dB) and at 40 dB above threshold (72dB) was only 0.93 msec. From Pollak et al. 1978.

Discharge Patterns Evoked by FM Signals

Brief downward sweeping FM signals evoke predominately phasic on discharges from collicular neurons in Mollosids (Pollak et al. 1978; Vater and Schlegel 1979). The phasic on patterns, however, are not homogeneous and can be divided into two subgroups. First, are neurons we have called phasic erratic latency responders (pELRs). The feature which characterizes these neurons is the large shifts in latency when an FM signal having a particular sweep, intensity and duration is repeatedly presented. (Fig. 1). The second type of phasic on pattern evoked by FM signals is quite dramatic and the significance of the response characteristics for echolocation is much better understood than for any other type of neuron. We call these phasic constant latency responders (pCLRs) (Suga 1970; Pollak et al. 1977c; Vater and Schlegel 1979) and they are distinguished by four features (Fig. 2): 1) a phasic on response pattern where the neuron fired at most two spikes to an FM burst; 2) a reliable firing to every or nearly every stimulus presentation for intensities 2-6 dB above threshold or greater; 3) a highly consistent latency of initial firing for a wide range of intensities; and 4) a relatively constant mean response latency where the average latency would change at most by 1.75 msec over a 40 dB intensity range.

It is of some interest to compare the response features of pCLRs when driven with tone bursts with the response features evoked by brief FM signals. Consider first that some pCLRs are unresponsive to any tonal stimulation while others will discharge to tone bursts (Pollak et al. 1978). Second, many, but not all, of the pCLRs that are responsive to tone bursts exhibit a pELR pattern to tone bursts but have a pCLR pattern to FM signals (Pollak et al. 1977c). In other words, in many units the discharges fall into tight registry with FM signals whereas when tone bursts are presented the firing latencies are erratic from stimulus to stimulus. Third, and most importantly, all pCLRs that can be driven with tonal stimulation have wide tuning curves (Fig. 3) where the Q values range from about 3-8 (Bodenhamer et al. 1979). The commonly accepted interpretation of such wide tuning is that these neurons are not frequency selective. However, and this point is critical, we can demonstrate that the tight registration of latencies is a direct consequence of an exceptionally high degree of frequency selectivity which only becomes apparent when the stimuli are FM bursts.

When FM signals are presented the pCLRs appear to respond to only one frequency or a very narrow band of frequencies in the FM signal. When such a brief FM burst is repeatedly presented, the neuron is fired by, and thus locks onto, the same frequency component or narrow frequency band each and every time and, as a consequence, the latencies fall into tight registration.

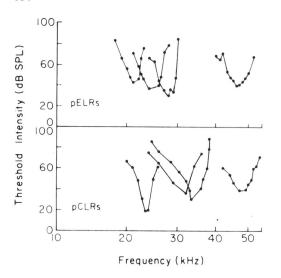

Representative tuning
curves of pELRs and pCLRs.

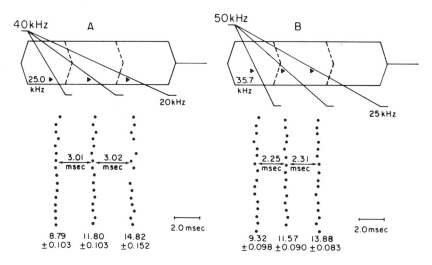

Fig. 3. pCLR sensitivity to a particular EF. Each dot column
represents spikes elicited by 16 presentations at a given duration. The dot columns have been aligned directly beneath the FM
signals to show how they line up with the proposed EF in each
sweep (arrow-heads). Each dot column's mean response latency and
standard deviation is shown below that column (in msec). In A,
the response latency increased by 3.01 msec. when signal duration
increased from 4.0 to 8.0 msec and by 3.02 msec when the duration
increased from 8.0 = 12.0 msec. The 25 kHz constant shifted by
3.01 msec for each change in duration & thus the neuron is
"following" the 25 kHz component of the FM pulse. A similar "following" is shown for another pCLR in B but this unit was following the 35.7 kHz component. From Bodenhamer et al. 1979.

This feature is illustrated by two pCLRs in Fig. 3. The rationale of this experiment is that if these neurons are synchronized to one frequency component in an FM burst, then the excitatory frequency (EF) will shift in time as the duration of the FM signal is increased and the discharge latency should shift by an equivalent amount of time. That this is indeed the case is demonstrated by the two units in Fig. 3.

The implications of these features for the processing of echolocation signals are considerable. These units are highly selective for both temporal and spectral attributes of the biosonar signals and are likely candidates for encoding both target distance or range as well as the spectral composition of the echo which could be important for target identification. Below, I shall discuss these features in greater detail.

The fact that pCLRs record the temporal event for a particular frequency in an FM burst means that these neurons can serve as precise time markers. That is to say, such neurons should encode the time of occurrence of a particular frequency component in the emitted pulse and the time of occurrence of the same frequency component in the returning echo. The period between the occurrences of a frequency in the pulse and echo is a function of the pulse-echo delay, i.e., target distance. As shown in Fig. 4, pCLRs can encode the temporal separation between two FM signals with remarkable precision and, as a consequence, are almost certainly conveying distance information to higher auditory centers.

ipi 6.0 msec

**FM 50-25 kHz
4 msec signals**

T, 62 db

T, 62db

6.0 msec

Fig. 4. An example of the precision with which pCLRs can encode the temporal separation between two FM signals.

The timing precision illustrated by the pCLR in Fig. 4 is a
necessary but not a sufficient demonstration to warrant a conclusion that pCLRs are encoding range information. The two signals in Fig. 5 were both of equal intensity and were presented
at only one time interval. Since range must be evaluated at a
variety of distances and with a wide variety of echo intensities,
it is important to demonstrate that discharge characteristics,
similar to those shown in Fig. 4, can be elicited with signals
simulating the spectrum of pulse-echo combinations that would
occur during echolocation.

Two features of the pCLRs are important in this regard.
First, is that the EF observed for a pCLR is invariant with intensity (Bodenhamer et al. 1979). Thus, the same frequency in both
the loud pulse and fainter echo will cause the neuron to fire and
this feature will serve to preserve the timing accuracy. The
second feature concerns the fast recovery times of pCLRs where
the presentation of a loud initial signal has almost no influence upon the response to a much fainter second signal following shortly thereafter (Pollak et al. 1977a,b,c). These

Fig. 5. Recovery in two collicular neurons. All signals were FM
bursts sweeping downward for 50-25 kHz. The intensity of the first
and second signals are shown as T_1 and T_2 respectively. Both neurons are phasic constant latency responders (pCLRs). Note the
fast recovery & the precision with which the response intervals
follow the inter-pulse-interval (IPI). Each pair of dot columns
was generated by 16 repetitions of a simulated pulse-echo-combina-
tion.

features, then, allow the pCLRs to encode the temporal period
between the pulse and echo under most of the conditions occuring during echolocation, as shown in Fig. 5.

The precision with which pCLRs encode the temporal interval
between two FM bursts, however, is influenced by the relative
intensities of the two signals. Louder signals evoke discharges
at shorter latencies than do fainter signals. It follows then
that a loud pulse followed shortly thereafter by a fainter echo
will, in any given pCLR, evoke two spikes but the discharge interval will be slightly longer than the temporal interval of the
signals. In this regard, it is most important to note that one
feature of the pCLRs is that their discharge latencies increase
on the average by only 0.75 msec over a 30 dB range of intensities
(Pollak et al. 1977c). In other words, if the bat emits a pulse
where the intensity reaching the cochlea is 82 db and receives
an echo 5 msec later having an intensity of 52 dB, the most
common pCLR will respond with discharges separated by 5.75 msec.
The unit, then, would encode a distance of 98 cm instead of the
true distance of 86 cm, an error of 12 cm. The error depends
only upon the pulse-echo intensity difference, the smaller the
difference the less the timing error. The important point is that
the possible range of errors is sharply limited because of the
small latency variation with intensity.

Such errors in an individual neuron are not difficult to re-
concile with the ranging precision required of bats. The argument is that the timing error depends only upon pulse-echo
intensity differences and since the bat encounters these errors
throughout life, it must somehow learn to make appropriate
corrections. In short, the bat could easily associate each pulse-
echo intensity difference with a given error and thereby correct
for it.

The spectrum of the biosonar signals could also be encoded by
the pCLRs. Consider that each pCLR discharges to only one frequency component in the FM sweep and that pCLRs have been found
with EFs from 24.5 to 45 kHz. Since the IC is tonotopically
organized, it follows that the pCLRs tuned to the various frequencies in the echolocation cries are topographically organized
with neurons discharging to the lower frequencies being situated
more dorsally while neurons discharging to the higher frequencies
in the biosonar signals being situated more ventrally.

The echoes bats receive from different targets vary in spectral
fine structure depending on the nature of the target. These echoes
may resemble filtered replicas of the emitted pulse with some targets acting primarily as high pass filters, others as low pass filters
and others as variable notch filters (Griffin, 1967). Such echo differences are important cues for target evaluation and since each pCLR is

triggered by only a small frequency band, a population of these cells, each tuned to a slightly different EF and arranged in a topographic manner within the IC, should be capable of encoding the fine structure of these echoes by a place code within the brain and hence be of value for encoding target features.

In summary, the pCLR's sensitivity to a particular EF is valuable to FM bats in at least two ways. First, it leads to a tight firing registration enabling the pCLRs to serve as pulse-echo time markers, and thereby encode target range information. Second, since each pCLR is fired by only a narrow frequency band of the FM signal, populations of pCLRs should be able to encode echo spectral characteristics and thus convey information about the physical nature of targets as well as target range.

Further Coding of Target Range

A very recent report by Feng et al. (1978) is highly suggestive of the way in which range information may be processed by higher centers. These workers recorded from individual neurons in the intercollicular nucleus, lying between the inferior and superior colliculi, of the loud FM bat, Eptesicus fuscus. They found neurons that are unresponsive to any tonal or FM stimuli but which would vigorously discharge when a pair of FM bursts were presented at a certain range of pulse-echo delays. The interpretation given to these results was that "target range may thus be encoded by a neural "place" mechanism incorporating different delay-tuned neurons in the central nervous system, presumably at levels above the inferior colliculus".

This is a most exciting hypothesis which, I believe, is likely to be the case. However, the preferred pulse-echo delay times exhibited by intercollicular neurons were quite broad. The width of the delay-tuning was typically 5-8 msec which would not satisfy the ranging requirements of echolocating bats.

Although I disagree with the authors' interpretation of their data, this should not distract from the importance of the finding. At the very least, neurons selective for a range of pulse-echo delays must be important for eliminating the clutter produced by echoes from nearer and more distant objects and may allow the bat to focus upon one, primary target of interest. Furthermore, the demonstration that a population of neurons discharge only to pairs of FM bursts having preferred temporal separations, opens the possibility that other neurons might have delay-tuning properties much sharper than the intercollicular neurons and these could be of value for the accurate encoding of target range, as Feng et al. originally proposed and Suga and O'Neill suggest in the following chapter.

LONG CF/FM BATS

Within the past several years considerable interest has been directed toward the long CF/FM bats. This interest has been generated by the demonstrations that these animals possess remarkable behavioral abilities as well as striking anatomical and physiological features which seem to correlate and provide a system designed to process the CF component of their biosonar cries.

The aspect of these bats that, I believe, has had a profound influence upon those who work with these animals is they emit a pure tone component (Novick 1963, 1971; Schnitzler 1968; Grinnell 1970; Grinnell & Hagiwara 1972). Following from this feature is the assumption that these bats are measuring the relative velocity between themselves and their targets by evaluating the amount of Doppler shift in the echo CF component of their echoes. When this is coupled with the fact that long CF/FM bats Doppler-compensate (Schnitzler 1968; Simmons 1973; Schuller et al. 1974), thereby holding echoes within a narrow band of frequencies only ± 50--100 Hz wide, one is led to conclude that they are interested in pure tones.

This is a most seductive idea. Tone bursts have traditionally been and continue to be the most widely used stimuli in studies of the auditory system and this is true of both vertebrates and invertebrates. Some of the historical reasons as well as the rationale for using tone bursts have been reviewed by Capranica (1972) and need not be repeated here. The point is that the long CF/FM bats emit pure tones and possess nervous systems which are thought by many investigators to be constructed for the evaluation of small frequency differences existing between two tonal signals. These bats, then, appear to be a mammalian subject for which tone bursts are the biologically relevant signals and hence are the stimuli of choice for neurophysiological investigations of the auditory system.

It seems to me, at least, that this is a major conceptual as well as a major strategic point, one which is of some interest to examine in greater detail. In many reports dealing with long CF/FM bats the argument is made that a CF signal is well suited for measuring Doppler shifts, i.e., relative velocity, due to the narrow bandwidth of the signal but, for the same reason, is poorly suited for target characterization, localization and target ranging (Simmons et al. 1975; Simmons 1979; Suga 1977, 1978). More recently, many workers have suggested that the filter system could also be important for prey recognition and identification by encoding the modulation patterns imposed on the echo CF component as it strikes and is reflected from the beating wings of a small insect (Schuller 1972, 1979a; Schuller & Pollak, 1979; Schnitzler 1970, 1978; Henson et al., 1974; Henson & Goldman 1976; Goldman & Henson 1977). Curiously, some investigators hold that the long

CF/FM bats are extracting both a precise estimate of relative velocity and modulation pattern information from the echo CF component
(Suga et al. 1975; Suga 1978). What makes this interpretation curious is that the two hypotheses are, in fact, incompatible. The
evaluation of relative velocity is based on the assumption that the
echo CF is a pure tone, and consequently of very narrow bandwidth,
whereas the prey recognition hypothesis is based on the fact that
echoes reflected from a small insect have periodic amplitude and
frequency modulations (AM and FM) which increase the echo bandwidth
and introduce a significant ambiguity into any velocity measurement.

Goldman and Henson (1977) have presented evidence strongly
indicating that the long CF/FM mustache bat chooses which among
a population of insects it will pursue and capture on the basis
of wing beat patterns. Furthermore, recent neurophysiological
studies have shown that neurons in the cochlear nucleus (Suga et
al. 1975; Suga and Jen 1977; Vater personal communication), inferior colliculus (Schuller 1972; 1979a; Pollak & Schuller 1978), and
auditory cortex (Ostwald, 1979) of long CF/FM bats are exceptionally well suited for encoding both AM and FM modulation patterns.
Considerations, then, both of the type of echoes the bats receive
during echolocation as well as neurophysiological features of
auditory neurons seem to support the idea that the primary function
of the filter system is for target characterization rather than for
relative velocity estimates.

In the following sections I shall review the results obtained
from recent single unit studies of the horseshoe bats' inferior
colliculus conducted in collaboration with Gerd Schuller. It is my
intention to show the advantage that accrues to Doppler-shift compensation for target characterization in light of the organizational features of the IC together with the manner in which collicular
neurons encode FM and AM modulation patterns.

Tuning Curves

Two general types of tuning curves are obtained from collicular
neurons in horseshoe bats: those that are broadly tuned and those
that are sharply tuned (Möller et al. 1978). The broadly tuned
neurons typically have BFs from 9 kHz to about 75-77 kHz. These
neurons rarely have Q values greater than 20 and in this respect
are similar to the collicular neurons in FM bats. In marked contrast, neurons tuned to the filter frequencies, i.e., 78-86 kHz,
are usually very sharply tuned and have Q values between 50-200
with some as sharp as 400-500. The demarcation point is quite
striking, as shown in Fig. 6, and it is this feature of the neurons
tuned between about 77-88 kHz that justifies the term "filter
neurons".

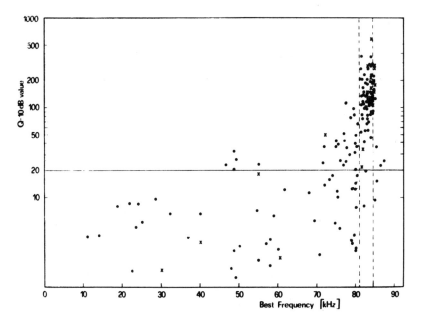

Fig. 6. Q values of collicular neurons recorded from horseshoe bats. From Möller et al. 1978.

Tonotopic Organization of the Horseshoe Bats' Inferior Colliculus

The finding that collicular neurons can be grouped into filter and non-filter neurons is also reflected in the tonotopic organization of the IC. The IC can be functionally divided into a dorsal region, where the BFs range from 9kHz to 75-76 kHz, and a ventral region where the filter frequencies are found (Fig. 7).

The tonotopic arrangement in the dorsal region is typically mammalian and is similar to that found in the IC of FM bats. Proceeding from the dorsal to the more ventral region, the neurons encountered have progressively higher BFs to a depth of approximately 800-1200 microns below the collicular surface. At that point there is an abrupt change in the topographic mapping of the cochlea upon the IC. For the final 1000-1200 microns of depth the tonotopy assumes a quasi-cortical appearance where almost all neurons along a particular vertical axis have the same or nearly the same BFs (Figs. 7 & 8).

Only filter frequencies are represented in the ventral, quasi-cortical region of the IC. For this reason, I shall hereafter refer to this region of the IC as the filter region. The tonotopic organization of the filter region is in the anterior-posterior axis

Fig. 7. Sagittal sections through the inferior colliculus of a
horseshoe bat showing the BFs of the neurons encountered in two
electrode penetrations. The regions, in the anterior-posterior axis,
from which these sections were obtained can be seen from lower sagittal section of one entire brain. Note the progressive increase in
BFs from dorsal surface to a depth of 700–900 microns and the relatively constant BFs for neurons at lower depths. Also note that the
BFs in the filter region were all 77–79.5 in the anterior region
whereas in the posterior region the BFs were 83.4–83.9. All electrode tracts were retracted from Alcian Blue Dye marks deposited from
the electrode. Some neurons were not driveable (ND) with acoustic
signals.

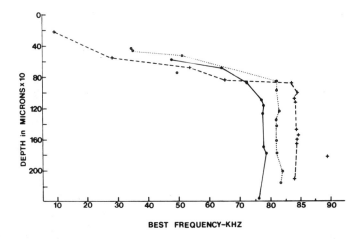

Fig. 8. Change in BFs with depth for three penetrations of the horseshoe bat's inferior colliculus. Note how similar the BFs in the filter region are within a given electrode penetration. These are different penetrations from those shown in Fig. 7.

where low frequencies (i.e., 77-80 kHz) are represented in the anterior portion while progressively higher frequencies (i.e., 81-86 or 87 kHz) are represented in the more posterior regions (Fig. 7).

It is important to point out that the arrangement described above holds true only for the central portion of the IC. A very different organization is found in both the medial and lateral margins of the IC. The tonotopy of the most medial edge of the IC is somewhat of a compromise between the dorsal and filter regions. The BFs in this region proceed from low to high with increasing depth but only the filter frequencies are represented. The BFs systematically increased with depth in this region but the range of BFs extends only from 78-89 kHz.

The most lateral region of the IC has yet a different organiza-tion. In this margin there is also a systematic progression of BFs with depth, from low to high, but the filter frequencies are only sparsely represented. This region, however, is truly striking but not for the tonotopy. Rather the neurons existing in vertical arrays within the lateral margin represent a homogeneous population in that they all have a high rate of spontaneous activity and, almost without exception, they respond to tone bursts with a pronounced inhibition of activity.

Fig. 9. (A) Histogram showing the number of units tuned to the frequencies comprising the audible range of horseshoe bats. The BFs are grouped and plotted in intervals of 1.0 kHz. (B) Histogram showing the distribution of units tuned to the filter frequencies where the BFs are grouped and plotted in intervals of 200 Hz. Note the precipitous increase in the number of neurons tuned between 83-84.5 kHz (cross-hatching). From Schuller and Pollak, 1979.

Proportional Representation of Audible Frequencies Within the IC

Previous studies have shown that approximately 50% of the neurons in any auditory center are tuned to the filter frequencies (Suga et al. 1976; Schlegel 1977; Möller et al. 1978). This result was verified in the IC by Schuller and Pollak (1978; 1979) (Fig. 9A). However, we had a systematic sample of the entire IC, having obtained the BFs of almost 600 neurons, and we therefore were able to plot the distribution of BFs for the filter frequencies in intervals of 200 Hz. The results, shown in Fig. 9B, are quite striking and we believe, are of fundamental importance. It is clear from this plot that the 83.0-84.5 kHz region has a much greater representation than do the other filter frequencies. This region of over-representation gains additional significance because the bats used in these studies had resting frequencies of 83.5-83.8 kHz. From studies of Doppler-shift compensation (Simmons 1973; Schuller et al. 1974) it is well established that horseshoe bats adjust their emitted CF components such that the echo frequencies are stabilized or held at the bat's reference frequency, a frequency 100-300 Hz above the emitted or resting frequency. Consequently, when Doppler-compensating these bats hold their echo CF component between 83.6–84.1 kHz. The advantage of DSC now becomes apparent. By manipulating the frequencies of their emitted CF components, horseshoe bats confine echoes to a very narrow frequency band and this behavior insures the bat that echoes will be processed by an exceptionally large number of sharply tuned filter neurons.

Doppler shift compensation then, can be thought of as, a mechanism for "foveation". In the visual system, eye movements are used to keep an image focused on the fovea. Horseshoe bats, on the other hand, move their voices, i.e., adjust the frequency of the CF component, in order to stabilize echoes on the "acoustic fovea". The "acoustic fovea" is the greatly expanded and highly specialized region of the cochlear partition devoted to the filter frequencies (Bruns 1976a,b) whereas the over-represented 83.0–84.5 kHz band is the "acoustic foveal" region of the brain.

The purpose of foveation, in vision, is for visual acuity. Similarly, it appears that DSC, or "acoustic foveation", is utilized for acoustic acuity; to enhance the resolving power for fine target features by devoting a disproportionately large number of neurons to the processing of the echo CF component.

The target features that the bat is attempting to resolve, most likely, are the frequency and amplitude modulation patterns which are created as the CF component is reflected from the beating wings of a small insect. These modulation patterns could well serve as a source of information for distinguishing flying insects from inanimate objects as well as for identifying which species the echo is reflected from.

Discharge Patterns to Frequency Modulated Signals

A few reports of evoked potentials (Schuller 1972) and recodings from cochlear nucleus (Suga et al. 1975; Suga and Jen 1977) neurons suggest that the nervous system of long CF/FM bats is well suited for encoding frequency modulation patterns. The ability to encode FM patterns was firmly established in a recent study by Schuller (1979a). He recorded the discharge patterns of neurons in the inferior colliculus of horseshoe bats when presented with sinusoidally frequency modulated signals (SFM). The truly spectacular synchronization of discharges to the modulation waveform, even when the depth of modulation is as small as $\pm$ 25 Hz, is illustrated in Fig. 14.

Neurons having firing patterns tightly locked to the modulating waveform are commonly encountered in the colliculus. However, most collicular neurons are quite selective with regard to some parameter of the modulating waveform to which they will synchronously discharge. This feature is perhaps best illustrated by considering how discharge synchrony is influenced by signal intensity.

Fig. 10.   Effect of stimulus intensity upon the locked discharges
to SFM signals in four collicular neurons. The neuron on the right
(unit 17-4-17) had tightly locked firings at all intensities above
threshold while the three other units each locked best to only a
small range of intensities.

        Intensity influenced locking behavior in two broad ways.
First are neurons in which the discharge registrations are about
equally secure at all intensities above threshold (Fig. 10).   The
second type of neuron exhibited the sharpest locking and most vigorous responding only at low or moderately low intensities and had a
significant decline or complete absence of locking at higher intensities (Fig. 10). Stated differently, these units discharged in
registry with the modulating waveform only over a preferréd range
of intensities. But notice from Fig. 10, that the preferred intensity
range differs slightly from neuron to neuron. The population, then,
forms a continuous gradient where some units encode the modulation
pattern equally well at all intensities above threshold whereas
others are more selective and only encode the modulation waveform
if the echo falls within a narrow intensity slot.

        Before considering how collicular neurons encode different mod-
ulation rates and depths, I should like to digress for a moment in

order to make an important point: the encoding features for SFM signals cannot always be predicted from the way in which the same neuron responds to tone bursts. In Fig. 11, for example, are shown
the responses to SFM signals centered at 85 kHz and the responses
to pure tones at 85 kHz, the BF of this neuron. Notice first that
the threshold for the 85 kHz tone burst was between 20-30 dB SPL
and that the discharges increased monotonically with intensity.
Both of these features were considerably different when SFM signals
were presented. The threshold for SFM was somewhat lower than 0 dB
and the firing rate, as seen in the amplitude of the histogram peaks,
first reached a maximum at 20 dB and then decreased, becoming neglig-
ably small at 50 dB.

The second parameter of the modulation waveform we varied was
the modulation rate which would simulate the various wing beat frequencies of different insects. As was the case for intensity, neurons exhibited a continuous gradation in their abilities to encode
modulation rate. Some neurons could accurately encode modulation
rates from 50 Hz to 500 Hz whereas most other units showed a preference for a range of rates (Fig. 12). In general the preferences were
expressed in a continuous shift among units in their capacity to encode progressively higher modulation rates. There is also evidence
suggestive of a synchronization in the opposite direction. In Fig.

Fig. 11. Differential behavior to SFM and tone burst stimulation.
Notice that threshold for SFM was about 25 dB lower than to tone
bursts at the neuron's BF. The neuron also responded monotonically
with intensity to tone bursts but was clearly non-monotonic to SFM.

Fig. 12.   Abilities of four collicular neurons to follow SFM
signals having different modulation rates.   Unit 36-5-16
(left) followed rates from 50-500 Hz with well synchronized
firings while unit 28-3-22 followed only up to 75 Hz.   Spike
scale for unit 28-3-22 is compressed for clarity.

13 for example, is a neuron which locked poorly to modulation rates
lower than 100 Hz but synchronized with reliable firings to modulation rates between 100 to about 250 Hz.   Notice that the modulation
depth was only ± 100 Hz.

      Another perturbation imposed upon the echo CF is the depth of
modulation.   The depth can vary considerably and depends upon the
size of the wings, wing beat frequency and the angle of the insect
with respect to the bat.   It is only fair to say that we have much
less data concerning modulation depth than we have about intensity
and rate.   However, the data obtained suggests that, in this case
also, neurons can be divided into two categories.   One category
discharges synchronously with the waveform at almost all depths
while the second category are more selective for a range of depths.
An example of the first type is shown in Fig. 14.

Fig. 13. Neuron exhibitory preference for higher modulation rates. The neuron responded poorly to 50 Hz rates but discharged with vigorous well synchronized firings at rates from 100-200 Hz and declined at higher rates. The modulation depth was + 100 Hz. From Schuller, 1979a.

Fig. 14. Example of a neuron able to respond with locked discharges for modulation depths ranging from + 2 kHz to + 25 Hz. From Schuller, 1979a.

The Effects due to Variations of the Carrier Frequency

All of the results presented above were obtained when the carrier frequency was set at the neuron's BF. However, the center frequency of the echo will not necessarily correspond to the BF of all neurons excited by the signal. Consider, for example, that upon Doppler compensating, a horseshoe bat holds the echo carrier around 84 kHz and that the FMs vary by $\pm$500 Hz around the carrier. Under these conditions, units having BFs of 84 kHz will be excited but so will other units having BFs both above and below 84 kHz. Consequently, it becomes of some interest to assess how filter neurons encode modulation patterns for a variety of different carrier frequencies. I shall refer to the locked responses across carrier frequency at a particular intensity as the SFM response area and the frequency range of tone bursts capable of evoking responses at a given intensity as the unit's response area (RA).

In Fig. 15 is shown how a typical neuron responds to SFMs having different carrier frequencies. The most striking feature of the SFM response areas of filter units is the very sharp boundaries on both the high and low frequency sides. This is especially evident in units which locked in a symmetrical manner around a carrier set at the unit's BF. In units having this type of symmetric SFM response area, the height of the locked discharge peaks correlated closely with the extent to which the SFM signals encroached upon the RA; the greater the encroachment the larger the response and if the signal failed to enter the RA, no discharges were evoked.

Other units have asymmetric SFM response areas where the locking, as judged by the amplitudes of the peaks in the histograms, was clearly superior on the low, or in other units, on the high frequency side of the BF. An asymmetric SFM response area where the neuron favored frequencies above the BF is shown in Fig. 16.

Response Patterns Evoked by Sinusoidally Amplitude Modulated (SAM) Signals

Neurons in the IC of horseshoe bats encode amplitude modulation patterns with discharges synchronized to the modulation waveform in a manner similar to the way in which they encode SFM patterns. An example of the degree of sensitivity for even very small modulation indices is shown in Fig. 17.

Intensity influenced the locking to SAM signals in a manner similar to that of SFM signals. Some units would lock to the SAM waveform over a wide range of intensities. However, most collicular units exhibited a preference for an intensity range

SFM CENTER FREQUENCY

MODULATING FREQUENCY 50 Hz   RF 212 UNIT 1
MODULATION ±2 kHz
CENTER FREQUENCY (kHz)

Fig. 15. Symmetric SFM response
area. The BF was 83.76 kHz
and the modulation depth was ±
500 Hz. When the carrier was
83.10 kHz, the SFM signal just
clipped the lower limit of the
RA (dashed lines) and elicited

Fig. 16.

some locked discharges. The locked firings became increasingly
vigorous as the carrier frequency progressively entered more of
the RA. As the carrier rose above the BF, the histogram peaks
decreased in amplitude and reached an upper limit for evoking responses when the carrier was 84.43 kHz. The unit failed to respond when the SFM carrier was 84.5 kHz, a signal that just failed
to enter any part of the RA.

Fig. 16. Asymmetric SFM response area. The bottom line describes
the time course of the modulation whereas the drawing to the left
of each histogram depicts the position of the modulation cycle
within the limits of the response area. Notice that when the
carrier was 78.0 kHz and crossed the neurons BF, hardly any firings
were evoked. L: lower limit of RA, U: upper limit of RA. From
Schuller, 1979a.

Fig. 17. Responses to sinusoidally amplitude modulated (SAM) signals. AM is indicated as the difference of the minimum and maximum intensities in decibels as modulation index in percent. Modulation
rate was 100Hz. From Schuller, 1979a.

Fig. 18. Neuron showing intensity preference for SAM stimulation.

Fig. 19. Interactive effect of intensity and modulation index
(percent modulation) upon locked firings evoked by SAM stimuli.
See text for further explanation.

over which they would discharge in registry with the SAM waveform. This intensity preference is illustrated by the neuron in Fig. 18.

In many units the ability to lock onto the modulation waveform at various depths was strongly influenced by the signal intensity. The interaction of intensity and modulation index is shown in Fig. 19. This unit responded with sharp, well synchronized discharges at all intensities, but only when the modulation index was 100%. At smaller modulation indices, of 60% and 40%, the locking was most secure and peaks of the histograms were highest at intensities of 30-50 dB, and a pronounced decline in responding was observed at 60 dB. At a modulation index of 20%, locking was seen only at 30-40dB and disappeared when the signal was 50 dB.

Response Selectivity

The majority of collicular neurons responded to tone bursts as well as SFM and SAM signals, in the case of the latter two signals with synchronized discharges, but there is a sizeable population that were more selective. The selectivity can be grouped into three categories. First, were neurons that were unresponsive to pure tones or SAM stimuli but responded to SFM with firings in tight registration with the modulation pattern. A few units were unresponsive to tone bursts and SFM stimuli but locked to the envelope of SAM signals.

Second were neurons that responded to tone bursts but were selective in the sense that they would respond with locked firings either to SFM stimuli or to SAM stimuli but not to both.

In the third category are neurons whose response properties, at least superficially, seem to be poorly suited for encoding modulation patterns. Included here are neurons that would respond to tone bursts, as well as SAM and SFM stimuli, but in no case were the discharges locked to the modulation waveform. Indeed, they responded to the SFM and SAM signals as they did to tone bursts. The other type of neurons that can be placed in the third category are those that could not be driven with any type of acoustic stimulation, including tone bursts, SFM, SAM nor by combinations of SFM-SAM, and were detected only because of their spontaneous activity. It might seem puzzling that neurons in the third category are considered as being highly selective for some stimulus but there is now evidence to suggest that some neurons of this type may be highly selective but the selectivity is only expressed while the animal is actively engaged in echolocation. I shall consider the evidence for this suggestion in the next section.

Single Unit Activity Recorded from Echolocating Bats

Recently, the activity of filter neurons has been monitored
while horseshoe bats were emitting pulses and receiving
electronically generated echoes played back from a loudspeaker
(Schuller 1979b; Schuller & Pollak unpublished observations). The
number of units recorded from behaving bats is quite limited and
the results should be considered to be of a preliminary nature.
The data, from at least some neurons, however, are of considerable interest and are instructive for guiding future studies.
Below I shall first discuss the activity evoked by the emitted
cries and correlate that activity with the discharge patterns
evoked by electronically generated tone bursts. Subsequently,
activity evoked by electronically generated echoes during periods
of pulse echo overlap will be presented.

The emission of an orientation cry typically evokes discharges from filter neurons, or at least from neurons in the
filter region of the IC. The discharge patterns can be tonic,
phasic or inhibitory and will often, but not always, be similar
to the pattern evoked by tone bursts at the BF or BIF. Of
particular interest is that some filter neurons that would not
discharge to tone bursts, SFM or SAM stimuli, fired bursts of
spikes to each emitted orientation cry. Whether these discharges resulted from some combination of internally generated
activity which primed or facilitated the responses to the
emitted sounds or whether the discharges were evoked solely by
internally generated activity is not known. On the other hand, a
few neurons of a similar type, i.e., not driveable with acoustic
signals, failed to respond when echolocation calls were emitted
and present what seems at the moment to be a paradox.

Most neurons would discharge upon pulse emission but the
events associated with pulse emission would also generate a
potent, underlying inhibition in the majority of cells studied.
The inhibition prevented the neuron from responding to echoes
that returned during periods of overlap. As soon as the pulse
had ended the normal response to the electronically generated
echo would be evoked. In other words, the response to an echo
which partially overlapped with the emitted cry would be
evoked only during the non-overlapping portion of the echo.

The above data seemingly indicate that the majority of
filter neurons do not respond to echoes during periods of pulse-
echo overlap which, if true, would be startling. However, it
should be stressed that the frequency relationship between the
emitted CF components and the echoes was substantially different
in experiments cited above from that occurring under most normal
conditions. While hunting insects, horseshoe bats Doppler
compensate and thereby listen to a cry having a frequency
substantially lower than the frequency of the echo CF

portion. Consequently, the sounds heard in the neurophysiological experiments represent a different set of stimulus
conditions than would normally occur.

In this context, a recent study by Möller (1978), of the
lateral inhibitory regions of filter neurons, seems to be highly
significant. Möller found that filter neurons tuned around
the bat's reference frequency had inhibitory surrounds flanking
and overlapping with the excitatory tuning curve. The important
point in this study is that a gap in the flanking inhibitory
region exists just below the bat's reference frequency and extends to frequencies 500-4000 Hz below the neurons BF. From
these data it follows, then, that two signals, an emitted
pulse at the resting frequency and an echo having roughly the
same frequency, would create a set of conditions whereby the
emitted cry would inhibit the responses to the echo. But the
surround inhibitory pattern also suggests that during Doppler-
compensation, when the emitted CF is lowered in frequency and
falls within the flanking gap of the inhibitory surround region,
no inhibition would be present and neurons would be free to respond to the echoes, even during periods of pulse echo overlap.

The most intriguing and surely the most important result
from the experiments with echolocating bats comes from a study
by Schuller (1979b). He showed that some neurons which do not
respond to SFM stimuli or others which do respond with discharges
that are not locked to the modulation waveform become synchronized
to the modulated signal but only if the bat is actively emitting
echolocation cries and only during periods of pulse-echo overlap.
These rather remarkable features are illustrated by the two
neurons shown in Fig. 20.

The unit in Fig. 20A responded to emitted cries (vocalizations) with a tonic firing pattern but was almost unresponsive to
SFM signals (AS(FM)). When the bat emitted orientation sounds and
the same SFM signal was simultaneously presented, the unit fired
with well defined discharges synchronized to the modulation waveform (Voc +AS(FM)). That active pulse emission was required
to evoke the locked firings is demonstrated by the control condition where a tone burst having the same frequency as the emitted
cry was presented simultaneously with the SFM signal (AS + AS
(FM)). Notice that almost no discharges were evoked with a simulated orientation cry. Finally, presenting an 83 kHz tone burst
during pulse emission resulted in a tonic firing pattern but one in
which the discharges were not in synchrony with the modulating waveform (VOC + AS), showing that the SFM stimulus was indeed necessary
to evoke locked firings.

A similar phenomenon is shown for another unit in Fig. 20B.

Fig. 20 Discharges of two single neurons evoked by emitted orientation cries (Voc) and electronically generated tone bursts (AS)
and frequency modulated signals (AS(FM)). (A): Neuron whose firings were locked to the modulation waveform only when orientation
cries were emitted. See text for further explanation. (B): Neuron illustrating the necessity of pulse-echo overlap for evoking
locked discharges. In top figure is the firing pattern to emitted
cries: The CF component was 83.3 kHz. The figure below shows response pattern to 84.3 kHz tone burst at 70 dB. In the subsequent
figure are shown the firing patterns evoked by vocalizations
coupled with SFM signals delayed by various times with respect to
the beginning of the emitted cry. Notice that locked discharges
were evoked only during the periods of pulse-echo overlap. From
Schuller 1979b.

This neuron also demonstrates that for locked firings to be evoked the echo must overlap with the emitted CF component.

What Makes Filter Neurons so Special?

The majority of reports concerned with the neurophysiology of long CF/FM bats have focused on those features associated with the processing of the CF component. The reasons, of course, derive from the highly dramatic nature of the behavior and the truly spectacular features of the nervous system that have evolved for processing this component of the biosonar signals. The descriptive term most often associated with the filter system is that it is specialized or even highly specialized. Implied in the usage of such a term is the idea that something unique derives from the specialization, that filter neurons have encoding properties so extraordinary as to set them apart from the neurons found in the nervous systems of other animals. To be sure, the system is "specialized," but I believe it worthwhile to define and to clarify a bit more carefully exactly what is meant by the term "specialized".

The three characteristics that have had an immediate and striking impact on all workers are the remarkably sharp tuning of the filter neurons, the great over-representation of the filter frequencies and the unusual tonotopic organization of the filter regions. What these features share in common is that they are basically structural in nature. The sharp tuning, for example, derives from the morphological features of that portion of the cochlea which responds to the filter frequencies (Bruns 1976a,b). But judging from a recent study by Bruns and Schmieszek (1979), there is little unique in the neural innervation of the cochlea. During evolution, apparently the most basal portion of the basilar membrane was, in essence, "stretched" out and fortified with a number of structural modifications. What this means is that a much longer strip of the basilar membrane is now devoted to each filter frequency than would have been the case if the membrane had not been "stretched". The mechanical properties of the "stretched" portion are sharpened by structural modifications. Within the elongated frequency strip, the density of the afferent neural innervation is reduced relative to that of the adjacent area but the decrease is more than compensated for by the increase in length devoted to each frequency. Thus, the absolute number of auditory nerve fibers devoted to each filter frequency is increased but the afferent neural innervation pattern is less dense than, but otherwise similar to that occurring in the nonfilter region. In short, it is the structural arrangement of the cochlear filter region that imparts the sharp tuning to a population of, otherwise, typically mammalian auditory neurons.

The results of recent experiments in the IC also lead me to
conclude that the specializations are largely architectural in nature. The filter region is somewhat amplified within the central
nervous system in that the proportion of neurons devoted to the
filter frequencies is greater in the central nervous system than in
the cochlea. Furthermore, the neurons deriving their synaptic inputs from the filter region of the cochlea assume their own domains
and tonotopic organization within the acoustic centers of the brain.
This arrangement serves to segregate the processing of the CF from
the processing of the FM component of the biosonar signals, as originally postulated by Simmons (1973) and more recently also demonstrated in the auditory cortex of the mustache bat (Suga and Jen,
1977). In this sense the system is indeed unique and highly
specialized but these features say little about the encoding
abilities of individual neurons.

It has been suggested in a number of reports that the sharp
tuning endows filter neurons with the ability to encode SFM modulation patterns (Suga et al. 1975; Suga & Jen 1977, Suga 1978). The
rationale is that an FM signal sweeping in and out of a very narrow response area would cause the neuron to discharge only when the
signal entered the RA, the result being discharges synchronized to
the period of the waveform. In this view, the ability to synchronize to the modulating envelope is a direct consequence of the
sharp tuning and thus is a specialized property of filter neurons.

This view, however, has not proven to be the case. Neurons
tuned to most of the audible frequency region of horseshoe bats
exhibit discharges to SFM signals that are as tightly locked to
the modulation waveform as are those of the filter neurons.
Furthermore, some sharply tuned filter neurons, as well as broadly
tuned nonfilter neurons, fail to lock to SFM stimulation. The
nonfilter neurons that do lock to the period of the SFM envelope
have broad tuning curves but the locking is still evoked when the
spectrum of the signals is confined entirely to, and sweeps back
and forth within, the neuron's RA. This feature is also observed
for filter neurons, as can be seen in Fig. 15[1]. The lock-

---

[1] The ability to lock to very small modulation depths, i.e., $\pm$10-40 Hz,
is seen only in filter neurons and has never been observed in neurons
having BFs below about 78 kHz (Schuller 1979a). A number of investigators consider the locking to very small modulation depths to be a
specialized property of filter neurons and have suggested that locking to such low depths is associated with the sharp tuning of the
filter neurons (Suga and Jen 1977; Schuller 1979a). However, it
should be pointed out that there does not appear to be a correlation
between the sharpness of tuning and the lowest modulation depth capable of evoking synchronized discharges. Apparently, neurons having

ing then is not so much a consequence of the sharp tuning but
rather seems much more to be due to the sequence and potency of
synaptic events evoked by the temporal and spectral features of
the signals. This, in principle, is the same argument advanced
for the pCLRs.

If, as I have argued, the sharp tuning does not confer extraordinary encoding abilities upon individual neurons, what advantage
accrues to having a population of sharply tuned elements? Perhaps
the functional significance of the sharp tuning is to confine and
spatially segregate the activity evoked by the emitted CF from
the activity evoked by the echo CF component (Suga et al. 1976;
Suga 1978). In conjunction with the pattern of the inhibitory
surround region observed for many filter neurons, which is a
specialized adaptation, the emitted CF would be prevented from
masking the responses to the echoes during periods of pulse-echo
overlap, a feature that is surely of primary importance to long
CF/FM bats.

The point I wish to make is simply this: there are a number of specialized structural and architectural features within
the auditory systems of long CF/FM bats which set these animals
apart from all other vertebrates. However, the response patterns
evoked by tone bursts are strictly mammalian (Möller et al., 1978
a; Pollak & Schuller, in prep.). Furthermore, except for the
locking to the very small modulation depths, the response features evoked by SFM and SAM signals are not fundamentally different in filter as compared to nonfilter neurons. This argument
is equally valid for the preferences of both filter and nonfilter neurons for the intensities, modulation depths and modulation rates to which each neuron will fire with locked discharges.

Although the encoding abilities of most individual filter
neurons appear not to be "specialized" to any extraordinary degree, the system with strong justification, can be said to be
functionally specialized. The functional specialization is imposed upon the system by the structural modifications in the
cochlea and by the architectural arrangement of elements within
the central nervous system. Each element receives input from a
sharp peripheral filter which severely restricts the extent of
the neuron's receptive field. Thus, a strip of basilar membrane
is projected upon a vertical array of collicular neurons with
each neuron in the array being selective for one or another signal

such encoding abilities must have a minimum Q value of about 40 (filter neurons encoding depths as low as $\pm$10 Hz can have Q values of 40
-45) but neurons having even sharper tuning curves, i.e., 100-300,
seem not to derive any additional advantage for encoding low modulation depths (Schuller 1979a; Pollak and Schuller, in prep).

parameter and with the extent of activity generated in each array
being sharply bounded in neural space. These features are then
expressed as a unique spatio-temporal pattern of activity which,
as a Gestalt, creates within the brain a neural image of a flying insect whose path happened to cross that of an echolocating
horseshoe bat.

Functional Organization of the Inferior Colliculus

There now exists sufficient information which can be fitted
together to provide a rough picture of how an acoustic image is
formed within the brain. Below I shall attempt to correlate the
known behavior modifications long CF/FM bats exhibit during echolocation with the architectural arrangement and the various encoding properties of collicular neurons and present what is likely
to be the form that the above mentioned spatiotemporal patterns of
activity might assume.

It seems reasonable to suppose that upon detecting a target,
the bat first needs to determine if the target is animate or is
inanimate, e.g., a falling leaf or background objects. A basis
for such a discrimination resides in the temporal patterns of the
responding elements; a flying insect can, presumably, be recognized
as such by the synchronized discharges locked to the modulation
pattern of the echo whereas the echo from an inanimate object will
be encoded in a manner similar to a tone burst.

After ascertaining the nature of the target, it may well
be that horseshoe bat seeks to determine whether or not the insect is one that is palatable and worthy of pursuit and capture.
In this context, the two manipulations the bat performs on the
emitted CF component during the goal directed portion of echolocation
seem to be important. The first type of manipulation is the
systematic lengthening of the emitted CF component that occurs
after target detection and continues for a considerable time
as the bat tracks its target (Novick & Vaisnys 1964; Novick
1971). One effect of the pulse lengthening is to increase the
number of modulation cycles present in each echo. The second
manipulation concerns Doppler-shift compensation where the bat
stabilizes the echoes within a narrow frequency band around the
reference frequency. The significance of this behavior is that
a disproportionately large number of filter neurons are tuned
to a 1.5 kHz band around the bat's reference frequency. It
appears, then, that horseshoe bats actively create a signal which
is designed to insure that an echo having a sufficient number of
modulation cycles will be received and processed by an exceptionally large number of sharply tuned neural elements, most of which
are well suited to encode some component of the modulation pattern.

Insights into the spatial extent of activity evoked by a
modulated echo can be obtained from considering the tonotopic organization together with the features of the SFM response areas. If
one assumes that filter units respond to frequency modulations in
a natural echo as they do to SFM signals, then the characteristics
of SFM response areas suggest that a modulated echo should produce
more or less equally vigorous activity over a large expanse of
neural tissue. That is, the discharge rate and synchrony should
be maximal in units tuned to the echo center frequency and having
symmetric SFM response areas. However, the activity should also be
maximal in other units having BFs above and yet in other units
having BFs below the center frequency of the echo because these
neurons have asymmetric SFM response areas. Furthermore, the
tuning properties of the filter neurons indicate that the active
region will have sharply confined boundaries in the anterioposterior
axis of neural space (Fig. 21).

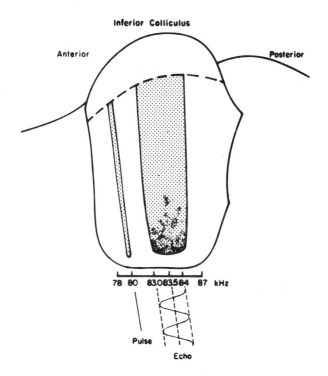

Fig. 21. Schematic diagram to visualize the spatial extent of
activity evoked by the CF of an emitted pulse and the CF of an echo
having frequency modulation of ± 500 Hz after the bat has compensated
for a 3.5 kHz Doppler shift in the echo CF component. The reference
frequency is assumed to be 83.5 kHz.

Within the active region, the echo will evoke discharges synchronized to the modulation waveform. Moreover, the frequency
changes occuring in the echo will, by virtue of their variations
with time, also cause a corresponding shifting, back and forth, of
the locus of activity as the spectrum enters and leaves the sharp
response areas of elements tuned to different frequencies. Thus,
the frequency modulation pattern is coded both by the temporal
sequence of discharges in a sharply bounded subpopulation of
filter neurons and in the spatial shifting of activity within
that subpopulation over time.

Due to changes in position, orientation and speed of either
the bat or its target, subsequent echoes will differ more or
less in carrier frequency, modulation pattern and intensity from
the previous echo(es). In principle, each echo will be encoded in
a similar manner. However, the preferences of many filter units
for selective ranges of intensity as well as modulation rate and
depth suggest that the changes in echo parameters will cause some
neural elements to drop out, new elements to be recruited while
others will simply change response vigor and/or firing registration to reflect the changes in echo characteristics. In short, the
properties of filter neurons endow the system with the ability to
encode the pertebations imposed on the echo CF component with the
sum total of spatio-temporal activity being a dynamic pattern
which differs from echo to echo.

Given the capacity to encode target features, the question
arises as to whether the information derived from one echo is
sufficient for target recognition or whether several echoes are
required. In this regard, it should be pointed out that there is
considerable variation of both wing length and wing beat frequency
among insect species. Both of these features have profound effects
on the frequency and amplitude modulations in the echo CF component and must result in substantial differences among the spatio-
temporal patterns of activity generated by echoes reflected from
different insects. Schuller (1979a) has discussed this issue in
considerable detail and from a number of considerations has concluded that the possibility for recognition from one echo, while
not impossible, is unlikely. On the other hand, the system encodes
modulation patterns with great precision and diversity. It, therefore, would be surprising if horseshoe bats could not associate
the acoustic image derived from the profile of activity integrated
over several echoes with a particular population of insect.

References

Aitkin, L.M. Tonotonic organization at higher levels of the auditory pathway. In: Internatural Review of Physiology: Neurophysiology II. R. Porter (Ed). University Park Press, London. Vol. 10, pp. 249-280 (1976).

Bruns, V. Peripheral auditory tuning for five frequency analysis by the CF-FM bat, Rhinolophus ferrumequinum. I. Mechanical specialization of the cochlea. J. Comp. Physiol. 106, 77-86 (1976a).

Bruns, V. Peripheral auditory tuning for fine frequency analysis by the CF-FM bat, Rhinolophus ferrumequinum: II Frequency mapping of the cochlea. J. Comp. Physiol. 106, 87-97 (1976b).

Bodenhamer, R., Pollak, G.D. and Marsh, D.S. Coding of fine frequency information by echoranging neurons in the inferior colliculus of the Mexican free-tailed bat. Brain Res. (in press).

Capranica, R. Why auditory neurophysiologists should be more interested in animal sound communication. Physiologist 15, 55-60 (1972).

Feng, A.S., Simmons, J.A., and Kick, S.A. Echodetection and target ranging neurons in the auditory system of the bat, Eptesicus fucus. Science 202, 645 (1978).

Friend, J.H., Suga, N. and Suthers, R.A. Neural responses in the inferior colliculus of echolocating bats to artificial orientation sounds and echoes. J. Cell. Physiol. 67, 319-332 (1966).

Goldman L.J. and Henson, O.W., Jr. Prey recognition and selection by the constant frequency bat, Pteronotus p. parnellii. Behav. Ecol. Sociobiol. 2, 411-419 (1977).

Griffin, D.R. Discriminative echolocation by bats. In Les Systems Sonars Animaux. R.G. Busnel (Ed.) INRA-CNRZ, Jouy-en-Josas, France, Vol. 1, pp. 273-300 (1967).

Grinnell, A.D. The neurophysiology of audition in bats; intensity and frequency parameters. J. Physiol., London 167, 38-66 (1963a).

Grinnell, A.D. The neurophysiology of audition in bats: temporal parameters. J. Physiol., London 167, 67-96 (1963b).

Grinnell, A.D. Comparative auditory neurophysiology of neotropical bats employing different echolocation sounds Z. Vergl. Physiol. 68, 117-153 (1970).

Grinnell, A.D. and Hagiuvara, S. Adaptations of the auditory system for echolocation: studies of New Guinea bats. Z. Vergl. Physiol. 76, 41-81 (1972).

Henson, M.M. The basilar membrane of the bat, Pteronotus p. parnellii. Am. J. Anat. 153, 143-158 (1978).

Henson, O.W. Jr., Pollak, G.D., Johnson, R.A., and Goldman, L.J. Specialized properties of the auditory system in the bat Pteronotus p. parnellii. Anat. Rec. 107, 373 (1974).

Henson, O.W., Jr. and Goldman, L.J. Prey detection and physiological aspects of the cochlea in the bat Pteronotus p. parnellii. Anat. Rec. 184, 425 (1976).

Jen, P.H.-S. Electrophysiological properties of auditory neurons in the superior olivary complex of echolocating bats. J. Comp. Physiol. 128, 47-56 (1978).

Möller, J., Neuweiler, G., and Zöller, H. Response characteristics of inferior colliculus neurons of the awake CF-FM bat, Rhinolophus ferrumequinum. I. Single tone stimulation. J. Comp. Physiol. 125, 217-225 (1978a).

Möller, J. Response characteristics of inferior colliculus neurons of the awake CF-FM bat, Rhinolophus ferrumequinum. II. Two tone stimulation. J. Comp. Physiol. 125, 227-236 (1978b).

Neuweiler, G. and Vater, M. Response patterns to pure tones of cochlear nucleus units in the CF-FM bat, Rhinolophus ferrumequinum. J. Comp. Physiol. 115, 119-133 (1977).

Novick, A. Orientation in neotropical bats. II. Phyllostomatidae and Desmodontidae. J. Mammal. 44, 44-56 (1963).

Novick, A. Echolocation in bats: Some aspects of pulse design. Amer. Sci. 59, 198-209 (1971).

Novick, A. and Vaisnys, R. Echolocation of flying insects by the bat, Chilonycteris parnellii. Biol. Bull. 128, 297-314 (1964).

Pollak, G.D., Marsh, D., Bodenhamer, R., and Souther, A. Echo-detecting characteristics of neurons in the inferior colliculus of unanesthetized bats. Science 196, 675-678 (1977a).

Pollak, G.D., Bodenhamer, R., Marsh, D.S., and Souther, A.

Recovery cycles of single neurons in the inferior colliculus of unanesthetized bats obtained with frequency modulated and constant
frequency sounds. J. Comp. Physiol. 120, 215-250 (1977b).

Pollak, G.D., Marsh, D.S., Bodenhamer, R. and Souther, A.
Characteristics of Phasic on neurons in the inferior colliculus
of unanesthetized bats with observations relating to mechanisms
for echo ranging. J. Neurophysiol. 40, 926-942 (1977c).

Pollak, G.D., Marsh, D.S., Bodenhamer, R., and Souther A.
A single unit analysis of inferior colliculus in unanesthetized
bats: response patterns and spike-count functions generated by
constant frequency and frequency modulated sounds. J. Neurophysiol. 41, 677-691 (1978).

Pollak, G.D., and Schuller, G. Tonotopic organization and
response patterns to frequency modulated signals in the inferior
colliculus of Horseshoe bats. Soc. Neurosci. Abstr. 4, 9 (1978).

Pollak, G.D., Henson, O.W. Jr. and Johnson, R.A. Multiple
specializations in the peripheral auditory system of the CF-FM
bat, Pteronotus parnellii. J. Comp. Physiol. (in press).

Schlegel, P. Directional coding of binaural brainstem units
of the CF-FM bat, Rhinolophus ferrumequinum. J. Comp. Physiol.
118, 327-352 (1977).

Schnitzler, H.-U. Die Ultraschall-Ortungslaute der Hufeisen-
Fledermäuse (Chiroptera, Rhinolophidae) in verschiedenen. Orientier-
ungesituationen. Z. Vergl. Physiol. 57, 376-408 (1968).

Schnitzler, H.-U. Echoortung bei der Fledermäus Chilonyteris
rubiginosa. Z. Vergl. Physiol. 68, 25-38 (1970)

Schnitzler, H.-U. Die Detektion von Bewegungen durch Echootung
bei Fledermäusen. Verh. Dtsche Zool. Gesc. 16-33 (1978).

Schnitzler, H.-U., Suga, N., and Simmons, J.A. Peripheral
auditory tuning for fine frequency analysis by the CF-FM bat,
Rhinolophus ferrumequinum. J. Comp. Physiol. 106, 99-110 (1976).

Schuller, G. Echoortung bei Rhinolophus ferrumequinum mit
frequenz-modulierten Lauten. Evoked potentials in colliculus inferior. J. Comp. Physiol. 77, 306-331 (1972).

Schuller, G. Coding of small sinusoidal frequency and amplitude modulations in the inferior colliculus of the "CF-FM" bat,
Rhinolophus ferrumequinum. Exp. Brain. Res. 34, 117-132 (1979a).

Schuller, G.   Vocalization alters responsivness of auditory neurons in CF-FM bat, Rhinolophus ferrumequinum.   J. Comp. Physiol. (1979b, in press).

Schuller, G., Beuter, K., and Schnitzler, H.-U.   Response to frequency shifted artificial echoes in the bat, Rhinolophus ferrumequinum.   J. Comp. Physiol. 89, 275-286 (1974).

Schuller, G. and Pollak, G.D.   Disproportionate frequency representation in the inferior colliculus of Horseshoe bats: Evidence for an "acoustic fovea".   J. Comp. Physiol. (in press).

Simmons, J.A.   The resolution of target range by echolocating bats.   J. Acoust. Soc. Amer. 54, 157-173 (1973).

Simmons, J.A., Howell, D.J., and Suga, N.   Information content of bat sonar echoes.   Amer. Sci. 63, 204-215 (1975).

Simmons, J.A., Fenton, M.B., and O'Farrell, M.J.   Echolocation and pursuit of prey by bats.   Science 203, 16-21 (1979).

Suga, N.   Single unit activity in the cochlear nucleus and inferior colliculus of echolocating bats.   J. Physiol. (London). 172, 449-474 (1964a).

Suga, N.   Recovery Cycles and responses to frequency modulated tone pulses in auditory neurons of echolocating bats.   J. Physiol. (London) 175, 50-80 (1964b).

Suga, N.   Echo-ranging neurons in the inferior colliculus of bats.   Science 170, 449-452 (1970).

Suga, N.   Feature extraction in the auditory system of bats. In: Basic Mechanisms in Hearing.   (Ed. A.R. Möller) pp. 675-742, Academic Press, New York (1973).

Suga, N.   Amplitude-spectrum representation in the Doppler-shifted CF processing area of the auditory cortex of the mustache bat.   Science 196, 64-67 (1977).

Suga, N.   Specialization of the auditory system for reception and processing of species-specific sounds.   Fed. Proc. 37, 2342-2354 (1978).

Suga, N., Simmons, J.A., and Jen, P.H.-S.   Peripheral control of acoustic signals in the auditory system of echolocating bats. J. Exp. Biol. 62, 277-311 (1975).

Suga, N. and Schlegel, P.   Coding and processing in the nervous system of FM signal producing bats.   J. Acoust. Soc. Amer. 84, 174-190 (1973).

Suga, N., Neuweiler, G., and Möller, J.   Peripheral auditory tuning for fine frequency analysis by the CF-FM bat, Rhinolophus ferrumequinum.   IV. Properties of peripheral auditory neurons. J. Comp. Physiol. 106, 111-125 (1976).

Suga, N. and Jen, P.H.-S.   Disproportionate tonotopic representation for processing CF-FM sonar signals in the mustache bat's auditory cortex.   Science 194, 542-544 (1976).

Vater, M., Schlegel, P. and Zöller, H.   Comparative auditory physiology of the inferior colliculus of two mollosid bats, Molossus ater and Molossus molossus.   I. Gross evoked potentials and single unit responses to pure tone stimulation.   J. Comp. Physiol. (1979a, in press).

Vater, M. and Schlegel, P.   Comparative auditory physiology of the inferior colliculus of two mollosid bats, Molossus ater and Molosus molosus.   II Single unit responses to frequency modulated signals and signal/noise combinations.   J. Comp. Physiol. (1979b, in press).

AUDITORY PROCESSING OF ECHOES: REPRESENTATION OF ACOUSTIC INFORMATION FROM THE ENVIRONMENT IN THE BAT CEREBRAL CORTEX

Nobuo Suga and William E. O'Neill

Department of Biology, Washington University

St. Louis, Mo., U.S.A.

Clues to understanding information processing by the auditory system are hidden in the differences in response properties of single neurons within and among individual auditory nuclei. In order to explore neural mechanisms for signal processing, we thus focus our research on differences in response properties among neurons as well as response properties per se. All the acoustic information of use to an animal is contained in the activity of its auditory nerve. In mammals, all nerve fibers with identical best frequencies show nearly the same response pattern and frequency-tuning curve. In bats, all these primary auditory neurons respond to both emitted orientation sounds (also called pulses) and echoes by discharging multiple action potentials to each. The role of the central auditory system is to process this activity and send appropriate information to regulate the motor system, if necessary.

If all neurons in a given central auditory nucleus responded to both emitted orientation sounds and echoes as do primary auditory neurons, they would be concerned with coding, but not encoding or decoding information. In the central auditory system, single neurons with identical best frequencies typically show wide variation in their response properties; this is true for characteristics such as (1) response pattern (e.g., Rose, Greenwood, Goldberg & Hind, 1963; Suga, 1971). (2) impulse-count function (e.g., Grinnell, 1963a; Suga, 1965a, 1977), (3) frequency-tuning curve (e.g., Grinnell, 1963a; Suga, 1965a, b, 1973), (4) response to FM sounds (e.g., Suga, 1965a, b, 1973), (5) recovery cycle (e.g., Grinnell, 1963b, 1970; Suga & Schlegel, 1973), and (6) directional sensitivity (e.g., Grinnell & Hagiwara, 1972; Knudsen & Konishi, 1978). Such wide variations for given response

properties are apparently related to encoding or decoding of
certain parameters of acoustic signals (for echolocating bats,
more specifically, echo patterns associated with certain physical
aspects of the environment).

The most recent research on the functional organization of
the auditory system clearly indicates that neurons with response
properties different from those at the periphery are the elements
which establish the various specialized and beautiful functional
organizations found at the higher levels of the auditory system:
that is, groups of neurons represent certain important aspects of
the acoustic environment in orderly patterns, usually along axes
or in coordinate systems. For instance, neurons with especially
narrow frequency-tuning curves contained within a population
showing a wide variation in the shape of the curves are associated
with the fine tonotopic representation found in the auditory
cortex of the mustached bat Pteronotus parnellii rubiginosus
(Suga & Jen, 1976; Suga, 1977). These same neurons show a wide
variation in impulse-count function and are arranged in an orderly
way to produce a systematic amplitopic representation in the same
area of the auditory cortex of this bat (Suga, 1977). Broad
variation in directional sensitivity curves is apparently essential
to form the neural map of auditory space discovered in the dorso-
lateral mesencephalic nucleus of the barn owl Tyto alba (Knudsen
& Konishi, 1978). Furthermore, neurons showing a broad range of
responses to FM sounds (in particular, FM-specialized neurons)
and a broad spectrum of recovery cycles are apparently responsible
for creating the various response properties of delay (or range)
sensitive neurons found in the mustached bat (Suga, O'Neill &
Manabe, 1978; O'Neill & Suga, 1979). Our most recent finding is
that these delay-sensitive neurons are systematically arranged
along a target-range axis in the cerebral cortical plane (Suga
& O'Neill, 1979). The two most important findings in acoustic
signal processing are that a broad distribution of response properties is created by excitatory and inhibitory neural interactions and that neurons with different response properties form
different aggregates and are arranged in specific ways in certain parts of the brain depending upon their function.

The following will summarize our recent findings on the
functional organization of the auditory cortex of the mustached
bat. We will minimize the detailed discussion of mustached bat
behavior because it will be covered in other parts of this volume,
and we will limit citation of the literature to those which are
directly related to understanding these findings.

For echolocation, the mustached bat produces biosonar signals
each consisting of a long constant-frequency (CF) component
followed by a short frequency-modulated (FM) component (Fig. 1).

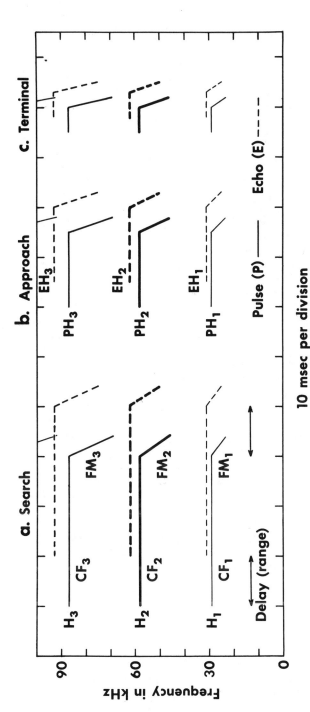

10 msec per division

Figure 1. Schematized sonagrams of the synthesized mustached bat biosonar pulses (P; solid lines) and echoes (E; dashed lines) mimicking three phases of target-oriented flight. The three harmonics of both the pulses (PH$_{1-3}$) and the echoes (EH$_{1-3}$) each contain a long CF component (CF$_{1-3}$) followed by a short component (FM$_{1-3}$). The fourth harmonic (H$_4$) is not shown in the figure. (a) search phase: CF and FM durations are 30 and 4 msec, respectively. (b) approach phase: CF and FM durations are 15 and 3 msec. (c) terminal phase: CF and FM durations are 5 and 2 msec. Repetition rates for (a), (b), and (c) were 10, 40, and 100 pairs per second, respectively. Thickness of the lines indicates the relative amplitudes of each harmonic in the pulses and echoes: H$_2$ is strongest, followed by H$_3$ (-6 dB) and H$_1$ (-12 dB). Echo delay is measured as the time interval between the onsets of corresponding components of the pulse and the echo in a stimulus pair.

Each component is composed of four harmonics (H_1-H_4), and therefore, there are eight components (CF_{1-4}, FM_{1-4}) in total. During the search phase, the frequency of the predominant second harmonic remains at about 61 kHz for 20 - 35 msec, then sweeps down to about 50 kHz within 2 - 4 msec. Since the reflected sound energy is highly concentrated at a single wavelength, the long CF tone is ideal for target detection and measurement of target velocity. The FM component, on the other hand, is more appropriate for ranging, localization, and characterization of a target, because of its short duration and wide bandwidth.

The mustached bat, like the horseshoe bat Rhinolophus ferrumequinum, performs a fascinating acoustic behavior called "Doppler-shift compensation" (Schnitzler, 1970). When there are no moving objects, the bat produces sounds with the second harmonic CF at about 61 kHz.[*] If the bat receives a Doppler-shifted echo, say at 63 kHz, from an approaching object, the bat reduces the frequency of the subsequent biosonar signals so as to stabilize the second harmonic echo CF at several hundred Hz above 61 kHz. Behaviorally the bat is extraordinarily sensitive to echoes from moving targets (Goldman & Henson, 1977).

To analyze the small shifts in the CF echo frequency produced by motion of the target, fine frequency resolution is required. Resolution at frequencies higher than 5 kHz is directly related to the sharpness of the frequency-tuning curves of auditory neurons. The cochlea of the Panamanian mustached bat is apparently specialized for fine frequency analysis of the CF components of the biosonar signal, because the cochlear microphonic response shows prominent ringing at frequencies at or near 61 kHz (Suga & Jen, 1977). As a consequence, single auditory nerve fibers most sensitive to 60 - 63 kHz, especially 61 kHz, show extremely sharp frequency-tuning. To a lesser extent, frequency-tuning curves of neurons sensitive to the 1st and 3rd harmonics are also sharper than others tuned to different frequencies (Suga, Simmons & Jen, 1976; Suga & Jen, 1977).

When the target is a flying insect, the returning echo CF may be modulated in amplitude and frequency by its wing beat. Sharply tuned peripheral neurons can code frequency modulation as low as $\pm$ 0.01%, i.e. $\pm$ 6.1 Hz for 61 kHz, by changing their discharge rate synchronously with the modulation (Suga & Jen,

---

[*] The resting frequency within a population of mustached bats may vary and 61 kHz is an approximate value for the Panamanian colony from which bats were taken. The average resting frequency of different populations may vary more widely (see Schnitzler, 1970).

1977). Since the Doppler-shift from the beating wing of a moth
can be as high as 800 Hz (Henson & Goldman, 1976), peripheral
auditory neurons code the wing beat of an insect by both synchronous
neural discharges of individual neurons and also synchronous
change in excitation from one group of neurons to another differing
in best frequency. The neurons sharply tuned at about 61 kHz are
potentially <u>flying-insect detectors</u>.

FUNCTIONAL ORGANIZATION OF THE CEREBRAL AUDITORY CORTEX OF
THE MUSTACHED BAT

     The auditory cortex reflects the remarkable specialization
of the peripheral auditory system (Fig. 2). There is a clear
tendency for high-frequency sensitive neurons to be located
anteriorly and low-frequency sensitive neurons to be located
posteriorly, as in the primary auditory cortex of other mammals
(Fig. 2B). However, in two respects this tonotopic organization
is quite unique: (1) neurons sensitive to 50 - 60 kHz, i.e.,
those processing the information carried by the dominant FM
component, FM_2, are not on the main tonotopic axis, but are
displaced anterodorsally (Fig. 2B,b). This <u>FM processing area</u> is
large and tonotopic representation in this area is vague and
complex; (2) 61-63 kHz sensitive neurons, i.e., those processing
the information carried by the main CF component, CF_2, occupy
about 30% of the primary auditory cortex, even though the range
of hearing is probably very broad, from a few kHz to 150 kHz.
Since this species' biosonar signals are quite distinct from its
communication sounds, the 61-63 kHz tuned area is undoubtedly
specialized for processing the main CF component of Doppler-
shifted echoes from moving targets. Accordingly, this area is
called the <u>Doppler-shifted-CF (or DSCF) processing area</u> (Fig. 2B,
c). This prominent disporportionate frequency representation of
the DSCF processing area was the first demonstration of a specific
functional organization in the auditory system and is comparable
to the over-representation of the fovea centralis of the eye in
the mammalian visual system and also of the hand areas in the
primate somatosensory and motor systems (Suga & Jen, 1976).

DSCF Processing Area

     In the DSCF processing area, each orthogonal microelectrode
penetration is characterized by neurons with nearly identical
<u>best frequencies</u> (<u>BF</u>'s), excitatory frequency-tuning curves, and
impulse-count functions (Suga & Jen, 1976; Suga, 1977). The peak
of the impulse-count function in response to a tone burst at the
BF uniquely defines the <u>best stimulus amplitude</u> (<u>BA</u>) for maximum
excitation. Since all the neurons arranged orthogonally to the
cortical surface are characterized by nearly identical BF's and

BA's, it is clear that for the above properties, the auditory cortex has a columnar organization.

In the DSCF processing area, the BF and BA of the columns systematically vary with the location on the cortical surface, so that there are systematic tonotopic and amplitopic representations: 61.0 kHz sensitive neurons are located at the center and 63.0 kHz sensitive neurons at the circumference, while neurons tuned to weak sounds of 30-40 dB SPL are located ventrally and those tuned to strong sounds of 80-90 dB SPL dorsally (Fig. 2C). Along the ventral margin of the DSCF processing area, neurons with BF's of 64-65 kHz are often found. Thus the amplitude spectrum of a signal is expressed in amplitude vs. frequency coordinates parallel to the cortical surface.* However, this amplitude spectrum representation is disproportionate, so that acoustic signals of 61.5 - 62.0 kHz and 30 - 50 dB SPL are projected to a much larger

---

* The tonotopic representation in the DSCF processing area correlates with the "resting frequency" of the CF_2 emitted by the individual mustached bat. When a bat's resting frequency is found to be about 60.5 kHz, its DSCF processing area represents sounds mainly between 60.5 and 62.5 kHz. On the other hand, if the resting frequency of another bat is about 61.5 kHz, its DSCF processing area represents sounds mainly between 61.5 to 63.5 kHz. Thus, differences between individual bats in the frequency of the biosonar signal are reflected by the DSCF processing area.

Figure 2. (A) The dorso-lateral view of the left cerebral hemisphere of the mustached bat. The auditory cortex is the area surrounded by the dashed lines. (a) CF/CF, (b) FM, and (c) Doppler-shifted-CF (or DSCF) processing area. (B) The DSCF processing area makes up the major portion of the main tonotopic axis of Al, which is illustrated by the iso-best frequency contour lines representing frequencies from about 100 kHz to 10 kHz antero-posteriorly. (C) Amplitopic (---) and tonotopic (———) representations in the DSCF processing area. Two-digit and three-digit numbers show the best amplitude in dB SPL and best frequency in kHz, respectively. (D) The amplitude and frequency axes represent, respectively, the subtended angular and velocity information about a target. The echo amplitude is the result of both the target cross-sectional area and the inverse 4th power of the target distance. Poorly directional neurons specialized for echo detection (E-E neurons) are mainly found in the shaded area, while neurons directionally sensitive and specialized for target localization (I-E neurons) are mainly found in the unshaded area. (Based upon Suga, 1977 and Manabe, Suga, and Ostwald, 1978).

area.   The frequency and amplitude of echoes from prey may predominantly be in this range from the moment of target detection through the early part of the target-oriented flight. It should be noted that the DSCF processing area is disproportionately large in the auditory cortex and that furthermore it itself is disproportionately organized to represent biologically important echoes by the activity of many neurons for fine echo processing. The functional significance of the tonotopic and amplitopic representation demonstrated in Fig. 2C is that the frequency of an echo, i.e., the target velocity information, is represented along the radial axis, while the amplitude, which is related to target range and size, is expressed along the circular axis (Fig. 2D).

In the area tuned to sound of 61.5 - 62.0 kHz and 30 - 50 dB SPL, BF changes at a rate of 1-2 Hz per micron along the frequency axis.   If the distance between neurons is assumed to be about 30 microns, BF would vary from neuron to neuron by only 30 - 60 Hz steps along the frequency axis. The frequency representation, therefore frequency resolution, is very fine.   Frequency resolution is directly related to the sharpness of the frequency-tuning curve.   Neural mechanisms exist in the mustached bat which further sharpen the previously mentioned narrow frequency-tuning curves of peripheral neurons whose BF's are near 61 kHz (Suga and Manabe, in preparation).   The narrowest curve obtained so far in the DSCF processing area has a level-tolerant bandwidth of 300 Hz, i.e., its bandwidth remains at 300 Hz over a broad amplitude range (Suga, 1977).   Because of lateral inhibition, the slope of the frequency-tuning curve is essentially infinite in a significant number of neurons.   Neurons with such properties act as narrow-band frequency detectors and as such would contribute to fine processing of frequency information.

When flying toward a stationary object, the speed of the mustached bat is expressed by the frequency difference between the emitted sound and the returning Doppler-shifted echo, not by the magnitude of the bat's Doppler-shift compensation.   The frequency information of the emitted sound is available to the bat in the form of vocal self-stimulation and perhaps efferent copy, about which little is currently known.   The frequency information in the Doppler-shifted echo is available regardless of whether Doppler-shift compensation is performed or not, but the measurement of its value becomes much more accurate with compensation because the echo frequency is then subjected to especially sharply tuned neural filtering near 61 kHz.

For velocity detection, a different mechanism is also conceivable in CF-FM bats.   The envelope modulation produced by the overlap of the Doppler-shifted echo with the self-vocalized sound may be coded by auditory neurons which can change their discharge

rate synchronously with envelope modulation as high as 3 kHz
(Suga, Simmons and Jen, 1975). However, this mechanism might be
unreliable for the measurement of relative velocity of a flying
insect, because of the ambiguity due to local motion produced by
the wing beat which evokes both amplitude and frequency modulation
of the echo.

When a bat pursues a flying insect, the resulting Doppler-
shifted echo consists of both D.C. and A.C. components: the
former is related to the velocity difference between the bat and
the insect, while the latter (periodic frequency modulation)
results from the insect's beating wings. In many mechano-receptor
systems, neurons have been found which are sensitive to either
D.C. or A.C. components of mechanical stimuli. Therefore, the
question may be raised whether cortical auditory neurons in the
DSCF processing area are suited for detection of either the D.C.
or the A.C. components in the echo from a flying insect. In the
DSCF processing area, some neurons are insensitive to the A.C.
component while still responding to the D.C. component. Others,
however, are sensitive to both components, and change their
discharge rate synchronously with frequency modulations as low as
0.05%, i.e., 31 Hz deviation from a 61 kHz carrier, easily evoked
by the beating wing. The A.C. component, which also is always
accompanied by amplitude modulation, is therefore represented by
these synchronous discharges and by synchronous alternation in
excitation from one group of neurons to another with different
best frequencies and best amplitudes. All peripheral neurons
tuned at about 61 kHz are sensitive to both the A.C. and D.C.
components in the Doppler-shifted echoes from flying insects
(Suga & Jen, 1977), while the auditory cortex contains some
neurons sensitive to both components and others sensitive to the
D.C. component only. There is, thus, some differentiation
between those neurons for signal processing.

The analysis of the D.C. component is obviously confounded
by the A.C. component. The D.C. component frequency is expressed
by the average BF of a group of activated neurons with different
best frequencies. If the positive and negative phases of the A.C.
component are not symmetrical, the measurement of the average
frequency may not accurately reflect the relative velocity of the
insect.

Doppler-shift compensation behavior has been discussed
extensively by Schnitzler (1968, 1970) and Simmons (1974). It
has also been discussed in relation to the neurophysiological
data (Grinnell, 1970; Johnson et al., 1974; Neuweiler, 1970;
Schnitzler, Suga & Simmons, 1976; Suga, Simmons & Jen, 1975).
Comprehensive physiological data on the peripheral auditory
system in mustached bats are reported by Suga and Jen (1977).

They also discuss the role of sharply tuned neurons in relation
to Doppler-shift compensation. Undoubtedly, other contributors
to this symposium will discuss this problem in detail, so that we
will end our discussion by pointing out the following. When the
bat pursues a flying insect, it initially lengthens the duration
of the orientation sound from about 20 to 28-37 msec, and then
quickly but systematically shortens the duration during the
approach and terminal phases, down to 6-8 msec (Novick & Vaisnys,
1964). This increase in duration during the initial approach
phase would increase the information carried by the CF component.
When the bat flies toward a landing platform, however, the initial
lengthening does not occur (Schnitzler, 1970; O'Neill, Kuriloff,
Berry & Suga, unpubl.). If the increase in duration takes place
only during insect pursuit, then perhaps it is related to maximizing
the information carried by the A.C. component, not the D.C. component.
Doppler-shift compensation also probably starts at the initial
part of the approach phase. Subsequently, as the duration of the
orientation sounds gradually shortens, the bat receives less and
less information about the beating wing during the major part of
Doppler-shift compensation. Furthermore, Doppler-shift compensation
occurs to stationary objects when the mustached bat is flying
(Schnitzler, 1970) or is passively moved in a "bat-mobile" (O'Neill,
Kuriloff, Berry & Suga, unpubl.). That is, the bat compensates
for the Doppler-shift even when there is no A.C. component.
These acoustic behaviors indicate that Doppler-shift compensation
functions in (i) target detection at long range, (ii) simple target identification, and (iii) target velocity estimation.

    Almost all neurons in the DSCF area are excited by acoustic
stimuli from the contralateral ear, and are either excited or
inhibited by stimuli to the ipsilateral ear. They are called E-E
and I-E neurons, respectively. Each cortical column is characterized not only by a BF and BA, but also by binaural interaction in
which (1) all neurons are E-E, (2) all are I-E, or (3) the interactions vary with depth. Thus in terms of aural representation,
this area consists of at least three different types of columns.
E-E neurons are poorly directional, but are tuned to weaker
echoes and integrate (or even multiply) signals from both ears
for effective target detection. On the other hand, I-E neurons
are directionally sensitive, and are tuned to stronger echoes.
E-E and I-E neurons form two functional subdivisions in the DSCF
processing area, one suited for target detection, the other for
target localization (Fig. 2D; Manabe, Suga & Ostwald, 1978).

FM Processing Area

    The functional organization of the FM processing area is
quite different from that of the DSCF processing area (Suga,
O'Neill & Manabe, 1978). Tonotopic organization is vague and

amplitopic representation has not been found, although the neurons
are tuned poorly to particular frequencies and amplitudes. This
area is composed of functional clusters of neurons which respond
to combinations of elements contained in the emitted biosonar
signals and Doppler-shifted echoes. Although our mapping of this
area is not yet complete, we have found separate clusters containing
neurons which respond strongly to either FM_2, FM_3, or FM_4 only
when it is delivered following FM_1 with particular delays. For
example, neurons in the largest cluster show facilitation of
response to an FM sound of the 2nd harmonic (FM_2) when it is
preceded by the FM component of the 1st harmonic (FM_1) and/or a
CF component of the 1st harmonic (CF_1). Some neurons in the
clusters do not show noticeable excitatory responses when stimulated
with any single component delivered alone, but respond strongly
when two sounds are combined in particular frequency, amplitude,
and time relationships. We call these neurons FM_1-FM_n facilitation
neurons (n designates the specific FM harmonic(s) of the second
sound to be combined with FM_1 of the first sound for facilitation.
n = 2, 3, or 4, or their combinations).

Since many facilitation neurons in the FM processing area do
not respond to pure tones, FM sounds or noise bursts delivered
alone, the measurement of their frequency-tuning curves can only
be accomplished by delivering pairs of acoustic stimuli. While
the frequency and amplitude of one member of the stimulus pair is
held constant at values producing the strongest facilitation, the
frequency and amplitude of the other stimulus can be varied to
determine the frequency-tuning curve for facilitation. This
process is then reversed for the measurement of the tuning curve
of the other, initially fixed stimulus. We call the curves
measured in this way facilitation-tuning curves.

Fig. 3A is an example of the frequency-tuning curves of an
H_1-FM_2 neuron obtained by the above method. There were no excitatory responses in this neuron to pure tones or FM sounds delivered
alone at any available frequency and amplitude. However, a CF
tone (30 msec duration) with a frequency within the first harmonic
FM frequency sweep (FM_1: 6 kHz sweep), followed by a FM sound
mimicking the FM_2 of the biosonar signal (12 kHz sweep), elicited
a vigorous response. By holding the initial CF tone burst constant
at its BF and BA for facilitation, we could measure the facilitation-
tuning curve for FM_2 (curve b, Fig. 3A). The BF and BA for FM_2
was 55.74 kHz (center frequency) and 41 dB SPL, respectively.
Conversely, by presenting a FM sound sweeping from 61 to 49 kHz,
we measured the facilitation-tuning curve for the CF_1 component
(curve a, Fig. 3A). Its BF and BA were 27.80 kHz and 56 dB SPL,
respectively. Similar facilitation could be evoked by replacing
the 27.80 kHz CF_1 tone burst with H_1 or FM_1 with the FM center
frequency at 27.80 kHz. In other words, the combination of the

first harmonic H_1 (or its elements) and the second harmonic FM
component was essential for the excitation of this neuron.
Therefore, we call it a H_1-FM_2 facilitation neuron.

Since these neurons were maximally facilitated only when
there was a delay between the FM_1 and FM_n, we hypothesized that
during echolocation they would be "conditioned" by self-stimulation
with the FM_1 component of the bat's emitted biosonar signal
(hereafter, the pulse, P) to respond to the higher harmonic FM
components in the time-delayed echo. In other words, these
neurons process FM components in an echo in a temporal relation
to the FM_1 of self-vocalized sound.

Since the primary cue for target range is the delay of an
echo from an emitted sound, (e.g. Simmons, Howell & Suga, 1975),
we tested the neurons in this area for delay sensitivity by
delivering pulse-echo stimulus pairs which could contain all of
the first three harmonics (H_1-H_3), or only the essential components
which elicited facilitation (Suga, O'Neill & Manabe, 1978; O'Neill
& Suga, 1979). These were delivered at three repetition rates
and durations, corresponding to biosonar signals and their echoes
in the search, mid-approach, and terminal phases (Fig. 1). By
holding the frequency and intensity of the pulse and echo components
constant at their BF's and BA's and then varying the delay of the
echo component at each repetition rate, we found that the strength

---

Figure 3. (A) Frequency-tuning curves of an H_1-FM_2 facilitation
neuron. This neuron did not respond at all to CF or FM sounds
presented alone. The "facilitation" tuning curves shown were
measured for the response when two sounds (CF_1 followed by FM_2)
were presented in a pair. "a" is the tuning curve for CF_1 which
caused facilitation in a combination with a 4 msec downward-
sweeping FM sound (61-49 kHz) at an amplitude of 41 dB SPL. "b"
is the tuning curve (center frequency) of an FM sound (12 kHz
sweep) measured while the first sound in the pair (CF_1) was held
at its BF (27.80 kHz) and 56 dB SPL (Suga, O'Neill, & Manabe,
1978; Copyright 1978 American Association for the Advancement of
Science). (B) Excitatory (dashed lines) and facilitation (solid
lines) tuning curves of a CF_1/CF_2 facilitation neuron. The
excitatory areas were measured by delivering a 34 msec long CF
tone. The facilitation areas were measured by delivering two CF
tones simultaneously, holding one at a constant frequency and
amplitude (x's) while measuring the other. The responses to CF
tones presented alone were poor and inconsistent. Presenting two
harmonically related CF tones at 29.75 and 59.39 kHz elicited a
strong facilitation accompanied by a dramatic decrease in the
threshold (Suga, O'Neill, & Manabe, 1979; Copyright 1978 American
Association for the Advancement of Science).

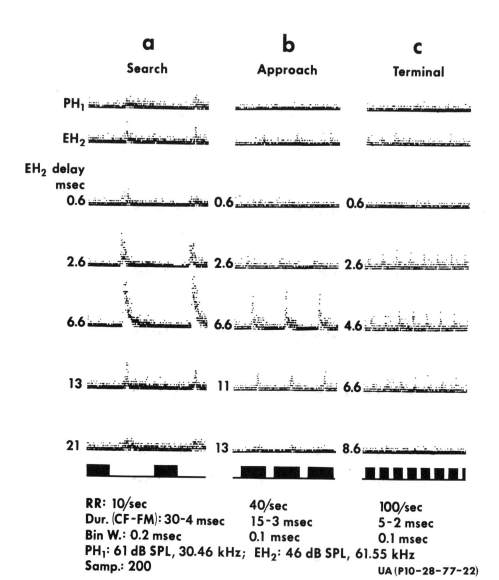

RR: 10/sec 40/sec 100/sec
Dur. (CF-FM): 30-4 msec 15-3 msec 5-2 msec
Bin W.: 0.2 msec 0.1 msec 0.1 msec
PH₁: 61 dB SPL, 30.46 kHz; EH₂: 46 dB SPL, 61.55 kHz
Samp.: 200

UA (P10-28-77-22)

of the facilitation, and therefore, the ability of the neuron to follow at the repetition rate of the stimulus pairs, was a function of the echo delay.

The PST histograms of Fig. 4 illustrate the responses of an FM_1-FM_2 neuron to stimulus pairs repeated at 10/sec (CF, 30 msec; FM, 4 msec duration), 40/sec (CF, 15 msec; FM 3 msec) and 100/sec (CF, 5 msec; FM, 2 msec). Note that the delay resulting in the best following to the stimuli (the <u>best delay</u>) at 40/second, for instance, is 6.6 msec. The neuron shows some following over the range from 2.6 to 13 msec delay. Also note that there is almost no response to presentation of the pulse or echo components alone.

In some neurons the best delay changed as a function of repetition rate and duration of stimuli, while it remained relatively constant in others. We call these two types <u>tracking</u> and <u>range-tuned neurons</u>, respectively. Their response properties are best illustrated by measuring their <u>delay-tuning curves</u> (Fig. 5). These curves are obtained by holding the amplitude of the pulse component (PH_1) constant at BA (71 $\pm$ 7 dB, N = 113) and measuring the threshold of response to the echo as a function of echo delay.

Range-tuned neurons have nearly the same delay-tuning curve regardless of repetition rate. In the neuron shown in Fig. 5A, the best delay (at the lowest threshold, 47 dB SPL) is 2.7 msec, corresponding to a 46 cm target range. Since a target would be

---

<u>Figure 4</u>. Peri-stimulus-time (PST) histograms of the responses of a tracking neuron (H_1-FM_2 specialized) to different echo delays during simulated search, approach, and terminal phases of target-oriented flight. The essential harmonics for facilitation of this neuron were H_1 of the pulse (CF_1, 30.46 kHz) and H_2 of the echo (CF_2, 61.55 kHz). PST histograms represent time on the abscissa and number of impulses per bin on the ordinate. The amplitudes of PH_1 and EH_2 for these responses were 61 and 46 dB SPL, respectively. The upper two histograms in each column show the poor response or lack of response of the neuron to PH_1 or EH_2 delivered alone at each repetition rate. The five histograms below them show the facilitation at various echo delays when PH_1 and EH_2 were delivered in a pair. Notice the decrease in the range of delays over which the neuron followed the stimulus pair, and also the shortening of the best delay as the repetition rate increased from 10 to 100 pairs per second (corresponding to a change in distance from 114 to 55 cm). Stimulus markers beneath the histograms indicate the pulse of the stimulus pair only (O'Neill & Suga, 1979; Copyright 1978 Am. Assoc. Adv. Sci.).

Target range in cm

at this range only during the late approach phase, this neuron is
apparently specialized to respond to the echo only around that
time. Some range-tuned neurons have upper thresholds in response
to an echo, and are then selective for both distance and target
subtended angle (Fig. 5B). The slope of the leading edge of the
delay tuning curve is steep, 24 dB/msec in Fig. 5A and infinite
in Fig. 5B. The width of the delay-tuning curve at 10 dB above
the minimum threshold is 2.8 msec for Fig. 5A. The tuning curve
does not appear to be terribly sharp because our criterion for
threshold was very low, i.e. about 0.1 response per stimulus
(just-noticeable response). If the criterion for threshold was
raised to 0.5 response per stimulus, the delay-tuning curve would
become much narrower and its minimum threshold higher. Thus, for
the neuron shown in Fig. 5B, the delay-tuning curve would become
a tiny circle centered at 5 msec delay and 70 dB SPL.

The best delay differs from neuron to neuron. We now have
data clearly indicating that target distance is expressed along an
axis, like velocity and subtended angular information are expressed
in the DSCF processing area (Suga, 1977), and also like azimuth
and elevation of a sound source are systematically mapped in the
midbrain auditory nucleus of the barn owl, Tyto alba (Knudsen &

Figure 5. Delay-tuning curves for three neurons in the FM processing area. (A) Range-tuned neuron (FM_1-FM_3 specialized). The
delay-tuning curves for the search (10 pairs per second), mid-
approach (40/second), and terminal (100/second) phases are very
sharp and nearly identical for the approach and terminal phases.
The best delay is also the same, 2.7 msec. This neuron is tuned
to a target 46 cm away – that is, during the late approach to
terminal phase. (B) Another range-tuned neuron (FM_1-FM_3 specialized) demonstrating upper thresholds for facilitation. The
curves are similar, and the neuron is tuned to a target that
returns an echo of about 65 dB SPL from a distance of about 85 cm.
The frequencies and amplitudes of the CF_1 components of the
pulses (PH_1) and the frequencies of the CF_1 components of the
echoes (EH_1) used for the measurements are shown at the right.
(O'Neill & Suga, 1979; Copyright 1978 Am. Assoc. Adv. Sci.). (C)
A tracking neuron (FM_1-FM_2) shows broad delay-tuning with echo
facilitation thresholds of 20-30 dB SPL and best delays of 3.8-5
msec (target range 70-85 cm) for search and early approach phases.
(▲, 2.5/sec; △, 5/sec; ●, 20/sec: these data were taken at an
early stage of the research before repetition rates were standard-
ized). During the terminal phase (○ , 100 pairs per second),
the best delay shortens dramatically to about 2.5 msec (50 cm
distance), the threshold increases to about 36 dB SPL, and the
tuning curve becomes much narrower. (Suga, O'Neill, & Manabe,
1978; Copyright 1978 Am. Assoc. Adv. Sci.).

Konishi, 1978). Neurons tuned to shorter delays and which follow better to repetitive stimuli at higher repetition rates are located more anteriorly in the FM processing area.

In contrast to range-tuned neurons, a tracking neuron changes both its best delay and the shape of the delay-tuning curve as the repetition rate changes (Fig. 5C; Fig. 4). These neurons respond to targets over a broad range during the search and early approach phases. But during the late approach and terminal phases, the best delay becomes shorter and the tuning curve narrower. Thus these neurons can follow the echo during all phases of target pursuit. The most interesting feature of these neurons is the narrowing of the delay-tuning curve as the bat approaches the target, so that the neuron effectively "ignores" possible secondary echo sources from objects farther away.

Behavioral experiments on range discrimination clearly indicate that "CF-FM" bats use the FM component for ranging just as "FM" bats do (Simmons, 1971; Simmons, Howell & Suga, 1975). The essential components in the pulse and echo for the excitation of delay-sensitive neurons which we found are particular combinations of FM components, so that our neurophysiological data parallel the behavioral results. Further, delay-sensitive neurons, apparently of this type, have been found in the intercollicular nucleus of an FM bat, Eptesicus fuscus (Feng, Simmons & Kick, 1978), and in the auditory cortex of another FM bat, Myotis lucifugus (W.E. Sullivan, unpublished). The importance of FM signals in ranging in both CF-FM and FM bats is thus demonstrated neurophysiologically.

In order to produce the response properties of range-tuned or tracking neurons, lower order neurons with a broad spectrum of recovery cycles (Friend, Suga & Suthers, 1966; Suga & Schlegel, 1973) and facilitation to echoes (Grinnell, 1963b, 1970) are necessary. The neural network ranging model proposed by Suga & Schlegel (1973) consists of three neural levels: (1) tonic neurons with short recovery cycles, (2) phasic latency-constant neurons with a broad spectrum of recovery cycles, and (3) neurons with different gates (or time slots) for echo detection and ranging. Modifying this model in the light of our cortical data, the second level should contain neurons with a recovery cycle which includes a facilitation period preceded and/or followed by an inhibitory period, and the third level should consist of neurons showing facilitation to the pulse-echo pair without a response to the pulse or echo alone. At the third level, range-tuned neurons express the extent to which echo delays match their best delays by the magnitude of their excitation. It is not necessary for this type of neuron to be very phasic and latency-constant, like those at the second level. The model contains the minimum neural elements for ranging. It is, of course, subject to further elaboration in the future. Neurons which respond to both pulse and echo even at short echo delays represent the lowest level of neurophysiological activity in that they simply reflect the tonic responses of peripheral neurons. The auditory system undoubtedly processes information by more elegant means than this.

CF/CF Processing Area

In addition to the FM processing area, we have found a third functional division of the auditory cortex, the CF/CF processing area (Fig. 2B, a; Suga, O'Neill & Manabe, 1979). Neurons here are facilitated by two (or more) harmonically related CF tones, one of which is always CF_1. These harmonic-sensitive neurons are labeled CF_1/CF_n (where n is the particular upper harmonic, n = 2, 3, or 4 or their combinations). The slash means that the harmonic components must be delivered simultaneously to elicit the best facilitation. CF_1/CF_2 neurons form one cluster, CF_1/CF_3 another, and so on. Between clusters are neurons facilitated by either of two upper harmonics combined with the first, e.g., $CF_1/CF_{2,3}$ neurons.

The responses of a CF_1/CF_2 neuron are shown by PST histograms in Fig. 6. This neuron showed some response to CF_1 alone, but no response to CF_2 alone. When CF_1 and CF_2 were delivered simultaneously, however, the neuron responded strongly. The best facilitation occurred when CF_1 and CF_2 were combined in an appropriate amplitude relationship: 50-70 dB SPL for CF_1 and 56 dB SPL for CF_2. The best amplitude relationships were different for different neurons. For facilitation, a precise harmonic relationship between the two CF signals was required. For the CF_1/CF_2 neuron in Fig. 3B, for example, the best-facilitation frequency was 29.75 kHz for CF_1 and 59.39 kHz for CF_2. The CF_2 frequency-tuning curve was always especially sharp.

In contrast to FM_1-FM_n neurons, CF_1/CF_n neurons are facilitated by combinations of CF signals rather accurately harmonically related. Furthermore, their best facilitation occurs when there is no delay between the two components. Therefore, these neurons are not processing echo components in relation to the emitted sounds, nor are they sensitive to target range in the sense that FM_1-FM_n neurons are. Most CF_1/CF_n neurons do not follow well to paired components over about 40 per second, although some do so up to 100 pairs per second. The importance of harmonic-sensitive neurons to the bat is not immediately clear: one notion is that they are used to detect the presence of nearby echolocating conspecifics who may be competing for insect prey in the same area.

Figure 6. PST histograms of responses of a single CF_1/CF_2 facilitation neuron (tuning curves shown in Fig. 3B). The response to the CF_1 alone is shown at the top of the right column, and the responses to the CF_2 alone and CF_2 with CF_1 are shown in the left and right columns respectively. The acoustic stimuli (a.s.) are 34 msec in duration, 0.5 msec rise-decay time. The CF_1 is 29.75 kHz and 46 dB SPL, while the CF_2 is 59.26 kHz and is either 26, 36, 46, 56, 66, 76, or 86 dB SPL. Each PST histogram consists of neural activity for 100 presentations of the same sound. (Suga, O'Neill, & Manabe, 1979; Copyright 1978 Am. Assoc. Adv. Sci.).

The importance of the third and fourth harmonic CF and FM components to the response of significant numbers of FM_1-FM_n and CF_1/CF_n neurons in the cortex of the mustached bat is rather surprising, because these harmonics are thought to be comparatively weak in the bat's sounds. As has been recently reported for the long-eared bat, Plecotus phyllotis (Simmons & Farrell, 1977), perhaps the mustached bat enhances the first harmonic while flying in the open and emphasizes the higher harmonics in confined environments. FM_1-FM_3 and FM_1-FM_4 neurons are better suited for the ranging of smaller targets than FM_1-FM_2 neurons, and the bats may exploit this capacity by actively shunting energy into the harmonics which are appropriate for the required resolution.

J. Ostwald (Poster session abstracts, this volume) reports that, analogous to the mustached bat, there is also a disproportionate cortical representation of frequencies associated with the predominant CF component in the biosonar signals and echoes of the CF-FM horseshoe bat, R. ferrumequinum. He has also found a separate and large area devoted to processing the frequencies associated with the FM component. There appears to be, then, a remarkable convergence in the nervous systems of these two unrelated species, apparently related to the processing of their similarly-structured biosonar signals.

CONCLUSION

In the central auditory system, there are neurons with response properties both similar to and different from those at the periphery. Primary-like neurons of the central auditory system may function to maintain intact the coding of signals by primary auditory neurons, while non-primary-like neurons would be concerned with encoding the signals to extract meaningful information. As we have demonstrated, the acoustic information received by the mustached bat during echolocation is segregated by the central nervous system into at least three separate, functional divisions of the auditory cortex. Neurons in each division are thus specialized to process particular subsets of the information reaching the bat's ears which have obvious relationships to successful orientation and prey capture. That the auditory system encodes and segregates sonic information in this systematic way is a very recent discovery: the amplitopic representation found in the DSCF processing area of the mustached bat (Suga, 1977), the neural map of target range found in the FM processing area of the mustached bat, and the neural map of auditory space found in the barn owl (Knudsen & Konishi, 1978) were the first significant departures from the classical view of vertebrate auditory neurophysiology. Our uncovering of cortical areas devoted to processing complex sounds is particularly germane to the general problem of understanding the neural mechanisms for the processing of meaningful acoustic patterns, and suggests that similar discoveries may be made about auditory processing in other species by employing biologically meaningful stimuli to analyze the auditory system.

EPILOGUE

The finding of neurons with different response properties in different aggregates in the brain is one of the most important recent advances in auditory physiology. This is true to some extent even in the cochlear nucleus, the lowest center of integration. Before this was understood, microelectrodes had been usually inserted into auditory nuclei in a standard orientation and in more or less the same part of the nucleus, so that sampling of single neurons was not random. In such studies, therefore, it is really not significant that a certain type of neuron comprises a particular percentage of the population in a certain nucleus. The discovery of different types of neurons in the nucleus and the detailed description of their response properties are, however, very important.

As a next step, we physiologists should study how neurons with particular response properties are arranged in a particular nucleus, from where and which type of neurons they receive signals, and to where and which type of neurons they send signals. We should go beyond the more classical studies which rely on quantitative descriptions of neural responses only to pure tones. According to our own experience, neurons which are not "interesting" in terms of responses to pure-tone stimuli are often "very interesting" in terms of responses to complex acoustic stimuli relevant to animal behavior. We should therefore use acoustic stimuli which are biologically significant.

With a lack of knowledge of the functional organization of auditory nuclei, lack of adequate control experiments with standardization in stimulus delivery and data processing, and lack of statistical treatment of these data, researchers sometimes reach hasty and inappropriate conclusions and criticize other data as incorrect. These researchers should determine whether two sets of data are adequate for comparison and whether a statistically significant difference between them, if any, is only due to a single factor. Since our knowledge of the system is still limited, hasty judgements do not promote scientific understanding and is somewhat analogous to those reached by "blind men touching an elephant". In order to better understand the auditory system, we must incorporate the data obtained by different scientists in reasonable and constructive ways.

One of the most neglected studies on the echolocation system concerns its fine functional neuroanatomy. New neuroanatomical techniques, such as labelled deoxyglucose, promise to be extremely useful for mapping the neural pathways which are active or are activated by echoes during echolocation. Other labelling techniques such as radioactive amino acids and/or horseradish peroxidase should also be used for detailed analysis of neural connectivity. The combined approach of neuroanatomy with electrophysiology and ethology is essential to promote our understanding of the echolocation system.

This work supported by NSF grant BMS 75-17077 and BNS 78-12987 to N.S. and NICDS (PHS) training grant 1-T32-NS07057-01 to W.E.O.

Supplementary References

Knudsen, E.I. & Konishi, M. (1978). A Neural Map of Auditory Space in the Owl. Science 200, 795-797.

O'Neill, W.E. & Suga, N. (1979). Target Range-Sensitive Neurons in the Auditory Cortex of the Mustache Bat. Science 203, 69-73.

Rose, J.E., Greenwood, D.D., Goldberg, J.M., & Hind, J.E. (1963). Some Discharge Characteristics of Single Neurons in the Inferior Colliculus of the Cat. I. Tonotopic Organization, Relation of Spike-Counts to Tone Intensity, and Firing Patterns of Single Elements. J. Neurophysiol. 26, 294-320.

Suga, N., O'Neill, W.E., & Manabe, T. (1979). Harmonic-Sensitive Neurons in the Auditory Cortex of the Mustache Bat. Science 203, 270-274.

Suga, N. & O'Neill, W.E. (1979). Neural Axis Representing Target Range in the Auditory Cortex of the Mustached Bat. Science (submitted).

Suga, N. (1979). Functional Organization of the Auditory Cortex Beyond Tonotopic Representation. Fed. Proc. Fed. Am. Soc. Exp. Biol. (submitted).

Chapter V
Theories and Models of Echolocation

Chairman : B. Escudié

co-chairmen
R.W. Floyd   and   J.A. Simmons

- Underwater
    Models of cetacean signal processing.
        R.W. Floyd

    Models for echolocation.
        R.A. Altes

    Energy spectrum analysis in echolocation.
        R.A. Johnson

- Airborne
    The processing of sonar echoes by bats.
        J.A. Simmons

    Signal processing and design related to bat sonar
    systems.
        B. Escudié

    Models of spatial information processing in biosonar
    systems, and methods suggested to validate them.
        P. Greguss

    A new concept of echo evaluation in the auditory system
    of bats.
        K.J. Beuter

MODELS OF CETACEAN SIGNAL PROCESSING

Robert W. Floyd

Naval Ocean Systems Center

Kailua, Hawaii 96734

INTRODUCTION

There are several reasons for studying signal processing in dolphins in addition to the pure delight in understanding this mechanism. The most immediate benefit is in being able to predict and understand the results of echolocation tasks assigned to the animal. This is especially important in psychophysical experiments where one wishes to know if a threshold is the result of the behavior of the animal or is actually determined by the stimuli and the ability of the animals' receiver to detect it. Another reason for studying signal processing is to discover what principals could be used by man in such areas as aids to the blind, sonar, geophysical exploration, etc. Finally, on a metaphysical level, knowledge of how the dolphin processes echoes helps to understand how the animal perceives his environment.

Although most of the signal processing theories can be applied conceptually to both bats and dolphins, there are differences in the medium that make the practical applications of techniques different. The first major difference is that in air, there is a great difference in impedances between the source (bat), the medium (air), and the target, while in water the differences in acoustic impedance are much less. The result of this is that while it is possible for an animal in water to produce an impulse-like sound at relatively high power density levels, it is much more difficult to do so in air. Dolphins are able to transmit signals of RMS amplitudes of 10^6 ergs/cm^2 sec and total energies of ~ 80 ergs/cm^2 (Au 79). Bats, as reported by Simmons (Simmons, 1979) emit signals on the order of 10^{-3} ergs/cm^2 sec. Even assuming that bats use much higher signals in the field than in the lab, it is obvious that dolphins still have peak intensities several orders of magnitude greater than bats. Thus bats are forced to use much longer pulses to get a reasonable amount of energy into the medium.

This difference in time-bandwidth product (TBW) has important implications as to the optimality of a given processor. For dolphin signals, whose TBW approaches the theoretical minimum (Au, 1979), it has been demonstrated that an energy detector is as effective as a matched filter (Urkowitz, 1967). However, as the TBW product increases, the number of degrees of freedom increase also, and the effectiveness of an energy detector decreases. Even without a matched filter, detection can be improved if one has a prior knowledge of the time frequency distribution of the expected signal, as happens in FM sweeps of fixed duration. Attempts to demonstrate this in humans has failed, but such an effect has been demonstrated for Rhinolophus ferrumequinum (Vater, 1979).

The next important difference relates to the relative impedances between the fluid and the potential target. In air, the differences are great enough so that there is very little penetration. For targets in water, however, there is considerable penetration, and internal reflections give important contributions to the target echo. Finally, the velocity of sound in water is approximately five times as great as it is in air. Since Doppler stretch is related to the target velocity divided by sound velocity, Doppler resolvency or tolerance questions are much less important to a dolphin using a short pulse.

The classic method of detecting a known signal in noise is to whiten and match. In other words, the noise spectrum is measured, a pre-filter is constructed that will flatten (or whiten) the spectrum, and a correlator matched to the whitened version of the expected signal makes the detection. Aside from the lack of physiological evidence of a matched filter, there are several theoretical reasons why the matched filter model may be inappropriate. First, the interferring noise in most cases is not gaussian or stationary, but actually reverberation from the reflection of the animal's outgoing signal with the clutter in its environment. This makes the whitening part very difficult because the reverberation has an average spectrum similar to the outgoing pulse. Secondly, the target echo is almost never known exactly a priori, so that the receiver would usually not have a stored echo for correlation, unless the receiver had the capability of storing nearly infinite numbers of echoes for purposes of comparisons.

A better way of approaching the target recognition/detection problem would be to choose some feature set that could describe the target and base the signal processing scheme on these features. For humans, using a visual modality, the natural way to classify

targets is by size and shape. Seville Chapman (Chapman, 1968)
gives an interesting discussion on how this might be approached
with a sonar system. His discussion was related to holography,
where one of the objectives was to measure the scattering centers
of the target. True holography is unlikely in the case of the
dolphin, because it requires a signal with extremely good phase
coherence whose length is large compared to the dimensions of the
target, neither of which is true for the dolphin (Au 1979).
However, there are other ways for measuring highlights. Given a
broadband signal, one way to accomplish this is with a replica
correlator. This detector passes the received signal through a
filter matched to the outgoing signal. The output of this filter
is a series of spikes in the time domain related to the time of
occurrences of the scattering centers, and whose amplitude is
related to the strength of the scatterer and to how well it is
correlated with the original signal. This model is attractive
because it approaches a matched filter in detecting signal in
noise and gives an intuitively useful description of the target.
The work of Au and Hammer (Au, Hammer 1979) indicates the dolphin's
ability to discriminate metal cylinders in both qualitatively and
quantitatively related to the output of a replica correlator.
Unfortunately, the physiological and neurological evidence suggests
no method by which true replica correlation could be accomplished.

   Another potentially useful feature set would be shape dependent
scattering functions. Although this technique had been applied to
radar signal analysis, R. A. Altes was the first to apply it to
animal sonar. The original model was the bionic coefficients
scheme, which represented the target transfer function as a power
series expansion in ω. This method was actually implemented on an
experimental sonar, the US Navy AN-SQQ27XN-2. The processor consisted of filters matched to a -5 to +5 expansion in ω of the
original waveform,

$$U(\omega) = \omega^{\nu} \exp \left\{ 2\pi jc \frac{\ln \omega}{\ln k} - \frac{(\ln(\omega))^2}{2 \ln (k)} \right\}$$

The arrival time of the echo was estimated and the filters sampled
at that instant. Since the filters overlapped, a matrix inversion
was performed to estimate the true value of each coefficient.
However, this system turned out to be impractical for several
reasons. First, the scattering model only applied to certain
idealized shapes and did not take into consideration the arrival
of components other than at one instant. Furthermore, the filters
were non-orthogonal. In other words, the energy in one filter
contributed to the output of other filters. The actual values
were separated by a matrix inversion, which left the system sensitive to noise. Some of these problems were alleviated by using
the geometric theory of diffraction (GTD) which used non-coherent

filters and a 1/2 power expansion in ω for the target transfer
function. When the outputs of the filters combined with a timing
parameter were used, this system performed reasonably well at
moderate to high signal-to-noise ratios. However, it still suffers
from not considering resonances and from having overlapping filters.
In addition, it is difficult to find a known biological mechanism
that could implement such a scheme.

Another group of theories are related to exploiting resonances
in the target. This is at first glance appealing, because the
modes of resonance are strictly related to the target structure.
Furthermore, any receiver that can perceive pitch would be an
effective detector. In actual fact, many resonances were stimulated by a broad band pulse and the echo spectrum depended on the
aspect of the target. This method by itself did not work very
well (Chestnut and Landsman, 1977). Another all pole model was
the use of linear prediction theory. This model assumes that the
output is the sum of the original signal plus delayed version
multiplied by same coefficient added to it. This method has been
used successfully in speech synthesis and recognition. The major
theoretical objections to this method is that the linear prediction coefficients are not unique, i.e., the same output can be
synthesized with many different combinations of coefficients. In
spite of this, the method worked well at moderate signal-to-noise
ratio (Chestnut and Landsman, 1977). These results are discussed
further in R. A. Johnson's review.

All of the above theories have started from target models
and then moved to signal processing methods based on exploiting
these models. Another approach is to start with what we know
about dolphin hearing and move from there. It is known, for
instance, that the dolphins auditory system can be modeled as a
bank of constant Q or proportional bandwidth filters (Johnson,
1968) which is functionally similar to the human auditory system.
Each of these filters is followed by a rectifier or square law
detector and then by an integrator whose time constant is approximately 1/BW (Jeffress, 1968). The output of such a bank of
filters would represent a spectrogram if all the energy occurred
during the integration period, or a time spectral history for
multiple highlights following outside this interval. Although
attractively simple, this method on the surface appears less than
optimal. Its performance is inferior to that of a replica correlator because both signal and noise are summed together incoherently
during the integration time of the filter. It also does not explain the demonstrably fine range resolution (Murchison, 1979)
of echolocating animals. However, spectrogram models have proved
the most successful of the models thus far at detecting and classifying echoes in introduced gaussian noise (Chestnut, Altes 1979).

In spite of the previously stated objections, there are ways
in which the spectrogram models can result in correlation equivalent processing gain and fine range resolution. If a spectrogram is taken of two wide band signals separated by some delay τ,
then ripples in the spectrum will appear at intervals $\Delta f = 1/\tau$
(Johnson et. al., 1976). These ripples are perceived by the auditory system as a pitch whose frequency is approximately Δf. This
phenomenon is known as time separation pitch (TSP) and has been
demonstrated both in humans (McClellan, 1967) and in the bottlenosed dolphin, Tursiops truncatus (Caine, 1976). One virtue of
this method is that it can be used to explain the results of many
dolphin echolocation experiments. For instance, the results of
the range resolution experiments (Murchison, 1979) were consistent
with what one would predict using the TSP model. As a method of
measuring range, the TSP model takes the spectrum, the outgoing
signal and returning echo, and produces a tone whose pitch is
inversely related to the distance. For very short distances, the
range resolution would depend on the auto correlation function of
the signal. As the range increased, resolution would be determined by the ability to resolve differences in the time difference

Figure 1. Spectra and replica correlator outputs for metal targets.

pitch. The results of Murchison fit this hypothesis if the porpoise's frequency resolution ($\Delta f/f$) equals .005. As mentioned
previously, the results of Hammer and Au indicate the classification performance closely follows a replica correlator. However,
the spectrogram model gives nearly equivalent results to the
replica correlator. Figure 1 shows the spectra and replica
correlator outputs of aluminum, bronze and glass cylinders. The
nulls in the spectrum occur at $1/\tau$ where τ is the time of arrival
of the secondary peak. The strength of this phenomena is
related to how well correlated the secondary peak is to the
primary. This implies that this model is correlation equivalent,
and may explain the success of animals in resisting jamming. One
problem this has as a complete model of echolocation is that it
cannot be used at longer ranges. First, as the integration
time increases between transmitted and received signal, the signal-
to-noise ratio becomes poorer. Secondly, the Δf decreases as
range increases until it is no longer discriminable.

SUGGESTIONS FOR FUTURE EXPERIMENTS

Briefly, there are four likely types of signal processing
models. The first, match filter or replica correlation, assumes
an ideal receiver and would give the best results for single or
multiple highlight targets in gaussian noise. Since the time
bandwidth (TBW) product of the dolphin is 1, another model that
would work as well for the single glint model is a simple energy
detector whose bandwidth equals that of the echo and whose integration time is 1/BW. However, this model would not perform as
well as a coherent receiver for multiple glint or greater TBW
product signals. A model based on the energy in a bank of constant
Q filters would require a 3 dB greater signal-to-noise ratio (S/N)
to have the same performance as the coherent models for either the
single or multiple glint echoes. The TSP model would predict also
a 3 dB poorer result for the single glint target, but for a
double highlight target the performance should approximate that
of a matched filter.

One problem in designing an experiment to test these hypotheses is that all of these are based on a single ping. If the
animal is allowed to emit multiple pings, one of three processes
could take place. First, the echoes could all be summed coherently, although this is likely only for the matched filter
hypothesis. In this case, the S/N would go up as log N, where N
is the number of echolocation clicks. A more likely case is that
the energy is summed noncoherently, in which case the S/N would
increase by $\log \sqrt{N}$. Finally, each echo could be treated as an
independent event, which would give approximately a $\log \sqrt[4]{N}$ for
an animal working near the minor diagonal on the Receiving
Operating Curve (Green and Swets, 1974). To avoid the problems

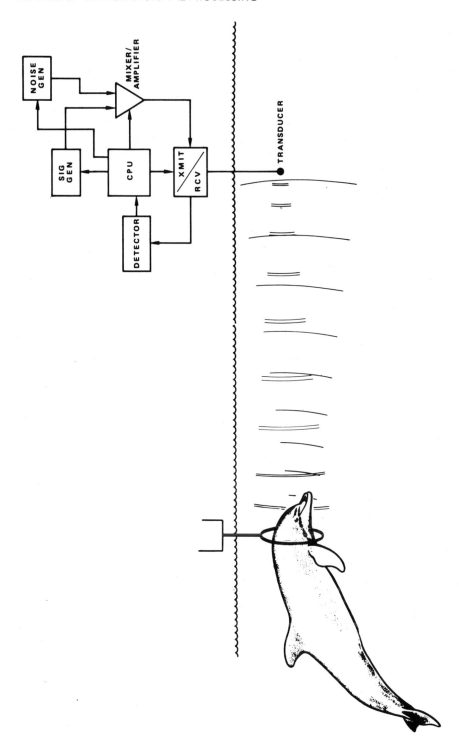

Figure 2. Experimental design for determining S/N at which dolphin can detect synthetic echoes.

of sorting out the multiple signal hypotheses, it would be preferable to have the animal make the decision on only one echolocation click. This could be accomplished by skillful training, or by using the following experimental set (Figure 2). The dolphin is trained to station in a hoop and to echolocate on a phantom target. This target is originally created by transmitting a synthetic echo each time the animal's echolocation pulse is detected. This is later changed so that a phantom echo is created only on the first pulse. This phantom echo always has the same energy, but may consist of either a single highlight or dual highlight. Gaussian noise is added to the stored echo, and the dolphin's performance is measured as a function of S/N. From this experiment, we would predict the following results:

If the matched filter hypothesis is true, then

	Single Glint	Double Glint
	$d^o = {}^{Eo}/No_2$	${}^{Eo}/No_2$
Energy detector, same Time-Bandwidth as transmitted signal	$d^1 = {}^{Eo}/No_2$	${}^{Eo}/{}_2 No_2$
Bank of Constant Q filters	$d^1 = {}^{E}/_2 No2$	${}^{Eo}/2No_2$
Time Separation Pitch	$d = {}^{E}/2No_2$	${}^{Eo}/No_2$

REFERENCES

Au, W. W. L., 1979, Echolocation signals of the Atlantic bottlenosed dolphin (Tursiops truncatus) in open waters, this volume.

Au, W. W. L., and Hammer, C., 1979, Target recognition via echolocation by Tursiops truncatus, this volume.

Caine, N. G., 1976, Time separation pitch and the dolphin's sonar discrimination of distance, Ph. D. Diss., San Diego State University, San Diego, California.

Chapman, S., 1968, Dolphins and multifrequency, multiangular images, Science, 160:208.

Chestnut, P., and Landsman, H., 1977, Sonar target recognition experiment, ESL Inc. Report N° ESL-ER178, 495 Java Dr., Sunnyvale, California.

Green, D. M., and Swets, J. A., 1966, "Signal Detection Theory and Psychophysics", J. Wiley, New York.

Jeffress, L. A., 1968, Mathematical and electrical models of auditory detection, J. Acoust. Soc. Amer., 44:187.

Johnson, C. S., 1968, Relation between absolute threshold and duration-of-tone pulses in the bottlenosed porpoise, J. Acoust. Soc. Amer., 43:757.

Johnson, C. S., 1968, Masked tonal thresholds in the bottlenosed porpoise, J. Acoust. Soc. Amer., 44:965.

Johnson, R. A., and Titlebaum, E. L., 1976, Energy spectrum analysis: a model of echolocation processing, J. Acoust. Soc. Amer., 60:484.

McClellan, M. E., and Small, A. M., 1967, Pitch perception of pulse pairs with random repetition rate, J. Acoust. Soc. Amer., 42:690.

Murchison, A. E., 1979, Detection range and range resolution of echolocating bottlenosed porpoise (Tursiops truncatus), this volume.

Simmons, J. A., 1979, Perception of echo phase information in bat sonar, in press.

Simmons, J. A., 1979, Processing of sonar echoes by bats, this volume.

Urkowitz, H., 1967, Energy detection of unknown deterministic signals, Proc. IEEE, 55:523.

Vater, M., 1979, Coding of sinusoidally frequency-modulated signals by single cochlear nucleus neurons of Rhinolophus ferrumequinum, this volume.

MODELS FOR ECHOLOCATION

Richard A. Altes

ORINCON Corporation

3366 N. Torrey Pines Ct., La Jolla, CA 92037, USA

I.   INTRODUCTION

Mathematical or engineering models of biological systems are viewed with extreme skepticism (if not with hostility) by some experimental biologists. A discussion about models should take account of this attitude, since the work may otherwise be doomed to obscurity, even if the results are correct. The skepticism about modelling can be attributed to one or more of the following arguments:

1. Models are unnecessary if experiments result in a good intuitive and verbal understanding of the functions that are performed by a biological system. If animal behavior and physiology can be understood in non-mathematical terms, the value of a mathematical model becomes questionable. "Physics is mathematical not because we know so much about the physical world, but because we know so little; it is only its mathematical properties that we can discover" (B. Russell, 1927). In this respect, biology may be fundamentally different from physics.

2. Models that are based upon insufficient data can be misleading, and it is not clear when the data base is sufficiently large to justify the formulation of a model. In A. C. Doyle's "A Scandal in Bohemia," Sherlock Holmes tells Dr. Watson: "It is a capital mistake to theorise before one has data. Insensibly one begins to twist facts to suit theories, instead of theories to suit facts." Even in engineering and physics, theories often rest upon assumptions that are "tractable" but overly simplistic.

3. Models should improve our comprehension of nature by using engineering concepts to interpret biological data, but the end result may be an overly restrictive or narrow viewpoint. Like philosophies and religions, models suggest a definite conceptual framework for the interpretation of nature, and the adoption of one such structure can preclude other interpretations. Engineers and physicists also have a tendency to favor theories that possess beauty, order, or simplicity. This attitude is nicely expressed by the physicist P. Dirac (1963). A predisposition toward mathematical elegance can again   result in a viewpoint that is too restrictive for biology.

4. Aside from possible improvement of man-made systems, there is sometimes no obvious benefit to be gained by constructing a model. Model makers often fail to suggest experiments to test their constructions, and these experiments are of primary importance to biologists. The result is an elaborate theory that is based largely upon conjecture--a house of cards without a foundation of check points or experiments that can be used to verify its relevance to nature.

The above list of objections to mathematical formulations can be used by the model maker to identify weaknesses in his exposition. Indeed, it will become apparent that items from the list apply to some past and present models for animal echolocation. In spite of the above objections, the following arguments seem to justify a modelling approach.

1. Models are a means for cross-fertilization between biology and engineering. The importance of this effect is described by A. Koestler in The Act of Creation (1964): "All decisive advances in the history of scientific thought can be described in terms of mental cross-fertilization between different disciplines." Examples in the study of animal echolocation:

(i) The mystification of L. Spallanzani concerning the interpretation of his bat navigation experiments, due to the fact that the wave theory of sound (and thus the concept of ultrasound) had not been discovered (Griffin, 1958).

(ii) Radar theory indicates that the E-E neurons of Menabe, Suga, and Ostwald (1978) can be used not only for detection, but also for estimation of echo strength (as discussed in Part VI of this section). This idea implies that detection and estimation may be performed by the same neuronal components.

2. Models provide a mechanism for expanding our conceptual framework and our vocabulary for describing biological data. For example:

(i) A study of angle estimation from the viewpoint of
binaural processing (Altes, 1978a) indicates that sound localization depends not only upon beamwidth of transmitter and
receiver (i.e., upon the amplitude of an array pattern) but
also upon the phase of this pattern as a function of frequency
and angle. (We are monaurally insensitive to a constant phase
shift, but not to a phase shift that varies with frequency.)
This concept suggests that many physical measurements of transmitted beam width and directional response of the pinna are
incomplete.

(ii) The discovery of excitatory-center, inhibitory-surround
receptive fields for sound localization in the midbrain of the
owl (Knudsen and Konishi, 1978) can be related to vision and
Von Bekesy's "neural sharpening" ideas. But such receptive
fields can also be interpreted as part of a principal component analysis for spatial pattern recognition (to be discussed in this section), or as part of a sequence of spatial
filters that are sensitive to the acoustic counterpart of
lines, edges, and more gradual contrast changes (Altes, 1976a;
Hauske, Wolf, and Lupp, 1976).

3. A model provides a method for evaluating the significance
of specific biological data. For example:

(i) Suppose that we have a set of neurons that fire in
response to a received echolocation pulse, but the firing
times exhibit a great deal of jitter. A model which forms
a best estimate of pulse arrival time reveals that a very accurate arrival time estimate can be obtained by averaging statistically independent neuronal responses. The variance of
the arrival time estimate is inversely proportional to the
number of independent neuronal responses that are averaged
together.

(ii) Green (1958) and Green, McKey, and Licklider (1959)
have deduced a summation law for auditory detection of multiple
sinusoids in humans. Detectability depends upon the sum of
the squares of the energies of the different components. This
behavior is found in a spectrogram correlation process, which
is a locally optimum detector for signals with known covariance
function (Altes, 1978b).   This detector and its relevance to
echolocation will be discussed further on.

(iii) An empirically determined equation for just-noticeable
difference in humans (Siebert, 1970) is commensurate
with the use of a spectrogram correlator for frequency
discrimination.

4. A definitive model can transform a conjecture into a hypothesis that can be experimentally tested, by predicting a particular aspect of animal behavior. If the prediction is supported by data, the conjecture may be correct. Models thus provide the experimenter with a feedback mechanism, such that conjectures from past data can be transformed into meaningful future experiments. Examples:

(i) An energy spectrum analyzer (Johnson and Titlebaum, 1976) is particularly susceptible to noise with a rippled power spectrum, e.g., noise added to a delayed version of itself (Johnson, 1972). A broadband matched filter is relatively unaffected by this kind of interference. This difference can be used as the basis for jamming experiments with bats and cetaceans, to differentiate between models.

(ii) If a spectrogram correlator is used for estimation of time delay, then we expect a just-noticeable delay difference $\Delta\tau$ that varies as $B^{-\frac{1}{2}}$, where B is the bandwidth of the echolocation signal (Altes, 1978b). For estimation with a matched filter, $\Delta\tau \sim B^{-1}$. Once again, one can construct experiments to test these hypotheses (spectrogram correlator versus matched filter).

5. A model provides a test of one's understanding if it can actually be built, or if its performance can be simulated on a computer. It can then be seen whether the model really "works" for realistic data, a necessary condition for the validity of the conceptual framework. For example, the geometrical theory of diffraction (Bechtel, 1976) along with concepts of Doppler tolerance (Altes and Reese, 1975) or simplicity of implementation (Altes, 1976b) suggest that target echoes should be divided into highlights, and that each highlight should be characterized by an echo $f_n(j\omega)^{n/2}U(\omega)$, where f_n is a constant, $U(\omega)$ is the transmitted signal in the frequency domain, and n is an integer between -2 and +2. A software implementation with real echo data, however, has revealed important deficiencies in this approach (Skinner, Altes, and Jones, 1977) and has necessitated a search for a different model of target discrimination capability in animal echolocation.

6. Although the concept of elegance or beauty has been listed as a disadvantage in applying mathematical models to biology, the quest for elegance and a unifying conceptual framework must also be listed as a fundamental motivation for much scientific inquiry (Koestler, 1964). An "inner beauty" is often found in a mathematical formulation that is truly descriptive of reality, "but sometimes the truth is discovered first and the beauty or 'necessity' of that truth seen only later" (Feinman, 1974). Mathematical models provide a pathway from a confusing melange of data to a "beautiful"--and hopefully simple--conceptual framework.

The remainder of this section describes some mathematical models that have recently been applied to animal echolocation.

Part II describes models that are associated with correlation or matched filtering (followed by envelope detection) and the point target assumption. Part III discusses the use of correlation processing for range-extended targets. Part IV involves a description of a target in terms of a scattering function or a time-frequency energy density function, and the use of spectrogram correlation to detect and classify such targets. Part V describes estimation and detection performance when spectrograms are used to describe an echo, and behavioral tests that can be used to distinguish between spectrogram correlation and matched filtering. Part VI discusses theories of binaural interaction and sound localization, and Part VII is concerned with the issue of phase sensitivity in echolocation. Part VIII discusses the results in terms of the objections to modelling that were expressed at the beginning of this introduction.

II.   CORRELATION OR MATCHED FILTERING FOLLOWED
      BY ENVELOPE DETECTION

a.   Elementary Target Models

     The simplest assumption concerning a radar/sonar reflector is that it behaves as a point target. The echo from a motionless point target is an exact replica of a signal that is transmitted from a motionless sonar, delayed in time. If the target is modelled as a linear filter, then the impulse response of this filter is assumed to be an impulse. The point target assumption can be applied to range-extended targets by assuming that the target echo is composed of a sequence of highlights or specular reflections (Freedman, 1962), and that each of these highlights behaves as a point target.

     A point target model implies that the echo from a target with zero range rate is exactly known except for delay, if the transmitted signal is known. In a background of white, Gaussian noise, the ideal detector for a point target echo is then a matched filter or correlator (Woodward, 1964; Cook and Bernfeld, 1967).

     A slightly more general point target model assumes that the target impulse response is a delta function multiplied by a complex constant with unknown phase. The echo is then known exactly except for delay and phase. In this case, a non-adaptive, locally optimum echo detector consists of a matched filter or correlator followed by an envelope detector (Altes, 1979b). A locally optimum detector is one that is optimized for low signal-to-noise ratio (SNR), where optimality is most important (Capon, 1961; Middleton, 1966). Such a detector may be non-ideal for higher SNR, but suboptimal behavior in such cases is often permissible. If a sequence of echoes is available, and if each echo has the same unknown phase, one can sequentially estimate the probability density function (pdf) that describes the phase, on the basis of accumulated echo data. This updated pdf can be used to adaptively improve the performance of the detector for each new echo (Roberts, 1965; Hodgkiss, 1978).

Matched filtering is also relevant to any extended target that can be modelled as a time-invariant linear filter. In this case, the echo is determined by convolving the transmitted signal with the target impulse response. A reasonable way to measure the impulse response is to transmit an approximation to an impulse. Because of transmitter power limitations, however, it is often preferable to transmit a broadband chirp. The response of a filter that is matched to the transmitted signal then approximates the target impulse response, if the signal is sufficiently broadband (Turin, 1957; Altes, 1977).

b. Doppler and Acceleration Tolerance

The maximum matched filter response to a point target or to a sequence of highlights is reduced when the signal is scaled because of target motion. A filter that is matched to the original signal is mismatched to the scaled signal. One can exploit this effect to both detect the echo and to measure target velocity by using a bank of filters, each of which is matched to a differently scaled version of the transmitted signal (Cook and Bernfeld, 1967; Rihaczek, 1969; Vakman, 1968). To detect a Doppler-scaled echo without resorting to a whole bank of filters, a Doppler-tolerant signal can be used. When a Doppler-tolerant signal is transmitted, the mismatch between the scaled signal and the original signal does not cause significant degradation of the envelope detected matched filter response.

A graphical derivation of such a signal (Altes and Titlebaum, 1975) is shown in Fig. 1. A picture of the transmitted signal is projected from P onto a screen at OT. A time-compressed picture of the signal (corresponding to an echo from a target that is getting closer to the sonar) is obtained by inserting a second screen closer to the projector. The last zero crossing of the unscaled signal occurs at time T on the far screen, and this zero appears at time T/s on the near screen. For Doppler tolerance, the zero at T/s should have a counterpart at time A = T/s on the far screen. The zero at time A projects back and intersects the near screen at time B. A zero should therefore be found at time C = B on the far screen. This process shows that the distance $T_i(t)/2$ between a zero crossing at time t and the next zero crossing is

$$T_i(t)/2 = (s-1)t \qquad (1)$$

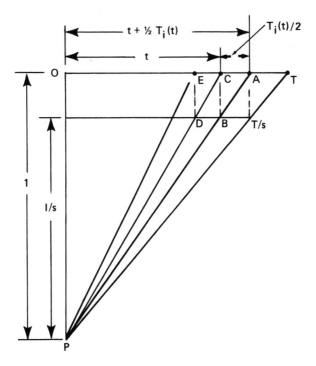

Fig. 1. Graphical derivation of zero crossings
for Doppler tolerance.

where s is the scale factor that is introduced by target motion,
i.e.,

$$s = (1+v/c) \ / \ (1-v/c) \approx 1 + 2v/c \ . \tag{2}$$

In (2), v is range rate and c is the speed of sound.

Suppose that a signal with zero crossings as in (1) is scaled
by a factor s_1 rather than s, where $s_1 = 1 + v/c$. The zero crossings of the scaled signal fall between those of the unscaled signal,
i.e., a third screen is inserted midway between the two screens in
Fig. 1. If the unscaled signal is a frequency modulated cosine
function, then the signal with scale factor s_1 can be written as a
frequency modulated sine function. A signal with an arbitrary scale
factor can be written as a linear combination of a cosine function
and a sine function. This observation implies that, for Doppler
tolerance, the receiver should use a signal with zero crossings
described by (1) together with two correlators, one of which corresponds to a frequency-modulated cosine function and the other to
a frequency modulated sine function. The sum of the squares of the
correlator outputs is relatively insensitive to Doppler-induced
scale changes. The response of each correlator, however, is

Doppler sensitive, and target velocity can be estimated by computing the arctangent of [the sine correlator response divided by the
cosine correlator response], i.e., by measuring the phase of the
echo (Altes and Skinner, 1977). This velocity estimate requires
only two matched filters, rather than the whole bank of filters
that was mentioned earlier.

In practice, a Doppler-tolerant receiver can be obtained by
passing the squared response of a single matched filter or correlator through a low pass filter or integrator. This envelope
detection process is approximately equivalent to summing the squares
of the outputs from the sine and cosine filters (Davenport and
Root, 1958).

A signal with zero crossings as in (1) has a period $T_i(t)$ that
is proportional to time, and it is called a linear-period modulated
(LPM) signal (Rihaczek, 1969; Kroszczynski, 1969). The instantaneous frequency is inversely proportional to $T_i(t)$, and it appears as
a hyperbola on a spectrogram plot. Such signals are employed by
many of the so-called FM bats (Cahlander, 1966; Griffin, 1958), and
they are used by other bats such as Rhinolophus when these bats
are close to the target of interest. One can also model some
cetacean echolocation clicks with LPM signals, although the clicks
have relatively few zero crossings (Altes and Reese, 1975; Altes,
1976b).

The technique in Fig. 1 can also be used to obtain signals that
are both Doppler and acceleration tolerant. By tilting the near
screen as in Fig. 2 and following the same procedure as before, we
obtain the counterpart of (1) for targets with variable velocity,
viz.,

$$T_i(t)/2 = [s(t)-1]t .\tag{3}$$

When the change in velocity over the duration of the pulse is much
less than the initial velocity, we obtain (Altes and Titlebaum,
1975)

$$T_i(t) = t/(C_0+C_1 t)\tag{4}$$

where C_0 and C_1 are constants. Eq. (4) describes an instantaneous
period that increases linearly with time and then levels off to a
constant as $C_1 t$ becomes much greater than C_0. The instantaneous
frequency first decreases in proportion to $1/t$ and then levels off
to a constant frequency toward the end of the pulse. Signals of
this type have been measured from Lasiurus borealis in the laboratory (Cahlander, 1966) and from Pipistrellus subflavus in its
natural environment (data obtained by Simmons and Griffin).

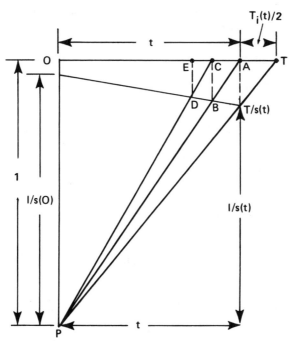

Fig. 2. Graphical derivation of zero crossings for
Doppler and acceleration tolerance.

c. Doppler Sensitivity

The above observations demonstrate a lack of uniqueness that
sometimes emerges from a theoretical interpretation of data. Having
found LPM signals in nature, we have been inclined to label the corresponding animal sonar system as Doppler-tolerant, although velocity can still be measured with more than one pulse. If a phase
measurement can be obtained, however, LPM signals can be used to
estimate radial velocity on the basis of a single echo.

Such a uniqueness problem can sometimes be solved by considering additional data and by careful analysis of theoretical assumptions. For example, a joint velocity and range measurement with
LPM signals is severely degraded unless range-Doppler coupling
errors are eliminated. The formulation of Altes and Skinner
requires LPM signals to be symmetric in time in order to eliminate
range-Doppler coupling, and bat signals do not display such symmetry. The original symmetry requirement, however, can be generalized,
and it can be shown that range-Doppler coupling is eliminated if an
LPM signal is preceded or followed by a constant frequency tone
with appropriate amplitude and duration. This situation occurs for
CW-FM bats, and for some "FM" bats such as Pipistrellus and

Lasiurus, which might better be labelled "FM-CW" bats. Signals that
are Doppler and acceleration tolerant when a matched filter is
followed by an envelope detector can thus also be Doppler sensitive,
with no range-Doppler coupling, if phase is measured.

For frequencies above 6 kHz, the auditory nerve cannot encode
the timing information that allows one to distinguish between a
sine and cosine function (Anderson, Rose, Hind, and Brugge, 1971).
It can therefore be argued that phase information is lost for ultrasonic signals, and a velocity measurement is impossible. Nevertheless, behavioral and neurophysiological evidence indicates that
humans and cats are sensitive to the relative phase between two
signals (Patterson and Green, 1970; Kotelenko and Radionova, 1975).
The energy spectrum of a signal with Fourier transform $U(f)\exp(j\phi)$
is $|U(f)\exp(j\phi)|^2 = |U(f)|^2$, and the phase ϕ is eliminated. For
the sum of two signals, however,

$$|U_1(f)\exp(j\phi_1) + U_2(f)\exp(j\phi_2)|^2$$

$$= |U_1(f)|^2 + |U_2(f)|^2 + 2\,\mathrm{Re}\{U_1(f)\,U_2^*(f)\exp[j(\phi_1-\phi_2)]\},$$

$$(5)$$

and the energy spectrum depends upon the difference $\phi_1-\phi_2$. Since
bat signals often include harmonics, the relative phase shift
between harmonics can be measured, and this quantity contains
velocity information. Two more methods of phase measurement that
do not require harmonics will be discussed in Part VII.

On the basis of past experimental data, it is not clear whether
velocity can be measured from a single echo by bats that use LPM
signals. This situation provides an example of the feedback from
models to future experiments that was listed in the introduction.
An appropriate behavioral measurement should be able to answer the
questions: (i) Are FM bats sensitive to the difference between
unscaled and scaled LPM signals, for very small scale changes?
(ii) Does this sensitivity (if it exists) depend upon the presence
of harmonics?

The possibility of phase measurement has been considered in
the context of hearing models, but what about the possibility of
matched filtering? Given that the peripheral auditory system
behaves as a bank of overlapping filters that are followed by envelope detectors (Siebert, 1968; Evans, 1977), how can one justify a
matched filter or correlator model? A matched filter followed by
an envelope detector is, in fact, theoretically feasible because the
original input signal can be reconstructed from its spectrogram,
aside from a constant, unknown phase shift (Altes, 1978b,d).

d. Signal-Filter Design for Maximization of
    Signal-to-Interference Ratio

It would seem that any mathematical model for a biological
system should be able to adapt to environmental changes. In
sonar, the environment is characterized by noise and reverberation
(clutter) with particular statistical characteristics. Detector
performance in a given clutter environment can be measured in
terms of the signal-to-interference ratio (SIR), where

$$
\text{SIR} = \cfrac{\text{Maximum detector output when input = target echo}}{\left\{ \begin{array}{l} \text{[Expected detector output when input = noise} \\ \text{(no signal transmitted)] + [Expected detector} \\ \text{output when input = clutter reflections} \\ \text{(reverberation)]} \end{array} \right\}} \tag{6}
$$

When the detector consists of a matched (or mismatched) filter
followed by an envelope detector, and when a point target model is
used, it is possible to find the best signal-filter pair for maximizing SIR. The resulting signal and filter functions depend upon
environmental clutter and noise conditions. The original work in this
area was performed in a radar context using narrowband signals (Delong
and Hofstetter, 1969; Stutt and Spafford, 1968; Rummler, 1966). The
radar formulation was then generalized (at the suggestion of E. L.
Titlebaum) to include wideband signals that apply to animal sonar
(Altes, 1971a; 1971b).

The SIR approach has recently been further extended by
M. Decouvelaère (this volume). Another extension, in which the
point target assumption is replaced by a distributed target with
known impulse response, has been considered by several authors
(Harger, 1970; Lee and Uhran, 1973; Altes and Chestnut, unpublished).

A problem with the SIR formulation is that it is not specifically adaptive, i.e., it does not specify a method for updating
the clutter or reverberation description and/or the estimate of
target impulse response. Recent adaptive detector formulations
(Picinbono, 1978; Brennan and Reed, 1973) allow the receiver configuration to change with the environment when the transmitted
waveform is fixed. When the signal is allowed to vary, however,
adaptive estimation of the clutter statistics (which are signal
dependent) becomes more complicated.

Bel'kovich and Dubrovskiy (1976) have argued that an adaptive
receiver without an adaptive waveform generator may be a reasonable
model for dolphin echolocation. Au, et al. (1974), however, have
found that Tursiops seems to use unusually high frequencies for
long range detection, apparently to minimize the effect of

low-frequency noise. Higher frequencies also result in a narrower
beam width, which implies spatial as well as temporal discrimination
against omnidirectional low-frequency interference.

Dziedzic and Alcuri (1977) have observed a target-dependent
signal change at close range in Tursiops. It would seem that waveform changes at close range should be associated with target characterization behavior, e.g., estimation of target impulse response
in the presence of clutter and noise (Altes, 1977), rather than
with a detection method such as SIR maximization.

III. CORRELATION PROCESSING FOR CHARACTERIZATION OF
      RANGE-EXTENDED TARGETS

The echo from a range-extended target can be parameterized
in many ways. A standard method for analyzing such an echo is to
correlate it with a sequence of orthonormal functions (Fig. 3).
When echo arrival time is unknown, the correlation operations can
be performed with linear filters (bandpass filters for Fourier
series analysis), and the sum of the squares of the filter outputs
(noncoherent detector response) can be used to detect the target
echo and to generate a sampling time, as shown in Fig. 3. The
desired parameters or expansion coefficients are the sampled filter
outputs, where samples are taken at the instant when the sum of
the squared responses is maximized. A matched filter for a known
signal can be synthesized by multiplying the output of each orthonormal filter by an appropriate coefficient and by summing the
weighted filter outputs (coherent detection). The process can be
made adaptive by using a weighted sum of coherent and noncoherent
detector outputs, where the weights depend upon the receiver's
uncertainty about the appropriate coefficients to use in the
coherent detector (Glaser, 1961).

Many different sets of orthonormal basis functions can be
used; sines and cosines are only one example. Which of these sets
is most relevant to animal echolocation? In fact, why should animals be restricted to an orthogonal basis set? Splines (Greville,
1969) and linear prediction coefficients (Makhoul, 1975), for
example, are useful non-orthogonal descriptions of a time function. An even more basic question is: Why should the echo be
parameterized at all? Why not just perform detection and classification with the echo waveform itself?

These questions will be considered in the following order:
First, "why parameterize?" Second, "which parameters or basis
functions are most relevant?" Third, an example will illustrate
the importance of orthogonality.

Fig. 3. Orthogonal decomposition of a signal with unknown
waveform and unknown epoch. SQ – square law device.
See Glaser (1961).

Parameterization can be justified for biological systems by equating the word "parameter" with the word "feature." Many experiments indicate that feature extraction is performed in animal echolocation. Bullock et al. (1968) found neural populations that were sensitive to rise time, intensity, frequency, frequency modulation, and sound direction in four dolphin species. Suga (1972) has described similar feature sensitivity in single bat neurons. In man-made systems, efficient parameterization lowers memory requirements and reduces the number of operations that are required for signal classification.

Given that echo parameterization is a reasonable operation for a biological system, what parameters or features are most meaningful or important for survival? In the study of echolocation, at least two different approaches to this problem have been pursued. One approach is to start with scattering theory and to obtain a set of parameters that are theoretically important for distinguishing between different sonar reflectors. Another approach is to empirically determine the important features for target classification by using pattern recognition methods. The first approach will be described in the following paragraphs, and the second method will be described in Part IV.

Research on the prediction of target scattering characteristics has resulted in two main theoretical approaches. Geometrical and physical optics have evolved into a technique that is known as the geometrical theory of diffraction (Bechtel, 1976). This theory involves the prediction of scattering characteristics from separate structures such as the tip of a cone or an edge, and it includes creeping wave effects (i.e., waves that propagate around the surface of the object). The echo components from various target structures are combined with proper delay, in order to form the composite echo. The geometrical theory of diffraction is primarily a high frequency approach for wavelengths that are smaller than the target (Kouyoumjian, 1965).

The second approach to target scattering prediction is the use of natural modes or resonances (Miller, et al., 1975). Natural modes tend to dominate the trailing edge of an echo, whereas specular components (mirror-like reflections or glints) with higher amplitudes are found at the beginning of the echo. The advantage of natural mode characterization is that the resonant frequencies are independent of aspect, and it would thus seem that the target echo need not be modelled as an aspect-dependent random process. Experimental results with various sonar targets, however, have shown that, although natural resonances may be independent of aspect, the excitation of a given resonance and the energy that is re-radiated toward the receiver is, in fact, aspect-dependent. Furthermore, the trailing edge of the echo tends to be attenuated

relative to the specular parts of the echo. Finally, the trailing
edge of an echo is difficult to analyze if the target is immersed
in clutter.

According to Bechtel and Ross (1966), the geometrical theory
of diffraction predicts that many scattering centers will have
cross-sections that vary as λ^n, where λ is wavelength and $-2 \leqslant n \leqslant 2$.
If cross section varies as λ^n, then the associated transfer func-
tion must vary as a half-power of frequency, and the echo spectrum
is $(j\omega)^{n/2} U(\omega)$ for $-2 \leqslant n \leqslant 2$, where $U(\omega)$ is the Fourier transform
of the transmitted signal. An idealized echo from a scattering center
or highlight can be synthesized by passing an impulse through a filter
with transfer function $(j\omega)^{n/2} U(\omega)$. The parameter n is range dependent,
and changes from one highlight to the next. A highlight causes a
local maximum in the echo envelope, and in Part II it was assumed
that all highlights corresponded to transfer functions which varied
as $(j\omega)^0 = 1$. Each scattering center or highlight echo was thus
assumed to be a delayed version of the transmitted signal.

In 1975, an attempt was made to generalize the point target
highlight model by applying a spline-like representation to measured target impulse responses (Altes and Reese, 1975; Altes,
1976b). Scattering centers in the spline model were assumed to
have transfer functions $(j\omega)^n$ rather than $(j\omega)^{n/2}$. This difference
has little effect upon sonar receiver configuration, and the following discussion of the spline model will use half-powers of $(j\omega)$.

The receiver consists of a detector/echo parameter estimator
that is similar to the one in Fig. 3. The frequency domain echo
from the mth scattering center can be written

$$E_m(\omega) = \sum_{n=-2}^{2} f_{nm}(j\omega)^{n/2} U(\omega) \exp(-j\omega\tau_m) \qquad (7)$$

where τ_m is the delay associated with highlight range, $U(\omega)$ is the
Fourier transform of the transmitted signal, and f_{nm} is an unknown
parameter. The problem is to estimate the parameters f_{nm}, $n=-2$,
-1, 0, 1, 2, and the delay τ_m for each highlight.

For the echo model in (7), the basis functions for echo analysis are the functions $(j\omega)^{n/2} U(\omega)$, and the matched filters in
Fig. 3 have transfer functions $[(j\omega)^{n/2} U(\omega)]*$, $n=-2$, -1, 0, 1, 2.
Since the basis functions are not orthogonal, minimum mean-square
error estimates of the parameters $\{f_{nm}\}$ are obtained by passing the
filter outputs through a matrix transformation that compensates for
this lack of orthogonality. This transformation can be implemented
by a set of interconnections between the filter outputs in Fig. 3.

    One additional assumption can be used to completely specify
the filter transfer functions and the signal spectrum $U(\omega)$. Assume
that the filters with transfer functions $[(j\omega)^{n/2}U(\omega)]*$, $n=-2,\ldots,2$
are all constrained to have the same time-bandwidth product. Time-
bandwidth product is a measure of filter complexity, and a constraint on the product of impulse response duration and transfer
function bandwidth is equivalent to a system complexity constraint.
Time-bandwidth product can be kept invariant by requiring all the
filter transfer functions to be scaled versions of one another, e.g.,

$$(j\omega)^{n/2} \; U(\omega) \;\; \propto \; U(\omega/k^{n/2}) \;. \tag{8}$$

For $k > 1$, the filter for $n=1$ is broader in the frequency domain
than the filter for $n=0$, but its impulse response is narrower in the
time domain. Equation (8) can be solved for $U(\omega)$. The solutions
correspond to linear-period-modulated signals that can be made to
resemble either bat pulses or dolphin echolocation clicks (Altes,
1976b). The constraint on time-bandwidth product can also be
related to a condition for Doppler tolerance (Altes and Reese,
1975; Altes and Titlebaum, 1970).

    If all the filter transfer functions $U*(\omega/k^{n/2})$, $n=-2,\ldots,2$,
are scaled versions of one another, then they all have the same
ratio of center frequency to bandwidth, i.e., the same Q. The
filter bank in Fig. 3 then becomes a set of overlapping proportional
bandwidth filters, as in models of the peripheral auditory system.

    The above theoretical construction seems to exhibit the kind
of "inner beauty" and self-consistency that was expressed by Dirac
and Feinman, as quoted in the introduction. Starting with scattering theory ideas, one obtains mathematical approximations for bat
and dolphin echolocation signals and an estimator/detector that is
similar to a cochlear spectrum analyzer. The theory seems to apply
to vision as well as to echolocation (Altes, 1976a; Hauske, Wolf,
and Lupp, 1976) and perhaps even to electrolocation in Mormyrids.

    Unfortunately, the beauty now appears to be only skin deep.
There are two main flaws in the theory, both of which provide important lessons for future modelling efforts:

(i) Non-separable highlights.

    The theory implicitly assumes that the echo from each scatter-
ing center is not contaminated by reflections from other parts of
the target. Computer processing of actual sonar echoes, however,
seems to indicate that a highlight is rarely isolated in range
(Skinner, Altes, and Jones, 1977). Sonar highlights are caused by
range-distributed impedance changes, and a large highlight is often
surrounded by many smaller highlights. The combined echoes from

these small highlights have an effect upon the receiver that is similar to noise, and when environmental noise is added, estimates of the parameters f_{nm}, n=-2, -1, 0, 1, 2, become unreliable.

(ii) Non-orthogonality.

It is not immediately obvious that a model for animal echo-location should employ orthogonal basis functions. The use of non-orthogonal basis functions, however, can severely degrade the performance of an estimator in a noisy environment. This effect is theoretically predicted by the matrix Cramér-Rao bound (Van Trees, 1968), which has been applied to the jointly estimated parameters f_{nm}, n=-2,...,2 by W. Hodgkiss (unpublished). Lower bounds on the variances of the estimated parameters are given by the diagonal elements of a matrix J^{-1}. The elements of J are proportional to the inner products (i.e., the degree of non-orthogonality or overlap) between basis functions. J is called the Fisher information matrix. For the transfer functions $U(\omega/k^{n/2})$ that are obtained from (8), the diagonal elements of J depend upon the areas under each of the curves in Fig. 4, and the off-diagonal elements depend upon the areas of overlap between the different functions. From Fig. 4, we see that the off-diagonal elements can be nearly as large as the diagonal elements. This observation means that the elements of J^{-1} can be very large, and estimation errors can be considerable even for a low level of environmental noise.

The matrix that is used to transform the filter outputs in Fig. 3, in order to estimate the parameters f_{nm} , is proportional to J^{-1}. The large off-diagonal elements of J make J^{-1} difficult to

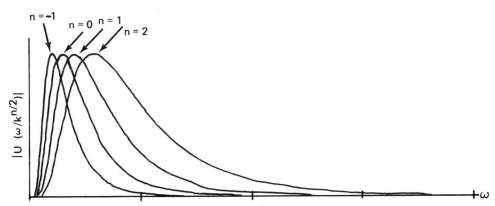

Fig. 4. Filter transfer functions used for estimation and detection of target parameters, shown with linear frequency and amplitude scales. On a log-log scale, the functions become a sequence of identical parabolas, uniformly spaced in logω.

implement, i.e., the elements of J^{-1} are large, and they must be exactly specified (with double precision computer arithmetic) or the system will not work. In mathematical terms, J is ill-conditioned (Skinner, Altes, and Jones, 1977). If orthogonal basis functions were used, J would be an identity matrix, and the receiver would be as shown in Fig. 3.

There may be some untried modifications of the system that will make it perform better. The use of an iterative subtraction method may help to separate highlights (Skinner, Altes, and Jones, 1977). An assumption that only one of the coefficients f_{nm}, n=2,...,2 is nonzero for each highlight may provide a way to circumvent the ill-conditioned matrix transformation J^{-1}. The estimator in this case can be a maximum likelihood hypothesis test, and only the maximum filter response is used.

In any case, the discussion has illustrated the power of a model, combined with computer processing of actual data, to test ideas and to reveal mistakes in assumptions. Insight into the target parameterization problem that is encountered in echolocation has also been obtained. Highlights are often difficult to analyze as separate echo waveforms, and orthogonal (or nearly orthogonal) basis functions provide better estimates of echo parameters in the presence of noise than can be obtained with non-orthogonal functions.

IV. SCATTERING FUNCTIONS, TIME-FREQUENCY ENERGY DENSITY
     FUNCTIONS, AND SPECTROGRAM CORRELATION

Another representation for a sonar target is the target's scattering function (Van Trees, 1971). The scattering function, $\bar{S}(t,f)$, describes the expected extent to which a sonar target (or time-varying communication channel) spreads out an input signal in time and frequency. By using a channel input-output relation derived by Bello (1963) and some spectrogram properties obtained by Ackroyd (1971), the spectrogram of a target echo $S_{zv}(t,f)$ can be related to the spectrogram of the input signal $S_{uv}(t,f)$ by the equation (Altes, 1978b)

$$E\left\{S_{zv}(t,f)\right\} = \bar{S}(t,f) * E\left\{S_{uv}(t,f)\right\} \qquad (9)$$

where (*) denotes a two-dimensional convolution operation in time and frequency, and $E\{\cdot\}$ denotes expected value or ensemble average. The spectrogram of the echo is thus a smeared version of the signal spectrogram, where the degree of smear is determined by the scattering function of the target.

A scattering function is useful for characterizing targets with
time-varying transfer functions, such as flying insects. If intra-
pulse time variations are neglected, then a time-frequency description of a target with transfer function H(f) and impulse response
h(t) is given by the time-frequency energy density function $e_{hh}(t,f)$
of the target impulse response (Rihaczek, 1968), where

$$e_{hh}(t,f) \equiv h(t) \; H^*(f) \; \exp(-j2\pi ft) \; . \tag{10}$$

It can be shown (Altes, 1979b) that for a time-invariant target,
the spectrogram of a target echo $S_{zv}(t,f)$ can be written

$$S_{zv}(t,f) \;=\; e_{hh}(t,f) \; \overset{t}{*} \; S_{uv}(t,f) \tag{11}$$

where $S_{uv}(t,f)$ is again the spectrogram of the input signal and where
$\overset{t}{*}$
* denotes one-dimensional convolution or smearing in the time
direction. Eq. (11) implies that the target time-frequency energy
density function can be estimated with a signal that has large
bandwidth and short duration, such as a dolphin echolocation pulse.

The time-frequency energy density function provides a useful
description of a target because it contains information about both
highlight distribution and energy spectrum. Local maxima of $|h(t)|^2$,
the time envelope of the target impulse response, provide highlight
locations. Using (10), one can demonstrate that

$$|h(t)|^2 \;=\; \int_{-\infty}^{\infty} e_{hh}(t,f) \; df \tag{12}$$

and highlight ranges can be obtained from $e_{hh}(t,f)$. Similarly,

$$|H(f)|^2 \;=\; \int_{-\infty}^{\infty} e_{hh}(t,f) \; dt \tag{13}$$

where $|H(f)|^2$ is the energy spectrum of the target impulse response.
Utilization of the energy spectrum for target classification is
discussed in the section by R. A. Johnson.

Now suppose that the scattering function or the time-frequency
energy density function of a target is known, and we wish to detect
the target when its echo is added to white, Gaussian noise. From
(9) and (11), knowledge of the target scattering function or t-f
energy density function is equivalent to knowledge of the expected
echo spectrogram. The locally optimum detector in this case is a
spectrogram correlator (Altes, 1978b), i.e., a device that computes

the quantity

$$\ell(\underline{z}) = \sum_{i=1}^{N} \sum_{j=1}^{M} z_{ij} \, E\{s_{ij}\} \tag{14}$$

where z_{ij} are samples of the data spectrogram and $E\{s_{ij}\}$ are samples of the expected spectrogram of the echo waveform.

A spectrogram correlator model for echolocation is especially attractive, since the mammalian peripheral auditory system functions like a bank of bandpass filters followed by envelope detectors (Siebert, 1968, 1970; Evans, 1977; de Boer, 1975). The resulting data representation is very similar to a spectrogram.

The spectrogram correlator can be used for classification as well as for detection by finding the maximum correlator response for a number of different reference spectrograms. This procedure can be very time consuming, however, since the number of samples NM in (14) is generally large. It would be more efficient to obtain meaningful features for distinguishing the spectrograms of different targets, and to decide whether or not these features are present in the data spectrogram.

An attempt to predict features from scattering theory, and to use non-orthogonal basis functions for feature extraction, has already been described. In the following analysis, features that are empirically determined from the data, along with orthogonal basis functions, will be used.

The desired number of features is limited in practice. A limited set of features is associated with a set of basis functions that is incomplete; one cannot exactly reconstruct every possible data waveform with a limited number of basis functions. For a given number of feature vectors (orthogonal filters) N, one can obtain a "best" incomplete N-dimensional basis set. A "best" basis corresponds to the N orthogonal functions that give the least average squared error between an echo that is reconstructed from N coefficients and the corresponding data echo. The average is taken over a representative set of data echoes. The resulting set of N feature vectors are called the "principal components" corresponding to the representative set of data echoes (Chien and Fu, 1968).

Since a limited number of basis functions are used to convey the maximum possible information about the data, one can assume (in the absence of a better approach) that these basis functions correspond to the most important features for classification.

   In order to obtain a representative set of data, echoes of
wideband, dolphin-like signals were obtained from targets with
different shape, size, and composition, and for many aspects of a
solid aluminum cylinder, from end-on to broadside. Two-dimensional
principal components were then determined from the echo spectrograms. The resulting orthogonal basis functions represent the most
important features for spectrogram echo classification. Seven of
these components were used in a target classification experiment
that will be described. The three most important components are
shown in Fig. 5 (Altes and Faust, 1978).

   The functions in Fig. 5 have interesting implications for
models of the acoustic and visual systems. Suppose that each of
the features in Fig. 5 corresponds to a neuron in the auditory cortex. The neuron that is associated with Fig. 5b, for example, has
an inhibitory center at 40 kHz, surrounded by an excitatory area in
frequency and time. The neuron in Fig. 5c is excited by spectrogram outputs that appear early in time, and it is inhibited by outputs that appear later in time. Such neurons display not only
lateral inhibition for neural "sharpening" in the frequency domain
(Nilsson, 1978), but also forward and/or backward masking in the
time domain (Sparks, 1976). The similarity to receptive fields in
vision (Hubel and Wiesel, 1968) is obvious. Our results suggest
that visual neurons with specialized receptive fields may form an
orthonormal basis for pattern recognition, and the well-known excitatory center/inhibitory surround cell is <u>one</u> of the principal
components for such a basis.

   The direction of the ridges in Fig. 5 seems to indicate that
temporal variations (highlight locations) are more important than
frequency variations (energy spectra) for classification. This
effect, however, is strongly dependent upon the spectrogram time
window (the impulse response duration of the filters that are used
to construct the spectrogram) and upon the "representative" set
of data echoes that are used to obtain the principal components.

   Vel'min and Dubrovskiy (1976) have found a "critical interval"
in dolphin audition. If a time-limited signal and time-limited
interference are separated in time by more than a critical interval,
the interference does not affect signal detectability. Just as a
critical band can be used to model the frequency window or bandwidth
of filters that are used to construct a spectrogram (Johnson, 1968b)
the critical interval can be used to model the duration of the time
window for spectrogram analysis. According to Vel'min and Dubrovskiy,
the critical interval is 200-300 μsec long. The time window corresponding to Fig. 5 is 128 μsec long.

   Before principal component analysis occurs, a data spectrogram
must be registered in time, i.e., the temporal position of the data

TOP VIEW

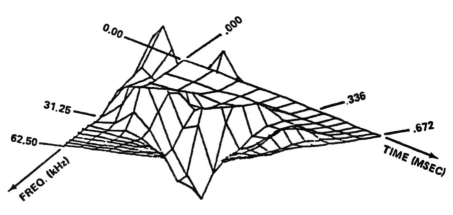

BOTTOM VIEW

Fig. 5a. Principal components or basis functions for the
analysis of echo spectrograms.

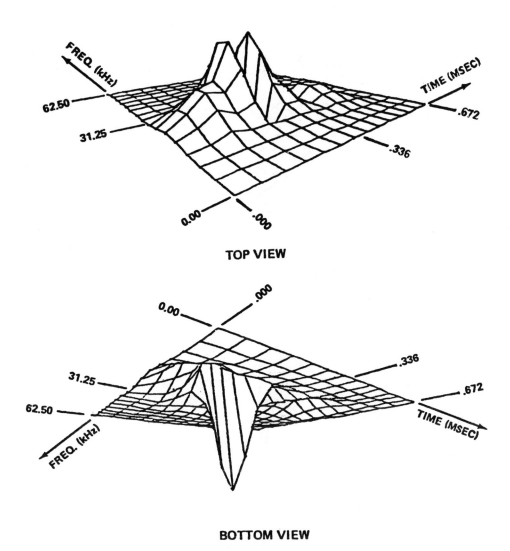

TOP VIEW

BOTTOM VIEW

Fig. 5b.   Principal components or basis functions for the
analysis of echo spectrograms.

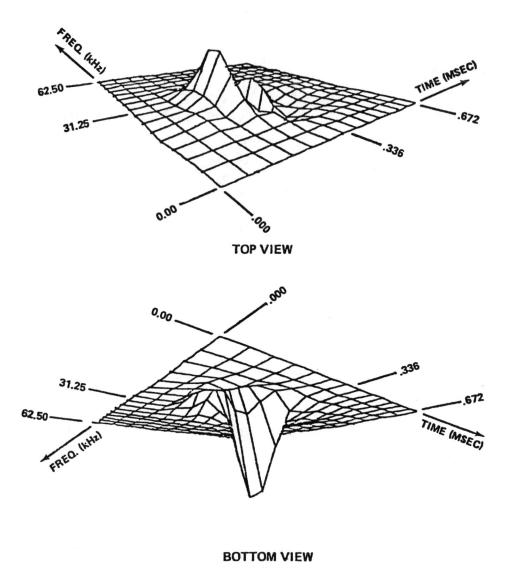

TOP VIEW

BOTTOM VIEW

Fig. 5c. Principal components or basis functions for the
analysis of echo spectrograms.

relative to the basis set must be determined. In the computer algorithm, this registration was accomplished by a noncoherent energy detector as in Fig. 3. Data energy was measured in a 128 μsec long time interval, and the energy in the interval as a function of interval starting time (the left-hand edge of the interval) was plotted. A maximum in the resulting energy versus time function was used to register the data spectrogram relative to the reference spectrograms. This registration process can be generalized by using Glaser's formulation (Glaser, 1961), but impressive classification performance can be obtained by using energy versus time as the only method of registration.

When the expected spectrogram of an echo can be predicted, a locally optimum echo detector is a spectrogram correlator. This detector implements a test for deciding between two alternative hypotheses, H_1 and H_0, where H_1 is the hypothesis that a target with specified scattering function plus white, Gaussian noise is present, and H_0 is the hypothesis that only noise is present. A classification problem can be viewed as a multiple hypothesis test that uses the same basic operations as the two-hypothesis case. For K different targets, a maximum likelihood classifier forms K two-hypothesis tests that decide between H_k and H_0, $k = 1, 2, \ldots, K$. The test statistics (in this case, the spectrogram correlator responses) for the K target hypotheses are compared, and the largest statistic is chosen. If this largest statistic exceeds a threshold, then the receiver decides that the corresponding target is present. If the largest correlator response is too small, then the receiver decides that only noise is present.

Principal component analysis does not change the basic structure of the classifier. Correlation between data spectrograms and reference spectrograms is performed in a new coordinate system with reduced dimensionality. Simplification occurs because of this reduced dimensionality, and because the reference functions for spectrogram correlation are the same for all hypotheses (the same features are extracted, regardless of hypothesis). When the coefficients of data and reference spectrograms in the new coordinate system are energy normalized, a nearest neighbor classifier is equivalent to a spectrogram correlator. A nearest neighbor classifier measures the Euclidean distance between a data point (i.e., a representation of the data spectrogram) and each reference point (i.e., a representation of each reference spectrogram). The reference point that is closest to the data point determines the classification of the data spectrogram. If there were three principal components, data points and reference points would appear in an ordinary three-dimensional x,y,z coordinate system, and energy normalization would cause all the points to fall on a sphere centered at the origin. The intersection of the sphere with the x, y, and z axes would correspond to the patterns in Figs. 5a, 5b, and 5c, respectively.

For the classification experiment, half the echoes in the
data set were obtained from a 6.35 cm x 17.8 cm aluminum cylinder
that was rotated from end-on aspect to broadside. The other half
of the echoes were from various other objects such as hollow,
open-ended metal cylinders with various wall thickness, styrofoam
cubes and cylinders, a metal sphere, etc. An interesting classification problem was to distinguish the solid aluminum cylinder from
the rest of the targets in a white noise background. Other problems
were to distinguish cylinders from non-cylinders and metal targets
from non-metal ones.

Reference vectors from all the targets were used in a nearest-
neighbor classifier with seven principal components. No attempt
was made to generalize the classifier by dividing the seven-
dimensional principal component space into decision regions. Such
a division is straightforward; a nearest-neighbor classifier is
replaced by a least-squares minimum distance classifier (Ahmed and
Rao, 1975). The least-squares classifier represents a method for
classifying future data on the basis of past experience, and it
constitutes a model for learning and generalization. An adaptive
version of such a classifier was first derived as a model for biological pattern recognition (Widrow, 1962).

Classification performance of a spectrogram correlator with
seven principal components is shown in Figs. 6a-6c. In these figures, chance classification (i.e., the result of random guessing)
occurs at different levels because of different numbers of target
echoes in each class. SNR is echo energy divided by average noise
power at the output of a bandpass filter with a 90 kHz wide bandwidth. This performance is compared with some suboptimum techniques
in Altes and Faust (1978).

In summary, it has been found that an empirical determination
of target features, combined with an orthogonal component analysis,
leads to impressive classification performance. The empirically
determined feature vectors are patterns on a time-frequency plane.
The correlation of these feature vectors with data spectrograms,
combined with nearest-neighbor classification, is a form of spectrogram correlation process. Spectrogram correlation is the basis
of a locally optimum test between a signal with known expected
spectrogram and white, Gaussian noise, so it is not surprising that
the classifier performs well in additive Gaussian noise. A signal
with known expected spectrogram corresponds to an echo from a target
with known scattering function or time-frequency energy density
function. The t-f energy density function can be related to the
highlight structure of the reflector and also to the energy spectrum of the target echo.

The analysis has illustrated a strong interconnection between
detection, estimation, and classification. If the data is classified

a.   Pr[misclassification, 2.5 x 7" Al. Cyl. versus other targets]

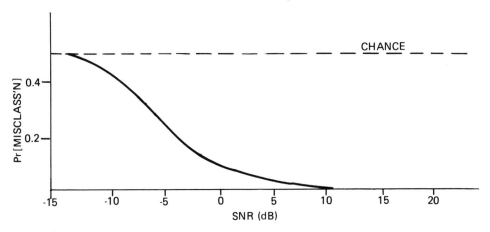

b.   Pr[misclassification, cylinder versus non-cylinder]

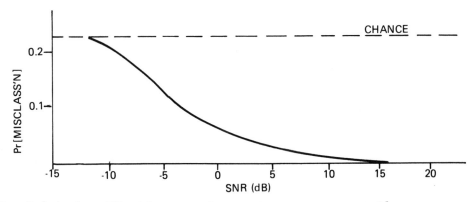

c.   Pr[misclassification, metal target versus non-metal]

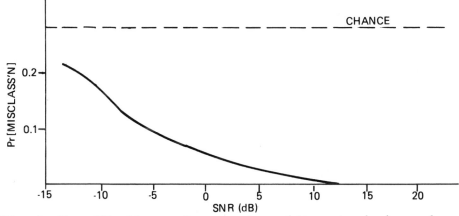

Fig. 6. Classification performance in a white noise background.

as noise, then it is rejected, and the <u>classifier</u> has functioned as
a <u>detector</u>. The coefficients that are <u>extracted</u> by correlation of
spectrograms with principal component feature vectors are minimum
mean-square error estimates of the feature values, and the
classifier/detector utilizes a parameter <u>estimator</u>. The relation
between detection, estimation, and classification is becoming an
increasingly popular theme in statistical communication theory
(Bangs and Schultheiss, 1973; Scharf and Nolte, 1977).

A spectrogram processor is an attractive model for echolocation because the encoding of sound by the peripheral auditory
system is similar to the formation of a spectrogram, and because
the underlying target model (the scattering function) takes account
of time-varying transfer functions. The next part of this review
considers behavioral tests to determine whether spectrogram correlation is a viable model for echolocation.

V. PERFORMANCE OF SPECTROGRAM PROCESSORS, AND TESTS FOR
THEIR UTILIZATION IN ANIMAL ECHLOCATION

The following paragraphs describe quantitative characteristics
of a spectrogram correlator model, tests that exploit these characteristics to determine the validity of the model, and the outcomes
of some of these tests on human subjects.

If a spectrogram correlator is to perform as well as a matched
filter, signal-to-noise ratio should be increased by approximately
5 dB (Altes, 1978b, 1979b). A five dB difference in detection performance, however, is difficult to measure experimentally, since
internal system noise is generally unknown.

For a two-alternative, forced choice (2AFC) experiment, the
probability of a correct choice is a monotone increasing function
of SNR, and this function is significantly steeper for a matched
filter than for an energy detector (Altes, 1978b). Data for a
2AFC experiment have been obtained for human subjects, using a sine
wave signal with a rectangular envelope. For such a signal, a spectrogram correlator is equivalent to an energy detector. The results
of this experiment (Green and Swets, 1966) favor a spectrogram correlator or energy detector over a matched filter.

For a sum of K sinusoids in white noise, the SNR at the output
of a spectrogram correlator is

$$\text{SNR(SC)} \propto \sum_{k=1}^{K} E_k^{\,2} \; / \; [\sigma^2 \sum_{k=1}^{K} E_k] \tag{15}$$

where σ^2 is the average noise power and E_k is the energy of the kth sinusoid. For a matched filter,

$$\text{SNR(MF)} \propto \sum_{k=1}^{K} E_k / \sigma^2 . \tag{16}$$

For a spectrogram correlator, signal detectability should vary as the sum of the squared energies of the sinusoidal components, if the signal is energy normalized. This behavior is similar to Green's summation law, which describes human detection of a sum of sinusoids (Green, 1958; Green, McKey, and Licklider, 1959). Energy normalization can be approximated by an automatic gain control (AGC), a function that may be accomplished by middle ear muscle contraction.

The frequency of a narrowband sinusoid with duration T can be estimated with a spectrogram correlator. The standard deviation or uncertainty Δf for the frequency measurement is proportional to $T^{-\frac{1}{2}}$ (Altes, 1978b, 1979b). For a matched filter, Δf is proportional to T^{-1} (Cook and Bernfeld, 1967). A number of experiments involving frequency discrimination in humans have been summarized by Siebert (1970). In Siebert's empirically determined equation, Δf varies as $T^{-\frac{1}{2}}$.

The arrival time of a wideband pulse with bandwidth B can also be estimated with a spectrogram correlator. The standard deviation or uncertainty $\Delta \tau$ for this epoch measurement is proportional to $B^{-\frac{1}{2}}$ (Altes, 1978b, 1979b). For a matched filter, $\Delta \tau$ is proportional to B^{-1} (Cook and Bernfeld, 1967).

Predictions of detection and estimation performance for a spectrogram correlator seem to be consistent with data from a number of behavioral experiments performed with human subjects. Data from similar experiments with echolocating animals have apparently not yet been compared with the predicted performance of a spectrogram correlator model. The predictions in the foregoing paragraphs, along with some of the techniques that have been used for human experimentation, provide a fertile area for future behavioral work with bats and dolphins. These quantitative predictions illustrate how a model (or two competing models) can produce a feedback mechanism for experimentation. Future experiments are obtained from models that are based upon past data.

VI. BINAURAL DETECTION AND LOCALIZATION

a. Theoretical Direction (Delay Vector) Estimators, and
    Associated Detector Configurations

Recent experiments that involve sound localization in echolocation will be described, but first it is worthwhile to examine theoretically optimum direction estimators and detectors that have
been developed for man-made radar/sonar systems. The configuration
of these estimators and detectors depends upon knowledge of the
echo waveform. We shall consider the two extreme cases: (i) echo
waveform unknown, (ii) echo waveform known exactly.

(i) Echo waveform unknown.

When the echo waveform that impinges upon a series of isotropic array elements is unknown, one can infer the direction of
the echo source from an estimate of the delay vector, i.e., the
delays in arrival time between each of the elements and the first
element. Hahn and Tretter (1973) and Hahn (1975) have shown that a
maximum likelihood estimate of the delay vector is obtained by pair-
wise cross-correlation of the signals that emerge from the different
array elements. For a binaural system, an ideal interaural delay
estimate involves cross-correlation of the signals from each ear,
when the echo waveform is unknown. The cross-correlator response
can be used for detection as well as for angle estimation (Wolff,
Thomas, and Williams, 1962).

(ii) Echo waveform known exactly.

When the echo waveform is known exactly, an ideal receiver
should synthesize a matched filter at the output of each array element. When the elements are anisotropic due to head shading, pinna
shape in bats, or multiple sound paths in the dolphin head, the
matched filters must incorporate a direction hypothesis. For each
direction hypothesis, a different matched filter is used. Monaural
direction hypotheses are also required when the transmitted signal
is direction-dependent, as in Tursiops (Evans, 1973). This complexity is offset by accurate monaural localization capability, which
exploits echo waveform and amplitude changes that are induced by
a direction change (Livshits, 1974; Altes, 1978a). A binaural
direction hypothesis is implemented by using the appropriate matched
filter for each ear and by summing delayed matched filter responses,
where the delay is determined from the direction hypothesis. The
resulting sum is a matched filter in space and time. The direction
hypothesis that gives the maximum space-time matched filter response
corresponds to a maximum likelihood direction estimate (Altes,
1978a). This filter response can also be used for estimation of echo
amplitude (Davis, Brennan, and Reed, 1976).

A sequence of space-time matched filters with different direction hypotheses can be used for detection as well as for estimation.
To detect a waveform that is known except for direction, the likelihood ratio for detection of an echo from a particular direction is
multiplied by the probability density function that describes the
direction uncertainty of the receiver, and the product is integrated
over all possible directions. To write this detection statistic in
the form of an equation, we let a column vector $\underline{r}$ denote the sampled
data from both ears, and we let $\underline{m}(\theta)$ be a noise-free version of $\underline{r}$
when an echo is received from direction θ. It follows that $\underline{r}^T\underline{m}(\theta)$
is the response of a space-time matched filter to a binaural data
vector $\underline{r}$, and $\underline{m}^T(\theta)\underline{m}(\theta)$ is the response of the filter to a noise-free
echo from direction θ. In Gaussian noise with variance σ^2, the
detection statistic is

$$\ell(\underline{r}) = \int_{-\pi}^{\pi} \exp\{[\underline{r}^T\underline{m}(\theta)- \frac{1}{2} \underline{m}^T(\theta) \underline{m}(\theta)]/\sigma^2 \} \, p(\theta) \, d\theta \ . \qquad (17)$$

According to (17), the signal is detected by forming a monotone
function of the matched filter response $\underline{r}^T\underline{m}(\theta)$ for each angle
hypothesis θ, and by summing a weighted version of this function
over all θ-values, where the weighting function is a probability
distribution that describes the receiver's uncertainty concerning
θ. An adaptive estimator/detector can be implemented by updating
$p(\theta)$ on the basis of accumulated echo data (Nolte and Hodgkiss,
1975). When θ is known exactly, e.g., $\theta=\theta_o$, then $p(\theta) = \delta(\theta-\theta_o)$,
and $\ell(\underline{r}) \propto \exp[\underline{r}^T\underline{m}(\theta_o)]$, the response of a filter that is matched
to both signal waveform and direction.

For a binaural system, an alternate form of maximum likelihood
direction estimate for a known signal is obtained by subtracting
the delayed matched filter responses from the two ears. If the
direction hypothesis is correct and if the matched filter response
of each ear is multiplied by an appropriate gain, a null will be
observed at the subtractor output in the absence of noise. The
subtractor implementation results in a "difference beam" at the
receiver (Davis, Brennan, and Reed, 1976), and the sign of a nonzero subtractor response can be used to track the target. A similar
estimation/tracking philosophy for delay, rather than direction,
results in a delay-lock discriminator (Spilker and Magill, 1961).
The delay-lock discriminator uses a time derivative of the reference
signal, and the subtractor or difference beam uses a spatial derivative of the reference signal (Davis, Brennan, and Reed, 1976).

b. Models for Binaural Interaction

    A detailed study of binaural interaction in humans has been
undertaken by H. S. Colburn and his associates (Colburn, 1973,
1977; Colburn and Latimer, 1978; Stern and Colburn, 1978). The
resulting model is shown in Fig. 7. According to this model, corresponding auditory nerve fibers from each ear (i.e., fibers tuned
to the same characteristic frequency) are combined after an internal,
interaural delay τ_m is applied to the output of one of the ears.
The combining operation is a coincidence counter for neural firings,
which depends upon a window function $f(t_i-t_j-\tau_m)$, where t_i-t_j is
the difference between firing times for the corresponding nerve
fibers. The coincidence counts L_m are multiplied by weights c_m
that depend upon interaural pressure differences. The weighted
sum of the coincidence counter responses is L_B, a binaural decision
variable that can be used for localization. Colburn has pointed
out that this model is similar to the models that were obtained
by Jeffress (1948), Sayers and Cherry (1957), and Licklider (1959).

    The model in Fig. 7 is similar to an ideal angle estimator/
detector for a signal with <u>unknown</u> waveform. A coincidence counter
seems to be a neural analog to a polarity coincidence correlator
(Wolff, Thomas, and Williams, 1962) or a dead-zone limiter coincidence correlator (Kassam and Thomas, 1976, 1977) if neural firings
are correlated with threshold crossings of an input signal. At
higher frequencies, however, neural firings may represent average
signal amplitude over a given time interval. Since spontaneous
action potentials can occur in the absence of an input signal, a
coincidence counter could then implement a summation operation. In
this case, the model resembles a spatial matched filter for a
<u>known</u> echo waveform.

    An ambiguity in the mathematical significance of a coincidence
counter can thus be associated with an ambiguous interpretation of
the corresponding estimation/detection problem. Could this ambiguity
be exploited by the biological system? By a subtle change in neural
coding or in window function operation, it may be possible to adapt
the behavior of the system to a particular estimation/detection
problem (waveform known, unknown, or partially known).

c. Neurophysiological Data

    If any localization system in nature fits into the "signal
with known waveform" category, it is the processor for the constant
frequency echo component of the mustache bat (<u>Pteronotus parnellii
rubiginosus</u>). The CF component of the mustache bat's signal is carefully tailored to furnish the receiver with a predictable waveform.
When the input to a linear filter is a sine wave, the output is a
sine wave at the same frequency (usually with different amplitude

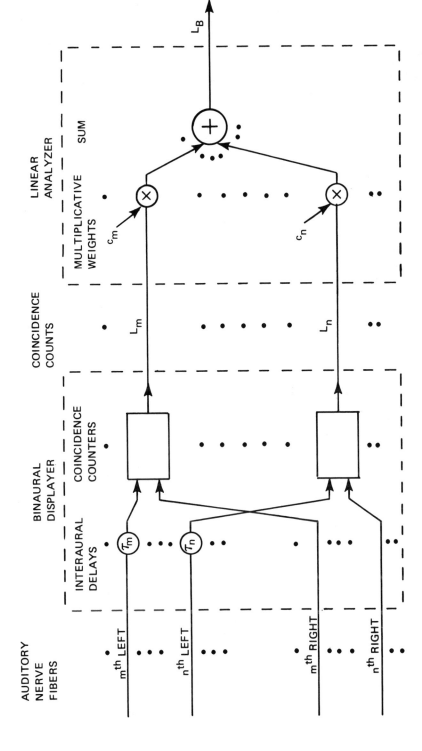

Fig. 7. Generation of binaural decision variable L_B from auditory-nerve firing patterns (From Colburn and Latimer, 1978).

and phase), so echoes from motionless targets are similar to the
transmitted signal. (This fact helps to explain the compatibility
of the point target assumption with a narrowband radar/sonar system.)
In Pteronotus, Doppler shift at the receiver is minimized by adjusting the transmitted signal frequency in order to compensate for
range rate. Similar behavior has been observed in Rhinolophus
(Schnitzler, 1973).

Menabe, Suga, and Ostwald (1978) have found that the CF processing area in the auditory cortex of the mustache bat has two
functional subdivisions. One population of neurons (the E-E neurons)
is excited by tone-burst stimulation of either ear, while another
population (the I-E neurons) are inhibited by stimulation of one
ear and excited by stimulation of the other. When both ears are
stimulated simultaneously by a sound source at various azimuth
angles, the responses of the E-E neurons are found to be angle
insensitive, while the I-E neurons are sensitive to angle. The
responses are also amplitude dependent. Most of the E-E neurons
respond best to weak tone bursts, while the I-E neurons respond
best to stronger stimuli.

Menabe, et al., have speculated that the I-E neurons are used
for accurate localization of echoes from targets at short range,
while the E-E neurons are used for long-range detection.

From (17), a direction-insensitive detector is ideal when the
target direction is not well known, i.e., when $p(\theta)$ is broad. As
more echo data becomes available, the detector should become more
angle sensitive, so that detection performance can be improved
(Nolte and Hodgkiss, 1975). One sign of such a capability would
be a gradual emergence of direction sensitivity in E-E neurons as
the stimulus is repeated. This conjecture implies that, in an
unanesthetized bat, repeated, weak tone pulses from a given direction should facilitate some E-E neurons and inhibit the responses
of others. Neurons that correspond to erroneous direction hypotheses should display decreased responses that could easily be misinterpreted as habituation to the stimulus.

Menabe, et al., state that the E-E neurons "integrate or sometimes even multiply faint signals originating from both ears for
effective detection of a target." The ambiguity in this observation
is similar to the functional ambiguity of the coincidence counter
in Fig. 7. If the outputs of the ears are multiplied and then integrated, the detector functions as though it lacks knowledge of the
echo waveform. If the outputs are passed through narrowband filters
and summed, the detector functions as though the echo waveform is
known. Better detection performance is expected in the latter case,
and the assumption that the echo waveform is known applies especially well to CF echolocation in the mustache bat.

The interpretation of I-E neurons as direction estimates is consistent with difference beam localization of targets when the echo waveform is known. The fact that this estimate is made for targets at relatively close range may indicate a capability to track moths during escape maneuvers (Roeder, 1970; Altes and Anderson, this volume). I-E neurons can also have an interference suppression function. This function will be discussed below under the heading of adaptive binaural interference suppression.

A close connection between detection and estimation has already been observed in target classification models, and (17) also illustrates such a connection. The matched filter response $\underline{r}^T\underline{m}(\theta)$ can be used for computing maximum likelihood direction and amplitude estimates, and the mean of an updated version of $p(\theta)$ can also be used as an angle estimate (Scharf and Nolte, 1977). These concepts may be exploited in animal echolocation. For example, each E-E neuron has a best amplitude as well as a detection capability, indicating a possible dual role as detector and amplitude estimator.

In summary:

(i)   E-E neurons can be used for amplitude estimation as well as for detection. The E-E neurons that were investigated by Menabe, et al., probably correspond to a detector that has knowledge of the echo waveform, but little prior knowledge of target direction. An adaptive detector would utilize E-E neurons that become increasingly angle sensitive as the stimulus is repeated.

(ii)   I-E neurons probably correspond to a direction estimator for a known echo waveform. An I-E response can also be used as an error signal for an angle tracker, and for interference suppression (spatial filtering).

Another interesting neurophysiological result is the discovery of neurons, in the midbrain of the owl, that are excited by a sound stimulus from a given direction and inhibited by sound from slightly different directions (Knudsen and Konishi, 1978). This excitatory center, inhibitory surround organization has been observed in the optical cortex of mammals when the stimulus is a point of light that emanates from various directions (Hubel and Wiesel, 1968). In vision, this neural point-spread function has been interpreted as a sharpening device for accurate estimation of line and edge position. In Part IV, an interpretation in terms of principal components was suggested.

An excitatory center, inhibitory surround function can be obtained from a pressure difference sensor by spatial differentiation (Altes, 1978a). This observation suggests a mathematical relation between the midbrain neurons of the owl and the cortical I-E neurons

of the mustache bat. The differentiation operation is a spatial
high-pass filter that tends to accentuate angular inhomogeneities
in pressure.

The visual analogy suggests that valuable insights may be
obtained by thinking of animal sonar as an acoustic imaging system
that attempts to "see with sound." The point spread function of an
imaging system determines its resolution capability. An acoustic
counterpart of the point spread function is the range-angle ambiguity
function, which determines sonar position resolution in range and
direction.

d. The Range-Angle Ambiguity Function

The range-angle ambiguity function (Altes, 1978c, 1979) is a
generalization of the usual range-Doppler ambiguity function of
radar theory (Woodward, 1964). For an array of filters (or neurons)
that correspond to different range and angle hypotheses, a point
target will give rise to a characteristic pattern of filter (neuronal) responses as its range and angle are varied. This pattern
(which sometimes resembles an excitatory center with inhibitory
surround) is described by the range-angle ambiguity function.

When the volume and sidelobe level of the ambiguity function
are reduced, range and angle resolution improves. The volume of the
range-angle ambiguity function is inversely proportional to the number of array elements multiplied by the bandwidth of the signal.
This volume dependence suggests a trade-off between array size and
signal bandwidth. Given the limited array size of echolocating animals, the use of very wide bandwidths (in comparison to many existing man-made systems) can be understood in terms of an acoustic
imaging device.

The range-angle ambiguity function also leads to a generalization of the SIR formulation (6) to include angle-dependent clutter.

e. Adaptive Binaural Interference Suppression

Ideal binaural estimator/detector configurations have been
described. These configurations assume white, Gaussian background
noise that is directionally uniform (isotropic). If noise from a
particular direction is introduced, the ideal receiver spatially
"whitens" the data, i.e., the receiver attempts to obtain constant
noise power as a function of direction. In order to make the noise
isotropic, a null must be placed in the direction of the noise source.

Adaptive beam-forming methods for obtaining the desired null
have been described by many authors, e.g., Widrow, et al. (1967),

Frost (1972), Griffiths (1969), and Reed, Mallett, and Brennan (1974). A biological counterpart of these beam-forming algorithms is indicated by a phenomenon called binaural masking level difference (Green and Yost, 1975). Durlach (1963) has modelled this phenomenon with an equalization-cancellation device. In Durlach's model (and in most adaptive beam formers) a null is obtained by delaying the output of one ear, weighting it by an appropriate factor, and subtracting it from the output of the other ear.

Neural analogs for this model of binaural interaction are the I-E neurons of the mustache bat. I-E neurons may thus be implicated in interference suppression as well as direction estimation. By placing a null in the direction of an interfering signal, a sonar effectively estimates the direction of the source of interference.

VII. PHASE SENSITIVITY IN ECHOLOCATION

a. Definition of Phase Sensitivity

Suppose that the Fourier transform of a signal $u(t)$ is a complex function

$$U(f) = A(f) \, \exp[j\phi(f)] \,. \tag{18}$$

$A(f)$ is called the magnitude of $U(f)$, and $\phi(f)$ is usually called the phase. When we speak of phase insensitivity, however, we shall mean that a signal $u_o(t)$ with Fourier transform

$$U_o(f) = A(f) \, \exp\{j[\phi(f) + \phi_o]\} \tag{19}$$

cannot be distinguished from $u(t)$, for any constant phase parameter ϕ_o. For example, if $\phi_o = 180°$, then $u_o(t) = -u(t)$. Such a sign reversal is apparently undetectable by a human listener. In this chapter, the word "phase" refers to ϕ_o in (19).

b. Why Measure Phase?

The phase shift that occurs when a sonar signal is reflected from a target is a useful feature in both air and water.

In air, all targets have a higher acoustic impedance than the propagation medium, and no information about target "hardness" is obtained from a phase measurement. In Fig. 1, however, we saw that a linear-period modulated (LPM) signal experiences a phase shift when it is reflected from a moving target. Although relative phase shifts between harmonics can be used for motion sensitivity, high frequencies can be rapidly attenuated by absorption in air (Griffin, 1971). The phase of the fundamental component of an

echo, measured with respect to the transmitted waveform, would
therefore be useful for moving target indication at long range.

In water, some targets (like rocks) have higher acoustic
impedance than the propagation medium, and echoes from such targets
experience no phase reversal. Other targets, like air bladders of
fish, have lower acoustic impedance than the medium, and echoes
from these objects are phase shifted by 180°. In order to measure
these impedance-induced phase shifts with linear-period modulated
signals, Doppler-induced phase shifts should be avoided by using
low chirp rates (only a few zero crossings). LPM dolphin clicks
exhibit this property.

Finally, phase sensitivity implies that the detailed structure
of a matched filter response can be measured, rather than the envelope of the response. This fine structure can yield superior range
resolution (Simmons, 1979) and can lead to coherent summation of
echoes for better long range detection (Vel'min, Titov, and
Yurkevich, 1975).

c. Possible Mechanisms for Phase Measurement in Echolocation

Mammalian echolocation signals are typically ultrasonic, or
at least above 6 kHz. For frequencies that are above 6 kHz, auditory nerve data ceases to contain phase information (Anderson, et
al., 1971). There are at least two possible mechanisms to circumvent this problem.

The first method is to calculate an energy spectrum from time
data that includes both the transmitted signal and the echo.
Although the 200-300 μsec critical interval of Vel'min and Dubrovskiy (1976) is too short for this application, the integration
times that were measured by C. S. Johnson (1968a) are ~ 0.10 second
long. If the echo $u_o(t-\tau)$ from a point target or highlight is
shifted by ϕ_o radians with respect to the transmitted signal $u(t)$,
the resulting energy spectrum is (Johnson and Titlebaum, 1976)

$$\left| U(f) + a\ U(f)\ \exp[j(\phi_o - 2\pi f\tau)] \right|^2 = (1+a^2)\left| U(f) \right|^2$$

$$+ 2a\left| U(f) \right|^2 \cos(\phi_o - 2\pi f\tau) \tag{20}$$

where a is an attenuation factor. Just as the energy spectrum is
sensitive to the relative phase between two harmonics, it is sensitive to the relative phase between two pulses that are separated
in time. The cosine term in (20) represents a ripple or amplitude
modulation, and the locations of peaks and nulls of the energy
spectrum depend upon ϕ_o.

The second method of phase measurement exploits low frequency noise components. If the data contains low frequency noise, then the phase of a noise component that is below 6 kHz can theoretically be measured. Since ϕ_O is independent of frequency, this phase measurement theoretically allows the receiver to completely reconstruct the input data (signal plus noise) from the spectrogram of the data (Altes, 1978b,d). Subsequent matched filtering eliminates low frequency noise components, but the phase of the echo can, in theory, be extracted from the completely reconstructed data waveform.

d. Suggested Experiments

The above conjectures about phase sensitivity can be experimentally investigated. By using a transponder that re-transmits a phase-shifted version of an animal's signal back to the animal, one can test for phase sensitivity in echolocation. If phase cannot be measured by passive listening but only via active echolocation, then the first phase measurement method, using an energy spectrum with long integration time, may be a relevant mechanism.

The second method of phase measurement, which exploits low frequency noise components, suggests that past phase sensitivity measurements may have been "too clean." Elimination of low frequency noise destroys a receiver's theoretical capability to completely reconstruct an ultrasonic pulse from a spectrogram.

Another experimental approach is to measure sensitivity to the fine structure of a signal's autocorrelation function, since phase insensitivity implies that only the envelope of the autocorrelation function can be used. This approach has been investigated by Simmons (1979), and his range resolution data seem to indicate phase sensitivity in the FM bat Eptesicus fuscus. Sensitivity to fine structure is also indicated by the ability to coherently add pulses. Coherent addition gives a faster increase in SNR than is obtained by summing envelope detected pulses. Vel'min, Titov, and Yurkevich (1975) claim to have observed coherent pulse summation in dolphins, for a pulse repetition frequency in the range 1-6 pulses per second.

VIII. SUMMARY AND CONCLUSION

Animal echolocation has been interpreted in terms of engineering problems such as estimation, detection, classification, pattern recognition, array processing, interference suppression, and description of range extended, time-varying sonar targets. Adaptive solutions to these problems have been accentuated because such solutions are becoming more important for man-made systems and because biological systems sometimes display adaptive behavior. Adaptivity is

also important because ideal receiver configurations depend upon
what is known or unknown (e.g., the echo waveform and target location), and this knowledge can change with time.

The three most challenging and important tasks have been
(i) to decide what problem an animal is actually solving (e.g.,
what target features should be estimated), (ii) to decide what
constraints exist on the animal's solution to the problem (e.g.,
monaural phase deafness), and (iii) to make quantitative performance predictions so that a model can be checked against experimental data.

Some objections to modelling were given in the introduction,
but counter-examples have been found for most of them:

The feeling that mathematics may be unnecessary for an appreciation of biology is belied by the need for statistical communication theory in order to understand (i) the relation of biological
data to behavior, i.e., to detection, estimation, and classification
functions, and (ii) the transfer of information within the auditory system via neural encoding. An impressive example of non-
mathematical insight can be found in the functional interpretation
of E-E and I-E neurons by Menabe, Suga, and Ostwald (1978). Even
in this case, however, theoretical models help to fully specify the
detection problem that is being solved by the bat and point to the
possible existence of an adaptive detector as well as a device for
interference suppression.

A premature theoretical construction, based upon sketchy data,
is perhaps illustrated by the model that uses the geometrical theory
of diffraction, along with a receiver complexity constraint. Quantitative analysis and experimental performance of the model have
indicated a need for revision. The theory has thus been twisted
to fit facts, rather than vice versa. In the course of finding
fault with the theory, previously unavailable echo data had to be
gathered, and these data have served as building blocks for an
improved model. Tests of the model have also led to some important
concepts, such as the use of near-orthogonal basis functions for
echo analysis, and the difficulty of analyzing separate scattering
centers that have overlapping echoes.

Although a model may be restrictive in its over-simplification
of reality, the simultaneous consideration of several models can
significantly increase our understanding. By comparing Colburn's
model for binaural interaction (Fig. 7) with theoretical systems
for detection and estimation, we have spotlighted the importance
of the operation that is performed by the coincidence counter. By
applying pattern recognition principles to spectrograms, we may
have obtained insight into the function of excitatory center, inhibitory surround neurons in vision and sound localization. Each of

these concepts corresponds to one restricted viewpoint of reality.
An analogous situation is found in the story of the blind men, each
of whom perceives an elephant as a different object because of his
own unique encounter with the animal (without the benefit of ultrasonic spectacles). A concerted effort by an organized, communicative group of blind men should eventually solve the problem.

Beauty and elegance appear in some of the models, e.g., in
the use of a scattering function and its relation to spectrograms.
But no matter how elegant a theory is, we are still obliged to search
for experiments to test its validity, since a theory is vacuous without such verification. We are thus unlikely to become overly attached
to a model that is fundamentally erroneous.

"Our objective is to abstract patterns from Nature..., but
many proposed patterns do not in fact correspond to the data. Thus
all proposed patterns must be subjected to the sieve of critical
analysis, and rigid skepticism without a search for patterns, are
the antipodes of incomplete science. The effective pursuit of knowledge requires both functions." (Sagan, 1977).

IX. REFERENCES

Ackroyd, M. H., 1971, Short-time spectra and time-frequency energy
        distributions, J. Acoust. Soc. Amer., 50:1229.
Ahmed, N., and Rao, K. R., 1975, "Orthogonal Transforms for Digital
        Signal Processing", Springer-Verlag, Berlin.
Altes, R. A., 1971a, Methods of wideband signal design for radar and
        sonar systems, Fed. Clearinghouse N° AD 732-494.
Altes, R. A., 1971b, Suppression of radar clutter and multipath
        effects for wideband signals, IEEE Trans. on Inform. Theory,
        IT-17:344.
Altes, R. A., 1976a, Bionic image analysis using lines and edges,
        Math. Biosciences, 31:317.
Altes, R. A., 1976b, Sonar for generalized target description and
        its similarity to animal echolocation systems, J. Acoust. Soc.
        Amer., 59:97.
Altes, R. A., 1977, Estimation of sonar target transfer functions
        in the presence of clutter and noise, J. Acoust. Soc. Amer.,
        61:1371.
Altes, R. A., 1978a, Angle estimation and binaural processing in
        animal echolocation, J. Acoust. Soc. Amer., 63:155.
Altes, R. A., 1978b, "Further Development and New Concepts for Bi-
        onic Sonar. Vol. 2. Spectrogram Correlation", Report OC-R-
        78-A004-1, ORINCON Corporation, 3366 N. Torrey Pines Ct.,
        La Jolla, Ca., Reprinted as NOSC TR-404, Naval Ocean Systems
        Center, San Diego, Ca.
Altes, R. A., 1978c, "Further Development and New Concepts for Bio-
        nic Sonar. Vol. 3. New Concepts and Experiments", see 1978b.

Altes, R. A., 1978d, Possible reconstruction of auditory signals
    by the central nervous system, J. Acoust. Soc. Amer., 64,
    Supp. 1, S137.
Altes, R. A., 1979a, Target position estimation in radar and sonar,
    and generalized ambiguity analysis for maximum likelihood
    parameter estimation, Proc. IEEE, 67.
Altes, R. A., 1979b, Utilization of spectrograms for detection and
    estimation, with applications to theories of hearing and
    animal echolocation. Part I. Spectrogram processing, sub-
    mitted for publication to J. Acoust. Soc. Amer.
Altes, R. A., and Anderson, G. M., Binaural estimation of cross-
    range velocity and optimum escape maneuvers by moths, this
    volume.
Altes, R. A., and Faust, J. W., 1978, "Further Development and New
    Concepts for Bionic Sonar. Vol. 1. Software Processors",
    see Altes, 1978b.
Altes, R. A., and Reese, W. D., 1975, Doppler-tolerant classification
    of distributed targets--a bionic sonar, IEEE Trans. on Aero-
    space and Electronic Systems, AES-11:708.
Altes, R. A., and Skinner, D. P., 1977, Sonar velocity resolution
    with a linear-period-modulated pulse, J. Acoust. Soc. Amer.,
    61:1019.
Altes, R. A., and Titlebaum, E. L., 1970, Bat signals as optimally
    Doppler tolerant waveforms, J. Acoust. Soc. Amer., 48:1014.
Altes, R. A., and Titlebaum, E. L., 1975, Graphical derivations
    of radar, sonar, and communication signals, IEEE Trans. on
    Aerospace and Electronic Systems, AES-11:38.
Anderson, D. J., Rose, J. E., Hind, J. E., and Brugge, J. F., 1971,
    Temporal position of discharges in single auditory nerve
    fibers within the cycle of a sine wave stimulus: frequency
    and intensity effects, J. Acoust. Soc. Amer., 49:1131.
Bangs, W. J., and Schultheiss, P. M., 1973, Space-time processing
    for optimal parameter estimation, in: "Signal Processing",
    Griffiths, stocklin, and Van Schooneveld, eds., Academic
    Press, London.
Bechtel, M. E., 1976, Short pulse target characteristics, in: "Atmos-
    pheric Effects on Radar Target Identification and Imaging",
    H. E. G. Jeske, ed., Reidel, Dordrect.
Bechtel, M. E., and Ross, R; A., 1966, Radar scattering analysis,
    CAL report N° ER/RIS-10, Cornell Aeronautical Laboratory,
    Buffalo, N.Y.
Bello, P.A., 1963, Characterization of randomly time-variant linear
    channels, IEEE Trans, on Comm. Sys., CS-11:360
Bel'kovich, V. M., and Dubrovskiy, N. A., 1976, "Sensory Bases of
    Cetacean Orientation", Chapter V. Echolocation, JPRS, L/7157.
de Boer, E., 1975, Synthetic whole-nerve action potentials for the
    cat, J. Acoust. Soc. Amer., 58:1030.
Brennan, L. E., and Reed, I. S., 1973, Theory of adaptive radar,
    IEEE Trans. Aerosp. and Electronic Systems, AES-9:237.

Bullock, T. H., Grinnell, A. D., Ikezono, E., Kamseda, K., Katsuki, Y., Nomoto, M., Sato, O., Suga, N., Yanagisawa, K., 1968, Electrophysiological studies of central auditory mechanisms in cetaceans, Zeitschrift für Vergleichende Physiologie, 59: 117.

Cahlander, D. A., 1966, Echolocation with wideband waveforms: bat sonar signals, Fed. Clearinghouse N° AD 605-322.

Capon, J., 1961, On the asymptotic efficiency of locally optimum detectors, IRE Trans. Inform. Theory, IT-7:67.

Chien, Y. T., and Fu, K. S., 1968, Selection and ordering of feature observations in a pattern recognition system, Inform. and Control, 12:395.

Colburn, H. S., 1973, Theory of binaural interaction based on auditory nerve data. I. General strategy and preliminary results on interaural discrimination, J. Acoust. Soc. Amer., 54:1458.

Colburn, H. S., 1977, Theory of binaural interaction based on auditory nerve data. II. Detection of tones in noise, J. Acoust. Soc. Amer., 61:525.

Colburn, H. S., and Latimer, J. S., 1978, Theory of binaural interaction based on auditory nerve data. III. Joint dependence on interaural time and amplitude differences, J. Acoust. Soc. Amer., 64:95.

Cook, C. E., and Bernfeld, M., 1967, "Radar Signals", Academic Press, New York.

Davenport, W. B., and Root, W. L., 1958, "Random Signals and Noise", McGraw-Hill, New York.

Davis, R. C., Brennan, L. E., and Reed, I. S., 1976, Angle estimation with adaptive arrays in external noise fields, IEEE Trans. on Aerospace and Electronic Systems, AES-12:179.

Decouvelaere, M., 1979, Signal design for matched filter detection in a reverberation-limited environment: application to cetacean echolocation signals, this volume.

De Long, D. F., and Hofstetter, E. M., 1969, The design of clutter-resistant radar waveforms with limited dynamic range, IEEE Trans. on Inform. Theory, IT-15:376.

Dirac, P. A. M., 1963, The evolution of the physicist's picture of nature, Scientific American, 208, N° 5:45.

Durlach, N. I., 1963, Equalization and cancellation theory of binaural masking-level differences, J. Acoust., Soc. Amer., 35:1206.

Dziedzic, A., and Alcuri, G., 1977, Reconnaissance acoustique des formes et caracteristiques des signaux sonars chez Tursiops truncatus, C. R. Acad. Sc. Paris, 285 D: 981.

Evans, E. F., 1977, Peripheral processing of complex sounds, in: "Recognition of Complex Acoustic Signals", T. H. Bullock,ed., Abakon Verlagsgesellschaft, Berlin.

Evans, W. E., 1973, Echolocation by marine dolphinids and one species of fresh-water dolphin, J. Acoust. Soc. Amer., 54:191.

Feinman, R., 1974, Structure of the proton, Science, 183:601.

Freedman, A., 1962, A mechanism of acoustic echo formation, Acustica, 12:10.

Frost. O. L., 1972, An algorithm for linearly constrained adaptive array processing, Proc. IEEE, 60:926.

Glaser, E. M., 1961, Signal detection by adaptive filters, IRE Trans. on Inform. Theory, IT-10:87.

Green, D. M., 1958, Detection of multiple component signals in noise, J. Acoust. Soc. Amer., 30:904.

Green, D. M., McKey, M. J., and Licklider, J. C. R., 1959, Detection of a pulsed sinusoid in noise as a function of frequency, J. Acoust. Soc. Amer., 31:1446.

Green, D. M., and Swets, J. A., 1966, "Signal Detection Theory and Psychophysics, Wiley",New York.

Green, D. M., and Yost, W. A., 1975, Binaural analysis, in: "Handbook of Sensory Physiology",Vol. 5, Part 2, Springer-Verlag, Berlin.

Greville, T. N. E., 1969, "Theory and Applications of Spline Functions", Academic Press, New York.

Griffin, D. R., 1958, "Listening in the Dark", Yale University Press, New Haven, Conn.

Griffin, D. R., 1971, The importance of atmospheric attenuation for the echolocation of bats (Chiroptera), Anim. Behav., 19:55.

Griffiths, L. J., 1969, A simple adaptive algorithm for real-time processing in antenna arrays, Proc. IEEE, 57:1696.

Hahn, W. R., and Tretter, S. A., 1973, Optimum processing for delay-vector estimation in passive signal arrays, IEEE Trans. Inform. Theory, IT-19:608.

Hahn, W. R., 1975, Optimum signal processing for passive sonar range and bearing estimation, J. Acoust. Soc. Amer., 58:201.

Harger, R. O., 1970, "Synthetic Aperture Radar Systems", Academic Press, New York.

Hauske, G., Wolf, W., and Lupp, U., 1976, Matched filters in human vision, Biol. Cybernetics, 22:181.

Hodgkiss, W. S., 1978, Detection of LPM signals with estimation of their velocity and time of arrival, J. Acoust. Soc. Amer., 64:177.

Hubel, D. H., and Wiesel, T. N., 1968, Receptive fields and functional architecture of monkey striate cortex, J. Physiol., 195:215.

Jeffress, L. A., 1948, A place theory of sound localization, J. Comp. Physiol. Psychol., 41:35.

Johnson, C. S., 1968a, Relation between absolute threshold and duration-of-tone pulses in the bottlenosed porpoise, J. Acoust. Soc. Amer., 43:757.

Johnson, C. S., 1968b, Masked tonal thresholds in the bottlenosed porpoise, J. Acoust. Soc. Amer., 44:965.

Johnson, R. A., 1972, "Energy spectrum analysis as a processing mechanism for echolocation", Ph. D. Diss., University of Rochester, New York.

Johnson, R. A., and Titlebaum, E. L., 1976, Energy spectrum analysis:
     a model of echolocation processing, J. Acoust. Soc., Amer.,
     60:484.
Kassam, S. A., and Thomas, J. B., 1976, Dead-zone limiters: an
     application of conditional tests in nonparametric detection,
     J. Acoust. Soc. Amer., 60:857.
Kassam, S. A., and Thomas, J. B., 1977, Improved nonparametric
     coincidence detectors, J. Franklin Inst., 303:75.
Knudsen, E. E., and Konishi, M., 1978, Space and frequency are re-
     presented separately in the auditory midbrain of the owl,
     J. Neurophysiol., 41:870.
Koestler, A., 1964, "The Act of Creation", Dell, New York.
Kotelenko, L. M., and Radionova, E. A., 1975, On the phase sensi-
     tivity of neurons in the cat's auditory system, J. Acoust.
     Soc. Amer., 57:979.
Kouyoumjian, R. G., 1965, Asymptotic high-frequency methods, Proc.
     IEEE, 53:864.
Kroszczynski, J. J., 1969, Pulse compression by means of linear-.
     period modulation, Proc. IEEE, 57:1260.
Lee, S. D., and Uhran, J. J., 1973, Optimum signal and filter
     design in underwater acoustic echo ranging systems, IEEE
     Trans. on Aerospace and Electronic Systems, AES-9:701.
Licklider, J. C. R., 1959, Three auditory theories, in: "Psychology:
     A Study of a Science", S. Koch, ed., McGraw-Hill, New York.
Livshits, M. S., 1974, Some properties of the dolphin hydrolocator
     from the viewpoint of a correlation hypothesis, Biofizika,
     19:916, JPRS 64329.
Makhoul, J., 1975, Linear prediction: a tutorial review, Proc. IEEE,
     63:561.
Menabe, T., Suga, N., and Ostwald, J., 1978, Aural representation
     in the Doppler-shifted-CF processing area of the auditory
     cortex of the mustache bat, Science, 200:339.
Middleton, D., 1966, Canonically optimum threshold detection, IEEE
     Trans. on Inform. Theory, IT-12:230.
Miller, E. K., Deadrick, F. J., Hudson, H. G., Poggio, A. J., and
     Landt, J. A., 1975, Radar target classification using temp-
     oral mode analysis, report UCRL-51825, Lawrence Livermore
     Laboratory, University of California, Livermore Ca.
Nilsson, H. G., 1978, A comparison of models for sharpening of the
     frequency selectivity in the cochlea, Biol. Cybernetics, 28:
     177.
Nolte, L. W., and Kodgkiss, W. S., 1975, Directivity or adaptivity?,
     EASCON '75, (35-A)-(35-H).
Patterson, J. H., and Green, D. M., 1970, Discrimination of trans-
     ient signals having identical energy spectra, J. Acoust. Soc.
     Amer., 48:894.
Picinbono, B., 1978, Adaptive signal processing for detection and
     communication, in: "Communication Systems and Random Process
     Theory",Skwirzynski, ed., Sijthoff and Hoordhoff, Alphen aan
     den Rijn.

Reed, I. S., Mallett, J. D., and Brennan, L. E., 1974, Rapid con-
    vergence rate in adaptive arrays, IEEE Trans. on Aerospace
    and Electronic Systems, AES-10:853.
Rihaczek, A. W., 1968, Signal energy distribution in time and fre-
    quency, IEEE Trans. on Inform. Theory, IT-14:369.
Rihaczek, A. W., 1969, "Principles of High-Resolution Radar",
    McGraw-Hill, New York.
Roberts, R. A., 1965, On the detection of a signal known except
    for phase, IEEE Trans. Inform. Theory, IT-11:76.
Roeder, K. D., 1970, Episodes in insect brains, Amer. Scientist,
    58:378.
Rummler, W. D., 1966, Clutter suppresssion by complex weighting of
    coherent pulse trains, IEEE Trans. on Aerospace and Electronic
    Systems, AES-2;689.
Russell, B., 1927, "An Outline of Philosophy", Allen and Unwin,
    London.
Sagan, C., 1977, "The Dragons of Eden", Ballantine Books, New York.
Sayers, B., McA., and Cherry, E. C., 1957, Mechanism of binaural
    fusion in the hearing of speech, J. Acoust. Soc. Amer., 29:
    973.
Scharf, L. L., and Nolte, L. W., 1977, Likelihood ratios for sequen-
    tial hypothesis testing on Markov sequences, IEEE Trans. on
    Inform. Theory, IT-23:101.
Schnitzler, H.-U., 1973, Control of Doppler shift compensation in
    the greater horseshoe bat Rhinolophus ferrumequinum, J. Comp.
    Physiol., 82:79.
Siebert, W. M., 1968, Stimulus transformations in the peripheral
    auditory system, in: "Recognizing Patterns", Kolers and Eden,
    eds., MIT Press, Cambridge, Mass.
Siebert, W. M., 1970, Frequency discrimination in the auditory sys-
    tem: place or periodicity mechanisms?, Proc. IEEE, 58:723.
Simmons, J. A., 1979a, Perception of echo phase information in bat
    sonar, in press.
Simmons, J. A., 1979b, Processing of sonar echoes by bats, this
    volume.
Skinner, D. P., Altes, R. A., and Jones, J. D., 1977, Broadband
    target classification using a bionic sonar, J. Acoust. Soc.
    Amer., 62:1239.
Sparks, D. W., 1976, Temporal recognition masking--or interference?,
    J. Acoust. Soc. Amer., 60:1347.
Spilker, J. J., and Magill, D. T., 1961, The delay-lock discrimina-
    tor--an optimum tracking device, Proc. IRE, 49;1403.
Stern, R. M., and Colburn, H. S., 1978, Theory of binaural inter-
    action based on auditory nerve data. IV. A model for sub-
    jective lateral position, J. Acoust. Soc. Amer., 64:127.
Stutt, C. A., and Spafford, L. J., 1968, A "best" mismatched filter
    response for radar clutter discrimination, IEEE Trans. on
    Inform. Theory, IT-14:280.

Suga, N., 1972, Analysis of information-bearing elements in complex sounds by auditory neurons of bats, Audiology, 11:58.

Turin, G. L., 1957, On the estimation in the presence of noise of the impulse response of a random, linear filter, IRE Trans. on Inform. Theory, IT-3:5.

Van Trees, H. L., 1968, "Detection, Estimation, and Modulation Theory, Part I", Wiley, New York.

Van Trees, H. L., 1971, "Detection, Estimation, and Modulation Theory, Part III", Wiley, New York.

Vakman, D. E., 1968, "Sophisticated Signals and the Uncertainty Principle in Radar", Springer-Verlag, New York.

Vel'min, V. A., and Dubrovskiy, N. A., 1976, The critical interval of active hearing in dolphins, Soc. Phys. Acoust., 22:351.

Vel'min, V. A., Titov, A. A., and Yurkevich, L. I., 1975, Temporal pulse summation in bottlenosed dolphins, in: "Kiev Morskiye Mlekopitayushchiye", Agarkov, ed.

Widrow, B., 1962, Generalization and information storage in networks of adaline "neurons", in: "Self-Organizing Systems-- 1962", Yovits, Jacobi, and Goldstein, eds., Spartan Books, Washington, D. C.

Widrow, B., Mantey, P. G., Griffiths, L. J., and Goode, B. B., 1967, Adaptive antenna systems, Proc. IEEE, 55:2143.

Wolff, S. S., Thomas, J. B., and Williams, T. R., 1962, The polarity-coincidence correlator: a non-parametric detection device, IRE Trans. on Inform. Theory, IT-8:5.

Woodward, P. M., 1964, "Probability and Information Theory, With Applications to Radar", Pergamon Press, Oxford.

ENERGY SPECTRUM ANALYSIS IN ECHOLOCATION

Richard A. Johnson

Naval Ocean Systems Center

San Diego, California   92152

INTRODUCTION

There are many approaches to the problem of understanding echolocation. Classically, observation and behavioral experimentation have been used to define echolocating animals' abilities, while more recently emphasis has shifted to neurophysiological investigations, particularly with certain bat species. However, details of the acoustic processing mechanisms involved remain obscure, although inferences can be made concerning the efficiency of extracting information from echoes.

Several theories have been proposed to explain the neurophysiological processing necessary for acoustic orientation. In order to infer the distance to objects for example, it is necessary to know the round trip travel time Δ associated with the acoustic propagation from the animal to the object and back. The distance d to the object is

$$d = (\Delta \cdot Vp)/2$$

where Vp is the velocity of acoustic propagation in the medium. Most simply, it has been theorized that Δ is directly measured neurophysiologically, roughly equivalent to neural potential spikes stimulated by acoustic pulses. However, many uncertainties such as noise, weak echo strength, and echo distortions make the direct determination of Δ by neural networks a rough estimate, at best. Slightly more complex theories, such as Kay's frequency modulation range-finder (Kay, 1967) or Pye's beat notes (Pye, 1967), are very restrictive in that they do not allow for the slight inaccuracies in the generation of signals which one finds

in biological systems. Nor do they account for non-FM signals.
However, at the same time, Pye recognized the utility of wide-
bandwidth pulses, spectral interference patterns, and the minimal
effect of Doppler on FM pulses. Consideration of the components
and functioning of man-made radar and sonar has led to more recent
theoretical emphasis on correlation and matched filter processing.
Encouragement was provided by signal derivation studies (Altes &
Titlebaum, 1970; Altes & Reese, 1975) and animal performance tests
(Simmons, 1970, 1971a, 1971b, 1973; Simmons, Howell & Suga, 1975;
Murchison, this volume).

     The main concerns of this paper are object detection, object
distance estimation, and object identification and how they may be
accomplished via energy spectrum analysis as an alternative to
correlation processing in the time domain sense. With the additional
cue of object direction, the task of acoustic orientation can be
satisfactorily explained.

INTERFERENCE PATTERNS IN SPECTRAL ANALYSIS

     Interference patterns are well-known wave phenomena. Examples
include Newton's rings (optics), two coherent sources (optics,
ripple tanks) and beat notes (acoustics). While the causes may
vary, the result is the same: alternating constructive and destructive interference phenomenon with applications to echolocation.

     Laboratory analysis of echolocation sounds can involve both
time domain techniques (oscilloscopes, period meters) and frequency
domain techniques (sonagraphs, FET's, spectral analyzers). The
latter represent implementations of Fourier analysis with considerations for integration time, frequency resolution, dynamic range,
etc. Also, usually only a representation of the magnitude squared
of the Fourier transform known as the energy spectrum is retained
from this analysis and often the time history of the energy spectrum
is displayed as a sonagram or waterfall spectrogram. The energy
spectrum can display interference patterns under certain conditions
(Johnson and Titlebaum, 1976):

     Let u(t) be the time domain representation of a finite
duration signal, and let U(f) be the Fourier transform of u(t) so
that $|U(f)|^2$ is the energy spectrum of u(t).

     Now the signal plus a time delayed replica of the signal is
represented by $u(t) + u(t-\Delta)$. Using well-known properties of
Fourier transforms, one finds that the transform of this signal
pair is represented by

$$|U(f)|^2 \, [1 + \exp \, (-2\pi f \Delta)]$$

and the energy spectrum is represented by

$$|U(f)|^2 \, [2 + 2 \, \cos \, (2\pi f \Delta)].$$

This indicates that the original energy spectrum is simply modified
by the multiplying term $2 + 2 \, \cos \, (2\pi f \Delta)$ which is an interference
pattern. The number of cycles present is a function of how wide
the original spectrum $|U(f)|^2$, is and the time delay, Δ. An
example illustrating this effect is depicted in figure 1. More
generally, this interference pattern, or rippled spectrum (Yost,
et al, 1978), can have a lesser amplitude as a function of the
ratio of the two pulse amplitudes and the coherence of the two
pulses. Also, several ripple patterns can be present simultaneously
as a function of the number of pulses present (Johnson and Titlebaum,
1976). As depicted in figure 1, the delay parameter, Δ, is represented by the reciprocal of the distance between peaks of the
ripple.

ECHOLOCATION WITH WIDEBAND PULSES

    This spectral interference phenomenon has application as a
model of echolocation processing. An animal's outgoing pulse
corresponds to the first signal, while the returning echo (or
echoes) corresponds to the delayed signal (or signals). A spectrum
analysis processor can detect objects (ripple), determine the
distance to the object (a function of the distance between ripple
peaks) and possibly identify the echo source (discussed in more
detail in a later section). All of these analyses will be facilitated
if the energy spectrum of the outgoing pulse is very broad (large
bandwidth). Signal theory has shown that two factors can contribute
to large signal bandwidth (Johnson and Titlebaum, 1976). The
first factor is signal duration; that is, the briefer a signal is,
the broader the bandwidth. The second factor is a large change in
the instantaneous frequency over the signal duration; that is,
frequency modulation or FM. Also, note that harmoic components by
themselves do not necessarily contribute to large bandwidths, but
when combined with either of the above two factors, can significantly
broaden the energy spectrum. Apparently without exception, all
echolocating animals are capable of using signals that meet this
large bandwidth criterion and for many, broadband signals are used
exclusively.
    Spectral analysis of broadband signals as an explanation for
echolocation processing has several advantages over alternative
hypotheses. First of all, the process is correlation-equivalent
(Johnson and Titlebaum, 1976). As such, signal derivations and
concepts relating to correlation processing or matched filtering
continue to be applicable. (Loosely speaking, one might envision
the interference pattern as a type of correlation.) For example,
Doppler-tolerant bat signals will have well-behaved ripple peak

(a)

(b)

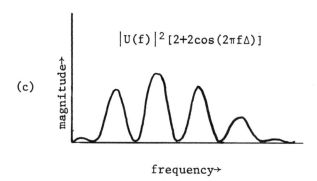

(c)

Figure 1.   (a) Signal spectrum, (b) Interference pattern,
            (c) Interference spectrum of signal pair

separations as well as well-behaved ambiguity functions (Johnson, 1972). However, the improvements over matched filtering or correlation as normally implemented in signal processing schemes include flexibility in the signal generation mechanism and not requiring an exact stored time replica of the transmitted signal as normally implemented in correlators.

A second benefit stems from the animal hearing its own outgoing pulse as well as the echoes for certain tested bats (Pollak et al., 1972) and since exclusion of the outgoing pulse from the auditory system would be anatomically difficult, if not impossible, in any animal, it should not be surprising to find that the exact nature of each pulse would be useful for echolocation. The extent to which the level of the outgoing pulse is controlled by the auditory system (Henson, 1970; Pollak and Henson, 1973) supports spectrum analysis in that it is the relative levels of the pulse and echoes that determine the level of the spectral ripple, and benefits will accrue to a processor that can more closely match the level of the outgoing pulse to the returning echoes.

A third advantage is that neurons are only called upon to measure the spectral peak spacings, so that there would appear to be no frequency limit beyond the physical responses of the middle ear and cochlea (Pye, 1967).

Finally, as a unifying theory for echolocation, it suggests that the general mammalian auditory system possesses the rudimentary ability to echolocate, and it should not be surprising to find that the echolocating animals have refined the processing on particular types of pulses contained in a large class of signals suitable for the basic mechanism. The pathways for evolutionary development are represented by the broad variety of signals, capabilities, and uses established for echolocation. This does not mean that other types of processing are excluded from use by echolocating animals. For example, the constant frequency portions of certain bat pulses must be processed in another manner. Also, long range echolocation where round-trip travel times exceed perhaps 20 msec is difficult to include in this theory since the interference patterns become difficult to resolve and integration times are extremely long for the spectrum analysis of the pulse and echoes.

It is important at this point to note that faulty interpretation of correlation theory can lead to troublesome interpretations of echolocation behavior. The example which often causes problems is the so-called range-Doppler coupling in echolocating signals. This is manifested as a "tilt" in the ambiguity function diagram ridge. The implicit assumption of constant relative velocity

between the animal and object allows a more plausible interpretation.
Instead of arbitrarily choosing a time origin for an echolocation
signal, choose an origin for which there is no range-Doppler
coupling. This can always be done for the constant relative
velocity assumption. This "uncoupled" origin will usually be
apart from the signal, and it is the round-trip travel time of
this origin which is measured. Hence, the true position of the
object is indicated for a time different from the actual time
during which the signal is reflected. Under the constant velocity
assumption, this will make no difference. It has been shown
(Johnson and Titlebaum, 1972) that for the linear period signals,
this "uncoupled" origin is maximally close to the signal itself.
The rate of frequency change will further improve this factor.
Thus, instead of measuring range-Doppler coupling, it is more
appropriate in echolocation to measure the time estimation error,
which is the distance from the "uncoupled" origin of a signal to
the signal itself. Note that for frequency downsweeps, this
origin always preceeds the signal. The opposite is true for
frequency upsweeps. Were an echolocating animal to use upsweeps,
it would be predicting (under the constant relative velocity
assumption) where an object will be. Since this is not generally
the case, the indication of where the object was (within a very
few milliseconds) would be the case for a correlation-type processor.

PITCH OF RIPPLED SPECTRA

     Evidence for this spectrum analysis of rippled spectra is
provided by the psychoacoustic phenomenon of repetition pitch or
time-separation pitch (TSP). While it has been associated with
echolocation previously (Nordmark, 1960; Pye, 1967), only more
recently has the link to a correlation-type processing been emphasized
In fact, the early conjecture that animals heard a "time-difference
tone", possibly in a frequency range below their hearing range,
was cause for much initial skepticism. However, psychoacoustically
the pitch is not maskable in the usual sense and therefore must be
arising centrally. Also, it is not necessary to conjecture that
echolocating animals hear a pitch, only that they process rippled
spectra in a similar manner and ultimately perceive information
about their environment. The pitch effect does present useful
analogies when applied to echolocation.

     Wilson (1967) summarized the psychoacoustics of delayed
stimuli in the first animal sonar symposium, and others have
elaborated on specific points since then (Bilsen and Ritsma, 1969;
Ritsma, 1967; Wightman, 1973). In a manner similar to Yost et al
(1978), I have also investigated the discriminability of TSP. The
motive for doing so was to get an analogy to range discriminability
in echolocation and to probe the processing mechanisms. Briefly,

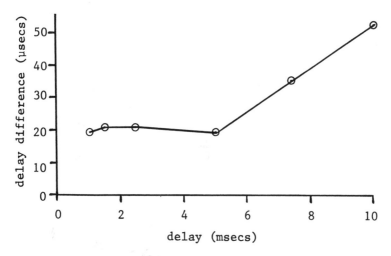

Figure 2. Performance of human subjects for time discrimination experiment.

the results (Johnson, 1977 and unpublished) were as follows:

For random pulse-pairs occuring at about 30 per second and for delays from 10 to 1 msec between pulse pairs, the pitch resolution varied from 0.5 Hz to 19 Hz respectively, which is considerably better than that found by Yost et al. The improvement may be attributed to the exact stimulus used and the technique for obtaining the threshold. Also, in a manner analogous to the range discrimination éxperiments in echolocation, the human subjects were allowed to make A-B comparisons repeatedly as they saw fit until they were ready to indicate their choice. All subjects felt that the discrimination thresholds reported by Yost et al were trivally easy to discriminate by this method. For comparison with animal experiments, time discrimination is plotted as a function of time delay in Figure 2.

As I have pointed out elsewhere (Johnson, 1977), the comparison to similar plots for echolocating animals is startling (see also Murchison, this volume). My subjects, under almost ideal conditions, could perform nearly as well as a dolphin and considerably better than bats. The time resolution of human subjects generally increase as a function of time, thus resembling the dolphin resolution function. It is not known why the bat data do not indicate similar function.

In an effort to probe the processing mechanism, pulses which exhibit the Doppler-tolerant behavior (Figure 3) were developed. The stimuli consisted of repetitions of either identical pulse pairs or pulse pairs where the second pulse of the pair was a Doppler-compressed version of the first pulse. The rather large

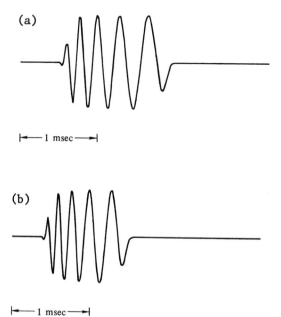

Figure 3.   (a) Doppler tolerant pulse, (b) Non-normalized,
           Doppler Compressed version of the pulse in (a)
           with exactly one cycle shifted over

amount of compression was chosen so that the cross-correlation function between the pulses was maximally like the autocorrelation function of the uncompressed pulse (Fig. 4). The time delays were chosen so that correlation processing would yield identical delay estimates for the two cases. The result was that the pitch of the two random pulse-pair trains was not identical. For a 2.5 msec delay between pairs of identical pulses, the pitch did not become the same for the non-identical pairs until the delay was increased to 2.9 msecs. Several conclusions can be drawn from this. First, time-domain correlation processing in the usual sense is unlikely to be involved, given the specially constructed cross-correlation function (Fig. 4). Second, other timing cues such as onset or envelope characteristics do not correspond to the interval predicted. Finally, since the corresponding spectra are different for the different delay (Fig. 5), it is not obvious that a spectral processing mechanism would produce the same pitch. Clearly not enough is known about this processing to predict what pitch will be produced.

Perhaps the most valuable aspect of TSP experiments is experiencing this perception. There is no single written treatise that best describes this complex auditory perception. In view of its continuing association with echolocation processing, the value of perceiving and even experimenting with TSP for oneself cannot be overemphasized.

OBJECT IDENTIFICATION

As indicated in section III, it is also possible to explain how spectrum analysis identifies objects. Before commenting on how interference spectra can aid in this task, it is necessary to review concepts in object identification, particularly as they have been investigated for the U. S. Navy Bionic Sonar program. The summary which follows is principally based on the work of Chestnut and Landsman (1977) from that program.

As compared to conventional man-made sonars, a sonar identification system must be wideband. The frequency band of some dolphin signals is 30 to 150 kHz, some bat signals cover 35 to 100 kHz, and as a comparison, human speech perception uses the band from 200 to 3000 Hz. Conventional sonars use a band of less than 1 kHz. A basic postulate is that in order to perform target identification, the system must analyze the frequency response of the target; hence, the system must have a wide frequency range.

Another postulate is that an identification system should have good resolution of frequency components. Animals and humans have good frequency resolution (Thompson and Herman, 1975), and this seems intuitively to be a useful characteristic for recognition.

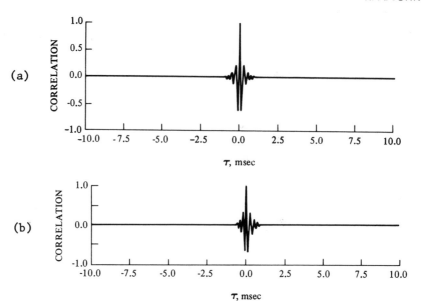

Figure 4. (a) Autocorrelation function for signal in Fig. 3a,
           (b) Crosscorrelation for signals of Figs. 3a and 3b.

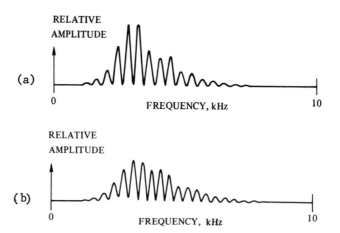

Figure 5. (a) Interference spectrum corresponding to uncom-
           pressed echo with 2.5 msec delay, (b) Interference
           spectrum corresponding to compressed echo with 2.9
           msec delay

Two types of features are used here to characterize the transfer functions of objects: energy in a bank of filters, and a mathematical model (the all-pole model) of the spectrum. While it may be impractical to store an accurate representation of a transfer function of an object, a sampled version may be adequate for object identification.

Suppose the frequency band of interest (f_a, f_b) is divided into subintervals defined by frequencies $f_1, f_2, ..., f_n$. The energy contained in the subinterval (f_{i-1}, f_i) is defined by

$$E_i = \int_{f_{i-1}}^{f_i} \left| \frac{S(f)}{U(f)} \right|^2 df$$

where $S(f)$ is the Fourier transform of the echo data and $U(f)$ is the Fourier transform of the signal. The values E_i, $= 1, 2, ...,$ N are a sampled version of the energy distribution in (f_o, f_n).

The subintervals here are chosen in two ways: uniform and constant Q. The uniform intervals are defined by

$$f_o = f_a$$
$$f_i = f_{i-1} + \frac{f_b - f_a}{N}, \quad c = 1, 2, ..., N.$$

"Constant Q" means that the ratio of the bandwidths to center frequencies is a constant. In this case, the frequency intervals are defined by

$$f_o = f_a$$
$$f_i = f_{i-1} \cdot \frac{2Q+1}{2Q-1}, \quad = 1, 2, ..., N$$

where Q is the value of the ratio of center frequency to bandwidth for each subinterval. The value of Q is given in terms of f_a, f_b, and N by

$$Q = \frac{1}{2} \cdot \frac{(f_b/f_a)^{1/N} + 1}{(f_b/f_a)^{1/N} - 1}$$

There are three parameters which can be varied in this presentation of the transfer function: the frequency range (f_a, f_b), the number of subintervals N, and the method of subdividing (constant Q versus uniform).

Spectral modeling connotes the fitting of a mathematical model to an experimentally derived spectrum. The spectrum is represented, not by a set of discrete values of the spectrum or its integral, but by a small number of parameters, such as the poles and/or zeroes of a rational function, or the coefficients of a polynomial. The model is defined as a continuous function of frequency throughout the domain of the function. The advantages of spectral modeling are that the amount of information required to be stored may be smaller, and template matching is easier. The information stored in the filter bank is highly redundant, because it represents samples of a continuous function. Template matching is easier because the comparison of model parameters replaces the comparison of the actual spectra.

The all-pole model was chosen for the transfer functions of objects. That is, the energy spectrum /S(f)/ of the data, when divided by the energy spectrum U(f) of the transmitted signal, is modeled by a function of the form

$$P(f) = \frac{G^2}{\left| 1 + \sum_{K=1}^{P} a_k e^{-2\pi j k f \Delta t} \right|^2}$$

where a_k, $k = 1, \ldots, p$ and G are the parameters of the model; Wt is the sampling period of the time data. The model is an all-pole model because, in terms of the variable

$$z = e^{2\pi j k \Delta t}$$

is a rational function with poles and no zeroes. The coefficients a_k of the model are obtained by the methods of linear prediction and are called linear prediction coefficients (LPC), or reflection coefficients.

The model is chosen because of its success in human speech applications, and because it performs well with respect to spectral maxima, perhaps caused by natural acoustic resonances of objects. Also, the coefficients which are used to characterize the all-pole model have the desirable properties of stability and independence; that is, as the order of the model is increased, the lower order coefficients do not change appreciably.

An all-pole model appears to be very good for delineating the narrow spikes which are usually observed in the energy spectra of echoes of common objects. The spikes arise from interference and

from natural object resonances. Other models require many more
coefficients to get equivalent approximations to these spectra.

Either type of feature extraction procedure results in a set
of numbers called a feature vector representing the transfer
function.

To evaluate these identifcation procedures, data were collected
from a set of objects, 13 cylinders of varying size and composition,
2 spheres and one bionic object. The signal used was a so-called
"bionic" signal introduced by Altes and Reese (1975) with parameters
chosen so that the center frequency was at 23 kHz, with a broad
bandwidth and short time duration (Fig. 6). Factors considered
important for this signal have no significance here, since the
effect of the specific signal is removed to obtain the transfer
function.

Both object identification techniques, one using energy in a
bank of filters, the other using the coefficients calculated from
the all-pole model of the transfer function, have been evaluated.
The performance criterion is the probability of misclassification
as a function of signal-to-noise ratio.

To establish the archetypal feature vectors for the set of
objects, a subset (called the design set) of the recorded echoes
was averaged, and the feature vector was calculated. This was
done for a variety of parameter settings for the two methods. In
each case, 4 echoes were used for this determination. The remaining
echoes (called the test set) were used to test the identification
method. Each echo not included in the design set was read, noise
was added, the feature vector for this noisy echo was calculated,
and the result was compared to the collection of archetypes. The
echo was associated with that object whose archetypal feature
vector was closest, in some metric, to the feature vector for that
object. If the minimum distance was too large, the sample was
rejected as not corresponding to any of the archetypal objects.
The probability of misclassification was estimated to be the
fraction of echoes that were associated with an incorrect object.
The probability of rejection was estimated to be that fraction of
the test echoes that were rejected. Of three metrics chosen, only
the data for the conventional Euclidean distance will be represented.
Other measures attempted were more sensitive to the added noise.
Parameters which were varied in the analysis were the frequency
band (15 to 45 kHz and 25 to 55 kHz) and the number of features
(5, 10, 15, and 30).

First, however, consider the results when no noise was added
to the test echoes (table 1). This table shows the fraction of
test echoes that were correctly identified using the various
methods indicated.

Figure 6.   Bionic Signal used for identification study.

The results show that good accuracy can be obtained with as few as 5 coefficients, and when 15 coefficients are used, some of the methods yield 100% correct identification. Energy detection achieves 100% when 15 or 30 filters are used, but this method degrades significantly when a smaller number of filters is used. The method based on the all-pole model is 100% accurate when 15 coefficients are used. The method based on the all-pole model is 100% accurate when 15 coefficient are used, and 99.4% correct when only 5 coefficients are used.

It is not advisable to make decisions about the relative merits of the various methods when the test is based upon data for which the signal-to-noise ratio is very high. A different picture is presented when noise is artificially created and added to the test echoes. The results of tests with noisy data are presented in figures 7 to 12. These figures show the probability of misclassification (dotted lines) and added to that the probability of rejection (solid lines) as a function of signal-to-noise ratio (SNR). If the signal power P after deconvolution is given by

$$P = \int_{f_1}^{f_2} \left| \frac{S(f)}{U(f)} \right|^2 \, df$$

and the expected noise power E(No) is given by

$$E\left\{No\right\} = s^2 \int_{f_1}^{f_2} \frac{1}{\left| U(f) \right|^2} \, df$$

Table I.   The Fraction of Test Echoes which were Correctly
           Identified Using Various Methods, as a Function
           of the Number of Parameters Used

	5	10	15	30
Reflection coefficients, 15–45 kHz	0.994	0.989	1.00	
Reflection coefficients, 25–55 kHz	0.977	0.983	0.994	
Bank of uniform filters, 15–45 kHz	0.960	0.977	1.00	1.00
Bank of uniform filters, 25–55 kHz	0.932	0.994	1.00	1.00
Bank of constant Q filters, 15–45 kHz	0.920	0.983	1.00	1.00
Bank of constant Q filters, 25–55 kHz	0.909	1.00	0.994	1.00

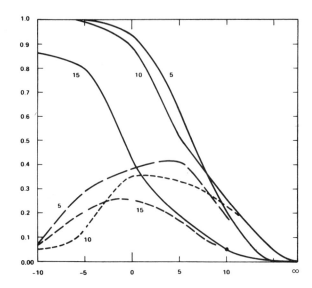

Figure 7.   Probability of Rejection and Misclassification as a
            Function of Signal-to-Noise Ratio for the All-Pole
            Model on 15-45 kilohertz Band Using the Euclidean
            Distance Measure

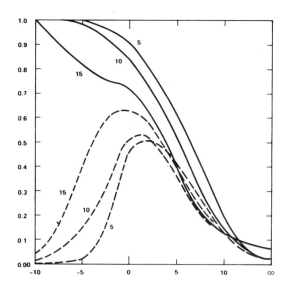

Figure 8.   Probability of Rejection and Misclassification as a
            Function of Signal-to-Noise Ratio for the All-Pole
            Model on 25-55 kilohertz Using the Euclidean Distance
            Measure

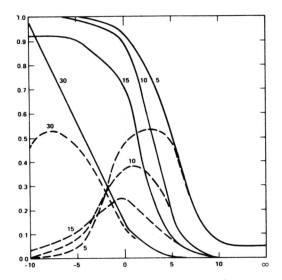

Figure 9. Probability of Rejection and Misclassification as a
           Function of Signal-to-Noise Ratio for a Bank of Constant
           Q Filters on the 15-45 kilohertz Band

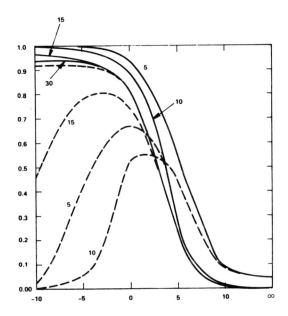

Figure 10. Probability of Rejection and Misclassification as a
            Function of Signal-to-Noise Ratio for a Bank of Constant
            Q Filters on the 25-55 kilohertz Band

Figure 11. Probability of Rejection and Misclassification as a
Function of Signal-to-Noise Ratio for a Bank of Filters
of Uniform Width on the 15-45 kilohertz Band

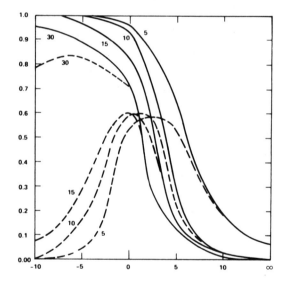

Figure 12. Probability of Rejection and Misclassification as a
Function of Signal-to-Noise Ratio for a Bank of Filters
of Uniform Width on the 25-55 kilohertz Band

where σ^2 is the expected value of the square of the magnitude of
the Fourier transform of the noise, $N(f)$, then the SNR is defined
to be

$$SNR = 10 \log_{10}(P/E\{N_0\})$$

Observe that as the SNR decreases, the probability of misclassification decreases and the probability of rejection tends to 1.
This is to be expected because as the noise increases, the distance
from the feature vector of an echo to every archetype increases.
The threshold for rejection was chosen by observing the set of
distances in the high SNR cases.

When noise is added to the echoes, the method which uses the
energy in a bank of filters is superior to the method which uses
the all-pole model. The best results are obtained using 30 filters
on the interval from 15 to 45 kHz for these objects. There are
only slight differences between the uniformly spaced filter bank
and the constant Q filter bank.

The reason for the superiority of the bank-of-filters method
needs to be analyzed. One factor may be the correlation of the
errors among the coefficients for the all-pole mode in the presence
of noise. The errors in the elements of a bank of filters are
uncorrelated if the noise is white, but this is not true of the
errors in the coefficients of a mathematical model. Another
factor may be the sensitivity of specific coefficients of the all-
pole model.

So, it has been demonstrated for one set of objects that at
least two methods for sonar object identification based on the
transfer function of the objects will yield a high percentage of
correct identifications with a low rate of rejection, provided the
signal-to-noise ratio is adequate. The superiority of the uniform
bank of filters lends some support to the theory of spectrum
analysis of interference patterns although the integration of
times are different. The transfer function of an object is
represented by the amplitude of the interference pattern across
the spectrum. Thus, if neurons can measure peak amplitudes as
well as peak spacings (section III), the analysis for object
identification is available as a bonus to this processor. The
analogy for TSP would be a coloration to the pitch much like
timbre in music.

SUMMARY AND CONCLUSIONS

The thrust of this paper has been to present a mechanism of
echolocation processing based on spectrum analysis of a signal
plus its echoes as well as to present possible methods of object

identification. Justification for this proposed mechanism is based on the mounting evidence that correlation-echolocation processing is involved in echolocation and on the need to emphasize an alternative to time-domain correlation techniques. The main strength of the proposal lies in its universal application to echolocating animals. This common link applied to the great variety of wide-band signals essentially satisfies the goal of an underlying generalization for echolocation that has long been sought. The actual mechanism of spectrum analysis of interference patterns remains to be demonstrated. Insight into this processing could aid in the explanation of known behaviors relating to depth-of-field adjustments and interference rejection, among others.

The object identification techniques presented will yield a high percentage of correct identifications with a low rate of rejection provided the signal-to-noise ratio is adequate. A method that is aspect-independent should be developed, although preliminary investigations suggest that the principle components transformation extracts invarient features of an object from data obtained at many aspects (Chestnut and Landsman, 1977). Extensions of these methods remain to be investigated.

REFERENCES

Altes, R. A., and Reese, W. D., 1975, Doppler-tolerant classification of distributed targets-- a bionic sonar, IEEE Trans. on Aerospace and Electronic Systems, AES-11:708.
Altes, R. A., and Titlebaum, E. L., 1970, Bat signals as optimally Doppler tolerant waveforms, J. Acoust. Soc. Amer., 48:1014.
Bilsen, F. A., and Ritsma, R. J., 1969, Repetition pitch and its implication for hearing theory, Acustica, 22:63.
Chestnut, P., and Landsman, H., 1977, Sonar target recognition experiment, ESL Inc., ReportN° ESL-ER178 (Address: 495 Java Dr., Sunnyvale, Ca.)
Henson, O. W., Jr., 1970, The ear and audition, in: "Biology of Bats", W. A. Wimsatt, ed., Academic Press, New York.
Johnson, R. A., 1972, "Energy Spectrum Analysis as a Processing Mechanism for Echolocation", Ph. D. Diss., University of Rochester, New York.
Johnson, R. A., 1977, Time difference pitch resolution in humans and animal echolocation capabilities, J. Acoust. Soc. Amer., 62 (Suppl)(Abstract).
Johnson, R. A., and Titlebaum, E. L., 1972, Range-Doppler uncoupling in the Doppler tolerant bat signal, in: "Proceedings of the 1972 IEEE Ultrasonics Symposium".
Johnson, R. A., and Titlebaum, E. L., 1976, Energy spectrum analysis a model of echolocation processing, J. Acoust. Soc. Amer., 60:484.

Kay, L., 1967, Discussion to: Neural processing involved in sonar, in: "Animal Sonar Systems: Biology and Bionics", R. G. Busnel, ed., Laboratoire de Physiologie Acoustique, INRA-CNRZ, Jouy-en-Josas, France.

Murchison, A. E., 1979, Detection range and range resolution of echolocating bottlenosed porpoise (Tursiops truncatus), this volume.

Nordmark, J., 1960, Perception of distance in animal echolocation, Nature, 183:1009.

Pollak, G., and Henson, O. W., Jr., 1972, Cochlear microphonic audiograms in the "pure tone" bat, Chilonycteris parnellii parnellii, Science, 176:66.

Pollak, G., and Henson, O. W., Jr., 1973, Specialized function aspects of the middle ear muscles in the bat, Chilonycteris parnellii, J. Comp. Physiol., 84:167.

Pye, J. D., 1967, Discussion to: Theories of sonar systems in relation to biological organisms, in: "Animal Sonar Systems: Biology and Bionics", R. G., Busnel, ed., Laboratoire de Physiologie Acoustique, INRA-CNRZ, Jouy-en-Josas, France.

Ritsma, R. J., 1967, Frequencies dominant in the perception of the pitch of complex sounds, J. Acoust. Soc. Amer., 42:191.

Simmons, J. A., 1970, Distance perception by echolocation: the nature of signal processing in the bat, Bydragen tot de Dierkunde, 40:87.

Simmons, J. A., 1971a, Echolocation in bats: signal processing of echoes for target range, Science, 171:925.

Simmons, J. A., 1971b, The sonar receiver of the bat, Annals of the New York Acad. of Sciences, 188:161.

Simmons, J. A., 1973, The resolution of target range by echolocating bats, J. Acoust. Soc. Amer., 54:157.

Simmons, J. A., Howell, D. J., and Suga, N., 1975, Information content of bat sonar echoes, Am. Scientist, 63:204.

Thompson, R. K., and Herman, L. M., 1975, Underwater frequency discrimination in the bottlenosed dolphin (1-140 kHz) and the human (1-8 kHz), J. Acoust. Soc. Amer., 57:943.

Wightman, F. L., 1973, The pattern-transformation model of pitch, J. Acoust. Soc. Amer., 54:407.

Wilson, J. P., 1967, Psychoacoustics of obstacle detection using ambient or self-generated noise, in: "Animal Sonar Systems: Biology and Bionics", R. G. Busnel, ed., Laboratoire de Physiologie Acoustique, INRA-CNRZ, Jouy-en-Josas, France.

Yost, W. A., Hill, R., and Perez-Falcon, T., 1978, Pitch and pitch discrimination of broadband signals with rippled power spectra, J. Acoust. Soc. Amer., 63:1166.

THE PROCESSING OF SONAR ECHOES BY BATS

James A. Simmons

Departments of Psychology and Biology

Washington University, St. Louis, Missouri 63130 U.S.A.

There are two different types of echolocation in bats, really the two extremes of a continuum of variations in the use of echo information by different species. One type is broadband echolocation, used for producing very acute, multidimensional acoustic images of targets, and it is used by all bats. The other is narrowband echolocation, used for target detection and for acoustic imaging of specific target features related primarily to movement. The results of behavioral (Griffin, 1958; Novick, 1977; Schnitzler, 1978; Simmons, 1977, in preparation; Simmons, Howell, Suga, 1975), physiological (Grinnell, 1973; Henson, 1970; Suga, 1973, 1978), and anatomical (Bruns, 1976; Henson, 1970) investigations reveal the existence of these two distinct modes of biological sonar. The dual nature of echolocation appears in the design of the orientation sounds, in the mechanisms for processing the signals, in the performance of echolocation and the kinds of information gathered, and, to the extent that it is understood, in the evolution of echolocation and bats (Simmons, 1979). Many species actually use a mixture of both extremes in different situations (Simmons, Fenton, O'Farrell, 1979).

These different types of echolocation reflect different emphases on the processing of signals either in the time domain or in the frequency domain. Within the auditory system the use of broadband sonar implies the use of temporal or periodicity representations of echoes, and the use of narrowband sonar implies some reliance on spatially-tuned or "place" representations. Such a duality is basic to any description of hearing (Nordmark, 1978; Siebert, 1973; Wever, 1949), and it is indeed a fundamental aspect of the nature of information in signals. This review attempts to

695

describe echolocation as a phenomenon in the context of hearing
and thus constitutes an auditory theory of echolocation mechanisms.

BROADBAND ECHOLOCATION

Peripheral Transformations of Signals

The broadband components of orientation sounds are short
(less than 5 msec), usually downward frequency-modulated (FM),
and often contain multiple harmonics (Novick, 1977; Simmons and
Stein, in preparation; Simmons, Howell, Suga, 1975). Initially
they are encoded by the auditory system as arrays of nerve discharges spread in space across the differently-tuned fibers of
the auditory nerve (which originate at different places along
the basilar membrane), and in time across the duration of the
signal (Pollak, Marsh, Bodenhamer, Souther, 1977; Suga, 1973).
The signals are first transformed by the external, middle, and
inner ears of bats (Henson, 1970; Suga and Jen, 1975). One part
of this peripheral transformation of FM signals is a slight
stretching due to the frequency-dependent traveling wave responses along the basilar membrane in the cochlea. In Myotis lucifugus the stretching amounts to about 60 μ sec per octave of frequency sweep (McCue, 1969).

Another part of the transformation is half-wave rectifica-
tion of the acoustic waveform at the hair-cell's interaction with
primary auditory neurons (Brugge, et al., 1969; Kiang, 1966;
Rose, et al., 1967; Siebert, 1973; Whitfield, 1978). The hair-
cells in the Organ of Corti, which convert mechanical into electrical and finally neural activity, are so connected to neurons
that displacement of the stimulus pressure in one direction causes
excitation of neurons and displacement in the other direction
causes inhibition. The activity of first-order auditory neurons
therefore is a half-wave rectified representation of the stimulus
at the hair-cells. The discharges of these neurons are locked
to the excitatory phases of stimulus waveforms. A certain amount
of internal noise is added in the form of jitter of the timing
of nerve discharges relative to stimulus cycles (see, for example, Goldstein and Srulovicz, 1977). At present no procedurally
definitive measure of this jitter is available; the smallest
estimate derived from single-unit recordings is about 40 μ sec
RMS jitter. New research suggests that it may be rather smaller,
of the order of 10 μsec (D. O. Kim, personal communication).

Central Neural Processing

The time of occurrence of discharges in first-order auditory neurons represents the time at which the FM transmission or

echo sweeps down into the excitatory areas of the tuning curves
for these neurons (Suga, 1973; Suga and Schlegel, 1973). The
intensity of the sound is represented by the number of discharges
following the initial discharge and probably also by the particular population of first-order neurons activated, if recent physiological data correctly suggests that there is an array of different excitatory thresholds for different neurons at any particular
characteristic tuned frequency (Liberman, 1978).

A variety of different temporal response patterns occur in
second-order and higher-order auditory neurons (Suga, 1973).
Two categories of neural processing have been identified that use
the timing of nerve discharges to represent stimulus characteristics. Binaural projections of timing and intensity information
to various midbrain auditory centers probably result in determination of the horizontal direction of a target. The timing of
nerve impulses relative to the occurrence of the stimulus is preserved in the auditory pathways of the brain up to the level of
the inferior colliculus (Pollak, Marsh, Bodenhamer, Souther, 1977;
Suga, 1970). The available evidence indicates that time intervals between sonar transmissions and echoes are recoded spatially,
by place, into delay-tuned neurons (Feng, Simmons, Kick, 1978;
Suga, 1978; Suga and Schlegel, 1973). The timing of nerve spikes
evoked by sonar sounds and echoes represents the target's range
and probably also target structure from timing relationships
within complex echoes (Simmons, in preparation).

The vertical direction of a target may also be represented
in the timing of nerve discharges. Another part of the peripheral transformation of acoustic waveforms is passage of the signal
through the pinna-tragus (external-ear) structures of the bat,
which probably introduce reflections with multiple path-lengths
into the waveform emerging at the tympanic membrane. The acuity
of detection of echo-timing shifts by the bat Eptesicus fuscus
(a broadband echolocator) is about 0.5 μsec, which would be adequate to determine vertical direction from the directionally-
dependent impulse response of the external ear (Simmons, in preparation).

At high levels of the auditory system (the auditory cortex,
for example) spatial arrays of excitation are processed as images
to display a variety of kinds of target information (Suga, 1978).
Some of the target features represented spatially in the bat's
auditory cortex were originally represented by the timing of
nerve discharges in peripheral parts of the auditory nervous system. Higher-order image information describing such target features as motion is also represented in the cortex. For example,
the target's range, which is represented temporally up to the
inferior colliculus and probably spatially thereafter, appears to

be displayed neurally in ways relevant to perception of approaching targets (Suga, O'Neill, Manabe, 1978; O'Neill and Suga, 1979).

Target Images and the Sonar Receiver

Echolocating bats behave as though broadband sonar sounds
and echoes are represented in the sonar receiver by the half-wave
rectified autocorrelation functions of the signals. Fig. 1
illustrates the results of an experiment on detection of jittered
echoes by Eptesicus fuscus which demonstrates both the fine temporal acuity of broadband echolocation and the presence of the
equivalent of phase information in the neural representation of
the signals. Ambiguities in perception of echo jitter correspond
to jitter times which are small integer multiples of the average
period (Simmons and Stein, in preparation) of the signals (Fig. 1).
The distinction between the actual autocorrelation function and
its envelope, if no significant smoothing takes place, is not
easily made if the autocorrelation function is first half-wave
rectified.

The results of behavioral studies of perception of target
range, horizontal direction, and the separation in depth of two
surfaces in complex targets (Peff and Simmons, 1972; Simmons,
1973; Simmons, et al., 1974) can be accounted for exclusively as
a discrimination of echo timing information, either between the
emitted sound and subsequent echoes or between samples of the
same echo at the two ears (Simmons, in preparation). Although
other acoustic cues may be present, for example, interaural intensity differences, the psychophysical data available at present
indicate that they are not used.

Fig. 1 shows results that are crucial to understanding the
mechanisms of broadband echolocation. The constellation of
physiological and behavioral data reviewed above suggests that
the broadband sonar receiver of the bat is a neural system operating on information in the time domain. The auditory system of
the bat extracts and displays periodicities in acoustic stimuli,
which for the bat represent such features of targets as range and
multi-planar structure. The data shown in Fig. 1 indicate that
nerve-discharge timing information is combined across the duration of the peripherally stretched, half-wave rectified stimulus
and across the frequencies (cochlear places) in the FM sweeps to
form an autocorrelation-like distribution of the signal's probable time of occurrence.

The bat's cochlea disperses successively-occurring frequencies
in the downward FM sequence to different tuned locations along
the basilar membrane. The result may be to apply each echo frequency to receptors (hair-cells) and primary neurons that are

Fig. 1. A graph comparing the performance (% errors; dotted curve with data points) of the bat _Eptesicus fuscus_ at detecting which of two simulated targets is jittering in echo delay (range) with the autocorrelation function of the bat's sonar signals (ACR; solid curve) and with the envelope of the full-wave-rectified ACR (ENV; dashed curve). The bat evidently perceives the half-wave-rectified ACR as the target's image along the delay (range) axis. The region of poor performance for echo jitters around 30 μsec indicates that echo delay changes corresponding to the average period of the signal are less easily detected than jitters of more than or less than one full average cycle (Simmons, in preparation). The curve showing the bat's performance corresponds with what is thought to be physiologically the auditory system's representation of a stimulus.

not very likely to have received much stimulation since the original transmission several milliseconds earlier. The cochlea thus spreads excitation from any one sonar sound or echo across many nerve fibers, setting up a spatial and temporal array of discharges within the volume of the auditory nerve bundle. The auditory nervous system integrates across time and frequency to specify from periodicities in discharges (transmission-echo intervals) an estimated time-of-occurrence for each echo. Fig. 1 shows what psychophysical studies reveal to be the approximate shape of this estimate as a range or delay image for the stimulus. The presence in hearing of periodicity-extracting, autocorrelation-like neural mechanisms has often been hypothesized (Licklider, 1959; Wever, 1949). The intricate psychoacoustical "periodicity" phenomena can justify theories about both time-domain, periodicity and frequency-domain, spectral processing operations (DeBoer, 1977; Wightman, 1973). The relatively unambiguous demonstration of time-domain mechanisms at high stimulus frequencies for broadband echolocation by bats should renew interest in general for time-domain models of auditory information processing.

It seems less probable that bats use a crosscorrelation receiver for echolocation than an autocorrelation receiver primarily because the sonar sounds of individual bats change from one situation to another (Johnson and Titlebaum, 1976; Simmons, 1973). The dimensions along which the signals change to adapt to different perceptual tasks are generally continuous (Simmons, Fenton, O'Farrell, 1979). Some species, such as Tadarida brasiliensis (Simmons, et al., 1978), even use radically different sounds in different situations. Crosscorrelation of transmitted and received signals requires storage of a replica of the transmission, and bats would require a very large number of replicas, including not only of the repertoire of emissions but of all likely Doppler-shifted echoes. An autocorrelation receiver could detect echoes without formal stored replicas by determining the presence of many pulse-echo periods of equal length across the range of frequencies in the FM sweeps of the sonar signal (see Fig. 1). Of course, the simultaneous presence in the auditory system of nerve discharges representing the previous sonar emission as well as presently received echoes constitutes a kind of stored replica. Two forms of the autocorrelation receiver have been proposed (Johnson and Titlebaum, 1976; Simmons, 1973), a time-domain autocorrelator and a frequency-domain analyzer of pulse-echo spectra. Interestingly, the similarity between stimuli encountered for echolocation of target distance by bats and for human perception of periodicity pitch suggested these models. The data reviewed here indicate that time-domain processing is probably used by bats.

The Periodic Structure of Echolocation Signals

It appears likely that one principal function of the bat's sonar receiver is to extract periodicities from acoustic stimuli arising from broadband echolocation signals and target echoes to estimate target features. One characteristic of the orientation sounds used by most species of bats that are primarily broadband echolocators is the presence of several harmonic FM sweeps. The total bandwidth of the signal is large due to the presence of several harmonic bands, but the signal's periodic structure is unusual and has interesting perceptual consequences (Simmons and Stein, in preparation).

Fig. 1 shows the autocorrelation function for an orientation sound used by Eptesicus fuscus, which we may take to represent in a general way the orientation sounds of all but the most specialized narrowband bat species (Simmons, 1979). The short, broadband sonar sounds of Eptesicus contain a strong first-harmonic sweep from 55 to 25 kHz, a strong second-harmonic sweep from 110 to 50 kHz, and sometimes a weaker third-harmonic in the region of 100 kHz. The harmonic relationship among the frequencies results in the waveform having an average repeating period equal to the average period of the fundamental frequency. (The average periodicity is only a useful concept when one is describing closely-spaced sections of the waveform of a broadband signal such as a bat sonar sound. The bandwidth is so large that there is substantial correlation within the waveform only over time intervals corresponding to a few periods.) The autocorrelation function in Fig. 1 illustrates the rapid decline in internal correlation by the sharp decrease in amplitude of the peak at 30 μsec compared with the central (zero) peak. The average period of the Eptesicus sound is 30 μsec, not the 14 μsec period corresponding to the average frequency across all harmonics, which is about 65 kHz.

The decline in autocorrelation for time lags greater than 50 to 100 μsec is almost complete for most broadband bat sonar sounds (Simmons and Stein, in preparation). This decline is represented by the width of the envelope of the autocorrelation function (Fig. 1), or approximately the reciprocal of the signal's bandwidth, which determines the acuity of target range perception in many practical situations (Simmons, 1973; Simmons, Howell, Suga, 1975). The ability of a bat to capture a flying insect in its wing or tail membrane (Webster and Griffin, 1962) would require determination of range to within about 2 to 4 cm, which seems well represented by the envelope of the autocorrelation function. If the bat must detect a small insect on a surface or resolve the distinction between different parts of an insect's body for identification of prey, the bat needs all of the time resolving power potentially available to it. As Fig. 1 shows, this resolving power corresponds to the half-wave rectified autocorrelation function of its orientation sound.

To minimize possible interference with perception of small time intervals by the secondary side peaks in the half-wave-rectified autocorrelation function (at 30 μsec time lag in Fig. 1), these side peaks should be kept as far away from the central peak as possible. The harmonic organization of the bat's sonar sounds accomplishes such separation by providing lower average periodicities than would be expected from the average frequency of the signal as a whole. The signals used by species of bats that hunt for prey in very cluttered environments are rich in harmonics, which provides for wide separation of positive-going peaks in the autocorrelation function. Some species even emit second and higher harmonics with the lower, first harmonic absent, often resulting in particularly wide spacing of half-wave peaks in the autocorrelation function.

NARROWBAND ECHOLOCATION

Bats universally use broadband orientation signals, even if only as components of sounds dominated by a narrowband component. When particular species are trained to discriminate target range or other features requiring fine temporal acuity, the time-resolving power of their broadband signals is related to performance (Simmons, 1973; Simmons, Howell, Suga, 1975). When confronted with perceptual tasks requiring target detection, bats use relatively narrowband signals, judging from observations of the addition of short CF components to signals by Eptesicus (Simmons, Lavender, Lavender, 1978), or the use of short, almost purely CF signals during the search stage of insect pursuit by Tadarida (Simmons, et al., 1978). Many other species emit even longer CF components and appear to place greater reliance on the kinds of information conveyed by these very narrowband signals (Schnitzler, 1978).

Peripheral Transformations of Signals

The auditory systems of bats appear to be tuned to the narrowband components of orientation sounds (Grinnell, 1973; Schnitzler, 1978; Simmons, Howell, Suga, 1975; Suga, Simmons, Shimozawa, 1974), indicating that echoes of narrowband emissions are emphasized. These CF signals are emitted in situations requiring great sensitivity. Neural responses specifically related to the initial detection of small targets at ranges likely to be encountered during hunting have been observed in Pteronotus gymnonotus (=suapurensis) (Grinnell and Brown, 1978).

The available evidence suggests that short-CF components (less than 5 msec), which are moderately narrowband signals, are emitted for detection of targets, and the auditory system is appropriately tuned to receive a range of biologically relevant Doppler-shifted echoes of these signals. It is also probable that detection of short-CF components in echoes involves the use of mechanisms similar to those involved in processing of broadband echoes, that is, periodicity detectors. The neural autocorrelation receiver "looks at" the narrowband elements in echoes represented by place in appropriate first-order neurons that are as sharply tuned as, or perhaps slightly more sharply tuned than, the first-order neurons tuned to the frequencies in the FM sweeps. The short-CF echo is also represented by the variety of temporal response patterns that occur in the auditory pathways to tone bursts, such as phasic, tonic, "on", and "off" responses (Suga, 1973). Frequency-domain preemphasis occurs in the peripheral auditory system at those frequencies present in the CF emissions.

Bats that emit very narrowband signals (long-CF components; more than 5 to 10 msec) extract a variety of kinds of information from CF echoes that are not available to bats using exclusively broadband signals or only short-CF sounds. In general terms long-CF signals convey information about target movements to bats (Schnitzler, 1978). The peripheral auditory systems of these bats contain particularly sharply-tuned neurons at frequencies present in echoes of long-CF transmissions. The frequency of the CF echo is in fact regulated to the sharply-tuned band by the bat's Doppler-compensation response. Mechanical tuning at the periphery in long-CF bats is an order of magnitude sharper than for other species (Suga, 1978), and resolution of changes in frequency is accordingly very acute (Long, 1977). This improved peripheral tuning and place representation may be sufficient to account for the accuracy of tracking at the reference frequency during Doppler compensation.

Modulations of the amplitude and frequency of long-CF echoes are represented in a dual manner. They appear as periodicities in neural discharges (Suga, Neuweiler, Möller, 1976; Suga and Jen, 1977; Suga, Simmons, Jen, 1975). They also appear as represented by place in the frequency domain since sidebands in the echo spectrum occur as a result of modulations. Beats produced by overlap of transmitted and received signals are a kind of modulation in this sense. The average frequency of the CF echo is represented by the place of activity in different neurons tuned to different closely-spaced frequencies around the bat's reference frequency, to which echoes are adjusted through the Doppler compensation response. These same neurons are carrying timing information about echo modulations. The FM components of long-CF/FM signals, which are the broadband transmissions of these bats, are represented in the manner described above for broadband signals in general.

Central Neural Processing

Many features of targets are represented topographically in the auditory cortex of long-CF/FM bats (Suga, 1978; Suga, O'Neill,

Manabe, 1978). Some of these features, such as average approach
velocity, were initially represented spatially in populations of
first-order neurons as well. Other features, such as target
range, are extracted from temporal information in first-order
neurons, and this presumably is true for periodicities in neural
discharges representing amplitude and frequency modulations of
CF echoes. The acoustically primitive features of targets represented by such straightforward echo parameters as overall intensity, echo delay, and interaural differences in echoes, are represented cortically, as are more complex features compounded from
patterns of information changing over time. Some neural response
patterns appear to be related to specific echo relationships from
naturally-occurring targets. For example, some neurons only
respond to weak echoes at appropriate delays, or to a series of
successive echoes from approaching targets.

Target Images and the Sonar Receiver

The broadband components of both long- and short-CF/FM sounds
probably are processed in the time domain to extract information
for high-resolution acoustic images of targets. The presence of
short-CF components does not so greatly degrade the signal's
bandwidth that these short-CF components must be isolated from
FM components before processing (Simmons, 1973; 1974). There are
no obvious reasons for concluding that the echolocation receiver
is radically different for species using purely broadband, FM
signals than for species using short-CF/FM or FM/short-CF signals
(Simmons, Howell, Suga, 1975). The same periodicity-detecting,
time-domain neural mechanisms presumably serve as the sonar receiver. It may be that the periodicity apparatus in hearing is
a more primitive or evolutionarily more basic auditory system
than the frequency-domain (tuned) or "place" system; comparative
studies of frequency discrimination in a broad range of vertebrate species support this conclusion (for example, Fay, 1978).

In species of bats where the durations of CF components
exceed 5 to 10 msec and the amount of peripheral tuning and preemphasis for the CF frequencies is great, a qualitatively distinctive frequency-domain mechanism must be present in the receiver.
The Doppler-compensation response of Rhinolophus appears to require
greater frequency resolution and specification of frequency than
can be obtained from beats occurring during pulse-echo overlap.
The physiology underlying fine tuning in the auditory periphery
and the explicit cortical display of average CF echo frequency
reveal the existence of this second, frequency-domain, narrowband
sonar receiver associated with the time-domain receiver in long-
CF bats. Since these bats can detect small modulations in the
amplitudes and frequencies of CF echoes, and since these modula-

tions are coded temporally and by place as spectral sidebands, some combination of time- and frequency-domain processing is likely to occur in the narrowband receiver system. It is possible for a purely frequency-domain (place) mechanism to account for the presently available behavioral data on narrowband echolocation (Schnitzler, 1978), so this question is an important one for future research.

THE ECHOLOCATION RECEIVER

The bat's sonar appears sufficiently different from many conventional sonar-receiver designs to warrant referring to it specifically as the echolocation receiver. Different species of bats seem to have different echolocation receivers, depending upon the extent to which they use narrowband echolocation with CF signals of suitable duration in addition to or in partial replacement of broadband FM echolocation. Several different categories of receivers already appear likely: for example, FM, short-CF/FM or FM/short-CF, and long-CF/FM (Grinnell, 1973; Simmons, Howell, Suga, 1975). Some short- or long-CF sonar systems incorporate the Doppler compensation response (Schnitzler, 1978), while others may not. Significant characteristics of the echolocation receivers of bats are summarized elsewhere in this volume (see reviews by Neuweiler, Pollak, Schnitzler and Henson, and Suga and O'Neill).

Signal Conditioning by the Auditory Periphery

Knowledge about the performance of echolocation and its neural basis suggest that the bat's sonar receiver is organized approximately as illustrated in Fig. 2. This diagram is an auditory model of the echolocation receiver. It shows in a general way the processing of sonar signals and echoes for the target information they contain. The dual, bilateral nature of the auditory system is represented in Fig. 2 (right half of auditory system at the top, left half at the bottom). Furthermore, the apparent unity of the bat's perceptions arrived at through echolocation, a unity revealed in echolocation behavior, is represented by convergence of bilaterally-acquired information upon a single neural display system for multiple dimensions of the acoustic images of targets (right side of diagram). Some means of resolving bilateral rivalry in perception must occur, perhaps through dominance of one of the halves of the auditory system over the other.

The external ears (left side of diagram) act as directional receiving antennae, separated by about 1 to 5 centimeters and

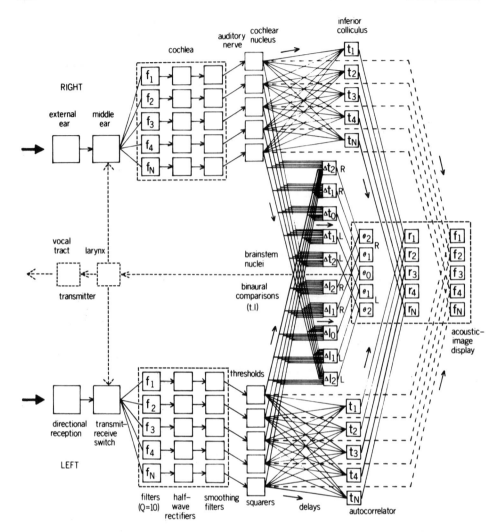

<u>Fig. 2.</u> A diagram of stages known or suspected to occur in the processing of sonar echoes within the auditory system of the bat. This diagram constitutes a model of the echolocation receiver expressed in terms of features of the auditory system (see text). This receiver decodes broadband FM transmission-echo delays and interaural echo time differences to represent target range, multi-planar target structure (as a "fine structure" to range), and the horizontal direction of a target by the place of activity within arrays of neurons. The receiver represents the average approach velocity of a target from average Doppler shifts of narrowband CF echoes by the place of activity in an array of neurons especially sharply tuned to frequencies across the narrow span of biologically relevant CF echo frequencies. The horizon-

pointing forward. The middle ears are the sites of mechanical
switches synchronized to the vocalizations and providing selective attenuation of the transmitted sonar sounds relative to
returning echoes at higher levels in the auditory system (Henson,
1970; Suga and Jen, 1975). Within the auditory system selective
neural attenuation of events representing the transmitted sounds
relative to returning echoes also occurs (Suga and Schlegel,
1973).

The cochlea disperses different sound frequencies to different locations along the Organ of Corti, and, hence, to different
fibers of the auditory nerve. Broadband sonar sounds and echoes
are thus represented in primary auditory neurons by activity in
many parallel channels, each tuned to its characteristic frequency and responding only in the vicinity of that frequency
(f_1 to f_N in Fig. 2). The mechanical and electrical signals
passing through the transduction process within the Organ of
Corti undergo half-wave-rectification and some low-pass filtering (smoothing) prior to triggering nerve impulses in primary
auditory neurons. These events appear to occur within each
tuned channel and are part of the response characteristics of
individual fibers of the auditory nerve. The amplitude response
of primary and many higher-order auditory neurons is similar to
a squaring operation; the neuron's response exhibits a rather
narrow dynamic range and is relatively unchanged (saturated)
over a wide range of higher stimulus amplitudes (Kiang, 1966;
Pollak, Marsh, Bodenhamer, Souther, 1977; Whitfield, 1978).

The frequencies in the FM orientation sounds of bats sweep
downward, in Eptesicus at rates of the order of an octave per
millisecond. In bats the tuning curves of primary auditory

---

tal direction of a target is decoded from interaural amplitude
differences in CF echoes. The temporal structure of CF echoes
(amplitude- and frequency-modulations induced by fluttering targets) is decoded from time intervals between nerve discharges
locked to the phase of the flutter modulations and represented
by the place of activity within other neural arrays. The same
neural mechanisms that decode transmission-echo delays to represent target range in channels tuned to FM echo frequencies decode CF echo flutter modulations in the more sharply-tuned channels tuned to CF echo frequencies. The receiver operates most
generally upon information concerning the timing of echoes and
transforms it into information concerning the location of targets
in space, displaying targets on multi-dimensional maps in the
brain. Other parts of the receiver convert echo amplitude or
frequency to spatial information for these same map displays.

neurons with characteristic frequencies anywhere along the broad
FM bandwidth have sharper high-frequency slopes (260-290 dB/octave)
than low-frequency slopes (110 dB/octave) (Suga and Jen, 1975).
Consequently, the sonar signals sweep down into the response
areas of individual neurons very abruptly. From the moment of
first entry into the neuron's response area, such signals reach
the neuron's characteristic frequency within the time span of
only a few cycles or average periods of the stimulus waveform,
and the signal effectively sweeps down and out of the response
area of the neuron within a short time thereafter. In most instances the discharge of the neuron marks the time-of-occurrence
of the first stimulus cycle to contain appreciable energy at frequencies at or just above the neuron's characteristic frequency.
To what extent does the smoothing filter just prior to the auditory nerve-fiber's threshold-trigger mechanism introduce jitter
or smear the registration of the first effective cycle of the
stimulus by the neuron's discharge? The psychophysical data
(Fig. 2) suggest that such jitter is probably smaller than the
average period (30 μsec) of the bat's FM signals. Measurements
of the true physiological jitter of discharges in peripheral
auditory afferent neurons in bats for FM stimuli should be made
as soon as possible.

Extraction of Target Information from Echoes

The information represented in peripheral auditory neurons
by the time-of-occurrence of successive frequencies in FM sonar
sounds and echoes probably supports a surprising proportion of
the bat's perceptions of targets. "Periodicities," or time-
intervals between separate occurrences of any one frequency, such
as in a sonar signal and its subsequent echo, may be extracted
from the neural representations of these sounds by higher brain
centers (Fig. 2). A neural network that displays transmission-
echo intervals spatially is shown as terminating in the bat's
inferior colliculus (t_1 to t_N). These time intervals and their
spatial display illustrate the fundamental role of the auditory
system in converting acoustic information into information describing targets. A target's range is represented acoustically
in the time delay between the sonar transmission and the echo,
and neurophysiological research clearly demonstrates that a
spatial map of range emerges from the first few levels of the
auditory nervous system. The complex of brainstem and midbrain
auditory nuclei must contain, among other things, the neural
time-to-place transformation processor, a kind of autocorrelator
for periodicities in discharge patterns in the auditory periphery
(Licklider, 1959). Echo delays represent target ranges (r_1 to r_N),
and the spatially presented range-image developed by the auditory system probably undergoes higher-order neural processing in

a manner similar to that observed in the visual cortex.

Interaural comparisons of time (Δt) and intensity (ΔI) are
likely to occur at multiple levels in the bat's brain, in the
brainstem, midbrain, and auditory cortex. Binaural differences
contribute to locating targets in the horizontal plane, and higher
neural centers within the auditory system would contain spatial
displays of horizontal angular directions (θ) for targets. The
neural display of target location in the bat's auditory cortex is
discussed elsewhere in this volume (see review by Suga and O'Neill),
and neurophysiological data concerning the extraction of information on interaural differences in echoes by the auditory system
are also described (see review by Neuweiler).

The relative contributions of interaural time and intensity
comparisons to localization of sonar targets are not yet made
clear from the results of experiments. The FM sonar sounds of
bats are sufficiently broadband that they do provide a basis for
discrimination of very small interaural time differences (overcoming the small interaural distances in bats), and the acuity
with which bats using broadband sounds actually can detect time
changes in echoes is correspondingly fine (Fig. 1). Echolocating
bats can in fact discriminate differences in the angles separating
targets (see review by Schnitzler and Henson, this volume). One
behavioral study of horizontal-angle discrimination by Eptesicus
(Peff and Simmons, 1972) yielded results compatible with the use
of interaural echo time cues (Simmons and Stein, 1979). Bats may
localize targets from broadband echoes by using their external
ears as the two antennae of an acoustic interferometer (see
Escudie, this volume). A spectral model for the localization of
targets by bats using FM sounds requires only interaural intensity
discriminations at different frequencies, not interaural time
discrimination (discussed comparatively by Grinnell, 1973). The
conventional distinction between interaural time (or phase) and
intensity cues in sound localization arises from psychoacoustical
research on localization of sounds, pure tones in most cases, by
humans. A demonstration that individual neurons in the bat's
auditory system can respond selectively to tone-bursts with particular interaural intensity differences may only be relevant to
understanding target localization with CF echoes. Bats using CF
sounds are directionally sensitive to echoes (Grinnell and
Schnitzler, 1977), and interaural intensity differences seem likely
to provide the basis for this directionality. Interaural time
differences have not yet been directly studied as cues for driving
the responses of neurons in the bat's auditory system. The distinction made between time and intensity cues in localization does
not imply the necessary existence of separate mechanisms for extracting each type of information; binaural crosscorrelation processes
may act upon both (see Altes, this volume).

Tonotopic (frequency) mapping is preserved from the cochlea through various levels of the bat's auditory system, and the frequency composition of sounds is displayed in central brain sites (f_1 to f_N in Fig. 2). The neural representation of stimulus frequencies, which originates in the tuning of primary auditory neurons, forms the basis for narrowband echolocation with long-CF signals (see reviews by Neuweiler, by Pollak, and by Suga and O'Neill in this volume). The Organ of Corti contains regions of especially sharp tuning of neurons to the frequencies occurring in CF echoes. The overall or average Doppler shift of echoes is represented by which of these neurons are activated by the echoes, and modulations of these same CF echoes produced by target fluttering motions are represented in part by a temporal or periodic structure to the discharges within these sharply-tuned channels. The same periodicity-extracting mechanisms (t_1 to t_N in Fig. 2) that process broadband echoes for their delay times to display target range (r_1 to r_N) may, in the sharply-tuned channels of "foveation" to CF echo frequencies (see reviews by Neuweiler and by Pollak in this volume), extract modulation rates and display wing-beat frequencies of insects.

The Bat's Description of Sonar Targets

The nature of the acoustic image of a target that bats derive from broadband FM sounds and echoes (Fig. 1) suggests that bats perceive complex targets as a series of "glints" (see reviews by Altes and by Floyd in this volume). A complex target would have a fine structure to its range; different parts of the target would be perceived at slightly different distances. Fig. 1 indicates that Eptesicus can reliably perceive separate glints if their ranges differ by more than 0.1 to 0.2 mm, corresponding to echo delay differences greater than 0.5 sec. Complex targets with angular widths greater than a few degrees would be represented by separate glints localized in slightly different directions. It seems likely that a complex target would appear as distributed across a segment of the bat's cortical spatial map. A target with a surface continuously sloping away in range would reflect a stretched or smeared echo not readily located at a discrete point in range (see Altes, this volume), and the smeared neural registration of the time-of-occurrence of such an echo would be a clue to the presence of the slope. Experiments with complex targets and then with simulated echoes that are acoustically simplified to study the bat's reactions to individual echo features are needed to explore the feature composition of complex perceptions and the perception of patterns of targets by echolocation.

The auditory model illustrated in Fig. 2 supposes that most of the bat's perceptions of target features arise from very acute resolution of the time-of-occurrence of sonar echoes. The bat

samples the echo sound field at two points in space corresponding
to the positions of the external ears, and assembles acoustic
images from the information thus obtained. The principal activity
of the broadband echolocation receiver once targets are detected
is the extraction of periodicities from transmission-echo delays,
with slightly different delays for each glint in a complex target
and slightly different delays and echo intensities at each ear.
It may prove useful to adopt so strongly a time-domain model
because it provides a clear reference point from which to state
alternatives. Purely frequency-domain models have been proposed
(see review by Johnson in this volume), and combinations of time-
domain and frequency-domain processing must also be considered
(Simmons, 1973; Simmons, et al., 1974; see review by Beuter in
this volume). Periodicity models of some auditory phenomena are
strongly favored by comparative auditory evidence (Fay, 1978) and
from many general considerations in auditory theory (Wever, 1949).

Acknowledgments

This research was supported by Grant No. BNS 76-23670 from
the U.S. National Science Foundation. I thank R. A. Altes,
K. Beuter, J. L. Goldstein, R. A. Johnson, and W. A. Lavender for
critical suggestions, and I thank C. Hale, S. A. Kick, and B. D.
Lawrence for assistance in research. I also thank R.-G. Busnel
and the organizers of the Animal Sonar Symposium for the opportunity to present this material.

REFERENCES

Brugge, J. F., Anderson, D. J., Hind, J. E., and Rose, J. E., 1969, Time structure of discharges in single auditory nerve fibers of the squirrel monkey in response to complex periodic sounds, J. Neurophysiol., 32:386.

Bruns, V., 1976, Peripheral auditory tuning for fine frequency analysis by the CF-FM bat, Rhinolophus ferrumequinum, II. frequency mapping in the cochlea, J. Comp. Physiol. 106:87.

DeBoer, E., 1977, Pitch theories unified, in: "Physiology and Biophysics of Hearing", E. F. Evans and J. P. Wilson, eds., Academic Press, New York.

Fay, R. R., 1978, Coding of information in single auditory-nerve fibers of the goldfish, J. Acoust. Soc. Amer., 63:136.

Feng, A. S., Simmons, J. A., and Kick, S. A., 1978, Echo detection and target-ranging neurons in the auditory system of the bat Eptesicus fuscus, Science, 202:645.

Goldstein, J. L., and Srulovicz, P., 1977, Auditory-nerve spike intervals as an adequate basis for aural frequency measurement, in: "Physiology and Biophysics of Hearing", E.F. Evans and J. P. Wilson, eds., Academic Press, New York.

Griffin, D. R., 1958, "Listening in the Dark", Yale University Press, New Haven, 1974, Dover Publications, New York.

Grinnell, A. D., 1973, Neural processing mechanisms in echolocating bats, correlated with differences in emitted sounds, J. Acoust. Soc. Amer., 54:147.

Grinnell, A. D. and Brown, P., 1978, Long-latency "subthreshold" collicular responses to the constant-frequency components emitted by a bat, Science 202:996.

Grinnell, A. D. and Schnitzler, H.-U., 1977, Directional sensitivity of echolocation in the horseshoe bat, Rhinolophus ferrumequinum. II. Behavioral directionality of hearing, J. Comp. Physiol., 116:63.

Henson, O. W., Jr., 1970, The ear and audition, in: "Biology of Bats, Vol. II", W. A. Wimsatt, ed., Academic Press, New York.

Johnson, R. A., and Titlebaum, E. L., 1976, Energy spectrum analysis; a model of echolocation processing, J. Acoust. Soc. Amer., 60:484.

Kiang, N. Y.-S., 1966, "Discharge Patterns of Single Fibers in the Cat's Auditory Nerve", M.I.T. Press, Cambridge, Mass.

Liberman, M. C., 1978, Auditory-nerve response from cats raised in a low-noise chamber, J. Acoust. Soc. Amer., 63:442.

Licklider, J. C. R., 1959, Three auditory theories, in: "Psychology, a Study of a Science", S. Koch, ed., McGraw-Hill, New York.

Long. G. R., 1977, Masked auditory thresholds from the bat, Rhinolophus ferrumequinum, J. Comp. Physiol., 116:247.

McCue, J. J. G., 1969, Signal processing by the bat, Myotis lucifugus, J. Aud. Res., 9:100.

Nordmark, J. O., 1978, Frequency and periodicity analysis, in: "Handbook of Perception, Vol. IV. Hearing", E. C. Carterette and

M. P. Friedman, Academic Press, New York.

Novick, A., 1977, Acoustic orientation, in: "Biology of Bats, Vol.
    III", W. A. Wimsatt, Academic Press, New York.

O'Neill, W. E., and Suga, N., 1979, Target range-sensitive neurons
    in the auditory cortex of the mustache bat, Science, 203:69.

Peff, T. C., and Simmons, J. A., 1972, Horizontal-angle resolution
    by echolocating bats, J. Acoust. Soc. Amer., 51:2063.

Pollak, G. D., Marsh, D. S., Bodenhamer, R., and Souther, A., 1977,
    Characteristics of phasic on neurons in inferior colliculus
    of unanesthetized bats with observations relating to mechan-
    isms for echo ranging, J. Neurophysiol., 40:926.

Rose, J. E., Brugge, J. F., Anderson, D. J., and Hind, J. E., 1967,
    Phase-locked responses to low-frequency tones in single audi-
    tory nerve fibers of the squirrel monkey, J. Neurophysiol.,
    30:679.

Schnitzler, H.-U., in press, Die Detektion von Bewegungen durch
    Echoortung bei Fledermäusen, Hand. Dtsch. Zool. Ges.

Siebert, W. M. 1973, Hearing and the ear, in: "Engineering Princ-
    iples in Physiology, Vol. I", J. H. U. Brown and D. S. Gann,
    eds., Academic Press, New York.

Simmons, J. A., 1973, The resolution of target range by echolocating
    bats, J. Acoust. Soc. Amer., 54:157.

Simmons, J. A., 1974, Response of the Doppler echolocation system
    in the bat, Rhinolophus ferrumequinum, J. Acoust. Soc. Amer.,
    56:672.

Simmons, J. A., 1977, Localization and identification of acoustic
    signals, with reference to echolocation, in: "Recognition of
    Complex Acoustic Signals", T. H. Bullock, ed., Dahlem Kon-
    ferenzen, Berlin.

Simmons, J. A., 1979, Phylogenetic adaptations and the evolution of
    echolocation in bats (Chiroptera), in: "Proceedings of the
    Fifth International Bat Research Conference", Texas Tech
    University Press, Lubbock, Texas.

Simmons, J. A., in preparation, Perception of echo phase information
    in bat sonar.

Simmons, J. A., Fenton, M. B., and O'Farrell, M. J., 1979, Echo-
    location and pursuit of prey by bats, Science, 203:16.

Simmons, J. A., Howell, D. J., and Suga, N.; 1975, Information con-
    tent of bat sonar echoes, Amer. Sci., 63:204.

Simmons, J. A., Lavender, W. A., and Lavender, B. A., 1978, Adapt-
    ation of echolocation to environmental noise by the bat,
    Eptesicus fuscus, in: "Proceedings of the Fourth International
    Bat Research Conference", Kenya National Academy for Advance-
    ment of Arts and Sciences, Nairobi, Kenya.

Simmons, J. A., Lavender, W. A., Lavender, B. A., Childs, J. E.,
    Hulebak, K., Rigden, M. R., Sherman, J., Woolman, B., and
    O'Farrell, M. J., 1978, Echolocation by free-tailed bats
    (Tadarida), J. Comp. Physiol., 125:291.

Simmons, J. A., Lavender, W. A., Lavender, B. A., Doroshow, C. F.,
    Kiefer, S. W., Livingston, R., Scallet, A. C., and Crowley,

D. E., 1974, Target structure and echo spectral discrimination by echolocating bats, Science, 186:1130.

Simmons, J. A., and Stein, R. A., in preparation, Acoustic imaging in bat sonar; the echolocation signals.

Suga, N., 1970, Echo-ranging neurons in the inferior colliculus of bats, Science, 170:449.

Suga, N., 1973, Feature extraction in the auditory system of bats, in: "Basic Mechanisms in Hearing", A. R. Møller, ed., Academic Press, New York.

Suga, N., 1978, Specialization of the auditory system for reception and processing of species-specific sounds, Federation Proceedings, 37:2342.

Suga, N., and Jen, P. H.-S., 1975, Peripheral control of acoustic signals in the auditory system of echolocating bats, J. Exp. Biol., 62:277.

Suga, N., and Jen, P.H.-S., 1977, Further studies on the peripheral auditory system of "CF-FM" bats specialized for the fine frequency analysis of Doppler-shifted echoes, J. Exp. Biol., 69:207.

Suga, N., Neuweiler, G., and Möller, J., 1976, Peripheral auditory tuning for fine frequency analysis by the CF-FM bat, Rhinolophus ferrumequinum. IV. Properties of peripheral auditory neurons, J. Comp. Physiol., 106:111.

Suga, N., O'Neill, W. E., and Manabe, T., 1978, Cortical neurons sensitive to combinations of information-bearing elements of bio-sonar signals in the mustache bat, Science, 200:778.

Suga, N., and Schlegel, P., 1973, Coding and processing in the nervous system of FM signal-producing bats, J. Acoust. Soc. Amer., 84:174.

Suga, N., Simmons, J. A., and Jen, P. H.-S., 1975, Peripheral specialization for fine frequency analysis of Doppler-shifted echoes in the auditory system of the "CF-FM" bat, Pteronotus parnellii, J. Exp. Biol., 63:161.

Suga, N., Simmons, J. A., and Shimozawa, T., 1974, Neurophysiological studies on echolocation systems in awake bats producing CF-FM orientation sounds, J. Exp. Biol., 61:379.

Webster, F. A., and Griffin, D. R., 1962, The role of the flight membranes in insect capture by bats, Anim. Behav., 10:332.

Wever, E. G., 1949, "Theory of Hearing", Wiley, New York.

Whitfield, I. C., 1978, The neural code, in: "Handbook of Perception, Vol. IV: Hearing", E. C. Carterette and M. P. Friedman, eds., Academic Press, New York.

Wightman, F. L., 1973, The pattern-transformation model of pitch, J. Acoust. Soc. Amer., 54:407.

SIGNAL PROCESSING AND DESIGN RELATED TO BAT SONAR SYSTEMS

Bernard Escudié

Signal Processing Laboratory
Institut de Chimie et Physique Industrielles de Lyon
25, rue du Plat   69288 Lyon   CEDEX 1   France

BAT SONAR SIGNALS: COHERENT RECEPTION AND RANGE ESTIMATION

Bat sonar signals cannot be correctly described if the information processor (or receiver) which extracts useful information
from the incoming echo added to noise is not taken into account.
Since the Frascati meeting, many investigators have assumed that
the bat has a coherent receiver, or a matched filter, or a correlator (Mermoz, 1967; Cahlander, 1967; Altes, 1971; Escudié and
Héllion, 1973).   Some results have suggested that the bat's receiver
is coherent and very near the ideal receiver (Van Trees, 1968).
Simmons (1971) and Simmons and Vernon (1971) measured range discrimination ability in bats.   Range resolution can be defined using
the performances of bats to discriminate between two targets very
close to each other.   The major results are shown in Fig. 1:

$$\frac{d}{C_o} = \tau_o \simeq \frac{1}{B} \qquad \begin{array}{l} d = \text{ target separation} \\ C_o = \text{ sound speed, } C_o \simeq 340 \text{ m/s} \end{array}$$

These agree with the theoretical results on coherent reception.
Many theoretical studies have been derived from a receiver model
(e.g., Johnson, 1972).   All of these models are very near the
matched filter or "correlator", even the "spectral analysis" model
(Johnson, 1972; Licklider, 1959).   These results now suggest that
the bat's receiver can be described (or modelled) as an ideal one
which processes the echoes as a matched filter (Bullock, 1976;
Altes, 1971).   Many of the theoretical results on optimal signals
described in this review resemble the actual signals emitted by
bats.   For the present, this seems to constitute proof.

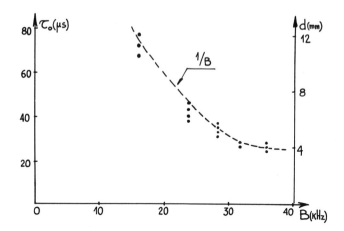

Fig. 1.   Range resolution in bats.

DOPPLER TOLERANT SIGNALS AND ACCELERATION TOLERANT SIGNALS

During cruise and pursuit of prey, bats are faced with the following problem: "How to detect and identify a moving target with constant performances?".

The relative radial speed of the target and the bat, V_R, distorts the signal by the so-called Doppler effect:

$$S(t) \longrightarrow S(\eta t), \quad \eta = \frac{1 - V_{R/C_o}}{1 + V_{R/C_o}}$$
$$t \longrightarrow \eta t$$

The Doppler rate is relatively high (Bard, 1975).

$$C_o \simeq 340 \text{ m/s} \qquad V_R \leq 20 \text{ m/s}$$

$$\frac{V_R}{C_o} \simeq \frac{1}{20} = 5.10^{-2}, \quad 1 \pm 2 \times 5.10^{-2} = 1 \pm 0.1 = \eta$$

Assuming coherent reception with an energy constraint, Altes (1971) derived an optimally Doppler tolerant signal (Fig. 2):

$$S(t) = \frac{A}{\sqrt{t}} J_7 (\beta t) \quad \sin\{2\pi\frac{\nu_0}{a}\log(1 + at)\}, \quad T \simeq 3.3 \text{ ms}$$

$$\beta = 2580 \quad J_n(x) n^{th} \quad \text{order Bessel function.}$$

A similar signal of <u>Myotis</u> <u>lucifugus</u> was recorded by Cahlander (1967) during cruise and pursuit of prey. Other optimally tolerant wave-forms have been recorded and analyzed by different authors during recent experiments (Tupinier et al., 1978). Altes (1977) has also shown that the hyperbolic frequency modulated signal processed by matched filter can estimate the relative speed of the target.

The ambiguity function (output of a matched filter) was de-rived for a high bandwidth duration product BT in hyperbolic fre-quency modulated signals emitted by bats $(20 \leq BT \leq 400)$ (Bard, 1975). Range estimation seems to be biased by a delay τ_m (Fig. 3).

$$\tau_m(\eta) = \frac{T}{\alpha_m - 1} \cdot \frac{1-\eta}{\eta}, \quad T \text{ duration}$$

α_m is the frequency modulation rate. Tupinier et al. (1978) pointed out that during pursuit:

$$\left| \begin{array}{l} T \text{ decreases} \\ \alpha_m \text{ increases} \end{array} \right. \longrightarrow \tau_m \text{ decreases}$$

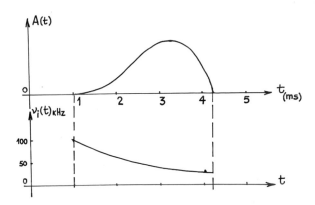

Fig. 2.   Amplitude and frequency modulation of a Doppler tolerant signal.

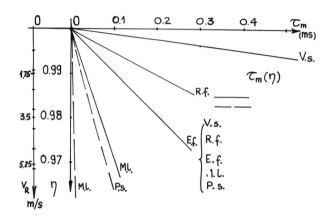

Fig. 3. Bias of range estimation as a function of Doppler effect.

This agrees with previous results reported by Cahlander (1967).

    Taking into account the fact that bats such as Lasiurus borealis
(Cahlander, 1967), Vampyrum spectrum (Bradbury, 1970), or Noctilio
leporinus (Suthers, 1965), emit signals of long duration, Altes
(1971) derived an optimally acceleration tolerant waveform assuming
a coherent receiver. Such a signal is similar to the hyperbolic
frequency modulated cruise signal recorded and analyzed by Cahlander
(1967). These optimally acceleration tolerant signals may be easily
compared with those emitted by different species (Simmons, 1971;
Bradbury, 1970; Suthers, 1965). Assuming a high BT value, the acceleration tolerance of a given signal can be derived using asymptotic
methods (Escudié and Héllion, 1973). Such acceleration tolerance
is very high. Even at acceleration values higher than 10 m/s^2, the
signal to noise ratio loss is less than -3 dB (Escudié and Héllion,
1973).

RANGE AND RANGE ESTIMATION IN BATS: COUPLED AND UNCOUPLED
ESTIMATIONS.

    During target identification and pursuit, the bat estimates
the range and range rate of the flying target. As the speed of the
bat is only two or three times higher than that of insects, the

"pursuit problem" is highly related to the major information: range and relative speed (this is not true for porpoises in hot pursuit of fish!). The incoming echo is mixed with noise; the problem is estimating two different interrelated parameters, range and speed $(r, v_R = \frac{dr}{dt})$. The estimates are correlated and the major problem is reducing the variance of estimation errors (Whalen, 1971):

$$\sigma_{\hat{\tau}}^2 = \frac{1}{E\{\frac{\sigma^2}{\sigma\tau^2} \log p(x|\tau)\}} \cdot \frac{1}{1 - \rho_{\hat{\tau}\hat{\eta}}^2} \qquad \begin{array}{l} \hat{\tau} \text{ range estimate} \\ \hat{\eta} \text{ Doppler rate estimate} \end{array}$$

$$X(t) = S(t) + N(t), \quad (N: \text{ noise})$$

$$\sigma_{\hat{\eta}}^2 \quad \frac{1}{\{\frac{\sigma^2}{\sigma\eta^2} \log p(x|\eta)\}} \cdot \frac{1}{1 - \rho_{\hat{\tau}\hat{\eta}}^2} \qquad \begin{array}{l} \rho_{\hat{\tau}\hat{\eta}} \text{ correlation rate of} \\ \text{the two estimates} \end{array}$$

The variance $\sigma_{\hat{\tau}}^2$ and $\sigma_{\hat{\eta}}^2$ may be minimized so as to reach the minimum Cramer-Rao limit (Whalen, 1971; Van Trees, 1968), if:

$$\left| \begin{array}{l} 1 - \rho_{\hat{\tau}\hat{\eta}}^2 = 1 \longrightarrow \rho_{\hat{\tau}\hat{\eta}} = 1 \\[2ex] 1 - \rho_{\hat{\tau}\hat{\eta}}^2 \approx 1 \longrightarrow \rho_{\hat{\tau}\hat{\eta}} \approx 1 \end{array} \right.$$

Such a solution implies that the designed signal provides uncorrelated estimates for range and range rate (Whalen, 1971; Van Trees, 1968). The signal design problem may be written as follows (Levine, 1970):

$$\lambda_L - <\omega> <t> = 0$$

$$<\omega> = \int_R \pi\nu|\gamma(\nu)|^2 d\nu, \qquad <t> = \int_R t|Z(t)|^2 dt, \quad |Z(t)| = A(t)$$

$$\left| \begin{array}{l} S(t) = A(t)\cos(t) \\ Z(t) = A(t)e^{i\Phi(t)}, \quad BT >> 1 \end{array} \right. \qquad \begin{array}{l} \rightleftharpoons \text{ Fourrier transform;} \\ Z(t): \text{ analytic signal} \\ \qquad \text{related to } S(t). \end{array}$$

$$\frac{1}{2\pi} \cdot \frac{d\Phi}{dt} = \nu_i(t): \text{ instantaneous frequency}$$

An easy solution is given by: $\qquad \left| \begin{array}{l} \nu_i(t) = \nu_i(-t) \\[1ex] A(t) = A(-t), \quad -\frac{T}{2} \leq t \leq +\frac{T}{2} \end{array} \right.$

A and ν_i are even functions!

Even hyperbolic frequency modulated signals have been derived by Altes (1977).

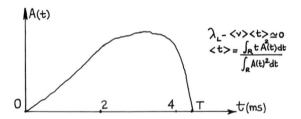

Fig. 4a.   Amplitude modulation of quasi-optimal estimation by bats.

Mamode (1979) claims that many bat signals have very low $\rho_{\hat{t}\hat{\eta}}$ coefficients such that

$$\rho_{\hat{t}\hat{\eta}} < 0.2 \longrightarrow 1 - \rho_{\hat{t}\hat{\eta}}^2 \simeq 0.96$$

and that some bat signals are very near an optimal solution (Tupinier et al., 1978) (Fig. 4a and 4b). An optimal derivation of such uncorrelated estimation signals may be performed under practical constraints such as those that the bats meet: Doppler tolerance, range resolution, energetic constraints.

ANGULAR ESTIMATION OF TARGETS, ANGULAR IDENTIFICATION AND
HARMONIC STRUCTURE

Experiments on Vampyrum spectrum by Bradbury (1970) have shown that bats are able to discriminate between spheres and prolate spheroids (ellipsoids). Some of the signals recorded during identification exhibit many harmonic components (Peff and Simmons, 1972). Simmons noted that under different conditions bats were emitting signals with different harmonic structures (Escudié and Hellion, 1973).   Peff and Simmons (1972) recorded signals and performances of Eptesicus fuscus during discrimination of two spheres in different directions.

If we assume that the bat is crosscorrelating the echoes arriving in the two pinnae, as described by many models of hearing (Bard and Simmons, 1976), we can describe the receiver as an "interferometric receiver" whose performance can explain the results obtained by Peff and Simmons, illustrated in Fig. 5 (Escudié and Héllion, 1973). The high frequency harmonic components increase the angular resolution as shown by the correlation function of the signal (Escudié and Héllion, 1973).

Fig. 4b. Frequency modulation of quasi-optimal estimation by bats.

Bard derived the angular response of such an interferometer without any narrow bandwidth approximation (Fig. 5). The results prove that angular resolution is primarily related to high frequency and low frequency components in the spectrum. Bard and Simmons (1976) have shown that the theoretical results agree with the previous results obtained by Peff during experiments with Eptesicus fuscus (Fig. 6) (Bard et al., 1975, 1977a, 1977b, 1978).

Bard has shown that the harmonic structure of a given signal (e.g., signals emitted by Vampyrum spectrum (Fig. 7)) is a major parameter related to angular response of the interferometric receiver. Furthermore, it may be proved that a "distance frequency equivalence" exists such that bats with small interferometric distances (between the pinnae) are able to emit "wide band" signals to obtain high angular resolution (Bard et al., 1977, 1978).

TARGET IDENTIFICATION AND PARAMETER ESTIMATION

Target Identification

Many experiments using different species (although particularly Eptesicus fuscus) have pointed out the ability of bats to extract target information from the echo. As discussed in the Life Science Research Report 5 (Bullock, 1976):

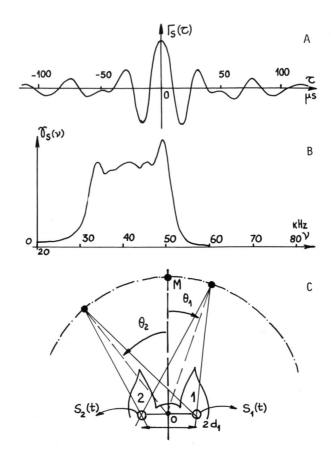

Fig. 5. Correlation function (A) and power spectrum density (B)
of an <u>Eptesicus</u> <u>fuscus</u> signal. Interferometric receiver
and targets (C).

Fig. 6.   Bard's Interferometric Receiver Model (A).   Interferometric angular response of _Eptesicus fuscus_ signals (B).

Fig. 7.   Interferometric angular response of a _Vampyrum spectrum_
signal.

Fig. 8.   Acoustic scattering function of a prolate spheroid.

- Bats can emit "wideband signals" to estimate target range and its size. May we point out the fact that the low frequency components may be used to estimate the shape or volume of the target, as suggested by the "scattering function" or "spectral cross section" defined in Acoustics (Fig. 8).

- Bats emit "wideband signals" with many harmonic components (two or three or more) to estimate small differences in the shape of targets, such as holes (Bullock, 1976). The information content may be described as the complex gain function related to the target. This information requires wideband signals to be extracted by time or frequency processing (Bullock, 1976).

Altes (1975) suggests that a target may be modelled by a set of parallel filters whose fransfer function is related to target parameters.

Assuming coherent reception, many wideband signals emitted by bats can estimate some of these parameters, as demonstrated by Altes and Reese (Bullock, 1976). Even angular estimation provides data on target parameters, as demonstrated using prolate spheroids (Bard, 1977).

Parameter estimation by "narrowband signals" emitted by different bats

Bats such as Rhinolophids or tropical bats (e.g., Pteronotus parnellii) emit "narrowband" signals, called "time diversity" signals by many authors (Escudié et al., 1976). Many examples of such signals have been studied and the frequency modulation may be divided into two parts (Fig. 9):

Fig. 9. Amplitude and frequency modulation of "time diversity" signals.

a) Constant frequency or pure tone signals (CF)
b) Frequency modulated signals (FM) which can
    be modulated in a linear or hyperbolic way
    (Suga et al., 1973).

Such signals can be described as "sequential estimation signals".
The first part involves speed or Doppler estimation and the second
range estimation assuming coherent reception (Suga et al., 1973).
Based on Schultheiss and Hill's theoretical approach to sequential
estimation, Escudié and Héllion (1973) pointed out that under optimal conditions (minimum value of σ_T^2, σ_R^2) the signals emitted by
Rhinolophus ferrumequinum provide very accurate estimations of target speed and range.

As the bat's receiver is coherent, such signals can estimate
target speed even in the presence of reverberation. The receiver
processes the signals in the same way as the so-called "Moving
Target Indicator" (MTI) in radar and sonar. Under severe reverberation conditions, the "sub-clutter visibility" (SCV)(which is
the ability to detect the useful echo against the "cluttering reverberation", may reach values close to -15 to -17 dB in the case
of pure tone signals, such as those emitted by Rhinolophus (Escudié
et al., 1976). FM signals whose BT product is close to 20-30 may
be processed in the same coherent way. The signal to reverberation
ratio at the matched filter output is:

$$\left| \frac{S}{R} \right| = 5.4 \, \frac{\sigma^t}{\sigma^c}, \quad BT \leq 40$$

where σ^t is the target cross section and σ^c the cross section of the
scatterers generating the reverberation (Escudié et al., 1976).

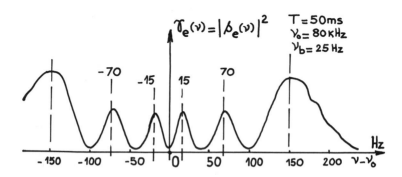

Fig. 10.   Power spectrum density of an amplitude and frequency
           modulated echo.

As suggested by Pye (1967) at the Frascati meeting, a theoretical approach of the echo from a flying moth takes into account:

- Periodic amplitude modulation,

- Periodic frequency modulation (highly related to amplitude modulation) of the echo resulting from wing movements (Escudié et al., 1976).

Such a model of the incoming signal demonstrates that the squared modulus of the echo spectrum exhibits "spectral peaks" related to frequency and amplitude modulation caused by target wing movements. These peaks may be analyzed by the bat's receiver (Simmons, 1972), (Fig. 10).

Some aspects of signal processing and signal design related to sonar signals emitted by bats and porpoises are described in a previous review paper (Escudié and Héllion, 1975).

REFERENCES

Altes, R. A., 1971, Methods of wideband signal design for radar and sonar systems, Fed. Clearinghouse N° AD 732-494.

Altes, R. A., 1975, Bionic signal processing for environmental sensing via sonar, electrostatic fields and vision, ESL. TM 637 Rept. Sept.

Altes, R. A., and Reese, W. D., 1975, Doppler-tolerant classification of distributed targets: a bionic sonar, IEEE Trans. on Aerospace and Electronic Systems, AES-11:708.

Altes, R. A., and Skinner, D. P., 1977, Sonar velocity resolution with a linear-period-modulated pulse, J. Acoust. Soc. Amer., 61:1019.

Altes, R. A., and Titlebaum, E. L., 1970, Bat signals as optimally Doppler tolerant waveforms, J. Acoust. Soc. Amer., 48:1014.

Bard, C., Beroujon, M., Escudié, B., Héllion, A., and Simmons, J. A., 1975, Performances de certains signaux sonar animaux à modulation homographique en présence d'harmoniques, in: "Compte-rendu du Cinquième Colloque GRETSI de Nice", B. Derouet, ed., Cagnes-sur-Mer.

Bard, C., Chiollaz, M., Simmons, J. A., and Tupinier, Y., 1977, Etude des performances angulaires de divers systèmes sonar biologiques aériens à l'étude d'un modèle de recepteur interférométrique: imagerie et identification des cibles, in: "Compte-rendu du Sixième Colloque GRETSI de Nice", B. Derouet, ed., Cagnes-sur-Mer.

Bard, C., and Simmons, J. A., 1976 (unpublished), mentioned in: "Recognition of Complex Acoustic Signals", T. H. Bullock, ed., Dahlem Konferenzen, Berlin.

Bard, C., et al., 1978, Acustica, 40, N°3:139.

Bradbury, J. W., 1970, Target discrimination by the echolocating
    bat, Vampyrum spectrum, J. Exp. Zool., 173:23.

Bullock, T. H., 1976, Life sciences research report, in: "Recogni-
    tion of Complex Acoustic Signals", T. H. Bullock, ed., Dahl-
    em Konferenzen, Berlin.

Cahlander, D., 1967, Discussion to: Theories of sonar systems and
    their application to biological organisms, in: "Animal Sonar
    Systems, Biology and Bionics, Vol. II", R. G. Busnel, ed.,
    Laboratoire de Physiologie Acoustique, INRA-CNRZ, Jouy-en-
    Josas, France.

Escudié, B., and Héllion, A., 1973, Comparaison entre certain sig-
    naux optimaux à grand BT et ceux utilisés par les chauve-
    souris, in: "Compte-rendu du Quatrième Colloque GRETSI de
    Nice", B. Derouet, ed., Cagnes-sur-Mer, France.

Escudié, B., and Héllion, A., 1975, Rev. CETHEDEC N° 44, Vol.12:77.

Escudié, B., et al., Rev. Acoustique, N° 38:216. (1976).

Johnson, R. A., 1972, "Energy Spectrum Analysis as a Processing
    Mechanism for Echolocation", Ph. D. Diss., Univ. Rochester.

Levine, B., 1970,"Radiotechnique Statistique, Vol. II", Mir, ed.,
    Moscow.

Licklider, J. C. R., 1959, "Three auditory theories, in: "Psychology:
    A Study of a Science", S. Koch, ed., McGraw-Hill, New York.

Mamode, M., in press, in: "Compte-rendu du Septième Colloque GRETSI
    de Nice", B. Derouet, ed., Cagnes-sur-Mer, France.

Mermoz, H., 1967, Discussion to: Grinnell, A. D., Mechanisms of
    overcoming interference in echolocating animals, in: "Animal
    Sonar Systems, Biology and Bionics, Vol. I", R. G. Busnel,
    ed., Laboratoire de Physiologie Acoustique, INRA-CNRZ, Jouy-
    en-Josas, France.

Peff, T. C., and Simmons, J. A., 1972, Horizontal-angle resolution
    by echolocating bats, J. Acoust. Soc. Amer., 51:2063.

Pye, J. D., 1967, Discussion to: Theories of sonar systems in
    relation to biological organisms, in: "Animal Sonar Systems,
    Biology and Bionics, Vol. II", R. G. Busnel, ed., Laboratoire
    de Physiologie Acoustique, INRA-CNRZ, Jouy-en-Josas, France.

Simmons, J. A., 1972, Third International Bat Conference, Plitvice,
    Yugoslavia.

Simmons, J. A., 1971, Echolocation in bats: signal processing of
    echoes for target range, Science, 171:925.

Simmons, J. A., and Vernon, J. A., 1971, Echolocation discrimination
    of targets by the bat, Eptesicus fuscus, J. Exp. Zool., 176:
    315.

Suga, N., Schlegel, P., Shimozawa, T., and Simmons, J. A., 1973,
    Orientation sounds evoked from echolocating bats by electrical
    stimulation of the brain, J. Acoust. Soc. Amer., 54:793.

Suthers, R. A., 1965, Acoustic orientation by fish-catching bats,
    J. Exp. Zool., 158:319.

Tupinier, Y., et al., 1978, Analysis of Verpertillionid sonar
        signals during cruise, pursuit, and prey capture, F.I.B.R.C.
        Meeting, New Mexico, U.S.A.
Van Trees, H. L., 1968, "Detection, Estimation, and Modulation
        Theory, Part I", Wiley, New York.
Whalen, A. D., 1971, "Detection of Signals in Noise", Academic Press,
        New York.

MODELS OF SPATIAL INFORMATION PROCESSING IN BIOSONAR SYSTEMS

AND METHODS SUGGESTED TO VALIDATE THEM

Pal Greguss

Applied Biophysics Laboratory
Technical University Budapest
H-1111 Budapest, Hungary

INTRODUCTION

The objective of this symposium is to review the efforts and results in the research on behavior, neurophysiology, ecology, signal processing and electronics engineering with respect to echolocation since the NATO conference on "Animal Sonar Systems - Biology and Bionics" held at Frascati in 1966. I think the Organizing Committee has made a very good decision by choosing 1979 for the year of this symposium. We can now commemorate the 250th anniversary of the man who first described an animal sonar system, that of a bat, the Italian Abbot Lazzaro Spalanzani, who was born on July 12, 1729, at Scandiano in the province of Modena.

Spalanzani discussed his findings first in his correspondence with the Abbot Vassali, as I learned from the first issue of the Archiv für die Physiologie, edited in 1796 by Prof. D. Joh. Christ. Reil, in Halle, Germany (Fig. 1). In Fig. 2, I present a copy of the first two pages of this essay from which we can learn that already at that time, i.e., more than 200 years ago, it was suggested that not only bats but also other animals, and even humans, may have the ability of "seeing" without eyes, if seeing means an information processing, the result of which is an appropriate knowledge of the three-dimensional environment around the species. The answer to this knowledge may then be twofold: orientation and/or target discrimination.

Basically, two types of orientation can be distinguished, depending on whether an organism uses self-generated information carrying waves, or the information carrying wave is independent of the organism. The same distinction holds for pattern recognition

731

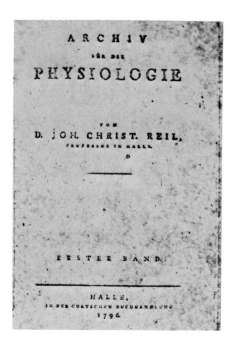

Fig. 1.   First page of the Archiv für die Physiologie published
          in 1796, Halle.

too. The difference between the two information processing mechanisms, the active and passive ones, lies in the strategy of the
information handling. In the first case, small time differences
are translated into spatial dimensions, while in the second case,
amplitude (echo intensity) and phase (time) bound information has
to be processed simultaneously.

     As a result of animal sonar research all around the world,
synthetic sonars are nowadays probably capable of achieving localization accuracies as good as, and perhaps better than, bats, for instance. However, not only the relative time needed to do so may
significantly exceed that required by these animals, but synthetic
sonars are far beyond the target (pattern) recognition abilities of
animal sonar systems. The reason for this is that we are still
rather far away from understanding the design philosophy which spec-
ifies the neural processing of the received wavefront.

     The general sense of conclusion drawn from Simmons' (1979)
review presented at this symposium on the relationship between
behavioral and physiological.data on bat sonar and signal processing
theory, Floyd's (1979) comprehensive discussion of the research data

Fig. 2. First pages of the essay "On a Presumable New Sense of Bats" by Brugnarelli.

on cetaceans, and the detailed theoretical work on echolocation by
Altes (1979), is that echolocating animals somehow retain information from the outgoing cry and compare it with information from
returning echoes. Based upon a variety of field and laboratory
results, scientists suggest that there may be a neural "template"
for storing the original emitted wavefront, and some sort of cross
correlation (or autocorrelation) is performed. A number of different hypothetical design philosophies have already been proposed,
each one involving some form of interaction between outgoing wavefront characteristics and the characteristics of the perceived
echoes.

    Although these models of animal sonars are capable of describ-
ing in a rather conceivable way the functioning of the neural network of echolocating animals where ranging or location is concerned,
they run into difficulty when target and/or shape discrimination
abilities of these animals have to be interpreted. The reason is
simple: the proposed performance strategies are generally not based
upon a simultaneous processing of the amplitude and phase bound
information content of the received wavefront. They presume a
neural activity which only retains the characteristics of the emitted
wavefront, but which is not coherent with the stimulus replica of
the incoming target-reflected wavefront. If, however, we assume
that the stimulus pattern resulting from perceived echo patterns
can interact with a coherent stimulus pattern replica of the emitted
wavefront, then it is evident that the neural network of the animal
is capable of recording the phase bound information of the target
as an intensity (amplitude^2) value simultaneously with its amplitude
bound information, because an interference pattern is formed which
has the properties of a hologram.

    The exact meaning of "hologram" is a controversial subject.
Without entering into that controversy, let us give a definition of
holography as an information processing strategy which includes what
is universally called "holography" as one subcase, and many other
methods as additional subcases. According to this definition, holography is an information processing method comprising two stages.
The encoding procedure involves the recording of amplitude and direction of the radiation coming from the target - the resulting interference pattern is the so-called hologram - and the second, decoding,
step involves the direct utilization of amplitude and direction
information it contains. This means that if the object were three-
dimensional, the decoded, reconstructed image will also be three-
dimensional.

    Applying this concept to describe information processing strat-
egies in biological systems means that the encoding of the spatial
pattern information is not as generally wave-optical in nature, but
stimulus-like. Both the author of this review (1967) and Dreher
(1967) arrived independently, and practically at the same time, at

the idea of applying the holographic concept to interpret the neuro-physiological passage of information in echolocating animals.

The scope of this presentation is twofold. First, it gives a survey of existing or proposed models of animal sonar systems based on concepts related to holography. Then, keeping in mind that it is far more difficult to validate models than to invalidate them, some methods will be suggested which may prove how far one or another of the models may match with reality.

MODELS OF AIRBORNE ANIMAL SONARS

Since the letters of Spalanzani, our knowledge about the mechanism underlying the bat's spectacular and highly successful mode of orientation has considerably increased. We now know that there are species which use ultrasound of continuous frequency (CF), and others which emit frequency modulated (FM) signals in which frequency always sweeps from high to low, while still others use both methods simultaneously: the starting CF signal is terminated with an FM sweep. It is interesting to note that CF bats are in general less agile flyers than FM species, which are able to hover in flight, allowing them to capture crawling or stationary prey.

Considering that the ultrasonic waves emitted by these animals can be regarded as coherent, it has been assumed (Greguss, 1967) that the nature of the mechanism that extracts arrival time information from echoes which are reflected and translated into spatial dimensions is such that the time intervals, i.e., phase bound information, are transformed via a "coherent reference background" in a reversible way into proportional intensity (amplitude^2) relations to which the acoustic receptors only are sensitive. It was therefore predicted that after the wavefront of the target echo enters the acoustic apparatus, the information about the target (and not only the range) is transferred into a stimulus pattern, which then proceeds towards the neural organization of the inferior colliculi, the place where the stimulus pattern replica of the outgoing sound pattern has just arrived, and interferes with it. In other words, both the amplitude and phase bound information of the target are encoded in a holographic form in the neural network of the animal. This prediction was partially based on a remark of Griffin (1950): "the bat's outgoing pulses can just barely be heard amid the random noise; the echoes are quite inaudible. Yet the bat is distinguishing and using these signals some 2000 times fainter than the background noise!"

This astonishing ability of the bats can be particularly well interpreted using the above bioholographic model. Fig. 3 shows the vector diagram of the formation of a stimulus hologram, where R denotes the amplitude of the reference stimulus, I is the amplitude of the stimulus representing the amplitude of the target echo, and θ

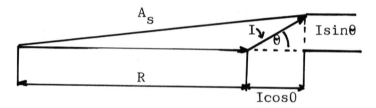

Fig. 3. Vector diagram of the formation of a stimulus hologram.

represents the phase bound information. The resultant stimulus
amplitude, A_s, recorded by the neural network will therefore be

1. $$A_s = (R^2 + I^2 + 2RI \cos \theta)^{1/2}$$

As seen from Equation 1, R^2 is uniform, I^2 is negligibly small,
and so the information is carried by the last term, which is amplified by the factor R. This, however, means that when the signal/
noise ratio gets worse, it is more appropriate to increase the value
of R, than to increase the signal strength. In other words, if the
described bioholographic model of echolocating holds, the bat in a
noisy environment will increase the intensity of its inner reference
stimulus pattern, and not the "loudness" of its cry.

Indeed, Airapetianz et al (1967) experimentally demonstrated
that at the time that bats – at least those belonging to the family
Rhinolophidae – emit an ultrasonic impulse, they also send a stimulus
pattern to that part of their brain where the information carried by
the wave reflected from the object is processed, and if the background noise increases, the intensity of this stimulus pattern increases too, without increasing the intensity of the emitted pulse.

To simulate the proposed neural information processing strategy
of Rhinolophidae we desined the model shown in Fig. 4. Although it
worked at higher frequencies than the frequencies ueed by bats, the
medium was water rather than air, which has a higher sound velocity.
The wavelength was therefore about the same order as used by bats.

A piezoelectric transducer P sends a CF wave into a solid block,
the wedge part of which deflects a portion of this wavefront. This
deflected wavefront simulates the emitted CF wave of the bat, which
in turn travels through the environment until it hits the target T,
from which it is reflected. The undeflected part of the wavefront
generated by the piezoelectric transducer represents the stimulus
replica of the outgoing CF wavefront which proceeds towards a definite neural organization of the inferior colliculi denoted by H in
Fig. 4, where it interferes with the stimulus pattern replica of the

Fig. 4.   Model of the neural information processing strategy
          of Rhinolophidae.

sound waves reflected from the target T.   To investigate how the
intensity distribution of the resulting stimulus pattern - the
biohologram - may look, a sonosensitized plate was placed in plane H
(Greguss, 1966).   Fig. 5a represents such an intensity pattern.
That it has the property of a hologram could be proved by reconstructing from it the image of the insonified target.   Fig. 5b shows such
a reconstruction of a structured metal block as a target.   The noisy
background is due to the not quite adequate reconstruction technique,
the only one available at that time.

Fig. 5.   a)   Resulting interference pattern in plane H.
          b)   Reconstructed image of a structured metal block.

This information processing strategy may, however, be suitable
as an explanation only in the case of CF bats. Bats belonging to
the family of Rhinolophidae terminate their CF signal by an FM sweep.
The horseshoe bat (Rhinolophus ferrumequinum), for example, emits a
CF sound of about 83 kHz for an average duration of 25-50 msec, which
is followed by an FM sweep of 1.5 msec duration from 83 kHz down to
about 56 kHz. Why does this change occur in the characteristics of
the emitted information carrier? Can the holographic concept be extended to that part of the bat's cry?

Our assumption is that in the first stage of their information
gathering, the Rhinolophidae are only interested in whether the space
volume covered by the emitted ultrasound is empty or whether there
are targets in it, and in the latter case, their shape and size,
since this knowledge is necessary to be able to avoid them. Those
targets which the bat may regard as prey, however, have a size roughly comparable to the wavelength used by the bat, so they cannot get
as good an image acoustically as they could optically. How can the
bat resolve this obvious inconsistency?

Perhaps by using an information processing strategy similar to
that used in optical pattern recognition procedures where the cross
correlational properties of the filtering of the Fourier series are
used. This, however, presumes that there are neural templates from
the prey to be recognized (not identical with the templates of the
outgoing cry!) and that these templates (engrams) are stored in a
holographic form (Greguss, 1967). If this holds true and the detected sound field contains information in features similar to one of
those stored in these templates, a stimulus will be generated, the
intensity of which will indicate the degree of correlation. However,
this information processing strategy is valid only if the size of the
target is the same as of that stored in the engram. Since there is
practically no chance that the size of a detected prey will fulfill
this requirement, the bat helps itself by terminating its CF signal
with an FM sweep.

For a better understanding of this strategy, let us first consider what happens when the bat starts to approach the target, i.e.,
when it uses CF waves. At this stage, the bat is in the far field
of the target, i.e., each point of the target manifests itself as a
plane wave propagating in an unique direction. As the bat gets
closer to the target, this is no longer valid, and to maintain the
former condition the target "has to be put" artificially into infin-
ity. The bat achieves this by changing the "focal length" of its
ear which we consider as acting as a varifocal acoustic mirror. If
the target gets into the focal plane of the bat's ear, each point on
the object is transformed into a plane wave, which is the equivalent
of moving the target into the far field.

Dealing now with plane waves, the situation in the receptor system of the bat is somewhat similar to that of an aperture illuminated by a coherent plane wave: a diffraction pattern is formed at a given distance Z from the aperture. Moving a detector along this diffraction pattern with a constant velocity, the intensity of the detector signal will be a function of time. Although no moving sound detector has yet been discovered in the sound receptor system of the bat, it may still experience a similar periodic signal, because sweeping the frequency causes the diffraction pattern to change. If the receptor is fixed at a distance x_0 from the center of the diffraction pattern, and the frequency ν varies with a sweep rate $m = \Delta\nu / \Delta t$, the intensity will vary with time according to:

2.
$$I(t) = \frac{K}{c} mt \left[\frac{\sin (2\pi d/cZ) \, x_0 mt}{(2\pi d/cZ) \, x_0 mt} \right]^2$$

where d is the width of the aperture, c is the sound velocity, and K is a constant which depends on the geometry (Goodman, 1968).

This means, however, that if a coherent receiver is displayed in the hearing system of the bat, it will register an output proportional to the complex field amplitude distribution associated with a portion of the diffracted pattern of the target. It is known from optics literature that a proper coherent processing of data obtained in this way may yield an image of the target, the size of which varies with frequency (Hidaka, 1975; Berbekar, 1975; Farhat, 1975). Since Berbekar and Tökes (1978) have shown for optical pattern recognition that by changing the frequency it is possible to recognize figures of different size, I am almost certain that this holds true for coherent acoustic pattern recognition as well. This then would be the reason why Rhinolophidae terminate their cries with an FM sweep.

I am, however, not unaware that this model is not the only information processing strategy that a bat using an FM signal can pursue, particularly if it is a "pure" FM species. The CF or the CF + FM bats initiate avoidance at around 3 to 10 m distance from the target, while the pure FM bats do so only at around 1 to 3 m distance. This means that pure FM bats have to process information leading to the decision "prey" practically at the same time, which requires most probably another information processing strategy.

MODELS OF UNDERWATER ANIMAL SONARS

Dreher (1967) was the first who considered the possibility that Odontocetes may use some sort of holographic principle in their information processing strategy. He came to this conclusion after analyzing the tegument of the melon of Tursiops. The tegument is anchored to the supporting tissue by structures called rete pegs, and a slice of it shows a remarkably regular array of tiny tubules,

most of them oval in cross section, with an average diameter of
0.01 mm, spaced in longitudinal collumns with an 0.01 mm separation
between columns. Considering the surface as a monostatic receptor
array, he calculated its approximate performance, and found that it
is about 100 lines per millimeter, i.e., in the range of a medium
resolution photographic emulsion. Although this resolution is of
about one order of magnitude less than needed for light holography,
the ultrasonic wavelengths used by dolphins are also several magni-
tudes larger than that of light, so that this resolution capability
matches fairly well the animal's requirement for targets of interest,
usually small to medium fish.

Apart from mentioning the possibility of a holographic information processing strategy, Dreher (1967) did not pursue this idea
systematically. He did, however, refer to an interesting observation
on the behavior of dolphins, quoting Purves (1966), which just may
be a proof of the validity of his idea: there is experimental evidence that a wild Globicephala scammoni may emit a double pulse, as
shown in Fig. 6.

In general, it is thought that by decoding a single hologram
all information on the object has been reconstructed. This, however,
is not quite true, as shown by Gabor (1965), because during reconstruction one component of the imaging amplitude is dropped, viz.,
the one in quadrature with the background (Equation 1). This loss
of information can nevertheless be avoided if a second hologram is

Fig. 6. Double pulse recorded from wild Globicephala scammoni,
         according to Purves (1966).

recorded from the target with a reference background delayed by
λ /4, since this will record the component which previously was
dropped. The reconstruction of these two superimposed holograms
will then completely restore the original wavefront, and all information on the target will be available.

The interval between the pulses in these particular signals
seems to correspond to a delay of λ /4, and so I believe that it is
fair to assume that the reason for such a double pulse is to get
information on the target or prey as complete as possible. My be-
lief is supported by the fact that most CF bats - which most probably use a similar holographic principle - radiate sound from a dipole
source separated so that they can use this principle to reconstruct
all information on targets reaching their processing mechanism.

That blind fish can swim around without touching obstacles
and find their prey was already described by Brugnarelli shortly
after Spalanzani's experiments became known in the late 1790s. Do
they use some sort of sonar system? To answer this question we have
to investigate the role of the lateral line organ which is found
along the sides of the body in all fish and in many amphibians. It
is a special sensory system that perceives vibrations, and which has
many anatomical features in common with mammalian sound receptors.
The mechanical waves generated during swimming are reflected from
different targets and are picked up by the lateral line organ which
produces a stimulus pattern to be processed. The undulating motion
of swimming fish is certainly coordinated by a kind of neural activity, the pattern of which may serve as a reference background in
the information processing strategy of the fish, since it can be
regarded as a stimulus replica of a plane wavefront perpendicular
to the swimming direction. As shown in the sketch of Fig. 7 (for
the sake of simplicity only one reflecting point is indicated), the
reflected waves are spherical, so that the interference of their
stimulus replica with the stimulus template may yield an interference
pattern equal to a one-dimensional zone plate. This means essentially that the information of the given point is coded in a holographic
form, somewhat similar to a "side-looking radar" technique. At
present, this model is only speculative, although Darwin's theory
of evolution allows us to infer the existence of a similar information processing strategy in the neural network of the fish.

Animals using sonar systems live in air or water, i.e., in an
environment in which more phase bound information has to be processed
than on the surface due to the higher degree of freedom of motion.
If an animal lives in an environment where it has to process a lot
of phase bound information, it will most probably use some sort of
holographic principle. If, however, this information is no longer
crucial, the animal may abandon the method. This is just what we
find when we study the sex life of Triturus vulgaris, an amphibian
species.

Fig. 7. "Side-looking radar" in fish?

For nine months of the year this amphibian moves on solid surfaces only, but during the three-month breeding season both males
and females return to water. And what happens? Their lateral line
system, which is totally nonfunctional during terrestrial life, re-
develops together with the fin. When returning to an environment
in which living requires more information processing, an organ
evolves which may be used to develop a holographic information processing strategy.

SUGGESTIONS ON HOW TO INVESTIGATE BIOSONAR MODELS

The holographic information processing strategy is only one
of the theoretical models of biological sonar phenomena, and it is
as heuristic as most of them, i.e., it is difficult to test. In
general, the validity of these theories is questioned by a demonstration of the failure of occurrence of requisite event features
at the level of measurement, as was correctly pointed out by Diercks
(1972) in his survey of biological sonar systems. Furthermore, these
models usually fail to predict an unexpected physiological event in
the neural system of the animal. Thus, claims of "solutions" to biological sonar problems are premature and pretentious, and the same
holds true for bioholographic models as well. These claims are
merely statements of the capabilities and limitations of one-dimensional signal form(s). The animal emits and receives patterns of
at least two dimensions, whose structure is barely known at present.
Furthermore, little is known about the acoustic pattern to stimulus
pattern conversion in the receptor apparatus of the animal, although

this knowledge is needed for an understanding of the neural process
resulting in a given kind of animal behavior. The purpose of the
closing part of my review is to suggest some measuring techniques
which may help to validate or invalidate animal sonar models, including of course the reviewed holographic models as well.

The works of Tonndorff and Khanna (1971) introduce the technique of holographic interferometry to study bioacoustical phenomena. From their method of real-time interferometry, researchers
have recently switched to the more sophisticated double-exposure
interferometry, and it appears now that this technique already
offers practical applications in otolaryngology diagnostics. Fig. 8,
recorded by von Bally (1977), shows for example the behavior of the
human acoustical-to-stimulus converter, the tympanic membrane, at
two different frequencies. Why not use a similar method to study
the performance of animal sonar systems?

The holographic analysis of the behavior of animal sonar
systems would require much care, however, not only because of the
size of these systems, but also because of their difficult access.
Those who are familiar with optical holographic methods know well
that two basic recording configurations exist: the original arrangement of Gabor (1948), where the reference and object wavefronts are
in line, and the arrangement introduced by Leith and Upatnieks (1962)
where these wavefronts are not collinear, the so-called off-axis
method. The merit of the in-line technique is its simple recording

Fig. 8.   Vibration mode of the human tympanic membrane
          at 1.5 kHz (left) and at 3 kHz (right).

configuration. During reconstruction of the hologram, however, the
virtual and real images may interfere with each other, and it is
rather difficult to separate them. The off-axis method, on the other
hand, yields separated images, but it would be rather difficult to
use it for testing the behavior of the melon of a dolphin.

To overcome these difficulties, our laboratory is presently
developing an easy-to-handle holographic camera based on a forgotten
paper of Vanderworker and Snow (1967). The recording begins in the
same way as in-line holography, i.e., the laser beam is diverged by
a micro-objective, whose optical axis coincides with the axis of the
object field. However, between the object and the film (which is
placed in the focal plane of the micro-objective) is a beam divider
which reflects a portion of the diverging laser beam to the film
plane, and this serves as a reference background. Depending on the
reflecting properties of the object, beam dividers of different reflecting properties can be used to produce a 2-to-1 reference-to-
subject energy ratio, said to be optimum for making holograms. To
avoid fogging of a large area of the film by the passage of the laser
beam through the photographic film, it is not enough to focus the
laser beam at the film plane. An area of about 2 to 3 mm^2 of the
emulsion (film) must be removed.

The advantage of this recording technique is that in spite
of being in-line during recording, it is off-axis during reconstruction, and rather good reconstructions can be obtained even with white
light.

Although the suggested holographic camera may help to furnish ideas on the performance of the acoustical-to-sensory converter
of the animal sonar system, it gives us practically no information
on the spatial pattern of the acoustic field acting on the acoustical-to-sensory converter (provided that the acting acoustic field
is visualized by a method which allows a holographic recording, i.e.,
the converter does not destroy the coherence of the laser beam).
From the growing literature on acoustical-to-optical converters
(Greguss, 1975; Wade, 1976; Greguss, in press), I wish to pick out
only one, which is based on the change in bi-refrigerant properties
induced by an ultrasonic field.

A thin layer of nematic liquid crystals (e.g., MBBA) is
sandwiched between two glass plates which are coated with a material
that is adherent to their surfaces and yet non-absorbant with respect
to the liquid crystals. This allows the crystals to be in a low
energy state, aligned on one axis but having no fixed alignments on
other axes (Greguss, 1973). An ultrasonic wavefront acting on such
a layer can easily produce a change in the predetermined order and
state of the layer and consequently alter the optical transmission
properties of the system. The liquid crystal molecules are bi-
refrigerant, and when they act as wave plates between polarizers,

the impinging acoustic wave pattern can be transformed in real time into an equivalent optical pattern. The ultrasonic sensitivity of such a display is at present on the order of several mW/cm^2, so that in theory they could be used to visualize sound fields of echolocating animals.

An advantage of such a display is that a laser can be used without losing its coherence to visualize the acoustic field. Thus, placing such an AOCC display before a holocamera, the spatial pattern of the echo to be produced by the animal sonar system could perhaps be recorded.

REFERENCES

von Bally, G., 1977, Holographic methods in biomedical sciences, Proc. Int. Conf. on Optical Computing in Res. and Dev., 68, Budapest.

Berbekar, G., 1975,"Hologram Aperturak Letapogatasa a Frekvencia Valtoztatasaval", Ph. D. Diss., BME, Budapest.

Berbekar, G., 1978, Hologram aperture synthesis with frequency sweeping, Ultrasonics, 16:251.

Farhat, N. A., 1975, A new imaging principle, in: Proc. IEEE (Lett).

Gabor, D., 1965, Imaging with coherent light, CBS Laboratories Seminar, Stanford.

Gabor, D., 1948, A new microscopic principle, Nature, 161:777.

Goodman, J. W., 1968, "Introduction to Fourier Optics", McGraw-Hill, New York.

Greguss, P., 1966, Pictures by sound, Perspective, 8:287.

Greguss, P., 1967, Bioholography, Lecture at the Society of Electronic Engineers, Budapest.

Greguss, P., 1975, Pictures by sound, Chimia, 29:273.

Greguss, P., in press,"Seeing by Sound", Focal Press, London.

Greguss, P., 1974, USA Pat. 3,831,434. Brevet français, 1973, 2.177.410.

Hidaka, T., 1975, Image reconstruction by using frequency sweep method, J. Appl. Phys., 46:786.

Leith, E. N., Upatnieks, J., 1962, Reconstructed wavefronts and communication theory, J. Opt. Soc. Amer., 52:1123.

Purves, P. E., 1967, Anatomical and experimental observations on the cetacean sonar system, in:"Animal Sonar Systems", R. G. Busnel, ed., Laboratoire de Physiologie Acoustique, INRA-CNRZ, Jouy-en-Josas, France.

Tonndorff, J., and Khanna, S. H., 1971, Validation of holographic observation on the displacement of the tympanic membrane in cats, J. Acoust. Soc. Amer., 49:120.

Vanderworker, R., and Snow, K., 1967, Low spatial frequency holograms of solid objects, Appl. Phys. Lett., 10:35.

Wade, G., 1976, "Acoustic Imaging: Cameras, Microscopes, Phase Arrays and Holographic Systems", Plenum Press, New York.

A NEW CONCEPT OF ECHO EVALUATION IN THE AUDITORY SYSTEM OF BATS

Karl J. Beuter[+]

Fachbereich Biologie, Johann Wolfgang Goethe Universität

Frankfurt am Main, Federal Republic of Germany

1. INTRODUCTION

Bats use their echolocation system very effectively for spatial orientation and prey hunting. Important tasks bats are able to perform are the detection of weak target echoes or echoes superimposed by clutter-interference, the precise localization of targets, and the identification of target structure.

In every echolocation system exist many different requirements which should be fulfilled simultaneously. Basic needs are for instance a high signal energy, a good directionality of transmitter and receiver, and a good time resolution for precise ranging. The analysis of echolocation sounds can give hints how bats have solved this problem in their bionic sonar.

Bat sounds can be recorded and qualitatively analyzed with a rather simple technical apparatus. As a consequence, a variety of different types of sounds has been described. A simple classification scheme according to basic frequency modulation patterns like constant frequency (cf) and frequency modulated (fm) appears more and more unsatisfactory. The goal of theoretical work is to find a generally accepted concept concerning the purpose of the different frequency modulation functions.

A combination of behavioural and systems theoretical methods is the most promising way to find an explanation for the different types of bat sounds. In a given behavioural situation, for example during target discrimination, bats emit very similar sequences of sounds during each approach. Sound sequences found in connection with different target configurations exhibit characteristics, which can be a basis for species specific sound classification.

[+]New address: Battelle-Institut e.V., Am Römerhof 35, Frankfurt am Main, Federal Republic of Germany

The purpose of this study is to derive a processing mechanism which
does not depend on smaller changes of frequency and amplitude modulation functions. This mechanism should enable the bat to make consistent estimates of target structure, even when the sound pattern
is drastically changed as for example during approaching a prey.

Any receiver model should be judged according to its biological
relevance. Therefore it has to take into account the basic properties of the neural elements concerned with coding the echo information. The output signals of two basic echolocation receiver types
will be dicussed in detail. Ambiguity diagrams and frequency spectra
of natural echolocation signals will form the basis for comparing
the properties of these receivers.

Since coding of distance and of structural differences in targets
has been studied in several behavioural experiments, the resolution
of time and time differences in echolocation receivers is of central interest. The echoes of ideal so called "point targets" and
simple structured targets will be simulated and analyzed.

The most simple structured target consists of two point targets
separated in range by a small distance. To avoid confusion, any
target combination will be called a structured target as long as
the partial echoes overlap. In this case only one wave train returns
to the bat's ear.

Discrimination between structured targets has been studied by Simmons et al. (1974), Vogler and Leimer (1979), Habersetzer and Vogler (1979) in different bat species. The structured targets consisted of massive plates with cylindrical holes of constant depth.
The bats were able to discriminate between plates, which had differences in depth as low as 1.5 mm. This corresponds to a difference
in sound propagation time of about 8.5 μsec.

Bats can discriminate between structured targets much better than
between separated targets. The lower limit for discrimination in
range between separated targets is in the order of 15 mm (Simmons,
1973).

2.   THE CONCEPT OF OPTIMUM RECEIVERS

Echolocation systems must derive target information from differen-
ces between the emitted signal s(t) and the echo s($\underline{a}$,t). $\underline{a}$ and t
represent a set of target parameters and time, respectively. A practically important measure of difference is the root mean square
deviation in the form

$$D(\underline{a}) = \int_{-\infty}^{\infty} (s(t) - s(\underline{a},t))^2 dt$$

In the case of energy normalized echoes it follows

$$D(\underline{a}) = const - 2\mathcal{X}(\underline{a})$$

where

$$\chi(\underline{a}) = \int_{-\infty}^{\infty} s(t)s(\underline{a},t)dt \tag{1}$$

In the expression for D, only χ depends on the target parameter $\underline{a}$. Therefore it is sufficient for the receiver to evaluate only χ, which is called the generalized ambiguity function (AF) of the signal s. The AF was introduced in a more special form by Woodward (1953). In a geometrical representation, where signals are described by vectors, the AF can be visualized as a scalar product between the signal- and the echo-vector.

Ambiguity diagrams can serve as a means to describe how a change in a certain parameter influences the deviation of the echo from the expected signal s(t). When the AF changes rapidly with small changes of the parameter, the signal is called "resolvent" for this parameter. If the AF changes only little, it is called "tolerant". Unfortunately it is not possible to make the AF resolvent for all parameters simultaneously. This is due to its so called volume invariance property (Altes, 1973).

Many problems in optimization theory lead to models, the properties of which can be described by the AF. A central problem in echolocation theory are echoes superimposed with interfering noise and ground clutter. Engineers are therefore forced to construct optimal receivers which give the best results according to the following criteria:
-Maximizing of the signal-to-noise ratio at the receiver output.
-Maximizing of the correct signal detection rate at a given "false alarm" rate.
-Optimization of parameter estimation according to a minimum mean square error criterion.
All solutions of this problem (Van Trees, 1968) lead to the same basic signal processing mechanism which is called the "optimum filter", but one has to keep in mind, that it is only optimal for a restricted class of problems. It can be shown that the AF simply describes the output of an optimum filter which is matched to the emitted signal. In technical systems the optimum receiver can be implemented in numerous ways such as correlators, matched filter banks, tapped delay lines or dispersive filters.

Let the target parameter $\underline{a}$ consist only of the Doppler factor $p = (1+v/c)/(1-v/c)$ and the time delay $\tau = x/v$, where x is the distance between bat and target, v the relative speed and c = 343 m/sec, i.e. the velocity of sound. This gives the normalized wideband AF in the form

$$\chi(\tau,p) = p^{1/2}\int_{-\infty}^{\infty} s(t)s(p(t-\tau))dt \tag{2}$$

This type of AF was used in the computer analyses of this study.

Up to this point, signals echoes and filters were only described in

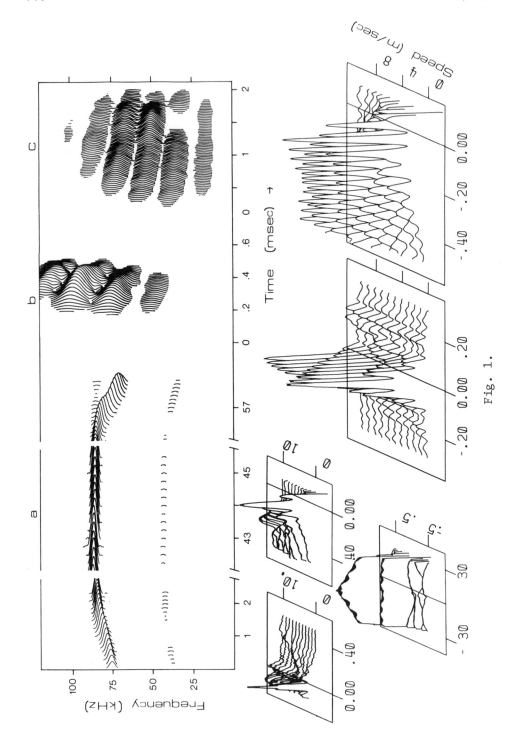

Fig. 1.

the time domain. The Fourier transform leads to an equivalent description in the frequency domain. Johnson and Titlebaum (1976)
pointed out that the energy spectrum contains the same information
as the AF, if the Fourier transform is taken from a time interval
which contains both the emitted signal and the echo.

Since the output of an optimum receiver gives the best estimate
of signal parameters in practically important cases, one can use
the AF to derive potential resolving capabilities of bat signals.
Energy spectra of signals and their echoes can serve the same purpose.

3.   SOUND RECORDING AND COMPUTATIONAL METHODS

3.1 Animals and Sound Recording

Standard equipment was used to record ultrasonic emissions. Echolocation sounds of four different bat species were recorded in the
laboratory when the bats performed defined tasks.

Rhinolophus ferrumequinum and Taphozous melanopogon were flying on
a straight path through an empty room of 7x3x2.5 m. Myotis myotis
had been trained to catch a mealworm from a stick (Habersetzer,
1978). The sounds were recorded along the whole flight path during
approach to the prey. Megaderma lyra had been trained to discriminate between two structured targets (Vogler and Leimer, 1979). The
sounds were recorded when the bat was flying 30 cm in front of the
targets.

3.2 Computer Analyses and Target Simulations

FORTRAN computer programs written by the author were used on a
PDP 11/40 laboratory computer for all signal analyses and for target simulations. After analog-to-digital conversion, bat signals
were stored in digital form on magnetic discs. AFs were calculated
according to Eq. 2. For graphical reasons and improved visibility,
their envelopes were derived by quadrature mixing and plotted in
the ambiguity diagrams.

Computer spectrograms which display the frequency modulation functions were constructed from a sequence of short time frequency
spectra. The partially overlapping cosine-tapered echolocation
sound segments, on which the Fourier analyses were performed, had
a duration of 200 µsec, each.

Figure 1 (opposite page): Spectrograms (above) and ambiguity functions (below) of typical echolocation sounds in different bat species. a) Rhinolophus ferrumequinum. b) Megaderma lyra. c) Taphozous
melanopogon.

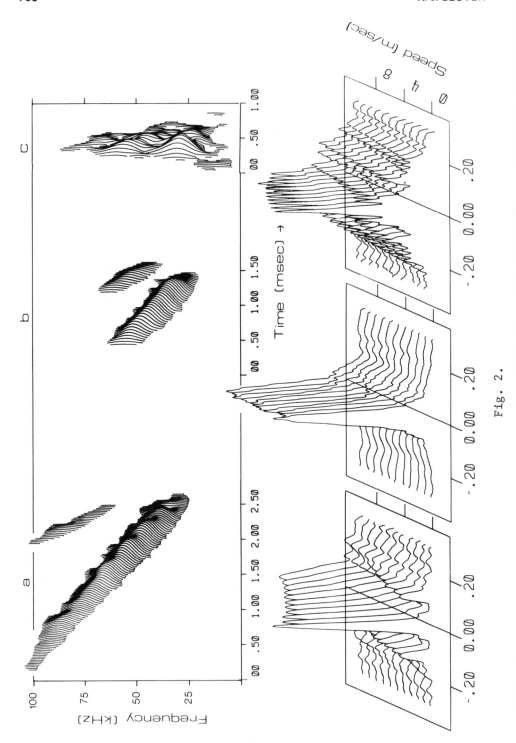

Fig. 2.

Simple structured targets reflecting the sound only in two planes
with limited size can be simulated by two point targets shifted in
range. The corresponding time shifted signals were added to their
original version. The sum was energy-normalized and stored in the
same way as normal echolocation sounds.

4.   ANALYSES OF ECHOES

4.1   Ambiguity Functions of Natural Bat Sounds

Typical echolocation sounds from three bat species are shown in
Fig.1. The Doppler parameter axis in the AFs is translated into
target speed. The shapes of the AFs corresponding to the basic frequency modulation functions cf, cf-fm, short harmonic fm and harmonic cf-fm demonstrate the following general properties.

Cf signals as in the sound of Rhinolophus ferrumequinum allow for
a very good speed resolution which can be as fine as 10 cm/sec
(Beuter, 1976). Harmonic structures as in the Megaderma lyra sound
and those of many other species lead to increased ambiguities,
since spurious sidelobes appear. They are caused by the different
frequencies simultaneously interfering in the harmonics. The sidelobes are closely spaced and have different hight. Structured targets would produce similar responses. If the bats would use an
optimum receiver working only in the time domain, a very precise
storage mechanism would be required to compensate for these ambiguities.

In Fig.1a (downward sweep) a cf component at the beginning causes
increased Doppler sensitivity, which is expressed in a rapid drift
of the maximum of the AF with increasing speed. This leads to an
overestimation of distance for approaching targets.

In contrast to the long cf-fm call of Rhinolophus ferrumequinum,
Taphozous melanopogon produces fm-sounds with varying sweep rates
and many harmonics. The slow sweep rates cause increased Doppler
sensitivity and simultaneously the harmonics produce many strong
sidelobes. Both effects would call for a very sophisticated evaluation in an optimum processor.

In Fig.2 sounds from three consecutive flight phases of Myotis
myotis were analyzed. These types of sound have a very small Doppler sensitivity at least in the biologically relevant speed range
below 10 m/sec. The shape of the AF in Fig.2b is very advantageous
for precise target ranging and target structure discrimination,
since it has only minor sidelobes. A very similar shape of AFs has
been found in the hyperbolic fm sweeps of Eptesicus fuscus (Beuter,
1978).

Figure 2 (opposite page): Spectrograms (above) and ambiguity functions of Myotis myotis echolocation sounds during approaching a
prey. a) Search phase. b) approach phase. c) Final phase.

When Myotis myotis is approaching a target, the time spread of the
AF increases. The harmonics in the final buzz sounds (Fig.2c) exhibit similar sidelobes as is typical for Megaderma lyra sounds.
From theoretical reasons, the width of the AF along the time axis
is for all signals in the order of magnitude of their inverse bandwidth. Therefore one can expect for all bats with similar sound
bandwidth a width of the AF from 20 μsec to 100 μsec. The high
bandwidth of the long linear fm sweep of Myotis myotis in Fig.1a
is the cause for the very narrowly peaked AF.

Linear or hyperbolic frequency modulations in bat sounds do not
severely influence the shape of the AFs within a biologically significant range of speed. Other unwanted influences such as harmonic structures or irregular sweep rates produce much stronger
effects. This may indicate that there exists no evolutionary
pressure in all bats to produce optimally Doppler tolerant hyperbolic fm sounds as has been speculated by Kroszynski (1968) and
Altes and Titlebaum (1970).

4.2   Echoes From Structured Targets

The output of optimum filters was simulated in the time domain by
caculating envelopes of cross-ambiguity functions between the
actually emitted echolocation sounds and their simulated echoes

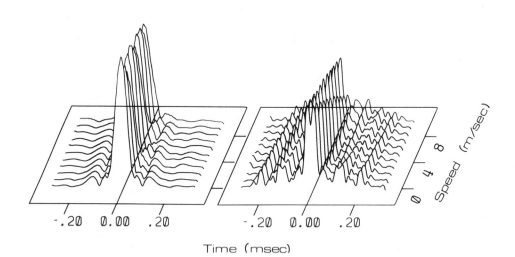

Time (msec)

Figure 3: Cross ambiguity function between signals from Figures.
2b and 2c and their simulated echoes from simple structured targets
formed by two point targets separated in depth by 2.8 mm.

from a simple structured (staggered) target with 2.8 mm depth.
Two examples are given in Fig.3 for the sounds from the approach-
and final buzz-phase in <u>Myotis myotis</u>. Separated peaks caused by
the two time-shifted partial echoes can be only observed in the
final buzz sound but the peak separation would lead to an erroneous
depth estimate of about 4 mm. Effects on the amplitude of the optimum filter outputs can also be seen, but they are of minor importance, since in natural situations the bat does not receive energy
normalized echoes and can evaluate the absolute amplitudes therefore only to a limited extent.

Four different representations of the echo of an <u>Eptesicus fuscus</u>
sound reflected from a simple structured target of 14 mm depth are

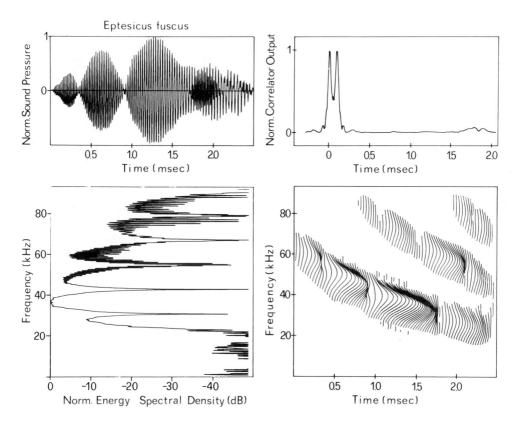

<u>Figure 4</u>: Simulated Echo from a simple structured target formed by
two point targets separated in depth by 14 mm. The original signal
was a natural echolocation sound of <u>Eptesicus fuscus</u>.
Upper left: Waveform. Upper right: Cross-AF between emitted sound
and echo for zero speed. Lower left: Energy spectrum. Lower right:
Spectrogram.

given in Fig. 4. The corresponding echo delay is 82 μsec.

Strong beats appear in the envelope of the echo. They have durations in the order of several hundred μsec. The cross AF for zero
Doppler correctly indicates the two point targets, separated by a
propagation delay of 82 μsec. But the two peaks are only separated
down to about 5o% of their maximal amplitude.

The most strikung effect appears in the energy spectrum, where the
originally smooth curve is periodically depressed. The distance between the periodic minima is 12.2 kHz. The depressions also appear
in 'the spectrogram at the same frequencies.

The observation can be qualitatively understood on the basis of
interfering waves which are shifted in space by a fixed propagation delay. Destructive interferenc is most effective for those
waves which are shifted by half a wavelength. The effect can be
also derived by simple mathematics. It follows, that the resulting
energy spectrum of the sum of the two echoes contains cosine-
shaped modulations, which are characteristic fot the time delay ΔT
between the echoes. The bandwidth or period of these modulations is

$$B = 1/\Delta T \qquad\qquad (3)$$

This explains the strong modulations in the energy spectrum in Fig.
4. It explains also the beats in the envelope of the echo waveform.
The beats are not regularly spaced as a consequence of the hyperbolic fm and the harmonics. This renders their direct evaluation
difficult.

The narrow oscillations in the energy spectrum are due to the higher
harmonics, which are similar to time-delayed versions of the first
harmonic. These "fast" oscillations in the spectrum are a good
example to visualize the inverse relationship between time delay
and the period of spectral oscillations. For time shifts of 3 msec,
the spectral oscillations have periods of only 33o Hz. To separate
these fine spectral structures, a receiver with its best frequency
at 6o kHz would require a frequency resolution better than o.3%.
At smaller echo delays however, the period of spectral oscillations
increases and is easier to resolve.

5.   A NEW CONCEPT FOR ECHO PROCESSING IN BATS

5.1   Time and Frequency Coding in the Auditory System of Bats

So far only the properties of theoretically derived receivers have
been discussed. We now have to evaluate the coding properties of
the auditory system represented by neural responses.

Suga and Schlegel (1973) have found fm-specialized neurons, which
only respond during a short time interval with very short latencies.

Their potential to dicriminate between a two-tone combination with very small time lags has not yet been measured. Phase-locked responses of single neurons to signal waveforms up to frequencies of 6 kHz have been found. At higher frequencies, the periods of the oscillations could not be distinguished from the time course of the neural responses. The time intervals which correspond to this upper limit of direct temporal coding are longer than 100 μsec.

Recently Vater and Schlegel (1979) have studied the responses of neurons in the inferior colliculus of molossid bats. These bats use strong fm components in their echolocation sounds. A high percentage of phasic on-neurons characterized by very short bursts of activity were found. They respond preferrentially to certain fm stimuli. The short time constants of these phasic neurons make them likely candidates for coding short time intervals down to about 100 μsec.

There is no neurophysiological evidence however, indicating that time intervals far below 100 μsec can be represented by corresponding temporal changes in neural activity. This means that direct coding of changes in the envelopes of echoes or receiver outputs, which are shorter than 100 μsec, is very unlikely.

Processing of information in the frequency domain requires an appropriate representation of signals. The main requirement therefore is a sufficient frequency resolution of the neural elements. Q-10 dB values in fm bats do not differ very significantly from those in other mammals and are typically in the order of 10. A typical neuron with its best frequency at 50 kHz would therefore exhibit a -10 dB bandwidth of about 5 kHz.

5.2 Proposal for a New Processing Mechanism

From the statements above, one can derive a general new concept for structure- and distance-discrimination mechanisms in bats. The model is based on the fact, that the resolution of target structures can be improved, when both temporal and spectral echo representations are analyzed.

The inverse relationship between echo delay and the resulting periods of modulations in the energy spectrum suggests, that the echo spectrum is used to derive fine structural information. Filtering in the time domain should be used for near optimum detection of weak echoes and for the estimation of larger distances. The proposed mechanism has the following properties.
1. Echo features caused by time shifts of more than a critical time resolution constant are evaluated by direct temporal coding.
2. Echo features which are caused by time shifts smaller than this critical time resolution constant are evaluated in the frequency domain.
3. The critical time resolution constant is in the order of

magnitude of 100 μsec and is similar in all bats.

These three conditions can be simultaneously fulfilled by the same
receiver type. Moreover, this receiver can be optimized for typical
sweep rates. For instance, the receiver can be constructed from a
broadband set of fm specialized bandpass filters with Q-values in
the order of 10. Each filter should be followed by an envelope

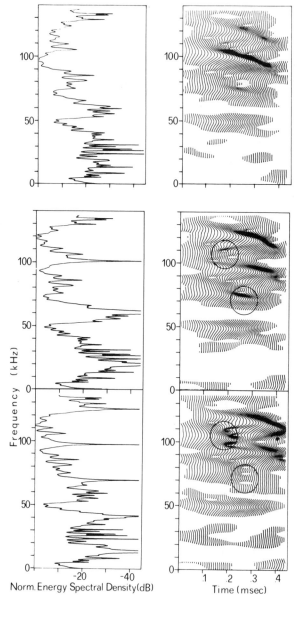

Figure 5: Energy spec-
tra (left) and spectro-
grams (right) of diffe-
rent computer simula-
ted targets based on a
natural echolocation
sound of Megaderma lyra.

a) Point target.

b) Simple structured
target formed by two
point targets separa-
ted in depth by 4.1 mm.

c) Target depth of
5.9 mm.

Circles indicate corres-
ponding areas with major
differences.

detector which transform the filter output into a sequence of action potentials.

Temporal features of the echo are coded in this model by the timing of activity in each channel. Spectral features are represented by the relative pulse rates in the channels. This coding gives a representation of the echoes in a frequency-time matrix. The distribution of activity in this neuronal matrix resembles closely to the distribution of the hills in the spectrograms of this study.

Target structure analysis in the frequency-time matrix of this model can be therefore simulated as in Fig. 5. The emitted sound of Megaderma lyra contains five harmonic downward modulated frequency bands. The echoes of the structured targets with 5.9 mm and 4.1 mm depth differ essentially from each other and from the emitted signal. These differences can serve the bat to discriminate between the two targets.

Frequency analysis of very short signal segments leads to reduced frequency resolution in the spectrograms. This inevitable result of the tradeoff between time and frequency resolution produces the smooth lines in the spectrograms of Fig. 5. In the proposed model this effect may be advantageous for processing structural information since random fluctuations in the energy spectra of the emitted sounds are averaged out. Elimination of variations in the emitted sounds may lead to improved detectability of target structures.

6. DISCUSSION

Simulated structured targets are well suited to describe the essential properties of echoes from real targets which have been used in behavioural experiments. This allows for finding out the shortcomings of optimum processors in their basic form which appear, when natural echoes are analyzed.

The shortcomings of conventional models can be solved by the proposed mixed model for frequency-time processing. The model can explain the coding of structural information on the basis of presently known neural properties. Extremly fine structural target information can be extracted from spectral echo patterns and range information can be evaluated by direct temporal coding. This method corresponds to the method for constructing spectrograms. The structural analysis of sonar targets is thus reduced to the recognition of frequency-time patterns or in the case of tonotopic neural organization a place-time pattern.

The proposed concept allows for matching the receiver to some given fm rate and makes optimization of signal-to noise ratio possible. Especially this receiver mechanism is well suited to evaluate all types of wideband sonar signals and does not require

receiver adaptations for changing signal structures. The distributed processing makes this model safe against partial failure
and gives the bat the opportunity to vary his echolocation call
according to external conditions.

7. REFERENCES

Altes, R. A. and Titlebaum, E. L.: Bat signals as optimally Doppler
tolerant waveforms. J. Acoust. Soc. Amer., 48, 1014-1020 (1970)

Altes, R. A.: Some invariance properties of the wide-band ambiguity function. J. Acoust. Soc. Amer., 53,1154-1160 (1973)

Beuter, K. J.: Systemtheoretische Untersuchungen zur Echoortung
der Fledermäuse. Doctoral Dissertation, University of Tübingen
(1976)

Beuter, K. J.: Optimalempfängertheorie und Informationsverarbeitung
im Echoortungssystem der Fledermäuse. In Kybernetik '77,
G. Hauske and E. Butenandt eds., Oldenbourg, München-Wien, 1978

Habersetzer, J.: Ortungslaute der Mausohrfledermäuse (Myotis myotis)
in verschiedenen Ortungssituationen. Diplomarbeit, Fachbereich
Biologie, University of Frankfurt (1978)

Habersetzer, J. and Vogler, B.: Discrimination between structured
plates in the bat Myotis myotis (in preparation)

Johnson, R. A. and Titlebaum, E. L.: Energy spectrum analysis: A
model of echolocation processing. J. Acoust. Soc. Amer., 60,
484-492 (1976)

Kroszynski, J. J.: Pulse compression by means of linear-period
modulation. Proc. IEEE, 57, 1260-1266 (1968)

Simmons, J. A.: The resolution of target range by echolocating
bats. J. Acoust. Soc. Amer., 54, 157-173 (1973)

Simmons, J. A., Lavender, W. A., Lavender, B. A., Doroshow, C. A.,
Kiefer, S. W., Livingston, R., Scallet, A. C. and Crowley, D. E.:
Target structure and echo spectral discrimination by echolocating
bats. Science, 186, 1130-1132 (1974)

Suga, N. and Schlegel, P.: Coding and Processing in the auditory
systems of FM-signal-producing bats. J. Acoust. Soc. Amer., 54,
174-190 (1973)

Van Trees, H.: Detection, estimation and modulation theory, Part I:

Wiley, New York, 1968

Vater, M., Schlegel, P.: Comparative auditory neurophysiology of
  the inferior colliculus of two molossid bats, Molossus ater
  and Molossus molossus. II: Single unit responses to frequency-
  modulated signals and signal and noise combinations. J. Comp.
  Physiol. (in press)

Vogler, B. and Leimer, U.: Discrimination between structured plates
  in the bat Megaderma lyra (in preparation)

Woodward, P. M.: Probability and information theory, with applications to radar. Pergamon press, Oxford, 1953.

Chapter VI
Sensing Designs

Chairman : R.S. MacKay

Sensing system design using knowledge of animal acoustic systems.
    R.S. MacKay

Air sonars with acoustical display of spatial information.
    L. Kay

Discrimination of complex information in an artificially generated auditory space using new auditory sensations.
    M.A. Do

SENSING SYSTEM DESIGN USING KNOWLEDGE OF ANIMAL ACOUSTIC SYSTEMS

R. Stuart Mackay

Biology Dept., Boston University

Boston, Mass., 02215 USA

By way of introduction to the topic and the following papers, the term "bionics" might be mentioned. A goal of this activity has been to use the slowly evolved adaptations and mechanisms of living organisms to guide the design of technical developments. There have been some interesting exceptions, but in general this has not often happened. More often the converse has taken place. That is, after a new development has been pro-duced it can be seen that some now understood related systems appear in nature, and in many cases technological progress has removed major mental blocks to understanding animal (and plant) functioning. However, increased learning from nature is becoming possible in some areas.

The natural sonar systems of some bats, birds and mammals, and the related water surface-wave systems of some insects, had come to be understood in many of their aspects largely by com-parison with laboriously developed commercial sonar navigation systems (which themselves in some cases are improved by consider-ing more advanced radar systems). Animal performance in these respects has often been excellent compared with what could be achieved by man-made systems as regards speed, reliability, classification capability, jamming freedom, and compact packag-ing. It should be emphasized that the mere existence of high performance in nature encourages the setting of higher but real-istic goals, even if no indication of mechanism is seen...watch-ing birds did not show us how to fly, but suggested we could.

Observation of unexpected performance also forces clarified thinking. For example, Tavolga has found swim bladders of the

catfish *Arius felis* to be direction sensitive to sound at frequencies as low as 100 cycles per second where the wavelength is
about 15 m. One might expect these to sense only a periodic
squeeze from all directions, independent of shape, with emphasis
on no direction. A possible explanation brings out interesting
concepts. Even at large distance, scattering of sound by a very
small low density object is not the same forward and backward,
which creates a "shadow". Thus there is unequal pressure on the
front and back of the object and so it vibrates in the *direction
of sound propagation*. Thus a swim bladder could tug differentially on its supports to give direction sensitivity. The effect
is inherent in equations of Lord Rayleigh from a century ago but
this observation suggests applications to many measuring systems.

Another consequence of related induced motions of gas filled
spaces is that small fish appear significantly larger than actual
to the sonars of whales, making them more noticeable. We use the
same effect to ultrasonically observe and measure the bubbles of
decompression sickness ranging in size down to 1 micrometer with
sound wavelengths of 200 micrometers in animal and human subjects
(Mackay & Rubissow, 1978). (The method mentioned in the next
paper could also be very useful for this application.) In principle dolphins are thereby able to sense things ranging from
dangerously expanding decompression bubbles to a baby requiring
burping.

Of course, the process of making animal observations is
itself instructive and there are reciprocal advantages. Thus
when one understands the observation that a typical hydrophone
records passing airplanes about the same when in or out of the
water of a dolphin tank, one is then better able to consider
animal hearing, and vice versa.

In recent years attempts to understand some aspects of biosonar performances have led to theoretical analyses of such
systems and predictions of their limiting performances. In-
corporation of such thinking into future technology should lead
to improved performance; things have now progressed to the point
that a reverse contribution can be made from understanding
animals to human innovation. Several of the other chapters in
this book contain information against which to measure device
performance, with the attempts to explain and analyze animal
sonars in theoretical terms clearly suggesting approaches to
waveform generation and analysis that could be incorporated into
future generations of man-made systems.

Some general comments can be made. Nature often seems to
employ parallel and peripheral processing of information. This
is widely observed in the visual system but has previously been

little exploited in primary optical systems, though recent
"optical processing" and holographic methods are of this sort.
Various workers contributing to this book observe it in the bat
auditory system. It could be employed with benefit in man-made
acoustic systems now that large scale integrated circuits are
available. (In this connection it is interesting to note that
what has come to be widely known as a "phased array" in the
medical and sonar literature received one of its earliest
mentions in the sound literature at Frascati: Vol 2 pg. 1190.)
In our laboratory we have long routinely sharpened ultrasonic and
other images by mixing with signals their own derivatives, and so
the useful effects of lateral inhibition in all sensory systems
are already understood from technology.

In studying animal systems for mechanisms one often sees
the use of alternative cues. But in attempting to understand
plasticity of function for possible application one must also
consider alternative interpretations of any given set of sensory
system inputs. Thus the ability of a bat to function properly
soon after leaving an altered sound velocity helium-oxygen atmosphere is not mysterious if one considers an analogous human
experience. Because the velocity of light in water is 3/4 that
in air, to a diver wearing a usual flat-fronted mask things
appear at 3/4 their actual distance, but the diver is able immediately to reach and grasp objects, probably through sensing
a series of fractional distances and noting closure. Similarly
the reasons for other performances may not follow our precon-
ceived ideas of assumed function. Thus the technology of certain
altimeters emphasized a possible use in distance estimation for
the frequency modulated portion of the cry of a bat. But in experiments with *Tursiops truncatus* I found that animals required
to produce a sustained pure tone of given frequency always preceded it by a sometimes quiet FM sweep, perhaps only as an adjustment (Mackay, 1975). In some cases there was also a transient apparently useless following frequency modulation which
training reduced in loudness but did not eliminate. To the
biologist and engineer alike this could emphasize an obvious
caution.

A real contribution of activities such as bionics, or
whatever they are named, is to bring together people from different disciplines who might not otherwise talk to each other.
Significant technical developments as well as biological understanding have come from interdisciplinary thinking. An area
where continuing work has been done that exploits the normal information processing of the animal or human brain, and which
perhaps somewhat mimics the functioning of a two-eared creature,
is in certain ultrasonic blind guidance devices that resolve
multiple object auditory space. There are extensions of applica-

tion ranging from ocean surveillance and fish finding to the
clinical exploration of the movement of the heart in the human
body.   The prosthesis application has been one of the most
vigorously pursued areas since our prior Frascati meeting.   Two
workers heavily involved in this contribute the next two papers.

References

Mackay, R.S. and Rubissow, G. Decompression Studies Using
    Ultrasonic Imaging of Bubbles.   Inst. Electrical and
    Electronics Engineers Trans. on Biomed. Eng. 25, pp. 537-544,
    1978.

Mackay, R.S. Acoustic Control of Environmental Events by
    Tursiops truncatus.   In: Conference on Biology and Con-
    servation of Marine Mammals (Abstract), pg. 55.   Santa Cruz
    1975.

AIR SONARS WITH ACOUSTICAL DISPLAY OF SPATIAL INFORMATION

Leslie Kay

University of Canterbury

Christchurch, New Zealand

The development of aids to spatial perception for blind persons has been actively pursued over a period in excess of two decades by a number of individuals or groups (Benham (6), Kay (23), Benjamin (7), Russell (42), Collins (15), Armstrong (2), Brindley (11) and Dobelle (16), to mention only a few who are well known researchers in the field. The primary goal in the late 50's and the early 60's was to improve mobility in the blind through the use of a sensory aid. Now most aids are designed to complement the Long Cane (36) which emerged during the 60's as a primary mobility aid, following its introduction in the U.S.A. in 1947. This was partly brought about through the strong influence of the teachers of orientation and mobility for the blind, whose teaching skills are based on Long Cane travel. This demands reliance on the natural sense of hearing as well as the probing of the cane ahead of the user. Thus, whilst the design of a sensory aid remains the prerogative of the electronic engineer it is most important that he pays adequate cognizance to the experience and views of these teachers before allowing his fertile imagination free rein. It is against this background that three distinctly different sensory aid philosophies have developed.

The most elementary philosophy calls for the "clear path indicator" or obstacle detector. Such an aid to travel by a blind person does not indicate where the path is. The aids of Benham, Benjamin, Russell, and Armstrong fit the philosophy. These aids indicate only the presence of the nearest object in the travel path and are intended to be easy to use. Clearly they can also be used to locate objects nearby. All are narrow beam devices and may be in the form of a torchlight, fitted into a cane, or mounted on the chest. Russell and Armstrong use ultrasonic waves, whereas Benham and Benjamin use light waves to gather environmental

FIG.1.   Sonic Spectacles for the Blind as used in the 1971-72
           Evaluation Programme

information.   The display of distance information is in either
auditory or tactile form.

The second philosophy due to Kay, proposes the use of the
auditory sense to its maximum capability in sensing the complex
environment in which we move.   This has led to devices in the form
of a hand-held torch (25), head-mounted in spectacle frames (26) or
a headband as shown in Figs.1 and 2.   The display may be either
monaural for a narrow high resolution field of view, or binaural for
a wide field of view.   This latter may be designed to match the
auditory system localisation function (46) or be used to give mainly
left, right or ahead indications of the direction of an object as
well as the distance to an object.   A special feature of this display
is the means to locate more than one object at a time and the audio
signal carries information about the nature of the object (24).
(This is missing in the "clear path indicators").

The third philosophy seeks to model vision by displaying frontal
images of the environment using either a matrax of tactile
stimulators on the body (Collins and Bach-y-Rita (4)), or a matrix
of neural stimulators implanted in the brain as proposed by

FIG.2.   Spatial Sensor for Blind Children Mounted in Cosmetically
           Acceptable Headband.

Brindley (11) and Dobelle (16). The information is collected by a
miniature camera using solid state microelectronics. At present
the displayed information is elemental (in a visual sense) resembling
at best spots of light (or phosphines), or a 20 x 20 array of
vibrators on the abdomen.

     This paper discusses the theory, design and use of the sonar
system due to Kay which aims to stimulate the auditory sense to its
maximum capability. As such, it may be said to most nearly model
the animal sonar. The "clear path indicators" cannot be considered
in this context except in a simple way, since the environmental
information is so processed as to be reduced to single object
distance only with a stimulus of predetermined form. A comparison
is made with echo-location used by animals and the spatial
perception they might enjoy.

     It has been found through a comparison between continuous

transmission frequency-modulation and tone-burst forms of radiated
ultrasonic energy that the former provides a means for matching to
the human auditory process (24). The linear frequency sweep can be
specially chosen to match optimally the auditory process so as to
provide a display of spatial information which is rich in object
character and sensitive to motion. User experience has shown that
the need for accurate judgment of distance and direction is less
important than perception of the flow of spatial sound patterns and
locomotor control in negotiating a complex environment can be
readily achieved - even by blind children - using only the acoustic
output information from an air sonar. It is also possible to
observe the motion of objects about their axis of rotation with an
appreciation of what is taking place which challenges some high
resolution pulse sonars with visual display

There are two possible reasons for this: a) the linear
frequency-modulation radiation which covers a frequency band of
approximately one octave has an ambiguity function which combines
time delay and Doppler shift in a way that a wide band "tone"
burst cannot, and b) auditory and visual perception of the display
of spatial information are different processes and the motion of
sonic reflectors which may be readily perceived auditorily may not
necessarily be readily perceived on a visual display.

BASIC CONCEPT OF AUDITORY SPATIAL PERCEPTION

In order to perceive objects in the environment and be aware
of their relative motion through the medium of an artificial sensing
system it is first necessary to establish spatial coordinates.
These are most readily understood as DISTANCE and DIRECTION and the
more precise the resolution the greater the environmental detail
which can be perceived. This requires a sensing wave energy with
as short a wavelength as practicable and a system bandwidth as wide
as possible. When vision is impaired as in blindness, the most
effective remaining spatial sense is hearing which requires an
acoustical form of coupling of the sensor to the ears and a coding
of the spatial information which is most readily interpreted.
It has been argued that the most effective spatial sense is the
tactile sense where spatial information is mapped in two dimensions
on the abdomen (15). However, it is later shown here that the
auditory sense can handle superior information.

The optimum spatial code appears to have two components.
Distance is best represented proportionally by the frequency of the
audible signal and directional angle by the proportional logrithmic
increasé in the interaural amplitude difference (IAD) of sound fed
to both ears. Left is indicated by a louder sound in the left ear
which causes a lateral shift of the binaural sound image to the

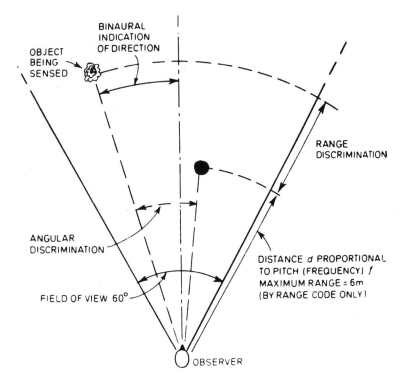

FIG.3.   Illustration of Distance and Direction Code for Spatial
          Sensor.

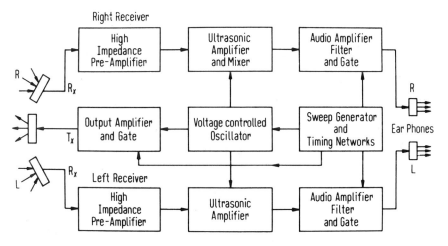

FIG.4.   Schematic of Binaural Spatial Sensor using Ultrasonic
          Radiation.

left of the midline. This code may be easily modelled in a computer
for simulation purposes (12) (17), but is not physically achievable
without introducing auditory artifacts which may be disturbing to
the user. Some apparent artifacts are nevertheless complex useful
spatial information which has been found difficult to simulate, as
shown later.

Ideally it would appear that an object acting as a point
reflector of wave energy in the sensing field would cause a
continuous tone to be heard which would be imaged in the head as a
pure tone, the frequency of which was proportional to the distance.
Its position would be shifted laterally in the head by an amount
corresponding to the direction of the object. If the object were
caused to move radially relative to the observer the tone would
decrease as the object came closer and the image would shift
laterally to the left as the object moved to the left.

The closest it is possible in practice to approach such a
spatial code is through the use of a linear ramp modulation on the
frequency of ultrasonic radiation and the reception of echos by
two receiving apertures having divergent polar receiving
sensitivities; each having a Gaussion variation in angular response
relative to the angles of divergence (42).

The code is illustrated in Fig.3 and a schematic of the system
is shown in Fig.4.

PHYSICAL REALISATION OF THE SPATIAL SENSING CODE

Distance Code

The radiated signal of a linear ramp frequency modulation
echo-location system may be represented by the relationship

$$S_T(t) = \exp j\theta_T(t) \tag{1}$$

where

$$\theta_T(t) = 2\pi \left| f_2 t_s - \frac{m}{2} t_s^2 \right|$$

and

$$t = t_n + t_s, \text{ shown in Fig.5.}$$

(t_n is the beginning of the nth cycle of sweep period T_s).

FIG.5.   Frequency Sweep and Audio Signal of Sonar.

Thus, while t varies between t_n and t_{n+1}, t_s varies from 0 to T_s, $T_s \gg \tau_{max} = 2r_{max}/c$

m is the sweep rate in Hz/s , and may be either positive or negative. In the system being described m is negative.

The instantaneous transmitted frequency is

$$f_T(t) = f_2 - mt_s \qquad (2)$$

f_2 is the frequency when $t_s = 0$.

When, due to motion, the received signal at t_s is a Doppler shifted "replica" of the transmitted wave radiated at $t_s - \tau$ the phase angle of the received signal at the time t_s is

$$\theta_R(t) = 2\pi \left\{ f_2 [t_s - \tau(t)] - \frac{m}{2} [t_s - \tau(t)]^2 \right\} \tag{3}$$

The received frequency is

$$f_R(t) = \frac{1}{2\pi} \frac{d\theta_R(t)}{dt}$$

$$= f - f_2 \dot{\tau}(t) - m [1 - \dot{\tau}(t)][t_s - \tau(t)] \tag{4}$$

where the dot represents the time derivatives. The distance to the reflecting structure is then indicated by the difference frequency obtained by simple multiplication between $S_T(t)$ and $S_R(t)$ giving

$$f_T(t) - f_R(t) = f_a(t)$$

$$= m \tau(t) [1 - \dot{\tau}(t)] - \dot{\tau}(t) (f_2 - mt_s) \tag{5}$$

The general expression for the time delay is

$$\frac{\tau(t)}{2} = \frac{1}{c} r \left(t - \frac{\tau(t)}{2}\right) \tag{6}$$

where $r(t)$ is the instantaneous distance and $\dot{r}(t)$ is positive for increasing $r(t)$.

Since the rate of change of $r(t)$ is much smaller than the velocity of propagation c, the following approximation may be used

$$\tau(t) = \frac{2r(t)}{c} \quad \text{for} \quad \left| \frac{2\dot{r}(t)}{c} \right| \ll 1$$

then

$$f_a(t) \doteq \frac{2m}{c} r(t) - f_T(t) \frac{2\dot{r}(t)}{c} \tag{7}$$

illustrated in Fig.6.

When the Doppler shift is zero for stationary reflections

$$f_a = \frac{2mr}{c} \qquad \frac{2m}{c} \quad \text{is called the distance code in Hz/m.}$$

FIG.6.   Doppler Shift Due to Motion.

It should be noted that the term $f_T(t) \, [2\dot{r}(t)/c]$ is not the Doppler shift induced by the moving structure. It is instead a function of the transmitted frequency at the instant $t = t_n + t_s$ (which is the instant of reception) and the instantaneous distance of the structure at the same time t. The relation is sufficiently general to include non-constant velocity and is in this form to facilitate ease of computation as well as ready appreciation of the range indicator $f_a(t)$.

When r(t) is increasing, i.e. the structure is moving away from the transducers thus making $\dot{r}(t)$ positive, the difference frequency is given by equation (7) as written

The audible signal

$$S_a(t) \; = \; 0 \qquad\qquad (t_n < t < t_{n+\tau})$$

$$\qquad = \; \beta(r,t) \, \cos \, 2\pi f_a t \; (t_{n+\tau} < t < t_{n+1}) \qquad\qquad (8)$$

$\beta(r,t)$ is both range and frequency sensitive (which is varying with t) carrying information about the reflection characteristics of the object during each frequency sweep.

$S_a(t)$ may thus be expressed as a Fourier Series (24)
for stationary conditions only by

$$\beta(r,t) \sum_{p=1}^{\infty} a_p \cos p \, \omega_s t$$

where $\omega_s = \dfrac{2\pi}{T_s}$

and $a_p = \dfrac{\sin(\omega_a + p\omega_s)k}{(\omega_a + p\omega_s)} + \dfrac{\sin(\omega_a - p\omega_s)k}{(\omega_a - p\omega_s)}$

and is illustrated in fig.7. With motion the line structure is
broken up. This echo signal is heard as an interrupted tone, the
quality of the tone being determined by the ratio T_s/τ which
should be as large as practicable. A high value of ω_s (say 25)
causes a faint motorbike sound and great care is needed in the
design to ensure this is minimised without destroying valuable
character in the signal induced by the time varying frequency
component of $\beta(r,t)$.

From Fig.5 it is clear that the system bandwidth and the
rate of change of radiation frequency m are independent variables
which allow a choice of T_s to suit the human ear. This determines
the resolution limit, that is the theoretical minimum difference
frequency between two tone signals which can be unambiguously
detected. From Fig.7 this is seen to be $2/(T_s - \tau)$ because of the
line structure of the signal spectrum.

If an object consists of a number of surfaces each
scattering energy back to the receiver and the distance to each
surface is $r + \Delta r_i$ the audible signal corresponding to the object
will be

$$S_a(t) = \sum_{i=1}^{n} \beta(r + \Delta r_i, t) \sum_{p=1}^{\infty} a_{p_i} \cos p_i \, \omega_s t \qquad (9)$$

where ω_a in a_{p_i} is given by $2\pi . \dfrac{2m}{c} (r + \Delta r_i)$

This narrow band of tones each representing a reflecting
elemental surface (like leaves on a tree) form a spectrum of
signals uniquely related to the geometry of the object relative
to the viewer. The sound therefore has a unique audible character

FIG.7.   Audio Signal Spectrum.

which provides a means for auditory object discrimination and even sometimes recognition. It should be noticed that if, as in the case of a tree with leaves fluttering in the breeze, the reflecting elements are in motion we have to write $(r + \Delta r_i)(t)$ when each element returns a Doppler shifted "replica" of the radiation frequency and $f_a(t)$ is given by equation 7.

There are thus several factors which uniquely determine the signal representing an object creating the maximum information transfer to the auditory system. It should be noted that the Doppler shift in radiated frequency is unmodified in the audio spectrum when a small fractional shift of the radiation frequency may become a very significant shift in the audio band.

Factors Affecting Choice of Distance Code

The distance to the object of interest and the rate of change of distance are represented in the audio frequency band by equation 7. When there is no motion the audio signal is simply $2mr/c$ and the variable m determines the frequency of f_a for any given distance r. This is illustrated in figure 8 for two values

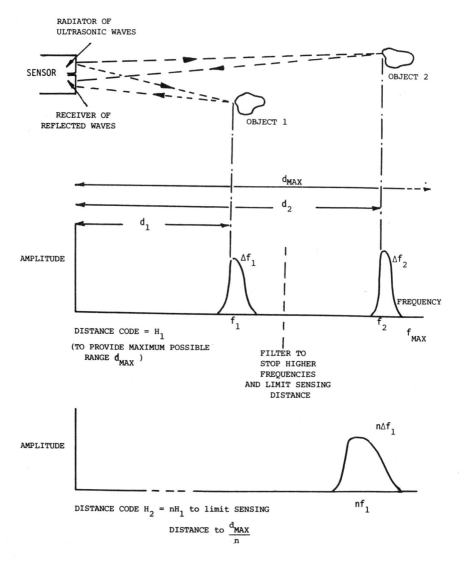

FIG.8.   Distance Code of Sensor.

of distance code H = 2m/c Hz/metre. The sensor as designed for
adult mobility on city streets uses a code H_1 of 900 Hz/m and the
maximum distance at which a wall can be sensed is arranged to be
approximately 6 metres producing an echo frequency of 5400 Hz.
When this distance code is changed to 10,800 Hz/m simply by
increasing the value of m, as is done in the aids used for babies,
the maximum distance which can be sensed is 0.5 m and the maximum
audible frequency thus remains 5400 Hz.

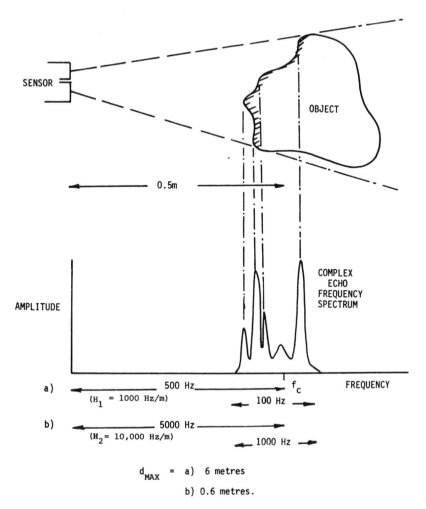

FIG.9.   Object Discrimination Using Two Extremes of the Distance
Code.

The spread in frequencies Δf_a representing the object
increases linearly as the distance code is increased and it might
be assumed that greater discrimination between tone complexes
would become possible as this spread in frequency Δf increased.
This is so only to a limited extent as shown by Plomp and
Steeneken (40); the ability to discriminate between sounds of
differing quality is superior above 1000 Hz to that below 1000 Hz.
From this work, as the mean frequency of the tone complex increases
so also must the spread in frequency in order to retain the
discrimination.   Beyond about 3000 Hz the spread in tonal
frequencies must increase more in proportion to the rise in mean

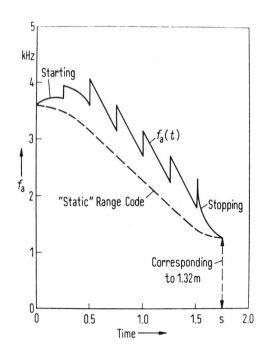

FIG.10.   Effect of User Motion Approaching an Object.

frequency. Consider the two codes shown in Fig.8 as used to
provide object discrimination at a distance of 0.5 m shown in
Fig.9. It will be evident that at these two extremes in the
usable distance code there will be a marked difference in object
discrimination. The long range code H_1 provides very poor
discrimination capability and the short range code H_2 may have
exceeded the optimum for an object at this distance. The actual
optimum has yet to be determined but using the variable range code
developed by Boys et al (10) it is now possible to "focus" the
sensor in distance so as to obtain the optimum for any distance
within wide limits.

When there is relative radial motion the audio signal $f_a(t)$
is determined both by the parameter m and the radiated frequency
$f_T(t)$ which is itself a time varying function. The effect of
walking up to a pole and stopping at a distance of 1.32 metres
(4 ft) is illustrated in Fig.10. The saw tooth in the distance
indication is due to the cyclic variation in $f_T(t)$. The error in
the indication caused by the Doppler shift is a significant fraction
of the instantaneous distance. It is explained later why this
does not cause any problem under user operating conditions.

TABLE 1

Sensory Aid Parameters

DESIGN

	A	B	C	D
Maximum sensing distance (r) in metres	2	1	1	0.5
Ultrasonic frequency band (kHz)	40–80	40–80	60–120	60–120
Sweep time T_s ms	96	48	72	36
Distance code H (kHz/m)	2.5	5	5	10
Maximum Doppler Shift (Hz) (maximum reaching velocity of 1 m/s)	500	500	750	750
Theoretical maximum resolution mm. (From Fig.7).	8.3	8.3	5.4	5.4

A more critical sensing situation is reaching for an object within arm's length where it is reasonable to expect that a blind child, for example, wearing a sensor would develop a degree of hand-ear coordination if it could sense the instantaneous position of its hand and the object of interest. A choice of four design parameters is shown in table 1 to illustrate the range of choice available to a designer of a sensory aid.

The auditory stimuli resulting from a reaching hand are shown in figs.11(a) and (b) where it will be seen that not only is there considerable stimuli variation but some of the stimuli are much less representative of the reach than others. The reaching indicator for designs A and C is very poor and for much of the reach the stimuli is inaudible (too low a frequency for the ear-phones to respond adequately). In designs B and D the indication of distance more nearly represents the reach: design D is thought to be the best. Quite evidently one design will not provide the optimum sensory stimuli for all situations and for a sensor to provide the best realisable spatial information it must have variable parameters under the control of the user.

Direction Code

The basic direction code is expressed in terms of the interaural amplitude difference (IAD) in decibels per degree

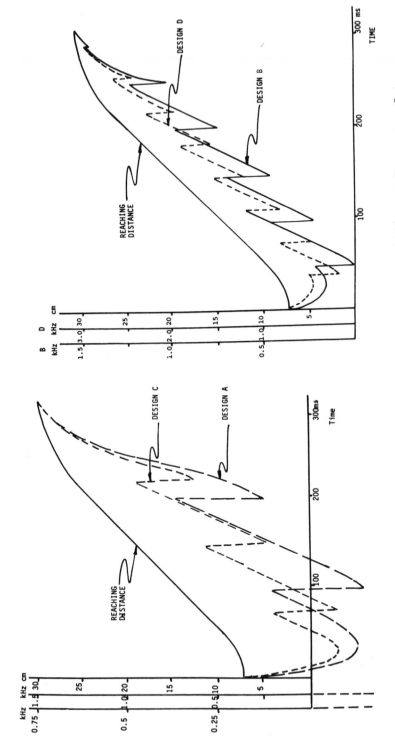

FIGS. 11(a) and (b). Audio Indication of a Reaching Hand for Four Different Parameter Sets.

azimuth angle and can be written in terms of the well known
relationship for the direction of a low frequency tonal image
reported by Stewart and Hovda (47). Thus the estimated angle of
an object is given by

$$\theta_E = 20 \ K \ \log \frac{A_L}{A_R} \tag{10}$$

where A_L and A_R are the sound levels of the stimulus at the
left and right ears respectively and K is the subject's constant
of lateralization of binaurally presented sounds. Ideally one
would expect the estimated angle θ_E arising from the lateralizing
sensation from the sensory aid should be the same as the actual
angle, or a known function of it. This is not easily arranged for
each individual aid user because the lateralizing constant K varies
between individuals and each one has to undergo special tests to
determine it (42,36). Even if the IAD could be readily adjusted to
suit individuals, the technical problems are such that it may be
suitable for only a portion of the distance being sensed. This is
partly because the lateralizing constant for each individual varies
with frequency, and partly because the technology for obtaining the
direction code has inherent physical limitations. An engineering
compromise is necessary, although it is not yet determined what this
should be.

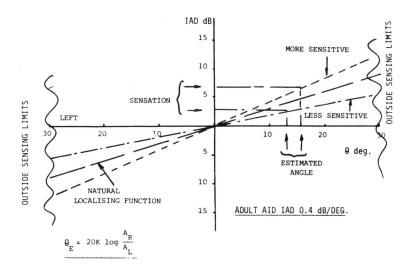

FIG.12.   Ideal Direction Code for Different Sensitivities of the
             Localising Function.

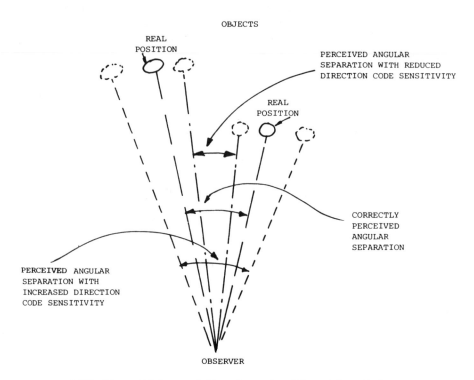

FIG.13.   Effect of Varying the Direction Code.

An important aspect of the direction code is the freedom to
choose the degree of angular resolution built into the sensory aid.
The direction code is shown in Fig.12 where it will be seen that
it can be set to have increased or decreased sensitivity relative
to that which corresponds to the natural lateralizing function.
In Fig.13 is shown the spatial effect of changing the direction
code. When the code is made more sensitive, objects close
together in azimuth angle may be perceived as being separate when
otherwise they may have been thought to be in the same direction.
A very close analogy may be drawn with changing optical lenses so
as to expand the visual image in order that more detail can be
seen.

There are technical difficulties in obtaining the ideal
direction code given by equation 8. To obtain this the azimuthal
field response of the two receivers as measured at the audio
output was shown by Rowell to be (42)

Left receiver $A_L(\theta) = C_L \exp\left[-(\theta + \alpha/2)^2\right]/2\alpha K$ (11)

Right receiver $A_R\theta = C_R \exp\left[-(\theta - \alpha/2)^2\right]/2\alpha K$ (12)

where C_L, C_R are constants and normally $C_L = C_R$

α is the angle of divergence between the polar axes of the two receivers.

Thus the polar plots follow a Gaussian function displaced by an angle $\alpha/2$ to the left and right of the median $\theta = 0°$ so as to create the interaural amplitude difference direction code in degrees/dB. To obtain the desired distance code a broadband radiation of the order of one octave is necessary. Thus the wavelength may change by a factor of 2:1 during the frequency sweep. It follows that without special time varying aperture shading the polar response of the ultrasonic transducers must also vary by a factor of approximately 2:1. The value of K in equations (11) and (12) is the desired localisation constant to match the user, but it will be seen that it is now time varying over the sweep period causing the estimated angle θ_E to vary as A_R and A_L vary for any given angle θ. It would seem that such an arrangement could not provide the desired direction cue. It was discovered however that during the sweep period T_S the auditory system did not respond to the instantaneous ratio A_L/A_R but to some integrated value corresponding to the energy difference in the period T_S.

Using solid dielectric transducers described in (26) the desired Gaussian response could be obtained to an adequate approximation over the field of view provided

$$\theta_E(\text{measured}) = 20 \; \text{K Log} \; \frac{\displaystyle\int_{t=t_n}^{t_{n+1}} A_L(t)\,dt}{\displaystyle\int_{t=t_n}^{t_{n+1}} A_R(t)\,dt}$$ (13)

It is here that the greatest technical difficulties lie in making the sensor. An approximately linear relationship between θ_E IAD (dB) can be obtained over the field of view provided this is restricted to an angle 3α by the radiation field of the transmitting transducer. Direction ambiguity will otherwise result as shown in Fig.14.

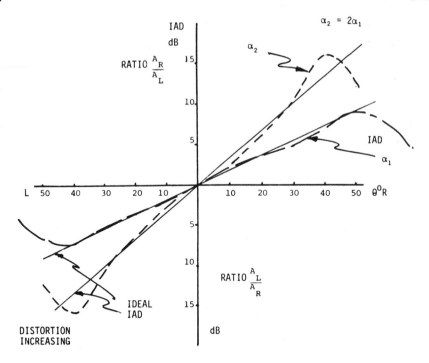

FIG.14.   Distortion in Direction Code Due to Physical Limitations
           of the System.

        The binaural direction cue thus relies entirely upon the
lateralizing sensation produced by the dichotic presentation of tone
complexes.   It cannot therefore immediately create an "out there"
or spatial sensation in a naive listener; only the equivalent of a
pina in the echo signal path can do this.   Nevertheless, long term
users do say they experience an "out there" sensation which is
probably induced by user motion.

                          Field of View

        Probably the most difficult of the several sensory parameters
to appreciate without personal experience is the field of view
through which objects may be perceived.   Imposed upon the cues to
distance and direction is the variable of object reflectivity which
determines the strength of the echo and the variation of echo
intensity caused by the change in system sensitivity with direction
and radiation frequency.   The dynamic range of the ear allows
perception of objects well outside the field of view expressed as
a cone of solid angle Ω representing the half power sensitivity and
greater.   Whilst this is a technical convenience and commonly used

FIG.15.   Illustration of Field of View of Binaural Sensor.

in sonars with visual displays it is very misleading in an air
sonar with auditory display.

An illustrative plot of a typical field of the Binaural
Sensory Aid is given in Fig.15 and derives from the transducer
sensitivity plots illustrated in Fig.16. (It should be
remembered that the transducer plots can be for only one frequency;
the field plot is taken to be the integrated audio output). The
"loudness" plot is entirely illustrative.

In the vertical plane the field sensitivity varies in a
complex way because of the splay angle α and has a significant
effect on the I.A.D. characteristic for angles off the horizontal.

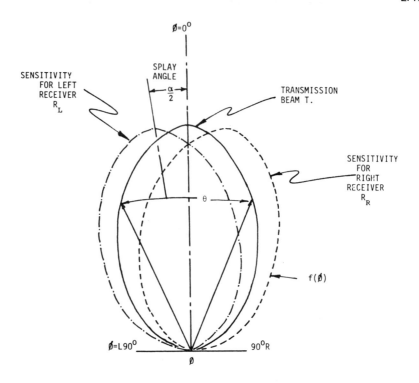

FIG.16.   Typical Transmitter and Receiver Field Plots in Azimuth.

     The user of an aid does not hear the left and right sounds
separately, but instead it is a fused signal which shifts to the
left or right according to the I.A.D.   As the head is turned
about the direction of an object, if the I.A.D. matches the
lateralization function, the real direction and the sensed
direction will coincide giving the impression that the object
remains still.

     However, because the instantaneous sensitivity of the field
of view varies with frequency "off axis" in a different way to
"on axis", the character of the object echo also varies as the head
is turned.   Object recognition is thus more reliable when "looking"
at the object than when "looking away" from it.

     Varying the splay angle has only a secondary effect on the
the field of view as plotted because of the radiation response but
it has a marked effect on the detection of an object entering the
field from the left or right when another object is already in the
field.   This is because with a large splay angle one ear hears the
echo more distinctly, thus increasing direction discrimination.
It is thus evident that whilst the distance code is independent of

the field of view the direction code and field of view are
dependent variables and must be jointly considered in any design.

OBJECT DISCRIMINATION AND MOTION PERCEPTION

Distinguishing one object from another when more than one
object is present in the field of view requires that two or more
objects can be separately resolved. This is so whether perception
of the objects is by visual or auditory means. When the objects
are widely spaced in direction so that only one can be in the field
of a sensory aid, all that is necessary so as to determine their
relative positions is turn the head. The viewer then seeks to
recognise one object from another by correlating the echo signal
with previous experience in sensing objects. This was tested by
Sharpe when he carried out an evaluation of the "Kay" Sonic Torch
in 1968 (46). It was established that, using the wide band CTFM
air sonar described in reference 25, subjects were able to
recognise a variety of objects and surface structures including
different trees, various forms of fencing and objects forming
landmarks. The study was based upon the training procedure designed
by Elliot et al (19). A special series of tape recordings provided
the basic training material prior to experience in the real world.

The binaural sensor in the form of spectacles or headband poses
a problem in defining resolution from which the degree of
discrimination can be determined. It was shown earlier that the
maximum theoretical resolution is given by $2(T_s - \tau)$Hz from which
the spatial resolution may be determined from $2/H(T_s - \tau)$ metres
where H is the range code. It was shown that for design D (Table 1)
this is 5.4 mm. In terms of auditory resolution it implies that
two echo tones representing two vertical fine strings at say 250 mm
in the field of view can be perceived as distinctly different from
one string at this distance. This is so; but only because the
sound is different, not because the two strings can be perceived as
being separate – as is the case for optics. In order that the
echos can be heard as distinctly separate the strings need to be
separate in distance by 100 mm when the farthest one is at 250 mm.
This is discussed by Kay and Do in references (27, 18) and by Do in
the next paper of this chapter. For less separation the two echo
tones become combination tones as described by Fletcher (20) and
confusion is experienced by subjects under test when questioned
about what they hear (42). In an attempt to describe the spatial
resolution capability of a user of the sensory aid as distinct from
determining the ability to discriminate between or recognise objects
Do and Kay (18) simulated the sensory aid auditory space using an
analogue computer to control the frequency of two or more tone
generators. One object in the field of view was simulated as moving
relative to another either in the same direction (same IAD) or in

FIG.17.   Simulated Moving and Stationary Objects to Test
           Auditory Resolution.

different directions (differing IAD). A simple simulation is
illustrated in Fig.17. $f_{a_1}(t)$ represents an object moving towards

the viewer. When the two objects are at the same distance the
combined auditory image lies midway between the two images if
perceived separately. As the changing frequency $f_{a_1}(t)$ decreases

the sound changes in character from a tone to increasing beats and
the image glides down in frequency until at a critical frequency
difference it must split into its two separate components. This
follows because at a difference frequency of 40% of the higher
frequency the two tones are perceived as separate even under
stationary conditions. It was discovered that the separation may
occur for as little difference as 10% of the initial frequency.

Do in the accompanying paper shows that for tone complexes this difference can be only 5%, almost 10 times better than the stationary state.

This has considerable relevance to a spatial sensor for the blind because motion of some kind is constantly taking place. Turning of the head also induces frequency glides which may influence the spatial resolution but this requires detailed study (17).

There are thus three forms of spatial perception determining how well (or poorly) a blind person can view his immediate environment.

a) Identification of an object by virtue of the sound character it produces determined in the limit by the theoretical resolution of tones.

b) Discrimination between objects due to the difference in sound from each which can be detected through sequential comparison.

c) Resolution between objects either stationary or in motion by virtue of the ability to listen to each separately even when they are simultaneously producing sounds.

These three forms have not been studied adequately but raise interesting theoretical issues as discussed by Do. Until they are well understood it will not be possible to determine the optimum design of aid.

BEHAVIOURAL STUDIES WITH THE BINAURAL SENSORY AID

A sonar design can be tested by well established measurement techniques to ensure that it meets the required engineering specifications. Evaluation of the sonar on the other hand in terms of usability - which seems to be the only valid criterion for a sensory aid to spatial perception - involves the total life function of the particular individual who will wear it. Only under day to day living conditions normally experienced by a blind person can an aid be assessed to see if it has user value. Generalisations about usability from an adequate sample of the user population may then be possible. Laboratory type experiments can be used to determine the level of attainment of a particular skill which may be acquired with an aid. There is however no way to determine the value of that skill to the user of an aid until many other skills have also been acquired and incorporated into a modified way of living. Those skills of little value then lapse into disuse. It is then that newly acquired skills of significant value can be tested. This process poses a

dilemma for both the designer of an aid, who is given no guidance
in the setting of system performance specifications which could be
inappropriate, and the agencies serving the user population who have
to determine the cost-effectiveness of their service in possibly
providing an aid. Two special conferences (39, 22) on this matter
were unable to resolve the many differences of opinion which existed
in the early 70's and indeed continue to exist today, if perhaps to
a lesser extent. A study undertaken by behavioural scientists to
determine the value of a sensor under controlled conditions
divorced from both the user's day to day emotional problems and the
designer's professional interest was thought to be the most
appropriate approach by several consultants: in the end, however,
general acceptance of the Binaural Sensory Aid as described by
Kay (26) came about through the hard school of user, professional
worker and designer interaction during an extensive evaluation
programme in the period 1970-72 (26, 28).

Adult Sensory Aid Use

To be of value a sensory aid to spatial perception should
provide the user with a useful percept of the immediate environment
over an adequate field of view. Since the Binaural Sensory Aid
output is sound, this can provide only a little of the information
sought by a blind person. On the other hand, the environment in
which we travel is so structured by human design that vision is not
essential in order that individuals may identify many features of
their environment. For example, many sound producing objects are
readily identified by their sounds alone although this should
not be construed as implying the stimulation of a mental image.
In a similar way, objects causing echo sounds in the sensory aid
can also be recognised by their sound character provided the
character is reliably repeatable and there is sufficient a-priori
information available for the recall of experience. For example,
walking along the street in a shopping area it is common to find
the pathway lined with parking meters. Blind people would normally
be unaware of their presence but when wearing the binaural sensor
it would produce a regular series of sound patterns every few steps
which seem to come from the side of the roadway.   These sounds can
quickly come to mean "parking meter". What this image constitutes
in the congenitally blind remains unknown.

In evaluating a sensory aid this kind of perception must be
carefully defined and the cognitive process employed by the aid
user considered in depth otherwise the wrong goals may be sought.
When a sensory aid only indicates the presence of a single object
at a time, as in the case of "clear path indicators", the user
information is trivial and evaluation relatively simple. Even so,
this information can reduce stress in the Long Cane traveller who

relies greatly on his natural hearing for oeientation. Evaluation
of the binaural sonar, instead, highlighted the need for a general
theory of mobility and an international seminar on the subject was
held at Oxford in 1973. The only paper to reach publication from
this seminar was that by Kay (29) expressing some thoughts on a
theory arising from the extensive observation of blind travellers
during the binaural sensor evaluation. This has now been carried
further by Brabyn (13). During the evaluation process some 25
Orientation and Mobility specialists whose primary role is that of
teaching mobility to the Blind were themselves trained to use the
sensor under blindfold over a period of four weeks. Although this
was thought to be an inadequate period of training, it was the
longest time likely to be made available and was planned accordingly.
These specialists then, between them, trained about 100 blind
persons, who had either Long Cane or Guide Dog travel skills, to
use the sensor in a variety of travel situations including downtown
travel. The pattern of training was built round the Long Cane
training programme so as to integrate the two aids to mobility.
In consequence, the binaural aid became a secondary aid to mobility
(the Long Cane being the primary aid) and its role as a spatial
sensor was largely suppressed in favour of obstacle detection and
"clear path indication".

On completion of the evaluation programme the 25 teachers and
96 users were sent confidential questionnaires by a professional
field service evaluator who designed the instrument (1). A number
of cross correlations were carried out from which it was found that
the responses from users cross correlated better than those from
the teachers, and there was often poor correlation between teacher
and user response. In Table II, for example, (from Kay, 30), the
responses from the Long Cane user and Guide Dog user are separated –
showing quite different responses to certain questions. It was to
be expected that in some situations a guide dog would provide more
assistance than could the information from a sensor with limited
distance sensing capability. Hence, from the question on "Crossing
a street in a downtown area" the guide dog user gained little
benefit from the aid since the guide dog is trained to give
assistance. The long cane user on the other hand did benefit as
is evident from the fact that 68% of the users reported improved
mobility. The teachers thought the aid had little impact on
mobility in this situation. Subsequent questioning of groups of
users revealed that the teachers had not taught the use of the aid
in this situation and therefore were making inexperienced judgments
compared with users who had tried the aid in this situation
subsequent to training. Little detailed analysis of the
questionnaire is available in the literature. Airasian discusses
a number of factors which may account for the discrepancies in the
72 user responses finally received and the 21 teacher responses:

TABLE II

The Sonic Glasses: Ratings of Usefulness in Mobility.

Skills	Percentage of Users *				Percentage of Trainers (N = 21)		
	Greater Mobility	Less Mobility	No Change	Never Encountered	Great Impact	Mod-erate Impact	Small Impact
Locating landmarks	93/100	2/0	5/0	0/0	58	42	0
Avoiding pedestrians in crowded areas	73/18	5/18	20/64	2/0	30	65	5
Avoiding obstacles in one's path	93/36	2/0	5/64	0/0	60	40	0
Following sidewalks bordered by grass	29/9	10/0	56/91	5/0	0	0	100
Locating doorways	80/70	2/0	17/30	0/0	58	37	5
Guidelining (shorelining)	80/36	5/0	13/64	2/0	37	42	21
Following paths with hedgerows	88/36	5/0	7/55	0/9	37	53	10
Following paths in hallways	61/27	12/18	17/55	10/0	35	55	10
Locating a clerk in a store	29/45	0/10	32/45	39/0	30	35	35
Taking one's place in a line	49/64	0/9	17/18	34/9	45	40	15
Locating a seat in a restaurant	17/9	7/0	17/64	59/27	15	40	45
Crossing a street in a downtown area	68/18	7/0	20/82	5/0	10	15	75
Traveling in moderate pedestrian traffic	76/45	2/9	20/45	2/0	21	63	16
Finding a specific store downtown	34/55	7/0	32/45	27/0	25	60	15
Seeking pedestrian assistance	46/30	5/0	32/60	17/10	40	30	30
Finding one's way in unfamiliar suburban area	59/55	7/0	12/36	22/9	15	55	30
Finding one's way in familiar suburban areas	71/36	2/0	22/55	2/9	20	60	20
Finding one's way in familiar shopping centers	71/45	2/0	12/45	15/10	60	25	18
Finding one's way in unfamiliar shopping centers	56/36	5/9	7/18	29/36	20	55	25
Finding one's way in familiar rural areas	30/40	7/0	8/30	55/30	26	42	32
Finding one's way in unfamiliar rural areas	27/50	7/0	2/20	63/30	11	47	42
Finding up curbs	38/27	10/0	50/73	2/0	0	10	90
Assessing the distance of stationary objects	98/82	0/18	2/0	0/0	70	30	0
Assessing the distance of moving objects	56/40	10/10	32/50	2/0	15	70	15
Squaring off to face objects or people	73/45	2/0	17/55	7/0	55	35	10
Finding one's way in large open spaces	34/9	2/0	49/91	9/0	15	5	80

*First figure represents percentage long cane users (N=43); second figure percentage dog guide users (N=11).

one of them relates simply to the basic social problem of
"diffusion of innovation".

In an attempt to obtain objective analysable data on sensory
aided skills, there was built into the "standard" training programme
a series of well defined exercises in the form of Lessons which
students were expected to complete before moving into everyday
travel situations. These were to provide the means to record
performance such as time to achieve specified levels of achievement.
An example of an exercise is shown in Fig.18. Further details may
be found in reference (31). The blind student was required to walk
centrally in a straight line between two rows of poles then slalom
back along one row. In terms of body control the task is not
trivial. Almost no records were kept by the teachers. However
five teachers under training were filmed executing this task and
the film was subsequently shown to 58 "judges" (a class of university
students) who were asked to rate the performance of each teacher on
a scale from 0 representing "blind" to 10 for "sighted" performance
(29). Their judgments are shown in Fig.19. There is considerable
variability in these showing how unreliable any individual judgment
may be for even a simple task. How much more variable then are

FIG.18.   Typical Pole Exercise for Testing Locomotion Skills
           Using Sensory Aids.

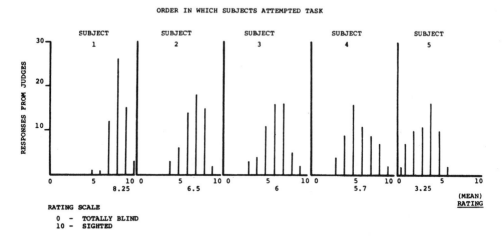

FIG.19.   Response of Judges After Viewing Film of Five Subjects
           Carrying Out Exercise in Fig.18.

judgments of mobility likely to be for real life situations and
the extent to which a sensory aid plays a part is exceedingly
difficult to determine.

    The skill shown by one blind subject in executing the slalom
is illustrated in Fig.20 which was produced from an anlaysis of
film of the action. Each footstep is plotted together with head
direction shown by an arrow. Brabyn,et.al.(12) reports an attempt
to produce a more objective measure still, by using a tracking
system in a specially treated room 20m x 13m. It was quickly
realised here that a meaningful definition of the ideal path and
deviation from this in order to find a figure of merit was difficult
to obtain except for walking a straight path at a specified
distance from a guideline - row of poles. For more complex tasks
such as the slalom a track of a subject's path was easily obtained
but an "ideal path" is less easily defined: a variation in path
between subjects did not mean a variation in the skill of executing
the task.

FIG.20.   Blind Subject Slaloming Round 10 Poles.

It became evident that accurate measurement of sensory aided
mobility was easiest for straight path travel where it was found
that blind sensory aided subjects could perform as well as sighted
in terms of the measurements taken. Variation in the direction cue
of the sensory aid had considerable effect on the subject's
performance, but changes in the distance cue had only small effect.
This is thought to be due to the generation of auditory flow
patterns which are more affected by the direction cue than the
distance cue (13). The detection threshold for a change in the
direction cue is important in determining a subject's path. When
the threshold is small, error correction can be made even during a
step as one foot is moving forward. It seems it is the auditory
flow pattern which becomes highly developed with sensory aid use,
that provides the primary guiding information for smooth locomotion
in a cluttered environment. The experiments of Do certainly show
that this kind of motion generates the greatest degree of object
resolution. It was observed that the visual looming effect of
Gibson (21) seems to apply equally well to spatial auditory
perception through an acoustic spatial sensor. It will be realised
that blind sensory aided performance approaching that of sighted

performance on one task does not lead to the general conclusion
that the Blind can now travel like the sighted. Such is the
evaluation problem.

Film records of totally blind persons travelling in a busy
downtown area using a spatial sensor do nevertheless show that
"near" sighted performance by a few is possible as judged by the
blending of the blind person into the pedestrian flow. Thornton
(52,53) describes his experience in using the sensor during a
period of extensive travel on foot but the individual personality
characteristics which lead to this proficiency have not been
determined. They are not restricted to an intellectually elite
group although such a group has more need of a spatial sensor in
their travels than do some others. Much too little is known about
the human characteristics which lead to good independent mobility
among the blind, and still, inadequate information is available
from which to design the optimum form of this aid. Two decades of
research have so far failed to quantify the problem sufficiently;
it is so complex.

Use of a Sensor by Children

The use of a spatial sensor which provides an entirely new
form of perception of the environment was often said to be more
suitable for children than adults. It was thought the children
would learn the spatial cues more quickly than adults who were
said to be slow in learning complex sound patterns and they could
be more interested in their surroundings. Bower (8) makes this
kind of point very strongly in his argument for using a sensor on
a baby although he has not produced any evidence in support of his
theory. Contrary evidence has now been obtained by other
experimenters (48), yet at the same time there does seem to be
reason to expect that babies and small children may come to adapt
more completely to a new sensor than older children and adults who
may already have established strong habits in their behaviour.
A well developed reliance on tactile information for example may
inhibit later adaption to auditory information. It was this kind of
argument which led to the study of sensory aid usage by children at
Canterbury (49). The acceptance of the Binaural Sensory Aid for
adults in 1972 following the evaluation opened the way to use by
children, but the sensory parameters used by adults were unsuitable.
They had been chosen for mobility out of doors in conjunction with
the long cane. Children have different needs, discussed by Strelow
and Hodgson (49) which called for considerable changes in the aid
design following a period of observation of blind children (48).

The choice of sensing distance is of prime importance for it
is a fundamental factor which must influence greatly the rate of
learning by a child. A long range sensor collects large amounts of

information requiring a high degree of object discrimination if the
many objects which can be simultaneously sensed are to be perceived
as separate entities. The binaural sensor, particularly with a
medium to long range, does not provide this discrimination in a
stationary environment as perceived by a baby. This has already
been shown to be so by Do. Close range objects which produce low
frequency sounds are also easily masked by the dominant low
frequency components of ambient noise contaminating our environment
(44). They are therefore less reliably perceived. For a developing
child these close objects represent the most important part of its
physical world.   On the other hand, as discussed by Warren (56),
a child which has reached some understanding of its immediate
surroundings and is ready to develop locomotor skills can be
expected to be more motivated to explore distant surroundings if it
can sense that these exist. It would seem reasonable then to
provide an aid which, initially, could be adjusted to provide
greatest detail about the close environment - which can be kept
uncluttered - and then later increase the sensing distance as the
need or interest develops in the child. With this approach the
complex effects of motion on the distance indicator can be most
readily accommodated.

An aid design evolved which provides a variable maximum sensing
distance and an automatic level controller which maintains an
almost constant signal to ambient noise ratio for an ambient noise
change in excess of 20 dB. It is arranged that the user is unaware
of the control taking place (22). It is now feasible to arrange
for the variable sensing distance to be under the control of
muscle potentials. The field of view may be chosen during the
design stage of an aid, but at present, once built it remains fixed.
This limits the research which can be carried out because the field
of view is an important parameter in object sensing and
discrimination. A narrow field of view of say 5°-10° provides
remarkably good auditory discrimination in azimuth but only radial
motion along the axis of the field may be perceived. Head
searching, no matter how slight this may be, is necessary to
determine the relative position of objects.

A wide field provides good motion perception but azimuthal
discrimination is poorer due to the multiple tone complexes
produced by several objects. Both fields have special value to a
child learning about its world but at present both are not
simultaneously available.   The requirement may be likened to
central and peripheral perception, but in an auditory sense. It
now seems feasible to create this artificially.

A small number of blind children have been exposed to sensory
aided perception and it is becoming evident that blind children in
general who are about to become mobile or who have become mobile
may learn to perceive their environment more effectively through

the sensor. The extent to which they do this must vary and is
undetermined at this time. The kind of experiments carried out
with animals to establish their use of a natural sonar cannot be
undertaken with children. Instead, use of the artificial sonar
forms part of a complex life style - even for a small child - and
development on a wide front is constantly taking place. Strelow,
et al. (48) discusses some operant training which was used but
explains why this lasted for only a few sessions before the research
procedure developed into case studies on development of two
children which is partly reproduced here.

Anophthalmic Boy. A totally blind 2½ year-old boy, whose
physical development was good, was provided with an experimental
aid to determine if the increased sensory stimulus would bring
about a change in his developmental pattern. The aid parameters
were the most appropriate that could be provided at that time.

While no other handicaps were assessed, he showed a serious
developmental delay both in walking and in acquisition of language.
He showed no form of walking or crawling on his own, although he
could walk when supported by parents and pull himself into a
standing position against a wall or piece of furniture. His
language was restricted to babbling and cooing sounds. Social
interaction was nonexistent; he showed no obvious awareness of
people around him unless touched or tickled, and did not reach out
for objects making sounds such as rattles and bells. He was not
even startled by loud noises.

It was decided to see whether he could be trained in a spatial
activity by teaching him to reach for the cup when it was brought
in front of him and sensed only through the aid. A small amount of
milk in the cup served as reinforcement. By training the boy to
reach to the sound cue, relating to his drinking cup, it was hoped
to teach him to attend to the device sounds and to relate these to
objects in space. The training task had elements of both classical
and operant conditioning. The response of reaching to the sound
of the cup being tapped, was transferred to the signal of the cup,
sensed through the aid, in a manner analogous to classical
conditioning. The reaching response was "shaped" to its final form
in a manner analogous to operant procedures.

The reaching study established that a spatial task involving
the aid could be learned by the boy. However, the extent to which
the boy perceived the aid sounds as spatial information meaning,
"An object is out there" as opposed to merely reaching on command
to sounds in the ear was not established by this comparatively
simple training procedure. More elaborate procedures in which
objects were placed in different spatial positions with the boy
required to respond differentially could have established this
point.

It was concluded by Strelow et al. after a considerable period
of training, that a developmentally delayed child who was initially
immobile and unresponsive to auditory stimuli, showed a major
improvement in locomotion skills and a marked improvement in
listening skills. While some of this was attributable to the
intervention activities per se, and some due to normal developmental
progress, the sensory aid appeared to help at critical stages of
development.

Congenitally Blind Girl. A 6½ year-old, congenitally blind
girl was in many ways a complete contrast to the boy. She was a
verbally precocious child of well above average intelligence (IQ
125-135), who attended an integrated school that had a visual
resource center. A month's regular observation indicated virtually
no independent locomotion. Requests for her to walk on her own
were met with stubborn refusal, a form of behaviour that was
particularly common when she was observed at home.

At first it was impossible to obtain cooperation for periods
longer than 2 or 3 minutes, but within a few sessions sustained
work for periods of 10-15 minutes was possible. The initial aim
of the work was to establish the extent of her ability to use the
aid in the Sensory Perception laboratory. A series of tasks was
tried in an effort to develop some skill in discrimination of aid
sounds.

Chasing the experimenter - The girl attempted to follow the
experimenter and was occasionally allowed to catch up. This was
felt to be particularly useful for encouraging quick response to
the spatial information provided by the device. The large Sensory
Perception laboratory was sound-deadened and carpeted, and ambient
cues were almost nonexistent.

Shoreling - The task required walking at arm's length along a
row of six poles spread one meter apart. At first she was allowed
to touch each pole as she went, but was later required to maintain
her distance by sound alone and walk up to the last pole.

Slaloming - This was performed with the same pole arrangement
as in shorelining.

Beacon Exercises - An omnidirectional ultrasonic radiator,
which could be detected only through the aid, was used as a homing
beacon. The direction could be sensed binaurally and an indication
of distance was gained from the strength of the signal. The device
was carried by the experimenter to provide a cue to his position
in games of hide and seek, even when the experimenter was hidden
among objects.

This regime was interspersed with occasional walks around the campus, holding hands, during which objects of interest were pointed out. The laboratory exercises were maintained until some level of skill was noted. The most successful task for maintaining interest was the chasing task. The pole exercises were at first the least interesting, although also the most demanding in requiring precise control of body motion in relation to objects.

The first travel task attempted was a walk between the girl's classroom and the visual resource center. There was an irregular concrete travel path between the two buildings requiring a number of 90° turns. At the beginning of the path there was grass on one side and buildings on the other, but it gradually opened up. The most difficult obstacles for an unaided blind traveller were metal poles of an outdoor gymnasium, although they were easily detected with the aid. A final open area led to a large tree in front of the resource center, a total travel distance of about 150m. There were no reliable ambient orientation cues and there were irregular gaps between the buildings so that a simple shorelining procedure was not possible. Some form of navigation skill had to be developed, and the child taught to recognise important objects. Since the school was for sighted children, little attention was given to keeping the travel routes free of unexpected obstacles, so the girl had to be able to cope with them.

The girl's fear of solo travel was at first hard to combat, but by walking with her over the route, taking her by the hand, she began to develop the confidence to move on her own. The experimenters travelled with her for about two weeks, monitoring her aid sounds and explaining what they signified. Finally, she was tested in her ability to handle the route entirely on her own, which she did, albeit slowly and awkwardly, but nevertheless safely.

It became important at this point to ensure that the girl was competent to look after the aid on her own, able to change batteries and bring it to school each day. To earn permission to keep the device, she had to pass a test, which included handling of the aid and batteries as well as performance on shoreline and slalom in the laboratory. The girl was surprisingly serious about the two practice sessions and the test itself. Her performance, while not outstanding, at least indicated a level of competence; two of the five slalom tasks were judged as good, as were three of the five shoreline trials. She was then allowed to keep the aid at home and bring it in herself each day.

To go to the visual center, it was necessary for the girl to put the aid on in class, pack up her work material, and proceed to the visual center on her own as required by the timetable. With the cooperation of teachers this soon became an established routine no longer worthy of comment by either teachers or the girl. Her performance was monitored daily by the research team but this soon

became an unnecessary precaution. She began to have a more
confident gait, and at no time was she observed to collide with or
trip over objects while using the aid. When occasionally she
wandered off course, she kept going until the next object was
sensed, walked up to feel it, and re-oriented herself.

Having established that the girl could use the device outdoors
as a primary aid, other settings were sought. Free use of the aid,
during recess periods and lunchtime, was not encouraged for a number
of reasons, primarily because the aid lacked an automatic level
control to cope with the very noisy environment during playtime.
The travel route to the visual center was taken during a quiet time
in the day when few children were about. However, it was decided
to attempt a more difficult travel route, which would allow her to
come into her classroom from the other side of the large schoolyard.

She arrived each day by taxi, which left her at the gate of
the schoolyard from which she was escorted to class by a teacher or
another student. The school environment was comparatively quiet at
these times (by comparison with the break periods), and a route was
charted out that allowed shorelining of fences and buildings. The
route was lengthy, required a good deal of turning, and there were
frequent unexpected obstacles. Considerable patience was required
by the experimenters and strong support from the teachers was again
helpful in convincing the girl that she could handle the task and
would soon be expected to do it on her own.

Again, daily training was given for two weeks, during which she
was brought to school by one of the team and accompanied on the
route. Her aid sounds were monitored by headphone so that objects
could be pointed out, and various travel strategies could be
explained. The time to travel the route dropped from 30 minutes
to about 9 minutes.

When she was able to perform the route successfully, she was
observed for another two weeks to make sure that she was coping
with the route. The precaution turned out to be unnecessary, as
her performance over the route soon became very good. Her gait
loosened up well and protective actions with her hands became
minimal. However, when unexpected obstacles were encountered, she
would walk up to them and place her hands very accurately to find
out what they were. She travelled quite fast on sections of the
route and generally ran the last part to the classroom, hitching
her school bags up under her arm and appearing little different
from a sighted child.

General Comment. It was discovered during the early stages of
training children in the use of the sensory aid that they learned
the apparently simple tasks less quickly than adults. The pole
tasks used in teaching adult mobility were eventually mastered and

performed proficiently by a very intelligent child but the time
scale of learning greatly exceeded that for adults. Outdoor
mobility in the playground of a school for sighted children also
came slowly but reached a higher level of attainment without the
additional aid of a cane than that seen for most adults. The child
occasionally broke into a jog even when there were other children
around. No blind adult did this during the reported studies.
Baird, Newcommer and Scionie (5,38,45) also report on sensory aid
use by children during lessons but these children do not get to
keep the aid and it is shared by the users. Little information is
provided on the levels of aid use each child reached.

Babies

The controversy over the use of a sensory aid by babies
resulting from a case study by Bower involving a blind baby raises
some fundamental questions (8,9,32). The binaural sensory aid has
been shown here to code the environmental information into an
acoustic form previously unknown to man: even the direction cue is
artificial in that the initial sensation is one of lateralization.
Hence the baby reported by Bower to have tracked an object with
its blind eyes through the sensory aid stimulation could also have
been induced to produce this response using the simple presentation
of dichotic sounds from standard earphones. Any lateral shift of
the fused auditory image should equally well have caused a
corresponding rotation of the eyes. Taking this further, the baby,
without prior knowledge, would have to attribute to the lateral
auditory shift an appropriate angular shift of the eyes so that it
appeared to look at the object represented by the sound. Film of
the baby as viewed by the writer was unconvincing. It could not
be established that the IAD direction cue was matched to the baby's
lateralization function and it is suggested here that any apparent
tracking of the object by the blind baby which was observed could
only have been fortuitous. Bower also argued that "placing" of
hands by the baby was sensory aid induced. This is quite feasible,
particularly since the auditory sensation when approaching a close
object is not unlike the visual looming discussed by Gibson (21).
This visual looming effect is very noticeable when viewing an
approaching light through frosted glass. Unfortunately, Bower in
his tests used two spaced specular reflecting batons as objects
and these were equidistant from the baby. As shown earlier in Do's
experiments and it is discussed elsewhere (34), two echos from
different directions but equal distances produce a single ghost
echo from midway between the objects. Hence, if the baby in a
placing position were stimulated to outstretch its hands by the
aid, it could only be the ghost which did this. How then did the
baby work out that there were two objects on either side of the
ghost and reach towards them as explained by the experimenter?
This, without turning its head to the extent that is necessary to

cause object separation (17). These apparent facts fit a theory proposed by Bower but the physics of aid performance do not fit the same facts. It is essential when experimenting with an aid such as this, that its physical performance be well understood and more should not be inferred from behaviour patterns than is physically reasonable to expect. What has to be realised is the difference between sensing the environment with light waves visually and acoustic waves audibly. The 1 to 10,000 ratio in wavelength itself processes out almost all the spatial information we are familiar with through the visual channel. Perhaps it is because some adult blind people have been able to demonstrate quite remarkable skills using the aid that some experimenters have come to believe much more was possible simply by exposing a very young receptive brain to the spatial information.

A wavelength of the order of 2.5 mm does in fact have the effect of creating mirrors in the environment because of the high proportion of object surfaces which are specular reflectors at this wavelength. Loudness of echo rather than structure becomes a dominant cue in many situations so care is needed in the design of any experiment. Bower, for example, used a cube suspended on a string as an object. This is highly specular as a reflector and as it freely turns, causes great changes in echo strength but, more importantly,it can cause strong left-right auditory sensations as the acoustic highlight of the radiator is reflected more to the left receiver than the right and then back again. A more appropriate object would be a sphere with a rough surface or, if large enough, a smooth sphere.

A number of babies are presently learning to perceive their environment through the medium of an aid: not all the aids have the same sensing parameters and most of the babies are multiply handicapped so each is a case study from which preliminary information about use may be obtained. Several researchers are involved and so far none of the work has been published with the exception of that by Bower. It can be said however that of four babies exposed to an aid none have exhibited the spontaneous responses reported by him. During a recent visit to one of the babies reported in the New Scientist (8) it was found that the aid being used by the baby could not be providing even the elementary information it was designed to provide. The experimenter reported failure in the use of the aid. The photograph of the sensory aid shows that this was a modified set of spectacle frames as used for adults and the side arms extended well behind the ears. The sound outlet from each of the side arms was then well behind the ears also and no acoustic coupler to the ears was used. Under these circumstances - as checked with the experimenter - the baby could only hear very loud sounds from specular reflectors; binaural direction cues must have been almost non-existent judging from the work of Keen (35); and the advantage of the wide dynamic range of the ear and its ability to discriminate subtle differences in sound quality was discarded. This gives rise to questions about other reports on babies who used aids which were not properly designed or fitted (50).

At this time experience suggests a child about to become mobile will appear to benefit most from a binaural sensory aid. Earlier than this the responses so far reported which are supported by the physics of the system do not yet support the hypothesis that a very young brain can readily respond to and assimilate acoustic environmental information. But it may be doing so – when later it may come to use this information more effectively than the older children who are now using aids. The aid in use by some researchers is shown in Fig.21. Special care has been taken to ensure a secure fit so that the sensors are well positioned (they are normally covered by special material as seen in Fig.2) and the sound feed to the ears of a baby remains correct for long periods of time during the day. Miniature phones as shown have the advantage of creating an "out-thereness" in the sound sensation which otherwise only comes from experience with the distance cue and is self induced. By this is meant that a mental projection of the echo sound comes about as experience is gained in moving with the sensor. This has been verbally reported by adult users. The ears of the baby are clear of sound shadowing caused by the phones which fit forward of the ears so that ambient sound cues can also be learned by a developing baby.

FIG.21.   Babies' Sensory Aid

RELATIONSHIP OF BINAURAL SENSORY AID TO
THE ANIMAL SONAR

It has now been reasonably well established that blind people
within the age bracket of approximately 2 to 55 years can learn to
interpret, in their own undefined rudimentary way, the rich echo
sounds from an ultrasonic sensory aid in which electronic signal
processing is minimal (multiplication of reception with radiation).
The incoming ultrasonic information is not removed, it is instead
transferred to a usable frequency band in the audible range for
processing by the human auditory neural system. Appropriate time
scaling is achieved in the radiation code. Blind people can learn
to perceive and identify certain objects as belonging to a limited
set of a given type of object and can negotiate their complex
environment with considerable skill. Some locomotion tasks
requiring a fine manipulation of auditory flow patterns can be
executed with a performance rating which is comparable with that
for sighted subjects, and even if only a few blind persons could
actually compete with a sighted subject in some of these tasks
this is nevertheless performance which two decades ago was not
thought likely to be possible. Some experiments with echo-locating
animals in the 50's and early 60's as reported at the first NATO
Animal Sonar Symposium (Busnel 1966) were really no more
sophisticated in relation to animal size and freedom of motion than
some of those which have now been carried out with blind humans.
Observation of sensory aided blind travellers in a down-town area
reveal locomotion skills greatly in excess of those measured in a
laboratory but are themselves unmeasurable at present. They cannot
even be defined. What then has been learned which can be related
to the use of sonar by animals?

The world as it is actually perceived by an echo-locating
animal remains unknown to us: up to now it has only been through
behaviour patterns that we have guessed at what it must be like.
The world as it is perceived by an echo-locating blind person using
audible sound bears only a trivial relationship to the real physical
world and object identification is negligible (41). Perception is
very unreliable. However the world as now perceived through a
binaural CTFM sonar does bear strong relationship to the real
physical world and anyone with fairly normal hearing can have the
same experience as a blind user. It is a unique experience, quite
different from that obtained using a high resolution scanning sonar.
It may not have been brought about had there not been evidence that
animals use echo-location to hunt their prey in complex environments
and must therefore be able to resolve multiple object auditory
space through two summing apertures.

A great deal has been learned about the ultrasonic reflection
characteristics of materials and surface structure of objects

forming our environment in a subjective way which no form of
mathematics or computer modelling could hope to convey; certainly
no simulation system as proposed by the Subcommittee on Sensory
Aids (51) could emulate this. The brain seems to be capable of
categorising a wide range of auditory signals representing objects
and relates these to the particular environment in which it is
operating. The parking meter, for example, can be identified only
if it is located in the environment for which it was designed and
forms an event in a travel sequence. It must be seen as one of
several in a complex pattern of events. A static situation
presented out of context is confusing unless the environmental
structure has been simplified. This is well illustrated in the
evaluation report by Thornton (54) and Armstrong (3). The
influence of ambient cues - sound, smell, etc. - may be considerable
at times.

It is therefore not just the measurable or specifiable
information about the spatial location of objects that the brain
is using, but instead it is the information from the several
sensory modalities which are integrated into a single perception
of the environment as it changes through motion. It is nevertheless
quite evident that the blind are being provided with a simple model
of the animal sonar which can be said to have commonality mainly in
terms of the frequency band of ultrasonic energy radiated into the
environment to be viewed and the reception of echos by two auditory
channels. The coding of the radiated energy has a similarity -
that of a near linear time varying frequency but, whereas the
period of sweep used by a bat may vary from 5 to 0.5 ms, that used
in the sensory aid varies from 250 to 25 ms. The difference arises
from the usable receiver bandwidth of man's sensory system being
only 5 kHz compared with up to 150 kHz enjoyed by bats.

The environmental information which can be extracted by the
two systems - that of the bat and that used by man - may not be
so very different however since the time-bandwidth products have
greater similarity: a 5 ms pulse of sweeping tone radiation
covering an octave frequency band from 100 to 50 kHz and emitted
every 50 ms as may be used by a bat, and a 50 ms pulse of sweeping
tone covering the same octave as may be used in the design of an
aid (see Table 1) have the same time-bandwidth product of 250
when man's limited receiver bandwidth of 5 kHz is used as the
effective handwidth of the system. With this latter system it is
possible to perceive a dead fly tied to a fine thread as it is
swung in and out of the branches of a small bushy fir tree which
may be compared with what the bat perceived in the photograph
shown by Webster (55) of a bat chasing its prey into a similar
shrub tree. The ambiguity function of the CTFM sonar having an
octave bandwidth as discussed by Do (17) helps to explain this
auditory acuity and it is evident that there is no known way in

which this same information can be displayed visually. The
discussion by Do in the next paper is highly relevant.

A similar comparison cannot be made using any other sonic
sensory aids for the Blind: they use a bandwidth of only 1 - 2 kHz
and have a time-bandwidth product of Unity.

Do (17) discusses an underwater sonar using a CTFM radiation
covering an octave bandwidth which has interesting features but
these do not relate to the echo-location by porpoises which employ
quite a different ultrasonic emission. Even so, it seems that
with a CTFM sonar it may be possible to recognise shoals of fish
from other underwater scatterers and it could become possible to
identify different species given adequate experience.

ON AID USE BY THE BLIND

There can be no doubt that blind people need and want a
sensory aid to help them in their everyday activities requiring
perception of their environment, whether this be considered to be
of primary or secondary importance to the Long Cane as a travel
aid matters little. So far, an aid has not been developed which
meets all their declared requirements and it is highly likely
that this goal is beyond man's technical capability. What has
been developed is a range of aids from the simplest - the Long
Cane -- to the most sophisticated conceptually - the Binaural
Sensory Aid - with a variety of sonic devices of a simple type,
such as the "Russell Path Sounder", the Nottingham Obstacle
Detector, the "Mowat Sensor" (57) and the Canterbury Single Object
Sensor (33). The Laser Cane also belongs to this latter category.
Of all the promising directions in which sensory aid development
could proceed employing imaging technology and visual implants or
high resolution ultrasonic scanning systems with tactile display,
none actually seem to offer the practical promise inherent in
making most use of the auditory system and its sensitive wide
dynamic range.

This reviewer remains convinced that, as for echo-locating
mammals, humans can make greater use of their auditory system for
sensing their environment. It seems unlikely that the man machine
interface of implanted microelectrodes or a matrix of tactile
stimulators can transfer to the brain the kind of information which
will allow greater perceptual detail for object identification and
motion perception under conditions of mobility. (This should not
be confused with the blind person's need to read or look at single
static objects and the like). But even so, man may not gain
sufficient environmental information from an auditory display to
motivate him to adapt to it regardless of the age of introduction.

The binaural sensor in its further developed form incorporating means to central and peripheral perception, giving greater object recognition in a central narrow field and motion perception in a wide peripheral field, and the means to "focus" on objects through mioelectric control seems to offer the possibility to overcome many of the current limitations of the sensor. This still leaves the need for intellectual stimulus which the lack of vision seriously reduces and spatial auditory perception may not be increased to above the critical useful level which motivates man.

The concept of modelling the bat's sonar so as to make available to man the kind of spatial information enjoyed by the bat fails to take into account man's intellectual needs. The bat is a creature of simple habits, either sleeping or hunting, and an auditory means to catch its prey on the wing in the dusk, remarkable though this may once have seemed to be, does not satisfy man who is already well cared for physically in a highly socialised community. The unknown way in which the bat recognises its environment and can return unerringly to its base poses a more appropriate question. Is the environmental detail much greater than man has available through a sensory aid? Does the pinna of the bat and the flexible use of these apertures give it added information of a kind man would come to use? Unless technological developments can move in these directions we wont learn the answer and this seems to be particularly difficult as shown by Bui (14).

However, the criticism of the Subcommittee on Sensory Aids (51) that the latest techniques in Radar and Sonar systems remain largely untapped in sensory aid development is refuted if it has ·been shown that the binaural sensor models, as well as it is presently possible technically to model, the bat's sonar and that this model is superior in sensing the environment of a blind man to any other system which can today be proposed. This, at least, would remove a major handicap to acceptance of an acoustic display for a sensory aid but would still leave social attitudes to blindness which determine the economic factors that do perhaps exert the ultimate control on sensory aid use.

REFERENCES

1.   Airasian, P.W. "Evaluation of the Binaural Sensory Aid",
      Research Bulletin No.26, pp.51-71, June 1973.

2.   Armstrong, J.D. "Summary Report of the Research Programme on
      Electronic Mobility Aids", Report, Psychology Dept, University
      of Nottingham, 1973.

3. Armstrong, J.D. "An Independent Evaluation of the Kay Binaural
    Sensor". Report to St.Dunstan's Scientific Committee.
    (St.Dunstan's, London, 1972).

4. Bach-y-Rita, P. "A Tactile Vision Substitution System Based
    on Sensory Plasticity", in (Sterling, T.D., et al. Eds) Visual
    Prosthesis, the Interdisciplinary Dialogue, (Academic Press,
    New York, 1971).

5. Baird, A.S. "Electronic Aids: Can They Help Children"?
    J.Visual Impairment and Blindness 71, No.3, pp.97-101, March
    1977.

6. Benham, T.A. "The Bionic Instruments Travel Aid" in
    (Dufton, R. Ed.), International Conference on Sensory Devices
    for the Blind. (St.Dunstan's, London, 1966).

7. Benjamin, J.M. "The Bionic Instruments C-4 Laser Cane" in
    (Nye, P.W. Ed.), Evaluation of Mobility Aids, (National
    Academy of Engineering, Washington, D.C. 1971).

8. Bower, T.G.R. "Blind Babies See with their Ears, New Scientist,
    73, pp.255-257, 1977.

9. Bower, T.G.R. "Babies are More Important than Machines",
    New Scientist, 74, pp.712-714, 1977.

10. Boys, J.T., Strelow, E.R., Clark, G.R.S. "A Prosthetic Aid
    for a Developing Blind Child", Ultrasonics, 17, pp.37-42, 1979.

11. Brindley, G.S., Lewin, W.S. "The Sensations Produced by
    Electrical Stimulation of the Visual Cortex". J.Physiol.
    196, pp. 479-493, 1968.

12. Brabyn, J.A., Strelow, E.R. "Computer-analysed measures of
    characteristics of human locomotion and Mobility", Behaviour
    Research Methods and Instrumentation, Vol.9(5), pp. 456-462,
    1977.

13. Brabyn, J.A. "Laboratory Studies of Aided Blind Mobility",
    Ph.D. Thesis, University of Canterbury, N.Z.,1978.

14. Bui, S.T. "A Single Object Sensor as a Mobility Aid for the
    Blind". Ph.D. Thesis, University of Canterbury, N.Z., 1979.

15. Collins, C.C. "Tactile Vision Synthesis" in (Sterling,T.D.,
    et al.), Visual Prosthesis, (Academic Press, New York, 1971).

16. Dobelle, W.H. "Current Status of Research on Providing Sight
    to the Blind by Electrical Stimulation of the Brain".
    J.Visual Impairment and Blindness, 71, pp.289-297, 1977.

17. Do, M.A. "Perception of Spatial Information in a Multiple Object Auditory Space", Ph.D. Thesis, University of Canterbury, N.Z., 1977.

18. Do, M.A., Kay, L. "Resolution in an Artificially Generated Multiple Object Auditory Space Using New Auditory Sensations", Acustica, 36, pp.9-15, 1976/77.

19. Elliot, E., Elliot, P.H., Roskilly, D. "Sonic Mobility Aid, St.Dunstan's Instruction Manual", (St.Dunstan's, London, 1969).

20. Fletcher, H. "Speech and Hearing in Communication" (D. Van Nostrand Co.Inc., N.Y. 1953).

21. Gibson, J.J. "Visually Controlled Locomotion and Visual Orientation in Animals", Brit.J.Psychology, 49, pp.182-194, 1958.

22. Guttman, N. (Ed.) "Evaluation of Sensory Aids for the Visually Handicapped",(National Academy of Sciences, Wash. D.C., 1972).

23. Kay, L. "Active Energy Radiating Systems: Ultrasonic Guidance for the Blind" in (Clark, L.L., Ed.) Proceedings of the International Congress on Technology and Blindness, (The American Foundation for the Blind, N.Y. 1963).

24. Kay, L. "Auditory Perception and its Relation to Ultrasonic Blind Guidance Aids". J. Brit. I.R.E., 24, pp.309-317, 1962.

25. Kay, L. "An Ultrasonic Sensing Probe as a Mobility Aid for the Blind", Ultrasonics, 2, pp.53-59, 1964.

26. Kay, L. "A Sonar Aid to Enhance Spatial Perception of the Blind: Engineering Design and Evaluation". The Radio and Electronic Engineer, 44, pp.605-627, 1974.

27. Kay, L., Do. M.A. "An Artificially Generated Multiple Object Auditory Space for Use Where Vision is Impaired", Acustica, 36, pp.1-8, 1976/77.

28. Kay, L. "Evaluation of the Ultrasonic Binaural Sensory Aid for the Blind" in (Guttman, N. Ed.), Evaluation of Sensory Aids for the Visually Handicapped, pp.85-94, (National Academy of Sciences, Wash.D.C. 1972).

29. Kay, L. "Towards Objective Mobility Evaluation: Some Thoughts on a Theory". (American Foundation for the Blind, N.Y.1974).

30. Kay, L. "The Sonic Glasses Evaluated", The New Outlook for the Blind, 67, pp.7-11, 1973.

31. Kay, L. "Evaluation of the Ultrasonic Binaural Sensory Aid for the Blind" in (Guttman,N. Ed.),Evaluation of Sensory Aids for the Visually Handicapped, Appendix C, pp.1-33, (National Academy of Sciences, Wash.D.C., 1972).

32. Kay, L., Strelow, E.R. "Blind Babies Need Specially Designed Aids", New Scientist, 74, pp.709-712, 1977.

33. Kay, L., Bui, S.T., Brabyn, J.A., Strelow, E.R. "Single Object Sensor: A Simplified Binaural Mobility Aid", J.Visual Impairment and Blindness, 71, pp.210-213, 1977.

34. Kay, L. "Enhanced Environmental Sensing by Ultrasonic Waves" in (Busnel, R.G., Ed.), Animal Sonar Systems, Tome II, pp.757-781, (Laboratoire De Physiologie Acoustique, Jouy-en-Josas 78, 1966).

35. Keen, K. "Coupling the Output of the "Sonicguide" to the Ear of the User", The New Outlook for the Blind, 70, pp.304-306, 1976.

36. Keith, W.J., "Adaption to Distortion of Auditory Space, M.Sc. Thesis, University of Auckland, 1971.

37. Long Cane; The Cane as a Mobility Aid for the Blind, (National Academy of Sciences, Wash.D.C., 1972).

38. Newcommer, J. "Sonicguide: Its Use with Public School Blind Children", J.Visual Impairment and Blindness, 77, pp.268-271.

39. Nye, P.W. (Ed.), Evaluation of Mobility Aids for the Blind, (National Academy of Engineering, Wash. D.C. 1971).

40. Plomp, R. Steeneken, A.J.M., "Interference Between Two Simple Tones". J.Acoustical Soc. Am. 43, pp.883-884, 1968.

41. Rice, C.E. "The Human Sonar System" in (Busnel, R.G., Ed.) Animal Sonar Systems, Tome II, pp.719-755, (Laboratoire De Physiologie Acoustique, Jouy-en-Josas 78, 1966).

42. Rowell, D. "Auditory Display of Spatial Information", Ph.D. Thesis, University of Canterbury, N.Z., 1970.

43. Russell, L. "Travel Path Sounder" in (R.Dufton, Ed.) International Conference on Sensory Devices for the Blind, pp.293-296 (St.Dunstan's, London, 1966).

44. Saunders, S.D.   "The Distortion of an Auditory Display with Noise", M.E. Thesis, University of Canterbury, N.Z., 1974.

45. Scionie, M.W.   "Electronic Sensory Aids in a Concept Development Program for Congenitally Blind Young Adults", J. of Visual Impairment and Blindness, 72, pp.88-93, 1978.

46. Sharpe, R.   "The Evaluation of the St.Dunstan's Manual of Instruction for the Kay Sonic Aid",   Report, Dept of Psychology, University of Nottingham.

47. Stewart, W.G., Hovda, O.   "The Intensity Factor in Sound Localization; an Extension of Weber's Law", Psychol. Rev. 25, pp.242-251, 1918.

48. Strelow, E.R., Kay, N., Kay, L.   "Binaural Sensory Aid: Case Studies of Its Use by Two Children", J.Visual Impairment and Blindness, 72, pp.1-9, 1978.

49. Strelow, E.R., Boys, J.T.   "The Canterbury Child's Aid: A Binaural Spatial Sensor for Research with Blind Children", Journal of Visual Impairment and Blindness [in press 1979].

50. Strelow, E.R., Kay, N., Kay, L.   "Binaural Sensory Aid: Response to Comments by Smith and Dailey", J. Visual Impairment and Blindness, 72, pp.320-321, 1978.

51. Subcommittee on Sensory Aids,   "Selected Research, Development and Organisational Needs to Aid the Visually Impaired". (National Academy of Engineering, Wash. D.C., 1973).

52. Thornton, W.   "The Binaural Sensor as a Mobility Aid",   The New Outlook for the Blind", 65, pp.324-326, 1971.

53. Thornton, W.   "The Binaural Sensor as a Mobility Aid: An Individual Evaluation of Its Use in Conjunction with the Long Cane", (St.Dunstan's, London, 1971).

54. Thornton, W.   "Evaluation of the Binaural Sensor or Sonic Glasses", (St.Dunstan's London, 1972).

55. Webster, F.A.   "Interception Performance of Echolocating Bats in the Presence of Interference", in (Busnel, R.G., Ed.), Animal Sonar Systems, Tome I, pp.673-712, (Laboratoire De Physiologie Acoustique, Jouy-en-Josas 78, 1966).

56. Warren, D.H.   "Blindness and Early Childhood Development", (American Foundation for the Blind, N.Y., 1977).

57. Wormald Sensory Aids International Ltd., "Mowat Sensor" a commercial product.   Christchurch, New Zealand.

DISCRIMINATION OF COMPLEX INFORMATION IN AN ARTIFICIALLY

GENERATED AUDITORY SPACE USING NEW AUDITORY SENSATIONS

Manh Anh Do

Radio Engineering Limited

P.O. Box 764, Dunedin, New Zealand

INTRODUCTION

Wide beam broad band CTFM (continuous transmission frequency modulated) sonar with auditory binaural display was first proposed to be used as a sensory device for the blind person by Kay [16] in 1959. In this system the combined distance and direction of an object is coded in the form of rising pitch with increasing distance (frequency proportional to distance) and the binaural differences which could include that of time, amplitude, and frequency. Subsequent research to optimise this display resulted in a reduction of the interaural differences to that of amplitude only, and this difference measured in dB was designed to be proportional to the azimuthal angle as far as was physically possible [17, 28]. Thus the CTFM sonar with the binaural display (or briefly the binaural sonar) basically performs a one-to-one transform of the two dimensional real space-distance (range) and direction (azimuthal angle) of an object into a two dimensional auditory space described by frequency and interaural amplitude difference (IAD) of a tone. This auditory space was readily produced by means of a sensory aid for the blind [24, 18], and a sonar to locate fish [33, 34]. The evaluations of these two systems were reported in [2, 19, 34, 35, 5]. The results indicated clearly the ability of human operators to use the new display and to perform their tasks successfully with relative ease, even though the input to the ears was rich in information.

The results obtained from field evaluation were however impossible to explain by the experimental results found by Rowell [28] and Anke [1]. The latter was from laboratory experiments

817

using system simulation designed to determine the ability of
subjects to resolve objects in space by auditory means. The
results of Rowell and Anke suggested a very poor auditory resolution.
The disagreement between these two results led the author and Kay
to the investigation of the role of Doppler effect in the binaural
sonar [20, 4, 5]. Rowell and Anke's experiments used only static
situations to facilitate controlled measurement. The temporal
effects and the complexity of the sounds produced by the binaural
sonar under dynamic conditions, which could possibly enhance the
resolution capability of subjects were neglected in their
experiments.

This paper reviews the experimental results reported in [5]
on the capability of subjects to discriminate complex information.
New auditory sensations were found to be perceived by subjects in
laboratory experiments using system simulation close to realistic
situations. This resulted in the high performances of the users of
the binaural sonars as described in [34, 2, 19]. The experimental
results were related to the theoretical research on the hearing
system and the sonar system to evaluate the performance of the
man-machine system as a whole.

THE AUDITORY FREQUENCY RESOLUTION OF TWO PURE TONES SIMULTANEOUSLY SOUNDED

Since the audible information displayed by the sonar, when a
multiple of objects is presented in the field of view, is a large
number of tones, each with its own characteristics, the first
approach is to determine the auditory resolution of two tones.
Several different psychological problems are however immediately
arisen under such simple conditions of simulation. Indeed, even
if the two tones are monaurally presented, depending on the
combinations of four variables (two amplitudes and two frequencies)
there will be different effects, namely, beats, combination tones
(CT's), masking, fusion, discrimination, etc., on subject's
perception.

In 1924, Wegel and Lane [41] designed a very elegant
experiment to measure the shift of the threshold of hearing of a
pure tone when masked by another. However, when using the 80 dB
1200 Hz tone as the masker and increasing the level of the masked
tone beyond its threshold of hearing up to the same level of the
masker, over the range from 400 to 4000 Hz, they could determine
the limits in intensity and frequency of the masked tone where
either beats, or combination tones, or one tone only, or two tones
only, etc., would occur (figure 1). Their results showed that a
tone can mask tones of higher frequency more effectively than those
of lower frequency than itself. Wever [42, 43] described the beat

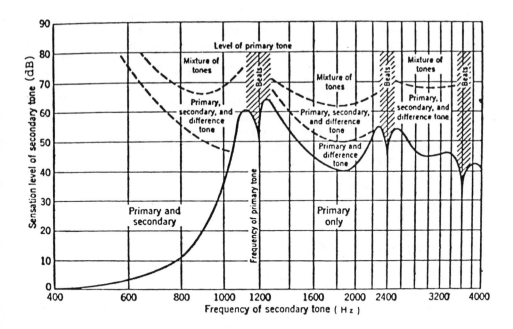

Figure 1.   Criteria of various sensations caused by the
             simultaneous sounding of two tones. The primary is the
             80 db, 1200 Hz tone. The frequency of the secondary
             varies from 400 to 4000 Hz and its level from 0 to 80
             db (Wegel and Lane 1924).

phenomenon by three clearly distinguishable stages, appearing
successively as the frequency difference of two tones is increased
from zero-(i). Noticeable oscillation: the loudness rises and
falls slowly; (ii). Intermittance or pulsation of tone: pulses of
tone separated by silence; (iii) Roughness without intermittance:
there appears a sort of whirr, and the sound is characteristically
rough. Observing two tones of equal sensation levels of 20 dB, one
was at 1024 Hz while the other was varied from that frequency upward,
Wever found the above three stages of the beat phenomenon at the
frequency differences of: (i) 0 to 6 Hz, (ii) 6 to 166 Hz, and
(iii) 166 to 356 Hz.   All these results [41-43] show that a
frequency difference of several hundreds Hz is required so that
subject can hear two tones, say, of 1000 Hz and 1000 Hz + Δf,
without being interfered by the combination tones. These
experiments were however not extended to different frequencies.

The purpose of this study is to seek for a definition of the auditory frequency resolution which is related to the spatial range resolution in a sense that: for two single objects presented in the field of view of the sonar system, the listener will hear as if only two tones are presented and the perceived tones must be the coding tones displayed by the sonar. Plomp et al [25-27] defined four criteria: (i) Frequency difference between partials of a complex of tones required to hear them separately, (ii) frequency difference between two tones required to distinguish two pitches, (iii) frequency difference between two tones for maximum harshness, and (iv) frequency difference for just absence of beats. Investigating the results shown in figure 2 and the experimental procedure described in [25-27] it is believed that the fourth criterion is equivalent to the end of the second stage of beats described by Wever [42, 43] while the third is the maximum perceptible intermittence of this stage. A forced choice procedure was used in the first and second criteria. Subject was asked to compare a complex stimulus with two different single tones successively, and to choose the single tone which belonged to the complex stimulus. This procedure does not testify to either the absence or the existence of the combination tones. So neither of these criteria are able to define the resolution of the auditory space generated by the binaural sonar. Using two tones having different IAD's, Rowell [28] found that subject requires a fractional frequency difference, $\Delta f/f$, of 40% to tell which tone, low or high, is in the left or in the right. This result seems not too dependent on the IAD's of the tones.

In [4,5] the author defined the frequency resolution as the least frequency difference between two tones so that they are perceived as being "distinctly separate" where the term "distinctly separate" implies the absence of the interference between two tones so that the stimuli are the only perceived tones. This ensures a one to one transformation from the real space to the auditory space generated by the binaural sonar. Two tests were used to verify the subject's judgments at the criterion of resolution.

(i)   Two and only two tones are perceived by subject, and

(ii) The perceived tones are the stimuli.

In test (i) subject is presented with a stimulus which randomly includes either two or three components, and has to answer the question "is this only two tones"?   Immediately after a correct "yes" answer is given, test (ii) is continued. Subject is presented with a single tone, and he has to answer either of the two questions, "is this the low tone"? or "is this the high tone"?, while the single tone may be either the low tone, the high

Figure 2.   Four Criteria of Sensations defined by Plomp et al.

● Frequency difference between the partials of complex
tones required to hear them separately.

▲ Frequency difference to distinguish two tones
simultaneously presented.

o Frequency difference for maximal roughness of two tones.

Δ Frequency difference for just absence of beats given
by two tones.

tone, the difference tone, or the cubic difference tone. Experiments with four subjects shows an average frequency resolution of $\Delta f/f = 40\%$. Below this limen, the perception of the fundamentals are interfered by the combination tones considerably.

ROLE OF DOPPLER EFFECT IN THE BINAURAL DISCRIMINATION OF AUDIBLE FREQUENCY PATTERNS PRODUCED BY WIDE BAND CTFM SONAR

The transmitted signal of a CTFM sonar system is cyclic and of the form:

$$S_T(t) = A \exp\left[2\pi j \left(f_2 t_s - \frac{m}{2} t_s^2\right)\right] \tag{1}$$

Where f_2 = upper limit frequency

   m = sweep slope

   t = $t_n + t_s$ at any instant, t_n is the beginning time of the n-th cycle of the modulating function, and t_s varies from 0 to T_s, the repetition period.

   A = constant amplitude of the transmitted signal.

Let $r(t)$ be the instantaneous range of an object. The received signal, $S_R(t)$, is then a Doppler shifted replica of the transmitted signal, $S_T(t-\tau(t))$, $\tau(t)$ seconds ago, which was reflected from the object at distance $r(t_r)$, where $t_r = t - \frac{\tau(t)}{2}$. Thus

$$\frac{\tau(t)}{2} = \frac{1}{c} r(t_r) = \frac{1}{c} r\left(t - \frac{\tau(t)}{2}\right) \tag{2}$$

where c is the propagation velocity of sound.

Substituting (2) in $S_T(t-\tau(t))$, the range coding frequency of an object is the frequency difference of the transmitted and received signals.

$$f_a(t) = m\tau(t)(1 - \dot{\tau}(t)) - \dot{\tau}(t)(f_2 - mt_s) \tag{3}$$

In perfectly stationary conditions, the range of an object is constant, thus the delay time is:

$$\tau = \frac{2r}{c} = \text{constant, and } \dot{\tau} = 0 \tag{4}$$

Hence $f_a = \dfrac{2m}{c} r = Kr$ $\hspace{4cm}$ (5)

The constant $K = \dfrac{2m}{c}$ is the range code of the sonar. With a suitable choice of the sweep slope m, the range coding frequency, f_a, can be made audible. Typically, the sensory aid for the blind has a range code of 941 Hz/metre, and the fishing sonar has four different range codes of 53.33, 26.67, 13.33, and 6.67 Hz/m for four different maximum operating ranges of 94, 188, 375 and 750m.

Under dynamic conditions, the range of an object is no longer simply proportional to the coding frequency of the displayed sound as described by relation (5). To solve equations (2) and (3), the simplest way is to approximate $\tau(t)$ by:

$$\tau(t) = \frac{2r(t)}{c} \hspace{4cm} (6)$$

this will meet an error of

$$\frac{\Delta\tau(t)}{\tau(t)} = \frac{1}{c} \frac{r(t) - r(t - \tau(t)/2)}{\tau(t)/2} = \frac{\dot{r}(t)}{c} \hspace{3cm} (7)$$

However, while the fishing sonar is being operated, the operator steering the boat does not travel faster than 3m/s (six knots), hence $\dot{r}(t)/c$ is smaller than 0.2%, where c = 1500 m/s is the sound velocity in water. A blind user of the sensory aid walks at a speed, at most, 5 km/hour. The average sound velocity in air is 340 m/s. Thus $\dot{r}(t)/c$ is smaller than 0.4%. The approximation in equation (6) is reasonable. Replacing (6) in (3) we have:

$$f_a(t) \doteq \frac{2m}{c} r(t) - \frac{2\dot{r}(t)}{c} f_T(t) \hspace{3cm} (8)$$

where $f_T(t) = f_2 - mt_s$ is the instantaneous transmitted frequency.

Hence the relative motion between object and sonar system basically changes the frequency of the sonic wave by a factor $(1 - 2\dot{r}(t)/c)$. The frequency change $\dfrac{2\dot{r}(t)}{c} f_T(t)$ is called the Doppler shift. Equation (8) shows that the Doppler shift is preserved in the range coding frequency of the CTFM sonar. An example of the audible frequency pattern produced by a fishing sonar due to constant radial velocity $\dot{r}(t) = -v$ is shown in figure 3.

$$f_a(t) = \frac{2m}{c} (r(0) - vt) + \frac{2v}{c} (f_2 - mt_s) \hspace{2cm} (9)$$

This sawtooth cyclic variation pattern is due both to the Doppler shift of frequency in the medium, and the wide variation in transmitted frequency.

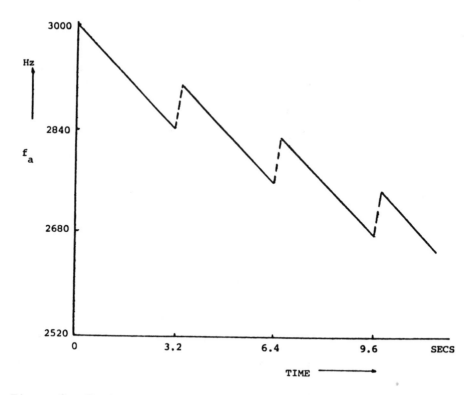

Figure 3.   Variation of Range Coding Frequency with Respect to
            Time.

$f_{a1}(t) = 3000 - 25t - 25\ t_s$, where $0 < t_s < 3.2$.
The broken lines correspond to ambiguous signals
blanked out.

Binaural Sound Pattern

        The discussion has thus far been about the range indication.
Under realistic conditions, the relative direction of the object
is also changing during locomotion of the blind aid user. Hence
the audible frequency pattern will flow left or right in
accordance with the changing spatial positions of the object
relative to the user. The realisation of the direction code is
shown by Kay in the same chapter.

        Two typical situations encountered in realistic conditions --
(i) object passed in one side as subject moves along a straight
course, and (ii) subject rotates his head to obtain more
information - were analysed by the writer in [5]. Figure 4 shows

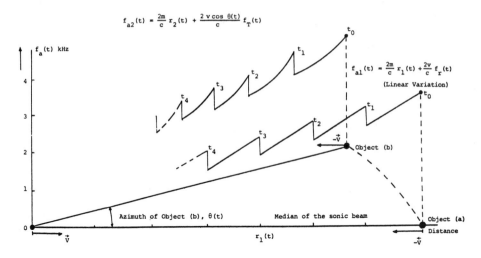

$$f_{a2}(t) = \frac{2m}{c} r_2(t) + \frac{2v\cos\theta(t)}{c} f_T(t)$$

$$f_{a1}(t) = \frac{2m}{c} r_1(t) + \frac{2v}{c} f_r(t)$$

(Linear Variation)

Figure 4.   Comparison between the sound patterns of a centre
             object and of a side object.

two different frequency patterns perceived by a user of the sensory
aid while moving along a straight course. One is produced by a
central object, and the other due to an object on the left side.
The possibility for a sensory aid user to discriminate two objects
different in direction but close in range, by rotating his head, is
illustrated in figure 5. The Doppler effect mainly creates a
negative frequency pulse when the user turns away from the object,
and a positive frequency pulse when he turns towards it.

Hence under real life conditions, for example, mobility by the
blind in a busy pedestrian area where people are moving with
varying acceleration from different directions, and the relative
velocity between user and fixed objects will also vary, the audio
signals from the sensory system will change in an indescribably
complex way, each signal having its own unique character.
Experience with blind people using the sensory aid suggests that
great resolution of these time varying tones is being enjoyed.
What they perceived are the patterns of sound, which flow towards
the left or the right in a unique way related to the changing
positions of the reflecting objects.

In the case of the fishing sonar, similar sound patterns
could be produced due to motion of either the boat or the targets.
Two kinds of motion of shoal target could be considered:
(i) translational motion of the entire shoal and, (ii) oscillatory
motion of individual fish within the shoal. The translation
velocity of the shoal is frequently smaller than 75 cm/s [13, 23],

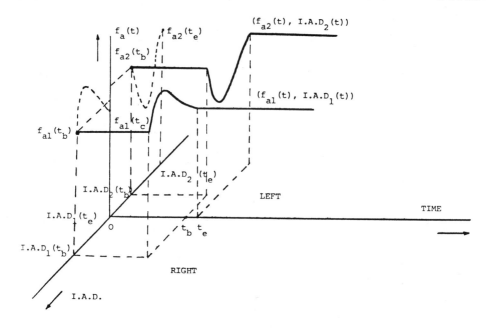

Figure 5.   The effect of head rotation on the binaural sound
             patterns of two objects initially close in range but
             different in azimuthal angle (head rotates to the
             right in this figure).

the equivalent frequency shift is less than 60 Hz. However the
sound pattern created by the shoal when it crosses the sonic beam
from left to right or vice-versa is not unlike that obtained in
the sensory aid when subject rotates his head. Bainbridge [3]
showed that average fish (length 75 cm) can reach a speed of 6 m/s
though only for a short period. Hence the oscillatory motion of
fish in the shoal can create the frequency shifts up to 480 Hz in
the echoes if 60 kHz signals are transmitted. Using 70 kHz
transmitted signal Hester [15] observed the frequency shifts of
about 200 Hz from individual fish. In addition to this effect,
amplitude modulations in the sound pattern (due to the changes in
scattering cross section of fish while they are swimming) are also
possibly heard. Fish of different sizes or different kinds have
different types of motion. Hence the sound patterns they produce
must be different. An experienced user of the binaural sonar will
be able to learn to recognise different species from the difference
in character of the sound patterns these species produce.

A DEFINITION OF FREQUENCY RESOLUTION USING NEW AUDITORY SENSATIONS

Consider two adjacent single objects, 1 and 2, described by their instantaneous ranges $r_i(t)$ and azimuthal angles $\theta_i(t)$ where $i = 1,2$ in the field of view of a binaural sonar. Correspondingly, in the auditory space the two time varying tones will be described by:

$$f_{ai}(t) = \frac{2m}{c} r_i(t) - \frac{2f_T(t)}{c} \dot{r}_i(t) \tag{12}$$

$$IAD_i(t) = 10 \log \frac{I_{Ri}(t)}{I_{Li}(t)} = K_1 \theta_i(t) \tag{13}$$

where $i = 1,2, I_{Ri}$ and I_{Li} are respectively the sound intensities in the right and left ears of the i-th tone. Constant K_1 is determined by the equation (11). Typically it is 0.4 to 0.5 dB/ degree in the sensory aid, and about 1.5 dB/degree in the fishing sonar.

Thus the capability of resolution must be a function of four dynamic variables $[f_{a1}(t), IAD_1(t), f_{a2}(t), IAD_2(t)]$. Clearly any function of four dynamic variables is too complicated to be studied and described by subjects in a psychophysical experiment. In all experiments described in (12) and (13) f_{a2}, IAD_2, and IAD_1 were held constant, only f_{a1} was varied as in equation (12). The procedure was based entirely upon the "psychological phenomenon" described below.

The Psychological Meaning of Auditory Resolution
Under Conditions of Change

It was known in Section 2 that under static conditions the assignment of different values of IADs to the two pure tones did not affect the frequency resolution, but it defined the psychological phenomenon of frequency resolution more clearly, since subjects were required to indicate the left or right directions of the tones. This procedure was adapted to the dynamic experiment as follows:

Initially at t=0, and $f_{a1}(t) \doteq f_{a2}$ the subject hears slow beats, and feels only one complex image in the auditory space. Its IAD is the combination of IAD_1 and IAD_2. As t increases $f_{a1}(t)$ decreases according to equation (12). The interference between two

Figure 6.   The Psychological Phenomenon of Resolution in a
             Changing Situation

tones changes in a unique way until a certain time $t=t_1$, the
subject feels the image of $f_{a1}(t)$ flow "left" or "right" in auditory
space according to the assigned values of IAD_1 and IAD_2 . The
attention appears to be captured by the "moving" tone which may be
thought of as forming a <u>flow pattern</u> in space. At this instant the
subject is required to respond by pressing a switch to store the
value of $f_{a1}(t) \cdot f_{a1}(t)$ is allowed to continue when the attention
can be transferred to f_{a2} which then appears in its appropriate
auditory spatial position. This is always secondary to the
initially perceived flow of $f_{a1}(t)$, but occurs almost

simultaneously. The psychological phenomenon is described in figure 6. The judgment of subjects can be verified by questioning them on the relative movement of the images. The frequency resolution can be defined here as the frequency difference between f_{a2} and $f_{a1}(t)$ at which the two tones separate and "slide" in auditory space to their respective positions.

Measurement of Frequency Resolution

The simulation related to a fishing sonar was chosen in the experiments because the rates of change encountered are such as to allow a subject to respond without introducing serious error due to the delay in response. The situation of constant radial velocity was simulated.

Typically v=1.5 m/s, c=1500 m/s, T_s= 3.2 secs, m=12.5 kHz/s, thus

$$f_{a1}(t) = f_{a1}(0) - 25(t + t_s) \qquad (14)$$

where $0 \leqslant t_s \leqslant 3.2s$.

The variation of $f_{a1}(t)$ with respect to time is shown in figure 3. Values of $f_{a1}(0) = f_{a2}$ were chosen to be 1000, 2000, and 3000 Hz, while the values of IAD were 0, +4, +8, + 12 dB for tone 1, and -10, -6, -2, +2, +6, +10 dB for tone 2. Figure 7 shows the results obtained from three subjects S_1, S_2, and S_3. The variation of $\Delta f_a = f_{a2} - f_{a1}(t_1)$ with respect to different values of IAD_1 and IAD_2 was found to be smaller than the possible error due to the delay of subject's reaction. Hence the dependence of frequency resolution on IAD could be considered as negligible.

From the above results it is obvious that the capability of frequency resolution, under dynamic conditions is significantly improved when compared with that under stationary conditions. Equation (14) shows that the Doppler effect results in an increase of the rate of change of the range coding frequency by a factor of two. This also takes part in the improvement of the frequency resolution. Indeed by removing the cyclic variation in the tone $f_{a1}(t)$ so that

$$f_{a1}(t) = f_{a1}(0) - 25t \qquad (15)$$

and repeating the above experiment with subject S_2, his capability of resolution was found to be worsened (see table 1).

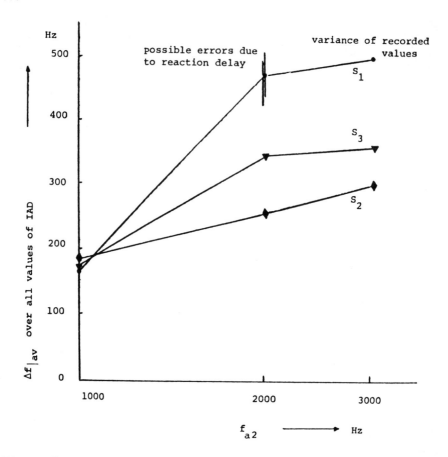

Figure 7.   Variation of the frequency resolution with respect to
frequency averaged over all values of IAD (IAD variable
has negligible effect).

Since the cyclic variation involving doubling the rate of
change of the audio frequency within each sweep, has a significant
effect on the resolution capability, a direct increase in the
object velocity, resulting in a higher rate of change of frequency
has to have certain effects on the resolution performance of the
subject. By increasing the object velocity to 3 m/s, so that

$$f_{a1}(t) = f_{a1}(0) - 50 (t + t_s) \tag{16}$$

and repeating the experiment with subject S_2, his resolution
capability was found to be improved quantitatively as well as
qualitatively (see table 1). The sensation of image slide in
auditory space is more pronounced and the separation of images is

very clear by all who experience the phenomenon. Subject S_2 was
unable to respond quickly to change greatly the measured resolution
capability obtained in the experiment using equation (14) for the
simulation. Hence in the conditions of high rates of change of
audio frequencies as in the sensory aid for the blind, signals
must be resolved cognitively long before reaction can be recorded.

TABLE 1.   Comparison of Frequency Resolution of Subject
S_2 in four different conditions.

f_{a2} Hz	$\Delta f(0)$ Hz	$\dfrac{\Delta f(0)}{f}$	$\Delta f(1)$ Hz	$\dfrac{\Delta f(1)}{f}$	$\Delta f(2)$ Hz	$\dfrac{\Delta f(2)}{f}$	$\Delta f(3)$ Hz	$\dfrac{\Delta f(3)}{f}$
1000	400	0.50	180	0.20	220	0.25	130	0.14
2000	700	0.42	250	0.13	440	0.25	230	0.12
3000	1100	0.45	300	0.11	630	0.23	240	0.08

$f(0)$: stationary tones

$f(1)$: saw-tooth variation of $f_{a1}(t)$

$f(2)$: Linear variation of $f_{a1}(t)$

$f(3)$: saw-tooth variation of $f_{a1}(t)$ at double rate

AUDITORY DISCRIMINATION OF TWO MULTIPLE COMPONENT TONES

The investigation on the multiple target resolution of the
binaural sonar so far has been carried out with single objects
only. Under realistic conditions a target always has a certain
size, and is characterised by its shape. Hence the corresponding
audible signals displayed by the binaural sonar at its two
channels must be the band limited tones whose spectra are combined
in a specific way in accordance with the shape of the target.
To describe a target consisting of, say, n "scattering centres"
(locations scattering most of the power back to the receiver)
n pairs of variables $(R_i(t), \theta_i(t))$ are required, where i=1, n,
and $R_i(t)$ and $\theta_i(t)$ are respectively the instantaneous range and
azimuth of the i-th scattering centre. Correspondingly in the
auditory space, the multiple component coding tone, or simply the
complex tone, must be described n pairs of variables
$(f_{ai}(t), IAD_i(t))$, where $f_{ai}(t)$ and $IAD_i(t)$ are determined by
equations (12) and (13), and i=1, n. However to simplify the
mathematical presentation of the problem, it will be assumed that

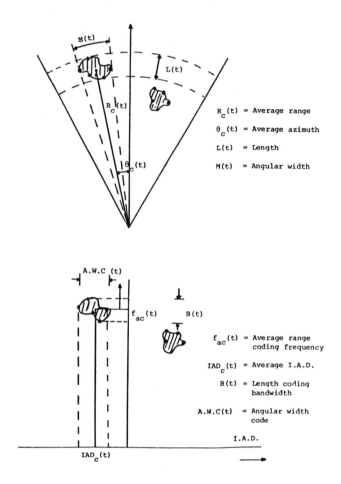

Figure 8.   Relationship between the variables in real space and
             in auditory space.

within a short duration of time, when a certain task such as
discrimination of two adjacent targets is carried out, the sizes
and shapes of the targets are unchanged. Consequently each target
can be simply described by two variables, its average range $R_c(t)$
and azimuth angle $\theta_c(t)$, and two parameters, the extent in range
(or length) L and the angular width M of the target. Figure 8
illustrates the one to one relationship between the above variables
and parameters in real space with the variables $f_{ac}(t)$ (average
range coding frequency) and $IAD_c(t)$ (average IAD), and parameters
B (length coding band-width) and A.W.C. (angular width code) of
the multiple component tone in the auditory space.

    A ten channel voltage controlled oscillator was especially
built to synthesize two complex tones, each of five equally spaced

components. The experimental procedure was based upon the
psychological phenomenon similar to that found in the experiment
with single tones described in Section 4. The most surprising
result is that the discrimination of two multiple component tones
under certain conditions is easier than the discrimination of two
single tones.

Initially the length coding bandwidths of the two complex
tones were set at 100 Hz, and the angular width codes were 2 dB.
The values of $g_{ac} = f_{ac}(0)$ were chosen to be 1000, 2000, and 3000
Hz while IAD_c were 0,4,8 dB, and JAD_c were -10, -6, -2, +2, +6 and
+10 dB. For each combination of g_{ac}, IAD_c, and JAD_c the recording
frequency discrimination in four times showed little deviation,
frequently less than 20 Hz. An average $f_{ac} \doteq 130$ Hz at $g_{ac} = 1000$
Hz at $g_{ac} = 2000$ Hz, and 240 Hz at $g_{ac} = 3000$ Hz. Figure 9 shows
very close results given by three subjects. Besides the better
discrimination performance obtained in this case of complex tone in
comparison to the performance with single tones, it was noted that
the frequency discrimination was slightly worsened when IAD_c and
JAD_c are close together. Although this increase of Δf_{ac} was not
very large, it was consistently found in the performance of all
three subjects at different frequencies, and IAD's.

In order to study the effect of the angular width code on the
frequency discrimination, the above experiment was repeated with
subject S_2 using complex tones having angular width code increased
to 4 dB. The bandwidths of the tones were still kept at 100 Hz.
Δf_{ac} is increased by up to 20% in this case. Also there is always
a small increase in Δf_{ac} as the difference between IAD_c and JAD_c
decreases.

The length coding bandwidths of the complex tones were then
increased to 200 Hz, while their angular width codes were restored
to 2 dB. Repeating the experiment with subject S_2, it was found
that in comparison to the results obtained in the experiment using
100 Hz bandwidth tones, Δf_{ac} increases 79% at $g_{ac} = 1000$ Hz,
65% at $g_{ac} = 2000$ Hz, and 69% at $g_{ac} = 3000$ Hz. At $g_{ac} = 1000$ Hz,
subject reported the difficulty in separating two tones.

DISCUSSION

From Basilar Membrane Filtering to Frequency Analysis
of Collicular Neurones

The "cochlea" in the inner ear is the primary frequency
selective part of the hearing organ. Von Békésy [6] discovered

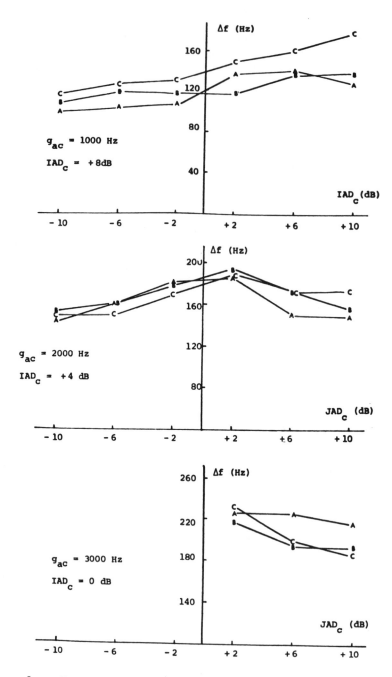

Figure 9.   Frequency discrimination of two complex tones as a
             function of I.A.D. performed by three subjects A, B,
             and C.

the "travelling waves" on the "basilar membrane" inside the cochlea and recognised it as a "non-uniform" transmission line where lower frequencies can travel further than higher frequencies from the "stapes" before being attenuated. This observation led to the "place theories" of pitch perception, according to which the position of maximum vibration of the basilar membrane determines the pitch of the tone. The frequency resolution of the basilar membrane as observed by Von Békésy was very low. Two tones simultaneously sounded must be separated by a fractional difference of $\Delta f/f \dot{=} 30\%$ before usual fatigue effects appear. He considered this as the limen for two tones to be heard distinctly. Another approach to study the frequency selectivity of the ear was the introduction of the concept of the "critical band" [6,8,12,14,29, 30,39,45,46] as : (i) either the limited band of noise effectively masking a tone, (ii) or the frequency band within which the energy of a multiple component tone is summated, etc. Widely diverse results were obtained from these authors. $\Delta f/f$ is from 10% to 25%. Several experimenters have had the tendency to compare these critical bandwidths with the frequency resolution capability [25-27], while it has been shown in section two that two single tones simultaneously sounded must differ considerably in frequency, $\Delta f/f \dot{=} 40\%$, to be perceived distinctly. However, if we consider the fact that the ear can easily detect a frequency change of $\Delta f/f$ less than 0.5% with respect to time, either abruptly or in a certain continual mode [7,35,44], can track a fast time varying tone (as experienced by the sensory aid users), and can resolve time varying tones better than stationary tones, then the assumption of a mechanical filtering process in the cochlea as the only frequency analysis is apparently inadequate.

To understand the above unconformable results, let us postulate that: The inner ear can perform a real time frequency analysis of the stimuli but sustains a high degree of distortion. The assumption of a real time frequency analyser mechanism explains the capability of the ear in tracking a fast time varying tone as well as detecting a very small frequency change with respect to time. However when the stimulus consists of two or more components, the perception of the fundamentals is considerably interfered by the distortion products, unless the fundamentals are well separated from each other.

Considering now three most prominent subjective components heard by a subject when he is presented with two adjacent tones, f_{a1} and f_{a2}, simultaneously in three different cases (Table 2):

It is obvious that while all three subjective components are stationary in case 1, in case 3 only the high tone is stationary, the cubic difference tone descends at a rate of change in frequency equal to twice that of the low fundamental tone. This can be

TABLE 2.   Three most prominent subjective components produced
           by two tones f_{a1} and f_{a2} simultaneously presented.

Subjective component	case 1 stationary tones	case 2 $f_{a1}(t)$, cyclic variation	case 3 $f_{a1}(t)$, linear variation
High tone	f_{a2}	$f_{a2}=f_{a1}(0)$	$f_{a2}=f_{a1}(0)$
Low tone	f_{a1}	$f_{a1}(0)-qt-qt_s$	$f_{a1}(0) - qt$
Cubic difference tone	$2f_{a1}-f_{a2}$	$f_{a1}(0)-2qt-2qt_s$	$f_{a1}(0) - 2qt$

where $q = \dfrac{2v}{c}$ m, t_s varies from 0 to T_s

generalised that under dynamic conditions, if the stimuli have
different rates of change in frequency, all the subjective
components will vary at the rates of change in frequency which are
entirely different from each other. Hence it is possible that at
higher levels of perception, following the filtering process of the
basilar membrane, the collicular neurones respond to these
subjective components differently, so the interference of the
combination tones in the perception of the fundamentals is less
significant than that in the static conditions when both the
fundamentals and the combination tones are of the same
characteristic: "STATIONARY". Accepting this hypothesis the
explanation for the higher resolution obtained in case 2 becomes
obvious. The Doppler effect in the CTFM sonar increases the rates
of change of the range coding frequencies, increasing also the
differences between the rates of change in frequency of the
subjective components. The same argument applies to the case of
direct increasing in the object velocity.

Little knowledge about this performance of human beings was
found in literature. Suga (1965, 1968) [36-38] studied the
responses of collicular neurones of bats to frequency modulated
(FM) and complex sounds, classified them in five main types:-
symmetrical, asymmetrical, FM sensitive, FM insensitive, and
upper threshold units - depending on the way they respond to their
best frequencies, the inhibitory areas which limit the ranges of
frequency modulation on one or both sides of the best frequencies,
the selective responses of the units to certain types of modulation

determined by the range, rate, direction and functional form of the frequency sweep, and the thresholds of intensity to activate the neurones. These principles were believed to apply to all kinds of mammals and human beings but of course with different time scales. The characters of selective responding of the collicular neurones explain the capability of the auditory system in analysing the structure of a complex sound. Hence under realistic conditions, the time varying tones produced by a binaural sonar are furnished with additional characters such as rates of change in frequency, functional forms, etc. which enable the auditory system to discriminate the fundamentals from the distortion products more easily.

The Improvement and the Worsening of Frequency Discrimination to Multiple Component Tones

It was known from place theory that if the stimulus consists of several frequency components which are not separately resolved, then an averaging mechanism does take place [21,22], a pitch of frequency approximately equal to that of the centre component will be heard. An explanation for the improvement or the worsening of the frequency discrimination of the multiple component tones in comparison to that of single tones can be based upon that averaging mechanism as follows.

Throughout the discussion in the previous section it was seen that the CT's, especially the cubic difference tone $2f_1 - f_2$, generated due to the non-linearity of the inner ear, interfere significantly in the discrimination of the fundamentals. Figure 10a shows the typical level of the $2f_1 - f_2$ tone when f_1 and f_2 are equal to 70 dB [9-11]. However if two complex tones of the same levels (70 dB) as the single tones, and having the same bandwidths B's, are now considered, the cubic difference tones generated by the combinations of various pairs of components will be spread out over a large frequency band as demonstrated in Figure 10b. This also applies to higher order CT's. Therefore if the bandwidths of the complex tones are narrow enough so that the averaging process can take place, two stimuli must raise two pitches of frequencies equal to the centre frequencies of the complex tones and of intensities equal to that stimulated, while all the CT's are spread out and create a low level wide band background. The interference of this background may not be as considerable as the interference of the CT's generated by a two pure tone stimulus.

Accepting the above explanation for the improvement of frequency discrimination of narrow band (100 Hz) tones leads to the only possible explanation for the worsening of frequency

Figure 10.   Explanation for the improvement of frequency
              discrimination of narrow band tones and the worsening
              of frequency discrimination of wide band tones.

discrimination of wide band (200 Hz) tones that the averaging
mechanism fails to summate the components of the complex tones
effectively in the latter case, the stimulus tones are diffused and
the CT's generated by the high complex tone interfere seriously
on the low tone (Figure 10c). Zwicker et al [45,46] and Sharf
[29,30] measured the critical band within which the power of a
complex tone is summated and found that this bandwidth could vary
from 80 Hz at f=100 Hz to 2500 Hz at f=10,000 Hz. According to
their results the critical handwidth is equal to 160 Hz at
f=1000 Hz, 300 Hz at f=2000 Hz, and about 500 Hz at f=3000 Hz.
Subject S_2 reported the diffiuclty in separating two tones at
g_{ac} = 1000 Hz because the bandwidths of the two complex tones were
larger than the critical bandwidths. At g_{ac} equal to 2000 Hz and
3000 Hz, the summation of the power of each complex tone may not
fail, but the CT's generated by the high tone still interfere on the
low tone considerably. The discrimination is worsened but the tones
are still easily perceived and separated.

CONCLUSION

Throughout this paper, results obtained from several
psychophysical experiments using system simulation techniques are
described. An investigation of the frequency discrimination
performance in a multiple object auditory space was approached from
the very simple situation of two single stationary tones to a much
more complicated situation of two multiple component tones whose
frequency difference varied with respect to time. The experimental
results have explained why a human operator was able to use the new
auditory display in realistic conditions with relative ease, as
observed in field evaluation, even though the inputs to the ears
were very rich in information.

This paper is a review of four chapters of the complete work
[5] whose purpose was to evaluate an auditory display for use with
the broad band CTFM sonar. The actual resolutions in range and in
velocity of the CTFM sonar itself are much higher than the
resolutions of the display. In [5] the ambiguity function of the
broad band CTFM signal was determined and applied in the example
of the fishing sonar described in section 4. In this approach, the
resolutions in range and velocity are defined by the limits where
the cross-correlation between the echoes returned from two adjacent
targets is equal to a quarter of the maximum value when the targets
are completely coincident. The maximum values of the resolutions
in range and velocity were found to be $\Delta r \doteq 0.375$ ft, and $\Delta v \doteq 4 \times 10^{-2}$
ft/s. These values correspond to a resolution in range coding
frequency of 2 Hz, and a resolution rate of change of the coding
frequency equal to 0.2 Hz/s. No physical display system has shown
to provide such high resolution. Simmons (1971, 1973) [31,32]
experimented with the FM bats, and found that the two species

Eptesicus and Phyllotomus could have an acuity in discriminating
target range of 1 to 2% at the absolute distances of 60 cm and
120 cm. One specimen of Eptesicus was found to be able to
discriminate two targets of 14 mm apart at the absolute distance
of 240 cm, i.e. $\Delta r/r \doteq 0.6\%$. It has been proposed that these species
have an ideal sonar receiver which cross correlates the transmitted
and received signals to extract target-range information. The
correlation process may take place at the higher level of
perception than the basilar membrane of the bat. This discovery
would be an impetus for researchers to look for a real time display
of high resolution and fast response for the CTFM sonar which could
emulate the performance of the bat.

REFERENCES

1.   Anke, Von D. "Blindenorientierungshielfen mit
      ultraschallortungs Verfahren und Horbarer Anzeige", Acustica,
      Vol.30, pp.67-80, 1974.

2.   Arasian, P. "Evaluation of the Binaural Sensory Aid for the
      Blind", A.F.B. Res. Bull., Vol.26, pp.35, 1973.

3.   Bainbridge, R. "The Speed of Swimming of Fish as Related to
      Size and to the Frequency and Amplitude of the Tail Beat",
      J.Exp. Biol., Vol.35, pp.109-133, 1958.

4.   Do, M.A. and Kay, L. "Resolution in an Artificially Generated
      Multiple Object Auditory Space Using New Auditory Sensations",
      Acustica, Vol.36, pp.9-15, 1976/77.

5.   Do, M.A. "Perception of Spatial Information in a Multiple
      Object Auditory Space", Ph.D. Thesis, University of Canterbury,
      1977.

6.   Fletcher, H. "Auditory Patterns", Review of Modern Physics,
      Vol.12, pp.47-65, 1940.

7.   Fletcher, H. "Speech and Hearing in Communication",
      D. Van Nostrand Co. Inc., New York, 1953.

8.   Greenwood, D.D. "Auditory Masking and the Critical Band",
      J. Acoust. Soc. Am., Vol.33, pp.248, 1961.

9.   Goldstein, J.L. "Auditory Nonlinearity", J.Acoust. Soc. Am.,
      Vol.41, pp.676-688, 1967.

10.   Hall, J.L. "Auditory Distortion Products, f_2-f_1, and $2f_1-f_2$",
      J.Acoust. Soc. Am.,Vol.51, pp.1863-1871,1972.

11. Hall, J.L. "Monaural Phase Effect: Cancellation and Reinforcement of Distortion Products f_2-f_1 and $2f_1-f_2$",J.Acoust. Soc.Am., Vol.51, pp.1872-1881, 1972.

12. Hamilton, P.M. "Noise Masked Threshold on a Function of Tonal Duration and Masking Noise Bandwidth", J.Acoust. Soc. Am., Vol.29, pp.506-511, 1957.

13. Harden Jones, F.R. "Fish Migration", Edward Arnold Publishers Ltd., London, pp.230-234, 1967.

14. Hawkins, J.E. and Steven, S.S. "The Masking of Pure Tone and Speech by White Noise", J.Acoust. Soc. Am., Vol.22, pp.6-13, 1950.

15. Hester, F.J. "Identification of Biological Sonar Targets from Body Motion Doppler Shifts", Symposium on Marine Bio-Acoustics, Pergamon Press, New York, Vol.2, pp.59-74, 1967.

16. Kay, L. "A New or Improved Apparatus for Furnishing Information as to Position of Objects", Patent Specification No.978741, The Patent Office, London, 1959.

17. Kay, L. "Blind Aid", Patent Specification No.3366922, United States Patent Office, Washington, D.C., 1965.

18. Kay, L. "A Sonar Aid to Enhance Spatial Perception of the Blind: Engineering Design and Evaluation", Radio and Electron. Eng., Vol.44, pp.605, 1974.

19. Kay, L. "Sonic Glasses for the Blind - Presentation of Evaluation Data", A.F.B. Res.Bull., Vol.26, pp.35, 1973.

20. Kay, L. and Do, M.A. "An Artificially Generated Multiple Object Auditory Space for Use where Vision is Impaired", Acustica, Vol.36, pp.1-8, 1976/77.

21. Lichte, W.H. and Gray, R.F. "The Influence of Overtone Structure on the Pitch of Complex Tones", J. Exp.Psychol., Vol.49, pp.431-436, 1955.

22. Licklider, J.C.R. "Auditory Frequency Analysis", Symposium on Information Theory, The Royal Institution, London, 1955, Butterworths Scientific Publications, London, 1956.

23. Madison, D.M. et al "Migratory Movements of Adult Sockeye Salmon (Oncorhyn Chus Verka) in Coastal British Columbia as revealed by Ultrasonic Tracking", J.Fish. Res.B. Can., Vol.29, pp.1025-1033, 1972.

24. Martin, G. "Electronics and Transducers for an Ultrasonic
     Blind Mobility Aid", M.E. Thesis, University of Canterbury,
     1969.

25. Plomb, R. "The Ear as a Frequency Analyser", J.Acoustic. Soc.
     Am., Vol.36, pp.1628-1636, 1964.

26. Plomb, R. and Levelt, W.J.M. "Tonal Consonance and Critical
     Bandwidth", J. Acoust. Soc.Am., Vol.38, pp.548-559, 1965.

27. Plomb, R. and Steeneken, H.J.M. "Interference Between Two
     Simple Tones", J. Acoust. Soc. Am., Vol.43, pp.883-884, 1968.

28. Rowell, D. "Auditory Display of Spatial Information", Ph.D.
     Thesis, University of Canterbury, 1970.

29. Scharf, B. "Critical Bands and the Loudness of Complex Sounds
     Near Threshold", J.Acoust.Soc.Am., Vol.31, pp.365-390, 1959.

30. Scharf, B. "Loudness of Complex Sounds as a Function of the
     Number of Components", J.Acoust.Soc.Am., Vol.31, pp.783-785,
     1959.

31. Simmons, James A. "Echolocation in Bats: Signal Processing
     of Echoes for Target Range", Science, Vol.171, pp.925-928, 1971.

32. Simmons, James A. "The resolution of Target Range by Echo-
     locating Bats", J.Acoust.Soc.Am., Vol.54, pp.157-173, 1973.

33. Smith, R.P. and Kay, L. "A Fish Finding Sonar Utilizing an
     Audio Information Display", Digest of Technical Papers,
     I.E.E.E., Ocean Conf., Panama City, Florida, pp.113, 1970.

34. Smith, R.P. "Transduction and Audible Displays for Broad
     Band Sonar Systems", Ph.D. Thesis, University of Canterbury,
     1973.

35. Steven, S.S. "Pitch Discrimination, Mels and Koch's
     Contention", J.Acoust.Soc.Am., Vol.26, pp.1075-1077, 1974.

36. Suga, N. "Analysis of Frequency Modulated Sounds by Auditory
     Neurones of Echolocating Bats", J.Physiology, Vol.179,
     pp.26-53, 1965.

37. Suga, N. "Responses of Critical Auditory Neurones to
     Frequency Modulated Sounds in Echolocating Bats", Nature,
     London, Vol.206, pp.890-891, 1965.

38. Suga, N. "Analysis of Frequency Modulated and Complex Sounds
     by Single Auditory Neurones of Bats", J. Physiology, Vol.198,
     pp.51-80, 1968.

39. Swets, J.A. et al "On the Width of Critical Bands",
    J.Acoust. Soc. Am., Vol.34, pp.108-113, 1962.

40. Von Békésy, G. "Experiments in Hearing", McGraw Hill,
    New York, 1960.

41. Wegel, R.L. and Lane, C.F. "The Auditory Masking of One Pure
    Tone by Another and its Probable Relation to the Dynamics of
    the Inner Ear", Phys. Rev., Vol.23, pp.266-285, 1924.

42. Wever, E.G. "Beats and Related Phenomena Resulting from the
    Simultaneous Sounding of Two Tones-I", Psychological Review,
    Vol.36, pp.402-418, 1929.

43. Wever, E.G. "Beats and Related Phenomena Resulting from the
    Simultaneous Sounding of Two Tones-II", Psychological Review,
    Vol.36, pp.512-523, 1929.

44. Wever, E.G. "Theory of Hearing", John Wiley, New York, 1949.

45. Zwicker, E. et al "Critical Band-width in Loudness Summation",
    J. Acoust. Soc. Am., Vol.29, pp.548-577, 1957.

46. Zwicker, E. "Subdivision of the Audible Frequency Range into
    Critical Bands", J.Acoust. Soc. Am., Vol.33, pp.248, 1961.

Chapter VII
Posters

THE ROLE OF CRANIAL STRUCTURES IN ODONTOCETE SONAR SIGNAL EMISSION

Gustavo Alcuri

Two independent hypotheses attempt to explain the sonar signal transmission mecanism on its path towards the outside environment: 1) focalization through reflection on the concave surface of the anterior cranial region, 2) transmission through the premaxilla.

In order to explain the sonar signal conduction system experimentally, thirty skulls were studied in 23 species from 7 different families, using two fundamental methods, deduction and direct visualization.

The deductive method was applied in several different ways:

1. Anterior cranial region analyses were performed from a geometric point of view. As the configuration of the surface determines the mode of reflection for a given source, the purpose of these analyses was to determine the focalization possibilities of the skull through reflection on the surface of the anterior cranial region.

2. A study was also made of the bone response to vibrations under pulse and sinosoidal excitation. Its purpose was to determine the signal transmission properties of the rostral bone structures. Electrical excitation was transmitted to the bone through a special piezoelectric ceramic whose sandwich-like design allowed simultaneous control of the attack signal. An effective comparison of emitted and received signals could therefore be made after their passage through the biological structure.

3. The propagation velocity of longitudinal and surface waves through the rostral bones was calculated.

4. Attenuation within normal sonar system frequencies was also studied.

Direct visualization of bone structure vibratory behavior was obtained using double exposure holographic interferometry (Cherbit and Alcuri, 1978). This technique permits interference in coherent light of diffraction spectra from excited and non-excited bone. Localized interference fringes can thus be observed on the bone which correspond to movements of the excited structure. The emitting ceramic is placed in the antero-posterior palatine cavity.

The results using these different methods lead to the following assertions:

1. Theorectically, the geometrical properties of the skull cannot explain focalization by reflection in all species.

2. Signal propagation along the rostrum is responsible for the significant reduction in the spectral bandwidth of the attack signal (Fig. 1).

3. The propagation velocity of longitudinal waves is not constant along the rostrum but this phenomenon is not connected to the constitutional differences in the bones.

4. Acoustic vibrations are highly attenuated during their propagation through the bone in the rostrum considered globally (Fig. 1).

5. From a vibratory point of view, the superior maxillary bones respresent a functional identity. In other words, the different rostral bones act together as a whole (Fig. 1 and 2).

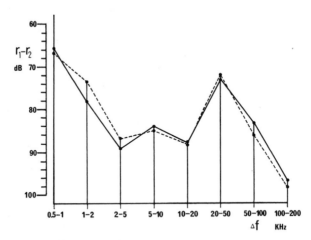

Fig. 1.   Frequency interval/reference level- response level $(r_1 - r_2)$ in D. delphis.

———————   Premaxillary

- - -   Maxillary

Fig. 2. Direct visualization of bone vibratory behavior by double
         exposure holographic interferometry. The result of this
         is an equivalence between the left and right sides of the
         rostrum and a functional continuity between the premaxilla
         and the maxilla (Sotalia teuzii).

         Excitation frequency: 20 kHz.

     Echolocation signal transmission by internal bone conduction
therefore seems unlikely. The results indicate that signal propagation takes place through the interface constituted by the premaxilla and the cephalic soft tissue. In some groups these transmission waves are complementary with the reflection method.

Cherbit, G. and Alcuri, G., 1978, Etude de la propagation des
         vibrations à travers le rostre de Sotalia teuzii (Cetacea)
         par interférométrie holographique.C.R. Acad. Sc. Paris,
         286: D.

BINAURAL ESTIMATION OF CROSS-RANGE VELOCITY AND OPTIMUM

ESCAPE MANEUVERS BY MOTHS

Richard A. Altes and Gerald M. Anderson

Roeder (1970) has shown that noctuid moths display erratic flight patterns during pursuit by bats, and that these moths are capable of sensing their approximate distance from the bat (from the loudness of echolocation signals) and the direction of the bat (from comparison of two sound sensors on opposite sides of the body). This behavior suggests the following two problems: (1) What is the best maneuver the moth can make?, and (2) Is the bat theoretically capable of sensing such a maneuver?

Problem (1) was solved by an application of zero-sum, perfect information, differential game theory (Isaacs, 1967). Assuming that the bat has perfect information about the moth's position and velocity, the theory yields a berrier which separates the bat's capture region (assuming the moth maneuvers optimally). From the data of Webster (1966), it would appear that a bat's velocity is ≈ 5 m/s, a moth's velocity is ≈ 1.5 m/s, a bat has a turn radius of ≈ 1 m, a moth has a turn radius of ≈ 0 m, and the bat can capture anything within a circle of diameter 5 cm (e.g., with a somersault maneuver). Using these parameters, the capture region has a minimum radius at a range of ≈ 30 cm from the bat, or $\approx .06$ sec before capture. The moth's best maneuver is to fly normal to the bat's velocity vector, and the best chance for escape occurs ≈ 30 cm from the bat.

Problem (2) was solved by assuming that a bat can measure a differential Doppler shift between the two ears. Using this differential Doppler shift, the standard deviation of an optimum estimate of tangential velocity v_t in the horizontal plane is:

$$\Delta v_t \;=\; rc/(\sqrt{2SNR}\ Tf_o d)$$

for a CW bat at range r with signal frequency f_o, pulse length T, and interaural distance d. If the bat first detects its prey at r = 3 m, and if the SNR at this distance is 3 dB, then at r = 30 cm, SNR is at least 2×10^4, and $\Delta v_t = (32T \times 10^2)^{-1}$ when c = 330 m/s, f_o = 80 kHz, and d = 2 cm. Typically, a CW bat pulse shortens to roughly T = 2 msec at 30 cm, but even with such a short pulse, $\Delta v_t \approx 0.2$ m/s. The bat's theoretical resolution capability is thus sufficient to detect an optimum escape maneuver with a horizontal

velocity component greater than 0.2 m/s, near the moth's best escape
position. If only interaural frequency shift were used to estimate
cross-range velocity, then the moth's best strategy would be a dive
in the bat's median plane, normal to the bat's velocity vector.

DOLPHIN WHISTLES AS VELOCITY-SENSITIVE SONAR/NAVIGATION SIGNALS

Richard Altes and Sam H. Ridgway

Certain dolphin whistles display a hyperbolic increase in frequency as a function of time, followed by a hyperbolic decrease. These whistles nearly always contain one or more harmonics. The signals were classified as distress calls by Lilly (1963), but they are also used under many other circumstances.

The above dolphin whistles are very similar to signals that can be used for accurate Doppler measurement (Altes and Skinner, 1977). Velocity is measured by observing the Doppler-induced phase shift in a signal $a(t) \exp(j \cdot b \cdot \log t)$. A Doppler-scaled echo for such a signal is:

$$s^{1/2} a\{s(t-\tau)\} \exp\{jb\log s(t-\tau)\} \simeq a(t-\tau)\exp\{jb\log(t-\tau) + jb\log s\}$$

where s is a velocity-dependent scale factor and τ is delay.

Objections to this theoretical conjecture are as follows:

1. Mammals are insensitive to a constant phase shift at high frequencies.

2. The actual dolphin whistles are often asymmetric in time, instead of symmetric, as specified by Altes and Skinner.

3. The whistles sometimes display rapid fluctuations in frequency and amplitude, relative to the theoretical versions.

These objections are satisfied by the following arguments:

1. Phase insensitivity in mammalian audition does <u>not</u> apply to <u>relative</u> phase shifts between signals <u>at</u> two different frequencies, and the relative phase shift of a signal and its harmonic, $2b\log s - b\log s$, should be audible.

2. The formulation in Altes and Skinner can be generalized to yield asymmetric signals with no range-Doppler coupling.

3.   Rapid, zero-mean frequency and amplitude fluctuations
     are "averaged out" by the velocity estimator, and
     have little effect upon the estimates.

The next step is to perform behavioral experiments to determine the jnd in scale factor for scaled, linear-period-modulated
signals with harmonics. Are these minimum perceptible scale differences small enough to be useful for sonar/navigation?

REFERENCES

Altes, R. A., and Skinner, D. P., 1977, Sonar velocity resolution
          with a linear-period-modulated pulse, J. Acoust. Soc. Amer.,
          61:1019.
Lilly, J. C., 1963, Distress call of the bottlenose dolphin:
          stimuli and evoked behavioral responses, Science, 139:116.

TARGET RECOGNITION VIA ECHOLOCATION BY <u>TURSIOPS</u> <u>TRUNCATUS</u>

Whitlow W. L. Au and Clifford E. Hammer, Jr.

Experiments were conducted to investigate the target echo recognition and discrimination behavior of <u>Tursiops</u> <u>truncatus</u>. Two specific hollow aluminum cylinders (3.81 cm and 7.62 cm O. D.) and two solid coral rock (3.81 cm and 7.62 cm diameter) cylinders, all 17.8 cm in length, were used as standard baseline targets. The coral rock targets were constructed of coral pebbles encapsulated in a degassed epoxy mix. Illustrated in Fig. 1 is the experimental pen with the two target funnels located at ranges of 6 and 16 m in front of the dolphin's pen. The funnels were used to present the targets into the water at approximately the same location at each range. The two-alternative, forced-choice technique was used with the targets presented successively in a random sequence. The animal was required to echolocate on the targets and respond to paddle A for the aluminum and paddle B for the coral rock standards. After near errorless performance with the standard targets was achieved, probe sessions were conducted to investigate differential reporting of probe targets varying in structure and composition. All of the probe targets were cylinders of 17.8 cm length. Two probe targets were used in each probe session and only 8 of the 64 trials of the session were used for probe trials, 4 for each probe target.

Fig. 1. Experimental Configuration.

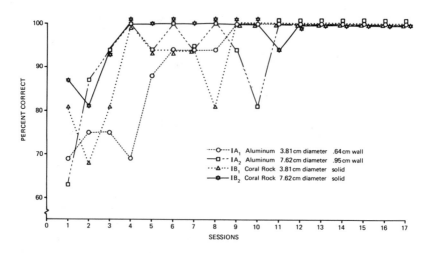

Fig. 2.   Acquisition curves for the baseline targets.

RESULTS

    The initial acquisition of the baseline standard targets are
shown in Figure 2.   The dolphin reached asymptotic performance with
near errorless behavior after session 11.   For experiment I, a general discrimination test was made with one rock, one PVC, two solid
and two hollow aluminum, and two corprene cylinders as probe targets.
Initially the dolphin classified 3 of the aluminum probes with the
standard aluminum targets but eventually classified all of the probes
with the coral rock standards, after the target range was moved from
6 to 16 m and then back to 6 m.   The dolphin began to focus in on
the salient features of the two aluminum standards and reject all
other targets as being "not A".   In other words, the dolphin became
a null detector and searched for the presence or absence of the aluminum standards.   In order to further test the dolphin's discrimination capability, hollow aluminum probes with the same outer diameters but different wall thicknesses as the aluminum standards were
used in experiment II.   The results showed that the dolphin could
reliably discriminate wall thickness difference of 0.16 cm for the
3.81 cm O. D. targets, and 0.32 cm for the 7.62 cm O. D. targets.
In experiment III, the dolphin's ability to discriminate material
composition was tested using bronze, steel and glass probes that
had the same dimensions as the aluminum standards.   It was found
that the animal could discriminate the bronze and steel targets from
the aluminum but classified the glass probes with the aluminum standards.

Fig. 3.   Acoustic measurement results.

DISCUSSION

     In order to determine the acoustic cues used by the dolphin,
all of the targets were examined acoustically, using a simulated
dolphin echolocation signal.   The results for the small cylinders
used in the material composition experiment are shown in Fig. 3.
The envelopes of the matched filter response with the reference
signal being the outgoing signal, are included in the figure.   The

Fig. 4.   Frequency spectra for the small aluminum standard.

matched filter model indicated that the arrival times of the secondary echo components for the aluminum and glass cylinders were very
similar, whereas the arrival times for the bronze and steel cylinders were quite different than for the aluminum. This indicates
that time of arrival cues may have been the primary cues used by
the animal in discriminating the targets. The arrival time difference between the first and second echo components probably resulted
in a time separation pitch, since these two echo components are
highly correlated. The spectrum for the first two echo components
is shown in Fig. 4, overlapping the total epectrum for the small
aluminum standard. Note how well the total spectrum is described
by the spectrum for the first two echo components. Furthermore,
such a rippled spectrum will introduce a time separation pitch that
should be perceived by the dolphin. The echoes were also analyzed
with a contiguous constant-Q filter bank of ten filters using a Q
of 13.8 to simulate the dolphin's auditory system. The results of
this analysis indicated that the bronze probe was more similar to
the aluminum than was the glass probe. However, the dolphin's performance indicated otherwise.

SPHERE–CYLINDER DISCRIMINATION VIA ECHOLOCATION

BY TURSIOPS TRUNCATUS

Whitlow W. L. Au, Ronald J. Schusterman,

and Deborah A. Kersting

An experiment was conducted to investigate the ability of
T. truncatus to discriminate between spherical and cylindrical targets by echolocation. The targets were constructed out of polyurethane foam to eliminate any internal reflections, and their dimensions were chosen so that the spheres and cylinders had overlapping target strengths. All targets were acoustically examined with
dolphin-like echolocation signals. The results of the acoustic
measurements performed under free-field conditions are shown in
Table 1, along with the correlation coefficients between the echoes.

Two spheres and two cylinders were used in each 64 trial session except during the first portion of the initial acquisition
phase (see first panel of Fig. 1). The targets were placed on an
assembly which allowed them to be submerged (1.12 m depth) at approximately the same location; 6 meters in front of the animal's
pen. The two-alternative forced-choice technique was used with the
targets presented successively in a random sequence. During the
initial acquisition phase, the dolphin was allowed to swim freely
while echolocating. Eventually, the animal was required to station

Table 1. Results of the Acoustic Examination of the Targets

| | Target | Target Strength Measurements | | | Correlation Coefficients Between Target Echoes | | |
		Diameter	Length	TS	S_1	S_2	S_3
Spheres	S_1	10.2 cm	–	–32.1 dB	1.00	0.98	0.99
	S_2	12.7 cm	–	–31.2 dB	0.98	1.00	0.99
	S_3	15.2 cm	–	–28.7 dB	0.99	0.99	1.00
Cylinders	C_1	1.9 cm	4.9 cm	–31.4 dB	0.94	0.97	0.96
	C_2	2.5 cm	3.8 cm	–32.3 dB	0.98	0.97	0.98
	C_3	2.5 cm	5.1 cm	–28.7 dB	0.91	0.93	0.92
	C_4	3.8 cm	3.8 cm	–30.1 dB	0.95	0.97	0.98
	C_5	3.8 cm	5.1 cm	–27.6 dB	0.98	0.98	0.99

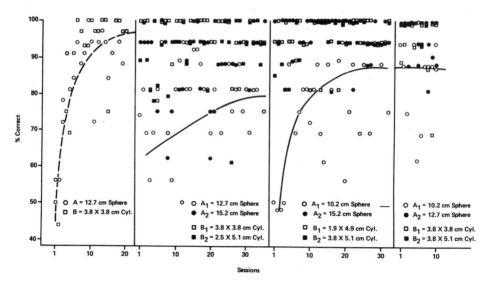

Fig. 1. Initial acquisition results with free-swimming animal.

in a hoop and only echolocate upon the introduction of an audio cue (see Schusterman, Kersting and Au, this volume). The hoop station was introduced to prevent the animal from receiving spatial cues by echolocating at different depths and thus respond on the basis of the aspect dependence of target strength for the cylindrical targets.

Results: The results of the four parts of the initial acquisition phase are shown in Fig. 1. The lines in the figure are curves fitted

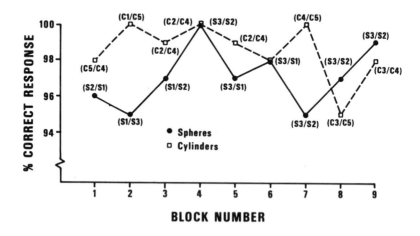

Fig. 2. Summary of the dolphin's in-hoop performance.

to the results of the smallest sphere. The curves show early acquisition with decrements in performance and reacquisition as a function of introducing different cylinders and spheres. Asymptotic
performance was considered to have been achieved as shown in the
last panel of Fig. 1. The results of the dolphin's performance
with the use of the hoop station are illustrated in Fig. 2. Each
block consisted of at least 5 sessions and at most 36 sessions.
A comparison between Fig. 1 and 2 indicates that the dolphin's
ability to differentiate between spheres and cylinders was superior
for the in-hoop situation than for the free-swimming situation.

Discussion: The acoustic examination of the targets showed that
the echoes were very similar, as can be seen in the high correlation
coefficient values of Table 1. It appears unlikely that the dolphin
made the discrimination based on differences in the measured reflective characteristics of the targets made in the free field. Further
acoustic measurements were made in which the planar transducer was
swept in azimuth to simulate the animal scanning across the targets.
These measurements did not provide any information as to how the
animal was making its discrimination. Finally, the transducer-target

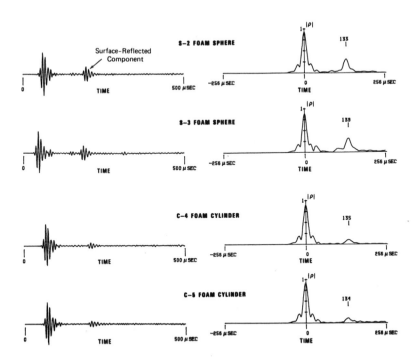

Fig. 3. Approximation of echoes received by the dolphin.

geometry was set up to approximate the animal's experimental condition.   Examples of the results of this acoustic measurement are
shown in Fig. 3.   Also shown in the figure are the envelopes of
the matched-filter responses, with the filter matched to the outgoing signal.   It was found that the spheres introduced a larger
surface reflected component to the echoes than did the cylinders.
Furthermore, the surface reflected component of the cylinder echoes
fluctuated in amplitude as wind and wave motion caused movements
in the targets.   The surface reflected component of the echoes
seemed to have provided the eventual cue for the dolphin's differentiation between the targets.   Superior performance in the hoop
station over the free-swimming situation is consistent with the
notion that the dolphin used the surface reflected component of
the echoes as the primary cue.

FREQUENCY SELECTIVITY OF CONSTANT LATENCY NEURONS IN THE INFERIOR COLLICULUS OF MEXICAN FREE-TAILED BATS

Robert Bodenhamer, George Pollak, and David Marsh

There are neurons in the inferior colliculus of the Mexican free-tailed bat, Tadarida brasiliensis mexicana, which are characterized by a highly stable discharge latency to frequency modulated (FM) signals. These phasic constant latency responders (pCLRs) are capable of accurately encoding the time interval between simulated orientation cries and echoes, and thus appear to be well suited for encoding target range.

Each pCLR is triggered by a particular frequency component of a downward sweeping FM burst. This can be shown by presenting bats with FM signals having the same modulation depth (typically sweeping from 50 kHz to 25 kHz, mimicking the orientation cries of Tadarida), but having different durations (Fig. 1). For each pCLR, the discharge latency increases as the signal duration is lengthened. The difference in latency from one duration to another consistently corresponds to the shift in the temporal position of a particular frequency component within the signal. Since a pCLR "follows", or is excited by, this component, we have labelled it the cell's excitatory frequency (EF). To date, units have been found with EFs ranging from 24.5 kHz up to 46 kHz.

The pCLRs' tight synchronization to very limited frequency ranges within a modulated signal might suggest that they are more sharply tuned to constant frequency signals than are other collicular neurons. That this is not the case is demonstrated by the representative tuning curves shown in Fig. 2. As a rule, tuning curves of pCLRs and of other collicular cells are indistinguishable with Q-10 dB values of from 3-11 being most common for each group.

Sensitivity to a particular EF represents a high degree of frequency selectivity, rivaling that of the sharply tuned filter neurons of the long-CF/FM bats such as Rhinolophus ferrumequinum. However, in Tadarida, this selectivity is apparently not due to mechanical specialization of the cochlea, but rather to complex neural circuitry.

In conclusion, the pCLRs' sensitivity to a particular EF is valuable to FM bats in at least two ways. First, it leads to the tight firing registration enabling the pCLRs to serve as pulse-echo time markers, and thereby encode target range information. Second, by restricting the response area of each pCLR to a small frequency band, it may enable a population of pCLRs to encode echo spectral

Fig. 1. Responses of two pCLRs which illustrate sensitivity to a
         particular EF. Top portion of both A and B show FM pulses
         of three durations. EAch dot column represents action
         potentials elicited by 16 stimulus presentations at a
         given duration. The first column in each half of the
         figure was produced in response to the 4.0 msec dura-
         tion signal, the second to the 8.0 msec duration sig-
         nal, and the third to the 12.0 msec duration signal.
         The dot columns have been aligned directly beneath the
         FM signals to show how they line up with the temporal
         position of the proposed EF in each sweep (indicated
         by arrows). Each dot column's actual mean response
         latency and standard deviation of mean latency is shown
         below that dot column (in msec). In A, the FM signals
         sweep from 40 kHz to 20 kHz and the calculated EF is
         25.0 kHz. Unit T10-12-88, 77 dB SPL. In B, the sig-
         nals sweep from 50 kHz to 25 kHz and the calculated EF
         is 35.7 kHz. Unit T82-5, 58 dB SPL.

characteristics, and thus to convey information about the physical
nature of targets as well as target range.

(Supported by NIH Research Grant NS-13276.)

Fig. 2.   Representative tuning curves of pCLRs (below) and
of other units for which response latency was not
constant (above). The important feature is the
similarity in the Q-10 dB values of the two groups.

STRUCTURAL ADAPTATION IN THE COCHLEA OF THE HORSESHOE BAT FOR THE

ANALYSIS OF LONG CF-FM ECHOLOCATING SIGNALS

Volkmar Bruns

The auditory system of the greater horseshoe bat, Rhinolophus ferrumequinum, is sharply tuned to the echo of the constant-frequency (CF) segment of the orientation signal of about 83 kHz (Neuweiler, 1970). The tuning is performed within the cochlea and is correlated with a number of specialized features (Bruns, 1976a and 1976b; Schnitzler et al., 1976; Suga et al., 1976).

The functional morphology of the cochlea of the horseshoe bat, studied last years in our Frankfurt group, is integrated in the review of G. Neuweiler, in this volume. This paper presents a mechanism creating the sharp cochlear tuning around 83 kHz by a transversal vibration component of the basilar membrane (BM) and its associated structures which is superimposed onto the longitudinal travelling wave.

The specialized basal region of the horseshoe bat cochlea where the CF-component is analyzed is shown in Fig. 1,B. In comparison to the cochlea of other "high frequency" mammals adapted to the analysis of a broad frequency band (e.g., FM-bats; see Fig. 1,A) the horseshoe bat demonstrated following specializations: (1) A displacement of the feet of the outer pillar from the thin center of the BM onto the inner thickening resulting in a pliable connection between the inner and outer segment of the BM. (2) The spiral lamina, with the exception of its inner margin, is almost free of radial fibers. Between the thickening of the secondary spiral lamina and the outer bony wall there is a small bone lamella. Thus the BM is only loosely attached to the outer bony wall. (3) The outer segment of the BM is connected with the thickening of the secondary spiral lamina by filaments of the inner margin of the spiral ligament. This part of the spiral ligament contains fibers in a longitudinally oriented bundle. This is interpreted that in between the loose structures in point (1) and (2) a rigid vibration unit exists enclosing the outer thickening of the BM and the secondary spiral lamina with the possibility of rotation around the axis of the longitudinal fibers in the spiral ligament.

In capacitive probe measurements of the vibration behavior of the BM and associated structures (joint studies with J. P. Wilson) two main results are found in the specialized region of the horseshoe bat cochlea. (1) The adjacent structures of the BM, the primary spiral lamina and the inner margin of the spiral ligament vibrate 35 and 20 dB lower than the BM. However, the thickening of

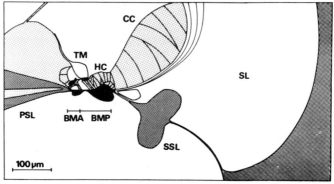

Fig. 1.   Cross sections of the organ of Corti of two high-frequency
          hearing mammals emphasizing the hydro-mechanically import-
          ant structures. Adaptations for the analysis of (A) a
          broad-band echolocating signal (e.g., FM bats) and (B)
          a small-band echolocating signal (CF bat, Rhinolophus
          ferrumequinum). BMA: Basilar membrane, arcuate zone;
          BMP: Basilar membrane, pectinate zone; C: Tunnel of
          Corti; CC: Claudius' cells; HC: Hensen's cells:
          PSL: Primary spiral lamina; SSL: Secondary spiral lamina;
          SL: Spiral ligament; SV: Stria vascularis; TM: Tector-
          ial membrane.

the secondary spiral lamina (see Fig. 1,B) vibrates with an unexpected high amplitude, only 5 dB lower than the BM. (2) The inner
and outer segments of the BM are in phase up to about 80 kHz, then
diverge and pass through an antiphase at the location of 83 kHz,
beyond which the phase difference increases further until it reaches
one full cycle for about 86 kHz.

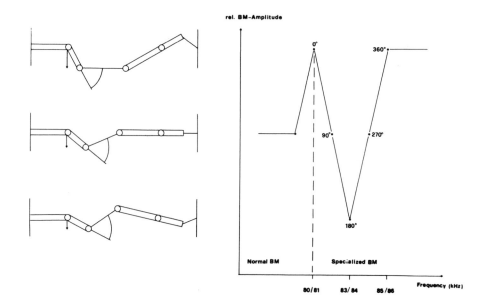

Fig. 2. (A) Model of the basilar membrane and associated structures of the horseshoe bat at different phase relations.
(B) The curve of mechanical frequency sharpening.

The fine structure and the vibration behavior are integrated
in a model of mechanical frequency sharpening (Fig. 2,A). In the
vibration modes of Fig. 2,A the angle β, a measure for the transversal vibration component of the receptor site of the organ of
Corti, is constant. The amount of vertical BM-amplitude (arrow)
required to achieve in the same angle depends on the phase relations
between the inner and outer vibrating units (double lined). In
phase (above) a high BM-amplitude is needed. The amplitude at 90°
phase divergence (middle) is medium and small at 180° (antiphase;
below). 270° corresponds to 90°, and 360° is in phase again. The
amplitude of the normal BM, in the cochlear region below 80 kHz,
is similar to the middle mode in the specialized BM. The curve
(Fig. 2,B) in which the relative BM-amplitude of the model in Fig.
2,A is correlated with the frequencies, fairly well matches to those
of the neuronal audiogram, around the frequency of the CF-component
of the echolocating signal.

SIMILARITIES IN DESIGN FEATURES OF ORIENTATION SOUNDS USED BY

SIMPLER, NONAQUATIC ECHOLOCATORS

Edward R. Buchler and Andrew R. Mitz

There are some remarkable similarities in the echolocation pulses of several species of nonaquatic echolocators that are relatively simple or conservative in terms of their echolocating abilities. I would like to illustrate this and offer speculations as to why this particular pattern of sounds has evolved repeatedly and, in many cases, independently. In the primary examples given, the echolocation signals consist of brief, unstructured sounds having a rapid onset and broad bandwidth. In addition, they often or always occur as paired pulses.

Supporting Data. Recently, Thurow and Gould (1977) presented strong circumstantial evidence that the broadband pulse pairs of a caecilian (probably Dermophis septentrionalis) are used for orientation. Each sound lasts ca. 20 ms and extends from <500 Hz to >25 kHz. The onsets of members of a pair are separated by ca. 200 ms.

The vagrant shrew, Sorex vagrans, emits paired pulses while echolocating (Buchler, 1976). Each pulse lasts from <1 ms to 2.5 ms. The members of each low intensity pulse pair are ca. 8-10 ms apart and span a frequency range of at least 20-60 kHz (Fig. 1). The broadband ultrasonic clicks of the short-tailed shrew, Blarina brevicauda, also are temporally associated most commonly as pairs (Tomasi, pers. comm.).

Three, and probably four, genera of tenrecs (Tenrecidae) echolocate using tongue clicks of 5-17 kHz which last only 0.1-3.6 ms. Again, the most common temporal configuration is pairs (Gould, 1965).

Chase (this volume) has recorded the very brief, low intensity clicks used by echolocating laboratory rats (Rattus norvegicus). They commonly are emitted in pairs or trains of pairs.

Three of the four most studied species of echolocating cave swiftlets (Aerodramus = Collocalia) use broadband clicks of uncertain origin. The frequencies range from 2-15+ kHz. There are ca. 15 ms between members of a pair (Griffin and Suthers, 1970; Fenton, 1975; Medway and Pye, 1977).

Finally, a few species of the megachiropteran, Rousettus, echolocate by using paired tongue clicks ranging from 10 kHz to as high as 100 kHz. The interclick interval is ca. 20-30 ms, each click being only a few milliseconds long (see Sales and Pye, 1974).

Fig. 1.   Two pairs of sonar pulses emitted by a vagrant shrew,
          Sorex vagrans.

Fig. 2.   Three pairs of FM sweeps emitted during the first flight
          of a little brown bat, Myotis lucifugus.

In addition, there are other species, some related to those above, that use broadband clicks that are not commonly paired; e.g., some shrews, a cave swiftlet and the oil bird, Steatornis. There appear to be correlates of this kind of broadband signal in some of the anatomically more primitive Microchiroptera. Megaderma, Rhinopoma and Nycteris use sounds of only a few milliseconds which have 3-5 harmonics and virtually no frequency modulation (Simmons, 1979). There might well also be ontogenetic correlates of the paired pulse phenomenon in the development of echolocation in some vespertilionids. The little brown bat, Myotis lucifugus, uses paired cruising and approach pulses during the first few nights of flight. The FM sweeps of a given pair are ca. 20 ms apart (Fig. 2) (Buchler, 1979). Although this pairing is quickly lost as the echolocation system of M. lucifugus matures, Woolf (1974) believes that a very similar form of pairing persists even into adulthood in the big brown bat, Eptesicus fuscus, and has its origin in a double note call. Pairing also occurs in the early flights of the cave bat, Myotis velifer (Brown, pers. comm.).

Discussion. It is obvious that very short pulses and their related wide bandwidth provide good range resolution and localization characteristics; but why the recurrence of pairing? One important adaptive advantage of multiple pulse sonar is that, by integration of successive similar pulses, the signal to noise ratio is increased. Regardless of the technique the individual uses to scan its environment, bounds can be set on the effectiveness of its integrating system. Consider a poor, but not worst, case near the lower bound. For a simple, or primitive, uniformly weighted, noncoherent integrator:

$$(S/N)_n \approx \sqrt{n}(S/N)_1$$

This illustrates that the integration of multiple pulses has rapidly diminishing returns associated with it (Skolnik, 1962). A second pulse increases the S/N by 3 dB, but a third by only 1.8 dB, etc. Thus, the paired pulse may be an energetically efficient strategy by which terrestrial insectivores, for example, with low intensity pulses can increase the S/N ratio while still retaining the advantages of short duration and broad bandwidth. Those species that use higher intensity pairs and move more quickly, such as the cave swiftlets, may be able to derive relative velocity information by comparing the interval between members of the echo pair with that of the emitted pair. The interval will be subject to the Doppler effect even though a given pulse is Doppler tolerant.

   It should be noted that the paired click structure is also used
commonly by several cetaceans; e.g., Tursiops (Norris et al., 1967),
Sotalia (Norris et al., 1972) and Delphinapterus (Evans, pers. comm.).
Although Mackay (1967) suggests that the second click may function
to greatly decrease ringing in the sending or receiving apparatus,
the explanations offered here may be more generally applicable.

REFERENCES

Buchler, E. R., 1979, The development of flight, foraging and echo-
       location in the little brown bat, Myotis lucifugus, Behav.
       Ecol. Sociobiol., (in press).
Simmons, J. A., 1979, Phylogenetic adaptations and the evolution of
       echolocation in bats (Chiroptera), Proc. 5th Internat. Bat
       Res. Conf., (in press).

RAT ECHOLOCATION: CORRELATIONS BETWEEN OBJECT DETECTION AND

CLICK PRODUCTION

Julia Chase

The laboratory rat Rattus norvegicus uses a variety of sensory cues to guide its nocturnal movements. Vision, olfaction, touch, and a well-developed spatial memory all aid the rat in orientation and obstacle avoidance. A half century ago, Shepard (1929), Lashley, (1929), and Honzik (1936) found that hearing is also important for orientation when a rat is deprived of visual cues. These experiments were published a decade before the discovery of echolocation in bats and 25 years before the detection of ultrasonic vocalizations in rats (Anderson, 1954) and therefore aroused little interest.

Riley and Rosenzwieg (Rosenzwieg et al., 1955; Riley and Rosenzwieg, 1955) demonstrated (1) that blind rats on an elevated Y-maze could determine which of two runways was obstructed by a barrier, and (2) that this discrimination broke down when the rat was deafened or alternatively if the barrier was rotated 45°. Bell et al., (1971) confirmed these findings and demonstrated that white noise of 20 dB also abolished the discrimination. Though both studies concluded the rats were echolocating, neither found evidence of sound production other than occasional scratches or sniffling noises. The results are thus open to two interpretations: (1) the rats are actively echolocating but emitting sounds too faint to be detected by the equipment, or (2) the rats are passively listening for differences in the ambient noise levels on the two runways. The published data remain as inconclusive as they were when Neuweiler reviewed the topic at the first of these symposia 13 years ago (Neuweiler, 1966).

Work currently in progress in this lab has confirmed Riley and Rosenzwieg's behavioral findings. With the improved equipment currently available, we were able in addition to detect brief, faint, broadband clicks that were not noted by the earlier investigators. The following progress report presents our results to date which suggest that these clicks are used by the rats as part of an active echolocation system.

Blind, water-deprived Long Evans rats were trained to seek water on an elevated Y-maze. One arm of the Y was open and the other was blocked by a 22 x 27 cm aluminum barrier suspended above the runway at a distance of about 35 cm from the rat's starting position. Sounds were monitored during testing with a Holgate ultrasonic detector tuned to 45 kHz. Preliminary analysis of the sounds was done in D. R. Griffin's laboratory using the more sensitive McCue and Bertolini ultrasound detector (1964) and a LockHeed high speed tape recorder. Taped sounds were photographed with a Grass Camera from a Tektronix oscilloscope.

Eleven rats learned to choose the open runway, scoring 75% or better on 50 consecutive trials. Five of these animals were then ear-plugged with cotton and vaseline and retested. Rats with their ears plugged were unable to select the open runway but when the plugs were removed after 5 days (50 trials) the discrimination was restored.

Detection of the sounds was difficult even with the Holgate and we believe that we were probably not detecting all of the rat's clicks. Nonetheless, there was a clear relationship between the production of clicks and the rat's performance in the maze. First, during training as the rat's performance improved, the rate of sound production increased. Second, while deafening sometimes caused an initial increase in clicks, they soon decreased or stopped altogether. Third, in all eleven animals, the rats were more successful on trials when clicks were detected than on silent trials.

Preliminary analysis of the clicks (Fig. 1.) shows that they are very brief, low intensity, broadband clicks of only 4-8 cycles. In the best recordings made with a carefully adjusted McCue-Bertolini detector, the unfiltered signals are about 3 times the noise level of 35 dB SPL. Filtering out all sounds below 40 kHz and above 50 kHz improved the signal-to-noise ratio.

Fig. 1.   Short train of clicks produced by a blind rat. Recorded with an unfiltered McCue-Bertolini detector and Lockheed tape recorder.

Clicks appeared both singly and in pairs and were sometimes recorded in trains of 30 or more. In these trains, paired clicks comprised about half the sounds produced and were reminiscent of the paired clicks of other echolocators such as shrews, cave swift-lets, tenrecs and Rousettus (see Henson, Buchler, and Mitz, this volume). Clicks also appeared more frequently (or louder?) when the rats were placed in a novel situation such as walking on a yardstick.

ACKNOWLEDGEMENTS

This research was supported by Barnard College and NIMH grant N° 1 RO3 MH 30169.

REFERENCES

Bell, R. W., Noble, M. E., and Davies, W. F., 1971, Echolocation in the blinded rat, Percep. Psychophysics, 10:112.
Dunning, D. C., 1975, Orientation by cave-dwelling pack rats, Neotoma floridana magister in the dark, paper presented at the XIV International Ethological Conference, Parma, Italy.

SIGNAL DESIGN FOR MATCHED FILTER DETECTION IN A REVERBERATION-

LIMITED ENVIRONMENT: APPLICATION TO CETACEAN ECHOLOCATION SIGNALS

Martine Decouvelaere

This study is concerned with cetacean echolocation signals as optimum signals for Doppler-tolerant, matched filter detection of targets in the presence of reverberation.

A numerical method is derived to find optimum signals in the following situation: a real, fixed-energy signal $s(t)$, $t\epsilon\{o,T\}$ is emitted. The echo at the receiver is the sum of the known target echo $c(t) = \sqrt{\eta_c}s\{\eta_c(t - \tau_c)\}$, white noise, and spurious echoes from scatterers. The reverberation echo is modeled as a gaussian random process of known expectation and covariance. The covariance is characterized by the response $r(t)$ of an elementary scatterer to $s(t)$ and by the reverberation density function. The receiver is a filter matched to $c(t)$.

The optimum signal, i.e., which leads to a maximum output signal-to-interference ratio, must then satisfy a nonlinear integral equation involving the wideband cross-ambiguity function of $s(t)$ and $r(t)$. This problem can be solved numerically (according to Altes, 1971) by writing $s(t)$ as a linear combination of M orthonormal basis functions of $L^2 \{o,T\}$. Conjectured optimum signals are then found with several choices of reverberation features and basis functions, including Doppler-invariant density functions (in order to yield Doppler-tolerant signals), Rayleigh approximation for $r(t)$, and basis functions derived from Angle Prolate Spheroidal functions.

The signals obtained compare well with experimental cetacean echolocation signals with respect to the above performance criterium as well as Doppler-tolerance and displayed form.

Altes, R. A., 1971, Methods of wideband signal design for radar and sonar systems, Fed. Clearinghouse N° AD 732-494.

VARIATIONS IN THE CHARACTERISTICS OF PULSE EMISSIONS OF A

TURSIOPS TRUNCATUS DURING THE APPROACH PROCESS AND THE ACOUSTIC

IDENTIFICATION OF DIFFERENT POLYGONAL SHAPES

Albin Dziedzic and Gustavo Alcuri

An alternative choice test was used to determine the ability of a 4-year-old Tursiops truncatus to discriminate acoustically a control ring (the positive stimulus) from a regular polygonal frame which was varied between tests. A study was then made of the physical characteristics of the acoustic emissions of the animal during the approach process and the identification of the positive stimulus which the animal was required to bring to the trainer in each trial.

The direct distance between the starting point of the animal and the objects was 8 m. Two acoustical arrays consisting of 10 hydrophones were used to record the signals. These were positioned so as to prevent interference between direct signals of <100 μs duration emitted during approach and their associated first order echoes. Signals were recorded during the complete approach process on a 14 track tape recorder and were synchronized with video recording during the last part of the trajectory (approximately 1.5 m). This allowed us to obtain projections of the trajectory at the end of the approach and to correlate the position of the subject at a specific time with its corresponding acoustic emissions. Discrimination performances are summarized in the following table:

N	3	4	5	6	8	10	12
%	100	100	78	72	77	70	85

where N represents the number of sides of the polygon tested and % the relative number of positive responses obtained with the corresponding polygon. The animal's discrimination performances are better than 70% chance level for all objects.

The video tapes showed three components in the movements of the animal: a) lateral sliding, b) rolling of the body around its longitudinal axis, and c) rotating movements of the head. For certain positions of the animal in its trajectory, an estimation was made of the angular parameters (azimuth θ, site ϕ), as well as of the rolling angle ψ, defining the orientation of the rostrum with respect to the hydrophone. Fig. 1 summarizes the trajectory and associated acoustic data obtained in a test using the pentagon.

881

Fig. 2.   Results of tests using
          a pentagon, an octagon,
          and a dodecagon. The
          resolution ($\simeq 1/\tau_o$) and
          the other three para-
          meters (F_c, B_e, F_{am})
          seem to increase with
          the number of sides of
          the polygon.

Fig. 1.   A summary of trajectory and associated acoustic data
          obtained during a test using a pentagon. Only four of
          the hydrophones are shown (H5, H7, H9, H10).
          Upper left: Actographs of the total group of signals
          emitted during the trial and capted by each of the
          hydrophones, showing the positions of the animal during
          selected portions of the trajectory.
          Upper right: Oscillographs of signals capted by each
          hydrophone at these points.
          Lower right: Associated angular parameters.

A study was made of the group of signals emitted during each
approach and of individual signals emitted at particular positions
in the trajectory.   For each group of signals, five parameters were
considered: 1) the total duration of signal emission,   2) the total
number of signals emitted during the trial,   3) the duration of the
part of the trial during which the frequency repetition rate (FRR)
increased approximately according to $e^{-\alpha t^2}$,   4) the number of signals
emitted during this portion of the trial,   5) the maximum frequency
repetition rate recorded in the trial.   The results indicate that
only the number of signals emitted during the first part of the
trial, prior to the characteristic increase in FRR, is proportional
to the number of sides of the polygon, which is probably related to
the difficulty of the discrimination task.

For each signal emitted at a specific point within a trial,
four parameters were considered: 1) the first null value of the
autocorrelation function,   2) the frequency ν_0 of the spectral
density function $\gamma(\nu)$, defined as:

$$F_c = \nu_0 = \frac{\int \nu \gamma(\nu) d\nu}{\int \gamma(\nu) d\nu}$$

3) the equivalent bandwidth of $\gamma(\nu)$, defined as:

$$B_e^{-1} = \frac{\int |\gamma(\nu)|^2 d\nu}{|\int \gamma(\nu) d\nu|^2}$$

4) the dominant frequency F_{am} of $\gamma(\nu)$.   Results are shown in Fig. 2.

Signals emitted during roughly the first half of the approach
(animal-to-object distance > 4 m) are independent of polygon shape.
They are similar for all tests showing high distance resolution and
a broadband spectrum.   Signals emitted during the second half of the
approach (animal-to-object distance < 4 m) differ according to object shape, showing in general lower values for the four parameters
than those emitted during the first half.

NEURAL MECHANISMS FOR TARGET RANGING IN AN ECHOLOCATING BAT

EPTESICUS FUSCUS

Albert Feng, James A. Simmons, Shelley A. Kick

and Beatrice D. Lawrence

We have concluded a series of behavioral, anatomical and physiological experiments in order to gain a better understanding of the neural basis of target ranging in Eptesicus fuscus. Behaviorally, we were interested in learning whether or not binaural integration is essential for ranging. We trained two bats to perform range discrimination using two-choice paradigms (Simmons, 1973) until they reached a criterion of 85% correct responses. We then observed their performance under intact conditions and when one ear was occluded. Occlusion was effected by inserting silicone grease to fill the ear canal of the selected ear. The performance of these two bats under monaural and binaural conditions is shown in Table 1. We found that the performance under monaural conditions was not significantly different from that of intact conditions (p<0.05, Hoel test for the difference of two means). Thus, bilateral interactions, which are essential for encoding directional information about the target, are not required for ranging in this species. Instead, time or intensity information from target echoes is more important for ranging.

Our physiological recordings from the central auditory structures of Eptesicus demonstrate that pulse-echo time is likely to be the essential information for these bats. We recorded single unit activity from the inferior colliculus, intercollicularis nucleus and the auditory cortex of lightly anesthetized or awake bats, studying the unit responses to paired acoustic stimuli in the free-field simulating pulse-echo sounds that the bat would have heard during prey-catching. We found that a number of neurons in the intercollicularis nucleus and the auditory cortex were selectively responsive to complex acoustic features of the simulated pulse-echo sounds, i.e., they responded only when the sounds were paired and were frequency-modulated, and the interval between the pulse and the echo was not so important in determining the neural responses. These neurons seemed to be "tuned" to a narrow range of pulse-echo delays (Feng, Simmons, and Kick, 1978). Different delay-tuned neurons showed response preferences for varying pulse-echo delays of from 4.5 to 24 msec as shown in Table 2. Thus, target range could be encoded by a place mechanism in the central nervous system as demonstrated most recently in the auditory cortex of Pteronotus parnellii rubiginosus (Suga and O'Neill, this volume).

Table 1.   Correct Responses for Target Range Discrimination

Bat W$_2$	Number of Trials	% Correct Response
Monaural	51	80%
Binaural	134	78%
Bat G		
Monaural	62	73%
Binaural	286	83%

Table 2.   Response Selectivity to Pulse-Echo Delays

Unit #	Optimal Echo Delay	Response Range	
12-1	11 msec	7-15	msec
12-2	9 msec	7.5-12	msec
12-6	10.5 msec	10-12	msec
12-8	12.5 msec	6-16.25	msec
12-9	6.75msec	4.7-10	msec
13-4	7 msec	4.4-12	msec
13-5	9 msec	5-15	msec
15-6	17 msec	15-20	msec
15-7	13 msec	10-15	msec
15-8	24 msec	20-30	msec

Fig. 1.   Uptake of ^{14}C-2-Deoxyglucose in the intercollicularis
nucleus following stimulation with paired pulse-echo
sound simulating a target at a distance of 2 meters.

The involvement of various neural centers in target ranging was investigated using the labed 2-Deoxyglucose functional mapping technique. In one immobilized bat, we injected about 12 μCi of C^{14}-2-Deoxyglucose intraperitoneally. Subsequently, the bat was exposed to paired acoustic stimuli simulating a target at a distance of 2 meters for 45 minutes. The bat was sacrificed and the brain extracted, frozen, and sectioned. The labeled areas in the brain were observed from X-ray films exposed to sectioned tissues. We found that the inferior colliculus, intercollicularis nucleus (Fig. 1), vast areas of cortex, and lower auditory centers were heavily labeled, indicating their involvement in processing these acoustic stimuli, and perhaps their involvement in processing events related to ranging.

In summary, the results from these combined studies suggest that temporal information may be utilized by Eptesicus for target ranging, and that it is processed along the ascending auditory pathway. Some neurons in the intercollicularis nucleus and the auditory cortex are specialized for processing the temporal delay between pulse and echo and are presumably independent of binaural processing of directional information.

Feng, A. S., Simmons, J. A., and Kick, S. A., 1978, Echo detection and target-ranging neurons in the auditory system of the bat Eptesicus fuscus, Science, 202:645.

Simmons, J.A., 1973, The resolution of target range by echolocating bats, J. Acous. Soc. Amer., 54:157.

Suga, N., and O'Neill, W. E., 1979, Auditory processing of echoes, part II. Representation of acoustic information from the environment in the bat cerebral cortex (this volume).

HUNTING STRATEGIES AND ECHOLOCATING PERFORMANCE OF BATS -

QUANTITATIVE BEHAVIORAL LABORATORY ANALYSIS

J. Fiedler, J. Habersetzer, and B. Vogler

Two bat species, Megaderma lyra and Myotis myotis, were tested behaviorally for the efficiency of their echolocation system. Bats were conditioned to discriminate between two targets, plastic plates with identical pattern of holes of different depths. M. lyra performed 75% correct choices, with hole depth of 8 mm (+) against 6 mm (-). M. myotis succeeded in 87% of flights with depth of 4 mm (+) against 3.2 mm (-). However, the combination of 8 mm and 7.2 mm yields only 68% correct flights. Since the discrimination threshold depends not only on the depth difference, but also on the absolute depths, it is suggested that the bat evaluates small differences in the power spectra of the echo.

How relevant are these performance tests for interpretations of the natural behavior of bats? Both bat species cease to echo-locate under certain hunting conditions. In about 46% of 35 trials M. lyra catches living mice without emitting any sound, just by passively listening to the noise generated by the moving prey (Fig. 1a). In the other 54% only sparse echolocation of 3 to 5 ultrasounds per approach is observed. On the other hand, dead mice are detected by means of echolocation (Fig. 1b). This situation is characterized by a typical "paired sound pattern", that was only found in situations where the bat has no other choice but to echolocate.

The echolocation flight of the insectivorous M. myotis consists of 1) search flight, 2) approach, and 3) final buzz. But M. myotis may also switch to passive orientation during hunting on the ground. The bat forages at a distance of 10 to 30 cm without emitting orientation calls. Dead mealworms are located in a 2 cm distance from a wall, and M. myotis crawls to the prey without emitting the final buzz, although the approach sound pattern is present. M. myotis can also rely on passive orientation in flight. It flies to a 6.5 x 6.5 mm mesh within a 40 x 40 cm wire grid if a scratching noise is produced behind the target. It is concluded that M. myotis abandons echolocation when the prey has to be spotted in front of a dense background. The natural prey of M. lyra are also targets on a dense background. For both species it is apparently easier to locate prey on the ground by its own sounds and noises.

Fig. 1. (a) Silent approach flight of M. lyra to a living mouse. (b) Unsuccessful approach flight of M. lyra towards a dead mouse. M₁ and M₂ indicate the positions of the microphones (M₂ is below the compartment). On the time-axis of the diagram the vertical bars (present only in Fig. 1(b)) indicate echolocation sounds recorded synchronously with the stroboscopic frames. Numbered arrows indicate the corresponding flight position shown above.

LOW-FREQUENCY RECEIVER OF THE MIDDLE EAR IN MYSTICETES AND ODONTOCETES

Gerald Fleischer

Mysticetes are known to emit sounds of low and even very low frequencies, down to the famous 20-Hz-signals. Because the mammalian ear is generally rather insensitive to low frequencies, the problem arises how these animals manage to hear them. Several mechanisms have been proposed, including the detection of resonances of the lungs. It is the purpose of this presentation, however, to show that the cetacean ear is structurally specialized for the detection of low frequencies.

Before dealing with the hearing organ itself, it is necessary to point out that mysticetes have structurally modified their entire skull in order to suppress resonances. This is mainly achieved by soft connections between the long and slender bony elements. The long rostrum is only loosely tied to the occipital part of the skull, but this is only the most prominent modification.

The entire hearing organ is not fused to the skull, but decoupled, which is a further step towards suppressing the influence of vibrations of the skull on the hearing organ. In order to stabilize the decoupled hearing organ, a lot of dense bony material has been added, much more than necessary to house the inner ear and to support the middle ear. A schematic cross section is shown on the left of Fig. 1. Although modified, the same holds true for dolphins and porpoises as well.

A biomechanical analysis of the mysticete middle ear revealed that it contains a stapes complex much like the one in other mammals. The malleus-incus complex is basically the same as in dolphins with a Tursiops-type middle ear, indicating that it is a torsional system. Although no detailed measurements have been made it is obvious from the stiffness and the mass of both subsystems that they are tuned to higher frequencies: probably the natural frequency of the stapes complex is in the range of a few kHz, and that of the malleus-incus unit at several hundred Hz. Both components are therefore not well suited as receivers for the very low frequencies mentioned.

The cetacean ear differs from that in other mammals in that it is clearly divided into two major components which are both bulky and massive. As shown on the right of Fig. 1, one is the periotic element which houses the cochlea, while the other is the tympanic bone to which the malleus is fused (a situation quite normal for most mammals). In mysticetes the Tympanic is fused to the Periotic

891

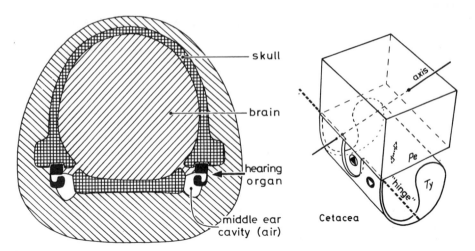

Fig. 1

left: schematic cross section through the cetacean head (lower
         jaw omitted) showing the isolation of the hearing organ.

right: schematic drawing illustrating the relation between the
         periotic bone (Pe) and the tympanic bone (Ty). The term
         "axis" refers to the rotational axis of the malleus-incus
         complex. Periotic and Tympanic can vibrate relative to
         each other around the "hinge" indicated. However, it is
         actually a flexional vibration.

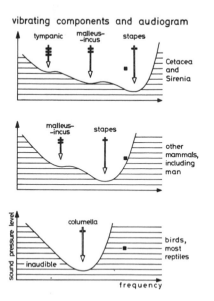

Fig. 2

simplified relation between the
vibrating elements of the middle
ear and the shape of the audiogram.
In man there are two vibrating
components: the malleus-incus
complex and the stapes. In
cetaceans the tympanic bone is
a third vibrating component. All
three subsystems combined increase
the frequency response of the ear.
(Sirenians have the same potential
but they do not seem to use it).

Fig. 1, right, and Fig. 2 are
reproduced from Fleischer (1978)
with the permission of Springer-
Verlag.

via two thin bony pedicles at the lateral side, while the medial
rim consists of a bulky mass of extremely dense bone. Because of
this configuration both elements, Periotic and Tympanic together,
vibrate relative to each other at low frequencies. The mode of
vibration can be visualized by the "hinge" indicated on the right
of Fig. 1, although it has to be emphasized that it is a flexional
vibration. (In dolphins and porpoises the situation is essentially
identical. A minor difference, which does not effect the functional
principle, is that in many of them there is no rigid fusion between
Periotic and Tympanic. Instead, this connection consists of a soft
coupling, via specialized contact areas and an intercalated layer
of cartilage).

An arrangement as the one just described can be simplified by
assuming that the Periotic is practically immobile during the vibration of the Tympanic, due to the inertia of its mass. The natural
frequency of the Tympanic then depends on the flexional stiffness
of the connection between Periotic and Tympanic, as well as on the
mass and the distribution of the mass of the Tympanic. It is obvious
that the development of the bulky medial rim of the Tympanic is a
measure to tune the system to low frequencies by shifting the center
of mass away from the "hinge", and by increasing the amount of mass.
The effect on the frequency response is the same as mass-loading a
tuning fork. (In odontocetes the flexional stiffness of the connection between Periotic and Tympanic is smallet, but the mass is also
smaller. However, basically the configuration is the same, making
it also a receiver for low frequencies, although it is very likely
not tuned to such extremely low frequencies as in mysticetes.

Preliminary experiments with a preparation of the hearing organ
from Balaenoptera physalus (fin whale) showed that the natural frequency of the Tympanic, relative to the Periotic, is at 29 Hz, thus
making it an effective receiver for very low frequencies. Because
the malleus is fused to the Tympanic vibrations of the latter will
be forced upon the malleus-incus unit, which in turn drives the
stapes as usual. Hence, vibrations of the Tympanic, relative to
the Periotic, will be transmitted to the cochlea via the ossicular
chain. The net effect of such a structural specialization is an
expansion of the frequency range of hearing, as shown in Fig. 2.
It enables the mysticetes to achieve a great sensitivity for very
low frequencies and it permits odontocetes to tune their ossicular
chain to high frequencies, while still retaining sensitivity for
low frequencies.

Fleischer, G., 1976, Uber die Beziehungen zwischen Hoervermoegen
    und Schaedelbau bei Walen. Saeugetierkundl. Mitteilungen
    (Munich), 24, 48-59.

Fleischer, G., 1978, Evolutionary Principles of the Mammian
    Middle Ear. Advances in Anatomy, Embrol., and Cell Biology,
    (Springer-Verlag; Berlin, Heidelberg, New York) 55:5.

MORPHOLOGICAL ADAPTATIONS OF THE SOUND CONDUCTING APPARATUS IN

ECHOLOCATING MAMMALS

Gerald Fleischer

Echolocating mammals, bats, porpoises and dolphins, have a sound conducting apparatus in their middle ear which is fundamentally similar. In both groups there is a tympanic membrane, or its equivalent, an ossicular chain, composed of malleus, incus, and stapes, as well as two well-developed middle ear muscles. Also in both groups the malleus is fused to the tympanic bone, the auditory ossicles are separated from each other by thin layers of soft tissue, and the middle ear cavity is filled with air. In appearance, however, the sound conducting apparatus in bats is very much different from that in odontocetes.

An analysis of the sound conducting apparatus across mammals revealed that echolocators are only specialized as far as the tuning of the middle ear to the frequency range of hearing, and to a lesser extent the development of the middle ear muscles is concerned. Of critical importance is the size of the entire ear (not the size of the animal) and whether it operates in air or underwater. Minute middle ears, such as in bats, need special design to overcome problems in detecting low frequencies. The relatively large ears of the odontocetes, on the other hand, are engineered to overcome problems with the conduction of high frequencies. Furthermore, especially in cetaceans, the ear is decoupled from the skull as a measure towards vibrational isolation of the ears.

In the following section an outline of some aspects of the biomechanics of the sound conducting apparatus in echolocators is presented, as part of the evolutionary radiation of the mammalian middle ear.

The innermost element of the ossicular chain is the stapes complex which is a mass-spring system, tuned to the frequency range of the greatest sensitivity of the ear. Mass is the bony material of the stapes, while the spring is represented by the annular ligament which anchors the stapes within the oval window. In porpoises and dolphins the stapes appears to be immobile, because the spring (annular ligament) has to be so stiff, in order to give the stapes the natural frequency of 60 kHz, or more. In bats the anchoring of the stapes is not so extremely firm because its mass is much smaller. In summary, the natural frequency of the stapes is determined by the stiffness of the annular ligament, and by the mass-load (stapes) attached to it. It is a translatory system, i.e., it moves hence and forth like a piston.

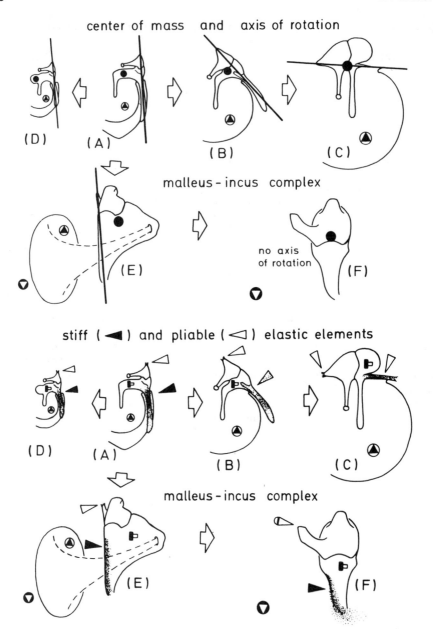

Fig. 1.   Evolutionary radiation of the malleus-incus complex in
          mammals. (A) – ancestral type; (B) – transitional type;
          (C) – freely mobile type; (D) – microtype; (E) – Tursiops-
          type; (F) – Kogia-type. Bats have the type (D), while
          dolphins have either type (E) or type (F). The black
          triangle symbolizes the tympanic membrane.

influence of the orbicular apophysis   vibrational modes of the
                                              microtype

Fig. 2

Analysis of the malleus-incus complex
in the microtype (bats). The orbicular
apophysis (X) is a bulky mass of bone
at the malleus. It determines the
frequency response of the malleus-incus
complex and its vibrational mode around
two rotational axes.

(Fig. 1 and Fig. 2 are reproduced from Fleischer (1978) with
the permission of Springer-Verlag; Berlin, Heidelberg, New York.)

      Besides the stapes there is a second semi-independent system
in the middle ear: the malleus-incus complex. It is a torsional
element, the axis of which is shown at the top of Fig. 1. Its
torsional stiffness is represented by the anchoring of the axis on
both ends, and it varies greatly (bottom of Fig. 1). Even more important is the center of mass, relative to the axis of rotation,
and it is this parameter which has the greatest influence on the
tuning of the malleus-incus complex, according to the parallel-axis
theorem. In bats (Fig.1 (D) and Fig. 2), this leads to the development of the orbicular apophysis, a bulky mass of bone at the base
of the manubrium which determines both, the natural frequencies as
well as the vibrational modes. Although different in appearance,
odontocetes have a malleus-incus complex not unlike the one in bats
(Fig. 1 (E) and (F)). In the Tursiops-type, a part of the tympanic
membrane is stretched into the tough tympanic ligament which in turn
rotates the malleus. In the Kogia-type, the tympanic membrane has

been replaced by a ceramic-like plate of thin bone (white triangle).
Without going into detail it has to be emphasized, however, that the
ossicular chain in odontocetes is functional and that it operates
according to the same principles as in bats or in man.

Fleischer, G., 1978, Evolutionary Principles of the Mammalian
    Middle Ear. Advances in Anatomy, Embrol., and Cell Biology,
    (Springer-Verlag; Berlin, Heidelberg, New York) 55:5.

ECHOLOCATION SIGNAL DESIGN AS A POTENTIAL COUNTER-COUNTERMEASURE AGAINST MOTH AUDITION

James Howard Fullard and M. Brock Fenton

Much of the work on the role of audition in tympanate moths indicates that they possess ears which are tuned to the predominant frequencies in the echolocation signals of sympatric, insectivorous bats (Roeder, 1970). Since moths are more sensitive to certain frequencies there may be selection for bats whose echolocation signals tend to minimize acoustic reception by the tympanate moths they are hunting.

Electrophysiological studies of a variety of tympanate moths captured at the Station Biologique de Lamto in Côte d'Ivoire, West Africa were used to prepare audiograms for moth species representing seven families. Using the notodontid moth, Desmeocraera graminosa, we tested the responses of the auditory nerve to the echolocation signals of four sympatric, insectivorous bats: Pipistrellus nanus (Vespertilionidae), Hipposideros ruber (Hipposideridae), Nycteris macrotis (Nycteridae) and Rhinolophus landeri (Rhinolophidae) as they flew up to 3 m from the ear preparation. The echolocation signals of hunting bats in the area were monitored in the field with a broad-band ultrasonic microphone and portable frequency analyzer (Simmons et al., 1979).

The moths that we tested in Côte d'Ivoire were most sensitive to frequencies between 20 and 45 kHz, matching the frequencies most commonly monitored for echolocating bats in the field (Fig. 1). The ear preparations of D. graminosa displayed a differential ability to detect the cries of the four species of bats that flew in the room. P. nanus and R. landeri, species that employ high intensity (+ 90 dB re. 2×10^{-5} N/m^2, 10 cm) and, in the case of R. landeri, long duration (+ 25 msec) echolocation cries were both easily detected by the moth as they flew up to 3 m from the moth (maximum distance possible for the trial). H. ruber, a high frequency (+ 100 kHz), directional signal emitter was less conspicuous to the moth and N. macrotis, which uses a low intensity (<50 dB) very short duration (2 msec) signal, was detected only when the bat was 10-15 cm from the moth and , as with H. ruber, only when flying directly towards the moth.

These data provide evidence that certain species of insectivorous bats may reduce the distance at which tympanate moths first detect them by emitting signals of a frequency and/or intensity that is difficult for the moth to hear. Coupled with the greater

899

Fig. 1.   The relationship between the echolocation frequencies of
          hunting bats (bar graph) monitored in the field and the
          neurological audiograms (•—•) of sympatric moths.   The
          moths tested are, 1 - Desmeocraera graminosa (n=10),
          2 - Miantochora interrupta (n=5), 3 - Rhodogastria vitrea
          (n=4).

flight speed of bats and a high degree of in-flight maneuverability
this reduced detection distance may provide some bats with a considerable advantage in catching tympanate moths.

     This research was supported by National Research Council of
Canada operating and equipment grants to MBF, and by a National
Research Council of Canada postgraduate scholarship to JHF.

REFERENCES

Roeder, K. D., 1970, Episodes in insect brains, Amer. Sci., 58:378.

Simmons, J. A., Fenton, M. B., Ferguson, W. R., Jutting, M., and
          Palin, J., 1979, Apparatus for research on animal ultra-
          sonic signals, Life Sci. Misc. Pub., R. Ontario Mus.
          (in press).

VOCALIZATIONS OF MALAYSIAN BATS (MICROCHIROPTERA AND MEGACHIROPTERA)

Edwin Gould

Repetitive, graded vocalizations occur in young bats during maternal-infant communication. In shrews, rodents, tenrecs, polar bears, ungulates, civet cats and some primates similar graded repetitive sounds are emitted and coupled to excitation levels. Occurrence of this coupling in such disparate groups suggests that phylogenetically the repetitive vocal emission system has been pervasive in placental mammals and that bats probably locked into an existing graded communication system for the development of echolocation.

During a visit to Malaysia I collected 10 species of bats that were either pregnant or nursing. Vocalizations of mothers and infants were recorded alone and during reunion. In general, calling rates of newborn infant bats are low; the rate increases from the first day (median=4 calls per sec; n=9 bats of 6 species) through the second week (median=6 calls per sec; n=11 bats of 6 species) and gradually wanes in the third week and near weaning (2 calls per sec; n=13 bats of 5 species) (Gould in press, Amer. Zool.).

In contrast to microchiropterans, Eonycteris spelaea (Megachiroptera) each night carries its infant out of the cave until the young infant is capable of flight. I netted nursing Eonycteris with infants attached to the nipple during their departure and return to the main entrance of Batu Cave. Infants detached from the mother at the cave were hand-held in front of the microphone and recorded with a Uher 4000 tape recorder. Sounds recorded later in the laboratory were indistinguishable from those recorded in the cave. Calling rates of Eonycteris infants were consistantly low compared to those of micro-chiropterans. Grinnell & Hagiwara (1972) reported that the temporal resolution and responsiveness to the second of a pair of sounds is the major receptive feature of the auditory system that is lacking in the non-echolocating bats. The inability of Eonycteris to emit calls of higher rate than about 2 per second may be another constraint that imposed limits on the ability of megachiropterans to echolocate with laryngeal sounds.

I recorded the vocalizations of mothers and infants of five Hipposideros species (Figure 1). Most infant Hipposideros sonar emissions emphasized harmonics that were lower than those of adult CF calls. As the infant matures, the second harmonic

is emphasized more and more; sounds higher than the second
harmonic of the typical CF call also occur. Infant non-sonar
calls emitted during reunion and isolation and in the absence
of any exploratory movements of the infants head or ears differ
from CF calls. Non-sonar calls of species such as H. armiger
and H. diadema differ in that H. diadema has a descending FM
sweep while H. armiger emits an ascending FM sweep (Figure 1).
I found nursing females of both species roosting in the same
cave. Likewise, about 4,000 nursing H. bicolor and more than
100 nursing H. cineraceus simultaneously occupied the same
roosting site; note their contrasting non-sonar calls.
Differences in neonatal calls presumably obviate confusion
during mother-infant reunion.

Figure 1.   Sonar and non-sonar calls of 10 species of Malaysian
bats compared.   No sonar calls were recorded for T. mops.
Eonycteris is a non-locater.

I recorded the adult sonar calls of five species of Hipposideros:
H. ridleyi, H. armiger, H. bicolor, H. cineraceus and H. diadema;
the first four had not been recorded previously. I plotted the
frequency of the CF portion of 16 species against forearm length
to determine whether vocal frequency and body size are related.
There is no linear relationship among the small species; it seems

that geographic distribution rather than size accounts for the
grouping of small species with similar frequencies. The larger
species that have rather widespread distribution form a second
group of low frequency emitters. Variable frequency of the CF
component occurs within and between populations in H. caffer
and H. commersoni (See p. 125, Novick, 1978). Several members
of this conference indicated their observation of variation in
other species occurring in Figure 2. To what degree this
tentative representation characterizes CF calls will depend on
subsequent reports for other localities and populations.

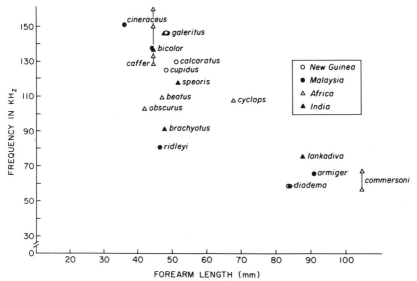

Figure 2. The CF portion of sixteen species of Hipposideros are
compared. Forearm length is a good measure of body size. For
details, see 1977.

Acknowledgements: NSF Grant BMS 7202337A01; NIH Development Award; Fulbright Hays Program.

REFERENCES

Novick, A. 1977.   Acoustic orientation, in: "Biology of bats", V.3, W. A. Wimsatt, ed., Academic Press, New York.

Grinnell, A. D. and S. Hagiwara 1972.   Adaptations of the auditory nervous system for echolocation.   Studies of New Guinea bats. Z. vergl. Physiol. 76, 41-81.

A RECONSTRUCTING TECHNIQUE FOR THE

NASAL AIR SACS SYSTEM IN TOOTHED WHALES

Vladimir S. Gurevich

Durable models of the nasal air sacs system of small delphinids were designed for three dimensional comparative anatomical study, and precise measurements and demonstrations were produced using a new technique. The few previous attempts to model the system were not completely successful (Gurevich, 1969; Mead, 1972; Schenkkan and Purves, 1973; Dormer, 1974). Latex, wax corrosion, and silicone rubber were used for reproducing the nasal air sacs system. Because the solvent in the previously used casting materials causes marked shrinkage (up to 70%) as it vaporizes, modeling of the small tubular and accessory sacs is unsatisfactory; the larger vestibular and premaxillary sacs may also be warped. Wax corrosion casts are very fragile, vinyl corrosions shrivel, and fusible alloy models are distorted.

In 1976, we began testing new materials. We were able to reproduce the entire nasal air sacs system and associated structures of Delphinus delphis, Stenella longirostris, and Tursiops gilli with Hasting's two-part epoxy resin, but could not obtain fully satisfactory results with fine details of muscle fibers, diagonal membranes, tubular and accessory sacs. Eventually, we discovered Batson's N 17 Anatomical Corrosion Compound, a product developed specifically for casting biological materials.

Batson's solution has many desirable characteristics for producing exact colored or colorless casts of any anatomical organ with an opening into which liquid plastic can be injected. The solution, a clear polyester resin, polymerizes at room temperature when mixed with one or more catalysts; shrinkage is about 15% and is easily injected by several means. Pigments are available if color is desired. Normal working time is about 20 minutes, but varies with amount of catalyst and color pigment used. The casts are relatively tough and easy to store and handle. The casting of the nasal air sacs system of nine species of delphinids Tursiops truncatus, T. gilli, Steno bredanensis, Delphinus delphis, Orcinus orca, Globicephala macrorhyncha, Phocoena phocoena, Phocoenoides dalli, Grampus griseus, and Kogia breviceps were produced with Batson's compound. Excellent details can be seen in the entire system including even such fine structures as the diagonal membrane, tubular and accessory sacs as well as their diverticuli and traces of the nasal plugs. Directions of the muscle fibres and their attachments also can be seen. The system of Kogia differs sufficiently from typical delphinids.

Figure 1

1. Width between external margins of vestibular sacs.
2. Length of vestibular sac.
3. Width of vestibular sac.
4. Height of vestibular sac.
5. Maximum width between external margins of nasal plugs.
6. Length of nasal plug.
7. Width of nasal plug.
8. Maximum width between external margins of premaxillary sacs.
9. Length of premaxillary sac.
10. Width of premaxillary sac.
11. Length of bony nàres.
12. Width of bony nares (top).
13. Width of bony nares (bottom).
14. Length of anterior horn of tubular sac.
15. Length of posterior horn of tubular sac.
16. Height of air sacs system.
17. Length of accessory sac.
18. Width of accessory sac.

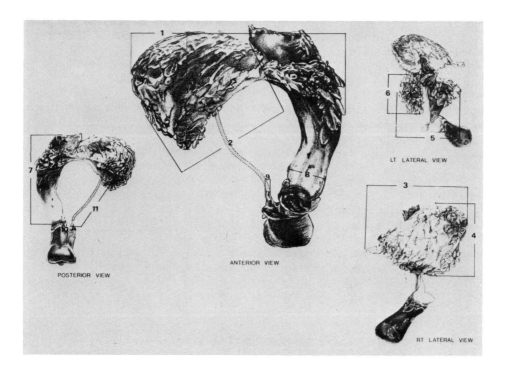

Figure 2

1. Width between external margins of vestibular sacs.
2. Length of vestibular sac.
3. Width of vestibular sac.
4. Height of vestibular sac.
5. Length of the left vestibular sac.
6. Width of left vestibular sac.
7. Length of left nasal bony passage.
8. Diameter of left nasal bony passage.
9. Diameter of right nasal bony passage.
10. Diameter of bony nares at bifurcation.
11. Projection of length of right nasal bony passage.

These anatomical structures are probably responsible for the sound producing in Odontoceti (Cetacea). Figures 1 and 2 show the casting models of the bottlenose dolphin (<u>Tursiops</u> <u>truncatus</u>) and pygmy sperm whale (<u>Kogia</u> <u>breviceps</u>) respectively. Numbers designate the area where the measurements were taken. The models, for the first time, allow us to conduct measurements of the nasal air sacs and associated structures which can be used for the future modeling of the dolphin's echolocator.

REFERENCES

Gurevich, V. S., 1969, Reconstruction technique of upper respiratory tract (nasal air sacs system) in Black Sea dolphins (unpublished).

Mead, J. G., 1972, Anatomy of the external nasal passages and facial complex in the Delphinidae (Mammalia: Cetacea). PH. D. Diss. University of Chicago.

Schenkkan, E. J., Purves, P. E., 1973, The comparative anatomy of the nasal tract and function of the spermaceti organ in the Physiteridae (Mammalia: Odontoceti), <u>Bijdragen tot de Dierkunde</u>, 43(1):93.

Dormer, K. J., 1974, Reconstruction technique for multiple castings of the sacs in silicone rubber. (Personal communication).

AN ANALOGUE DEVICE FOR THE GENERATION OF SONAR AMBIGUITY DIAGRAMS

Justin A. T. Halls

The ambiguity diagram is a measure of the sonar or radar properties of a given echolocation signal. It consists of a three-dimensional graph in which the x and y axes represent range and velocity, and the height at any point is the probability of a target having the corresponding range/velocity co-ordinates. Each section through the diagram parallel to the range axis represents the output of an ideal, range measuring receiver (eg. matched filter or correlation receiver) in response to an undistorted, energy normalised echo from a target moving at the velocity corresponding to that section; this output is the cross-correlation function of the transmitted signal and a doppler shifted copy of the transmitted signal. Each section through the diagram parallel to the velocity axis represents the output of an ideal velocity measuring receiver (eg. Fourier Transform receiver) in response to an undistorted, energy normalised echo from a target at the range corresponding to that section.

The ambiguity function was first defined by Woodward (1953) as:-

$$|\chi(\tau,\phi)|^2 = |\int u(t). u*(t+\tau). e^{j\phi t}dt|^2$$

in which $u(t)$ is the transmitted signal; τ = negative of time delay; $\phi = -(2\omega_0 v/c)$ = frequency shift caused by the Doppler effect; ω_0 is the carrier frequency in radians; c is the speed of signal propagation; v is the radial velocity of the target, taken to be positive for motion towards the receiver and negative away from it. This form of the ambiguity function is referred to as the narrowband approximation, since it was formulated for use with radar signals; it assumes that the bandwidth of the signal is small compared to the carrier frequency, and that the velocity of the target is small compared to the speed of propagation of the signal; the doppler effect can then be considered to be a simple spectral shift or frequency offset. In this case the height of the function at the origin, and the volume of the function are fixed and proportional to the energy contained in the original signal. Since the diagram is antisymmetrical about the origin it is only necessary to calculate half the function, the other half being a mirror image of this.

For wideband sonar signals the generalised ambiguity function must be used, although for convenience χ may be plotted rather than $|\chi|^2$.

$$|\chi(\tau,s)|^2 = |s^{1/2}\int u(t). u*(s(t+\tau))dt|^2$$

where s is the doppler stretch or compression factor and equals

$(1+\beta)/(1-\beta)$, and $\beta=v/c$. $s^{1/2}$ normalises the energy in the 'echo'. This is the form of the ambiguity function used by Cahlander (1967) to generate ambiguity diagrams of a number of echolocation signals from Myotis lucifugus and Lasiurus borealis. In this wideband case, the ambiguity diagram is not necessarily antisymmetrical, and its volume is not fixed.

The ambiguity function computer described here uses analogue techniques, and is being used to generate the ambiguity diagrams of real bat signals from over 100 species, recorded under a variety of conditions, by J. D. Pye. $\chi(\tau,s)$ is calculated for 1000 successive values of τ, and the resulting correlation function is plotted. This process is repeated for as many values of s as are required. Up to 16 values of s, which may be +ve or -ve, may be calculated in one run, although for most signals it is only necessary to compute 5 different values.

The signal is stored at 1/8 or 1/10 speed on the rotating magnetic drum of a Kay sonagraph, and is replayed continually into the computer. The computer directs one copy of the signal into an analogue delay line consisting of charge coupled analogue shift registers. A second copy of the signal goes to a similar delay line in which the delay varies continuously throughout the duration of the signal. This results in a stretching or compression of the signal in time, thus imposing a doppler shift proportional to the rate of change of delay. The outputs of the two delay lines are then filtered and multiplied in a balanced mixer. The product is integrated and the result stored in the memory of a Biomac 1000 averaging computer. The delay in the first delay line is then incremented and the process is repeated until a full 1000 point correlogram has been generated. This is then plotted on an x-y or y-t plotter, the doppler factor s is incremented, and the process is repeated. For convenience the correlation functions themselves are then assembled into the full ambiguity diagram, although a squaring circuit and envelope detector may be inserted between the Biomac 1000 and the plotter if desired.

The computer can also be used for generating cross-correlation functions or cross-ambiguity functions between two different signals, or between a signal and a real echo of that signal. A simple modification also permits the generation of range/acceleration as well as range/velocity ambiguity diagrams.

It should not be assumed that the bat is making use of all the information that the ambiguity diagram indicates is available from a given signal. The ambiguity function represents an upper bound on the precision with which certain target parameters can be measured, with a single pulse. In a real system the available information will always be less than this. Also the bat is not

relying on a single pulse to provide a complete picture of his
environment, and many improvements can be made by using pulse
trains and by making a priori assumptions about the nature of the
target.   However, the ambiguity function does tell us something
of the sonar properties of a given pulse, enabling us to estimate
the type of system in which it would be of most use.   As such the
ambiguity diagram is an essential supplement to the oscillogram
and the sonagram when attempting to describe signals.   The computer
described here provides an economical way of producing ambiguity
diagrams for real or simulated pulses on a routine basis.

REFERENCES
Cahlander, D.A., 1967, Discussion of Theory of Sonar Systems, in
     "Animal Sonar Systems: Biology and Bionics", Vol. II, ed.
     Busnel, R.-G., Jouy-en-Josas, pp. 1052-1081.
Woodward, P.M., 1953, "Probability and Information Theory, with
     Applications to Radar", Pergamon Press, Oxford.

THE CONSTANT FREQUENCY COMPONENT OF THE BIOSONAR SIGNALS OF

THE BAT, PTERONOTUS PARNELLII PARNELLII

O. W. Henson, Jr., M. M. Henson,

J. B. Kobler and G. D. Pollak

The auditory system of Pteronotus parnellii has been the subject of numerous studies but detailed knowledge of the emitted signals is limited. Previous studies utilizing tape recorded signals and sound spectrographs have shown that each emitted pulse consists of a brief upward sweeping initial FM component, a relatively long CF component and a brief downward sweeping terminal FM component. In addition there is a series of harmonics: the second harmonic (ca. 61 kHz) is the strongest. A study of the fine structure of this second harmonic CF component was undertaken to gain more critical information on pulse design, pulse emission patterns and the relationship of the signals to the sharply tuned cochlear microphonic (CM) audiogram.

CF component frequency measurements were made from filmed oscillographic displays of the output of a phase-locked-loop device which produced a dc voltage proportional to frequency (Smith et al., 1977) (Fig. 1). The CF component comprised from 70-83% of the total pulse. When the bats were not flying, the CF component usually reached a peak (maximum) frequency in the middle portion of each pulse and this frequency was maintained within 100 Hz for 33-65% of the total pulse duration. The beginning and terminal ends of the CF component represent transitions from the beginning and terminal FM components of the pulse and they can best be described as slow FM components, with frequency changes ca. 300-500 Hz; these slow FM changes constitute 15-40% of each pulse.

Records from flying bats were obtained by attaching the animals and a microphone to a pendulum with known velocity characteristics. Under these conditions the bats beat their wings, emitted pulses in characteristic way, and adjusted the peak frequency of the CF to compensate for Doppler shifts created by movement relative to a stationary target (Fig. 2). When a barrier was attached to the pendulum and positioned in front of the bats so that the echoes were not Doppler shifted, the bats did not adjust their pulse frequencies. Thus, Doppler compensation is entirely dependent on the perception of echoes.

Experiments with seven bats with chronically implanted electrodes showed that they consistently held the peak frequencies of the echoes at or below the tuned peak (best frequency) of the CM

microphone

Fig. 1. Oscillographic display of the pulses of Pteronotus p. parnellii. Upper trace shows the pulses as detected by a condenser microphone. The middle and lower traces show the output of a phase-locked-loop device where the CF pulse frequency has been converted into a dc voltage with gain settings of 100 Hz/div. and 1,000 Hz/div., respectively. Frequency calibration applies to middle trace. From Smith et al., 1977, J. Acoust. Soc. Amer. 61, 1092.

Fig. 2. Graph of the changes in the peak frequencies of the emitted pulses in one bat before, during and after three forward swings (flights) on the pendulum device. Data for the different swings are indicated by the open and closed circles and triangles. The horizontal line marked "best frequency of the CM audiogram" corresponds to the tuned peak or threshold minimum and was determined for the same bat with the aid of chronically implanted electrodes. Note the close correspondence of the decrease in peak (pulse) frequencies with the increase in the peak frequencies of the echoes that should occur as a result of Doppler shifts imposed by the movements of the pendulum.

Fig. 3. Cochlear microphonic (CM) audiogram for frequencies near the 2nd harmonic of the CF component of the emitted pulses and histogram of pulse peak frequencies emitted by the same bat when not flying.

Fig. 4. Cochlear microphonic responses recorded from flying <u>Pteronotus parnellii</u>. Upper trace (MIKE) shows pulses detected by a microphone; middle trace shows CM potentials to emitted pulses (p) and Doppler shifted echoes (e); horizontal lines under CM responses indicate the exact time of pulse emission relative to CM responses. Dot display is output of period meter with frequency scale indicated on right. Note the small responses to emitted cries compared to echo-evoked potentials and the beats created during periods of pulse-echo overlap.

audiogram; in most cases the peak frequencies were 200-400 Hz below the tuned peak (Fig. 3). In this frequency band the maximum amplitude of the CM potentials is higher and the intensity function curves are much more linear than at the best frequency of the CM audiogram. These data explain how the ear of flying Pteronotus may show a relatively small response to the pulse CF component and an enormous response to Doppler shifted echoes which overlap the outgoing cry (Fig. 4).

(Supported by NIH Grants NS 12445 and NS 13276.)

THE AVOIDANCE OF STATIONARY AND MOVING OBSTACLES BY

LITTLE BROWN BATS, MYOTIS LUCIFUGUS

Philip H.-S. Jen, Yung H. Lee and R. Kelman Wieder

Echolocating bats sense their environments by emitting ultra-sonic signals and listening to the echoes. The simplest way to measure this ability is by the obstacle-avoidance test in which a bat flies through an array of stationary wires and the numbers of hits and misses are recorded. By extending this test and making their bats fly between both stationary and moving obstacles, Jen and McCarty (1978) reported that bets avoid moving objects more successfully than stationary ones. Using the same experimental method and more bats, we report here more quantitative data on the sensitivity of the echolocation system of bats.

Two groups of little brown bats, Myotis lucifugus, caught in early summer (20 bats) and early fall (9 bats) of 1978 were tested in a room which is 4.6 m long, 2.2 m wide and 2.1 m high. An aluminum frame containing six vertically strung nylon monofilament lines (1 mm in diameter) was placed in the center of the room. The distance between the lines was always 30 cm. The total opening of this frame through which a bat flew was 1.4 m wide and 2.0 m high. By means of an electric motor, the frame could be made to oscillate with an amplitude of 14.5 cm in the horizontal direction parallel to the alignment of the nylon lines. Any effect of the movement of the frame itself on the bat's performance was eliminated by surrounding the frame with large wooden panels which were attached to the walls, ceiling and floor. Thus a bat was confronted only with the movement of the nylon lines, but not that of the frame. All surfaces of the wooden panels, ceiling and walls were covered with convoluted polyurethane foam to reduce echoes. During the test, the numbers of misses and hits were recorded when the obstacles were stationary or oscillating at different rates. The average oscillating rate used for testing the bats caught in early summer ranged from 0 to 28.7 cm/sec.

Assuming a bat could express its detection of moving obstacles by increasing its score of misses, we have studied the sensitivity of its echolocation system by statistically calculating the just-noticeable moving speed of the obstacles which can be perceived by the bat. That is, the minimum average moving speed of obstacles which can be detected by a bat and behaviorally expressed by sig-nificantly increasing its score of misses relative to that obtained from stationary obstacles. Nine bats caught in early fall were made to fly through the array of obstacles which was oscillated at an average speed of either 0, 2, 4, 6, 8, or 13.2 cm/sec.

917

The difference in scores of misses obtained under each test condition were then statistically compared (X^2 test, p = 0.05).

The results of these tests are summarized in Fig. 1. Bats can
avoid moving obstacles with higher scores of misses than stationary
ones. In both the fall and summer plots, all the data points obtained from moving obstacles are significantly higher than those
obtained from stationary ones with only one exception. That is, in
the fall plot, the data point obtained from obstacles moving at an
average speed of 2 cm/sec is not significantly higher than that obtained from stationary ones. Interestingly, bats caught in early
fall were able to avoid obstacles more successfully than bats caught
in early summer.

Fig. 1.   The average percentage of misses obtained under different
          test conditions from two groups of Myotis lucifugus
          caught in early summer (20 bats) and early fall (9 bats)
          of 1978. Respectively, the ordinate and abscissa represent
          percentage of misses and average moving speed of the ob-
          stacles in cm/sec. Vertical bars at each point represent
          one standard deviation. Respectively, the numbers without
          and within parentheses under each point represent the total
          number of bats tested and the total number of flights
          under each test condition.

Three of the nine bats caught in early fall avoided moving obstacles more successfully than stationary ones, but the increases in their scores of misses were not significant. In the remaining six bats the scores of misses when presented with moving obstacles were significantly greater than when presented with stationary ones. The just-noticeable moving speed obtained from these six bats ranged from 2 to 6 cm/sec with an average of 3.7 $\pm$ 1.5 cm/sec.

That little brown bats can avoid moving obstacles more successfully than stationary ones may be associated with the fact that they rely upon moving prey for survival and they apparently concentrate more intently on moving objects while hunting. Since the flight speeds of insects preyed upon by bats are generally higher than 4.0 cm/sec (Johnson, 1969), bats apparently have no difficulty in detecting them. We suggest that a whole summer's practice of echolocation and the high activity of bats during the mating season may contribute to the superior performance of fall bats in avoiding obstacles.

Acknowledgements: Work supported by NSF and University of Missouri Research Council.

REFERENCES

Jen, P.H.-S., and McCarty, J.K., 1978, Nature, 275:743-744.

Johnson, C. G., 1969,"Migration and dispersal of insects by flight", Methuan, London.

ECHOLOCATION AND BEHAVIOR

Jean-Claude Lévy

An echolocating animal can adjust its signal in such a way as to receive a constant frequency for which the ear shows maximum sensitivity and ability to recognize echo characteristics. This is known as an "acoustic fovea". Is this adjustment the effect of automatic or purposeful behavior? Our assumption is that it is an automatic process involving not the echo but the original signal, which is automatically adjusted to obtain an optimal echo whose characteristics are therefore determined and known in advance.

Let $S(t)$ be the signal defined as a time function and $s(t)$ be the echo received after emission of the signal $S(t)$:

$$s(t) = O\{S(t)\}$$

where O is a transformation operator containing all target characteristics including range, size, shape, elasticity, and composition. To determine the operator O, both signal and echo must be known, but it is first necessary to determine which of the two is given and which is observed.

To illustrate the basic premises of the model, an analogy can perhaps be useful: we know the weight of an object that we lift from the strength of our muscles, which adjust automatically according to the value of this weight. In the same way, a bat could observe: "I am emitting loud signals, therefore my echo is weak" and not "my echo is weak, therefore I must emit loud signals". As frequency correction compensates for the Doppler effect and acts to cancel it, a bat could also observe: "my signal has a high frequency component, therefore my target is flying away from me" and not "my target is flying away from me, therefore I must emit a high frequency component".

Ranging would be an exception to this, as even if the optimal echo $s(t)$ is known, the time origin is not. For this reason, the signal alone is insufficient and the set, signal-echo, must be taken into account:

$$s(t + \tau) = O^*\{S(t)\}$$

where the unknowns are $S(t)$ and τ; $s(t)$ is given as the optimal echo which can be adjusted by the animal to obtain the most useful information about the target, e.g., frequency-time ambiguity.

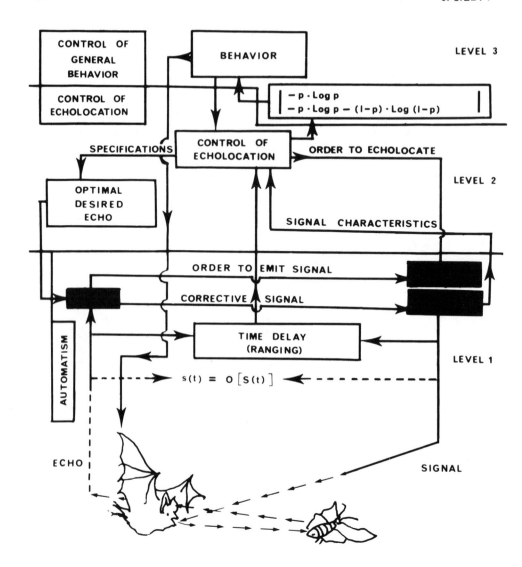

Fig. 1. Model of hypothetical neural integration.

Level 1:   Automatism of the echolocation signal.
Level 2:   Control of echolocation.
Level 3:   Control of general behavior.

The principle of the operation is illustrated in Fig. 1 which presents different levels of hypothetical neural integration:

1. Automatism of the echolocation signal.
2. Control of echolocation.
3. Control of general behavior.

An observation made by Griffin (1958) lends support to the hypothesis of automatism and at the same time allows an additional element, the concept of probability relations, to be introduced between the second and third levels of the neural integration model. An Eptesicus fuscus, habituated to a room containing obstacles which it deftly avoided in its flights, began to collide with them when they were placed in different positions, in spite of the fact that it was emitting orientation pulses at the time. Apparently it was navigating from memory and did not bother to extract any acoustic information from its automatically emitted signals or their echoes.

This additional element can be described mathematically using the relationship given by Shannon's entropy. Let p be the a priori probability of finding an unobstructed passage in the room. The amount of attention that the bat pays to its echolocation pulses is proportional to the entropy term which, according to different hypotheses, can be given by one of the two formulae:

$$h = - p \cdot \log p$$

$$h^* = - p \cdot \log p - (1-p) \cdot \log (1-p)$$

If, on the basis of past experience, the bat is sure of finding the passage unobstructed, $p = 1$, h (and h^*) $= 0$, and the animal would pay no attention to its echolocation pulses which are automatically emitted. The phenomenon is not surprising as much of our own behavior is also based on probability relations and not on immediate perception of our surroundings. Consider, for example, the act of walking. When a particular route is familiar to us, we don't rivet our attention to the ground to make sure that it is always there when we take each step. Based on accumulated past experience, we are sure of finding the ground under our feet. But if one day a hole appears in the sidewalk, we will fall in it just as stupidly as the poor bat colliding with a newly moved and therefore unexpected obstacle in its familiar flight path.

THE FUNCTION OF LATERAL INHIBITION IN THE RECOGNITION OF FREQUENCY SPECTRA

Jean-Claude Lévy

A biosonar system can recognize the nature of targets through an interpretation of echo characteristics. Humans are able to do the same thing only by striking an object and listening to the sound produced. The basic difference in the two procedures is that humans cannot do this at a distance.

An approximate estimation of frequencies is performed in the inner ear by the basilar membrane, whose movements are detected by the organ of Corti. The ear shows high precision in frequency discrimination and yet the response curve of the basilar membrane is very broad. We suggest that this precision is increased through the phenomenon of lateral inhibition.

The principle of lateral inhibition is well known in the study of the visual centers. Its function is to detect the boundaries of spots or illuminated fields by increasing the contast between bright and dark areas. It is also used in the auditory system but only for one dimension. Retinal cells which are fired by a field of brightness are replaced here by a segment of the organ of Corti. By increasing the contrasts and consequently the slopes of the response curves, lateral inhibition allows detection of boundaries with a precision greater than that required for a mere definition of the broad peak position.

Fig. 1 illustrates two frequency spectra, each representing 10 frequencies which are equidistant on a logarithmic scale. One spectrum is slightly higher than the other for upper frequencies, allowing theoretical differentiation of the two spectra. The global response of the organ of Corti to the two spectra (Fig. 2) is given by the summation of these curves, each with a weighting factor proportional to the log of the amplitude of the corresponding frequency. If $v(f,x)$ is the excitation of the organ of Corti for the frequency f, and $a(f)$ is the amplitude of that frequency, then the global response curve is:

$$r(x) = \Sigma a(f_i) \cdot v(f_x, x).$$

The organ of Corti cannot resolve the two spectra, and only one broad peak is observed.

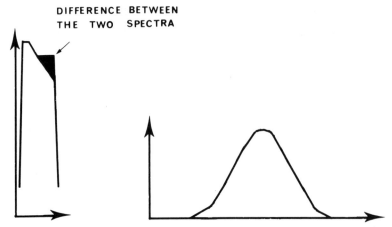

Fig. 1. Frequency spectra.   Fig. 2. Response of the organ of corti
                                   (no difference between the two
                                   spectra).

      The principle of lateral inhibition is illustrated schemat-
ically in Fig. 3.   Columns of interconnected neurons whose relationships are excitatory extend from the organ of Corti.   The neurons
of each line are connected by inhibitory relationships to the
neurons in adjacent columns of the following line.   With such inhibitory effects integrated over broad areas of the organ of Corti,
stimulation of one area would directly inhibit activity in adjacent
less stimulated areas, thus leading to a sharpening and tuning of
the normally wide and flattened response curve of the basilar
membrane.

Fig. 3. Schematic representation of lateral inhibition.

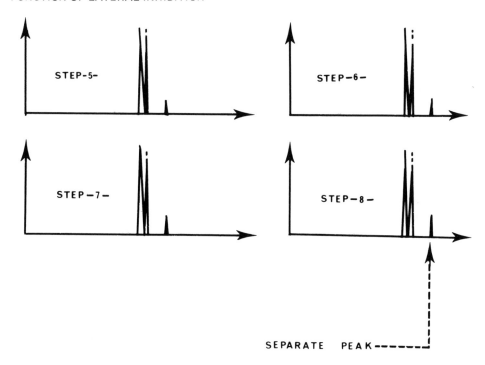

Fig. 4. The last four steps of lateral inhibition.

This procedure was simulated by means of an HF Mark 30 computer connected to a plotter (Fig. 4). The simplest possible curve
was taken for the characteristic response of each neuron: for a
positive input value, output equals input, while for a negative
input value, output is null. Following the action of lateral inhibition, the two spectra shown in Fig. 1 can be distinguished.
The higher spectrum is now characterized by a separate peak which
could theoretically be easily discriminated. No assumptions are
made about the meaning of these patterns. The only condition
stipulated is that the brain could easily distinguish and interpret
the two.

FURTHER STUDIES OF MASKING IN THE GREATER HORSESHOE BAT,

RHINOLOPHUS FERRUMEQUINUM

Glenis R. Long

Psychophysical studies have shown that the auditory system of
man and other mammals seem to act like a series of bandpass filters
which, with the exception of the lowest frequencies, increase
linearly with log frequency. This author (Long, 1977) obtained an
indirect measure of these bandwidths (critical ratios) using a
classically conditioned response to shock and found that, while the
critical ratios from Rhinolophus ferrumequinum were consistent with
those from other mammals below 75 kHz, they decreased dramatically
near the reference frequency. At that time it was not possible to
specify the relation of these results to other modifications in
these bats (see Neuweiler and Pollak, this volume). Further research
was conducted using essentially the same method except that the noise
was generated by multiplying 9.9 kHz low-pass noise by a pure tone
giving rise to 19.8 kHz bandwidth 18 dB SPL/Hz noise centered at
the frequency of the tone.

Fig. 1 shows that the tuning of the area of decreased thresh-
hold associated with the reference frequency of these bats is not as
pronounced in the masked thresholds as in the quiet but shows the
same frequency dependency. The right-hand axis gives the estimated
bandwidths based on Fletcher's formula (see Long, 1977). It should
be noticed that the calculated bandwidths for frequencies around
81 kHz are greater than the 19.8 kHz noise available. Bruns (this
volume and reviewed by Neuweiler, this volume) has developed a
model of basilar membrane motion in this bat in which the relative
motion of the basilar membrane is smaller at these frequencies due
to a discontinuity in the nature of the membrane vibration. Neuweiler and Water (reviewed by Neuweiler, this volume) found 8th
nerve fibers which responded to frequencies above and below these
frequencies but not to these frequencies. Möller (reviewed by
Neuweiler, this volume) found that some collicular neurons show an
absence of inhibition at these frequencies. The unusually large
critical ratios at these frequencies may be related to these
results and indicate that at least at the frequencies within the
transition area of the basilar membrane this indirect measure is
not a good measure at the frequency processing capabilities of the
animal.

Direct measurements of the critical bandwidth were obtained
using two different techniques: (a) Band Narrowing: pure tone thresholds were measured in noise of constant energy but decreasing bandwidth. The critical band was taken as the point where the threshold

929

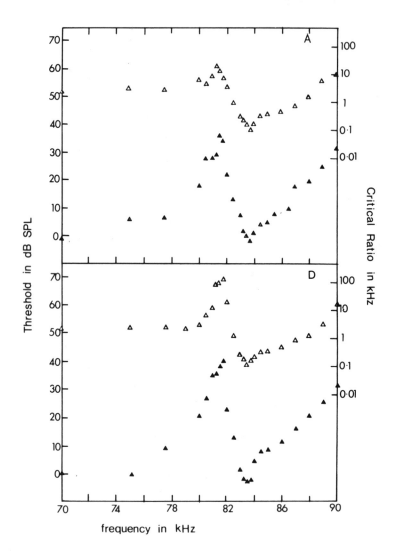

Fig. 1. Quiet thresholds (solid triangles) and masked thresholds
(open triangles) as a function of frequency for two animals. The
right-hand axis gives the calculated critical ratio bandwidths
from the masked thresholds.

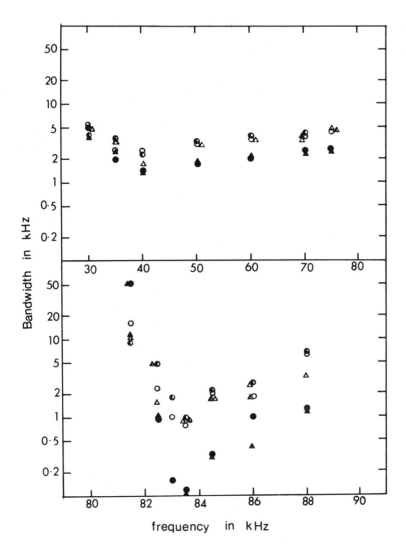

Fig. 2. Critical ratios (solid symbols) and two different measures of critical bands: band narrowing (open symbols) and two-tone masking of noise (half-filled symbols) as a function of frequency for two animals (D triangles, E circles).

stops increasing, (b) <u>Two-tone Masking of Noise</u>; the threshold of a
100 Hz bandwidth of noise is measured as the separation of two pure
tones flanking it increases. The critical band is taken as the
point at which the thresholds begin to decrease as the tones move
outside the critical band. The results are presented in Fig. 2
along with the critical ratios from Fig. 1. Below 78 kHz the different measures of critical band give similar results that are approximately twice as large as the estimates of critical ratio and maintain a linear relation to log frequency except at low frequencies,
i.e., in this area this bat does not differ significantly from
other mammals, despite the increased innervation of the organ of
Corti from 70-80 kHz (see Neuweiler's review, this volume). Above
80 kHz the measurements no longer maintain a constant relation to
each other to the extent that near 81 kHz the critical ratios are
larger than the critical bands and at the reference frequency the
critical bands are six times the critical ratio. This may be due
to unusual patterns at vibration at the modified basilar membrane
and to unusual patterns of interaction between sounds as reviewed
by Neuweiler and Pollak (this volume). Additional data from this
author (in preparation) indicate that the bandwidths are asymmetrical
around the reference frequencies and approximately 2 kHz below it
(815 kHz).

Schnitzler (this volume) suggests that these bats have frequency resolution capacities of 30-60 Hz at the reference frequency.
The small critical ratios and critical bands at these frequencies
are consistent with these claims. As these results show this fine
frequency resolution capacity not only means that the bat is able
to detect small frequency changes but that he can also detect and
process these frequencies in high noise conditions. It also appears
that a Doppler shift compensating bat would not easily detect his
emitted sounds in a noisy environment.

Long, G. R., 1977, Masked auditory thresholds from the bat, <u>Rhino-
       lophus</u> <u>ferrumequinum</u>, <u>J. Comp. Physiol.</u>, 116:247.

This work was conducted at the Polytechnic of Central London,
England. Now at Central Institute for the Deaf, St. Louis, Mo.,
U.S.A.

DOLPHIN AIR SAC MOTION MEASUREMENTS DURING VOCALIZATION BY TWO

NONINVASIVE ULTRASONIC METHODS

R. Stuart Mackay

Method one uses short pulses of 2 MHz sound in a sonar system.
A probe on the subject's head projects impulses downward or inward.
Echoes from successively deeper structures return to the probe to
brighten a downward moving spot on a cathod ray tube. The spot
jumps back to the top of the display and moves slightly to the right
for the next pulse. Traced from left to right (Fig. 1) is the time
pattern of movement of a line of internal reflecting structures, all
correctly laid out in relative depth within the animal. Instead of
being held stationary, the probe can be angulated on a small water-
filled balloon (flexible coupling) to outline internal structures
(form image) and thus achieve the desired aim for time recording.

The 6 Tursiops truncatus observed seemed undisturbed, though
nonlinearities in their hearing mechanism in principle could allow
them to sense the modulation envelope of this high frequency sound.
Approximately 2000 distance measurements per second are performed
(the pulse rate). Reflections from tissue-air interfaces are very
strong but usually not directed back to the probe; what is seen is
cessation of tissue echoes beyond the gas interface. Lower frequency sound gives deeper penetration and less resolution. Low
intensity is harmless.

One can see which sacs move, when and how much, during various
vocal activity. Fig. 1 shows motion of the anterior vestibular sac
during a whistle. Distance from the probe on the skin surface to
the intersection of sonic beam with sac is recorded. This diverticulum seems more than just a place to trap water leaks. Aiming
in from down on the head allowed looking deeper under that sac at
the nasofrontal sac.

There would be less possibility of animal disturbance if a
steady input tone could give position information. It can. Movement velocity is recorded by measuring Doppler frequency shifts in
the continuous signal returned from a steady outgoing 2 MHz. If
desired, signal passage through an integrator records position
(relative to initial position) as a function of time. In Fig. 2,
special circuits recorded maximum sensed instantaneous advancing
speed prior to and during clicks by an animal. Microphonics reduce
if a bandpass filter follows the receiving transducer. All probes
(Fig. 3) work in or out of water, and they can be fronted with a
concave plastic lens to concentrate sensitivity on a smaller region.
Fig. 4 suggests that signals from highly tilted surfaces might

933

Fig 1 Echograph trace of vestibular
      sac motion.

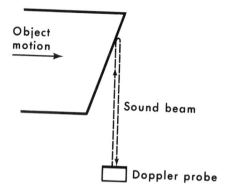

Fig 3 Probe on head of dolphin.

Fig 4 Exaggerated velocity is more
problem with pulsed than Doppler
units.

Fig 2 Doppler trace during
      clicks.

Fig 5 Similar patterns are
made with blowhole open or
closed.

exaggerate velocity if the geometry was unknown; in practice coherent echoes do not return in such a direction and one records component of motion along the beam, though in a pulsed system when
noting cessation of tissue echoes it can be misleading. Doppler
methods are especially useful for measuring the frequency of very
rapid motions since they can respond faster than periodically sampling methods such as x-ray movies or pulsed ultrasound systems.
Thus they might also have application to dolphin hearing mechanism
studies where gas also seems involved.

Interpretation is aided by supplementary facts. Air interfaces in the vicinity of a primary sound source will be forced to
vibrate and modify the overall radiated pattern just as does a
piece of metal in the vicinity of a radio antenna (either transmitting or receiving). In a human whistle air vibrates rather than
the lips and vortices are shed. I find it difficult to whistle at
pressures corresponding to below 25 m depth but it remains easy to
talk or to make "Bronx cheer" sounds. (A hyperbaric symphony orch-
estra has not been studied but mechanical whistles might be expected
to employ a different stimulation of resonances more related to
their click sequence generation.

Based upon the above, H. M. Liaw has begun his doctoral research with the methods. We find in Tursiops truncatus that before
or at the start of clicks the vestibular sac inflates on at least
one side and a nasal plug then vibrates. This sac acts as a reservoir with air coming and going but sound generates when air moves
up giving inflation. A "cry" such as in Fig. 5 when analyzed has
a low pulsatile frequency falling in the calculated resonance of
the vestibular sac which itself moves as the frequency is varied
and thus also probably acts as a resonator. Gross motion of the
other sacs has not yet been seen during clicks.

Some generalities on the nondestructive study of cetaceans,
including other details on ultrasound, have been given (Mackay,
1966).

Mackay, R. S., 1966, Telemetering physiological information from
        within cetaceans, and the applicability of ultrasound to
        understanding in vivo structure and performance, in: "Whales,
        Dolphins, and Porpoises", K. Norris, ed., U.C. Press,
        Berkeley.
Mackay, R. S., Rumage, W. T., and Becker, A., 1977, Sound velocity
        in spermaceti organ of a young sperm whale, in: "Proc. Second
        Conference on Biology of Marine Mammals", San Diego.

A THEORY OF THE SPERMACETI ORGAN IN SPERM WHALE SOUND PRODUCTION

R. Stuart Mackay

At a depth of 2500 m from where William Whitney recorded sperm whales sounding normal (unpublished), the density of air is about 1/4 that of water but the velocity of sound is approximately unchanged from the surface. (Air is not quite a perfect gas: at 250 atmospheres volume is 20% too large.) Acoustic impedance (velocity x density) thus is still less than that of water, and an air film remains a good sound reflector. Reflection of a sound pulse from the frontal air sac of a sperm whale showed excellent reflection with a phase change of 180° (inversion as in Fig. 1) suggesting the spermaceti organ might serve as a frequency controlling resonator or delay line analogous to an open ended rather than closed pipe. A reverberation model (Norris and Harvey, 1972) and a bugle model (Mackay, 1972) could be relevant. The energy release mechanism can involve air but frequency is depth independent if controlled by a liquid or solid resonator (or air if geometry could be fixed, viscosity changing little).

The exceptional anatomy of the sperm whale head was described by Raben and Gregory (1933) and by Norris and Harvey (1972). The cylindrical spermaceti organ has an air sac at each end, with a pair of lips in the forward one (Fig. 2-4), and is about 1/4 the length of the whale. The frontal sac of Physeter catodon provides a stable astigmatic concave-forward air film that focuses some sound in the vicinity of the Museau du Singe in the anterior sac, according to our measurements on one animal. The rear sac has a knobby back wall that can act as a spacer to stabilize the air film from mechanically distorting closed or dissolving even while pressurized to ambient by the lungs.

Vibrations of the lips could send sound down the organ where reflection would return the sound to shake the anterior sac and move air in synchronism through the lips, causing the lips to vibrate at the preferred frequency being generated just as do the lips of a trumpet player. In such modes of operation the lip end acts as a closed pipe with the sound period being 4 one-way transit times. The anterior sac would act as an air reservoir for recycling, as the analog of a trumpet mouthpiece, and as the coupling to the water and spermaceti. The organ is longer than wide and hence for the fundamental tone the proximal mirror is less than a wavelength across, making aim noncritical. The acoustic properties of the organ are rather like the surroundings and post mortem observations of channeling of sound at higher frequency can be misleading. However, if

937

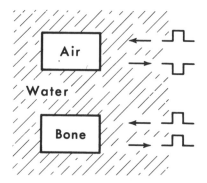

Fig 1   Fate of compression pulse after reflection
        from rather good reflectors, one acousti-
        cally hard and the other soft.

Fig 2   Spermaceti organ removed from over upper
        jaw has the Museau lips on front end and
        at the back had pressed against knobby
        tissue here held in place against front
        of skull by the hand.

Fig 3   Covering of the back wall of the frontal
        (posterior) sac.

Fig 4   Museau du Singe lips after removal of
        overlying front wall of anterior sac.

ducting of sound in the organ were involved, its conical taper would double the resonant frequency to one with a period of 2 transit times as if both ends were "open" or "soft", such as in a flute. (An oboe plays twice the frequency of the same length clarinet and overblows the octave rather than the twelfth.) Predicted frequencies are in the correct range, while variable lip tension, starting transients and short pulse length could cause the observed spread. Overtone emphasis depends on sizes and absorptions of all structures. Further observations will sort out alternatives, but the organ can control the frequency content of the clicks.

For the action to be like a trumpet blown for a few cycles, the natural frequency of the lips must be lower than the chamber resonant frequency, which seems true. If the column instead of interacting back on the lips merely emphasized certain frequency ranges (as with our use of vocal cords) then the start of each click would best communicate whale length. Another less likely model is that amplitude of a few cycles could be gradually built up in the column and then the stored cycles suddenly released as in a "Q switched" laser. Actual patterns seem not to favor models of passive axial reverberation of an initial click between two air sacs. While vortex shedding can take place in water (though not with water blown into air), I find it difficult to whistle at elevated pressure where "Bronx cheering" and speech with vocal cords remain easy; this may relate to why deep diving Physeter only clicks.

The velocity of sound in the spermaceti organ of a young animal at 35° C (a possible temperature) was V = 1.4 km/sec with a temperature coefficient from 20°-35° (while softening) of -5×10^{-3} km/sec/C°, versus the literature value of V = 2.6 km/sec. This velocity value relates a shorter whale to a given vocal frequency. If spermaceti density were very different from sea water, whales would have trim problems swimming level. Specific gravity was 0.86 here. Thus with the velocity of sound in tissue also being similar to that of water there is a fair impedance match for the release of sound, with sound being released in all directions from the forward sac (as is observed) and the monopole pattern being slightly modified toward a dipole pattern from the two sacs. The layered "junk" might help couple sound forward.

Norris, K.S., and Harvey, G. W., 1972, A theory for the function of the spermaceti organ of the sperm whale, in: "Animal Orientation and Navigation", Galler, Schmidt-Koenig, Jacobs and Belleville, ed., NASA SP-262.
Mackay, R. S., 1972, Discussion to above paper, 415-416.

HOW THE GREEN LACEWING AVOIDS BATS:

BEHAVIOR AND PHYSIOLOGY

Lee A. Miller

Bat sonar systems are primarily used for avoiding obstacles and detecting prey. But, not all prey are deaf to the ultrasonic cries of bats. Members of two orders of insects, namely the moths (Lepidoptera) and the green lacewings (Neuroptera), have evolved mechanisms for detecting and avoiding bats. In this poster I will present the results of studies designed to investigate the avoidance behavior of freely flying green lacewings (Chrysopa carnea) to hunting bats and the physiological mechanisms underlying this behavior. Detailed accounts are found in Miller and Olesen (1979) and Olesen and Miller (1979).

Interactions between hunting bats and freely flying green lacewings were studied in a large room using stop-motion photography and simultaneous recordings of emitted orientational cries. Pulsed ultrasound was also used. Electromyograms were taken from individual flight muscles while insects were in stationary flight.

Freely flying green lacewings often respond to hunting cries of bats by folding their wings and passively nosediving. Nosediving is an "early warning" response that takes the insect out of the bat's acoustic field. In the figure the insect begins its nosedive just prior to the bat's buzz (flash 5; the dots and numbers indicate stroboscopic flashes. L shows the bat's cries (repetition rate and relative cry duration) recorded by an ultrasonic detector to the left of the picture field). In this case the bat detected the insect too late to catch it. Bats caught less than 10% of the green lacewings they attacked and only 27% of the insects "buzzed" (see (1) in Table). Could green lacewings have other strategies for escaping bats after they had been detected? Should a falling green lacewing be detected, it can momentarily extend its wings and fold them again, continuing the fall. This "wing flip" constitutes a "last chance" response, which can be triggered by the high cry repetition rate in the buzz. The last chance response was studied in more detail using pulsed ultrasound. Green lacewings show other responses to ultrasonic stimuli such as changes in flight path and pattern. In addition the responses are variable, making them unpredictable. Hearing and responding to ultrasound gives intact green lacewings at least a 47% selective advantage over deaf green lacewings (see Table; destroying the ears does not interfere with flight). Green lacewings enjoy about the same selective advantage as tympanate moths with respect to bat predation (see Table) even though moth strategies differ from those of green lacewings.

941

Fig. 1.

Table: Encounters between insectivorous bats and insects.

CATEGORY	(A) ATTACKS	(B) CATCHES	(C) MISSES	(D) BUZZES	(B/A) SELECTION PRESSURE	(B/D) CATCHING EFFICIENCY	SELECTIVE ADVANTAGE- REACTORS OVER NON-REACTORS
(1) REACTING GREEN LACE-WINGS	185[x]	18	167	67	0.10	0.27	$1-\dfrac{(2c/2a)}{(1c/1a)}$
(2) NON-REACTING GREEN LACE-WINGS	63 [x]	33	30	37	0.52	0.89	0.47
(3) REACTING MOTHS	174	12	162	-	0.07	-	$1-\dfrac{(4c/4a)}{(3c/3a)}$
(4) NON-REACTING MOTHS	172	82	90	-	0.48	-	0.44

x : BASED ON THE NUMBER OF OCCURRENCES OF BUZZES AND HIGH CRY REPETITION RATES.

1 AND 2: DATA BASED ON ENCOUNTERS BETWEEN 3 BATS (TWO P. PIPISTRELLUS AND ONE M. BRANDTII) AND 507 INSECTS (C. CARNEA). MAXIMUM ERROR IN THE DATA IS 10%. NON-REACTING GREEN LACEWINGS WERE EARLESS.

3 AND 4: DATA BASED ON FIELD STUDIES OF ENCOUNTERS BETWEEN 408 MEDIUM SIZED MOTHS AND UNIDENTIFIED BATS. COMPUTATION OF SELECTIVE ADVANTAGES IS FROM ROEDER & TREAT (1962), PROC. XI ENTOMOL. CONGR. 3: 7-11.

The activity patterns recorded from flight muscles correlated well with the behavioral responses from freely flying insects. Restraint has little influence on physiological responses. Green lacewings in stationary flight stop some or all of their flight muscles in response to pulsed ultrasound. If all muscles stop, the elevators stop last and start first upon the resumption of flight. This response would produce a nosedive with the wings in a dorsal position. The stopping of some muscles, but not others, would produce, irregular flight patterns. Physiological responses show variability both in which muscles stop and when they stop. Phase measurements between antagonistic muscles indicate that the stopping is produced by a temporary uncoupling of motor neurons from the continuously active flight pattern generator. The uncoupling mechanism is sufficient to explain the behavioral responses and variability in freely flying green lacewings. Variability, it seems, may increase the survival advantage of the prey by making its behavior more unpredictable for the predator.

I acknowledge the Danish Natural Sciences Research Council.

REFERENCES

Miller, L. A., and Olesen, J., 1979, Avoidance behavior in green
        lacewings. I. Behavior of free flying green lacewings to
        hunting bats and ultrasound, J. Comp. Physiol., 131:113.
Olesen, J., and Miller, L. A., 1979, Avoidance behavior in green
        lacewings. II. Flight muscle activity, J. Comp. Physiol.,
        131:121.

CYLINDER AND CUBE SHAPE DISCRIMINATION BY AN ECHOLOCATING

BLINDFOLDED BOTTLENOSED DOLPHIN

Paul E. Nachtigall, A. Earl Murchison

and Whitlow W. L. Au

An Atlantic bottlenosed dolphin was trained to wear rubber eyecup blindfolds and to choose a cylinder as compared to a cube when the two targets were simultaneously presented. The animal was housed and tested within a circular fiberglass tank filled with filtered sea water. As may be seen in Fig. 1, the animal was trained to station within the tank and echolocate down a water-filled trough.

The targets were presented 2 m away from the animal, 40 cm below the surface, and 38 cm apart. The targets were placed between four 30 lb test monofilament lines attached, with stretchable elastic cord, to a wood frame.

On each of the 63 trials per session the animal was required to indicate the randomly predetermined right or left position of the cylinder by pressing one of two adjacent manipulanda. Each of the three different sized cylinders (with heights and diameters of

Fig. 1. Dolphin tank and apparatus.

4 x 4, 5 x 5, and 6 x 6 cm) was repeatedly paired with each of three different sized cubes (with heights, depths, and widths of 4 x 4 x 4, 5 x 5 x 5, and 6 x 6 x 6 cm). This variation in target sizes controlled for possible differences in overall reflectivity between cubes and cylinders. The order of presentation of the target pairs was randomly predetermined within each nine trials. That is, the 63 trials were broken down into seven nine-trial blocks. Each possible combination of targets was presented once within each nine-trial block.

Data in terms of the overall percent correct choice of the cylinder in each of the nine possible target combinations is presented in Fig. 2. The data indicate that the animal differentiated cylinders from cubes and that discriminability was influenced by target size. Comparisons of cubes and cylinders of the same size show increased discriminability of shape as target sizes increase.

Once the ability of the animal to discriminate cylinders from cubes was well-established, a probe technique was used to examine the effects of changing target aspect. Baseline performance trials were continued on 56 of the 63 trials per session but on the other seven trials one of the targets was either rotated or layed down horizontally. One probe trial was randomly presented within each

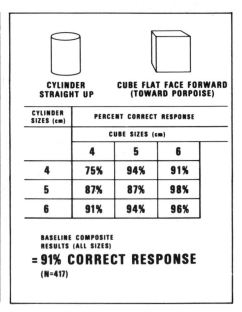

Fig. 2.   Percent correct choice of cylinder.

nine trial block. A fish reinforcer was provided regardless of the animal's choice. As illustrated in Fig. 2, during the first two probe sessions the cylinder was placed straight up but the cube was rotated to present the edge forward. This change did not disrupt the animal's performance. During the next two sessions probe trials were presented with the cube edge forward and the cylinder layed down on its side. Performance declined to 71%. During the final two sessions both cubes and cylinders were presented on these probe trials with the faces forward. Performance declined to near chance level.

Following the completion of the experiment the targets were examined acoustically with a monostatic sonar measurement system. The echo returns for the various targets were processed, and re-sulting frequency spectra were obtained, by using a 1024 point fast Fourier transform. Comparisons of the target spectra failed to reveal consistent and obvious cues for the discrimination of shape.

Fifteen measures of target strength for each target, however, revealed an interesting possible cue for shape discrimination. The standard deviation of the target strength measures of the cubes straight up were 1.22, 1.01, and 1.01. The standard deviations of the cubes with the flat faces forward were much higher at 2.71, 5.92, and 6.18. Measures similar to the cubes were obtained from the cylinder presented with the flat face forward: 2.58, 4.93, and 5.86. These differences in variability paralleled the performance given by the animal. The animal most likely received repeated pulse echoes varying in amplitude when scanning across the flat surfaces of cubes or the tops of solid cylinders and relatively uniform pulse echoes when scanning across the curved portion of cylinders.

"RANGE" INFORMATION PROCESSING AT THE HIGHEST LEVELS OF THE
AUDITORY SYSTEM.

William E. O'Neill and Nobuo Suga

Echolocating bats determine target range by measuring the time
delay between the emitted biosonar pulse and the returning echo.
We have found two types of neurons in a part of the auditory cortex
of the mustached bat, Pteronotus parnellii rubiginosus, which
demonstrate selective sensitivity to echo delays, and therefore,
target range.

The biosonar signals consist of up to four harmonics, each of
which contains a long constant-frequency (CF) component followed
by a short frequency-modulated (FM) component. The duration and
repetition rate of these signals become shorter and higher as the
bat pursues and captures prey. In order to mimic as much as
possible the actual biosonar signals and echoes for stimulation,
we built two harmonic generators, which could produce up to three
harmonics from a single fundamental frequency. We could, therefore,
stimulate using some or all of these harmonics by switching them
in or out at will.

Neurons in the FM processing area of the auditory cortex respond vigorously to a weak FM sound at frequencies near one of the
non-fundamental harmonics of the biosonar signal, only if it is
preceded by an FM sound mimicking the fundamental FM component of
the emitted pulse in both amplitude and frequency. The responses
remain essentially the same whether the pulse and echo contain all
the harmonic components or just the essential FM components, or
whether the CF components are produced continuously (no silent
periods) or not at all (just FM). The time delay between the
fundamental FM component (emitted sound) and the second, non-
fundamental FM sound strongly influences the sensitivity of the
neuron to the second FM sound (echo), and one can determine a best-
delay for maximum excitation of a single neuron. Since in reality
the delay is related to the distance to the target, these neurons
are range-sensitive.

When presented with pulse-echo paired stimuli, tracking
neurons respond to echo FM's from a wide range of delays (target
distances) at repetition rates mimicking the search and approach
phases of target-oriented flight, but the best delay decreases
dramatically during the terminal phase, just prior to target interception. Correspondingly, their sensitivity to targets at other
distances decreases sharply. Thus they appear to follow, or track,
the target as the bat approaches it.

PST histograms illustrating a tracking neuron are shown in the accompanying Figure. Note that as the repetition rate (RR) of the stimulus pairs increases, both the range of echo delays over which the neuron responds (EH_2 delay) and the best delay decreases. No responses were elicited to pulses or echoes delivered alone, or to pulse-echo pairs with delays shorter than two msec.

Range-tuned neurons, on the other hand, do not change their best delays with changes in repetition rate and duration of the stimuli. They respond to targets located only at certain distances, regardless of the stage of echolocation. Although best delay remains similar in any one cortical column, it differs among columns in the FM processing area. We have found that the best delay increases systematically in recordings from oblique electrode tracks moving rostro-caudally through the FM processing area. Range-tuned neurons are therefore arranged on an "odotopic" axis, in which target range is systematically represented by place on the cortical surface. The average change in best delay from neuron to neuron along this target range axis is 0.116 msec, corresponding to an average resolution of 2.0 cm. This correlates well to behavioral evidence for the acuity of target range discrimination in bats. The bats Eptesicus fuscus, Phyllostomus hastatus, Pteronotus suapurensis, and R. ferrumequinum are able to discriminate 1.2 to 2.5 cm range differences at an absolute distance of 30 and/ or 60 cm (Simons, 1971; Simmons, Howell and Suga, 1975). This would also correspond to our finding, if we assume that the rate of change in best range, 2.0 cm per neuron, is the theoretical limit of just noticeable difference in distance.

It is puzzling that the fundamental (H_1, in particular FM_1) of the orientation sound is so critical to the response of range-tuned neurons in spite of the fact that H_1 is always much weaker than the second and third harmonics and is sometimes barely detectable in recordings in our laboratory. We propose that this is an adaptation to reduce the probability of jamming by the sounds of echolocating conspecifics. When we delivered pairs of orientation sounds and echoes without H_1, we found no range-sensitive neurons in the FM processing area. This means that range-sensitive neurons are probably not excited by combinations of orientation sounds and/or echoes produced by conspecifics flying nearby. To excite range-sensitive neurons, H_1 must stimulate the ears prior to an echo, in spite of its weakness in the emitted sound. This suggests that H_1 produced by the vocal cords stimulates the animal's own ears by bone conduction, while it is not emitted at a significant amplitude possibly because of suppression by vocal-tract anti-resonance. In nature, range-sensitive neurons would be selectively excited only when the animal itself emits orientation sounds and echoes return with relatively short delays. Without this property, target ranging would be frequently impaired by conspecifics. Thus these neurons are jamming-tolerant in most situations.

Supported by NSF grant BMS 75-17077 and BNS 78-12987 to Nobuo Suga, and NINCDS (PHS) training grant 1-T32-N507057-01 to William E. O'Neill.

References

Simons, J.A. (1971). The sonar receiver of the bat. Ann. N.Y. Acad. Sci. 188, 161-174.
Simmons, J.A., Howell, D.J., and Suga, N. (1975). Information content of bat sonar echoes. Am. Sci. 63, 204-215.

THE FUNCTIONAL ORGANISATION OF THE AUDITORY CORTEX IN THE

CF - FM BAT RHINOLOPHUS FERRUMEQUINUM

Joachim Ostwald

Greater Horseshoe Bats, Rhinolophus ferrumequinum, use orientation sounds consisting of a long constant frequency (cf) part, followed by a short, downward frequency modulated (fm) part. By lowering their emission frequency during flight they compensate Dopplershifts and keep the cf-part of their echoes in a very small frequency band that is characteristic for each individual animal. Thereby they create a constant carrier frequency on which the fast, complex amplitude and frequency modulations are superimposed that are produced by the movements of targets, for instance the wingbeat of flying insects. Those modulations are characteristic for each insect species and size and may therefore be used to classify or even identify prey and to distinguish them from background clutter.

Multi-unit recordings show that the auditory cortex of Rhinolophus is especially adapted to the processing of echoes. Units with best frequencies (BF's) in the frequency band of the orientation sounds are strongly overrepresented in the tonotopical organisation (Fig. 1). There are two large areas that are concerned with the processing of the cf- and the fm-part of the Dopplershifted echoes respectively (shaded in Fig. 1). In both areas the frequency axis is stretched compared with other areas. In the fm-processing-area and in areas not concerned with echo processing neurons with equal BF's are organised in dorsoventral slabs with raising frequency from posterior to anterior. In the cf-processing-area frequencies between resting frequency and 2 kHz above are represented which are the frequencies occuring mostly in the Doppler-shifted cf-parts of the echoes. Here equal-BF-contour-lines are bent around each other in a semicircle arrangement with the resting frequency at the outside. When frequencies are normalized to the individual resting frequency of the animal, this pattern stays constant even for large variations in resting frequency. A similar disproportionate representation of echo frequencies was found in the auditory cortex of Pteronotus parnellii (summarized in: Suga & O'Neill, review at this conference). In contrast to Rhinolophus equal-BF-contour-lines in the Doppler-shifted-CF processing area of Pteronotus are organised in closed circles and with a reversed frequency arrangement.

The tuning curves of neurons with BF's in a frequency band of ± 2 kHz around resting frequency are extremely sharp (Q 10-dB values up to 400). They have high frequency selectivity and therefore are well suited for coding even small frequency modulations. With de-

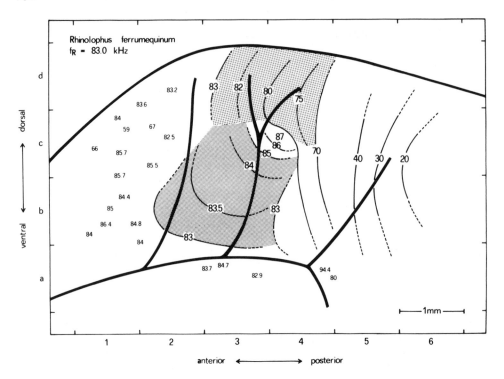

Fig. 1. Tonotopical organisation of multi-unit best frequencies
recorded in 5 bats. Measured frequencies have been normalized to the
resting frequency of the individual animal. For plotting the resting
frequency was set to a mean value of 83.0 kHz. The shaded areas represent the cf- and the fm-processing-area respectively.

creasing BF the Q 10-dB value becomes smaller. For BF's in the frequency band below 70 kHz, that is not used for echolocation, Q 10-dB
values are smaller than 20 which is the maximum value reached in
other mammals. Sharp tuning of neurons with BF's around resting frequency is found in all centres of the auditory pathway of Rhinolophus
studied so far and is probably due to a sharp mechanical filter in
the cochlea (summarized in: Neuweiler, review at this conference).

The response pattern of these neurons changes when sinussoidally frequency modulated stimuli (SFM) down to a frequency shift
of 0.05 % are used. With larger modulation depth phase locked responses occur. The degree of phase locking depends not only on the
modulation depth but also on the position of the carrier frequency
within the pure tone tuning curve. The response to complex frequency
and amplitude modulations caused by the wingbeat of flying moth is
phase locked to the wingbeat and differs from the response to SFM
(Fig. 2 B, C). Both responses can occur at carrier frequencies where
there is no response to the unmodulated carrier itself (Fig. 2 A - C).

Fig. 2. Response pattern of cortical neurons to a pure tone, 82.0
kHz, 80 dB SPL (A), to a sinussoidally frequency modulated stimulus
(SFM), modulation depth ± 1 kHz (B), and to the reproduced echo of
a flying moth (C) on the same carrier frequency. In the complex moth
echo frequency and amplitude modulations are present. The response
is phase locked to the wingbeat. (D) Threshold curves of the same
unit with (a) pure tone, (c) SFM, and (b) moth echo stimulation.
Plotted are frequency and intensity of the carrier.

Compared to the pure tone tuning curve (Fig. 2 D.a) the threshold
curve for SFM stimuli, modulation depth ± 1 kHz, (Fig. 2 D.c) is
1 kHz wider on both the high and low frequency side. For moth echo
modulations there seems to be a special mechanism of processing in
the unit shown in Fig. 2. Although in this stimulus there is an
asymetrical widening of the frequency band present with each wingbeat of −2 to +1 kHz (see Schnitzler & Henson, review at this conference, Fig. 16), carrier frequencies below the pure tone BF are
more effective in eliciting a response (Fig. 2 D.b).

TURSIOPS BIOSONAR DETECTION IN NOISE

Ralph H. Penner and James Kadane

Biosonar detection experiments conducted in the ocean are confounded as the range from the animal to the target is increased because added interference, such as fish schools, occlude the "field of view". This experiment was designed to test deduction performance as a function of noise at a fixed range of 16.5 meters, thus reducing the range variability problem. Echolocation data were collected from two Tursiops truncatus, named Ehiku and Heptuna, in Kaneohe Bay, Hawaii, during November 1978. They were conditioned to respond discretely in a two alternative forced choice detection problem of 0.5 a priori random presence or absence of a 7.65 cm steel water-filled sphere (Ts. -32 dB). The sphere, displayed one meter below the surface and 16.5 meters from the porpoise at. scan station, assured stable low effort baseline detections in the absence of added noise. Detection accuracy, pulse train duration,, and response latency data were recorded in response-terminated trials. Five levels of white noise, 67, 72, 77, 82, 87 dB rel μPa in a 1 Hz band, were presented in 10 trial blocks, each level occurring twice in each 100 trial session. There were a minimum of 220 trials at each sound level. Ambient noise, which is extremely high in Kaneohe Bay, was about the same as the lowest white noise level used in the experiment.

A white noise source was located 4 meters from the porpoise and directly in line with the sphere or the position occupied by the sphere. It was activated before an acoustically opaque screen was lowered which permitted an acoustic view of the stimulus area, and deactivated after a response on one of the manipulanda.

Although there was no significant change in response accuracy over baseline with the 67 dB noise, response bias became neutral (Fig. 1) and there was an abrupt increase in the number of clicks per click train (from 25 to 50) emitted by the two animals (Fig. 2). Overall performance degraded with both animals as the noise level increased. Ehiku's detection/rejection ratio remained neutral, whereas Heptuna showed a liberal bias in his decision process as the noise level increased. It was interesting to find that the click count ("echolocation effort") (Fig. 2 and 3) and response latency (Fig. 4) both increased with both animals until the noise exceeded 77 dB. At 82 and 87 dB, where the overall performance was near chance, the click trains got shorter and latencies longer.

Fig. 1.   Response Accuracy Ogive and Signal Detection Parameters,
          Sensitivity d[1] and Response Bias β.

Fig. 2. $\overline{X}$ Clicks per Trial

Fig. 3. X̄ Seconds Click Train per Trial

Fig. 4. X̄ Seconds Response Latency, Time from End of Click Train
to Activation of Response Manipulanda.

CORRELATION ANALYSIS OF ECHOLOCATION PULSES

J. David Pye

Following the lead of D. A. Cahlander (1967) and more recently of K. J. Beuter (this conference), radar ambiguity diagrams of bat pulses must now be regarded as an important extension to merely descriptive or acoustic analyses. The sonar properties of such signals are not intuitively obvious for several reasons:

(1) Their frequency and amplitude patterns differ from those of 'worked examples' given in radar texts and, indeed, often defy the engineers' design constraints (eg. envelope shaping, unusual f.m. patterns and the presence of harmonics); such signals may be difficult to define for purely mathematical analysis. (2) Several of the assumptions of radar theory (or even under-water sonar theory) are not valid for ultrasound in air: the fact that band-width is large compared with the centre frequency means that Doppler shift must be treated as a factor (Cahlander) and not approximated by an incremental offset, so that for example, Doppler tolerance is achieved by linear period sweeps (Altes and Titlebaum, 1970), not by linear f.m. (3) Propagation attenuation in air is strongly dispersive, the atmosphere acting as a low-pass filter (Griffin, 1971), and this must affect many long-range 'cruising' pulses (Fig. 1). (4) The back-scattering cross-sectional area of many targets is dispersive (eg. small insects are high-pass Rayleigh scatterers, Fig. 2). At certain ranges, depending on frequency and water vapour concentration, these two effects cancel out to give a 'flat' echo spectrum (Fig. 3). Bats presumably allow for all these effects in the evolution of their signals and in their methods of signal processing.

Our hybrid, true-Doppler correlator, described by J.A.T. Halls (poster, this conference) is therefore being used to examine:

(1) Pulse structures from our own recordings of 118 species of echolocating birds and bats. (2) The changes of pulse structure shown by many species under different conditions. (3) Artificial signals which simulate real pulses (Pye, 1967), so that noise is minimised and each of the pulse parameters can be varied independently. (4) Signals propagated over different distances. (5) Real echoes reflected from various targets. Correlation also offers the possibility of locating sources from short baselines and so of tracking individual bats in flight. Clearly such methods have great potential for the study of echolocation mechanisms in both theory and practice.

References: please see author's Review chapter.

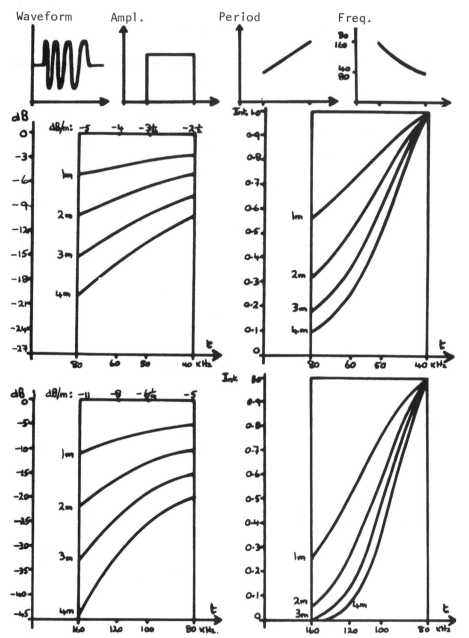

Figure 1. High frequency attenuation in air. Top: two model pulses
with square envelopes and linear period modulation over one octave.
Middle: the effects of propagation over 0-4 m in air with 1% H$_2$O,
at 80-40 kHz (dB left, normalised intensity right). Bottom: the
same at 160-80 kHz. Higher water vapour content gives greatly
increased effects. Data from Griffin (1971).

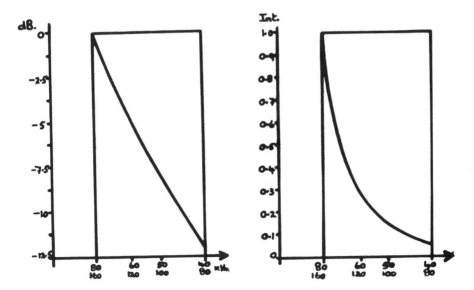

Figure 2.   Rayleigh back-scatter from a small sphere ($\lambda \gg \pi d$) for
the model pulses of Fig. 1.

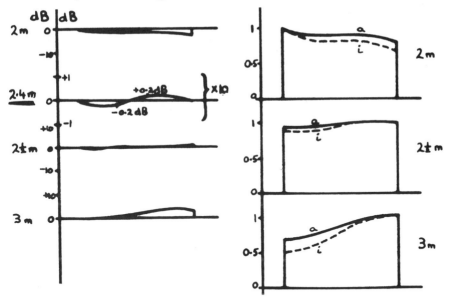

Figure 3.   Low-pass attenuation and high-pass Rayleigh scattering
can almost cancel out at certain ranges to give a 'flat' echo
spectrum.   For 80-40 kHz in 1% H_2O this occurs at 2.4 m;   left:
the echo in dB (at 2.4 m expanded x 10 to show ± 0.2 dB);   right:
intensity.   Similar 'flat' returns occur at 0.9 m for 160-80 kHz in
1% H_2O and at 1.5 m for 80-40 kHz and 0.6 m for 160-80 kHz in
3% H_2O.

THE COCHLEA IN PTERONOTUS PARNELLII

Ade Pye

Pteronotus parnellii provides an exceptional example among
the New World bats so far studied in being able to compensate for
Doppler shifts (Schnitzler 1970) and in possessing a cochlea which
is sharply tuned to just above emitted frequency at rest (Pollak
et al 1972). Structurally the cochlea exhibits a number of
specialized features (Pye,A. 1967 and 1978). For example, the
basal turn is much enlarged, including an enormous second half-turn
which exhibits the largest scala vestibuli found in any of the bats
studied. It has recently been reported by Pye,A. (F.I.B.R.C.,
Albuquerque 1978) that another special feature of the basal turn
is the presence of an un-identified substance in the scala tympani,
see Fig.1. It is located mainly at the medial and posterior
walls of this scala and has also now been found to reach around
the round window and the cochlear aqueduct. A smaller area is
situated anteriorly, see Fig. 2. This substance has been found
in every cochlea of Pteronotus parnellii studied (24 in all) from
various locations in Trinidad and Panama. But it is completely
absent in Pteronotus davyi and all other bats examined to date
(68 species from 14 families).

Fig.1. Horizontal section through the cochlea of Pteronotus
parnellii (x 29). The arrows indicate the position of the
substance. Note lysis of the cartilage situated in the
bony cochlear wall on the left side.

Fig.2. Micro-dissected cochlea of <u>Pteronotus parnellii</u> (x 16)
       S indicates the substance. a) shows the posterior part of
       the substance around the round window and onto the medial
       walls, b) shows the anterior part of the substance. Notice
       the enormous basal turn when compared with the second turn.

Under light microscopy it appears amorphous in structure, measuring
between 30μm and 40μm in thickness and being surrounded by a
distinct membrane in places. Further histological staining has
shown it to stain a) negative with Gram, b) positive with Amido
Black (specific for haemoglobin), c) red with Masson's Trichrome
(red for blood corpuscles and muscles), d) with Congo Red and
Van Gieson's (Elastin and Haemotoxylin) the same as for red blood
corpuscles in the surrounding vessels. Electron microscopic studies
also indicated it to be amorphous, but now new observations have
shown that there are distorted red blood corpuscles on the surface,
with massed haemoglobin underneath, see Fig.3. This is a surprising
finding and yet it agrees with the results obtained from histological
staining viewed under light microscopy. The massed haemoglobin
is obviously not randomly distributed, since it appears in precise
locations only. But as lysis of the cartilage situated in the bony
cochlear wall (see Fig.1.) also seems at present to be confined
to this species, the two phenomena could possibly be related.
Since this symposium only one foetus and one baby bat of this
species were available for histological study and neither showed
the presence of the massed haemoglobin nor lysis of the cartilage.

Fig.4. Electron micrograph of the substance, showing distorted red
       blood corpuscles on the left, with massed haemoglobin
       underneath on the right. (x 9250) (Courtesy of Dr.A.Forge)

It is obviously of great importance to trace the evolution of this
massed haemoglobin through foetal and young development to the
adult cochlea. Further investigations are being carried out, but
at present the definite origin and possible functional significance
of the massed haemoglobin remains unknown.

Pye, A. 1967. The structure of the cochlea in Chiroptera. III
       Microchiroptera: Phyllostomatoidea. J. Morph. 121, 241-254

Pye, A. 1978. Aspects of cochlea structure and function in bats.
       Proceedings of the Fourth International Bat Research
       Conference, Kenya National Academy for Advancement of
       Arts and Sciences, Kenya Literature Bureau.

ACTIVITY OF THE RECURRENT LARYNGEAL NERVE DUE TO THE PRODUCTION

OF ULTRASONIC ECHOLOCATION SOUNDS IN THE CF-FM BAT,

RHINOLOPHUS FERRUMEQUINUM

Rudolf Rübsamen

The Greater Horseshoe bat, Rhinolophus ferrumequinum emits a stereotyped echolocation sound consisting of a long constant-frequency portion (CF) terminated by a short frequency-modulated part (FM). During echolocation the frequency of the orientation pulse is controlled with high accuracy (Dopplershift compensation) and the duration of the call is shortened while the bat is approaching its targets.

The orientation pulses are produced in the larynx, which is innervated by two branches of the vagal nerve, the superior laryngeal nerve and the inferior laryngeal nerve. The superior laryngeal nerve, innervating exclusively the cricothyroid muscle, controls the frequency of the orientation sound. The inferior laryngeal nerve innervating the rest of the intrinsic laryngeal muscles, seems to be responsible for the temporal control of sound emission.

Simultanously recorded respiration signals and nerve activity of the recurrent laryngeal nerve show that this nerve is active during respiration as well as during vocalization (Fig. 1 and 2). The respiration is highly adapted to the demand of vocalization. Even in the quiet respiration without vocalization the expiration phase is a short pulse of air lasting about the length of an orientation call (50 msec). The following inspiration phase is longer and lasts about 150 msec. In the experimental situation, where vocalization was elicited by electrical stimulation, the vocalization does not change the rhythm of respiration and is strictly correlated with the expiration phase.

During vocalization a stereotyped activity pattern superimposes the respiratory activity (Fig. 3 and 4). 20 msec prior to the onset of vocalization the activity of the inferior laryngeal nerve reaches its first maximum. This level of activity is roughly maintained until 10 msec before the sound terminates. A short notch of reduced activity is then followed by a second strong activity maximum that parallels the termination of vocalization. The subglottic pressure is build up shortly before the onset of vocalization, it is maintained during the emission of the sound and drops at the end of the outgoing vocalization.

Viewing subgottical pressure and inferior laryngeal nerve activity
together suggests that the activity of the inferior laryngeal nerve
might be responsible for the closure of the larynx to build up
subglottic pressure prior to the vocalization and for the opening
of the larynx to reach the respiratory state at the end of the sound.

Fig. 1   Registration of respiration activity of the recurrent
          laryngeal nerve.

          In the upper trace respiration is recorded. The middle
          trace represents the registration of the nerve activity
          and the lower trace shows the comparator signal of the
          nerve registration.

Fig. 2   Recording of vocalization activity in the recurrent
          laryngeal nerve.

          Continuous recordings of the envelopes of the orientation
          calls, their CF-frequency, the gross activity of the nerve
          and the comparator signal for histogram-formation are
          shown.

a. Nerve activity prior to cutting of the N.recurrens

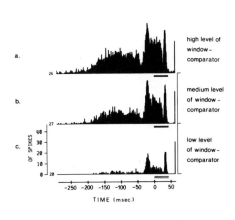

a.

high level of
window –
comparator

b.

medium level
of window –
comparator

c.

low level
of window –
comparator

b. Nerve activity after unilateral cutting of the
N.recurrens

Fig. 3.   Fig. 4.

Fig. 3   Ante-stimulus-time-histogram of nerve activity of the
         recurrent laryngeal nerve during respiration without
         concurrent vocalization.

         a. Nerve activity before the transection of the recurrent
         laryngeal nerve.
         b. Nerve activity after unilateral transection of the
         recurrent laryngeal nerve.
         The start of inspiratory phase is marked by an arrow.
         Each histogram is composed of 40 sweeps.

Fig. 4   Ante-stimulus-time-histogram of nerve activity of the
         recurrent laryngeal nerve during vocalization. The
         histograms are triggered by the onset of vocalization. The
         position of the orientation calls is represented by
         double-bars. All three histograms are summed up from the
         same sample of orientation calls using different discri-
         minator levels (a,b,c). The histograms show that indepen-
         dent of the discriminator level the same discharge-pattern
         of the recurrent laryngeal nerve during vocalization is
         obtained.

SINGLE BRAIN STEM UNIT RESPONSES TO BINAURAL STIMULI SIMULATING

MOVING SOUNDS IN RHINOLOPHUS FERRUMEQUINUM

Peter A. Schlegel

Echolocating bats are essentially dependent on the exact determination of target locations and their changes relative to the bat.
Echoes reflected from any space location are determined by the distance and the vertical and horizontal (azimuth) angle to the bat's
longitudinal axis.

The generally accepted theory of directional hearing assumes
neural mechanisms that integrate inputs from both ears (binaural
hearing). The most significant stimulus cue concerning the azimuth
angle of a sound source, i.e., echoes returning from targets, for
bats is the intensity difference between the two ears.

Directionality of Rhinolophus' hearing system

From neurophysiological data on the directionality of the ears
(Schlegel, 1977a) one can derive "interaural pressure difference"
curves (IPD) coding azimuth angles between ±40° from straight ahead
direction, probably the biologically most significant range. This
convertability offers the advantage of replacing free field stimulation via loudspeakers by binaural intensity combinations applied
through earphones (Schlegel, 1977b).

Neurophysiological properties of binaural neurons

Single unit recordings from binaural brain stem levels revealed
that in Rhinolophus, similar to other mammals studied, the most
prominent binaural mechanism consists in a subtraction of the neural inputs between right and left ears: stimuli delivered to one
channel provide excitation, whereas stimuli into the other inhibit
excitation so that ideally the resulting activity is a directly
proportional measure of the azimuth angle. This general relationship holds for the whole excitatory/inhibitory dynamic range of a
unit but is quantitatively not independent of absolute intensity
levels; yet one can state that brain stem units are suited to even
precisely code angular positions (best incremental limen estimated
as to ± 1°: Schlegel, 1977a).

Coding of moving sounds

Since bats hunt while flying one has to consider relative movements between bat and target as biologically highly relevant features

fig. 1-3

A. Time representation

Time course (abscissae, 1s time bar; 2oo bins, binwidth 46.7 ms, 5 runs averaged) of single unit activity (ordina- tes, number of impulses/bin) when trapezoidal intensity changes, indicated by thin solid lines, are applied simu- lating rotating sound sources. Histograms gained with diffe- rent sweep speeds are display- ed vertically (up to down, in- creasing velocity as marked); absolute intensity used as in- dicated. The histograms are aligned one above the other to the reference of o° appa- rent azimuth angle (dashed vertical lines). Curves were measured binaurally unless otherwise labeled. Wordings "contra" and "ipsi" stand for virtual sound source locations refered to the recording side.

B. Angle representation

Same data replotted in angular representation (abscissae, azi- muth angle, °; ordinates as in A) for the two opposite move- ment directions as indicated by arrows. Note the amount of shifts ("hysteresis") as sweep speed increases!

fig. 1, proportionalist type

fig. 2, velocity and angle coder type

fig. 3, movement direction coder type

(for more description, see text)

(changes of the dynamic cue, sound motion, its velocity and direction). Therefore neurons were stimulated with electronically produced linear intensity increases in one ear and simultanous decreases in the other. This device simulates horizontal sound movements of constant velocity which was changed systematically (maximally $\pm$ 4o$^{\circ}$ around straight ahead direction corresponding to 4o dB shifts).

The neurons can be classified according to their basical binaural properties but moreover according to their response types to dynamic stimuli as:
1. "Proportionalists" (most common type) code instantanous azimuth angles within a certain working range. The best of these coders indicate the value directly proportionally even during fast movements and independently of movement direction without "hysteresis". The slope of the response increase and decrease is directly proportional to the movement velocity (fig. 1).

2. "Velocity coders" respond with as more activity as faster is the angular velocity, but almost independently of movement direction. Some of these units are activated only when the stimulus sweeps through a certain angle range and sometimes indicate a particular angle by the maximum of excitation independently of velocity and movement direction (fig. 2).

3. "Movement directionalists" respond to motions into one direction with excitation, into the opposite direction with inhibition whereby the amount of the reaction is directly related to velocity (fig. 3). Information on absolute azimuth angles is lost as these units extract the dynamic features movement velocity and direction.

Discussion and conclusion

1. On the brain stem levels studied, neurons with binaural coding of azimuth angles were found as postulated which showed almost no specificity for dynamic cues ("proportionalists"; fig. 1).
2. Although less frequently, some kinds of "feature extracting" neurons were encountered that preferrentially analysed the dynamic cues, sound movement velocity (fig. 2, 3) and direction (fig. 3). Extreme "feature detectors" were not found -; a few rare units, however, only responded significantly with a modulation of a yet stimulus correlated, overall activity (transient excitation and depression; example presented by Neuweíler, this volume). One may ask whether on higher auditory brain levels still higher specializations take place at all.
3. Some but not all dynamic neural response characteristics can already be understood on the base of stationary monaural and binaural input/output functions (comp. with Schlegel, 1977a).
4. The biological relevance of the informations contained in the more or less specialized units' responses for the echolocating bat while hunting in free flight, i.e. for its motor control, might be obvious and should be studied neurophysiologically and behaviourally in more detail.

ALTERATIONS OF AUDITORY RESPONSIVENESS BY THE ACTIVE EMISSION OF

ECHOLOCATION SOUNDS IN THE BAT, RHINOLOPHUS FERRUMEQUINUM

Gerd Schuller

During echolocation, the greater horseshoe bat (long CF-FM calls) faces the problem of processing the faint echo during strong acoustic self stimulation by the emitted vocalization. The question arises whether the vocalization acts mainly as a masking stimulus on the echo processing or whether the behavioral state of vocalization prepares the auditory system for the analysis of the echo.

Behavioral findings that the bat only processes Doppler shift information if it is received during vocalization supports the latter hypothesis (Schuller, 1974).

Consequently, single neuron recordings from the inferior colliculus were made in the actively vocalizing greater horseshoe bat. Most neurons with best frequencies between 70 and 86 kHz responded to the electrically elicited vocalizations with response patterns commonly found in collicular neurons. However, if the response to the emitted sound was compared to the response of the same neuron to a pure tone at the CF frequency of the vocalization about 50% of the cells showed differences in the responses to the two stimuli (Fig. 1).

If the vocalization was combined with an artificial echo even more striking effects of vocalization on the collicular neurons were revealed. Some neurons only responded to sinusoidally frequency modulated stimuli (SFM) with synchronized discharges, if they were presented during vocalization. Replacing the vocalization by a pure tone mimicking the CF portion of the vocalization did not result in a synchronized response to the SFM stimulus (Fig. 2).

Other neurons that were stimulated with SFM sounds delayed to the onset of the vocalization only responded with synchronized discharges as long as the echo temporally overlapped with the vocalization.

Using stimulus sets combining the vocalization with frequency-shifted echoes (simulating the Doppler shift compensation system) neurons were found that encoded the threshold for Doppler shift compensation and the frequency shifts relative to the resting frequency by response pattern changes or latency changes respectively (Fig. 3).

977

Fig. 1: Responses of 2 collicular neurons to electrically elicited vocalizations and to artificial pure tones. A) Response of the neuron during vocalization (upper histogram) and to artificial pure tones at the frequency of the CF-portion of the echolocation signal at different intensities. B) Response of a neuron during vocalization and subsequent pure tone stimulation at different frequencies (82.0-83.4 kHz) at 80 dB SPL. The CF-frequency of the echolocation signal lies at 83.4 kHz.

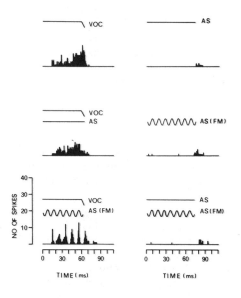

Fig. 2: Discharge patterns of a collicular neuron during vocalization (VOC), to a pure tone at the CF-frequency of the echolocation sound (AS), to a combination of vocalization and the same pure tone (VOC+AS), to a sinusoidally frequency modulated tone (AS(FM)) (carrier frequency: CF-frequency), to a combination of vocalization and the frequency modulated echo (VOC+AS(FM)) and to a combination of a pure tone at the CF-frequency and the frequency modulated echo (AS+AS(FM)). The intensity of the artificial echoes was 80 dB SPL. Note especially the different responses to the two acoustically identical stimulus situations in the lower two graphs.

Fig. 3.   Responses of 2 collicular neurons to vocalization presented together with frequency-shifted playback echoes, simulating the Doppler shift situation encountered by the
bat during flight towards a target.   The echoes were
played back with a delay of 1 ms after the onset of the
vocalization and the frequency shifts are indicated to
the right of each histogram.

These effects of vocalization on the collicular responses to
echoes cannot be explained by mere acoustic interferences between
emitted echolocation pulse and echo, but show that the behavioral
state of active vocalization neuronally affects auditory processing
in collicular neurons when analyzing behaviorally relevant echoes
(SFM signals-wing beat reflections) or Doppler shift information
in echoes.

Supported by the Deutsche Forschungsgemeinschaft, grant number
Schu 390/1,2 and Br 593/2 and Stiftung Volkswagenwerk.

STIMULUS CONTROL OF ECHOLOCATION PULSES IN <u>TURSIOPS</u> <u>TRUNCATUS</u>

Ronald J. Schusterman, Deborah A. Kersting,

and Whitlow W. L. Au

Until recently, it was extremely difficult to determine what cue or set of cues a dolphin used in detecting or differentiating between targets by means of its sonar system. One of the major problems was the ambiguous nature of the echo return in relation to the position of the dolphin during pulse emission and prior to the animal's choice response. In this experiment, we solved the problem by acquiring unambiguous control over a dolphin's position and pulse emission while it was actively engaged in a discriminative echolocation task.

Fig. 1. Diagram of an Atlantic bottlenose dolphin (<u>T. truncatus</u>) maintaining a fixed position in a water-filled plastic hoop. The dolphin remained in the hoop up to its flippers and when given an audio signal, it emitted a burst of echolocating clicks at a target 6 m directly in front of its melon. Following the termination of the last pulse within a burst of pulses, the dolphin backed out of the hoop and reported the presence of a sphere or cylinder by hitting the "A" or "B" response paddle or manipulandum with its rostrum.

An adult male Atlantic bottlenose dolphin (Ekahi), originally signalled by a tone to echolocate and discriminate spheres from cylinders while swimming across his pen, was trained to continue the same task from a fixed position. The animal was first trained to place its head(including blowhole) in a sixteen inch diameter hoop. The hoop was submerged to a depth that placed the dolphin's rostrum directly in line with the center of the target located six meters away (see Fig. 1). When the dolphin maintained the appropriate position in the hoop and emitted no echolocation pulses, the tone cue was activated. The tone signalled the dolphin to echolocate from its position in the hoop. If the animal left the stationary hoop to continue echolocation, the tone was terminated before a choice response was made. If it left the hoop only to surface and make a choice response, the tone remained activated until a response was made. With continued training, control over both the position of the dolphin and the time that it emitted echolocation pulses was exerted in a precise fashion. Control over echolocation pulses was so refined that termination of the tone invariably resulted in the termination of echolocation pulses even though the animal was stationed correctly and the target was present in front of him.

In order to determine if the dolphin was echolocating before the tone was given, a simultaneous measurement of the animal's signals and the tone cue was performed. A B and K 8103 hydrophone was placed one meter behind the target to measure the animal's echolocation signals. The hydrophone signal was amplified by 20 dB and recorded on one channel of an Ampex FR-1300 instrumentation tape recorder. The audio cue was recorded on another channel. Both channels were examined simultaneously using a dual-trace storage oscilloscope in the chop-mode. Analysis of tape recordings made during testing showed that the porpoise did not emit any echolocation signals while stationary in the hoop before the onset of the tone cue.

A second dolphin (Sven) was trained in essentially the same way as we trained Ekahi. The training took approximately the same amount of time (two weeks) and again resulted in controlling not only the postural orientation, but also the signal emissions of an echolocating dolphin in a materials discrimination task.

Supported by the Naval Ocean Systems Center and ONR contract N0014-77-C-0185.

RESPONSE BIAS AND ATTENTION IN DISCRIMINATIVE ECHOLOCATION

BY TURSIOPS TRUNCATUS

Ronald J. Schusterman, Deborah Kersting

and Whitlow W. L. Au

In the Einstellung experiment, stimulus correlated response set or bias can be induced which renders an animal in a maze "functionally blind" to a subsequent simple solution. Until the bias is extinguished, the animal remains insensitive to cues leading to the goal object. Observations on echolocating dolphins suggest that they too may become subject to an Einstellung phenomenon. The present experiment, which grew out of a series of biosonar studies on underwater material discrimination, tested the notion that a response bias acquired in an insolvable discriminative echolocation task will strongly influence the attention of a dolphin on a solvable discriminative echolocation task.

Method. In a two-alternative forced-choice procedure with targets presented on successive trials at a distance of 6 m (see Fig. 3 in Schusterman, this volume), the task of an adult male T. truncatus (Sven) was to differentiate echoes from 17.8 long hollow aluminum and hollow glass targets—each having two different sized outer diamters (OD) and wall thicknesses (see Fig. 1). Targets were submerged in a vertical orientation about 114 cm below the water surface in approximately the same location. Both large and small aluminum cylinders were to be reported on the A or left manipulandum and the large and small glass cylinders were to be reported on the B or right manipulandum. Results from synthesized dolphin-like clicks reflected from these targets in water are shown in Fig. 1. Target reflection characteristics are virtually identical for large aluminum and glass targets (7.62 cm OD), but the differences between small aluminum and glass(3.81 cm OD) are different in terms of the relative strength of the second component of the envelopes of the matched filter output and also with regard to frequency spectra—particularly between 100 and 130 kHz.

In Phase 1, each test session consisted of all four targets being presented 16 times in a quasi-random sequence. In Phase 2 (the Einstellung phase), a test session consisted of either a solvable discrimination problem (small aluminum and glass targets) or an insolvable problem (large aluminum and glass targets). Test sessions in Phase 2 were sequenced in a quasi-random fashion so that the dolphin would have difficulty "predicting" on which test session a solvable or insolvable problem would be given. In Phase 3, the dolphin was given the solvable task on each test session,

TARGET REFLECTION CHARACTERISTICS

Fig. 1.   Results of backscatter measurements for hollow aluminum
           cylinders and hollow glass cylinders. Wall thickness of
           the smaller OD's was 0.32 cm and that of larger OD's was
           0.40 cm. Comparison of echoes from the same sized OD's
           can be made on the basis of the time-domain waveforms,
           the frequency spectra (superimposition of echoes from
           aluminum and glass) and the envelopes of the matched
           filter output (superimposition of the echoes from
           aluminum and glass).

i.e., the differentiation between small glass and small aluminum.
Following the original experiment, the three phases were replicated
in the same sequence.

    Results and Discussion. After 30 test sessions of Phase 1 of
the original experiment, the dolphin perfected the differentiation
between the small aluminum and small glass targets. However, as
predicted from the target reflection characteristics shown in Fig. 1,
the animal did not distinguish between the large aluminum and glass
targets. Instead, the dolphin developed a response bias and consistently reported both large glass and large aluminum on the B
response manipulandum. In Phase 2, response bias not only increased

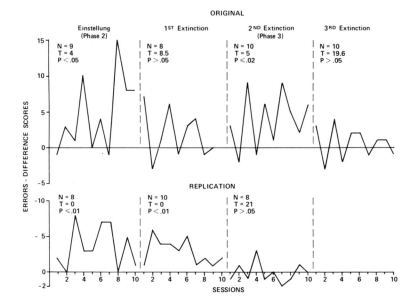

Fig. 2. Comparison of the dolphin's errors (reflecting response bias) during 1st and 2nd halves of each test session during Phases 2 and 3 of the original experiment and the replication. Values on the ordinate were obtained by subtracting errors made during the 2nd half of the test session from errors made during the 1st half of the test session. Positive values indicate response bias during the early part of a test session and 0 values indicate no difference in bias between the 1st and 2nd halves of a test session.

during insolvable large-target sessions, but bias also occurred in the first half of the solvable small-target sessions, decreasing significantly (Wilcoxin matched-pairs signed ranks test) during the second half of the sessions (see Fig. 2).

In Phase 3, when the solvable discrimination was repeatedly given, the initial bias or the Einstellung phenomenon was extinguished. When the entire experiment was replicated, the results were very similar to the first experiment except that the Einstellung phenomenon was extinguished more rapidly. Analysis of click trains from a single test session of Phase 2 showed that the dolphin emitted essentially the same echolocation signals during the solvable discrimination problem regardless of whether or not it had a response bias. The results are interpreted as indicating that persistant spatial responses during an insolvable echolocation task predisposed the dolphin to diminish its attention for listening to previously established distinctive echoes. This interpretation is similar to one suggesting that bats relying on spatial memory pay less attention to echoes in an obstacle avoidance task. (This research was supported by the Naval Ocean Systems Center and ONR contract N00014-77-C-0185.)

THE AUDITORY PATHWAY OF THE GREATER HORSESHOE BAT,

RHINOLOPHUS FERRUMEQUINUM

Hermann Schweizer and Susanne Radtke

Although the central auditory pathway of mammals has been the subject of numerous anatomical studies, only a few anatomical investigations exist on the auditory pathway in echolocating bats. In contrast to this shortage of anatomical knowlege, a lot of neurophysiological data exist from different auditory nuclei. To bridge the gap between anatomical and neurophysiological data we investigated the central auditory pathway of the greater horseshoe bat.

The auditory nuclei were studied with classical histological techniques (Nissl, Klüver-Barrera, Bodian, Golgi). The fiber connections of the inferior colliculus (IC) and the auditory cortex were investigated by using the anterograde and retrograde horse-radish peroxidase (HRP) method. The extension of the auditory cortex and the best frequency range at the HRP-injection sites were determined by using evoked potential recordings.

Auditory centers: No significant differences between the organization of the auditory pathway of this bat species and other mammals were found. This is also true for the composition of the superior olivary complex. In Rhinolophus as well as in Molossus and other investigated bats, a more or less prominent medial superior olivary nucleus was found. The only difference to acoustically non-specialized mammals is, that most of the auditory nuclei are hypertrophied, especially the anteroventral and posteroventral cochlear nucleus, the medial nucleus of the trapezoid body, the lateral superior olivary nucleus, the ventral nucleus of the lateral lemniscus, the central nucleus of the inferior colliculus, the medial geniculate body and the auditory cortex.

The structure and connections of the inferior colliculus (IC): The inferior colliculus of mammals represents an obligatory synapse for lemniscal fibers directly ascending from the lower auditory nuclei. Therefore, the auditory system has no direct pathway to the diencephalic thalamus.

As in other mammals, the IC of Rhinolophus can be divided in three subnuclei: the central nucleus, the pericentral nucleus and the external nucleus. Within the central nucleus a dorsolateral laminated part can be distinguished from a ventromedial non-laminated part. In Golgi-preparations, the same neuron types were found

987

as in other mammals. A comparison of anatomical and electrophysiological data shows that the frequency band from 10 to about 78 KHz
is represented in the dorsolateral part. The frequencies near the
constant frequency part of the echolocation call, where the audiogram of Rhinolophus is sharply tuned ("filter frequencies"), are
represented in the ventromedial part of the central nucleus of the
IC. The "filter frequencies" are overrepresented in the central
nucleus of the IC as well as in all other lower and higher auditory
centers. Therefore, the distribution of frequencies, within all
auditory centers, directly reflects the distribution of frequencies
along the basilar membrane of the cochlea.

The IC gets direct ipsilateral and/or contralateral inputs in
a tonotopic order from all auditory centers with the exception of
the medial nucleus of the trapezoid body and the medial geniculate
body. Ipsilateral projecting nuclei: dorsal cochlear nucleus,
lateral and ventral nucleus of the trapezoid body, medial superior
olivary nucleus, lateral superior olivary nucleus, dorsal periolivary nuclei, ventral and dorsal nucleus of the lateral lemniscus,
auditory cortex. Contralateral projecting nuclei: dorsal, anteroventral and posteroventral cochlear nucleus, ventral nucleus of
the trapezoid body, lateral superior olivary nucleus, dorsal nucleus
of the lateral lemniscus and the inferior colliculus. These results
show there is a strong convergence of auditory input into the IC;
this points to a very important function of the IC in auditory
processing.

Descending fibers of the IC project to the nuclei of the lateral
lemniscus and to the nuclei of the superior olivary complex. The
ascending fibers project to the medial geniculate body.

Furthermore, the IC is connected to nuclei not belonging to the
classical auditory pathway. These regions are: 1.) The deep layers
of the superior colliculus (they are commonly viewed to have multi-
sensory and motor functions); 2.) The central gray matter (electrical stimulation of the central gray matter elicits vocalization);
3.) The reticular formation of the medulla oblongata (through this
connection an alerting response mediated by auditory signals might
be possible); 4.) The lateral pontine nuclei (these project to the
cerebellum by relaying auditory information to the motor system).

The structure and connections of the auditory cortex: Studies
employing evoked potential recordings and the HRP-technique have
shown that both the auditory cortex and the medial geniculate body
(MGB) of Rhinolophus as the lower auditory centers are vastly
hypertrophied.

The auditory cortex shows the typical six-layered construction
of the mammalian sensory cortex. The anatomical neuron types
correspond to those in other mammals and are layer-specific. The
auditory cortex is tonotopically arranged. The lower frequencies
are located caudally and the higher frequencies rostrally. The

"filter frequencies" are overrepresented (For more detailed information on the tonotopical and functional organization of the auditory cortex, see the contributions of J. Ostwald and N. Suga, this volume).

The auditory cortex gets its acoustical inputs exclusively from the MGB. A detailed analysis of the organization of the MGB and its tonotopical and functional projections is still in progress. But, on the basis of Nissl-stained material, a posterior dorsal cell group can be distinguished from the remaining principal part. Preliminary studies, using relatively large HRP-injections into different frequency regions of the auditory cortex, have shown that in the dorsal cell group as well as in the principal part the entire frequency range of the audiogram is represented. In the principal part, the lower frequencies are localized ventrolaterally and the higher frequencies dorsomedially. The "filter frequencies" are overrepresented and occupy about 50% of the nucleus.

Projections to the auditory cortex also come from other dien-cephalic regions: from the medial and lateral posterior thalamic nuclei and from the zona incerta. The auditory cortex is recipro-cally connected with the contralateral cortex and with the MGB. It also projects back to the IC. Therefore, the auditory cortex could control its own auditory inputs from the diencephalic and mesencephalic levels.

With the techniques used, connections between the auditory cortex and other forebrain centers, the cingular area, the amygdala and a distinct region in the frontal cortex were found. In all of this nuclei, JÜRGENS and PLOOG (1970) were able to elicit vocalization with electrical stimulation in the squirrel monkey. Electrophysiological evidence in Rhinolophus is still lacking, but it is suggested that this forebrain centers might represent vocalization centers, which are presently unknown in bats. (Supported by Deutsche Forschungsgemeinschaft, grant No. Schn.136/6, Br. 593/2.

REFERENCES

JÜRGENS, U., PLOOG, D.: Cerebral representation of vocalization in the squirrel monkey. Exp. Brain Res. 10, 532-554 (1970).

GREY SEAL, HALICHOERUS: ECHOLOCATION NOT DEMONSTRATED

Bill L. Scronce and Sam H. Ridgway

Active echolocation by pinnipeds has been claimed by at least
one investigator (Poulter, 1963), but disputed by others
(Schusterman, 1967). The grey seal produces clicks. Its underwater hearing is fairly sensitive in the ultrasonic regions important for echolocation in bats and porpoises, and considering the
natural environment of the species, we suspected that it used active
echolocation.

Our subject was a four-year-old female from Iceland. The
seal was trained to wear an opaque elastic band about 15 cm wide
that fit over its head blocking vision (Fig. 1). The use of echolocation by this seal was evaluated in two experiments. The first
required the seal to retrieve an air-filled plastic ring 20 cm in
diameter (D) placed at random locations in a 5 x 1 m section of a
10 m D redwood tank. The second was a forced choice test requiring
the seal to detect the flat surface of a 25 cm D styrofoam disk as
opposed to the edge of the same disk (-30 dB target strength),
placed randomly on either side of a divider.

Without the blindfold the seal's ring retrieval rate was
100% with a latency of 3.8 sec. In 427 blindfolded trials there
were 99% correct responses, but latency increased to 6.5 sec. When
the 5 x 1 m area, where the ring was randomly placed, was ensonified
with echolocation clicks like those employed by dolphins, the seal
made 95% correct responses with a latency of 6.1 sec. Head scanning movements were observed on about half the blindfolded trials
and click trains were recorded on about 10% of the trials observed.

In the second experiment the blindfolded seal approached the
divider and made a choice about target location (right or left) at
least 1.5 m away (Fig. 2). Without the blindfold the seal's performance was almost 100%, but with the blindfold correct responses
never exceeded 65% and average correct responses for 617 trials was
46%. No click-trains were detected.

The strongest evidence for echolocation was the observation
of head scanning movements concurrent with click-trains on some of
the blindfolded trials. The animal, however, retrieved the ring
just as well when there were no click-trains or head scans. The
air-filled ring floating in the water may have been a good passive
target for the seal. Since the seal performed the discrimination

Fig. 1. Halichoerus grypus with blindfold in place, vibrissae
         extended.

Fig. 2. Halichoerus grypus discrimination experiment setup.

task at chance level, we concluded that the grey seal orients primarily by vision with good capability for passive detection by audition or by tactile senses (vibrissae). The grey seal makes sounds underwater but does not appear to use these sounds for echolocation.

REFERENCES

Poulter, T. C., Sonar signal of the sea lion, Science 139:753.
Schusterman, R. J. 1967, Perception and determinants of underwater vocalization in the California sea lion. In: R. G. Busnel (Ed.), Animal Sonar Systems, Vol. 1. Laboratoire de Physiologie Acoustique, Jouy-en-Josas, France, 535-617.

THE ROLE OF THE ANTERIOR AND POSTERIOR CRICOTHYROID MUSCLES IN

THE PRODUCTION OF ECHOLOCATIVE PULSES BY MORMOOPIDAE

Roderick A. Suthers and Gary E. Durrant

Pteronotus parnellii in the family Mormoopidae emits pulses having a long initial CF portion followed by a short downward FM sweep. Most of the acoustic energy is in the second harmonic at 60 kHz, but the fundamental and higher harmonics can also be introduced when greater bandwidth is needed (Fig. 1). Griffiths (1978) reported that the cricothyroid muscle (CTL) of Mormoopidae has two distinct bodies--an anterior longitudinal portion (aCTM) and a posterior transverse portion (pCTM). The latter attaches to the flexible lateral walls of the laryngeal cavity. On the basis of its anatomy, Griffiths suggested that it might enable the bat to tune its larynx, in the manner of a variable Helmholtz resonating amplifier, by altering the cross-sectional area of the laryngeal cavity anterior to a larger post-laryngeal chamber formed by the enlarged trachea. We have examined the effect of a light gas mixture (20% oxygen, 80% helium) on the harmonic content and frequency structure of orientation pulses emitted before and after selective denervation of particular parts of the CTM in an effort to determine if the pCTM is responsible for varying the harmonic content of the emitted pulse by changing the vocal tract resonance as suggested by Griffiths. By increasing the speed of sound, without altering the vibration of the laryngeal generator or changing the dimensions of the airway, a light gas mixture enables one to evaluate the extent to which the emitted frequencies depend on resonance or filtering by the vocal tract.

Denervation of aCTM. Bilateral denervation of the aCTM causes no pronounced change in harmonic emphasis. Pulse frequency rises slightly during the first few msec before becoming CF and the frequency of the CF portion of the second harmonic is about 2 kHz below normal. Light gas mixtures suppress the second harmonic and enhance the fundamental and third harmonic as they do in the case of a normal bat. The aCTM appears not to play a major essential role in controlling the frequency composition of emitted pulses.

Denervation of pCTM. Bilateral denervation of the pCTM also results in a 2 to 3 kHz drop in the frequency of the second harmonic (Fig. 2). The terminal FM sweep begins earlier in the pulse and proceeds at a more gradual rate than is normal. The intensity of the third harmonic is increased to approximately equal that of the second harmonic. The fundamental, fourth and sometimes higher harmonics, are usually also present at a lower intensity. A light gas mixture shifts the emphasis to the fundamental and third

Fig. 1.   Typical orientation pulses produced by an intact bat
in a plastic bag filled with air or with a light gas
mixture (He). Wide band sonagrams are above narrow
band sonagrams.

Fig. 2.   Pulses emitted in air and light gas mixture after
bilateral denervation of the pCTM.

Fig. 3.   Pulses emitted in air and light gas mixture after
          bilateral denervation of the pCTM and aCTM.

harmonic, but the second and fourth harmonics are still present.

       Denervation of aCTM and pCTM.   Bilateral denervation of both
aCTM and pCTM causes a dramatic drop in generator frequency (Fig. 3).
The fundamental drops to 10 or 12 kHz and as many as 9 or 10
harmonics may be present.   Harmonics in the range between about 40
and 65 kHz often tend to be emphasized in air, but suppressed in
the light gas mixture, indicating a significant amount of filtering
by the vocal cavities remains.

       Conclusions.   The aCTM and pCTM together determine the fre-
quency of the vocal generator.   If either portion is paralyzed, the
remaining part can largely compensate for its absence.   The aCTM
is particularly important in establishing the frequency at the
beginning of the pulse, while the pCTM is also important in timing
the start and controlling the rate of the terminal FM sweep.   The
pCTM may play a role in controlling the harmonic content of the
emitted pulse, possibly by varying the tuning of the vocal tract.
However, a significant amount of vocal tract tuning remains after
denervation of both the a CTM and pCTM, indicating the existence of
a second independent filter.

(Supported by NSF grant BNS 76-01716).

CODING OF SINUSOIDALLY FREQUENCY-MODULATED SIGNALS BY SINGLE

COCHLEAR NUCLEUS NEURONS OF RHINOLOPHUS FERRUMEQUINUM

Marianne Vater

The Greater Horseshoe bat, Rhinolophus ferrumequinum, emitts a
long constant-frequency (CF-) component of about 83 kHz during echolocation. The echoes are Doppler-shifted upward in frequency due to
the realative movement between bat and target. The bat compensates
for these deviations in the echo-frequency by lowering the frequency
of the emitted call so that the echo-frequency remains constant
within a narrow frequency band (Schnitzler, H.-U., 1968; Schuller et
al., 1974). The wing beat of prey insects results in periodic frequency and amplitude modulations of this carrier frequency which can
be used by the bats as clues for detection and probably identification of moving prey objects. Neurophysiological investigations of
the inferior colliculus (IC) and auditory cortex of Rhinolophus
(Schuller, G., 1979; Ostwald, J., this volume) showed that the sharply tuned filterneurons around 83 kHz specifically analyse the information contained in periodic modulations.

In order to examine the neuronal encoding characteristics of
these complex time-varying signals within the ascending auditory
pathway of Rhinolophus, recordings were made from single cochlear
nucleus (CN) neurons. The sample comprises neurons from all subdivisions of the CN-complex, localized with Alcainblue spots. Various
parameters of sinusoidally frequency-modulated (SFM-) signals were
presented.

The neurons synchronize their discharge activity to the period
of the SFM-stimulus. Tonic neurons were found that faithfully reproduced the time course of the periodic modulation. It can therefore
be expected that signals with more complex temporal patterns, as for
example generated by the wing beat of prey insects are also reproduced with only little distortion and relayed as input information
to higher auditory centers. Other tonic cells asymmetrically enhanced
the upward or downward component of the modulation cycle in their
discharge pattern. Phasic-on neurons did not preserve the stimulus
time structure: they responded with a transient discharge activity to
distinct portions of the modulation cycle.

The modulation height of the periodic modulation of the echo-
frequency depends on the size of the prey insect. CN-neurons code a
wide range of modulation heights thereby fulfilling the criteria for
processing echo-information of a variety of prey sizes. The threshold for modulation height was found to be especially low in CN-

filterneurons (minimum ± 20 to ± 50 Hz) as compared with neurons of
other frequency ranges. These minimum modulation heights are comparable to the values found by Schuller, G. (1979) for IC-neurons,
therefore no further neuronal interaction is required to account for
this response property of the higher auditory neurons. Rather, the
narrow bandwidth and the steep tuning curve flanks of filterneurons,
which are the result of specialized mechanical properties of the
cochlea (Bruns, V., 1976a, b) seem to be responsible for this response characteristic. Other factors governing the response to SFM
are discussed later.

SFM-stimuli centered at the best frequency of a neuron are not
necessarily the optimal stimulus. Shifting the carrier frequency towards the tuning curve flanks can lead to a better synchronization
and to a lowering of the modulation threshold.

The echo intensity depends on the distance between bat and target. CN-neurons clearly synchronize their discharges to the modulation
cycle over a wide range of stimulus intensities.

The response to SFM-stimuli of different modulation frequencies
correlates with the neurons´ response patterns to CF-stimuli and
therefore depends on the localization within the different subdivisions of the CN. Primary-like neurons of the anteroventral CN (Fig.
1 A) are able to synchronize their discharges to modulation frequencies up to 900 Hz. Functions, relating the magnitude of the neurons
response to modulation frequency typically possess an optimum around
400 Hz. The stimulus coding in the time domain therefore far exceeds
the frequencies of prey insect wing beat (i.e. between 30 and 100 Hz).
In contrast, build up and complex neurons of the dorsal CN, even if
they are sharply tuned, were unable to follow the periodic modulations in the modulation frequency range tested (minimum modulation
frequency 20 Hz). Their response patterns to SFM-stimuli resembles
the patterns to CF-stimuli. Therefore, the structure of the synaptic
connections within the subnuclei of the CN also plays a role in the
neurons´ responses to periodic stimuli. Phasic-on neurons in the
dorsal CN (Fig. 1 B) typically lock only to modulation frequencies
below 300-400 Hz with an optimum response below 100 Hz, a range
largely restricted to the biologically important parameters. These
response properties were also reported for IC-neurons of Rhinolophus
(Schuller, G., 1979).

These results show that this peripheral stage of the ascending auditory pathway can not be treated as a uniform population of neurons
simply relaying information to higher auditory centers for further
processing, but that the extraction of biologically important parameters has already begun.

Fig. 1. Response of anteroventral (A) and dorsal (B) CN-neurons to
SFM-stimuli of varying modulation frequency. Lower histogramms: response to CF-signals at best frequency (BF; A) or to CF-signals at the
carrier frequency (MF; B) of the SFM. df: modulation height; stimulus
intensity as indicated. Binwidth 0.5 ms, 50 stimuli. bars: duration.

SINGLE UNIT RESPONSES TO FREQUENCY-MODULATED SOUNDS AND

SIGNAL-NOISE COMBINATIONS. A COMPARATIVE STUDY OF THE

INFERIOR COLLICULUS OF MOLOSSID BATS

Marianne Vater and Peter Schlegel

Recordings were made from single inferior colliculus neurons of two closely related bat species, Molossus ater and Molossus molossus. Their short CF-FM echolocation calls differ only in frequency range.

The audiograms of both species derived from evoked potential measurements and threshold distributions of single neurons are broadly tuned with maximum auditory sensitivity reached at different frequency ranges according to the different spectral content of the orientation calls. Tuning curves, Q_{10} dB values, response patterns and spike count functions of single units are very similar in the samples obtained from both molossid species and closely resemble data from bats using FM orientation calls (Pollak et al., 1978). Contrary to long CF-FM bats such as Rhinolophus ferrumequinum, no conspicuous neuronal adaptations to the short CF component were found in molossids.

However, the use of FM stimuli revealed a high degree of specialization to the FM component of the orientation call. Artificial FM pulses of different durations, modulation heights and directions centered at the neurons' best frequency were tested. In about 30% of the neurons, thresholds to FM pulses mimicking the natural FM part of the echolocation call in sweep rate, were lower than thresholds to pure tone pulses of equal duration (2 ms), resulting in a better detectability of FM.

Response patterns of single units depended on the frequency time course of the stimulus. Neuronal response activity to slowly sweeping FM pulses could be greater than to any other stimulus configuration. Stabilization of initial latency was frequently observed with fast sweeping FM, resulting in increased accuracy of time information.

Threshold signal/noise ratios, measured by masking the neuron's response to the signal by noise, were usually equal for CF and FM stimuli or lower for the FM signal as compared to the CF signal. This increased detectability of FM sweeps in noise is advantageous for the bat usually hunting under nonideal acoustical conditions.

ASCENDING AUDITORY PATHWAYS IN THE BRAIN STEM OF THE BAT,

PTERONOTUS PARNELLII

J. M. Zook and J. H. Casseday

Although mammalian echolocation has been extensively studied in recent years, little is known concerning the connections of neural centers which might be involved in the special auditory functions of echolocation. We have mapped the ascending auditory pathways from cochlear nucleus in the bat, Pteronotus parnellii parnellii (Gray), using techniques for tracing axonally trans-ported molecules (^{3}H leucine or horseradish peroxidase) in the anterograde or retrograde directions. Our principal observations concern (1) direct pathways from cochlear nucleus to the central nucleus of the inferior colliculus and (2) an indirect pathway which relays within the lateral lemniscus enroute from cochlear nucleus to the central nucleus of the inferior colliculus.

The results show first that the direct projections of the three subdivisions of the cochlear nucleus, the anteroventral, posteroventral and dorsal cochlear nuclei, do not completely over-lap within the central nucleus of the inferior colliculus. The anteroventral cochlear nucleus projects only to the ventral and lateral parts of the central nucleus. From the posteroventral cochlear nucleus, the region of octopus cells projects to an even more limited area of the ventral part of the central nucleus of the inferior colliculus. Only the dorsal cochlear nucleus pro-jects to all parts of the central nucleus of the inferior col-liculus, but this projection is most dense in the dorsal and lat-eral aspects of the central nucleus. We conclude that each of the major subdivisions of the cochlear nucleus has a specific target in the central nucleus of the inferior colliculus but that there is partial overlap of one target with another. Additional obser-vations reveal topographic projections from the anteroventral, posteroventral and dorsal cochlear nuclei to their respective targets.

In addition to the direct pathway from the subdivisions of the cochlear nucleus to the central nucleus of the inferior col-liculus, Pteronotus, as other mammals, has several indirect path-ways which relay in nuclei in the superior olivary complex and the lateral lemniscus before terminating in the central nucleus of the inferior colliculus. One of these indirect pathways may be unique to Pteronotus in that the pathway relays within the lateral lemniscus at a large nucleus which has no homologue in mammals that do not echolocate. This "intermediate" nucleus of the lat-eral lemniscus receives its input primarily from the anteroventral cochlear nucleus and sends its main projections in turn to the central nucleus of the inferior colliculus. It is quite possible that this indirect pathway from cochlear nucleus to the central nucleus of the inferior colliculus is an elaboration unique to echolocating bats.

(Supported by NSF grant BNS 77-°6796)

Chapter VIII
General Bibliography

- Bibliography on echolocation in bats.
      compiled by G. Neuweiler

- Bibliography on echolocation in odontocetes.
      compiled by P. Nachtigall

- Bibliography on echolocation in non-bat non-cetacean species.
      compiled by E. Gould

BIBLIOGRAPHY ON ECHOLOCATION IN BATS

(Literature cited: articles published from 1965 to 1. Febr. 1979.
Russian literature is not included.)

1. Aidley, D. J. Echo intensity in range estimation by bats. Nature, 224, 1330-1331. 1969.
2. Altes, R. A. Optimum waveforms for sonar velocity discrimination, in: Proc. IEEE, 59, 1615-1617. 1971.
3. ——— Suppression of radar clutter and multipath effects for wide-band signals. IEEE Transactions on Information Theory, IT-17, 344-346. 1971.
4. ——— Methods of wideband signal design for radar and sonar systems. Federal Clearinghouse Number AD 732 494. 1971.
5. ——— Some invariance properties of the wide-band ambiguity function. J. Acoust. Soc. Am., 53, 1154-1160. 1973.
6. ——— Study of animal systems with application to sonar. Progress Report No. ESL PR 144, Electromagnetic Systems Labs., Sunnyvale, Cal. 1974.
7. ——— Mechanism for aural pulse compression in mammals. J. Acoust. Soc. Am., 57, 513-515. 1975.
8. ——— Sonar for generalized target description and its similarity to animal echolocation systems. J. Acoust. Soc. Am., 59, 97-105. 1976.
9. ——— The Fourier-Mellin transform and mammalian hearing. J. Acoust. Soc. Am., 63, 174-183. 1978.
10. ——— Angle estimation and binaural processing in animal echolocation. J. Acoust. Soc. Am., 63, 155-173. 1978.
11. ——— Further development and new concept for bionic sonars, Vols. II and III. Report OC-R-78-A 004-1, ORINCON Corp., 3366 N. Torrey Pines Ct., La Jolla, CA. 92037. 1978.
12. ———, and Faust, W. I. Further development and new concept for bionic sonar, Vol. I. Report OC-R-78-A 004-1, ORINCON Corp., 3366 N. Torrey Pines Ct., La Jolla, CA. 92037. 1978.
13. ———, and Reese, W. D. Doppler-tolerant classification of distributed targets - a bionic sonar. IEEE Transactions Aerospace and Electronic Systems, Vol. AES-11 (5), 708-724. 1975.
14. ———, and Skinner, D. P. Sonar velocity resolution with a linear-period-modulated pulse. J. Acoust. Soc. Am., 61, 1019-1030. 1977.
15. ———, and Titlebaum, E. L. Bat signals as optimally Doppler tolerant waveforms. J. Acoust. Soc. Am., 48, 1014-1020. 1970.
16. ———, ——— Graphical derivations of radar, sonar, and communication signals. IEEE Transactions Aerospace and Electronic Systems, Vol. AES-11 (1), 38-44. 1975.
17. Anderson, G. M. A model for the bat versus moth pursuit-evasion problem. J. Acoust. Soc. Am., Suppl. No. 1, S88. 1978.

18. Bach Andersen, B., and Miller, L. A. A portable ultrasonic detection system for recording bat cries in the field. J. Mammal., 58, 226-229. 1977.

19. Barclay, R. M. R. Vocal communication and social behavior of the little brown bat, Myotis lucifugus (Chiroptera: Vespertilionidae). M. Sc. Thesis, Dept. of Biology, Carleton University, Ottawa, Canada, 1978.

20. Baron, G. Morphologie comparative des relais auditifs chez les chiroptères. Dissertation, Montreal, 1972.

21. ———— Differential phylogenetic development of the acoustic nuclei among chiroptera. Brain Behav. Evol., 9, 7-40. 1974.

22. Beuter, K. J. Systemtheoretische Untersuchungen zur Echoortung der Fledermäuse. Dissertation, Universität Tübingen, 1976.

23. ———— Maskierung von konstantfrequenten Echoortungslauten mit Bandpaßrauschen bei der Fledermaus Rhinolophus ferrumequinum. Verh. Dtsch. Zool. Ges., 1978, 195, G. Fischer Verl., Stuttgart, 1978.

24. ———— Optimalempfängertheorie und Informationsverarbeitung im Echoortungssystem der Fledermäuse, in: "Kybernetik 77," Oldenburg-Verlag, München, 1978.

25. Bradbury, J. W. Target discrimination by the echolocating bat Vampyrum spectrum. J. Exp. Zool., 173, 23-46. 1970.

26. ————, and Vehrencamp, S. L. Social organization and foraging in emballonurid bats. I. Field studies. Behav. Ecol. Sociobiol., 1, 337-381. 1976.

27. Brosset, A. "La Biologie des Chiroptères," Masson et Cie., Paris, 1966.

28. Brown, A. M. An investigation of the cochlear microphonic response of two species of echolocating bats, Rousettus aegyptiacus (Geoffroy) and Pipistrellus pipistrellus. J. comp. Physiol., 83, 407-413. 1973.

29. ————, and Pye, J. D. Auditory sensitivity at high frequencies in animals. Adv. Comp. Physiol. Biochem., 6, 1-73. 1975.

30. Brown, P. E. Vocal communication and the development of hearing in the pallid bat, Antrozous pallidus. Ph. D. Thesis, University of California Los Angeles, 1973.

31. ———— Vocal communication in the pallid bat, Antrozous pallidus. Z. Tierpsychol., 41, 34-54. 1976.

32. ————, Grinnell, A. D., and Harrison, J. B. Developing of hearing in pallid bat, Antrozous pallidus. J. comp. Physiol. A, 126, 169-182. 1978.

33. Bruns, V. Die Anpassung des Hörsystems der Hufeisennase an die Echoortung mit konstantfrequenten Ortungslauten. Verh. Dtsch. Zool. Ges., 1976, 268, G. Fischer Verl., Stuttgart, 1976.

34. ———— Peripheral auditory tuning for fine frequency analysis by the CF-FM bat, Rhinolophus ferrumequinum. I. Mechanical specializations of the cochlea. J. comp. Physiol. A, 106, 77-86. 1976.

35. ———— Peripheral auditory tuning for fine frequency analysis of the CF-FM bat, Rhinolophus ferrumequinum. II. Frequency mapping in the cochlea. J. comp. Physiol. A, 106, 87-97. 1976.

36. Bruns, V. Die Feinstruktur der für die Hydrodynamik wesentlichen Elemente der Cochlea der Großen Hufeisennase, Rhinolophus ferrumequinum. Verh. Dtsch. Zool. Ges., 1978, 154, G. Fischer Verl., Stuttgart, 1978.

37. ————, Neuweiler, G., and Suga, N. Specialized frequency processing in the auditory system of the echolocating horseshoe-bat. Exp. Brain Res., 23, Suppl., 30. 1975.

38. Burikova, N. N. Efferent links in the cerebral auditory cortex of the bat Myotis oxygnathus. J. Evol. Biochem. Physiol., 7, 450-455. 1971.

39. Busnel, R. G., ed. "Animal Sonar Systems," Vol. I and Vol. II. Lab. Physiol. Acoust., C. N. R. Z., Jouy-en-Josas, France, 1967.

40. Cahlander, D. A., McCue, J. J. G., and Webster, F. A. The determination of distance by echolocating bats. Nature, 201, 544-546. 1965.

41. Dalland, J. I. Auditory thresholds in bats: a behavioral technique. J. Aud. Res., 5, 95-108. 1965.

42. ———— Hearing sensitivity in bats. Science, 150, 1185-1186. ·1965.

43. ————, Vernon, J. A., and Peterson, E. A. Hearing and cochlear microphonic potentials in the bat Eptesicus fuscus. J. Neurophysiol., 30, 697-709. 1967.

44. David, A. Can young bats communicate with their parents at a distance? J. Bombay Natur. Hist. Soc., 65, 210. 1968.

45. Dunning, D. C., and Roeder, K. D. Moth sounds and the insect catching behaviour of bats. Science, 147, 173-174. 1965.

46. Dziedzic, A. Quelques performances des systèmes de detection par echos des chauve-souris et de delphinidae. Rev. Acoust., 1, 23-28. 1968.

47. Engelstätter, R. Einzelzellantworten aus dem Colliculus inferior von Rhinolophus ferrumequinum auf Reizung mit Rauschen unterschiedlicher Bandbreite und Kombinationen von Rauschen und Reintönen. Diplomarbeit, Fachbereich Biologie, Universität Frankfurt, 1978.

48. Escudie, B. Etat actuel des travaux sur les "sonars biologique", comparaison des propriétés de ces systemes avec les systèmes artificiels. I. C. P. I. de Lyon, Report TS 72/02. 1972.

49. ————, and Hellion, A. Comparison entre certains signaux optimaux a grand Bt et ceux utilisés par les chauve-souris, in: C. R. du contrat DRME 72419, "Modèle de reception acoustique biologique ...", 1973.

50. Fairbanks, J. T. Detection and triggering device for research on bat-echolocation systems. Mass. Inst. Techn. Res. Lab. Electr. Quart. Prog. Rep., 86, 337-338. 1967.

51. Feng, A. S., Simmons, J. A., and Kick, S. A. Echo detection and target-ranging neurons in the auditory system of the bat Eptesicus fuscus. Science, 202, 645-648. 1978.

52. Fenton, M. B. The role of echolocation in the evolution of bats. American Naturalist, 108, 386-388. 1974.

53. ———— Acuity of echolocation in Collocalia hirundinacea (Aves: Apodidae), with comments on the distribution of echolocating swiftlets and Molossid bats. Biotropica, 7, 1-7. 1975.

54. Fenton, M. B. Observations on the biology of some Rhodesian bats, including a key to the Chiroptera of Rhodesia. Life Sciences Contribution, Royal Ontario Museum, No. 104, 1-27. 1975.

55. ————, Boyle, N. G. H., Harrison, T. M., and Oxley, D. J. Activity patterns, habitat use and prey selection by some African insectivorous bats. Biotropica, 9, 73-85. 1977.

56. ————, Cumming, D. H. M., and Oxley, D. J. Prey of bat hawks and availability of bats. Condor, 79. 495-497. 1977.

57. ————, Jacobson, S. L., and Stone, R. N. An automatic ultrasonic sensing system for monitoring the activity of some bats. Can. J. Zool., 51, 291-299. 1973.

58. Firbas, W. The number of neurons in the cochlea of some bats. J. Mammal., 51, 809-810. 1970.

59. ———— Über anatomische Anpassungen des Hörorgans an die Aufnahme hoher Frequenzen. (Eine vergleichend-anatomische Untersuchung an Fledermäusen.) Monatsschr. Ohrheilk. Laryng. Rhin., 106, 105-156. 1972.

60. ————, and Einzinger, H. Über das Ganglion spirale der Chiroptera. Z. Säugetierkunde, 37, 321-326. 1972.

61. ————, Welleschik, B. Über die Verteilung der Acetylcholinesterase-Aktivität im Cortischen Organ von Fledermäusen. Acta Oto-Laryng., 70, 329-335. 1970.

62. Flieger, E., and Schnitzler, H.-U. Ortungsleistungen der Fledermaus Rhinolophus ferrumequinum bei ein- und beidseitiger Ohrverstopfung. J. comp. Physiol., 82, 93-102. 1973.

63. Friend, J. H., Suga, N., and Suthers, R. A. Neural responses in the inferior colliculus of echolocating bats to artificial orientation sounds and echoes. J. Cell. Comp. Physiol., 67, 319-332. 1966.

64. Fullard, J. H., and Fenton, M. B. Acoustic and behavioural analysis of the sounds produced by some species of Neoarctic Arctiidae (Lepidoptera). Can. J. Zool., 55, 1213-1224. 1977.

65. Funakoshi, K., and Uchida, T. A. Studies on the physiological and ecological adaptations of temperate insectivorous bats. I. Feeding activities in the Japanese long-fingered bats. Japanese J. Ecol., 25, 217-234. 1975.

66. Glaser, W. Eine systemtheoretische Interpretation der Fledermausortung. Studia Biophysica, 27, 103-110. 1971.

67. ———— Zur Fledermausortung aus dem Gesichtspunkt der Theorie gestörter Systeme. Zool. Jb. Physiol., 76, 209-229. 1971.

68. ———— Zur Hypothese des Optimalempfangs bei der Fledermausortung. J. comp. Physiol., 94, 227-248. 1974.

69. Goldman, L. J., Henson, O. W. Prey recognition and selection by the constant frequency bat, Pteronotus p. parnellii. Behav. Ecol. Sociobiol., 2, 411-420. 1977.

70. Gould, E. "Echolocation and communication in bats." Southern Methodist University Press, Dallas, Texas 75222, 1970.

71. ———— Individual recognition by ultrasonic communication between mother and infant bats (Myotis), in: "Animal Orientation and Navigation," S. R. Galler, K. Schmidt-König, G. J. Jacobs, R. E. Belleville, eds. NASA Press, Virginia, USA, 1970.

72. Gould, E. Neonatal vocalizations in bats of eight genera. J. Mammal., 56, 15-29. 1975.

73. ———— Experimental studies on the ontogeny of ultrasonic vocalizations in bats. Dev. Psychobiol., 8, 333-346. 1975.

74. ————, Lemkau, P. V., and Hume, J. C. "Ontogeny of Echolocation in Bats." J. Hopkins Univ., Dept. Mental Hygiene, Baltimore, MD. 21218, 1972. 105 pp.

75. Greguss, P. Bioholography - a new model of information processing. Nature, 219, 482. 1968.

76. ———— Variable holography. Laser in der Forschung, (2), 43-46, (3), 57-60. 1969.

77. Griffin, D. R. Discriminative echolocation by bats, in: "Animal Sonar Systems," R. G. Busnel, ed., pp. 273-306. Lab. Physiol. Acoust., C. N. R. Z., Jouy-en-Josas, France, 1967.

78. ———— Migration and homing of bats, in: "Biology of Bats," Vol. I, W. A. Wimsatt, ed. Academic Press, New York, 1970.

79. ———— The importance of atmospheric attenuation for the echolocation of bats. Anim. Behav., 19, 55-61. 1971.

80. ———— Comments on animal sonar symposium. J. Acoust. Soc. Am., 54, 137-138. 1973.

81. ———— Echolocation, in: "Basic Mechanisms in Hearing," A. Møller, ed. Academic Press, New York, 1973.

82. ———— "The Sensory Physiology of Animal Orientation." Harvey Lectures Series 71, Academic Press, New York, 1978.

83. ————, Friend, J. H., and Webster, F. A. Target discrimination by echolocation in bats. J. Exp. Zool., 158, 155-168. 1965.

84. ————, and Simmons, J. A. Echolocation of insects by horseshoebats. Nature, 250, 731-732. 1974.

85. Grinnell, A. D. Mechanisms of overcoming interference in echolocating animals, in: "Animal Sonar Systems," R. G. Busnel, ed., pp. 451-481. Lab. Physiol. Acoust., C. N. R. Z., Jouy-en-Josas, France, 1967.

86. ———— Comparative physiology of hearing. Ann. Rev. Physiol., 31, 545-580. 1969.

87. ———— Comparative auditory neurophysiology of neotropical bats employing different echolocation signals. Z. vergl. Physiol., 68, 117-153. 1970.

88. ———— Neural processing mechanisms in echolocating bats, correlated with differences in emitted sounds. J. Acoust. Soc. Am., 54, 147-156. 1973.

89. ———— Rebound excitation (off-responses) following non neural suppression in the cochleas of echolocating bats. J. comp. Physiol., 82, 172-194. 1973.

90. ————, and Brown, P. Long-latency "subthreshold" collicular responses to the constant-frequency components emitted by a bat. Science, 202, 996-999. 1978.

91. ————, and Grinnell, V. S. Neural correlates of vertical localization by echolocating bats. J. Physiol., 181, 830-851. 1965.

92. ————, and Hagiwara, S. Adaptation of the auditory system for echolocation: studies of New Guinea bats. Z. vergl. Physiol., 76, 41-81. 1972.

93. Grinnell, A. D., and Hagiwara, S. Studies of auditory neuro-physiology in non-echolocating bats, and adaptations for echo-location in one genus Rousettus. Z. vergl. Physiol., 76, 82-96. 1972.

94. ————, and Schnitzler, H.-U. Directional sensitivity of echo-location in the horseshoe bat Rhinolophus ferrumequinum. II. Behavioral directionality of hearing. J. comp. Physiol., 116, 63-76. 1977.

95. Gunier, W. Experimental homing of gray bats to a maternity colony in a Missouri barn. Am. Midl. Nat., 86, 502-506. 1971.

96. Habersetzer, J. Ortungslaute der Mausohrfledermaus (Myotis myotis) in verschiedenen Ortungssituationen. Verh. Dtsch. Zool. Ges., 1978, 196, G. Fischer Verl., Stuttgart, 1978.

97. Hall, J. G. The cochlea and the cochlear nuclei in the bat. Acta Otolaryngol., 67, 490-500. 1969.

98. Halls, J. A. T. Radar studies of the bat sonar. Proc. 4th. Int. Bat Res. Conf. Kenya Literat. Bureau, Nairobi (eds. Olembo als.), 137-142. 1978.

99. Harrison, J. B. Temperature effects on responses in the auditory system of the little brown bat Myotis l. lucifugus. Physiol. Zool., 38, 34-48. 1965.

100. Henson, M. M. Unusual nerve-fibre distribution in the cochlea of the bat Pteronotus p. parnellii (Gray). J. Acoust. Soc. Am., 53, 1739-1740. 1973.

101. ———— The basilar membrane of the bat, Pteronotus p. parnel-lii. American J. Anat., 153, 143-158. 1978.

102. ———— Structural changes in the inner ear of the bat, Pteronotus p. parnellii after overstimulation with constant frequency signals. Amer. J. Anat., 190, 610 (Abst.). 1978.

103. ————, and Lay, D. M. Scanning electron microscopy of the basilar membrane. Anat. Rec., 184, 425 (Abst.). 1976.

104. Henson, O. W. The activity and function of the middle-ear muscles in echolocating bats. J. Physiol., 180, 871-887. 1965.

105. ———— Fluctuations of middle ear air pressure in bats. Anat. Rec., 151, 456. 1965.

106. ———— Comparative physiology of middle ear muscle activity during echolocation in bat. Amer. Zool., 6, 603. 1966.

107. ———— Auditory sensitivity in Molossidae. Anat. Rec., 157, 363-364. 1967.

108. ———— The perception and analysis of biosonar signals by bats, in: "Animal Sonar Systems," R. G. Busnel, ed., II., pp. 949-1003. Lab. Physiol. Acoust., C.N.R.Z., Jouy-en-Josas, France, 1967.

109. ———— Neuroanatomical and physiological correlates of acoustic orientation. Anat. Rec., 163, 305 (Abst.). 1969.

110. ———— The central nervous system of Chiroptera, in: "Biology of Bats," W. A. Wimsatt, ed., chap. 2, 57-152. Academic Press, New York, 1970.

111. ———— The ear and audition, in: "Biology of Bats," W. A. Wimsatt, ed., vol. II, chap. 4. Academic Press, New York, 1971.

112. Henson, O. W.   Comparative anatomy of the middle ear, in:
     "Handb. Sens. Physiol.", W. D. Keidel and W. D. Neff, eds.,
     vol. V, pp. 39-100. Springer Verlag, Berlin, Heidelberg, New York,
     1974.

113. ————, Eager, R. P., and Henson, M. M.   Tecto-cerebellar pro-
     jections in the bat Tadarida brasiliensis. Anat. Rec., 160, 473-
     474 (Abst.). 1968.

114. ————, and Henson, M. M.   Middle ear muscle contractions and
     their relation to pulse- and echo-evoked potentials in the bat,
     in: "Animal Orientation and Navigation," S. R. Galler et al., eds.,
     NASA Press, Washington, 1972.

115. ————, and Pollak, G. D.   A technique for chronic implanta-
     tion of electrodes in the cochlea of bats. Physiol. Behav. 8,
     1185-1187. 1972.

116. ————, ————, and Novick, A.   Continuous monitoring of
     cochlear potentials and middle ear muscle activity in echolocating
     bats. Anat. Rec., 172. 328. 1972.

117. Hinchcliffe, R., and Pye, A.   The cochlea in Chiroptera: a
     quantitative approach. Int. Audiol., 7, 259-266. 1968.

118. ————, ————   Variations in the middle ear of the mammalia.
     J. Zool., 157, 277-288. 1969.

119. Hooper, J. H. D.   Potential use of a portable ultrasonic receiver
     for the field identification of flying bats. Ultrasonics, 7, 177-
     181. 1969.

120. Howell, D. J.   Acoustic behavior and feeding in Glossophagine
     bats. J. Mammal., 55, 293-308. 1974.

121. Jen, P. H.-S.   Coding of the directional information of single
     neurons in the superior olivary complex of echolocating bats.
     Ph. D. Thesis, Dept. of Biology, Washington Univ., St. Louis,
     Missouri, 1974.

122. ————   Electrophysiological properties of auditory neurons
     in the superior olivary complex of echolocating bats. J. comp.
     Physiol., 128, 47-56. 1978.

123. ————, Alster, H., and Suga, N.   Coordinated activity of the
     middle ear and laryngeal muscle in echolocating bats. J. Acoust.
     Soc. Am., 57, S42, VII. 1975.

124. ————, Lee, Y. H.   The just-noticeable movement speed of ob-
     stacles perceived by echolocating bats. J. Acoust. Soc. Am., 64,
     S88, HH21. 1978.

125. ————, and McCarty, J. K.   Bats avoid moving objects more
     successfully than stationary ones. Nature, London, 275, 743-744.
     1978.

126. ————, ————, and Lee, Y. H.   The avoidance of obstacles
     by little brown bats, Myotis lucifugus. Proc. Fifth Int. Bat.
     Res. Conf., (in press). 1979.

127. ————, and Ostwald, D. J.   Response of cricothyroid muscle
     to frequency-modulated sounds in FM-bats, Myotis lucifugus.
     Nature, 265, 77-78. 1977.

128. ————, ————   Further study of the electrophysiology of the
     middle-ear and cricothyroid muscles of FM bats. Nation. Sci. Coun.
     Mon. (China), V, 1014-1028. 1977.

129. Jen, P. H.-S., Ostwald, J., and Suga, N. Acoustic reflexes of middle-ear and laryngeal muscles in the FM bat Myotis lucifugus. J. Acoust. Soc. Am., 60, S4, B3. 1976.

130. ———, ———, ——— Electrophysiological properties of the acoustic middle ear and laryngeal muscle reflexes in the awake echolocating FM-bats, Myotis lucifugus. J. comp. Physiol., 124, 61-73. 1978.

131. ———, and Suga, N. Coding of directional information by neurons in the superior olivary complex of bats. J. Acoust. Soc. Am., 55, S52. 1974.

132. ———, ——— Properties of acoustic middle-ear muscle reflex in echolocating bats. J. Acoust. Soc. Am., 56, S4, B7. 1974.

133. ———, ——— Coordinated activities of middle ear and laryngeal muscles in echolocating bats. Science, 191, 950-952. 1976.

134. ———, ——— Difference between CM and N1 tuning curves in the CF-FM bats, Pteronotus parnellii rubiginosus. J. Acoust. Soc. Am., 59, S19, H13. 1976.

135. ———, ——— Functional properties of the acoustic laryngeal muscle reflex in echolocating bats. J. Acoust. Soc. Am., 59, S1, OO10. 1976.

136. ———, ——— Electrophysiological properties of middle ear and laryngeal muscle in little brown bats. Neurosci. Abst., II, 20. 1976.

137. ———, ——— Physiological properties of middle-ear and laryngeal muscles of echolocating bats. Nation. Sci. Coun. Mon. (China), IV, 2606-2617. 1976.

138. Johnson, R. A. Energy spectrum analysis as a processing mechanism for echolocation. Ph. D. Thesis, Dept. Electr. Engn., Univ. Rochester, N. Y., 1972. xv + 175 pp.

139. ———, Henson, O. W., and Goldman, L. J. Detection of insect wing beats by the bat Pteronotus parnellii. J. Acoust. Soc. Am., 55, 53. 1974.

140. ———, and Titlebaum, E. L. Energy spectrum analysis a model of echolocation processing. J. Acoust. Soc. Am. (Suppl.), 56, S39. 1974.

141. Kämper, R., and Schmidt, U. Die Morphologie der Nasenhöhle bei einigen neotropischen Chiropteren. Zoomorphologie, 87, 3-19. 1977.

142. Khajuria, H. Courtship and mating in Rhinopoma h. hardwickei Gray (Chiroptera: Rhinopomatidae). Mammalia, 36, 307-309. 1972.

143. Kindlman, P. J., Berman, L. B., Johnson, R. A., Pollak, G. D., Henson, O. W., and Novick, A. Measurement of "instantaneous" carrier frequency of bat pulses. J. Acoust. Soc. Am., 54, 1380-1382. 1973.

144. Kolb, A. Ortungsprinzip bei Fledermäusen. Z. Säugetk., 35, 306-320. 1970.

145. ———, Funktion und Wirkungsweise der Riechlaute der Mausohrfledermaus, Myotis myotis. Z. Säugetk., 41, 226-236. 1976.

146. Kuhn, H. J. Über die Innervation des Kehlkopfes einiger Flughunde (Pteropodidae, Megachiroptera, Mammalia). Zool. Anz., 181, 168-181. 1968.

147. Kulzer, E., and Weigold, H. Das Verhalten der Großen Hufeisennase (Rhinolophus ferrum-equinum) bei einer Flugdressur. Z. Tierpsychol., 47, 268-280. 1978.
148. Ladik, J., and Greguss, P. Possible molecular mechanisms of
information storage in the long term memory (bat). Symp. Biol.
Hungar., 10, 343-355. 1971.
149. Long, G.R. Masked auditory thresholds from the bat, Rhinolophus ferrumequinum. J. comp. Physiol., 116, 247-255. 1977.
150. ————, and Schnitzler, H.-U. Behavioural audiograms from the
bat, Rhinolophus ferrumequinum. J. comp. Physiol., 100, 211-219.
1975.
151. Machmert, H., Theiss, D., and Schnitzler, H.-U. Konstruktion
eines Luftultraschallgebers mit konstantem Frequenzgang im Bereich
von 15 kHz bis 130 kHz. Acustica, 34, 81-85. 1975.
152. Manabe, T., Suga, N., and Ostwald, J. Aural representation in
the Doppler-shifted-CF processing area of the auditory cortex of
the mustache bat. Science, 200, 339-342. 1978.
153. Manley, G. A., Irvine, D. R. F., and Johnstone, B. M. Frequency
response of bat tympanic membrane. Nature, 237, 112-113. 1972.
154. Manley, J. A., and Johnstone, B. M. A comparison of cochlear
summating potentials in the bat and Guinea pig, including temperature effects. J. comp. Physiol., 88, 43-66. 1974.
155. McCue, J. J. G. Signal processing by the bat Myotis lucifugus.
J. Audit. Res., 9, 100-107. 1969.
156. Miline, R., Devecenski, V., and Krstic, R. Effects of auditory
stimuli on the pineal gland of hibernating bats. Acta Anat., 73
(Suppl. 56), 293-300. 1969.
157. Möhres, F. P., and Neuweiler, G. Die Ultraschallorientierung
der Großblatt-Fledermäuse (Chiroptera - Megadermatidae). Z. vergl.
Physiol., 53, 195-227. 1966.
158. Möller, J. Die Funktion von Hemmeinflüssen bei Neuronen der
lemniscalen Hörbahn bei der Echoortung von Rhinolophus ferrumequinum. Dissertation, Universität Stuttgart, 1977.
159. ———— Response characteristics of inferior colliculus neurons
of the awake CF-FM bat Rhinolophus ferrumequinum. II. Two-tone
stimulation. J. comp. Physiol. A, 125, 227-236. 1978.
160. ———— Anpassung von Neuronen der Hörbahn an die Echoortung
fliegender Hufeisennasen. Verh. Dtsch. Zool. Ges., 1978, 164,
G. Fischer Verl., Stuttgart, 1978.
161. ————, Neuweiler, G., and Zöller, H. Response characteristics
of inferior colliculus neurons of the awake CF-FM bat Rhinolophus
ferrumequinum. I. Single-tone stimulation. J. comp. Physiol. A,
125, 217-225. 1978.
162. Mogus, M. A. A theoretical approach to bat echolocation. Ph. D.
Dissertation, PA. State Univ., 1970.
163. Müller, H. C. Homing and distance-orientation in bats. Z. Tierpsychol., 23, 403-421. 1966.
164. Nakajima, Y. Fine structure of the medial nucleus of the trapezoid body of the bat with special reference to two types of synaptic
endings. J. Cell Biol., 50, 121-134. 1971.

165. Neuweiler, G.   Interaction of other sensory systems with the sonar system, in: "Animal Sonar Systems," R. G. Busnel, ed., vol. I, pp. 509-533. Lab. Physiol. Acoust., C. N. R. Z., Jouy-en-Josas, France, 1967.

166. ————— Neurophysiological investigations in the colliculus inferior of Rhinolophus ferrumequinum. Bijdr. Dierk., 40, 59-61, 1970.

167. ————— Neurophysiologische Untersuchungen zum Echoortungssystem der Großen Hufeisennase Rhinolophus ferrum equinum Schreber, 1774. Z. vergl. Physiol., 67, 273-306. 1970.

168. ————— Frequenzdiskriminierung in der Hörbahn von Säugern. Verh. Dtsch. Zool. Ges., 66, 168-176. 1973.

169. ————— Echoortung, in: "Biophysik," W. Hoppe, W. Lohmann, H. Markl, H. Ziegler, eds. Springer Verlag, Berlin, Heidelberg, New York, 1977.

170. ————— Recognition mechanisms in echolocation of bats, in: Life Sci. Res. Rep. 5, "Recognition of Complex Acoustic Signals," Th. H. Bullock, ed. Dahlem Konferenzen, Berlin, 1977.

171. ————— Die Echoortung der Fledermäuse. Rhein. Westf. Akad. Wiss., Vorträge N 272, 57-82. 1978.

172. —————, and Möhres, F. P.   The role of spacial memory in the orientation, in: "Animal Sonar Systems," R. G. Busnel, ed., vol. I, pp. 129-140. Lab. Physiol. Acoust., C.N.R.Z., Jouy-en-Josas, France, 1967.

173. —————, ————— Die Rolle des Ortsgedächtnisses bei der Orientierung der Großblatt-Fledermaus Megaderma lyra. Z. vergl. Physiol. 57, 147-171. 1967.

174. —————, Schuller, G., and Schnitzler, H.-U.   On- and off-responses in the inferior colliculus of the greater horseshoe bat to pure tones. Z. vergl. Physiol., 74, 57-63. 1971.

175. —————, and Vater, M.   Response patterns to pure tones of cochlear nucleus units in the CF-FM bat, Rhinolophus ferrumequinum. J. comp. Physiol. A, 115, 119-133. 1977.

176. Novick, A.   Echolocation of flying insects by the bat (Chilonycteris psilotis). Biol. Bull., 128, 297-314. 1965.

177. ————— Echolocation in bats: some aspects of pulse design. Amer. Sci., 59, 198-209. 1971.

178. ————— Echolocation in bats: a zoologist's view. J. Acoust. Soc. Am., 54, 139-146. 1973.

179. ————— Acoustic orientation, in: "Biology of Bats," W. A. Wimsatt, ed., vol. III. Academic Press, New York, 1977.

180. —————, and Dale, B. A.   Foraging behavior in fishing bats and their insectivorous relatives. J. Mammal., 52, 817-818. 1971.

181. —————, ————— Bats aren't all bad. National Geographic, 143, 614-637. 1973.

182. —————, and Leen, N.   "The World of Bats." Holt, Rinehart and Winston, New York, Chicago, San Francisco, 1970.

183. O'Neill, W. E., and Suga, N.   Target range-sensitive neurons in the auditory cortex of the mustache bat. Science, 203, 69-73. 1979.

184. Ostwald, J.   Tonotope Organisation des Hörcortex der CF-FM-
Fledermaus Rhinolophus ferrumequinum. Verh. Dtsch. Zool. Ges.,
1978, 198, G. Fischer Verl., Stuttgart, 1978.

185. Patterson, A. P., and Hardin, J. W.   Flight speed of five
species of vespertilionid bats. J. Mamm., 50, 152-153. 1969.

186. Peff, T. C., and Simmons, J. A.   Horizontal-angle resolution by
echolocating bats. J. Acoust. Soc. Am., 51, 2063-2065. 1972.

187. Pirlot, P.   Perspectives in sensory-ecological studies of
higher vertebrates. Rev. Canad. Biol., 36, 337-350. 1977.

188. ————, and Pottier, J.   Encephalization and quantitative brain
composition in bats in relation to their life habits. Rev. Canad.
Biol., 36, 321-336. 1977.

189. Pollak, G. D., Bodenhamer, R., Marsh, D. S., and Souther, A.
Recovery cycles of single neurons in the inferior colliculus of
unanesthetized bats obtained with frequency-modulated and constant-
frequency sounds. J. comp. Physiol. A, 120, 215-250. 1977.

190. ————, and Henson, O. W.   Specialized functional aspects of
·the middle ear muscle in the bat, Chilonycteris parnellii. J. comp.
Physiol., 84, 167-174. 1973.

191. ————, ————, and Novick, A.   Cochlear microphonic audiograms in the pure tone bat Chilonycteris parnellii parnellii.
Science, 176, 66-68. 1972.

192. ————, Marsh, D., Bodenhamer, R., and Souther, A.   Echo-detecting characteristics of neurons in inferior colliculus of unanesthetized bats. Science, 196, 675-678. 1977.

193. ————, ————, ————, ———— Characteristics of phasic
"on" neurons in the inferior colliculus of unanesthetized bats
with observations relating to mechanisms for echo-ranging. J.
Neurophysiol., 40, 926-942. 1977.

194. ————, ————, ————, ———— A single-unit analysis of
inferior colliculus in unanesthetized bats: Response patterns
and spike count functions generated by CF and FM sounds. J. Neurophysiol., 41, 677-691. 1978.

195. Prince, J. H.   The use of echo-locating by bats. Australian
Nat. Hist., 16, 367-370. 1970.

196. Pye, A.   The structure of the cochlea in Chiroptera. I. Microchiroptera: Emballonuroidea and Rhinolophoidea. J. Morph., 118,
495-510. 1966.

197. ———— The structure of the cochlea in Chiroptera. II. The
Megachiroptera and Vespertilionidea of the Microchiroptera.
J. Morph., 119, 101-120. 1966.

198. ———— The structure of the cochlea in Chiroptera. III. Microchiroptera: Phyllostomatidae. J. Morph., 121, 241-254. 1967.

199. ———— The structure of the cochlea in Chiroptera. A selection
of Microchiroptera from Africa. J. Zool., 162, 335-343. 1970.

200. ———— The aural anatomy of bats. Bijdr. Dierk., 40, 67-70,
1970.

201. ———— Some aspects of cochlear structure and function in bats.
Proc. 4th Int. Bat Res. Conf., Kenya Literature Bureau, Nairobi,
73-83. 1971.

202. Pye, A.   Comparison of histological changes in the cochlea of
the guinea pig and the bat Rousettus after acoustic trauma. Symp.
Zool. Soc. London, 37, 97-105. Academic Press, London, 1975.

203. ――――   The structure of the cochlea in Chiroptera. Periodicum
Biologorum 75, 83-87. 1973.

204. Pye, J. D.   Theories of sonar systems and their application to
biological organisms (Discussion), in: "Animal Sonar Systems,"
R. G. Busnel, ed., pp. 1121-1136. Lab. Physiol. Acoust., C.N.R.Z.,
Jouy-en-Josas, France, 1967.

205. ――――   Synthesizing the waveforms of bats' pulses, in: "Animal
Sonar Systems," R. G. Busnel, ed., pp. 43-65. Lab. Physiol. Acoust.
C. N. R. Z., Jouy-en-Josas, France, 1967.

206. ――――   Bats, in: "UFAW Handbook on the Care and Maintenance of
Laboratory Animals," 3rd edn., chap. 31, pp. 491-501. Livingstone,
London, 1967,

207. ――――   "Bats" - a natural science picture book illustrated by
Colin Threadgall. Bodley Head, 1968.

208. ――――   Animal sonar in air. Ultrasonics, 6, 32-38. 1968.

209. ――――   Hearing in bats, in: "Hearing Mechanisms in Vertebrates"
A. V. S. de Reuck and J. Knight, eds., pp. 66-84. Churchill,
London, 1968,

210. ――――   The diversity of bats. Sci. J., 5, 47-52. 1969.

211. ――――   Bats and fog. Nature, 229, 572-574. 1971.

212. ――――   Bimodal distribution of constant frequencies in some
hipposiderid bats (Mammalia: Hipposideridae). J. Zool. (Lond.),
166, 323-335. 1972.

213. ――――   Echolocation by constant frequency in bats. Period.
Biol., 75, 21-26. (Proc. 3rd Int. Bat Res. Conf., Yugoslavia, 1972)
1973.

214. ――――, and Roberts, L. H.   Ear movement in a hipposiderid bat.
Nature, 225, 285-286. 1970.

215. Radtke, S.   Struktur und Verschaltung des Hörcortex der Großen
Hufeisennase (Rhinolophus ferrumequinum). Staatsexamensarbeit,
Fachber. Biologie, Univ. Frankfurt/M., 1979.

216. Ramprashad, F., Money, K. E., Landolt, J. P., and Laufer, J.
A neuro-anatomical study of the cochlea of the little brown bat,
Myotis lucifugus. J. Comp. Neurol., 178, 347-364. 1978.

217. Roberts, L. H.   Variable resonance in constant frequency bats.
J. Zool., 166, 337-348. 1972.

218. ――――   Correlation of respiration and ultrasound production
in rodents and bats. J. Zool., 168, 439-449. 1972.

219. ――――   Cavity resonance in the production of orientation cries.
Period. Biol., 75, 27-32. (Proc. 3rd Int. Bat Res. Conf.) 1973.

220. ――――   Confirmation of the echolocation pulse production
mechanisms of Rousettus. J. Mamm., 56, 218-220. 1975.

221. Roeder, K. D.   Acoustic sensory responses and possible bat
evasion tactics of certain moths. Can. Soc. Zool. Ann. Meeting
Proc., Univ. New Brunswick, Fredericton, pp. 71-78. 1974.

222. Sales, G., and Pye, D. "Ultrasonic Communication by Animals."
Chapman and Hall, London, 1974.

223. Schlegel, P. Directional coding by binaural brainstem units
     of the CF-FM bat, Rhinolophus ferrumequinum. J. comp. Physiol.,
     118, 327-352. 1977.

224. ―――――― Calibrated earphones for the echolocating bat, Rhino-
     lophus ferrumequinum. J. comp. Physiol., 118, 353-356. 1977.

225. ――――――, and Suga, N. Recovery cycles of single neurons in the
     lateral lemniscus and inferior colliculus of echolocating bats.
     Psychologist, 14 (3). 1971.

226. ――――――, and Vater, M. Vergleichende hörphysiologische Unter-
     suchungen an zwei echo-ortenden Fledermausarten, Molossus ater
     und Molossus molossus (Molossidae). Verh. Dtsch. Zool. Ges., 1978,
     165, G. Fischer Verl., Stuttgart, 1978.

227. Schmidt, U., and Schmidt, Ch. Echolocation performance of the
     vampire bat (Desmodus rotundus). Z. Tierpsychol., 45, 349-358.
     1977.

228. Schneider, R. Das Gehirn von Rousettus aegyptiacus (E. Geoffroy
     1810) (Megachiroptera, Chiroptera, Mammalia). Abh. senckenb.
     naturf. Ges. (Frankfurt/M.), 513, 1-160. 1966.

229. Schnitzler, H.-U. Discrimination of thin wires by flying horse-
     shoe bats (Rhinolophidae), in: "Animal Sonar Systems," R. G.
     Busnel, ed., pp. 69-87. Lab. Physiol. Acoust., C. N. R. Z.,
     Jouy-en-Josas, France, 1967.

230. ―――――― Kompensation von Doppler-Effekten bei Hufeisen-Fleder-
     mäusen. Naturwissensch., 54, 523. 1967.

231. ―――――― Die Ultraschall-Ortungslaute der Hufeisen-Fledermäuse
     (Chiroptera-Rhinolophidae) in verschiedenen Orientierungssitua-
     tionen. Z. vergl. Physiol., 57, 376-408. 1968.

232. ―――――― Comparison of the echolocation behavior in Rhinolophus
     ferrum-quinum and chilonycteris rubiginosa. Bijdr. Dierk., 40,
     77-80. 1970.

233. ―――――― Echoortung bei der Fledermaus Chilonycteris rubiginosa.
     Z. vergl. Physiol., 68, 25-38. 1970.

234. ―――――― Fledermäuse im Windkanal. Z. vergl. Physiol., 73, 209-
     221. 1971.

235. ―――――― Die Echoortung der Fledermäuse und ihre hörphysiologi-
     schen Grundlagen. Fortschr. Zool., 21, 136-189. 1973.

236. ―――――― Control of Doppler shift compensation in the greater
     horseshoe bat, Rhinolophus ferrumequinum. J. comp. Physiol., 82.
     79-92. 1973.

237. ―――――― Die Echoortung der Fledermäuse, in:"Grzimeks Tierleben",
     Ergänzungsbd.,"Verhaltensforschung", 189-204. Kindler Verlag,
     Zürich, 1974.

238. ―――――― Die Richtwirkung der Echoortung bei der Großen Hufeisen-
     nase. Verh. Dtsch. Zool. Ges. 1976, 267, G. Fischer Verl.,
     Stuttgart, 1976.

239. ―――――― Die Detektion von Bewegung durch Echoortung bei Fleder-
     mäusen. Verh. Dtsch. Zool. Ges. 1978, 16-33, G. Fischer Verl.,
     Stuttgart, 1978.

240. ――――――, and Grinnell, A. D. Directional sensitivity of echo-
     location in the horseshoe bat Rhinolophus ferrumequinum. I. Direc-
     tionality of sound emission. J. comp. Physiol., 116, 51-61. 1977.

241. Schnitzler, H.-U., Schuller, G., and Neuweiler, G. Antworten des Colliculus inferior der Fledermaus Rhinolophus eureale auf tonale Reizung. Naturw., 58, 627. 1971.

242. ————, Suga, N., and Simmons, J. A. Peripheral auditory tuning for fine frequency analysis by the CF-FM bat, Rhinolophus ferrumequinum. III. Cochlear microphonics and auditory nerve responses. J. comp. Physiol. A, 106, 99-110. 1976.

243. Schuller, G. Echoortung bei Rhinolophus ferrumequinum mit frequenzmodulierten Lauten. Evoked potentials im Colliculus inferior. J. comp. Physiol., 77. 306-331. 1972.

244. ———— The role of overlap of echo with outgoing echolocation sound in the bat Rhinolophus ferrumequinum. Naturw., 61, 171-172. 1974.

245. ———— Echo delay and overlap with emitted orientation sounds and Doppler-shift compensation in the bat, Rhinolophus ferrumequinum. J. comp. Physiol. A, 114, 103-114. 1977.

246. ———— Coding of small sinusoidal frequency and amplitude modulations in the inferior colliculus of "CF-FM" bat, Rhinolophus ferrumequinum. Exp. Brain Res., 34, 117-132. 1979.

247. ————, Beuter, K., and Rübsamen, R. Dynamic properties of the compensation system of Doppler shifts in the bat Rhinolophus ferrumequinum. J. comp. Physiol., 97, 113-125. 1975.

248. ————, ————, and Schnitzler, H.-U. Response to frequency shifted artificial echoes in the bat Rhinolophus ferrumequinum. J. comp. Physiol., 89, 275-286. 1974.

249. ————, Neuweiler, G., and Schnitzler, H.-U. Collicular response to the frequency modulated final part of echolocation sounds in Rhinolophus ferrumequinum. Z. vergl. Physiol., 74, 153-155. 1971.

250. ————, and Suga, N. Storage of Doppler-shift information in the echolocation system of the "CF-FM"-bat, Rhinolophus ferrumequinum. J. comp. Physiol. A, 105, 9-14. 1976.

251. ————, ———— Laryngeal mechanisms for the emission of CF-FM sounds in the Doppler-shift compensating bat, Rhinolophus ferrumequinum. J. comp. Physiol., 107, 253-262. 1976.

252. Schweizer, H. Struktur und Verschaltung des Colliculus inferior der Grossen Hufeisennase (Rhinolophus ferrumequinum). Dissertation, Universität Frankfurt am Main, 1978.

253. Segall, W. Auditory region in bats including Icaronycteris index. Fieldiana Zool., 58, 103-108. 1971.

254. ———— The external morphology of the inner ear in bats from the phophorites of Quercy. Fieldiana Geol., 33, 59-81. 1974.

255. Shimozawa, T., Suga, N., Hendler, P., and Schuetze, S. Directional sensitivity of echolocation system in bats producing frequency modulated signals. J. Exp. Biol., 60, 53-69. 1974.

256. Shirley, D. J., and Diercks, K. J. Analysis of the frequency response of simple geometric targets. J. Acoust. Soc. Am., 48, 1275-1282. 1971.

257. Simmons, J. A. The sonar sight of bats. Psychol. Today, 2, 50-57. 1968.

258. Simmons, J. A. Depth perception by sonar in the bat Eptesicus fuscus. Ph. D. Dissertation, Princeton Univ., (x + 208 pp.), 1968.

259. ————— Acoustic radiation patterns for the echolocating bats, Chilonycteris rubiginosa and Eptesicus fuscus. J. Acoust. Soc. Am., 46, 1054-1056. 1969.

260. ————— The sonar receiver of the bat. Ann. N. Y. Acad. Sci., 188, 161-174. 1971.

261. ————— Narrow band CF-FM echolocation in bats (Summary). J. Acoust. Soc. Am., 50, 148. 1971.

262. ————— Echolocation in bats: Signal processing of echoes for target range. Science, 171, 925-928. 1971.

263. ————— The resolution of target range by echolocating bats. J. Acoust. Soc. Am., 54, 157-173. 1973.

264. ————— Response of the Doppler echolocation system in the bat, Rhinolophus ferrumequinum. J. Acoust. Soc. Am., 56, 672-682. 1974.

265. ————— (Rapporteur), et al. Localization and identification of acoustic signals, with reference to echolocation. Group report in: Life Sci. Res. Rep. 5, "Recognition of Complex Acoustic Signals," Th. H. Bullock, ed. Dahlem Konferenzen, Berlin, 1977.

266. —————, Fenton, M. B., and O'Farrell, M. J. Echolocation and pursuit of prey by bats. Science, 203, 16-21. 1979.

267. —————, Howell, D. J., and Suga, N. Information content of bat sonar echoes. Amer. Sci., 63, 204-215. 1975.

268. —————, Lavender, W. A., Lavender, B. A., et al. Target structure and echo spectral discrimination by echolocating bats. Science, 186, 1130-1132. 1974.

269. —————, —————, —————, et al. Echolocation by free-tailed bats (Tadarida). J. comp. Physiol., 125, 291-299. 1978.

270. —————, and O'Farrell, M. J. Echolocation by the long eared bat, Plecotus phyllotis. J. comp. Physiol., 122, 201-214. 1977.

271. —————, and Vernon, J. A. Echolocation discrimination of targets by the bat Eptesicus fuscus. J. Exp. Zool., 176, 315-328. 1971.

272. Skinner, D. P., Altes, R. A., and Jones, J. D. Broadband target classification using a bionic sonar. J. Acoust. Soc. Am., 62, 1239-1246, 1977.

273. Small, R., and Levine, R. R. Interspecific communication between bats and moth, in: "Course in Animal Communication," R. G. Busnel, ed. The City College, Paris, 1971.

274. Smith, D. M., Mercer, R. M., Goldman, L. H., Henson, W. W., and Henson, M. M. Phase-locked loop device for the fine frequency analysis of the biosonar signals of bats. J. Acoust. Soc. Am., 61, 1092-1093. 1977,

275. Stewart, J. L., and Kasson, J. M. Simulation mechanisms in animal echo ranging. U. S. Gov. Research Dev. Rep., 70. 1970.

276. Stones, R. C., and Branick, L. P. Use of hearing in homing by two species of Myotis bats. J. Mamm., 50, 157-160. 1969.

277. Strother, G. K. Comments on aural pulse compression in bats and humans. J. Acoust. Soc. Amer., 41, 529. 1967.

278. Strother, G. K., and Mogus, M. Acoustical beam patterns for bats: Some theoretical considerations. J. Acoust. Soc. Am., 48, 1430-1432. 1970.

279. Suga, N. Analysis of frequency-modulated sounds by auditory neurons of echolocating bats. J. Physiol., 179, 26-53. 1965.

280. ———— Functional properties of auditory neurons in the cortex of echolocating bats. J. Physiol., 181, 671-700. 1965.

281. ———— Responses of cortical auditory neurons to frequency modulated sounds in echo-locating bats. Nature, 206, 890-891. 1965.

282. ———— Analysis of frequency-modulated and complex sounds by single auditory neurons of bats. J. Physiol., 198, 51-80. 1968.

283. ———— Classification of inferior collicular neurons of bats in terms of responses to pure tones, FM sounds, and noise bursts. J. Physiol., 200, 555-574. 1969.

284. ———— Echo-location and evoked potentials of bats after ablation of inferior colliculus. J. Physiol., 203, 707-728. 1969.

285. ———— Echo-location of bats after ablation of auditory cortex. J. Physiol., 203, 729-739. 1969.

286. ———— Echo-ranging neurons in the inferior colliculus of bats. Science, 170, 449-451. 1970.

287. ———— Responses of inferior collicular neurons of bats to tone bursts with different rise times. J. Physiol., 217, 159-177. 1971.

288. ———— Analysis of information-bearing elements in complex sounds by auditory neurons of bats. Audiol., 11, 58-72, 1972.

289. ———— Neurophysiological analysis of echolocation in bats, in: "Animal Orientation and Navigation," S. R. Galler et al., eds. NASA Press, Washington, D. C., 1972.

290. ———— Feature extraction in the auditory system of bats, in: "Basic Mechanisms in Hearing," A. R. Møller, ed., pp. 675-744. Academic Press, New York and London, 1973.

291. ———— Amplitude spectrum representation in the Doppler-shifted-CF processing area of the auditory cortex of the mustache bat. Science, 196, 64-67. 1977.

292. ———— Specialization of the auditory system for reception and processing of species-specific sounds. Federation Proc., 37, 2342-2354. 1978.

293. ————, and Jen, P. H.-S. Muscular control of incoming signals in the auditory system of echolocating bats. J. Acoust. Soc. Am., 55, S52, AA9. 1974.

294. ————, ———— Peripheral control of acoustic signals in the auditory system of echolocating bats. J. Exp. Biol., 62, 277-311. 1975.

295. ————, ———— Disproportionate tonotopic representation for processing CF-FM sonar signals in the mustache bat auditory cortex. Science, 194, 542-544. 1976.

296. ————, ———— Neural vs. mechanical tuning curves in the CF-FM bats Pteronotus parnellii rubiginosus. J. Acoust. Soc. Am., 59, Suppl. No. 1, S18. 1976.

297. Suga, N., and Jen, P. H.-S. Further studies on the peripheral
     auditory system of CF-FM bats specialized for fine frequency
     analysis of Doppler-shifted echoes. J. Exp. Biol., 69, 207-232.
     1977.
298. ———, Neuweiler, G., and Möller, J. Peripheral auditory
     tuning for fine frequency analysis by the CF-FM bat, Rhinolophus
     ferrumequinum. IV. Properties of peripheral auditory neurons.
     J. comp. Physiol. A, 106, 111-125. 1976.
299. ———, ———, Jen, P. H.-S., and Möller, J. Adaptation of
     the peripheral auditory system of CF-FM bats for reception and
     analysis of predominant components in orientation sounds and
     echoes. J. Acoust. Soc. Am., 59, Suppl. No. 1, S46. 1976.
300. ———, and O'Neill, W. E. Mechanisms of echolocation in
     bats - comments on the neuroethology of the biosonar system of
     CF-FM bats. Trends Neurosc., 1, 35-38. 1978.
301. ———, ———, and Manabe, T. Cortical neurons sensitive
     to combinations of information-bearing elements of biosonar signals
     in the mustache bat. Science, 200, 778-781. 1978.
302. ———, ———, ——— Harmonic-sensitive neurons in the
     auditory cortex of the mustache bat. Science, 203, 270-274. 1979.
303. ———, and Schlegel, P. Neural attenuation of responses to
     emitted sounds in echolocating bats. Science, 177, 82-84. 1972.
304. ———, ——— Coding and processing in the auditory systems
     of FM-Signal producing bats. J. Acoust. Soc. Am., 54, 174-190.
     1973.
305. ———, ———, Shimozawa, T., and Simmons, J. A. Orientation
     sounds evoked from echolocating bats by electrical stimulation of
     the brain. J. Acoust. Soc. Am., 54, 793-797. 1973.
306. ———, Schuller, G., and Jen, P. H.-S. Neurophysiological
     studies on the echolocation system of bats sensitive to Doppler-
     shifted echoes. J. Acoust. Soc. Am., 60, S4, B2. 1976.
307. ———, and Shimozawa, T. Site of neural attenuation of re-
     sponses to self-vocalized sounds in echolocating bats. Science,
     183, 1211-1213. 1974.
308. ———, Simmons, J. A., and Jen, P. H.-S. Peripheral special-
     ization for fine analysis of Doppler-shifted echoes in the audi-
     tory system of the "CF-FM" bat Pteronotus parnellii. J. Exp. Biol.,
     63, 161-192. 1975.
309. ———, ———, ——— Peripheral specialization for anal-
     ysis of Doppler-shifted echoes in the auditory system of the "CF-
     FM" bat, Pteronotus parnellii. I. Cochlear microphonic. J. Acoust.
     Soc. Am., 57, S42. 1975.
310. ———, ———, ——— Peripheral specialization for anal-
     ysis of Doppler-shifted echoes in the auditory system of the "CF-
     FM" bat, Pteronotus parnellii. II. Properties of peripheral audi-
     tory neurons. J. Acoust. Soc. Am., 57, S42. 1975.
311. ———, ———, Shimozawa, T. Neurophysiological studies on
     echolocation systems in awake bats producing CF-FM orientation
     sounds. J. Exp. Biol., 61, 379-399. 1974.

312. Suthers, R. A. Acoustic orientation by fish-catching bats.
J. Exp. Zool, 158, 319-348. 1965.
313. ———— Comparative echolocation by fishing bats. J. Mamm.,
48, 79-87. 1967.
314. ———— Wie orten fischfangende Fledermäuse ihre Beute?
Umschau i. W. u. T., 67, 693-696. 1967.
315. ———— A comment on the role of choroidal papillae in the
fruit bat retina. Vision Res., 10, 921-922. 1970.
316. ———— Vision, olfaction, taste, <u>in</u>: "Biology of Bats,"
W. Wimsatt, Ed., vol. II. Academic Press, New York, 1970.
317. ———— Responses of neurons in the visual cortex of the echolocating bat, Rousettus aegypticus. Proc. East Afr. Acad.

318. ————, and Chase, J. Visual pattern discrimination by an
echolocating bat. Amer. Zool., 6, 573. 1966.
319. ————, ————, and Braford, B. Visual form discrimination
by echolocating bats. Biol. Bull, 137, 535-546. 1969.
320. ————, and Fattu, J. M. Fishing behaviour and acoustic orientation by the bat (Noctilio labialis). Anim. Behav., 21, 61-66.
1973.
321. ————, ———— Mechanisms of sound production by echolocating
bats. Amer. Zool., 13, 1215-1226. 1973.
322. ————, Thomas, St. P., and Suthers, B. J. Respiration, wingbeat and ultrasonic pulse emission in an echo-locating bat. J. Exp.
Biol., 56, 37-48. 1972.
323. ————, and Wallis, N. E. Optics of the eye of some echolocating bats. Vision Res., 10, 1165-1173. 1970.
324. Tepaske, E. R. Some morphological aspects and taxonomic relationship of the middle ear in bats. Diss. Abstr., 26, 65-8784 (155 pp.)
1966.
325. Vater, M., and Schlegel, P. Antwortcharakteristika einzelner
Neurone im Colliculus inferior der echoortenden Fledermausarten
Molossus ater und Molossus molossus auf tonale und FM Signale.
Verh. Dtsch. Zool. Ges., 1978, 197, G. Fischer Verl., Stuttgart,
1978.
326. Vernon, J. A., Dalland, J. S., and Wever, E. G. Further studies
of hearing in the bat Myotis lucifugus, by means of cochlear
potentials. J. Aud. Res., 6, 153-163. 1966.
327. ————, and Peterson, E. A. Echolocation signals in the free-
tailed bat, Tadarida mexicana. J. Aud. Res., 5, 317-330. 1965.
328. ————, ———— Hearing in the vampire bat, Desmodus rotundus,
as shown by cochlear potentials. J. Aud. Res., 6, 181-187. 1966.
329. Wassif, K., and Madkour, G. M. The anatomy of the hyoid bone,
larynx, and upper part of the trachea in some Egyptian bats.
Bull. Zool. Soc. Egypt., 22, 15-26. 1969.
330. Webster, F. A. Some acoustical differences between bats and men.
Int. Conf. on Sens. Devices for the Blind, St. Dunstan's, London,
63-87. 1967.

331. Webster, F. A., and Brazier, O. G. Experimental studies on target detection, evaluation, and interception by echolocating bats. T. D. R. No. AMRL-TR-65-172, Aerospace Medical Div., U. S. A. F. Systems Command. 1965.

332. ———, ——— Experimental studies on echolocation mechanisms in bats. T. D. R. No. AMRL-TR-67-192, Aerospace Medical Div., U. S. A. F. Systems Command. 1968.

333. ———, ——— Echolocation investigations on bats and humans: Target localization evaluation. Sensory Sys. Lab., Tucson Ariz., AD-697 070, pp. 80. 1968.

334. Wickler, W., and Seibt, U. Doppelklick-Orientierungslaute bei einem Epauletten-Flughund. Naturw., 61, 367. 1974.

335. Williams, T. C. Nocturnal orientation techniques of a neotropical bat. Ph. D. Thesis, Rockefeller Univ., New York, 1968.

336. ———, and Williams, J. M. Radio tracking of homing bats. Science, 155, 1435-1436. 1967.

337. ———, ——— Radio tracking of homing and feeding flights of a neotropical bat Phyllostomus hastatus. Anim. Behav., 18, 302-309. 1970.

338. ———, ———, and Griffin, D. R. The homing ability of the neotropical bat Phyllostomus hastatus with evidence for visual orientation. Anim. Behav., 14, 468-473. 1966.

339. Wilson, J. P. Towards a model for cochlear frequency analysis, in: "Psychophysics and Physiol. Hearing," E. F. Evans and J. P. Wilson, eds. Academic Press, New York, 1977.

340. Wimsatt, W. A., ed. "Biology of Bats," vol. I - III. Academic Press, New York, 1970.

341. Woolf, N. The ontogeny of bat sonar sounds: With special emphasis on sensory deprivation. Sc. D. Thesis, J. Hopkins Univ., Baltimore, 1974.

342. Yalden, D. W., and Morris, P. A. "The Lives of Bats." Quadrangle/The New York Times Book Co., New York, 1975.

BIBLIOGRAPHY OF ECHOLOCATION PAPERS ON AQUATIC MAMMALS PUBLISHED
BETWEEN 1966 AND 1978

Abramov, A. P., Golubkov, A. G., Yershova, I. V., Fradkin, V. G.
    and Korolev, V. I., 1972, Investigation of the dolphin's abil-
    ity to differentiate the volume of objects according to
    dimensions and materials, in: "Problemy Biologicheskey
    Kibernetiki", A. I. Berg, ed., Isdatelstvo Nauka, Leningrad.
Abramov, A. P., Ayrapet'yants, E. Sh., Burdin, V. I., Golubkov,
    A. G., Yershova, I. V., Zhezherin, A. R., Korolev, V. I.,
    Malyshev, Yu. A., Ul'yanov, G. K., and Fradkin, V. G., 1971,
    Investigations of delphinid capacity to differentiate between
    three-dimensional objects according to linear size and
    material, "Report from the 7th All-Union Acoustical Confer-
    ence", Leningrad, as cited in: Bel'kovich, V. M. and Dubrovskiy,
    N.A., 1976, "Sensory Bases of Cetacean Orientation", Nauka,
    Leningrad (English translation JPRS L/7157).
Ackman, R. G., Eaton, C. A., and Litchfield, C., 1971, Composition
    of wax esters, triglycerides and diacyl glyceryl ethers
    in the jaw and blubber fats of the Amazon river dolphin
    (Inia geoffrensis), Lipids, 6:69.
Ackman, R. G., Sipos, J. C., Litchfield, C., and Hilaman, B.,
    1972, Characterization of unusual wax esters from the jaw
    fat of the Atlantic bottle-nosed dolphin (Tursiops truncatus),
    Jour. Am. Oil Chem. Soc., 49:305.
Ackman, R. G., Sipos, J. C., Eaton, C. A., Hilaman, B. L. and
    Litchfield, C., 1973, Molecular species of wax esters within
    jaw fat of the Atlantic bottlenosed dolphin, Tursiops
    truncatus, Lipids, 8:661.
Agarkov, G. B., 1969, Soviet cetacean morphology research outlined,
    Vestnik Akademii Nauk, 38(8):58.
Agarkov, G. B. and Khomenko, B. G., 1973, Morphological and func-
    tional analysis of certain sensory areas of the dolphin's
    head, IV-oe Vsesoyuznoye soveshchaniye po Bionike, 4:1.
Agarkov, G. B. and Valiulina, F. G., 1974, The issue of the inner-
    vation of the mandibular region of the common dolphin,
    Bionika, 8:117.
Agarkov, G. B., Khomenko, B. G., and Khadzhinskiy, V. G., 1974,
    in: "Morphology of Delphinidae", Izdatel'stvo Naukova
    Dumka, Kiev.
Agarkov, G. B., Solukha, B. V., and Khomenko, B. G., 1971, Echo-
    location capacity of dolphins, Bionika, 5:52.
Agarkov, G. B., Solukha, B. V., and Khomenko, B. G., 1973, On the
    hydrolocation capability of dolphins, National Translation
    Center, 75:7.
Agarkov, G. B., Solukha, B. V., and Zakletskiy, A. V., 1975, Con-
    vergence in structure and function of the auditory system of
    fish, amphibians, reptiles, birds and mammals, in: "Morskiye
    Mlekopitayushchiye", G. B. Agarkov, ed., Izdatel'stvo
    Naukova Dumka, Kiev.

Alexeyeva, T. V., Golubkov, A. G., and Ershova, I. V., 1971, On
     the problem of the active width of the spectrum of dolphins'
     echolocation signals, Trudy Akusticheskogo Instituta 17:99.
Altes, R. A., 1971, Computer derivation of some dolphin echoloca-
     tion signals, Science, 173:912.
Altes, R. A., 1973, Study of animal signals and neural processing
     with application to advanced sonar systems, ESL Inc. Rpt. No.
     115.
Altes, R. A., 1974, Study of Animal Sensory Systems with Applica-
     tion to Sonar, ESL Inc. Rpt. No. 144.
Altes, R. A., 1976, Sonar for generalized target description and
     its similarity to animal echolocation systems, Jour. Acous.
     Soc. Amer., 59:97.
Altes, R. A., 1977, Localization in cetacean echolocation, in:
     "Proceedings (Abstracts) Second Conference On The Biology of
     Marine Mammals", San Diego.
Altes, R. A., and Skinner, D. P., 1977, Sonar velocity resolution
     with a linear-period-modulated pulse, Jour. Acoust. Soc.
     Amer., 61:1019.
Altes, R. A., Evans, W. E., and Johnson, C. S., 1975, Cetacean
     echolocation signals and a new model for the human glottal
     pulse, Jour. Acoust. Soc. Amer., 57:1221.
Andersen, H. T. (ed.), 1969, "The Biology of Marine Mammals",
     Academic Press, New York.
Andersen, S., 1970, Auditory sensitivity of the harbour porpoise
     Phocoena phocoena, in: "Investigations on Cetacea", G. Pilleri,
     ed., Bentelli AG., Berne.
Andersen, S., 1970, Directional hearing in the harbour porpoise
     Phocoena phocoena, in: "Investigations on Cetacea Vol. 2", G.
     Pilleri, ed., Bentelli AG., Berne.
Andersen, S., 1971, Orientierung hos delphiner, Nat. Verden, 5:181.
Andersen, S. and Pilleri, G., 1970, Audible sound production in
     captive Platanista gangetica., in: "Investigations on
     Cetacea Vol. 2", G. Pelleri, ed., Bentelli AG., Berne.
Asa-Dorian, P. V. and Perkins, P. J., 1967, The controversial
     production of sound by the California gray whale, Eschrichtius
     robustus Norsk Hvalfangsttid, 56:74.
Au, W. W. L., 1977, Target analysis using simulated dolphin echo-
     location signals, in: "Proceedings (Abstracts) Second Conference
     on the Biology of Marine Mammals, San Diego.
Au, W. W. L. and Hammer, C., 1978, Analysis of target recognition
     via echolocation by an Atlantic bottlenose porpoise
     (Tursiops truncatus), Jour. Acoust. Soc. Amer., 60:587.
Au, W. W. L., Floyd, R. W., and Haun, J. E., 1978, Propagation of
     Atlantic bottlenose dolphin echolocation signals, Jour.
     Acoust. Soc. Amer., 64:411.

Au, W. W. L., Floyd, R. W., Penner, R. H. and Murchison, A. E., 1974, Measurement of echolocation signals of the Atlantic bottle-nose dolphin, Tursiops truncatus (Montagu), in open waters, Jour. Acoust. Soc. Amer., 56:1280.

Audouin, F., 1967, Etude bionique de l'auto-information par echoacoustiques dans le regne animal, Rapport techn.du Marche DRME n°66-34179-00-480-75-05, Thomson-Houston, Vol. II.

Awbrey, F. T., 1979, Background study of acoustical and bioacoustical factors in tuna fishing productivity and associated porpoise mortality, Fish. Bull. (in press).

Ayrapet'yants, E. Sh. and Konstantinov, A. I., 1970, "Echolocation in Nature", Nauka, Leningrad (English translation JPRS 63328).

Ayrapet'yants, E. Sh. and Konstantinov, A. I., 1976, Physiological investigations of ultrasonic echolocation in animals, paper presented at the 25th International Congress of Physiol. Sciences, Munich.

Ayrapet'yants, E. Sh., Konstantinov, A. I., and Matjushkin, D. P., 1969, Brain echolocation mechanisms and bionics, Acta Physiol. Acad. Sci. Hung., 35:1.

Ayrapet'yants, E. Sh., Golubkov, A. G., Yershova, I. V., Zhezherin, A. R., Zvorkin, V. N., and Korolev, V. I., 1969, Echolocation differentiation and characteristics of radiated pulses in dolphins, Report of the Academy of Science of the USSR, 188:1197 (English translation JPRS 49479).

Ayrapet'yants, E. Sh., Voronov, V. A., Ivanenko, Y. V., Ivanov, M. P., Ordovskii, D. L., Sergeev, B. F., and Chilingiris, V. I., 1973, The physiology of the sonar system of Black Sea dolphins, Jour. Evol. Biochem. Physiol., 9:364 (English translation JPRS 60298).

Babkin, V. P., and Dubrovskiy, N.A., 1971, Range of action and noise stability of the echolocation system of the bottle-nose dolphin in detection of various targets, Tr. Akust. Inst., Moscow, 17:29, as cited in: Bel'kovich, V. M. and Dubrovskiy, N. A., 1976, "Sensory Bases of Cetacean Orientation", Nauka, Leningrad (English translation JPRS L/7157).

Babkin, V. P., Dubrovskiy, N. A., Krasnov, P. S, and Titov, A. A., 1971, Discrimination of material of spherical targets by the bottlenose Dolphin, in: "Report from the 7th All-Union Acoustical Conference, Leningrad".

Backus, R. H. and Schevill, W. E., 1966, Physeter clicks, in: "Whales, Dolphins, and Porpoises", K. S. Norris, ed., Univ. California Press, Berkeley.

Bagdonas, A., Bel'kovich, V. M., and Krushinskaya, N. L., 1970, Interaction of analyzers in dolphins during discrimination of geometrical figures under water, Jour. Higher Neural Act., 20:1070 (English translation in: "A Collection of Translations of Foreign Language Papers on the Subject of Biological Sonar Systems", K. J. Diercks, 1974, ed., Applied Research Lab, U. of Texas, Austin, Tech. Rept. 74-9).

Barham, E. G., 1973, Whales' respiratory volume as a possible resonant receiver for 20 Hz signals, Nature, 245:220.

Barta, R. E., 1969, Acoustical pattern discrimination by an Atlantic bottlenosed dolphin, unpublished manuscript, Naval Undersea Center, San Diego, CA.

Bastian, J., 1967, The transmission of arbitrary environmental information between bottle-nose dolphins, U. S. Naval Ordnance Test Stat., China Lake, Calif., NOTS TP 4117.

Bastian, J., Wall, C., and Anderson, C. L., 1968, Further investigations of the transmission of arbitrary information between bottlenose dolphins, NUWC TP 109.

Bateson, G., 1966, Problems in cetacean and other mammalian communication, in: "Whales, Dolphins, and Porpoises", K. S. Norris, ed., Univ. of Calif., Berkeley.

Beach, F. A. III, and Herman, L. M., 1972, Preliminary studies of auditory problem solving and intertask transfer by the bottlenose dolphin, Psychological Record, 22:49.

Beach, F. A. III, Pepper, R. L., Nachtigall, P. E., Simmons, J. V., and Siri, P. A., 1974, Spatial habit reversal in two species of marine mammals, Psychological Record, 24:385.

Beamish, P., 1971, Biological sonar of diving mammals, an analytical model, in: "Proc. 7th Annu. Conf. Biol. Sonar and Diving Mamm., Stanf. Res. Inst.", Menlo Park, Calif.

Beamish, P., 1972, Whale sonar, in: "Proc. 10th Annu. Meeting Canadian Committee on Acoustics", R. J. Donato, ed., Nation. Res. Council, Ottawa.

Beamish, P., and Mitchell, E., 1971, Ultrasonic sounds recorded in the presence of a blue whale Balaenoptera musculus, Deep-Sea Res., 18:803.

Beamish, P., and Mitchell, E., 1973, Short pulse length audio frequency sounds recorded in the presence of a Minke whale Balaenoptera acutorostrate, Deep-Sea Res., 20:375.

Bel'kovich, V. M., 1970, About "Acoustical Vision" in the dolphin. Experimental matters. Ekologia, 6:89.

Bel'kovich, V. M. and Borisov, V. I., 1971, Locational discrimination of figures of complex configuration by dolphins, Trudy Akusticheskogo Institute, 17:19.

Bel'kovich, V. M. and Dubrovskiy, N. A.,1976, "Sensory Bases of Cetacean Orientation", Nauka, Leningrad (English translation JPRS L/7157).

Bel'kovich, V. M. and Gurevich, V. S. 1969, The whales (Cetacea, Odontoceti) are the experimental animals, "Scientific Conference on Problems of the World Ocean", M:16.

Bel'kovich, V. M. and Gurevich, V. S., 1970, Problems on keeping of the dolphins in captivity and performing experimental work with them, in: "Collection Material of the Conference 'Ocean'", Moscow:108-112.

Bel'kovich, V. M. and Nesterenko, I., 1971, How the dolphin's locator operates, Priroda., 7:71 (German translation by W. Petri, 1974, Naturwissenschaftliche Rundschau, 0004:0143).

Bel'kovich, V. M., and Resnikov, A. Ye., 1971, What's new in dolphin sonar, Priroda, 11:84.

Bel'kovich, V. M. and Solntzeva, G. N., 1970, Anatomy and function of the ear in the dolphin, Zoolischeskiy Zhurnal, 49(2):275. (English translation JPRS 50253).

Bel'kovich, V. M., Borisov, V. I., Gurevich, V. S., 1970, Angular resolution by echolocation in Delphinus delphis, in: "Proceedings of Scientific Technical Conference, Ministry of Higher and Secondary Specialized Education RSFSR, Leningrad" (English translation in: "A Collection of Translations of Foreign Language Papers on the Subject of Biological Sonar Systems", K. J. Diercks, 1974, ed., Applied Research Lab, Univ. of Texas, Austin, Tech. Rept. 74-9).

Bel'kovich, V. M., Gurevich, V. S., and Borisov, V. I., 1970, Discrimination by the dolphins of the geometrical figures according to the angle, in: "The 23rd Scientific-Technical Conference of the LIAP", Leningrad.

Bel'kovich, V. M., Andreyev, F. V., Vronskaya, S. D., and Cherdantsev, A. I., 1975, Investigation of orientation and behavior of wild bottlenose dolphins (Tursiops truncatus), in: "Morskiye Mlekopitayushchiye", G. B. Agarkov, ed., Isdatel'stvo Naukova Dumka Kiev.

Bel'kovich, V. M., Bagdonas, A. P., Gurevich, V. S., and Krushinskaya, N. L., 1969, Echolocation of the toothed whales (Cetacea), "Scientific Conference of Problems of the World Ocean".

Bel'kovich, V. M., Borisov, V. I., Gurevich, V. S., and Krushinskaya, N. L., 1969, Echolocating capabilities of the common dolphin (Delphinus delphis), Zoologicheskiy Zhurnal, 48:876 (English translation JPRS 48780).

Bel'kovich, V. M., Borisov, V. I., Gurevich, V. S., Krushinskaya, N. L., and Dmitrieva, I. L., 1972, Echolocation discrimination of distance between the targets (angular resolution) by dolphins, in: "Proceedings of the 5th All-Union Conference on Marine Mammals", Makhachkala, Part 2:30.

Bel'kovich, V. M., Gurevich, V. S., Dmitrieva, I. L., Borisov, V. I., and Krushinskaya, N. L., 1973, Echolocation capabilities of the dolphins discriminating the geometrical figures by the square, in: "Material of the 6th All-Union Conference on Studying of Marine Mammals, Makhachkala: 57.

Berzin, A. A., 1971, "The Sperm Whale". Pacific Scientific Research Institute of Fisheries and Oceanography (USSR), (Trans. by Israel Program for Scientific Translations, publ. by U. S. National Technical Information Service, Springfield, Va.).

Blevins, C. E., and Parkins, B., 1973, Functional anatomy of the porpoise larynx, Amer. J. Anat., 138:151.

Blomberg, J., 1972, Pilot whale head oil: lipid analysis and ultrasonic studies, in: "Program of the XIth Congr. Int. Soc. for Fat Res.", Goteborg, Abstract 223.

Bloome, K. A., 1969, An electron microscopic study of the organ
    of corti of the porpoise (Abstract), Am. Zool., 9:1148.
Bogoslovskaya, L. S., 1974, Distinctions of neuronal structure
    of primary acoustic centers in the dolphins, in: "Morfologiya
    Fiziologiya i Akustika Morskikh Mlekopitayushchikh",
    V. Ye. Sokolov, ed., Izdatel'stvo Nauka, Moscow.
Bogoslovskaya, L. S., 1975, Ultrastructure of the spiral ganglion
    and acoustic nerve in bottlenose dolphins (Tursiops truncatus),
    in: "Morskiye Mlekopitayushchiye", G. B. Agarkov, ed.,
    Izdatl'stvo Naukova Dumka, Kiev.
Bourcart, 1970, le dauphin, Rev. des Corps de Sante, 11:311.
Bounton, K. L., 1966, The remarkable dolphin (ear, sonar, echos),
    III: Sound senders and receivers, Can. Audubon., 28:24.
Brown, A. M. and Pye, J. D., 1975, Auditory sensitivity at high
    frequencies in mammals, in: "Advances in Comparative Physiol-
    ogy and Biochemistry, Vol. 6", O. Lowenstein, ed., Academic
    Press, New York.
Bullock, T. H. and Gurevich, V. S., 1979, Soviet Literature on
    the nervous system and psychobiology of cetaceans, Interna-
    tional Review of Neurobiology, (In Press).
Bullock, T. H. and Ridgway, S. H., 1972, Evoked potentials in the
    central auditory system of alert porpoises to their own and
    artificial sounds, J. Neurobiol., 3:79.
Bullock, T. H. and Ridgway, S. H., 1972, Neurophysiological
    findings relevant to echolocation in marine mammals, in:
    "Animal Orientation and Navigation, S. R. Galles, K.,
    Schmidt-Konig, G. J. Jacobs, and R. E. Belleville, eds., NASA,
    Washington, D. C.
Bullock, T. H., Ridgway, S. H., and Suga, N., 1971, Acoustically
    evoked potentials in midbrain auditory structures in sea
    lions (Pinnipedia), Z. vergl. Physiol., 74:732.
Bullock, T. H., Grinnell, A. D., Ikezono, E., Kameda, K.,
    Katsuki, Y., Nomoto, M., Sato, O., Suga, N., and Yanagisawa, K.,
    1968, Electrophysiological studies of central auditory
    mechanisms in cetaceans, Zeitschrift fur vergieichende
    Physiologie, 59:117.
Burdin, V. I., Reznik, A. M., Skornyakov, V. M. and Chupakov, A. G.
    1974, Study of communicative signals in Black Sea dolphins,
    Akusticheskiy Zhurnal, 20(4):518.
Burdin, V. I., Reznik, A. M., Skornyakov, V. M. and Chupakov,
    A. G., 1975, Communication signals of the black sea bottle-
    nose dolphin, Sov. Phys. Acoust., 20:314.
Burdin, V. I., Markov, V. I., Reznik, A. M., Skornyakov, V. M. and
    Chupakov, A. G., 1971, Ability of Tursiops truncatus (Ponticus
    barabash) to distinguish a useful signal against a noise back-
    ground, in: "Morfologiya i Ekologiya Morskikh
    Mlekopitayushchikh", V. Ye. Sokolov, ed., Izdatl'stvo
    Nauka, Moscow.

Burdin, V. I., Markov, V. I., Reznik, A. M., Skornyakov, V. M., and Chupakov, A. G., 1971, Determination of the just noticeable intensity difference for white noise in the Black Sea dolphin, in: "Morfologiya i Ekologiya Morskikh Mlekopitayushchikh", V. Ye. Sokolov, ed., Izdatl'stvo Nauka Moscow.

Burdin, V. I., Markov, V. I., Reznik, A. M., Skornyakov, V. M., and Chupakov, A. G., 1971, Determination of the just noticeable intensity difference for white noise in the bottlenose dolphin (Tursiops truncatus Barabash), in: "Morphology and Ecology of Marine Mammals", K. K. Chapskii and V. Y. Sokolov, eds., Wiley, New York.

Busnel, R.-G., 1966, Information in the human whistled language and sea mammal whistling, in: "Whales, Dolphins and Porpoises", K. S. Norris, ed., Univ. Calif. Press, Berkeley.

Busnel, R.-G., 1970, Themes de reflexions a propos de l'echolocation biologique et de la bionique, in: "Principles and Practice of Bionics", H. E. Von Gierke, W. D. Keidel, H. L. Oestreicher, eds., Technivision Serv., Slough - G. B.

Busnel, R.-G., 1973, Symbiotic relationship between man and dolphins Trans. N. Y. Acad. Sci., Series II, 25:112.

Busnel, R.-G., and Dziedzic, A., 1966, Acoustic physiologique - caracteristiques physiques de certains signaux acoustiques due Delphinide Steno bredansis LESSON, C. R. Acad. Sci., 262:143.

Busnel, R.-G., Dziedzic, A., 1966, Acoustic signals of the pilot whale Glogicephala melaena and the porpoises Delphinus delphis and Phocoena phocoena, in: "Whales, Dolphins and Porpoises", K. S. Norris, ed., U. of Calif. Press, Berkeley.

Busnel, R.-G. and Dziedzic, A., 1967, Observations sur le comportement et les emissions acoustiques du cachalot lors de la chasse, Bocagiana, Mus. Municipal do Funchal, Madeira, 14:1.

Busnel, R.-G. and Dziedzic, A., 1967, Resultats metrologigues experimentaux de l'echolocation chez le Phocaena phocaena at leur comparison avec ceux de certaines chauves - souris, in: "Animal Sonar Systems, Biology and Bionics", R.-G. Busnel, ed., Laboratoire de Physiologie Acoustique, Jouy-en-Josas, France.

Busnel, R.-G. and Dziedzic, A., 1968, Caracteristiques physiques des signaux acoustiques de Pseudorca crassidens OWEN (Cetace odontocete), Extr. d. Mammal, 32:1.

Busnel, R.-G., and Dziedzic, A., 1968, Etude des signaux acoustiques associes a des situations de detresse chez certains cetaces odontocetes, Ann. de L'Inst. Oceanographique, Nouvelle Serie, 46:109.

Busnel, R.-G., Dziedzic, A., and Alcuri, A., 1974, Etudes preliminaires de signaux acoustiques du Pontoporia blainvillei CERVAIS et d' ORGIBNY (Cetacea, Platanistidae), Mammalia, 38:449.

Busnel, R.-G, Dziedzic, A., and Escudie, B., 1969, Autocorrelation et analyse spectrale des signauz "sonar" de deux especes Cetaces Odontocetes utilisant les basses frequences, C. R. Acad. Sci. Paris, 269:365.

Busnel, R.-G., Pilleri, G., and Fraser, F. C., 1968, Notes concerant le dauphin Stenella styx Gray 1846, Extrait de Mammalia, 32:192.

Busnel, R.-G., Escudie, B., Dziedzic, A. and Hellion, A., 1971, Structure des clics doubles d'echolocation du Globicephale (Cetace Odontocete), C. R. Acad. Sci., Serie D, 272:2459.

Caine, N. G., 1976, "Time Separation Pitch and the Dolphin's Sonar Discrimination of Distance", Master's thesis San Diego State University.

Caldwell, D. K., and Caldwell, M. C., 1966, Observation on the distribution, coloration, behavior and audible sound production of the spotted dolphin, Stenella plagiodon COPE, Los Angeles County Mus., Contrib. in Sci., 104:1.

Caldwell, D. K., and Caldwell, M. C., 1970, Echolocation-type signals by two dolphins genus Sotalia, Quart. J. Fla. Acad. Sci., 33:124.

Caldwell, D. K., and Caldwell, M. C., 1971, Sounds produced by two rare cetaceans stranded in Florida, Cetology, 4:1.

Caldwell, D. K., and Caldwell, M. C., 1971, Underwater pulsed sounds produced by captive spotted dolphins, Stenella plagiodon, Cetology, 1:1.

Caldwell, D. K., and Caldwell, M.C., 1972, Senses and communication, in: "Mammals of the Sea, Biology and Medicine", S. H. Ridgway, ed., C. C. Thomas, Springfield, Ill.

Caldwell, M.C., and Caldwell, D. K., 1967, Intraspecific transfer of information via pulsed sound in captive Odontocete Cetaceans, in: "Animal Sonar Systems: Biology and Bionics", R.-G. Busnel, ed., Lab de Physiol. Acoust. Jouy-en-Josas, France.

Caldwell, M. C. & Caldwell, D. K., 1968, Vocalization of naive captive dophins in small groups, Science, 159:1121.

Caldwell, M. C., and Caldwell, D. K., 1969, Simultaneous but different narrow-band sound emissions by a captive Eastern Pacific pilot whale, Globicephala scammoni, Mammalia, 33:505.

Caldwell, M. C., and Caldwell, D. K., 1970, Echolocation-type signals by dolphins genus Inia, Sea Frontiers, 15:343.

Caldwell, M. C., and Caldwell, D. K., 1970, Etiology of the chirp sounds emitted by the atlantic bottlenosed dolphins: a controversial issue, Underwater Naturalist, 6:6.

Caldwell, M. C., and Caldwell, D. K., 1970, Further studies on audible vocalizations of the Amazon freshwater dolphin, Inia geoffrensis, Contrib. in Sci., 187:1.

Caldwell, M. C., and Caldwell, D. K., 1971, Statistical evidence for individual signature whistles in Pacific whitesided dolphins, Lagenorhynchus obliquidens, Cetology, 3:1.

Caldwell, M. C., and Caldwell, D. K., 1972, Vocal mimicry in the whistle mode by an Atlantic bottlenosed dolphin, Cetology, 9:1.

Caldwell, M. C., Caldwell, D. K., and Evans, W. E., 1966, Preliminary results of studies on the sounds and associated behavior of captive Amazon freshwater dolphins, Inia geoffrensis, in: "Proc. 3rd Annu. Conf. Biol. Sonar & Diving Mammals", Stanford Res. Inst., Menlo Park, Calif.

Caldwell, M. C., Caldwell, D. K., and Evans, W. E., 1966, Sounds and behavior of captive amazon freshwater dolphins, Inia geoffrensis, Contr. in Science, L. A. County Museum, 108:1.

Caldwell, M. C., Caldwell, D. K., and Hall, N. R., 1969, An experimental demonstration of the ability of an Atlantic bottlenosed dolphin to discriminate between whistles of other individuals of the same species, Los Angeles County Mus. Nat. Hist. Found., 6:35.

Caldwell, M. C., Caldwell, D. K., and Hall, N. R., 1970, An experimental demonstration of the ability of an Atlantic bottlenosed dolphin to discriminate between playbacks of recorded whistles of conspecifics, in: "Proc. of the 7th Annu. Conf. on Biol. Sonar & Diving Mammals", Menlo Park, Calif.

Caldwell, D. K., Caldwell, M. C., and Miller, J. F., 1969, Three brief narrow-band sound emissions by a captive male Risso's dolphin, Grampus griseus, Bull. S. Calif. Acad. Sci., 68:252.

Caldwell, M. C., Caldwell, D. K. and Miller, J. F., 1970, Statistical evidence for individual signature whistles in the spotted dolphin, Stenella plagiodon, Los Angeles County Mus. Nat. Hist. Found., Tech Rept., 7:45

Caldwell, M. C., Caldwell, D. K., and Miller, J. F., 1973, Statistical evidence for individual signature whistles in the spotted dolphin, Stenella plagiodon, Cetology, 16:1.

Caldwell, D. K., Caldwell, M. C., and Rice, D. W., 1966, Behavior of the sperm whale, Physeter catodon L., in: "Whales Dolphins, and Porpoises", K. S. Norris, ed., Univ. of California Press, Berkeley.

Caldwell, M. C., Caldwell, D. K., and Turner, R. H., 1970, Statistical analysis of the signature whistle of an Atlantic bottlenose dolphin with correlations between vocal changes and level arousal, Los Angeles Co., Calif. Mus. Nat. Hist., 8:1.

Caldwell, M. C., Hall, N. R., and Caldwell, D. K., 1971, Ability of an atlantic bottlenosed dolphin to discriminate between, and potentially identify to individual, the whistles of another species, the spotted dolphin, Cetology, 6:1.

Caldwell, M. C., Hall, N. R., and Caldwell, D. K., 1971, Ability of an Atlantic bottlenosed dolphin to discriminate between, and respond differentially to whistles of eight conspecifics, in: "Proc. 8th Annu. Conf. Biol. Sonar & Diving Mammals", Menlo Park, Calif.

Caldwell, D. K., Prescott, J. H., and Caldwell, M. C., 1966, Production of pulsed sound by the pigmy sperm whale, Kogia breviceps, Bull. S. Calif. Acad. Sci., 65:245.

Caldwell, M. C., Hall, N. R., Caldwell, D. K., and Hall, H. I., 1971, A preliminary investigation of the ability of an Atlantic bottlenosed dolphin to localize underwater sound sources, Marineland Res. Lab., Techn. Rept. 6.

Chapman, S., 1968, Dolphins and multifrequency, multiangular images, Science, 160:208.

Cherbit, G., and Alcuri, G., 1978, Etude de la propogation des vibrations a travers le rostre de Sotelia teuzii (cetacea) par inferometric holographique, C. R. Acad. Sci. Paris, 286:607.

Corcella, A. T., and Green, M., 1968, Investigation of impulsive deep-sea noise resembling sounds produced by a whale, Jour. Acoust. Soc. Amer., 44:483.

Cummings, W. C., and Philippi, L. A., 1970, Whale phonations in repetitive stanzas, NUC Techn. Publ. 196, San Diego, Calif.

Cummings, W. C., and Thompson, P. O., 1968, Self-noise of a deep submersible in a bioacoustic investigation off Catalina Island, California, Jour. Acoust. Soc. Amer., 44:1742.

Cummings, W. C. and Thompson, P. O., 1971, Gray whales, Eschrichtus robustus, avoid the underwater sounds of killer whales, Orcinus orca, Fishery Bull., 69:525.

Cummings, W. C., and Thompson, P. O., 1971, Underwater sounds from the blue whale, Balaenoptera musculus, Jour. Acoust. Soc. Amer., 50:1193.

Cummings, W. C. and Thompson, P. O., 1971, Bioacoustics of marine mammals, R. V. Hero Cruise 70-3, Antarctic J. U. S., 6:158.

Cummings, W. C., Fish, J. F., and Thompson, P. O., 1972, Sound production and other behavior of southern right whale, Eubalena glacialis, Transact. of the San Diego Soc. Natur. Hist., 17:1.

Cummings, W. C., Thompson, P. O., and Cook, R., 1967, Sound production of migrating gray whales, Eschrichtius gibbosus ERXLEBEN, 74th Meeting of the Acoustical Soc. of America.

Cummings, W. C., Thompson, P. O., and Cook, R., 1968, Underwater sounds of migrating gray whales, Eschrichtius glaucus COPE, Jour. Acoust. Soc. Amer., 44:1278.

Cummings, W. C., Fish, J. F., Thompson, P. O., and Jehl, J. R., 1971, Bioacoustics of marine mammals off Argentina, R. V. Hero Cruise 71-3, Antarct. J. U. S., 6:266.

Dailly, M., 1971, The primary cochlear nuclei in the Amazon dolphin Inia geoffrensis, in: "Investigations on Cetacea: Vol. III", G. Pilleri, ed., Bentelli, Berne.

Dailly, M., 1972, Contribution to the study of the cochlear apparatus in dolphins, in: "Investigations on Cetacea, Vol. II", G. Pilleri, ed., Bentelli, Berne.

Defran, R. H., and Caine, N. G., 1976, Periodicity pitch difference
    limens in the bottlenose dolphin (Tursiops truncatus), Jour.
    Acoust. Soc. Amer., 60: (Supplement 1) S5 (Abstract).
Diercks, K. J., 1972, Biological sonar systems: a bionics survey,
    Applied Research Laboratories, ARL-TR-72-34, Austin, Texas.
Diercks, K. J., 1974, A collection of translations of foreign
    language papers on the subject of biological sonar systems,
    Applied Res. Lab. ARL-TR-74-9, Austin, Texas.
Diercks, K. J., and Evans, W. E., 1975, Delphinid sonar: pulse
    wave and simulation studies, NUC Techn. Publ. 175, San Diego.
Diercks, K. J., and Hickling, R., 1967, Echoes from hollow aluminum
    spheres in water, Jour. Acoust. Soc. Amer., 41:380.
Diercks, K. J., and Trochta, R. T., 1972, Animal sonar: measurements
    and meaning, Jour. Acoust. Soc. Amer, 51:133.
Diercks, K. J., Trochta, R. T., and Evans, W. E., 1973, Delphinid
    sonar: measurement and analysis, Jour. Acoust. Soc. Amer.,
    54:200.
Diercks, K. J., Trochta, R. T., Greenlaw, R. L. and Evans, W. E.,
    1971, Recording and analysis of dolphin echolocation signals,
    Jour. Acoust. Soc. Amer., 49:1729.
Dmitrieva, I. L., Krushinskaya, N. L., Bel'kovich, V. M., and
    Shurkhal, A. V., 1972, Echolocational discrimination of
    the shape of different subjects by dolphins, in: "Proc. of the
    Abstracts of the 5th All-Union Conference on Marine Mammals",
    Makhachkala, Part 2:76.
Dormer, K. J., 1974, The mechanism of sound production and measure-
    ment of sound processing in delphinid cetaceans, Ph.d.
    dissertation, University of California, Los Angeles.
Dormer, K. J., 1979, Mechanisms of sound production and air
    recylcing in delphinids: Cineradiographic evidence, Jour.
    Acoust. Soc. Amer., 65:229.
Dral, A. D. G., 1972, Aquatic and aerial vision in the bottlenosed
    dolphin Neth. J. Sea Res, 5:510.
Dreher, J. J., 1966, Cetacean communication: small-group
    experiment, in: "Whales, Dolphins, and Porpoises", K. S.
    Norris, ed., Univ. Calif. Press, Berkeley.
Dreher, J. J., 1967, Bistatic target signatures and their acoustic
    recognition: a suggested animal model, in: "Marine Bio
    Acoustics", W. N. Tavolga, ed., Pergamon Press, N. Y.
Dreher, J. J., 1969, Acoustic holographic model of cetacean
    echolocation, in: "Acoustical Holography, Vol. I", A. E.
    Metherell, ed., Plenum Press, New York.
Dubrovskiy, N. A., 1972, Discrimination of objects by dolphins
    using echolocation, "Report of the 5th All-Union Conference
    on Studies of Marine Mammals, Part 2, Makhachkala", as cited
    in: Ayrapet'yants, E. Sh. and Konstantinov, A. I., 1974,
    "Echolocation in Nature", Nauka, Leningrad (English translation
    JPRS 63328-2).

Dubrovskiy, N. A., and Krasnov, P. S., 1971, Discrimination of
    elastic spheres according to material and size by the
    bottlenose dolphin, Trudy Akusticheskogo Institute, 17,
    as cited in: Bel'kovich, V. M., and Dubrovskiy, N. A.,
    1976, "Sensory Bases of Cetacean Orientation", Nauka,
    Leningrad (English translation JPRS L/7157).
Dubrovskiy, N. A., and Titov, A. A., 1975, Echolocation
    discrimination by bottlenose dolphins (Tursiops
    truncatus) of spherical targets which differ simul-
    taneously in dimensions and materials, Akusticheskiy
    Zhurnal, 21(3):469.
Dubrovskiy, N. A., and Zaslavsky, G. L., Dolphin echolocation
    Akusticheskiy Zhurnal, 3 (English Translation JPRS 65777).
Dubrovskiy, N. A., and Zaslavsky, G. L., 1973, Temporal structure
    and directivity of sound emission by the bottlenosed
    dolphin, in: "Proceedings of the 8th All-Union
    Acoustical Conference".
Dubrovskiy, N. A., and Zaslavski, G. L., 1975, Role of the skull
    bones in the space-time development of the dolphin echo-
    location signal, Sov. Phys. Acoust., 21:409 (Translation:
    American Institute of Physics).
Dubrovskiy, N. A., Krasnov, P. S., and Titov, A. A., 1970, On the
    emission of echo-location signals by the Azov Sea harbor
    porpoise, Soviet Physics-Acoustics, 16:444.
Dubrovskiy, N. A., Krasnov, P. S. and Titov, A. A., 1970, On
    the question of the emission of ultrasonic ranging signals
    by the common porpoise, Akusticheskiy Zhurnal, 16:521
    (English translation JPRS 52291).
Dubrovskiy, N. A., Krasnov, P. S., and Titov, A. A., 1971,
    Discrimination of solid elastic spheres by an echolocating
    porpoise, Tursiops truncatus, in: "Proc. 7th International
    Acoust.", Budapest, 25(3):533.
Dubrovskiy, N. A., Titov, A. A., Krasnov, P. S., Babkin, V. P.
    Lekomtsev, V. M., and Nikolenko, G. V., 1970, Investigation
    of the emission capacity of the Black Sea Tursiops
    truncatus echolocation apparatus, Trudy Akusticheskogo
    Instituta, 10:163.
Dudok van Heel, W. H., 1966, Navigation in Cetacea, in: "Whales
    Dolphins and Porpoises", K. S. Norris, ed., U. of Calif.
    Press, Berkeley.
Dunn, J. L., 1969, Airborn measurements of the acoustic charac-
    teristics of a sperm whale, Jour. Acoust. Soc. Amer., 46:1052.
Dziedzic, A., 1968, Quelques performances des systemes de detection
    par echos des chauves souris et des delphinidae, Rev. Acoust.,
    1:23.
Dziedzic, A., 1971, Performances dans la detection de cibles par
    echolocation par le Marsouin Phocoena phocoena et etude
    analytique des signaux mis en oeuvre dans ce processus,
    These, Ecole Pratique des Hautes Etudes.

Dziedzic, A., 1971, Theorie et applications de l'acoustique sousmarine Ecole Super. des Techn. avancees, 16:1.

Dziedzic, A., 1972, Les sonars biologiques, La Recherche, 3:315.

Dziedzic, A., and Alcuri, G., 1977, Acoustic recognition of forms and characteristics of the sonar signals of Tursiops truncatus, C. R. Acad. Sci. Paris, 285:981.

Dziedzic, A., and Levy, J. C., 1973, Description et mise en oeuvre d'une base de triangulation destinee a l'etude en mer des signaux de Dauphins, 4°Coll. Traitement du signal et ses applications, Nice, 4:585.

Dziedzic, A., Escudie, B., and Hellion, A., 1975, Realistic methods for analysis of biological sonar signals, Ann. Telecomm., 30:270.

Dziedzic, A., Jeanny, M., and Levy, J. C., 1975, Resultats d'une etude sur les caracteristiques spatio-temporelles et spectrales des signaux acoustiques emis par un dauphin l'approche sonar d'une cible, in: "Colloque Nat. Sur le Traitement du Signal et ses Applications", Nice.

Dziedzic, A., Chiollaz, M., Escudie, B., and Hellion, A., 1977, Some properties of low frequency sonar signals of the dolphin Phoecena phoecena, Acustica, 37:258.

Dziedzic, A., Escudie, B., Guillard, P., and Hellion, A., 1974, Evidence of tolerance to the effect of Doppler in the sonar emissions of Delphinus delphis, C. R. Acad. Sci. Paris, 279:1313.

Dziedzic, A., Escudie, B., Hellion, A., and Vial, C., 1969, Resultats preliminaires d'une etude de certains sonars biologiques - Analyse spectrale de signauz de 7 especes de Cetaces odontocetes, 2°Coll. traitement du signal et ses applications, Nice, 2:785.

Eberhardt, R. L. and Evans, W. E., 1962, Sound activity of the California gray whale, Eschrichtius glaucus, Jour. Aud. Eng. Soc., 10:324.

Erulkar, S. D., 1972, Comparative aspects of spatial localization of sound, Physiological Reviews, 52:237.

Escudie, B., 1972, Etat actuel des travaux sur les "sonars biologiques", Comparaison des proprietes de ces systemes avec les artificiels, Perspectives, I. C. P. de Lyon, 2:27.

Escudie, B., Hellion, A., Dziedzic, A., 1971, Results from studies of air and marine sonars through signal processing and spectral analysis, in: "Proc. Troisieme Colloque sur le Traitement Signal et ses Applications", 3:533.

Evans, W. E., 1967, Discussion, in: "Animal Sonar Systems, Biology and Bionics", R.-G. Busnel, ed., Laboratoire de Physiology Acoustique, Jouy-en-Josas, France.

Evans, W. E., 1967, Vocalizations among marine mammals, in: "Marine Bio-Acoustic II", W. N. Tavolga, ed., Pergamon Press, New York.

Evans, W. E., 1969, Marine mammal communication: Social and ecolo-
    gical factors, in: "The Biology of Marine Mammals", H. T.
    Andersen, ed., Academic Press, New York.
Evans, W. E., 1973, Echolocation by marine delphinids and one
    species of fresh-water dolphin, Jour. Acoust. Soc. Amer.,
    54:191.
Evans, W. E. and Bastian, J., 1969, Marine mammal communication:
    social and ecological factors, in: "The Biology of Marine
    Mammals", H. T. Anderson, ed., Academic Press, N. Y.
Evans, W. E., and Herald, E. S., 1970, Underwater calls of a
    captive Amazon manatee, Trichechus inunguis, J. Mammal,
    51:820.
Evans, W. E., and Maderson, P. F. A., 1973, Mechanisms of sound
    production in delphinid cetaceans: A review and some
    anatomical considerations, Amer. Zool., 13:1205.
Evans, W. E., and Powell, B. A., 1967, Discrimination of different
    metallic plates by an echolocating delphinid, in: "Animal
    Sonar Systems, Biology and Bionics", R.-G. Busnel, ed.,
    Laboratoire de Physiologie Acoustique, Jouy-en-Josas, France.
Fadeyeva, L. M., 1973, Discrimination of spheres with various
    echosignal structure by dolphins, in: "Proceedings of the
    8th All-Union Acoustical Conference", 1:134.
Fay, R. R., 1974, Auditory frequency discrimination in vertebrates,
    Jour. Acoust. Soc. Amer., 56:206.
Filimonoff, I. N., 1966, On the so-called rhinencephalon in the
    dolphin, J. Hirnforsch, 8:1.
Fish, J. F., and Lingle, G. E., 1977, Responses of spotted
    porpoises (Stenella attenuata) to playback of distress (?)
    sounds of one of their kind, in: "Proceedings (Abstracts)
    Second Conference on the Biology of Marine Mammals", San Diego.
Fish, J. F., and Turl, C. W., 1976, "Acoustic source levels of four
    species of small whales", NUC Techn. Publ. 547, San Diego.
Fish, J. F., and Vania, J. S., 1971, Killer whale, Orcinus orca,
    sounds repel white whales, Delphinapterus leucas, Fishery
    Bull., 69:531.
Fish, J. F., and Winn, H. E., Sounds of marine animals, in:
    "Encyclopedia of Marine Resources", F. E. Firth, ed.,
    Van Nostrand Reinhold Co.
Fish, J. F., Johnson, C. S., and Ljungblad, D. K., 1976, Sonar
    target discrimination by instrumented human divers, Jour.
    Acoust Soc. Amer., 59:602.
Fish, J. F., Sumich, J. L. and Lingle, G. E., 1974, Sounds
    produced by the gray whale, Eschrichtius robustus, Mar. Fish
    Rev., 36:38.
Fitzgerald, J. W., 1973, "A qualitative model of the dolphin
    sonar, Part II: The high-frequency echo-location apparatus",
    Technical Report 117, Office of Naval Res., Acoustics Progr.
    Off., Arlington, Virginia.

Flanigan, N. J., 1972, The central nervous system, in: "Mammals of the Sea: Biology and Medicine", S. H. Ridgway, ed., C. C. Thomas, Springfield, Ill.

Fleischer, G., 1973, On structure and function of the middle ear in the bottlenosed dolphin (Tursiops truncatus), in: "Proc. 9th Ann. Conf. Biol. Sonar and Diving Mammals", Stanford Res. Inst. Press, Menlo Park, Calif.

Fleischer, G., 1973, Structural analysis of the tympanicum complex in the bottle-nosed dolphin (Tursiops truncatus), Jour. Aud. Res., 13:178.

Fleischer, G., 1975, Uber das spezialisierte Gehororgan von Kogia Breviceps (odontoceti), Z.f. Saugetierkunde Bd. 40, H.2, S:89.

Fleischer, G., 1976, Hearing in extinct cetaceans as determined by cochlear structure, Journal of Paleontology, 50:133.

Fleischer, G., 1976, On bony microstructures in the dolphin cochlea, related to hearing, N. Jb. Geol. Palaont., 151:166.

Fleischer, G., 1976, Uber Beziehungen Zwischen Hovermogen und Schadelbau bie Walen, Saugetierkundliche Mitteilungen: 48.

Fleischer, G., 1976, Uber Die Verankerung des Stapes im Ohr der Cetacea und Sirenia, Z.f. Saugetierkunde Bd. 41, H.5, S:304.

Fleischer, G., 1978, Evolutionary principles of the mammalian middle ear, in: "Advances in Anatomy, Embryology and Cell Biology; Springer Verlag, Berlin.

Ford, J. K. B., and Fischer, H. D., 1978, Underwater acoustic signals of the narwhal (Monodon monoceros), Can. J. Zool., 56:552.

Fraser, F. C., 1973, Record of a dolphin (Sousa teuszii) from the coast of Mauritania, Transact. N. Y. Acad. Sci., 35:132.

Gales, R. S., 1966, Pickup analysis and interpretation of underwater acoustic data, in: "Whales, Dolphins and Porpoises", K. S. Norris, ed., U. of Calif. Press, Berkeley.

Gallien, C. L., Chalumeau-le-Foulgoc, M. T., and Fine, J. M., 1967, Les proteines serigues de Delphinus delphis Linne (Cetace Odontocete) C. R. Acad. Sci. Paris, 264:1359.

Gallien, C. L., Chalumeau-le-Foulgoc, M. T., and Fine, J. M., 1970, Comparative study of serum proteins in four dolphin species, Comp. Biochem. Physiol., 37:375.

Gapich, L. I. and Supin, Y. A., 1974, The activity of single neurons of the acoustical region of the cerebral cortex of the porpoise Phocoena phocoena, J. Evol. Biochem. Physiol., 10:182.

Gentry, R. L., 1967, Underwater auditory localization in the California sea lion (Zalophus californianus), Jour. Aud. Res., 7:187.

Giraud-Sauveur, D., 1969, Biophysical research on the ossicles of cetaceans, Extrait de Mammalia, 33:285.

Giraud-Sauveur, D., and Miloche, M., 1968, Sur la structure particuliere des os de l'oreille moyenne des cetaces odontocetes, J. Microscop., 7:1098.

Giro, L. R. and Dubrovskiy, N. A., 1972, Correlation between recurrence frequency of delphinid echolocation signals and difficulty of echoranging problem, in: "Marine Instrument Building Series, 2:84, as cited in: Vel'min V. A., 1975, Target detection by the bottlenose dolphin under artificial reverberation conditions, in: "Marine Mammals, Proceedings of the Sixth All-Union Conference on the Study of Marine Mammals", G. B. Agarkov, ed., Nauka Dumka, Kiev (English translation JPRS L/6049).

Giro, L. R. and Dubrovskiy, N. A., 1973, On the origin of low-frequency component of the echoranging pulsed signal of dolphins in: "VIII-aya Vsesoyuznaya Akusticheskaya Konferentsiya", (Eighth All-Union Acoustical Conference) Moscow.

Giro, L. R., and Dubrovskiy, N. A., 1974, The possible role of supracranial air sacs in the formation of echo-ranging signals in dolphins, Akusticheskiy Zhurnal, 20(5):706.

Golubkov, A. G., and Ivanenko, Yu. V., 1970, Study of the range resolution ability of the dolphin's echolocator, in: "Proc. of the 23rd Scientific-Technical Conference of LIAP",

Golubkov, A. G., Ershova, I. V., Korolev, I. V. and Malyshev, Yu. A., 1972, On energetical parameters of the Black Sea bottlenosed dolphin echolocational apparatus, Trudy LIAP 76:9.

Golubkov, A. G., Korolev, V. I., Antonov, V. A., and Ignat'eva, E. A. 1975, Comparison of dolphin echolocation signals with results of optimal signal calculations, Doklady Akad. Nauk SSSR 223:1251.

Golubkov, A. G., Zworykhin, V. N., Ershova, I. V., Korolev, V. I., Burdin, V. I., and Malyshev, Ju. A., 1969, Some experimental data and prospects for investigation of the analyzing capacity in dolphins, Trudij Leningr. Inst. Aviats., 64:128.

Gracheva, M. S., 1971, A contribution to the structure of Larynx in Tursiops truncatus, Rev. de Zool., Moscow, 50:1539.

Grinnel, A. D., 1967, Mechanisms of overcoming interference in echolocating animals, in: "Animal Sonar Systems, Biology and Bionics", R.-G. Busnel, ed., Laboratoire de Physiologie Acoustique, Jouy-en-Josas, France.

Gruenberger, H. B., 1970, On the cerebral anatomy of the Amazon dolphin, Inia geoffrensis, in: "Investigations on Cetacea V. II", G. Pilleri, Bentelli AG., Berne.

Gurevich, V. S., 1968, Experience of study of dolphin's capabilities (Delphinus delphis) for echolocation discrimination of geometrical figures, (Theses of report), "Scientific Conference of Moscow State Lomonosov University, Moscow.

Gurevich, V. S., 1969, Echolocation discrimination of geometric figures in the dolphin, Delphinus delphis, Moscow, Vestnik Moskovskoga Universiteta, Biologiya, Pochovedeniye, 3:109 (English Translation JPRS 49281).

Gurevich, V. S., 1972, Morpho-functional studying of the upper
     skull respiratory tracts of the Common dolphin (Delphinus
     delphis), in: "Conference on using Mathematical Methods
     and Computing in Mathematics and Bionics", Leningrad.
Gurevich, V. S., 1973, Experience of the plastical reconstruction
     of the upper skull respiratory tract of the Common dolphin
     (Delphinus delphis), Journal Ontogenez: 78.
Gurevich, V. S., 1973, Histological study of the sound
     production organs in the Lagenorhynchus obliquidens, in:
     "Proceedings of the Eighth All-Union Acoustical Conference",
     Leningrad.
Gurevich, V. S., 1973, Possibilities of interaction of the larynx
     and upper skull respiratory tracts in the process of sound
     production in the dolphins, in: "Material of the Fourth
     All-Union Conference on Marine Mammals", Makhachkala: 23-25.
Gurevich, V. S., and Evans, W. E., 1976, Echolocation discrimina-
     tion of complex planar targets by the Beluga whale
     (Delphinapterus leucas), Jour. Acoust. Soc. Amer., 60(1):5.
Gurevich, V. S., and Evdokimov, V. N., 1973, Rentgenological
     investigation of the lungs and respiratory tracts of the
     Common dolphin (Delphinus delphis), in: "Materials of the
     Fourth All-Union Conference Marine Mammals", Makhachkala, 6:14.
Gurevich, V. S., and Korol'kov, Yu. I., 1973, A rentgenological
     study of respirative act in Delphinus delphis, Zool. Jour.,
     52(5):786.
Hall, J. G., 1967, Hearing and primary auditory centers of the
     whales, Acta Otolaryngol., Suppl., 224:244.
Hall, J. D., and Johnson, C. S. 1972, Auditory thresholds of a
     killer whale Orcinus orca (Linnaeus), Jour. Acoust.
     Soc. Amer., 51:515.
Hammer, C. E., 1977, Biosonar report: An experimental analysis
     of salient target characteristics for the echo-locating
     porpoise (Tursiops truncatus) (Technical Report 7709-1),
     SEACO, Incorporated, Kailua, Hawaii.
Hammer, C. E., 1978, Echo-recognition in the porpoise (Tursiops
     truncatus): An experimental analysis of salient target
     characteristics, Naval Ocean Systems Center, San Diego,
     Tech. Rep. 192.
Hammer, C., and Au, W. W. L., 1978, Target recognition via
     echolocation by an Atlantic bottlenose porpoise (Tursiops
     truncatus), Jour. Acoust. Soc. Amer., 64(1):587.
Hansen, I. A., and Cheah, C. C., 1969, Related dietary and
     tissue lipids of the sperm whale, Comp. Biochem. Physiol.,
     31:757.
Hellion, A., Guillard, P., Escudie, B., and Dziedzic, A.,
     1974, Etude des signaux impulsifs utilises par les sonars
     biologiques, Coll. s. les methodes d'etudes et simulation
     des chocs, Lyon.

Herman, L. M., 1975, Interference and auditory short-term memory in the bottlenosed dolphin, Animal Learning & Behavior., 3:43.

Herman, L. M., and Arbeit, W. R., 1971, Auditory frequency discrimination from 1-36 kHz in Tursiops truncatus, in: "Proc. 8th Annu. Conf. on Biol. Sonar and Diving Mammals", Stanford Res. Inst., Menlo Park, Calif.

Herman, L. M. and Arbeit, W. R., 1972, Frequency difference limens in the bottlenose dolphin: 1-70 kHz, Jour. Aud. Res., 12:109.

Herman, L. M., and Arbeit, W. R., 1973, Stimulus control and auditory discrimination learning sets in the bottlenose dolphin, J. exp, Analys. Behav., 19:379.

Herman, L. M., and Gordon, J. A., 1974, Auditory delayed matching in the bottlenose dolphin, J. Exp. Analys. Behav., 21:19.

Herman, L. M., Beach, F. A., Pepper, R. L., and Stalling, R. B., 1969, Learning-set formation in the bottlenose dolphin, Psychon. Sci., 14:98.

Herman, L. M., Peacock, M. F., Yunker, M. P., and Madsen, C. J., 1975, Bottlenosed dolphin: Double split pupil yields equivalent aerial and underwater duirnal acuity, Science, 189:650.

Herald, E. S., Brownell, R. L., Frye, R. L., Morris, E. J., Evans, W. E., and Scott, A. B., 1969, Blind river dolphin: First side-swimming cetacean, Science, 166:1408.

Hollien, H., Hollien, P., Caldwell, D. K., and Caldwell, M. C., 1976, Sound production by the Atlantic bottlenosed dolphin, Tursiops truncatus, Cetology, 26:1.

Ivanova, M. P., and Kurganskiy, N. A., 1973, A dolphin's trajectory of motion in the process of discriminating spherical and cylindrical objects, IV-oe Vsesoyuznoye soveshchaniye po Bionike, Moscow, 2:106.

Jacobs, D., 1972, Auditory frequency discrimination in the Atlantic bottlenose dolphin, Tursiops truncatus, Montague: A preliminary report, Jour. Acoust. Soc. Amer., 52:696.

Jacobs, D. W., and Hall, J. D., 1972, Auditory thresholds of a freshwater dolphin, Inia geoffrensis Blainville, Jour. Acoust. Soc. Amer., 51:530.

Jacobs, M., 1971, "Talking" through the nose. Sound and communication in whales and dolphins, Anim. Kingdom, 74:2.

Jansen, J., and Jansen, J. K. S,. 1969, The nervous system of cetecea, in: "The Biology of Marine Mammals", H. T. Anderson, ed., Acad. Press, New York.

Jerison, H. J., 1973, "Evolution of the Brain and Intelligence", Acad. Press, New York.

Johnson, C. S., 1966, Auditory thresholds of the bottlenosed porpoise, Tursiops truncatus (Montague), N. O. T. S. TP 4178.

Johnson, C. S., 1967, Discussion to paper by Evans and Powell,
    in: "Animal Sonar Systems, Biology and Bionics", R.-G.
    Busnel, ed., Laboratoire de Physiologie Acoustique, Jouy-en-
    Josas, France.

Johnson, C. S., 1967, Sound detection thresholds in marine mammals,
    in: "Marine Bio-Acoustics, Proc. of the Second Symposium on
    Marine Bio-Acoustics, New York", W. N. Tavolga, ed., Pergamon
    Press, New York.

Johnson, C. S., 1968, Masked pure tone thresholds in the bottle-
    nosed porpoise, Jour. Acoust. Soc. Amer., 44:965.

Johnson, C. S., 1971, Auditory masking of one pure tone by
    another in the bottlenosed porpoise, Jour. Acoust. Soc.
    Amer., 49:1317.

Johnson, R. A., and Titlebaum, E. L., 1976, Energy spectrum
    analysis: A model of echolocation processing, Jour. Acoust.
    Soc. Amer. 60:484.

Kadnadzey, V. V., Kreychi, S. A., Kakhalkina, E. N., Nikolenko,
    G. V., and Titov, A. A., 1975, Morskiye Mlekopitayushchiye,
    in: "Sixth All-Union Conference on the Study of Marine
    Mammals", G. B. Agarkov, ed., (English translation JPRS
    L6049-1).

Kanazu, R., Morii, H., and Fukuhara, T., 1969, Studies on the
    little toothed whales in the west sea area of Kyushu - XVII,
    about higher branched chain fatty acids in head oil of the
    little toothed whales - I., Bull. Fac. Fish. Nagasaki Univ.
    28:161.

Karol, R., Litchfield, C., Caldwell, D. K., and Caldwell, M. C.,
    1978, Compositional topography of the melon and spermaceti
    organ lipids in the pygmy sperm whale Kogia breviceps:
    Implications for echolocation, Mar. Biol., 47:115.

Kasuya, T., 1973, Systematic consideration of recent toothed whales
    based on the morphology of tympano-periotic bone, Sci, Rept.
    Whales Res. Inst., 25:103.

Kaufman, B. W., Siniff, D. B., and Reichle, R. A., 1972, Colony
    behavior of Weddell seals, Leptonychotes weddelli, at Hutton
    Cliffs, Antartica, in: "Symposium on the Biology of the
    Seal", G. B. Farquhar, ed., Univ. of Guelph, Ontario.

Kenshalo, D. R., 1967, Discussion, in: "Animal Sonar Systems,
    Biology and Bionics", R.-G. Busnel, ed., Laboratoire de
    Physiologie Acoustique, Jouy-en-Josas, France.

Kesarev, V. S., Trykova, O. V., and Malofeyeva, L. I., 1975,
    Structural prerequisites for corticalization of the dolphin's
    acoustic analyzer, in: "Morskiye Mlekopitayushchiye", G. B.
    Agarkov, ed., Izdatel'stvo Naukova Dumka, Kiev.

Khakhalina, E. N., 1975, Emotional signals in the dolphin's
    communication system, in: "Morskiye Mlekopitayushchiye",
    G. B. Agarkov, ed., Isdatel'stvo Naukova Dumka, Kiev.

Khomenko, B. G., 1969, Distinctions in structure of the peripheral
    nervous system of the dolphin's melon (Tursiops truncatus
    Ponticus Barabash), in: "Tezisy II vsesoyuznogo simpoziuma
    mododykh uchenykh", Sevastopol.
Khomenko, B. G., 1970, Some distinctions of histological structure
    and innvervation of the rostrum of the Black Sea dolphins,
    Bionika, 4:70.
Khomenko, B. G., 1970, The histostructure and innvervation of the
    sound apparatus –nasal sacs – in dolphins, Akad. Nauk. URSS,
    Dopovid, 32:83.
Khomenko, B. G., 1973, A contribution to the morphology of I-VIII
    pairs of cranial nerves in dolphins, Zoologicheskiy Zhurnal,
    52(3):407.
Khomenko, B. G., 1973, Morphological basis for the echolocation
    properties of dolphins, Bionika, 7:60.
Khomenko, B. G., 1974, A morphological and functional analysis of
    the structure of the supercranial respiratory passage as a
    possible echolocation signal generator in the dolphin,
    Bionika, 8 (English translation JPRS 63492).
Khomenko, B. G., 1975, Comparative anatomical analysis of the
    structure and innervation of nasolabial region in certain
    mammals and man, Vestnik Zoologii, 5:39.
Khomenko, B. G., 1975, Morphological characteristics of the
    respiratory and suckling reflexes in dolphin embryogenesis,
    in: "Morskiye Mlekopitayushchiye", G. B. Agarkov, ed.,
    Izdatel'stvo Naukova Dumka, Kiev.
Khomenko, B. G., and Khadzhinskiy, V. G., 1974, Morphological
    and functional principles underlying cutaneous reception in
    dolphins, Bionika, 8:106.
Kibblewhite, A. C., Denham, R. N., and Barnes, D. J., 1967, Unusual
    low-frequency signals observed in New Zealand waters, Jour.
    Acoust. Soc., Amer., 41:644.
Kinne, O., 1975, Orientation in space: Animals, marine, in:
    "Marine Ecology Vol. 2 – Physiological Mechanisms",
    O. Kinne, ed., Wiley, London.
Kiselev, V. I., and Mrochkov, K. A., 1975, Products of dissolu-
    tion of cetacean connective tissue collagen and use thereof,
    in: "Morskiye Mlekopitayushchiye", G. B. Agarkov, ed.,
    Izdatel'stvo Naukova Dumka, Kiev.
Konstantinov, A. I., Mel'nikov, N. F., and Titov, A. V., 1968,
    On dolphin ability of identifying objects, in: "Tezisy
    Dokladov na II-oy Ukrainskoy Respubliknaskoy Konferentsii
    po Bionike", Kiev.
Korolev, L. D., Lipatov, N. V., and Resvov, R. N., Savel'ev, M.A.,
    and Flenov, A. B., 1973, Investigation of the spatial
    directivity diagram of the dolphin sound emitter, in:
    "Proceedings of the 8th All-Union Acoustical Conference",
    Leningrad.

Koshovoy, V. V., and Mykhaylovs'kyy, V. M., 1972, The operating principle of the echolocating system of marine mammals, Dopovidi Akad. Nauk Ukrayins'kayi RSR Ser. A Fizyko-tekhn. ta matemat. Nauky, Ukranian 12:1097 (English translation JPRS 58344).

Kozak, V. A., 1974, Receptor zone of the video-acoustic system of the sperm whale (physeter Catodon L., 1758), Fiziologichnyy Zhurnal Academiy Nauk Ukraynskoy RSR 120 (English translation: JPRS 65017).

Kruger, Lawrence, 1966, Specialized features of the cetacean brain, in: "Whales, Dolphins, and Porpoises", K. S. Norris, ed., Univ. of California Press, Berkeley.

Krushinskii, L. V., Dashevskii, B. A., Krushinskaya, N. L., and Dmitrieva, I. L., 1972, A study of the ability of the dolphin Tursiops truncatus (Montague) to operate with the empirical dimensionality of geometric figures, Doklady Akademii Nauk SSR, 204:755 (Translation: Plenum Publishing Company, New York, 1972).

Ladygina, T. F., and Supin, A. Ya., 1970, The acoustic projection in the dolphin cerebral cortex, Fiziol. Zh. SSSR im. I. M. Sechenova, 56:1554.

Ladygina, T. F., and Supin, A. Ya., 1977, Localization of sensory projection zones in the cerebral cortex of the bottlenosed dolphin, Tursiops truncatus, Zurnal Evolyutsionnoy Biokhimii i Fiziologii, 6:712.

Leatherwood, J. S., Johnson, R. A., Ljungblad, D. K., and Evans, W. E., 1977, "Broadband measurements of underwater acoustic target strengths of panels of tuna nets", NOSC TR 126, San Diego.

Lekomtsev, V. M., and Titov, A. A., 1974, Procedures for studying the echolocation apparatus of dolphins, Bionika, 8:83 (English Translation JPRS 63492).

Lende, R. A., and Walker, W. I., 1972, An unusual sensory area in the cerebral neocortex of the bottlenose dolphin, Tursiops truncatus, Brain Res., 45:555.

Levenson, C., 1974, Source level and bistatic target strength of the sperm whale (Physeter catodon) measured from an oceanographic aircraft, Jour. Acoust. Soc. Amer., 55:1100.

Lilly, J. C., 1966, Sonic-ultrasonic emissions of the bottlenose dolphin, in: "Whales, Dolphins, and Porpoises", K. S. Norris, ed., Univ. Calif. Press, Berkeley.

Lilly, J. C., 1967, Dolphin vocalization, in: "Brain Mechanisms Underlying Speech and Language", Grune and Stratton, New York.

Lilly, J. C., 1967, Dolphin's vocal mimicry as a unique ability and a step toward understanding, in: "Research in Verbal Behavior and Some Neurophysiological Implications", Academic Press, New York.

Lilly, J. C., 1967, Mind of the dolphin: a nonhuman intelligence, Doubleday and Co., New York.

Lilly, J. C., 1968, Sound production in Tursiops truncatus
    (bottlenose dolphin), Ann. N. Y. Acad. Sci, 155:320.
Lilly, J. C., Miller, A. M., and Truby, U. M., 1968, Reprogram-
    ming of the sonic output of the dolphin sonic burst count
    matching, Jour. Acous. Soc. Amer., 43:1412.
Lipatov, N. V., and Solntseva, G. N., 1974, Morphological and
    functional features of the external auditory canal of
    the common and bottlenose dolphin, Bionika, 8 (English
    translation JPRS 63492).
Litchfield, C., and Greenberg, A. J., 1974, Comparative lipid
    patterns in the melon fats of dolphins, porpoises, and
    toothed whales, Comp. Biochem. Physiol., 478:401.
Litchfield, C., Karol, R., and Greenberg, A. J., 1973, Composi-
    tional topography of melon lipids in the Atlantic bottle-
    nose dolphin Tursiops truncatus: implications for
    echolocation Mar. Biol., 23:165.
Litchfield, C., Ackman, R. G., Sipos, J. C. and Eaton, C. A.,
    1971, Isovaleroyl triglycerides from the blubber and melon
    oils of the beluga whale (Delphinapterus leucas), Lipids,
    6:674.
Litchfield, C., Kinneman, J., Ackman, R. G., and Eaton, C. A.,
    1971, Comparative lipid patterns in two freshwater dolphins
    Inia geoffrensis and Sotalia fluviatilis, Jour. Amer. Oil
    Chem. Soc., 48:91.
Litchfield, C., Greenberg, A. J., Caldwell, D. K., Caldwell, M. C.,
    Sipos, J. C., and Ackman, R. G., 1975, Comparative lipid
    patterns in acoustical and nonacoustical fatty tissues
    of dolphins, porpoises, and toothed whales, Comp. Biochem.
    Physiol., 508:591.
Livshits, M. S., 1974, Some properties of dolphin hydrolocator
    from the viewpoint of correlation hypothesis, Biofizika,
    19:916 (English translation JPRS 64329).
Livshits, M. S., 1975, Correlation model for the recognition
    of objects by echolocating animals, Biofizika, 20(5):920
    (English translation JPRS L/5744).
Ljungblad, D. K., Leatherwood, S., and Johnson, R. A., 1977,
    Echolocation signals of wild Pacific bottlenosed dolphins
    Tursiops Sp, in: "Proceedings (Abstracts) Second Conference
    on the Biology of Marine Mammals", San Diego.
Lockyer, C., 1977, Observations on diving behaviour of the sperm
    whale Physeter catodon, in: "A Voyage of Discovery", M.
    Angel, ed., Pergamon Press, Oxford.
Love, R. H., 1971, Measurements of fish target strength: a review,
    Fish. Bull., 69:703.
Love, R. H., 1973, Target strengths of humpback whales, Megaptera
    novaeangliae, Jour. Acous. Soc. Amer., 54:1312.

Mackay, R. S., 1966, Telemetering physiological information from within cetaceans, and the applicability of ultrasound to understanding in vivo structure and performance, in: "Whales, Dolphins and Porpoises", K. S. Norris, ed., Univ. Calif. Press, Berkeley.

MacKay, R. S., Rumage, W. T., and Becker, A., 1977, Sound velocity in spermaceti organ of a young sperm whale, in: "Proceedings (Abstracts) Second Conference on the Biology of Marine Mammals", San Diego.

Malins, D. C., and Varanasi, V., 1977, Acoustic pathways in the cetacean head: Assessment of sound properties through the use of a new microtechnique, in: "Proceedings (Abstracts) Second Conference on the Biology of Marine Mammals", San Diego.

Markov, V. I., and Ostrovskaya, V. M., 1975, On the "identification" signal of dolphins, in: "Morskiye Mlekopitayushchiye", G. B. Agarkov, ed., Izdatel'stov Naukova Dumka, Kiev.

Markov, V. I., Tanchevskaya, V. A., and Ostrovskaya, V. M., 1974, Organization of acoustic signals in the Black Sea bottlenose dolphins (Tursiops truncatus), in: "Morfologiya, Fiziologiya i Akustika Morskikh Mlekopitayushchikh", V. Ye. Sokolov, ed., Izdatel'stvo Nauka, Moscow.

McCormick, J. G., 1968, "Theory of hearing for delphinids", Doctoral Dissertation, Princeton University, New Jersey.

McCormick, J. G., 1972, The physiology of hearing in the porpoise, in: "Marine Mammals: Biology and Medicine", S. H. Ridgway, ed., Thomas, Springfield, Ill.

McCormick, J. G., Wever, E. G., Palin, J. and Ridgway, S. H., 1970, Sound conduction in the dolphin ear, Jour. Acoust. Soc. Amer., 48:1418.

McCormick, J. G., Wever, E. G., Ridgway, S., and Palin, J., 1970, Function of the porpoise ear as shown by its electrical potentials (Abstract), Jour. Acoust. Soc. Amer., 47:67.

McDonald-Renaud, D. L., 1974, Sound localization in the bottlenose porpoise, Tursiops truncatus (Montague), Doctoral Dissertation, Univ. of Hawaii, Honolulu.

Mead, J. G., 1972, On the anatomy of the external nasal passages and facial complex in the family delphinidae of the order cetacea, Doctoral Dissertation, Univ. Chicago, Chicago.

Mermoz, H., 1967, Discussion, in: "Animal Sonar Systems, Biology and Bionics", R.-G. Busnel, ed., Laboratoire de Physiologie Acoustique, Jouy-en-Josas, France.

Metsaveer, J., 1971, Algorithm for calculation of echo-pulses from elastic spherical shells in fluid by summation of wave groups, Inst. of Cybernetics, Acad. Sci. Eston. SSR Tallinn (Preprint), 3:1.

Miyasnikov, V. S., and Titov, M. S., 1975, On the identification of marine animals by their acoustic signals,
Izvestiya TINRO, 94:38.

Mizue, T., Nishiwaki, M., and Takemura, A., 1971, The underwater
sound of Ganges river dolphins (Platanista gangetica),
Sci. Repts. Whales Res. Inst., 23:123.

Mizue, T., Takemura, A., and Nakasai, K., 1967, Studies on
the little toothed whales in the west sea area of Kyushu,
XIII, Mating calls and others of the bottlenose dolphin
caught at Arikawa in Goto Is., Nagasaki Pref. Bull. Fac.
Fish. Nagasaki Univ., 23:197.

Mizue, T., Takemura, A., Nakasai, K., 1968, Studies on the
little toothed whales in the west sea area of Kyushu,
XV, Underwater sound of the Chinese finless porpoise
caught in the Japanese coastal sea, Bull. Fac. Fish.
Nagasaki Univ., 25:25.

Mohl, B., 1967, Frequency discrimination in the common seal
and a discussion of the concept of upper hearing limit,
in: "Underwater Acoustics", V. M. Albers, ed., Plenum
Press, New York.

Mohl, B., 1968, Auditory sensitivity of the common seal in
air and water, Jour. Aud. Res., 8:27.

Mohl, B., 1968, Hearing in seals, in: "The Behavior and
Physiology of Pinnipeds", R. J. Harrison, R. C. Hubbard,
R. S. Peterson, C. E. Rice, and R. J. Schusterman, eds.,
Appleton-Century-Crofts, New York.

Mohl, B., and Andersen, S., 1973, Echolocation: high-frequency
component in the click of the harbour porpoise (Phocoena
Ph. L), Jour. Acoust. Soc. Amer, 54:1368.

Mohl, B., Ronald, K., and Terhune, J. M., 1972, The harp seal,
Pagophilus groenlandicus (Erxleben, 1777), XVIII, Under-
water calls, in: "Symposium on the Biology of the
Seal", G. B. Farquhar, ed., Univ. of Guelph, Ontario.

Moore, J. C., 1968, Relationships among the living genera of
beaked whales, Fieldiana:Zoology, 53(4):209.

Moore, P. 1975, Underwater localization of click and pulsed
pure tone signals by the California sea lion (Zalophus
californianus), Jour. Acoust. Soc. Amer., 57:406.

Moore, P., and Au, W., 1975, Underwater localization of pulsed
pure tones by the California sea lion (Zalophus
californianus), Jour. Acoust. Soc. Amer., 58:721.

Moore, P. W. B., and Schusterman, R. J., 1977, The upper limit
of underwater auditory frequency discrimination in the
California sea lion, in: "Proceedings (Abstracts) Second
Conference on the Biology of Marine Mammals", San Diego.

Morgan, D. W., 1971, The reactions of belugas to natural
sound playbacks, in: "Proceedings of the Seventh Annual
Conference on Biological Sonar and Diving Mammals",
Stanford Research Institute, Menlo Park, Calif.

Morgane, P. J., and Jacobs, N.S., 1972, Comparative anatomy
    of the cetacean nervous system, in: "Functional Anatomy
    of Marine Mammals", R. J. Harrison, ed., Academic Press,
    New York.
Morii, H., and Kanazu, R., 1972, Fatty acids in the foetus,
    nurseling and adult of a kind of dolphin, Stenella
    attenuata, Bull. Jap. Soc. Scient. Fish., 38:599.
Morozov, V. P., Akopian, A. I., Zaytseva, K. A. and Titov, A. A.,
    1975, On the characteristics of spacial directivity of the
    dolphin acoustic system in signal perception against the
    background of noise, in: "Marine Mammals, Proceedings of the
    Sixth All-Union Conference on the Study of Marine Mammals",
    G. B. Agarkov, ed., Naukova Dumka, Kiev (English translation
    JPRS L/6049).
Morozov, V. P., Akopian, A. I., Burdin, V. I., Zaytseva, K. A.
    and Sokovykh, Yu. A., 1972, Tracking frequency of the loca-
    tion signals of dolphins as a function of distance to the
    target, Biofizika, 17:139 (English translation JPRS 55729).
Morozov, V. P., Akopian, A. I., Burdin, V. I., Donskov, A. A.,
    Zaytseva, K. A., and Sokovykh, Yu. A., 1971, Audiogram
    of the dolphin, Tursiops truncatus, Fiziologicheskiy
    Zhurnal imeni I. M. Sechenova, 57(6):843.
Morris, R. J., 1973, The lipid structure of the spermaceti organ
    of the sperm whale (Physeter catodon), Deep-Sea Res., 20:911.
Murchison, A. E., 1976, Range Resolution by an Echolocating Dolphin
    (Tursiops truncatus), Jour. Acoust. Soc. Amer., 60:S5.
Murchison, A. E., 1979, "Maximum Detection Range and Range
    Resolution in Echolocating Tursiops truncatus (Montague)",
    Doctoral Dissertation U. of Calif. at Santa Cruz, Calif.
    (manuscript).
Murchison, A. E., and Nachtigall, P. E., 1977, Three dimensional
    shape discrimination by an echolocating bottlenose porpoise
    (Tursiops truncatus), in: "Proceedings (Abstracts) of the
    Second Conference on the Biology of Marine Mammals",
    San Diego.
Murchison, A. E., and Penner, R. H., 1975, Open water echolocation
    in the bottlenose dolphin (Tursiops truncatus): Metallic
    sphere detection thresholds as a function of distance, in:
    "Conference on the Biology and Conservation of Marine
    Mammals, Abstracts", 2:42.
Nachtigall, P. E., 1969, Visual size discrimination in the
    East Asian clawless otter (Amblyonyx cineria) in air and
    under water, in: "Proceedings of the Sixth Annual Conference
    on Biological Sonar and Diving Mammals", Stanford Research
    Institute Press, Menlo Park, Calif.
Nachtigall, P. E., 1971, Spatial discrimination and reversal
    based on differential magnitude of reward in the dolphin
    Tursiops truncatus, in: "Proceedings of the Eighth Annual
    Conference on Biological Sonar and Diving Mammals",
    Stanford Research Institute, Menlo Park, Calif.

Nachtigall, P. E., 1976, "Food-Intake and food-rewarded instru-
    mental performance in dolphins as a function of feeding
    schedule", Doctoral Dissertation, Univ. of Hawaii, Honolulu.
Nachtigall, P. E., 1977, A comparison of porpoise (Tursiops
    truncatus) performance with fish and a prepared food,
    in: "Proceedings (Abstracts) Second Conference on the
    Biology of Marine Mammals", San Diego.
Nachtigall, P. E., Murchison, A. E., and Au, W. W. L., 1978,
    Discrimination of solid cylinders and cubes by a blind-
    folded echolocating bottlenose dolphin (Tursiops truncatus),
    Jour. Acoust. Soc. Amer., 64(1):587.
Naumov, N. P., 1973, Signal (biological) fields and their
    significance for animals, Zhurnal Obshchey Biologii, 6:808.
Naumov, N. P., 1975, Biological (signal) fields and their
    significance in mammalian life, Vestnik Akademii Nauk
    SSSR, 45(2):55.
Naumov, N. P., 1975, Bionics and acoustics, Berliner Zeitung,
    (E. Germany) Lomonosov St., Univ., 4.
Neproshin, A. Yu., 1975, Sounds as part of the behavior of the
    Pacific beluga, Priroda, 4:99 (English translation: JPRS
    65658).
Ness, A. R., 1967, A measure of asymmetry of the skulls of odonto-
    cete whales, Jour. Zool., 153:209.
Norris, K. S., 1968, The evolution of acoustic mechanisms in
    odontocete cetaceans, in: "Evolution and Environment", E. T.
    Drake, ed., Yale Univ. Press, New Haven.
Norris, K. S., 1969, The echolocation of marine mammals, in:
    "The Biology of Marine Mammals", H. T. Andersen, ed.,
    Academic Press, New York.
Norris, K. S., 1974, "The Porpoise Watcher", George J. McLeod
    Limited, Toronto.
Norris, K. S. and Evans, W. E., 1967, Directionality of echolocation
    clicks in the rough-tooth porpoise Steno bredanensis (Lesson),
    in: "Marine Bio-Acoustics, Proceedings Second Symposium on
    Marine Bio-Acoustics, New York", W. N. Tavolga, ed., Pergamon
    Press, New York.
Norris, K. S., and Harvey, G. W., 1972, A theory for the function
    of the spermaceti organ of the sperm whale, in: "Animal
    Orientation and Navigation, S. R. Galler et al., eds., National
    Aeronautics and Space Administration, Washington, D. C.
Norris, K. S., and Harvey, G. W., 1974, Sound transmission in
    the porpoise head, Jour. Acoust. Soc. Amer., 56:659.
Norris, K. S., and Watkins, W. A., 1971, Underwater sounds of
    Arctocephalus philippii, the Juan Fernandez fur seal,
    Antarct. Res. Ser., 18:169.
Norris, K. S., Evans, W. E. and Turner, R. N., 1967, Echolocation
    of an Atlantic bottlenose porpoise during discrimination,
    in: "Animal Sonar Systems, Biology and Bionics", R.-G.
    Busnel, ed., Laboratoire de Physiologie Acoustique, Jouy-en-
    Josas, France.

Norris, K. S., Dormer, K. J., Pegg, J., and Liese, G. J., 1971,
    The mechanisms of sound production and air recyling in
    porpoises: A preliminary report, in: "Proc. 8th Annual
    Conf. Biol. Sonar and Diving Mammals", Stanford Research
    Institute, Menlo Park, Calif.
Norris, K. S., Harvey, G. W., Burzell, L. A., and Krishna Kartha,
    T. D., 1972, Sound production in the freshwater porpoises
    Sotalia and Inia, in Rio Negro, Brazil, in: "Investigations
    on Cetacea Vol. 4., G. Pilleri, ed., Benetelli AG, Berne.
Northrop, J., Cummings, W. C., and Morrison, M. F., 1971,
    Underwater 20-Hz signals recorded near Midway Island,
    Jour. Acoust. Soc. Amer., 49:1909.
Northrop, J., Cummings, W. C., and Thompson, P. O., 1968, 20-Hz
    signals observed in the central pacific, Jour. Acoust. Soc.
    Amer., 43:383.
Ostroumov, G. A., 1968, At what distances could marine animals
    communicate by means of electromagnetic waves (in problems
    of wave propagation), Izdatel'stvo Leningradskogo
    Universiteta, 8:3.
Paul, L. J., 1969, Dolphin noises recorded by echosounder, New
    Zealand Jour. Marine and Freshwater Res., 3:343.
Payne, R. S., and McVay, S., 1971, Songs of humpback whales,
    Science, 173:587.
Payne, R. S., and Payne, K., 1972, Underwater sounds of southern
    right whales, Zoologica, 56:159.
Payne, R., and Webb, D., Orientation by means of long-range
    acoustic signaling in Baleen whales, in: "Orientation:
    Sensory Basis", H. E. Adler, ed., Ann. N. Y. Acad. Sci.,
    188:110.
Penner, R. H., 1977, Paired simultaneous echo ranging by Tursiops
    truncatus, in: "Proc. (Abstracts) Second Conference on the
    Biology of Marine Mammals", San Diego.
Penner, R., and Murchison, A. E., 1970, Experimentally demonstrated
    echolocation in the Amazon river porpoise Inia geoffrensis
    (Blainville), in: "Proceedings of the Seventh Annual
    Conference on Biological Sonar and Diving Mammals", Stanford
    Research Institute, Menlo Park, Calif.
Penner, R. H., and Murchison, A. E., 1970, Experimentally
    demonstrated echolocation in the Amazon River porpoise,
    Inia geoffrensis (Blainville), Naval Undersea Center,
    Technical Publication 187, San Diego, Calif.
Perkins, P. J., 1966, Communication sounds of finback whales,
    Norsk Hvalfangst-Tidend, 55:199.
Perkins, P. J., Fish, M. P., and Mowbray, W. H., 1966, Underwater
    communication sounds of sperm whale Physeter catodon,
    Norsk Hvalfangst-Tidende, 55:225.
Perrin, W. F., and Hunter, J. R., 1972, Escape behavior of the
    Hawaiian spinner porpoise (Stenella cf. S. longirostris), U.
    S. Nat. Mar. Fish. Bull., 70:49.

Perryman, W. and Au, D., 1977, Aerial observations of evasive
    behavior of dolphin schools, in: "Proceedings 2nd Conference
    Bio. Mar. Mammals", San Diego.
Pilleri, G., 1970, Records of cetaceans off the Italian and
    Dalmatian coasts, in: "Investigations on Cetacea, Vol. 2",
    G. Pilleri, ed., Bentelli AG, Berne.
Pilleri, G., and Busnel, R.-G., 1969, Brain/body weight ratios
    in Delphinidae, Acta. Anat. 73:92.
Pilleri, G., and Gihr, M., 1970, Brain-body weight ratio of
    Platanista gangetica, in: "Investigations on Cetacea, Vol.
    2", G. Pilleri, ed., Bentelli AG, Berne.
Pilleri, G., Kraus, C., and Gihr, M., 1970, Frequenzanalyse
    der Laute von Platanista indi (Cetacea), Rev. Suisse
    Zool., 77:922.
Pilleri, G., Kraus, C., and Gihr, M., 1971, Physical analysis
    of the sounds emitted by Platanista indi, in: "Investigation
    on Cetacea, Vol. 3", G. Pilleri, ed., Bentelli AG, Berne.
Popov, V. V., and Supin, A. Yu., 1976, Determination of the
    hearing characteristics of dolphins by measuring induced
    potentials, Fiziol. Zh. SSSR im I. M. Sechenova, 62:550.
Poulter, T. C., 1966, Biosonar, in: "McGraw-Hill Yearbook
    of Science and Technology", McGraw-Hill, New York.
Poulter, T. C., 1966, The use of active sonar by the California
    sea lion Zalophus californianus (Lesson), Jour. Aud. Res.,
    6:165.
Poulter, T. C., 1967, Systems of echolocation, in: "Animal
    Sonar Systems: Biology and Bionics", R.-G. Busnel, ed.,
    Laboratoire de Physiologie Acoustique, Jouy-en-Josas,
    France.
Poulter, T. C., 1968, Marine mammals, in: "Animal Communication",
    T. A. Sebeok, ed., Indiana University Press, Bloomington.
Poulter, T. C., 1968, Underwater vocalization and behavior of
    pinnipeds, in: "The Behavior and Physiology of Pinnipeds",
    R. J. Harrison, et. al., eds., Appleton-Century-Crofts,
    New York.
Poulter, T. C., 1968, Vocalization of the gray whales in Laguna
    Oja de Liebre (Scammon's Lagoon), Baja, California,
    Mexico, Norsk Hvalfangsttid, 57:53.
Poutler, T. C., 1969, Conditioning marine mammals for testing
    echolocation abilities, in: "Proceedings of the Sixth
    Annual Conference on Biological Sonar and Diving Mammals",
    Stanford Research Institute, Menlo Park, Calif.
Poulter, T. C., 1969, Sonar discrimination ability of the
    California sea lion, Proceedings California Acad. Sci.
    36:381.
Poulter, T. C., 1969, Sonar of penguins and fur seals, Proceedings
    California Academy Science ", 36:363.

Poulter, T. C., 1970, Ultrasonic frequencies recorded from three captive blind dolphins, Platanista gangetica, in: "Proceedings of the Seventh Annual Conference on Biological Sonar and Diving Mammals", Stanford Research Institute, Menlo Park, Calif.

Poulter, T. C., 1972, Sea lion vibrissae - an acoustic sensor, in: "Proceedings of the Ninth Annual Conference on Biological Sonar and Diving Mammals", Stanford Research Institute, Menlo Park, Calif.

Powell, B. A., 1966, Periodicity of vocal activity of captive Atlantic bottlenose dolphins, Tursiops truncatus, Bull. So. Calif. Acad. Sci., 65:237.

Pryor, K. and Norris, K. S., 1978, The tuna porpoise problem: Behavioral aspects, Oceanus, 21:31.

Purves, P. E., 1966, Anatomy and Physiology of the outer middle ear in cetaceans, in: "Whales, Dolphins and Porpoises", K. S. Norris, ed., Univ. Calif. Press, Berkeley.

Purves, P. E., 1967, Anatomical and experimental observations on the cetacean sonar systems, in: "Animal Sonar Systems, Biology and Bionics", R.-G. Busnel, ed., Laboratoire de Physiologie Acoustique, Jouy-en-Josas, France.

Purves, P. E., and Pilleri, G., 1973, Observations on the ear, nose, throat and eye of Platanista indi., in: "Investigations on Cetacea", G. Pilleri, ed., Bentelli AG, Berne.

Ray, C., 1967, Social behavior and acoustics of the Weddell seal, Antarct. Jour. U. S., 2(4):105.

Ray, G. C., 1972, "Marine mammals: The biological basis of productivity and conservation", Smithsonian Inst., Wash. D. C.

Ray, G. C., and Watkins, W. A., 1972, Social function of phonation in walrus, in: "Symposium on the Biology of the Seal", G. B. Farquhar, ed., Univ. of Guelph, Ontario.

Ray, G. C., Watkins, W. A., and Burns, J. J., 1969, The underwater song of Erignathus (bearded seal), Zoologica 54:79.

Renaud, D. L., and Popper, A. N., 1975, Sound localization by the bottlenose porpoise, Tursiops truncatus, Jour. Exp. Biol. 63:569.

Repenning, C. A., Underwater hearing in seals: Functional morphology, in: "Functional Anatomy of Marine Mammals", R. J. Harrison, ed., Academic Press, New York.

Resvov, R. N., Savel'ev, Ya. A., and Flenov, A. B., 1973, The study of dolphin's echolocator ability in detection of different targets, difference in range and angle on vertical and horizontal planes, in: "Proceedings of the 8th All-Union Acoustical Conference", Leningrad.

Reysenbach de Haan, F. W., 1966, Listening underwater: Thoughts on sound cetacean hearing, in: "Whales, Dolphins, and Porpoises", K. S. Norris, ed., Univ. of Calif. Press, Berkeley.

Reznik, A. M., Skornyakov, V. M., and Chupakov, A. G., 1970,
    Echolocation activity of Black Sea Tursiops truncatus
    being presented targets, Trudy Akusticheskogo Instituta,
    12:116.

Reznik, A. M. Skornyakov, V. M., Chernyshev, O. B., and
    Chupakov, A. G., 1968, On some peculiarities of dolphin
    communication system, in: "Trudy VIoy Vsesoijuznoyi
    Akusticheskoi Konferentsii", Moscow.

Reznikov, A. E., 1970, On sound-vision in dolphins, in: "Tezisy
    Dokladov 23ey Nauchno-tekhnicheskoy. Konferentsii Lenin
    gradskogo Institute Aviatsionnogo Priborostroeyniya",
    Leningrad.

Reznikov, A. E., 1971, Mechanisms of delphinid hydroacoustic
    ranging, Priroda, 9:171.

Rice, D. W., 1978, Beaked whales, in: "Marine Mammals", D.
    Haley, ed., Pacific Search Press.

Ridgway, S. H., 1966, Dall porpoise, Phocoenoides dalli (True):
    Observations in captivity and at sea, Norsk Hvalfangst-
    Tidende, 5:97.

Ridgway, S. H., 1972, (ed.) "Mammals of the Sea, Biology and
    Medicine", Charles C. Thomas, Springfield, Ill.

Ridgway, S. H., and Brownson, R. W., 1979, Relative brain
    sizes and cortical surface areas in odontocetes,
    (In preparation).

Ridgway, S. H., and McCormick, J. G., 1967, Anesthetization
    of porpoises for major surgery, Science, 158:510.

Ridgway, S. H., Flanagan, N. J., and McCormick, J. G., 1976,
    Brain-spinal cord ratios in porpoises: possible correlation
    with intelligence and ecology, Psychon. Sci., 6:491.

Ridgway, S. H., McCormick, J. G., and Wever, E. G., 1974, Surgical
    approach to dolphin's ear, Jour. Exp. Zool., 188:265.

Ridgway, S. H., Scronce, B. L., and Kanwisher, J., 1969,
    Respiration and deep diving in the bottlenosed porpoise,
    Science, 166:1651.

Ridgway, S. H., Carder, D. A., Green, R. F., Gaunt, A. S.,
    Gaunt, S. L., and Evans, W. E., 1977, Electromyography
    and dolphin sound production, in: "Proceedings (Abstracts)
    Second Conference on the Biology of Marine Mammals, San
    Diego.

Rieger, M. F. P., and Johnson, R. A., 1977, Chorusing patterns
    of pilot whales, Globicephala macrorynchus, in: "Proceedings
    (Abstracts) Second Conference on the Biology of Marine
    Mammals", San Diego.

Robisch, P. A., Malins, D. C., Best, R., and Varanasi, V., 1972,
    Differences in triacylglycerols from acoustic tissues and
    posterior cranial blubber of the narwhal (Monodon monoceros),
    Biochem. Jour., 130:33.

Rodionov, V. A., 1975, Morpho-functional features of the
    respiratory musculature of dolphins, Zoologicheskiy
    Zhurnal, 53(6):919.

Romanenko, Ye. V., 1972, Near acoustic field of the bottlenose dolphin (Tursiops truncatus), Voe Vsesoyuznoye Soveshchaniye po izucheniyu morskikh mlekopitayushchikh, Makhachkala, Part II:200.

Romanenko, Ye. V., 1973, Mechanism of emission and formation of echolocating signals of the bottlenose dolphin (Tursiops truncatus), in: "Eighth Annual All-Union Acoustical Conference", Moscow.

Romanenko, Ye. V., 1973, Sound signals in the near field of the Atlantic bottlenose dolphin (Tursiops truncatus, Delphinidae), Zoologicheskiy Zhurnal, 52(11):1698 (English translation JPRS 61553).

Romanenko, Ye. V., 1974, On the mechanism of the dolphin emission of some pulse and whistle signals, in: "Morfologiya, Fiziologiya i Akustika Morskikh Mlekopitayushchikh", V. Ye. Sokolov, ed., Izdatel'stvo Nauka, Moscow.

Romanenko, Ye. V., 1974, "Physical fundamentals of bioacoustics", Fizichekiye Osnovy Bioakustik, Moscow (English translation JPRS 63923).

Romanenko, Ye. V., 1975, On the hearing of dolphins, in: "Morskiye Mlekopitayushchiye", G. B. Agarkov, ed., Izdatel'stvo Naukova Dumka, Kiev.

Romanenko, Ye. V., 1975, Recording sounds in the dolphin's respiratory system, in: "Morskiye Mlekopitayushchiye", G. B. Agarkov, ed., Izdatel'stvo Naukova Dumka, Kiev.

Romanenko, Ye. V., Tomilin, A. G. and Artemenko, B. A., 1965, in: "Bionics", M. G. Gaaze-Rapoport and V. E. Yakobi, eds., Nauka, Moscow (English translation JPRS 35125).

Romanenko, Ye., V., Yanov, V. G., and Akopian, A. I., 1974, A method of studying the echolocating system of the dolphin using a radiotelemetry system, in: "Morfologiya, Fiziologiya i Akustika Morskikh Mlekopitayushchikh", V. Ye. Sokolov, ed., Izdatel'stvo Nauka, Moscow.

Ryan, W. W., Jr., 1978, Acoustical reflections from aluminum cylindrical shells immersed in water, Jour. Acous. Soc. Amer., 64:1159.

Sales, G., and Pye, D., 1974, "Ultrasonic Communication by Animals," Chapman and Hall, London.

Saprykin, V. A., Korolev, V. I., Kovtuneko, S. V., and Dmitriyeva, Ye. S., 1973, An investigation of spatial frequency characteristics of the Tursiops echolocator in object identification, Doklady Akademii Nauk SSSR, 214(3):727.

Saprykin, V. A., Korolev, V. I., Kovtuneko, S. V., Dmitriyeva, Ye. S., and Sokovykh, Yu. A., 1975, Spatial analysis of the tonal supra threshold signals in dolphins, Biofizika, 20:319.

Saprykin, V. A., Kovtuneko, S. V., Ekker, I. V., Dmitriyeva, Ye. S., and Korolev, V. I., 1975, Invariance of the sensing properties of the auditory analyser of the dolphin in differentiating pulse tone signal, Doklady Akademii Nauk SSSR, 221(4):999.

Saprykin, V. A., Kovtuneko, S. V., Korolev, V. I., Dmitriyeva,
    Ye. S., and Belov, B. I., 1975, Study of the auditory percep-
    tion of dolphins as a function of signal characteristics in
    the time domain, Biofizika, 20(4):720.
Saprykin, V. A., Kovtuneko, S. V., Korolev, V. I., Dmitriyeva,
    Ye. S., Belov, B. I., and Mar'yasin, V. G., 1975, Study of the
    auditory perception of dolphins as a function of signal char-
    acteristics in the time domain, Biofizika, 20 (English
    translation JPRS 65877).
Saprykin, V. A., Kovtuneko, S. V., Korolev, V. I., Dmitriyeva,
    Ye. S., Ol'shanskiy, V. I., and Bekker, I. V., 1976, Invari-
    ance of the auditory perception relative to frequency-time
    signal transformations in the dolphin Tursiops truncatus,
    Zhurnal Evolyutsionnoy Biokhimii I Fiziologii (English
    translation JPRS L/6275).
Schenkkan, E. J., 1971, The occurrence and position of the "con-
    necting sac" in the nasal tract complex of small odontocetes
    (Mammalia, Cetacea), Beaufortia, 19:37.
Schenkkan, E. J., 1972, On the nasal tract complex of Pontoporia
    blainvillei GERVAIS and d'ORBIGNY, 1884 (Cetacea, Plata-
    nistidae), in: "Investigations on Cetacea Vol 4",
    G. Pilleri, ed., Bentelli AG, Berne.
Schenkkan, E. J., 1973, On the comparative anatomy and function of
    the nasal tract in odontocetes (Mammalia, Cetacea), Bijdragen
    Tot de Dierkunde, 43(3):127.
Schenkkan, E. J., and Purves, P. E., 1973, The comparative anatomy
    of the nasal tract and the function of the spermaceti organ
    in the Physeteridae (Mammalia, Odontoceti), Bijdragen Tot
    de Dierkunde, 43(1):93.
Schevill, W. E., 1968, Sea lion echo ranging?, Jour. Acoust. Soc.
    Am., 43:1458.
Schevill, W. E., and Watkins, W. A., 1966, Sound structure and
    directionality in Orcinus (Killer Whale), Zoologica, 51:71.
Schevill, W. E., and Watkins, W. A., 1971, Directionality of the
    sound beam in Leptonychotes weddelli (Mammalia, Pinnipedia),
    Antarct. Res. Ser., 18:163.
Schevill, W. E., and Watkins, W. A., 1971, Pulsed sounds of the
    porpoise Lagenorhynchus australis, Breviora, 366:1.
Schevill, W. E., and Watkins, W. A., 1972, Intense low-frequency
    sounds from an Antarctic minke whale, Balaenoptera acutoro-
    strata, Breviora, 388:1.
Schevill, W. E., Watkins, W. A., and Ray, G. C., 1966, Analysis of
    underwater Odobenus calls with remarks on the development and
    function of the pharyngeal pouches, Zoologica, 51:103.
Schevill, W. E., Watkins, W. A., and Ray, G. C., 1969, Click struc-
    ture in the porpoise, Phocoena phocoena, Jour. Mamm., 50:721.
Schusterman, R. J., 1966, Serial discrimination-reversal learning
    with and without errors by the California sea lion, J. Exp.
    Analysis Behav., 9(5):97.

Schusterman, R. J., 1966, Underwater click vocalizations by a
    California sea lion: effects of visibility, Psychol. Rec.,
    16:129.
Schusterman, R. J., 1967, Perception and determinants of under-
    water vocalization in the California sea lion, in: "Animal
    Sonar Systems, Biology and Bionics", R.-G. Busnel, ed.,
    Laboratoire de Physiologie Acoustique, Jouy-en-Josas, France.
Schusterman, R. J., 1968, Experimental laboratory studies of pin-
    niped behaviour, in: "The Behaviour and Physiology of Pin-
    nipeds", R. J. Harrison, R. C. Hubbard, R. S. Peterson, C. E.
    Rice and R. J. Schusterman, eds., Appleton-Century-Crofts,
    New York.
Schusterman, R. J., 1969, Aerial and underwater visual acuity in
    the California sea lion as a function of luminance, Naval
    Undersea Research and Development Center, San Diego, Calif.,
    Final Report.
Schusterman, R. J., 1972, Visual acuity in pinnipeds, in: "Beha-
    vior of Marine Animals Vol. II", H. E. Winn and B. Olla,
    eds., Plenum Press, New York.
Schusterman, R. J., 1973, A note comparing the visual acuity of
    dolphins with that of sea lions, Cetology, 15:1.
Schusterman, R. J., 1974, Low false-alarm rates in signal detec-
    tion by marine mammals, Jour. Acoust. Soc. Amer., 55:845.
Schusterman, R. J., 1976, California sea lion auditory detection
    and variation of reinforcement schedules, Jour. Acoust. Soc.
    Amer., 59:997.
Schusterman, R. J., 1977, Temporal patterning in sea lion barking
    (Zalophus californianus), Behavioral Biology, 20:404.
Schusterman, R. J., and Balliet, R. F., 1969, Underwater barking by
    male sea lions (Zalophus californianus), Nature 222:1179.
Schusterman, R. J., and Balliet, R. F., 1970, Conditioned vocaliza-
    tion as a technique for determining visual acuity thresholds
    in the sea lion, Science, 169:498.
Schusterman, R. J., and Balliet, R. F, 1970, Visual acuity of the
    harbor seal and the Steller sea lion under water, Nature,
    226:563.
Schusterman, R. J., and Balliet, R. F., 1971, Aerial and underwater
    visual acuity in the California sea lion (Zalophus californi-
    anus) as a function of luminance, Ann. N.Y. Acad. Sci.,
    188:37.
Schusterman, R. J., and Barrett, B., 1973, Amphibious nature of
    visual acuity in the Asian 'clawless' otter, Nature, 244:518.
Schusterman, R. J., and Dawson, R. G., 1968, Barking, dominance,
    and territoriality in male sea lions, Science, 160:434.
Schusterman, R. J., and Johnson, B. W., 1975, Signal probability
    and response bias in California sea lions, Psych. Rec.,
    25:39.

Schusterman, R. J., and Kersting, D., 1978, Selective attention in discriminative echolocation by the porpoise (Tursiops truncatus), Paper read at the Animal Behavior Society Annual Meeting, June 19-23, U. of Wash., Seattle.

Schusterman, R. J., Balliet, R. F., and Nixon, J., 1972, Underwater audiogram of the California sea lion by the conditioned vocalization technique, J. exp. Analysis Behav., 17:339.

Schusterman, R. J., Balliet, R. F., and St. John, S., 1970, Vocal displays by the gray seal, the harbor seal and the Steller sea lion, Psychon. Sci., 18:303.

Schusterman, R. J., Barrett, B., and Moore, P. W. B., 1975, Detection of underwater signals by a California sea lion and a bottlenose porpoise: Variation in the payoff matrix, Jour. Acoust. Soc. Amer., 57:1526.

Scronce, B. L., and Johnson, C. S., 1975, Bistatic target detection by a bottlenosed porpoise, Jour. Acoust. Soc. Am., 59(4):1001.

Seeley, R. L., Flanigan, W. F., and Ridgway, S. H., 1976, A technique for rapidly assessing the hearing of the bottlenosed porpoise Tursiops truncatus, Naval Undersea Center TP 522, San Diego, Calif.

Sergeant, D. E., 1978, Ecological isolation in some cetacea, in: "Recent Advances in the Study of Whales and Seals", A. N. Severtsov, ed., Nauka, Moskow.

Shevalev, A. Ye,. and Flerov, A. I., 1975, Study of the communicative signal of dolphins connected with food-finding, Bionika, 9:119.

Shirley, D. J., and Diercks, K. J., 1970, Analysis of the frequency response of simple geometric targets, Jour. Acoust. Soc. Am., 48:1275.

Sokolov, V., 1971, Cetacean research in the USSR, in: "Investigations on Cetacea Vol. 3", G. Pilleri, ed., Bentelli AG, Berne.

Sokolov, V., and Kalashnikova, M., 1971, The ultrastructure of epidermal cells in Phocoena phocoena, in: "Investigations on Cetacea Vol. 3", G. Pilleri, ed., Bentelli AG, Berne.

Sokolov, V. Ye., Ladygina, T. F., and Supin, A. Ya., 1972, Localization of sensory areas in the dolphin's cerebral cortex, Doklady Akademii Nauk SSSR, 200(2).

Solntseva, G. N., 1971, The comparative anatomical and histological characteristics of the structure of the external and internal ear of certain dolphins, Issledovaniya Morskikh Mlekopitayushchikh, Trudy Atlantnipo, 39:369.

Solntseva, G. N., 1974, Function of the hearing organ underwater, Priroda, 10:21 (English translation JPRS 65061).

Solntseva, G. N., 1975, Objectives and prospects of research on routes of sound transmission in cetaceans, in: "Morskiye Mlekopitayushchiye", G. B. Agarkov, ed., Izdatel'stvo Naukova Dumka, Kiev.

Solntseva, G. N., 1975, Morpho-functional peculiarities of the
    hearing organ in terrestrial, semiaquatic and aquatic
    mammals, Zoologicheskiy Zhurnal, 54(10):1529.
Solukha, B. V., and Mantulo, A. P., 1973, Mechanism of sound trans-
    mission in dolphins, in: "Nekotorye voprosy ekologii i
    morfologii zhivotnykh" (Materially Nauchnoy Konferentsii),
    Izdatel'stvo Naukova Dumka, Kiev.
Stewart, J. L., 1968, Analog simulation studies in echoranging,
    T. D. R. No. AMRL-TR-68-40, Aerospace Medical Div., U.S.A.F.
    Systems Command, Washington, D.C.
Sokolov, V. Ye. (ed.), 1978, Electrophysiological studies of the
    dolphin's brain, Izdatel'stvo, Nauka, Moscow.
Spong, P., and White, D., 1971, Visual acuity and discrimination
    learning in the dolphin (Lagenorhynchus obliquidens), Exp.
    Neurol., 31:341.
Spong, P., Bradford, J., and White, D., 1970, Field studies of the
    behaviour of the killer whale (Orcinus orca), in: "Pro-
    ceedings of the Seventh Annual Conference on Biological Sonar
    and Diving Mammals", Stanford Research Institute, Menlo Park,
    Calif.
Spong, P., Michaels, H., and Spong. L., 1972, Field studies of the
    behaviour of the killer whale (Orcinus orca) II, in: "Pro-
    ceedings of the Ninth Annual Conference on Biological Sonar
    and Diving Mammals", Stanford Research Institute, Menlo Park,
    Calif.
Spong, P., Spong, L., and Spong, Y., 1972, Field studies of
    the behaviour of the killer whale (Orcinus orca) III, in:
    "Proceedings of the Ninth Annual Conference on Biological
    Sonar and Diving Mammals", Stanford Research Institute,
    Menlo Park, Calif.
Sukhoruchenko, M. N., 1971, Upper limit of hearing of dolphins
    with reference to frequency, Trudy Akusticheskogo Instituta,
    12:194.
Sukhoruchenko, M. N., 1973, Frequency discrimination of dolphin
    (Phocoena phocoena), Fiziologicheskiy zhurnal imeni I. M.
    Sechenova, 59(8):1205.
Supin, A. Ya., and Sukhoruchenko, M. N., 1970, The determination
    of auditory thresholds in Phocoena phocoena by the method
    of skin-galvanic reaction, Trudy Akusticheskogo Instituta,
    12:194.
Supin, A. Ya., and Sukhoruchenko, M. N., 1974, Characteristics of
    acoustic analyzer of the Phocoena phocoena L. dolphin, in:
    "Morphology, Physiology, and Acoustics of Marine Mammals",
    V. Ye. Sokolov, ed., Nauka, Moscow (English translation
    JPRS 65139).
Tavolga, W. N., 1977, Mechanisms for directional hearing in the
    sea catfish (Arius felis), Jour. Exp. Biol., 67:97.
Terhune, J. M, 1974, Directional hearing of a harbor seal in air
    and water, Jour. Acoust. Soc. Amer., 56:1862.

Terhune, J. M., and Ronald, K., 1972, The harp seal, Pagophilus groenlandicus (Erxleben, 1777), X. The underwater audiogram, Can. J. Zool., 50:565.

Terhune, J. M., and Ronald, K., 1975, Underwater hearing sensitivity of two ringed seals (Pusa hispida), Can. J. Zool., 53:227.

Terhune, J. M., and Ronald, K., 1975, Masked hearing thresholds of ringed seals, Jour. Acoust. Soc., Amer., 58:515.

Thompson, P. O., and Cummings, W. C., 1969, Sound production of the finback whale, Balaeoptera physalus, and Eden's whale, B. edeni, in the Gulf of California (abstract), in: "Proceedings of the Sixth Annual Conference on Biological Sonar and Diving Mammals", Stanford Research Institute, Menlo Park, Calif.

Thompson, R. K. R., and Herman, L. M., 1975, Underwater Frequency Discrimination in the Bottlenose Dolphin (1-140 kHz) and Human (1-8 kHz), Jour. Acoust. Soc. Amer., 57:943.

Titov, A. A., 1971, Characteristics of sonic signaling of common dolphins (Delphinus delphis) under new conditions, Bionika, 5:62.

Titov, A. A., 1972, Investigation of sonic activity and phenomenological characteristics of the echolocation analyzer of Black Sea delphinids. Canditorial dissertation, Karadag, as cited in: Bel'kovich, V. M. and Dubrovskiy, N. A., 1976, "Sensory Bases of Cetacean Orientation", Nauka, Leningrad (English translation JPRS L/7157).

Titov, A. A., 1975, Recognition of spherical targets by the bottlenose dolphin in the presence of sonic interference, in: "Marine Mammals, Proceedings of the Sixth All-Union Conference on the Study of Marine Mammals", G. B. Agarkov, ed., Naukova Dumka, Kiev (English translation JPRS L/6049).

Titov, A. A., and Nikolenko, G. V., 1975, Quantitative evaluation of the sounds in three species of Black Sea dolphins, Bionika, 9:115.

Titov, A. A., and Tomilin, A. G., 1970, Sonic activity of the common dolphin (Delphinus delphis) and harbor porpoise (Phocoena phocoena) in various situations, Bionika, 4:88.

Titov, A. A., and Yurkevich, L. I., 1971, Physical characteristics of nonecholocation sounds of Black Sea dolphins, Bionika, 5:57.

Titov, A. A., and Yurkevich, L. I., 1975, Temporary pulse summation in the bottlenose dolphin, Tursiops truncatus, in: "Morskiye Mlekopitayushchiye", G. B. Agarkov, ed., Izdatel'stvo Naukova Dumka, Kiev.

Titov, A. A., Tomilin, A. G., Baryshnikov, N. S., Yurkevich, L. I., and Lekomtsev, V. M., 1971, Communication-emotional signals of Black Sea dolphins, Bionika, 5:67.

Tomilin, A. G., 1968, Factors promoting powerful development of the brain in Odontoceti, Trudy Vsesoyuznogo Sel'sko-Khozyaystvennogo Instituta Zaochnogo Obrasovaniya, 31:191 (in Russian JPRS 49777).

Tomilin, A. G., 1969, Present status of study of the acoustical vocalizations in Cetacea, "IVoe Vsesoyuznoye Soveshchaniye po izucheniyu Morskikh Mlekopitayushchikh", Kaliningrad.

Tsuyuki, H., and Itoh, S., 1973, Fatty acid component of blubber oil of Amazon river dolphin, Sci. Rept. of the Whales Res. Inst., 25:293.

Turner, R. N.,, and Norris, K. S., 1966, Discriminative echolocation in a porpoise, Jour. of the Exp. Anal. of Beh., 9:535.

Valiulina, F. G., 1975, The role of the mandible in conduction of sonic oscillations in man and dolphins, in: "Morskiye Mlekopitayushchiye", G. B. Agarkov, ed., Izdatel'stvo Naukova Dumka, Kiev.

Varanasi, V., and Malins, D. G., 1970, Unusual wax esters from the mandibular canal of the porpoise (Tursiops gilli), Biochemistry, 9:3629.

Varansai, V., and Malins, D. G., 1970, Ester and ether-linked lipids in the mandibular canal of a porpoise (Phocoena phocoena). Occurrence of isonaleric acid in glycerolipids, Biochemistry, 9:4576.

Varanasi, V., and Malins, D. G., 1971, Unique lipids of the porpoise (Tursiops gilli): difference in triacyl glycerols and wax esters of acoustic (mandibular canal and melon) and blubber tissues, Biochem. Biophys. Acta., 231:415.

Varanasi, V., and Malins, D. G., 1972, Triacylglycerols characteristics of porpoise acoustic tissues: Molecular structure of diisolvaleroyl-glycerides, Science, 176:926.

Varanasi, V., Everitt, M., and Malins, D. G., 1973, The isomeric composition of the diisovaleroyl-glycerides: a specificity for the biosynthesis of the 1,3-Diisolvaleryl-Glycerides structures, Int. J. Biochem., 4:373.

Varanasi, V., Feldman, H. R., and Malins, D. G., 1975, Molecular basis for formation of lipid sound lens in echolocating cetaceans, Nature, 255:340.

Vel'min, V. A., 1975, Target detection by the bottlenose dolphin under artifical reverberation conditions, in: "Marine Mammals, Proceedings of the Sixth All-Union Conference on the Study of Marine Mammals", G. B. Agarkov, ed., Naukova Dumka Kiev, (English translation JPRS L/6049).

Vel'min, V. A., and Dubrovskiy, N. A., 1975, On the auditory analysis of pulsed sounds by dolphins, Doklady Akad. Nauk. SSSR, 225(2):470.

Vel'min, V. A., and Dubrovskiy, N. A., 1976, The critical interval of active hearing in dolphins, Sov. Phys. Acoust., 22(4):351, (English translation, J. S. Wood, Amer. Inst. Physics).

Vel'min, V. A., and Titov, A. A., 1975, Auditory discrimination of interpulse intervals by bottlenose dolphin, in: "Marine Mammals, Proceedings of the Sixth All-Union Conference on the Study of Marine Mammals", G. B. Agarkov, Ed., Naukova Dumka, Kiev (English translation JPRS L/6049).

Vel'min, V. A., Titov, A. A., and Yurkevich, L. I., 1975, Differential intensity thresholds for short-pulsed signals in bottlenose dolphins, in: "Morskiye Mlekopitayushchiye", (Proceedings of the Sixth All-Union Conference on the Study of Marine Mammals, G. B. Agarkov, ed., Izdatel'stvo Naukova Dumka, Kiev (Translation JPRS L/6049-1).

Vel'min V. A., Titov, A. A., and Yurkevich, L. I., 1975, Temporary pulse summation in bottlenose dolphins, in: "Morskiye Mlekopitayushchiye", (Proceedings of the Sixth All-Union Conference on the Study of Marine Mammals), G. B. Agarkov, ed., Naukova Dumka, Kiev (English Translation JPRS L/6049-1).

Voronov, V. A., and Stosman, I. M., 1977, Frequency-threshold characteristics of subcortical elements of the auditory analyzer of the Phocoena phocoena porpoise, Zhurnal Evolyutsionnoy Biokhimii I Fiziologii, 6:719.

Vronskiy, A. A., 1975, Some data on extraorgan innervation of dolphin pharynx, Doklady Akademii Nauk SSSR, 6:545.

Vronskiy, A. A., and Manger, A. P., 1975, Extraorganic innervation of the larynx and pharynx of some delphinids, in: "Morskiye Mlekopitayushchiye", G. B. Agarkov, ed., Izdatel'stvo Naukova Dumka, Kiev.

Watkins, W. A., 1966, Listening to Cetaceans, in: "Whales, Dolphins and Porpoises", K. S. Norris, ed., Univ. of Calif. Press, Berkeley.

Watkins, W. A., 1967, Air-borne sounds of the humpback whale Megaptera novaengliae, Jour. Mammal, 48:573.

Watkins, W. A., 1967, The harmonic interval: fact or artifact in spectral analysis of pulse trains, in: "Marine Bio-Acoustics Vol. II", W. N. Tavolga, ed., Pergamon Press, New York.

Watkins, W. A., 1968, Comments on "Spectral Analysis of the calls of the Male Killer Whale", IEEE Transact. on Audio and Electroacoustics, 16:523.

Watkins, W. A., 1974, Computer measurement of biological sound-source locations from four hydrophone array data, Technical Report, Off. Naval Res., Washington, D.C.

Watkins, W. A., 1974, Bandwidth limitations and analysis of cetacean sounds, with comments on "Delphinid sonar measurements and analysis", (K. J. Diercks, R. T. Trochta and W. E. Evans, J. Acoust. Soc. Amer. 54, 200-204 (1973)), Jour. Acoust. Soc. Amer., 55:849.

Watkins, W. A., 1976, Biological sound-source location by computer analysis of underwater array data, Deep-Sea Research, 23:175.

Watkins, W. A., 1977, Acoustic behavior of sperm whales, Oceanus, 20:50.

Watkins, W. A., and Ray, G. C., 1977, Underwater sounds from ribbon seal, Phoca (Histriophoca) fasciata, Fishery Bulletin, 75:450.

Watkins, W. A., and Schevill, W. E., 1968, Underwater playback of their own sounds to Leptonychotes (Weddell seals), Jour. Mammal, 49:287.

Watkins, W. A., and Schevill, W. E., 1971, Four hydrophone array for acoustic three-dimensional location, Technical Report 71-60, Woods Hole Oceanographic Inst., Massachusetts.

Watkins, W. A., and Schevill, W. E., 1971, Underwater sounds of Monodon (Narwhal), Jour. Acoust. Soc. Amer., 49(2):595.

Watkins, W. A., and Schevill, W. E., 1972, Sound source location by arrival-times on a non-rigid three-dimensional hydrophone array, Deep-Sea Research, 19:691.

Watkins, W. A., and Schevill, W. E., 1974, Listening to Hawaiian spinner porpoises, Stenella cf. longirostris, with a three-dimensional hydrophone array, Jour. Mammal., 55:319.

Watkins, W. A., and Schevill, W. E., 1975, Sperm whales (Physeter catodon) react to pingers, Deep-Sea Research, 22:123.

Watkins, W. A., and Schevill, W. E., 1976, Right whale feeding and baleen rattle, Jour. Mammal., 57:58.

Watkins, W. A., and Schevill, W. E., 1977, Spatial distribution of Physeter catodon (sperm whales) underwater, Deep-Sea Research, 24:693.

Watkins, W. A., and Schevill, W. E., 1977, Sperm whale codas, Jour. Acoust. Soc. Amer., 62:1485.

Watkins, W. A., Schevill, W. E., and Best, P. B., 1977, Underwater sounds of Cephalorhynchus heavisidii (Mammalia: Cetacea), Jour. Mammal., 58:316.

Watkins, W. A., Schevill, W. E., and Ray, G. C., 1971, Underwater sounds of Monodon (Narhwal), Jour. Acous. Soc. Amer., 49:595.

Wedmid, G., Litchfield, C., Ackman, R. G., and Sipos, J. C., 1971, Heterogeneity of lipid composition within the cephalic melon tissue of the pilot whale (Globicephala melaena), Jour. Amer. Oil Chem. Soc., 48:332.

Wedmid, G., Litchfield, C. Ackman, R. G., Sipos, J. C., Eaton, C. A., and Mitchell, E. D., 1973, Heterogeneity of lipid composition within the cephalic melon tissue of the pilot whale (Globicephala melaena), Biochemica & Biophysica Acta, 326:439.

Wever, E. G., McCormick, J. G., Palin, J., and Ridgway, S. H., 1971, The cochlea of the dolphin Tursiops truncatus: general morphology, Proc. Nation. Acad. Sci., USA, 68:2381.

Wever, E. G., McCormick, J. G., Palin, J., and Ridgway, S. H., 1971, The cochlea of the dolphin Tursiops truncatus: the basilar membrane, Proc. Nation. Acad. Sci., USA, 68:2708.

Wever, E. G., McCormick, J. G., Palin, J., and Ridgway, S. H., 1971, The cochlea of the dolphin, Tursiops truncatus: hair cells and ganglion cells, Proc. Nation. Acad. Sci., USA, 68:2908.

Wever, E. G., McCormick, J. G., Palin, J., and Ridgway, S. H.,
    1972, Cochlear structure in the dolphin Lagenorhynchus
    obliquidens, Proc. Nation. Acad. Sci., USA, 69:657.
White, M .J., Ljungblad, D., Norris, J. and Baron, K., 1977,
    Auditory thresholds of two Beluga whales, Proceedings
    (abstracts) Second Conference on the Biology of Marine
    Mammals, San Diego, Calif.
Whitmore, F. C., Jr., and Sanders, A. E., 1976, Review of the
    Oligocene Cetacea, Syst. Zool., 25(4):304.
Winn, H. E., Perkins, P. J., and Poulter, T. C., 1970, Sounds of
    the humpback whale, in: "Proceedings of the Seventh Annual
    Conference on Biological Sonar and Diving Mammals", Stanford
    Research Institute, Menlo Park, Calif.
Winn, H. E., Perkins, P. J., and Winn, L., 1970, Sounds and behavi-
    our of the northern bottle-nosed whale, in: "Proceedings of
    the Seventh Annual Conference on Biological Sonar and Diving
    Mammals", Menlo Park, Calif.
Wood, F. G., 1973, "Marine Mammals and Man", R. B. Luce, Inc.,
    New York.
Wood, F. G., 1978, The cetacean stranding phenomenon: An hypothe-
    sis, in: "Report on the Marine Mammal Stranding Workshop",
    N. T. I. S., Washington, D.C.
Yablokov, A. V., Bel'kovich, V. M., and Borisov, V. I., 1972,
    Whales and Dolphins, Parts I and II, Kity i Del'fini, Izd-vo
    Nauka, Moscow (English translation JPRS 62150-1).
Yanagisawa, K., Sato, O., Nomoto, M., Katsuki, Y., Ikeono, E.,
    Grinnel, A. D., and Bullock, T. H., 1966, Auditory evoked
    potentials from brain stem in cetaceans, Fed. Proc.
    25:464.
Yunker, M. P., and Herman, L. M., 1974, Discrimination of auditory
    temporal differences by the bottlenose dolphin and by the
    human, Jour. Acous. Soc. Amer., 56:1870.
Zaslavskiy, G. L., 1971, On directivity of sound emission in the
    Black Sea bottlenosed dolphin, Trudy Akusticheskogo Instituta,
    17:60.
Zaslavskiy, G. L., 1972, Study of echolocation signals of bottle-
    nose dolphin (Tursiops truncatus) two channel system of
    registration, Biofizika, 17:717.
Zaslavskiy, G. L., 1974, Experimental study of time and space
    structure of the dolphin's echolocating signals, candida-
    torial dissertation, Karadag, as cited in: Bel'kovich, V. M.
    and Dubrovskiy, N. A., 1976, "Sensory Bases of Cetacean
    Orientation", Nauka, Leningrad (English translation JPRS
    L/7157).
Zaslavskiy, G. L., Titov, A. A., and Lekomtsev, V. M., 1969,
    Investigation of hydrolocation capabilities of the Azov
    dolphin (Phocoena phocoena), Trudy Akusticheskogo Instituta,
    8:134.

Zaytseva, K. A., Akopian, A. I., and Morozov, V. P., 1975, Noise
    resistance of the dolphin auditory analyzer as a function of
    noise direction, Biofizika, 20(3):519.
Zlatoustova, L. V., and Nizova, A. B., 1971, An acoustical analysis
    of some whistles of the bottlenose dolphin (Tursiops
    truncatus Montagu), "Morfologiya i Ekologiya Morskikh
    Mlekopitayushchikh", V. Ye. Sokolov, ed., Izdatel'stvo
    Nauka, Moscow.

This bibliography was compiled and prepared by Paul E. Nachtigall.

The assistance of Vladimir Gurevich and Mike Fielding in the
acquisition of Russian references is gratefully acknowledged.
Jeff Haun and Earl Murchison graciously assisted in proofreading,
while Phyllis Johnson expertly typed the entire list.

Bibliography on echolocation in non-bat, non-cetacean species

1. Buchler, E. R. The use of echolocation by the wandering shrew Sorex vagrans. Anim. Behav., 24, 858-873, 1976.
2. Cranbrook, E. and Medway, L. Lack of ultrasonic frequencies in the calls of swiftlets. Ibis, 107, 258. 1965.
3. Eisenberg, J. F. and Gould, E. The behavior of Solenodon paradoxus in captivity with comments on the behavior of other insectivora. Zool. 51, 49, 1966.
4. Eisenberg, J. F. and Gould, E. The tenrecs: a study in mammalian behavior and evolution. Smithsonian Contributions to Zoology No. 27, 138 p.
5. Fenton, M. B. The role of echolocation in the evolution of bats. Am. Nat., 108, 386-388. 1974.
6. ----- Acuity of echolocation in Collocalia hirundinacea (Aves: Apodidae), with comments on the distribution of echolocating swiftlets and molossid bats. Biotropica, 7, 1-7. 1975.
7. Gans, C. and Maderson, P. F. A. Sound producing mechanisms in recent reptiles: review and comment. Amer. Zool., 13, 1195-1203. 1973.
8. Gehlback, F. R. and Walker, B. Acoustic behavior of the aquatic salamander, Siren intermedia. BioScience, 20, 1107-1108. 1970.
9. Griffin, D. R. and Suthers, R. A. Sensitivity of echolocation in cave swiftlets. Biol. Bull., 139, 495-501. 1970.
10. Griffin, D. R. and Buchler, E. Echolocation of extended surfaces. In: "Animal migration, navigation and homing." Ed. by K. Schmidt-Koenig, New York, Springer-Verlag, 1978. p. 201-08.
11. Grunwald, A. Untersuchungen zur orientiergun der weisszahnspitzmaus (Soricidae-Crocidurinae). Z. vergl. Physiol., 65, 191-217. 1969.
12. Harrisson, T. Onset of echolocation clicking in Collocalia swiftlets. Nature, 212, 530-531, 1966.
13. Holyoak, D. T. Undescribed land birds from the Cook Islands, Pacific Ocean. Bull. Br. Ornith. Club, 94, 145-150. 1974.
14. Hutterer, V. R. Beobachtungen zur geburt and jugendentwicklung der zwergspitzmaus, Sorex minutus L. (Soricidae - Insectivora). Z. Saugetierkunde, 41, 1-22. 1976.
15. Ilichev, V. D. Certain problems in studying the orientation of birds. U. S. Gov. Res. Dev. Rep. 70, 173. 1970.
16. Lore, R., Kam, B. and Newby, V. Visual and nonvisual depth avoidance in young and adult rats. J. Comp. and Physiol. Psychol., 65, 525-528. 1967.
17. Komarov, V. T. Underwater sounds of the muskrat, Ondatra zibethica and the water vole Arvicola terrestris. Zool. Zh., 55, 632-633. 1976.

18. Konishi, M. and Knudsen, E. I. The oilbird: hearing and echolocation. Sci. 204, 425–427. 1979.

19. Konstantinov, A.I., Akhmarova, N.I. and Golovina, S.S. The possibility of small rodents using ultrasonic location. In: "Echolocation in animals." Ed. by E. Sh. Airapetyants, A. I. Konstantinov, Springfield, Virginia, U.S. Department of Commerce, National Tech. Info. Serv., 1966. 390 pp.

20. Medway, L. The swiftlets (Collocalia) of Java, and their relationships. J. Bombay Nat. Hist. Soc., 59, 146–153. 1962.

21. -----Field characters as a guide to the specific relations of swiftlets. Proc. Linn. Soc. Lond., 177, 151–172. 1966.

22. -----The function of echonavigation among swiftlets. Anim. Behav., 15, 416–420. 1967.

23. -----Studies on the biology of the edible nest swiftlets of southeast Asia. Malay, Nat. J., 22, 57–63. 1969.

24. -----The nest of Collocalia v. vanikorensis, and taxonomic implications. Emu, 75, 154–155. 1975.

25. Medway, L. and Wells, D. R. Dark orientation by the giant swiftlet Collocalia gigas. Ibis, III, 609–11. 1969.

26. -----The Birds of Malay Peninsula, Vol. 5, London, Witherby. 1976.

27. Medway, L. and Pye, J. D. Echolocation and the systematics of swiftlets. In: "Evolutionary ecology." Ed. by B. Stonehouse and C. Perrins, Baltimore, Maryland, University Park Press, 1977. p. 225–238.

28. Pecotich, L. Grey swiftlets in the Tully River gorge and Chillagoe Caves. Sunbird, 5, 16–21. 1974.

29. Penny, M. The birds of Seychelles and the outlying islands. London, Collins. 1976.

30. Pernetta, J. C. Anatomical and behavioral specializations of shrews in relation to their diet. Can. J. Zool. 55, 1442–1453. 1977.

31. Player, J. In search of Guianas oilbirds. Animals, 15, 74–79. 1973.

32. Sato, Y. Visual sense organs of soricoidea. Zool. Mag., 81, 419. 1972.

33. Schwartzkopff, J. Auditory communication in lower animals role of auditory physiology. In: "Annual review of psychology", Vol. 28. Ed. by M. R. Rosenzweig and L. W. Porter, Palo Alto, California, Annual Reviews, Inc., 1977. p. 61–84.

34. Sergeev, V. E. Characteristics of the orientation of shrews in water. Ekologiya, 4, 87–90. 1973.

35. Simmons, J. A. Echolocation for perception of the environment. J. Acoust. Soc. Am., 58, S18. 1975.

36. Smythies, B. E. A note on the swiftlets Collocalia found in Burma. J. Bombay Nat. Hist. Soc., 72, 847–851. 1975.

37. Snow, D. W. Oilbirds, cave-living birds of South America. Stud. Speleol. 2, 257–264. 1975.

38. Tavolga, W. N. Acoustic obstacle detection in the sea catfish (<u>Arius</u> <u>felis</u>). In: "Sound reception in fish." Ed. by A. Schuijf and A. D. Hawkins, Amsterdam, Elsevier Scientific Publ. Co., 1976. p. 185–203.

39. ----- Acoustic orientation in the sea catfish, <u>Galeichthys</u> <u>felis</u>. In: "Orientation: sensory basis." Ed. by H. Adler, Ann. N.Y. Acad. Sci., 188, 80–97, 1971.

40. ----- Mechanisms for directional hearing in the sea catfish (<u>Arius</u> <u>felis</u>). J. Exp. Biol. 67, 97–115. 1977.

41. Thurow, G. R. and Gould, H. J. Sound production in a caecilian. Herpetologica, 33, 234–237. 1977.

42. Vaughan, T. A. Mammalogy. Philadelphia, W. B. Saunders Co., 1972. 463 pp.

43. Vogel, P. Comparative investigations of the mode of ontogenesis of domestic Soricidae <u>Crocidura</u> <u>russula</u>, <u>Sorex</u> <u>araneus</u> and <u>Neomys</u> <u>fodiens</u>. Rev. Suisse Zool. 79, 1201–1332. 1972.

Compiled by Edwin Gould

SYMPOSIUM PARTICIPANTS

ALCURI (Gustavo)
Laboratoire d'Acoustique Animale
E.P.H.E. - I.N.R.A. - C.N.R.Z.
78350 Jouy-en-Josas   France

AU (Whitlow W.L.)
Naval Ocean Systems Center
P.O. Box 997
Kailua, Hawaii 96734   U.S.A.

BEUTER (Karl J.)
Battelle-Institut e.V.
Am Römerhof 35   F.R.
Frankfurt/Main   Germany

BODANHAMER (Robert)
Department of Zoology
University of Texas
Austin, Tx. 78712   U.S.A.

BOWERS (Clark A.)
Naval Ocean Systems Center
Code 5111
San Diego, Ca. 92152   U.S.A.

BROWN (Patricia)
Department of Biology
University of California
Los Angeles, Ca. 90024   U.S.A.

BRUNS (Volkmar)
Fachbereich Zoologie
J.W. Goethe-Universität   F.R.
D-6000 Frankfurt/Main   Germany

BUCHLER (Edward)
Department of Zoology
University of Maryland
College Park, Md. 10742   U.S.A.

BULLOCK (Theodore H.)   Department of Neurosciences
School of Medicine
University of California
San Diego, Ca. 92093   U.S.A.

BUSNEL (René-Guy)   Laboratoire d'Acoustique Animale
E.P.H.E. - I.N.R.A. - C.N.R.Z.
78350 Jouy-en-Josas   France

CHASE (Julia)   Biology Department
Columbia University
Barnard College
New York, N.Y. 10027   U.S.A.

DECOUVELAERE (Martine)   Ecole Nationale Supérieure
des Télécommunications
46, rue Barrault
75634 Paris Cedex 13   France

DIERCKS (K. Jerome)   Applied Research Laboratories
University of Texas
P.O. Box 8029
Austin, Tx. 78712   U.S.A.

DO (Manh Anh)   Radio Engineering Limited
P.O. Box 764
Dunedin   New Zealand

DZIEDZIC (Albin)   Laboratoire d'Acoustique Animale
E.P.H.E. - I.N.R.A. - C.N.R.Z.
78350 Jouy-en-Josas   France

ESCUDIE (Bernard)   Laboratoire du Traitement du
Signal - I.C.P.I.
25, rue du Plat
69288 Lyon   France

EVANS (William E.)   Hubbs-Sea World Research Institute
1700 South Shores Drive
San Diego, Ca. 92109   U.S.A.

FENG (Albert S.)   Department of Physiology and
Biophysics
University of Illinois
Urbana, Il. 61801   U.S.A.

FENTON (M. Brock)   Department of Biology
Carleton University
Ottawa, KIS 5B6   Canada

FIEDLER (Joachim)   Fachbereich Biologie-Zoologie
                           J.W. Goethe Universität   F.R.
                           D-6000 Frankfurt/Main   Germany

FISH (James F.)   Sonatech. Inc.
                           700 Francis Botello Road
                           Goleta, Ca. 93017   U.S.A.

FLEISCHER (Gerald)   c/o Umweltbundesant
                           Bismarckplatz 1   F.R.
                           D-1000 Berlin 33   Germany

FLOYD (Robert W.)   Naval Oceans Systems Center
                           P.O. Box 997
                           Kailua, Hawaii 96744   U.S.A.

FULLARD (James)   Department of Biology
                           Carleton University
                           Ottawa K1S 5B6   Canada

Von GIERKE (Henning E.)   Biodynamics and Bioengineering
                           Division
                           Aerospace Medical Research
                           Laboratory
                           Wright Patterson Air Force Base
                           Dayton, Oh. 45433   U.S.A.

GOULD (Edwin)   Johns Hopkins University
                           School of Hygiene and Public Health
                           615 North Wolfe Street
                           Baltimore, Md. 21205   U.S.A.

GREGUSS (Pal)   Applied Biophysics Laboratory
                           Technical University Budapest
                           H-1111 Budapest   Hungary

GRIFFIN (Donald R.)   Rockefeller University
                           New York, N.Y. 10021   U.S.A.

GRINNELL (Alan D.)   Department of Biology and
                           Physiology
                           University of California
                           Los Angeles, Ca. 90024   U.S.A.

GUREVICH (Vladimir Sol)   Hubbs-Sea World Research Institute
                           1700 South Shores Road
                           San Diego, Ca. 92109   U.S.A.

HALLS (Justin A.T.)   Department of Zoology and
                             Comparative Physiology
                             Queen Mary College
                             Mile End Road
                             London E1 4NS   England

HENSON (O'Dell W.)   Department of Anatomy
                             School of Medicine
                             University of North Carolina
                             Chapel Hill, N.C. 27514   U.S.A.

JEN (Philip H.S.)   Division of Biological Sciences
                             University of Missouri
                             Columbia, Mo. 65201   U.S.A.

JOHNSON (C. Scott)   Naval Ocean System Center
                             Code 5102
                             San Diego, Ca. 92152   U.S.A.

JOHNSON (Richard A.)   Naval Ocean Systems Center
                             Code 5122
                             San Diego, Ca. 92152   U.S.A.

KAY (Leslie)   Department of Electrical Engineering
                             University of Canterbury
                             Christchurch   New Zealand

KICK (Shelley)   Department of Psychology
                             Washington University
                             St. Louis, Mo. 63130   U.S.A.

LEVY (Jean-Claude)   Bureau de Psychologie Appliquée
                             de la Direction du Personnel
                             Militaire de la Marine
                             3, Avenue Octave Gréard
                             75007 Paris   France

LONG (Glenis R.)   Central Institute for the Deaf
                             818 South Euclid
                             St. Louis, Mo. 63110   U.S.A.

Mc.CORMICK (James G.)   Department of Otolaryngology
                             Bowman Gray School of Medicine
                             Wake Forest University
                             Winston-Salem, N.C. 27103   U.S.A.

McKAY (R. Stuart)   Biology Department
                             Boston University
                             Boston, Mass. 02215   U.S.A.

MARCUS (Stanley R.)   Naval Sea Systems Command (Sea 038)
                           Washington, D.C. 20363   U.S.A.

MILLER (Lee A.)   Department of Psychology
                           Stanford University
                           Stanford, Ca. 94305   U.S.A.

MOORE (Patrick W.B.)   SEACO, Inc.
                           146 Hekili St.
                           Kailua, Hawaii 96734   U.S.A.

MURCHISON (Earl A.)   Naval Ocean Systems Center
                           P.O. Box 997
                           Kailua, Hawaii 96734   U.S.A.

NACHTIGALL (Paul)   Naval Ocean Systems Center
                           P.O. Box 997
                           Kailua, Hawaii 96734   U.S.A.

NEUWEILER (Gerhard)   Fachbereich Biologie-Zoologie
                           Zoologisches Institut
                           J.W. Goethe-Universität   F.R.
                           D-6000 Frankfurt/Main   Germany

NORRIS (Kenneth S.)   Coastal Marine Laboratory
                           Environmental Studies
                           University of California
                           Santa Cruz, Ca. 95064   U.S.A.

O'NEILL (William E.)   Department of Biology
                           Washington University
                           St. Louis, Mo. 63130   U.S.A.

OSTWALD (Joachim)   Fachbereich Biologie-Zoologie
                           Philipps-Universität Marburg
                           Lahnberge   F.R.
                           3550 Marburg/Lahn   Germany

PENNER (Ralph H.)   Naval Ocean Systems Center
                           P.O. Box 997
                           Kailua, Hawaii 96734   U.S.A.

POLLACK (George)   Department of Zoology
                           University of Texas
                           Austin, Tx. 78712   U.S.A.

PORTER (Homer O.)   Biosciences Department
                           Naval Ocean Systems Center
                           Code 51
                           San Diego, Ca. 92152   U.S.A.

POWELL (William A.)   Naval Ocean Systems Center
                               Code 05
                               San Diego, Ca. 92152   U.S.A.

PYE (Ade)   Institute of Laryngology and
                               Otology
                               330, Gray's Inn Road
                               London WCIX 8EE   England

PYE (J. David)   Department of Zoology and
                               Comparative Physiology
                               Queen Mary College
                               Mile End Road
                               London E1 4NS   England

RADTKE (Susanne)   Fachbereich Biologie-Zoologie
                               J.W. Goethe-Universität   F.R.
                               D-6000 Frankfurt/Main   Germany

RIDGWAY (Sam H.)   Naval Ocean Systems Center
                               Code 5103 Seaside
                               San Diego, Ca. 92152   U.S.A.

RÜBSAMEN (Rudolf)   Fachbereich Biologie-Zoologie
                               J.W. Goethe-Universität   F.R.
                               D-6000 Frankfurt/Main   Germany

SCHLEGEL (Peter)   Fachbereich Biologie-Zoologie
                               J.W. Goethe Universität   F.R.
                               D-6000 Frankfurt/Main   Germany

SCHNITZLER (Hans-Ulrich)   Fachbereich Biologie-Zoologie
                               Philipps-Universität
                               Lahnberge   F.R.
                               3550 Marburg/Lahn   Germany

SCHULLER (Gerd)   Fachbereich Biologie-Zoologie
                               J.W. Goethe-Universität   F.R.
                               D-6000 Frankfurt/Main   Germany

SCHUSTERMAN (Ron)   Biology Department
                               California State University
                               Hayward, Ca. 94542   U.S.A.

SCHWEIZER (Hermann)   Fachbereich Biologie-Zoologie
                               J.W. Goethe-Universität   F.R.
                               D-6000 Frankfurt/Main   Germany

SIMMONS (James A.) Departments of Psychology and
          Biology
          Washington University
          St. Louis, Mo. 63130   U.S.A.

SUGA (Nobuo)   Department of Biology
          Washington University
          St. Louis, Mo. 63130   U.S.A.

SUTHERS (Roderick A.) Medical Sciences Program -
          Myers Hall
          Indiana University
          School of Medicine
          Bloomington, Ind. 47401   U.S.A.

STONE (Harris B.)   Office of Chief of Naval
          Operations (OP 987)
          Washington, D.C. 20350   U.S.A.

TIPPER (Ronald C.)   Office of Naval Research
          Oceanic Biology Branch
          Code 484 - NSTL Station
          Bay St. Louis, Miss. 39529 U.S.A.

VATER (Marianne)   Fachbereich Biology
          J.W. Goethe-Universität   F.R.
          D-6000 Frankfurt/Main   Germany

WATKINS (William A.) Woods Hole Oceanographic Institute
          Woods Hole, Mass. 02543   U.S.A.

WOOD (Forrest G.)   Biosciences Department
          Naval Ocean Systems Center
          Code 5104
          San Diego, Ca. 92152   U.S.A.

ZOOK (John M.)   Department of Psychology
          Duke University
          Durham, N.C. 27706   U.S.A.

OBSERVERS

HARNISCHEFEGER (Günther)   Fachbereich Biologie-Zoologie
                           J.W. Goethe-Universität   F.R.
                           D-6000 Frankfurt/Main   Germany

HERVIEU (René)   20, rue du Calvaire
                           44000 Nantes   France

MENNE (Dieter)   Fachbereich Biologie
                           Philipps-Universität
                           Lahnberge   F.R.
                           3550 Marburg/Lahn   Germany

REISS (Diana)   Laboratoire d'Acoustique Animale
                           E.P.H.E. - I.N.R.A. - C.N.R.Z.
                           78350 Jouy-en-Josas   France

WHEELER (Leslie A.)   Laboratoire d'Acoustique Animale
                           E.P.H.E. - I.N.R.A. - C.N.R.Z.
                           78350 Jouy-en-Josas   France

ABSENT CONTRIBUTORS

ALTES (Richard A.)   Orincon Corporation
                           3366 N. Torrey Pines Ct.
                           La Jolla, Ca. 92037   U.S.A.

POPPER (Arthur N.)   Department of Anatomy
                           Georgetown University
                           Washington, D.C. 20007   U.S.A.

INDEXES

- Author Index

- Species Index

- Subject Index

Compiled by L. Wheeler

Author Index

Underlined numbers refer to pages on which the complete references are listed.

1085

Species Index

PHYLUM ARTHROPODA

CLASS INSECTA, 670

Order Orthoptera
Super-family Tettigonioidea
Family Tettigoniidae
  Pflugus, 7

Order Neuroptera, 343-344, 435-
    436
Sub-order Hemeroboidea
Family Chrysopidae, 348
  Chrysopa carnea, 348, 435-436,
    445, 941-944

Order Lepidoptera, 343, 435-438,
    444-446, 666, 1009,
    1011, 1020, 1023
Super-family Pyralidoidea
Family Pyralidae, 436
Family Thyretidae, 436
Super-family Geometroidea
Family Drepanidae, 436
Family Geometridae, 343, 350,
    436
  Gorua apicata, 438
Family Thyatiridae, 436
Super-family Noctuoidea
Family Agaristidae, 436
Family Arctiidae, 343, 346, 436-
    438, 444, 1012
  Cycnia tenera Hubner, 346
  Rhodogastria vitrea, 900
Family Ctenuchidae, 343, 436
Family Lymantriidae, 436

Family Noctuidae, 343-347, 349-
    350, 436, 851
  Feltia subgothica, 346
  Heliothis spp., 345
  Heliothis zea, 345
  Miantochora interrupta, 900
  Ostrinia nubilalis, 345
  Prodenia cridania, 345, 443
  Scoliopteryx libatrix L., 351
  bollworm moth, 345
  tobacco budworm, 345
Family Notodontidae, 436
  Desmeocraera graminosa, 899-900
  Tmetopteryx bisecta, 437
Family Thaumetopoeidae, 436
Super-family Sphingoidea
Family Sphingidae, 436
  Celerio euphorbiae, 343, 350-351
  Celerio lineata, 343, 350-351
  Daphnis nerii, 157-158
Super-family Papilionoidea
Family Nymphalidae
  peacock butterfly, 438, 445

PHYLUM CHORDATA

CLASS OSTEICHTHYES, 825-826, 840-
    842, 1029

Super-order Teleostei

Order Clupeiformes
Sub-order Mormyroidei
Family Mormyridae, 406-407, 640
  Gnathonemus petersii, 406-407,
    421

1105